D0709443

Renewable and Efficient Electric Power Systems

Renewable and Efficient Electric Power Systems

Gilbert M. Masters
Stanford University

WILEY-INTERSCIENCE
A JOHN WILEY & SONS, INC., PUBLICATION

Published by John Wiley & Sons, Inc., Hoboken, New Jersey.
Published simultaneously in Canada.

Library of Congress Cataloging-in-Publication Data

Masters, Gilbert M.
Renewable and efficient electric power systems / Gilbert M. Masters.
p. cm.
Includes bibliographical references and index.
ISBN 0-471-28060-7 (cloth)
1. Electric power systems–Energy conservation. 2. Electric power systems–Electric losses. I. Title

TK1005.M33 2004
621.31–dc22

2003062035

Printed in the United States of America.

10 9 8 7 6 5 4

To the memory of my father,
Gilbert S. Masters
1910–2004

CONTENTS

PREFACE

Engineering for sustainability is an emerging theme for the twenty-first century, and the need for more environmentally benign electric power systems is a critical part of this new thrust. Renewable energy systems that take advantage of energy sources that won't diminish over time and are independent of fluctuations in price and availability are playing an ever-increasing role in modern power systems. Wind farms in the United States and Europe have become the fastest growing source of electric power; solar-powered photovoltaic systems are entering the marketplace; fuel cells that will generate electricity without pollution are on the horizon. Moreover, the newest fossil-fueled power plants approach twice the efficiency of the old coal burners that they are replacing while emitting only a tiny fraction of the pollution.

There are compelling reasons to believe that the traditional system of large, central power stations connected to their customers by hundreds or thousands of miles of transmission lines will likely be supplemented and eventually replaced with cleaner, smaller plants located closer to their loads. Not only do such distributed generation systems reduce transmission line losses and costs, but the potential to capture and utilize waste heat on site greatly increases their overall efficiency and economic advantages. Moreover, distributed generation systems offer increased reliability and reduced threat of massive and widespread power failures of the sort that blacked out much of the northeastern United States in the summer of 2003.

It is an exciting time in the electric power industry, worldwide. New technologies on both sides of the meter leading to structural changes in the way that power is provided and used, an emerging demand for electricity in the developing countries where some two billion people now live without any access to power, and increased attention being paid to the environmental impacts of power production are all leading to the need for new books, new courses, and a new generation of engineers who will find satisfying, productive careers in this newly transformed industry.

This book has been written primarily as a textbook for new courses on renewable and efficient electric power systems. It has been designed to encourage self-teaching by providing numerous completely worked examples throughout. Virtually every topic that lends itself to quantitative analysis is illustrated with such examples. Each chapter ends with a set of problems that provide added practice for the student and that should facilitate the preparation of homework assignments by the instructor.

While the book has been written with upper division engineering students in mind, it could easily be moved up or down in the curriculum as necessary. Since courses covering this subject are initially likely to have to stand more or less on their own, the book has been written to be quite self-sufficient. That is, it includes some historical, regulatory, and utility industry context as well as most of the electricity, thermodynamics, and engineering economy background needed to understand these new power technologies.

Engineering students want to use their quantitative skills, and they want to design things. This text goes well beyond just introducing how energy technologies work; it also provides enough technical background to be able to do first-order calculations on how well such systems will actually perform. That is, for example, given certain windspeed characteristics, how can we estimate the energy delivered from a wind turbine? How can we predict solar insolation and from that estimate the size of a photovoltaic system needed to deliver the energy needed by a water pump, a house, or an isolated communication relay station? How would we size a fuel cell to provide both electricity and heat for a building, and at what rate would hydrogen have to be supplied to be able to do so? How would we evaluate whether investments in these systems are rational economic decisions? That is, the book is quantitative and applications oriented with an emphasis on resource estimation, system sizing, and economic evaluation.

Since some students may not have had any electrical engineering background, the first chapter introduces the basic concepts of electricity and magnetism needed to understand electric circuits. And, since most students, including many who have had a good first course in electrical engineering, have not been exposed to anything related to electric power, a practical orientation to such topics as power factor, transmission lines, three-phase power, power supplies, and power quality is given in the second chapter.

Chapter 3 gives an overview of the development of today's electric power industry, including the regulatory and historical evolution of the industry as well as the technical side of power generation. Included is enough thermodynamics to understand basic heat engines and how that all relates to modern steam-cycle, gas-turbine, combined-cycle, and cogeneration power plants. A first-cut at evaluating the most cost-effective combination of these various types of power plants in an electric utility system is also presented.

The transition from large, central power stations to smaller distributed generation systems is described in Chapter 4. The chapter emphasizes combined heat and power systems and introduces an array of small, efficient technologies, including reciprocating internal combustion engines, microturbines, Stirling engines, concentrating solar power dish and trough systems, micro-hydropower, and biomass systems for electricity generation. Special attention is given to understanding the physics of fuel cells and their potential to become major power conversion systems for the future.

The concept of distributed resources, on both sides of the electric meter, is introduced in Chapter 5 with a special emphasis on techniques for evaluating the economic attributes of the technologies that can most efficiently utilize these resources. Energy conservation supply curves on the demand side, along with the economics of cogeneration on the supply side, are presented. Careful attention is given to assessing the economic and environmental benefits of utilizing waste heat and the technologies for converting it to useful energy services such as air conditioning.

Chapter 6 is entirely on wind power. Wind turbines have become the most cost-effective renewable energy systems available today and are now completely competitive with essentially all conventional generation systems. The chapter develops techniques for evaluating the power available in the wind and how efficiently it can be captured and converted to electricity in modern wind turbines. Combining wind statistics with turbine characteristics makes it possible to estimate the energy and economics of systems ranging from a single, home-size wind turbine to large wind farms of the sort that are being rapidly built across the United States, Europe, and Asia.

Given the importance of the sun as a source of renewable energy, Chapter 7 develops a rather complete set of equations that can be used to estimate the solar resource available on a clear day at any location and time on earth. Data for actual solar energy at sites across the United States are also presented, and techniques for utilizing that data for preliminary solar systems design are presented.

Chapters 8 and 9 provide a large block of material on the conversion of solar energy into electricity using photovoltaics (PVs). Chapter 8 describes the basic physics of PVs and develops equivalent circuit models that are useful for understanding their electrical behavior. Chapter 9 is a very heavily design-oriented approach to PV systems, with an emphasis on grid-connected, rooftop designs, off-grid stand-alone systems, and PV water-pumping systems.

I think it is reasonable to say this book has been in the making for over three and one-half decades, beginning with the impact that Denis Hayes and Earth Day 1970 had in shifting my career from semiconductors and computer logic to environmental engineering. Then it was Amory Lovins' groundbreaking paper "The Soft Energy Path: The Road Not Taken?" (Foreign Affairs, 1976) that focused my attention on the relationship between energy and environment and the important roles that renewables and efficiency must play in meeting the coming challenges. The penetrating analyses of Art Rosenfeld at the University of California, Berkeley, and the astute political perspectives of Ralph Cavanagh at the Natural Resources Defense Council have been constant sources of guidance and inspiration. These and other trailblazers have illuminated the path, but it has been the challenging, committed, enthusiastic students in my Stanford classes who have kept me invigorated, excited and energized over the years, and I am deeply indebted to them for their stimulation and friendship.

I specifically want to thank Joel Swisher at the Rocky Mountain Institute for help with the material on distributed generation, Jon Koomey at Lawrence Berkeley National Laboratory for reviewing the sections on combined heat and power and Eric Youngren of Rainshadow Solar for his demonstrations of microhydro power and photovoltaic systems in the field. I especially want to thank Bryan Palmintier for his careful reading of the manuscript and the many suggestions he made to improve its readability and accuracy. Finally, I raise my glass, as we do each evening, to my wife, Mary, who helps the sun rise every day of my life.

GILBERT M. MASTERS

Orcas, Washington
April 2004

Renewable and Efficient Electric Power Systems

CHAPTER 1

BASIC ELECTRIC AND MAGNETIC CIRCUITS

1.1 INTRODUCTION TO ELECTRIC CIRCUITS

In elementary physics classes you undoubtedly have been introduced to the fundamental concepts of electricity and how real components can be put together to form an electrical circuit. A very simple circuit, for example, might consist of a battery, some wire, a switch, and an incandescent lightbulb as shown in Fig. 1.1. The battery supplies the energy required to force electrons around the loop, heating the filament of the bulb and causing the bulb to radiate a lot of heat and some light. Energy is transferred from a *source*, the battery, to a *load*, the bulb. You probably already know that the *voltage* of the battery and the electrical *resistance* of the bulb have something to do with the amount of *current* that will flow in the circuit. From your own practical experience you also know that no current will flow until the switch is closed. That is, for a circuit to do anything, the loop has to be completed so that electrons can flow from the battery to the bulb and then back again to the battery. And finally, you probably realize that it doesn't much matter whether there is one foot or two feet of wire connecting the battery to the bulb, but that it probably would matter if there is a mile of wire between it and the bulb.

Also shown in Fig. 1.1 is a model made up of *idealized* components. The battery is modeled as an ideal source that puts out a constant voltage, V_B, no matter what amount of current, i, is drawn. The wires are considered to be perfect

Renewable and Efficient Electric Power Systems. By Gilbert M. Masters
ISBN 0-471-28060-7

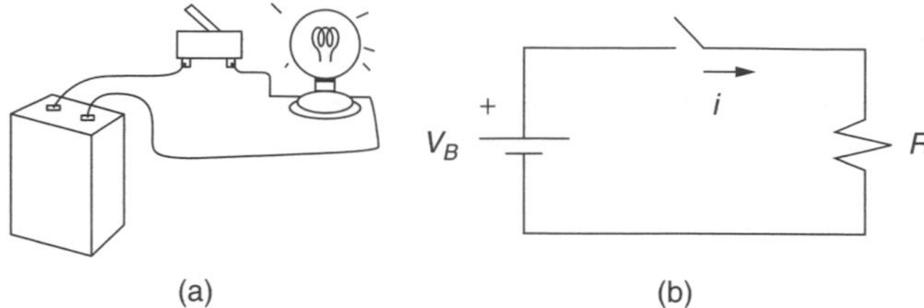

Figure 1.1 (a) A simple circuit. (b) An idealized representation of the circuit.

conductors that offer no resistance to current flow. The switch is assumed to be open or closed. There is no arcing of current across the gap when the switch is opened, nor is there any bounce to the switch as it makes contact on closure. The lightbulb is modeled as a simple resistor, R, that never changes its value, no matter how hot it becomes or how much current is flowing through it.

For most purposes, the idealized model shown in Fig. 1.1b is an adequate representation of the circuit; that is, our prediction of the current that will flow through the bulb whenever the switch is closed will be sufficiently accurate that we can consider the problem solved. There may be times, however, when the model is inadequate. The battery voltage, for example, may drop as more and more current is drawn, or as the battery ages. The lightbulb's resistance may change as it heats up, and the filament may have a bit of inductance and capacitance associated with it as well as resistance so that when the switch is closed, the current may not jump instantaneously from zero to some final, steady-state value. The wires may be undersized, and some of the power delivered by the battery may be lost in the wires before it reaches the load. These subtle effects may or may not be important, depending on what we are trying to find out and how accurately we must be able to predict the performance of the circuit. If we decide they are important, we can always change the model as necessary and then proceed with the analysis.

The point here is simple. The combinations of resistors, capacitors, inductors, voltage sources, current sources, and so forth, that you see in a circuit diagram are merely models of real components that comprise a real circuit, and a certain amount of judgment is required to decide how complicated the model must be before sufficiently accurate results can be obtained. For our purposes, we will be using very simple models in general, leaving many of the complications to more advanced textbooks.

1.2 DEFINITIONS OF KEY ELECTRICAL QUANTITIES

We shall begin by introducing the fundamental electrical quantities that form the basis for the study of electric circuits.

1.2.1 Charge

An atom consists of a positively charged nucleus surrounded by a swarm of negatively charged electrons. The charge associated with one electron has been found to be 1.602×10^{-19} coulombs; or, stated the other way around, one coulomb can be defined as the charge on 6.242×10^{18} electrons. While most of the electrons associated with an atom are tightly bound to the nucleus, good conductors, like copper, have *free electrons* that are sufficiently distant from their nuclei that their attraction to any particular nucleus is easily overcome. These conduction electrons are free to wander from atom to atom, and their movement constitutes an electric current.

1.2.2 Current

In a wire, when one coulomb's worth of charge passes a given spot in one second, the current is defined to be one *ampere* (abbreviated A), named after the nineteenth-century physicist André Marie Ampère. That is, current i is the net rate of flow of charge q past a point, or through an area:

$$i = \frac{dq}{dt} \tag{1.1}$$

In general, charges can be negative or positive. For example, in a neon light, positive ions move in one direction and negative electrons move in the other. Each contributes to current, and the total current is their sum. By convention, the direction of current flow is taken to be the direction that positive charges would move, whether or not positive charges happen to be in the picture. Thus, in a wire, electrons moving to the right constitute a current that flows to the left, as shown in Fig. 1.2.

When charge flows at a steady rate in one direction only, the current is said to be *direct current*, or *dc*. A battery, for example, supplies direct current. When charge flows back and forth sinusoidally, it is said to be *alternating current*, or *ac*. In the United States the ac electricity delivered by the power company has a frequency of 60 cycles per second, or 60 hertz (abbreviated Hz). Examples of ac and dc are shown in Fig. 1.3.

1.2.3 Kirchhoff's Current Law

Two of the most fundamental properties of circuits were established experimentally a century and a half ago by a German professor, Gustav Robert Kirchhoff (1824–1887). The first property, known as Kirchhoff's current law (abbreviated

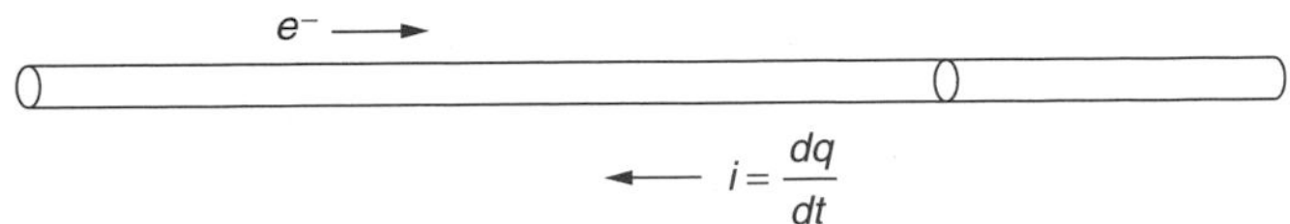

Figure 1.2 By convention, negative charges moving in one direction constitute a positive current flow in the opposite direction.

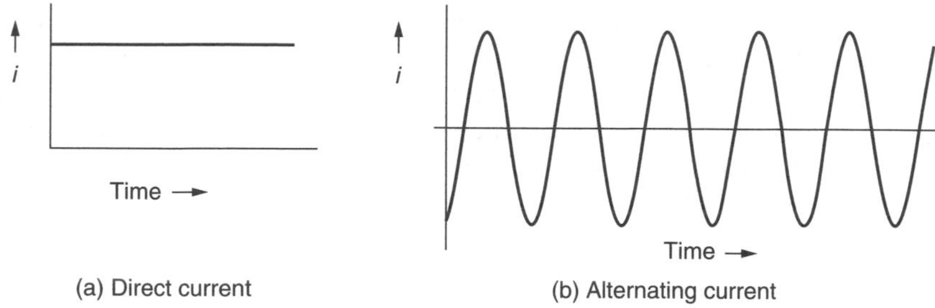

Figure 1.3 (a) Steady-state direct current (dc). (b) Alternating current (ac).

KCL), states that *at every instant of time the sum of the currents flowing into any node of a circuit must equal the sum of the currents leaving the node*, where a node is any spot where two or more wires are joined. This is a very simple, but powerful concept. It is intuitively obvious once you assert that current is the flow of charge, and that charge is conservative—neither being created nor destroyed as it enters a node. Unless charge somehow builds up at a node, which it does not, then the rate at which charge enters a node must equal the rate at which charge leaves the node.

There are several alternative ways to state Kirchhoff's current law. The most commonly used statement says that the sum of the currents into a node is zero as shown in Fig. 1.4a, in which case some of those currents must have negative values while some have positive values. Equally valid would be the statement that the sum of the currents leaving a node must be zero as shown in Fig. 1.4b (again some of these currents need to have positive values and some negative). Finally, we could say that the sum of the currents entering a node equals the sum of the currents leaving a node (Fig. 1.4c). These are all equivalent as long as we understand what is meant about the direction of current flow when we indicate it with an arrow on a circuit diagram. Current that actually flows in the direction shown by the arrow is given a positive sign. Currents that actually flow in the opposite direction have negative values.

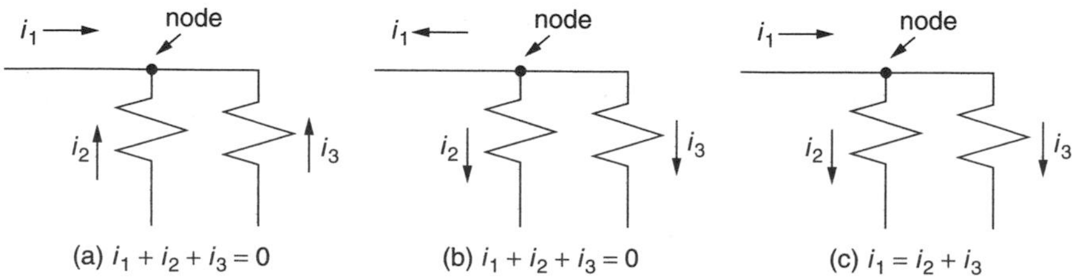

Figure 1.4 Illustrating various ways that Kirchhoff's current law can be stated. (a) The sum of the currents into a node equals zero. (b) The sum of the currents leaving the node is zero. (c) The sum of the currents entering a node equals the sum of the currents leaving the node.

Note that you can draw current arrows in any direction that you want—that much is arbitrary—but once having drawn the arrows, you must then write Kirchhoff's current law in a manner that is consistent with your arrows, as has been done in Fig. 1.4. The algebraic solution to the circuit problem will automatically determine whether or not your arbitrarily determined directions for currents were correct.

Example 1.1 Using Kirchhoff's Current Law. A node of a circuit is shown with current direction arrows chosen arbitrarily. Having picked those directions, $i_1 = -5$ A, $i_2 = 3$ A, and $i_3 = -1$ A. Write an expression for Kirchhoff's current law and solve for i_4.

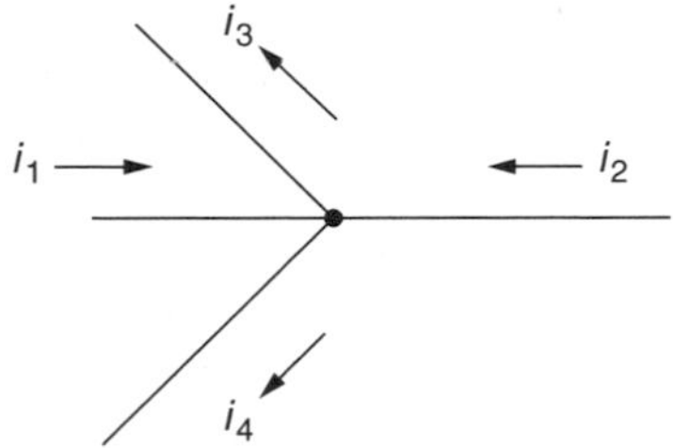

Solution. By Kirchhoff's current law,

$$i_1 + i_2 = i_3 + i_4$$

$$-5 + 3 = -1 + i_4$$

so that

$$i_4 = -1 \text{ A}$$

That is, i_4 is actually 1 A flowing into the node. Note that i_2, i_3, and i_4 are all entering the node, and i_1 is the only current that is leaving the node.

1.2.4 Voltage

Electrons won't flow through a circuit unless they are given some energy to help send them on their way. That "push" is measured in volts, where voltage is defined to be the amount of energy (w, joules) given to a unit of charge,

$$v = \frac{dw}{dq} \tag{1.2}$$

A 12-V battery therefore gives 12 joules of energy to each coulomb of charge that it stores. Note that the charge does not actually have to move for voltage to have meaning. Voltage describes the potential for charge to do work.

While currents are measured *through* a circuit component, voltages are measured *across* components. Thus, for example, it is correct to say that current through a battery is 10 A, while the voltage across that battery is 12 V. Other ways to describe the voltage across a component include whether the voltage rises across the component or drops. Thus, for example, for the simple circuit in Fig. 1.1, there is a voltage rise across the battery and voltage drop across the lightbulb.

Voltages are always measured with respect to something. That is, the voltage of the positive terminal of the battery is "so many volts" with respect to the negative terminal; or, the voltage at a point in a circuit is some amount with respect to some other point. In Fig. 1.5, current through a resistor results in a voltage drop from point A to point B of V_{AB} volts. V_A and V_B are the voltages at each end of the resistor, measured with respect to some other point.

The reference point for voltages in a circuit is usually designated with a *ground* symbol. While many circuits are actually grounded—that is, there is a path for current to flow directly into the earth—some are not (such as the battery, wires, switch, and bulb in a flashlight). When a ground symbol is shown on a circuit diagram, you should consider it to be merely a reference point at which the voltage is defined to be zero. Figure 1.6 points out how changing the node labeled as ground changes the voltages at each node in the circuit, but does not change the voltage drop across each component.

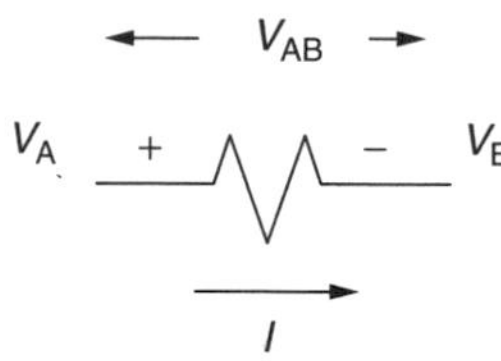

Figure 1.5 The voltage drop from point A to point B is V_{AB}, where $V_{AB} = V_A - V_B$.

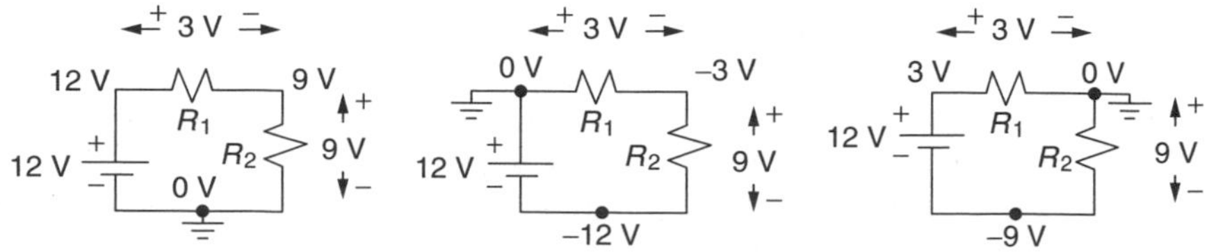

Figure 1.6 Moving the reference node around (ground) changes the voltages at each node, but doesn't change the voltage drop across each component.

1.2.5 Kirchhoff's Voltage Law

The second of Kirchhoff's fundamental laws states that *the sum of the voltages around any loop of a circuit at any instant is zero*. This is known as Kirchhoff's voltage law (KVL). Just as was the case for Kirchhoff's current law, there are alternative, but equivalent, ways of stating KVL. We can, for example, say that the sum of the voltage rises in any loop equals the sum of the voltage drops around the loop. Thus in Fig. 1.6, there is a voltage rise of 12 V across the battery and a voltage drop of 3 V across R_1 and a drop of 9 V across R_2. Notice that it doesn't matter which node was labeled ground for this to be true. Just as was the case with Kirchhoff's current law, we must be careful about labeling and interpreting the signs of voltages in a circuit diagram in order to write the proper version of KVL. A plus (+) sign on a circuit component indicates a reference direction under the assumption that the potential at that end of the component is higher than the voltage at the other end. Again, as long as we are consistent in writing Kirchhoff's voltage law, the algebraic solution for the circuit will automatically take care of signs.

Kirchhoff's voltage law has a simple mechanical analog in which weight is analogous to charge and elevation is analogous to voltage. If a weight is raised from one elevation to another, it acquires potential energy equal to the change in elevation times the weight. Similarly, the potential energy given to charge is equal to the amount of charge times the voltage to which it is raised. If you decide to take a day hike, in which you start and finish the hike at the same spot, you know that no matter what path was taken, when you finish the hike the sum of the increases in elevation has to have been equal to the sum of the decreases in elevation. Similarly, in an electrical circuit, no matter what path is taken, as long as you return to the same node at which you started, KVL provides assurance that the sum of voltage rises in that loop will equal the sum of the voltage drops in the loop.

1.2.6 Power

Power and *energy* are two terms that are often misused. Energy can be thought of as the ability to do work, and it has units such as joules or Btu. Power, on the other hand, is the *rate* at which energy is generated or used, and therefore it has rate units such as joules/s or Btu/h. There is often confusion about the units for electrical power and energy. Electrical power is measured in watts, which is a rate (1 J/s = 1 watt), so electrical energy is watts multiplied by time—for example, watt-hours. Be careful not to say "watts per hour," which is incorrect (even though you will see this all too often in newspapers or magazines).

When a battery delivers current to a load, power is generated by the battery and is dissipated by the load. We can combine (1.1) and (1.2) to find an expression for instantaneous power supplied, or consumed, by a component of a circuit:

$$p = \frac{dw}{dt} = \frac{dw}{dq} \cdot \frac{dq}{dt} = vi \tag{1.3}$$

Equation (1.3) tells us that the power supplied at any instant by a source, or consumed by a load, is given by the current through the component times the voltage across the component. When current is given in amperes, and voltage in volts, the units of power are *watts* (W). Thus, a 12-V battery delivering 10 A to a load is supplying 120 W of power.

1.2.7 Energy

Since power is the rate at which work is being done, and energy is the total amount of work done, energy is just the integral of power:

$$w = \int p\,dt \tag{1.4}$$

In an electrical circuit, energy can be expressed in terms of joules (J), where 1 watt-second = 1 joule. In the electric power industry the units of electrical energy are more often given in watt-hours, or for larger quantities kilowatt-hours (kWh) or megawatt-hours (MWh). Thus, for example, a 100-W computer that is operated for 10 hours will consume 1000 Wh, or 1 kWh of energy. A typical household in the United States uses approximately 750 kWh per month.

1.2.8 Summary of Principal Electrical Quantities

The key electrical quantities already introduced and the relevant relationships between these quantities are summarized in Table 1.1.

Since electrical quantities vary over such a large range of magnitudes, you will often find yourself working with very small quantities or very large quantities. For example, the voltage created by your TV antenna may be measured in millionths of a volt (microvolts, μV), while the power generated by a large power station may be measured in billions of watts, or gigawatts (GW). To describe quantities that may take on such extreme values, it is useful to have a system of prefixes that accompany the units. The most commonly used prefixes in electrical engineering are given in Table 1.2.

TABLE 1.1 Key Electrical Quantities and Relationships

Electrical Quantity	Symbol	Unit	Abbreviation	Relationship
Charge	q	coulomb	C	$q = \int i\,dt$
Current	i	ampere	A	$i = dq/dt$
Voltage	v	volt	V	$v = dw/dq$
Power	p	joule/second	J/s	$p = dw/dt$
		or watt	W	
Energy	w	joule	J	$w = \int p\,dt$
		or watt-hour	Wh	

TABLE 1.2 Common Prefixes

Small Quantities			Large Quantities		
Quantity	Prefix	Symbol	Quantity	Prefix	Symbol
10^{-3}	milli	m	10^3	kilo	k
10^{-6}	micro	μ	10^6	mega	M
10^{-9}	nano	n	10^9	giga	G
10^{-12}	pico	p	10^{12}	tera	T

1.3 IDEALIZED VOLTAGE AND CURRENT SOURCES

Electric circuits are made up of a relatively small number of different kinds of circuit *elements*, or *components*, which can be interconnected in an extraordinarily large number of ways. At this point in our discussion, we will concentrate on idealized characteristics of these circuit elements, realizing that real components resemble, but do not exactly duplicate, the characteristics that we describe here.

1.3.1 Ideal Voltage Source

An ideal voltage source is one that provides a given, known voltage v_s, no matter what sort of load it is connected to. That is, regardless of the current drawn from the ideal voltage source, it will always provide the same voltage. Note that an ideal voltage source does not have to deliver a constant voltage; for example, it may produce a sinusoidally varying voltage—the key is that that voltage is not a function of the amount of current drawn. A symbol for an ideal voltage source is shown in Fig. 1.7.

A special case of an ideal voltage source is an ideal battery that provides a constant dc output, as shown in Fig. 1.8. A real battery approximates the ideal source; but as current increases, the output drops somewhat. To account for that drop, quite often the model used for a real battery is an ideal voltage source in series with the internal resistance of the battery.

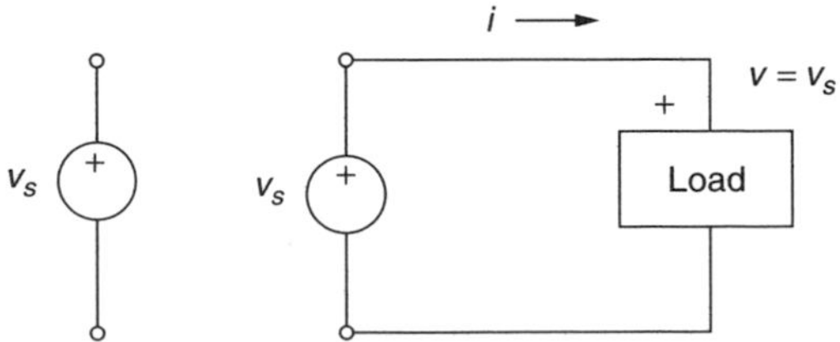

Figure 1.7 A constant voltage source delivers v_s no matter what current the load draws. The quantity v_s can vary with time and still be ideal.

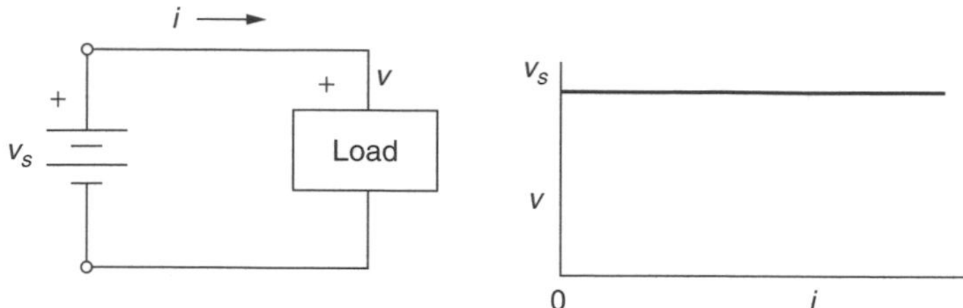

Figure 1.8 An ideal dc voltage.

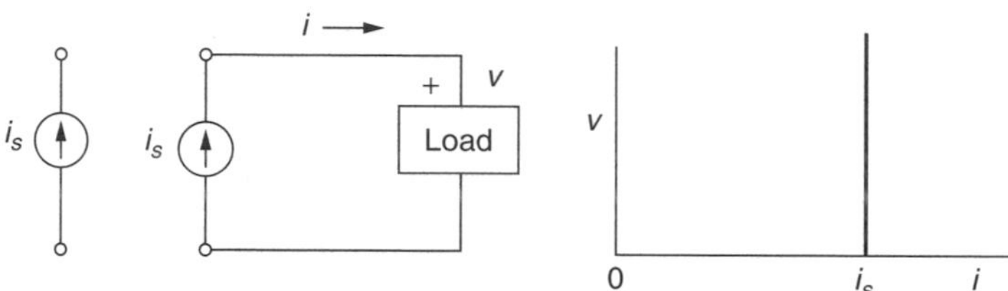

Figure 1.9 The current produced by an ideal current source does not depend on the voltage across the source.

1.3.2 Ideal Current Source

An ideal current source produces a given amount of current i_s no matter what load it sees. As shown in Fig. 1.9, a commonly used symbol for such a device is circle with an arrow indicating the direction of current flow. While a battery is a good approximation to an ideal voltage source, there is nothing quite so familiar that approximates an ideal current source. Some transistor circuits come close to this ideal and are often modeled with idealized current sources.

1.4 ELECTRICAL RESISTANCE

For an ideal *resistance* element the current through it is directly proportional to the voltage drop across it, as shown in Fig. 1.10.

1.4.1 Ohm's Law

The equation for an ideal resistor is given in (1.5) in which v is in volts, i is in amps, and the constant of proportionality is resistance R measured in ohms (Ω). This simple formula is known as *Ohm's law* in honor of the German physicist, Georg Ohm, whose original experiments led to this incredibly useful and important relationship.

$$v = Ri \tag{1.5}$$

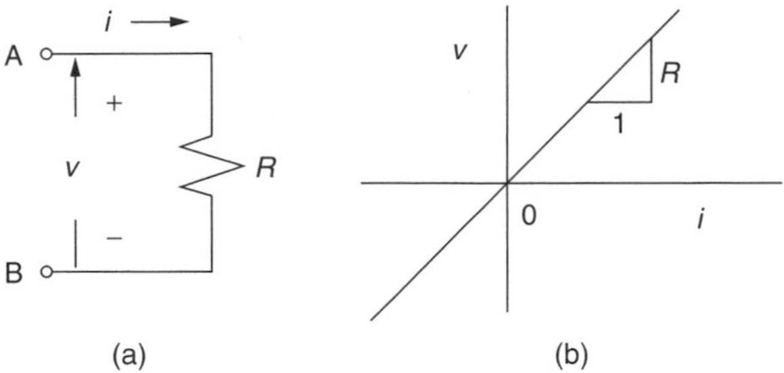

Figure 1.10 (a) An ideal resistor symbol. (b) voltage–current relationship.

Notice that voltage v is measured across the resistor. That is, it is the voltage at point A with respect to the voltage at point B. When current is in the direction shown, the voltage at A with respect to B is positive, so it is quite common to say that there is a *voltage drop* across the resistor.

An equivalent relationship for a resistor is given in (1.6), where current is given in terms of voltage and the proportionality constant is conductance G with units of siemens (S). In older literature, the unit of conductance was mhos.

$$i = Gv \tag{1.6}$$

By combining Eqs. (1.3) and (1.5), we can easily derive the following equivalent relationships for power dissipated by the resistor:

$$p = vi = i^2 R = \frac{v^2}{R} \tag{1.7}$$

Example 1.2 Power to an Incandescent Lamp. The current–voltage relationship for an incandescent lamp is nearly linear, so it can quite reasonably be modeled as a simple resistor. Suppose such a lamp has been designed to consume 60 W when it is connected to a 12-V power source. What is the resistance of the filament, and what amount of current will flow? If the actual voltage is only 11 V, how much energy would it consume over a 100-h period?

Solution. From Eq. (1.7),

$$R = \frac{v^2}{p} = \frac{12^2}{60} = 2.4\ \Omega$$

and from Ohm's law,

$$i = v/R = 12/2.4 = 5\ \text{A}$$

Connected to an 11-V source, the power consumed would be

$$p = \frac{v^2}{R} = \frac{11^2}{2.4} = 50.4 \text{ W}$$

Over a 100-h period, it would consume

$$w = pt = 50.4 \text{ W} \times 100 \text{ h} = 5040 \text{ Wh} = 5.04 \text{ kWh}$$

1.4.2 Resistors in Series

We can use Ohm's law and Kirchhoff's voltage law to determine the equivalent resistance of resistors wired in *series* (so the same current flows through each one) as shown in Fig. 1.11.

For R_s to be equivalent to the two series resistors, R_1 and R_2, the voltage–current relationships must be the same. That is, for the circuit in Fig. 1.11a,

$$v = v_1 + v_2 \tag{1.8}$$

and from Ohm's law,

$$v = iR_1 + iR_2 \tag{1.9}$$

For the circuit in Fig. 1.11b to be equivalent, the voltage and current must be the same:

$$v = iR_s \tag{1.10}$$

By equating Eqs. (1.9) and (1.10), we conclude that

$$R_s = R_1 + R_2 \tag{1.11}$$

And, in general, for n-resistances in series the equivalent resistance is

$$R_s = R_1 + R_2 + \cdots + R_n \tag{1.12}$$

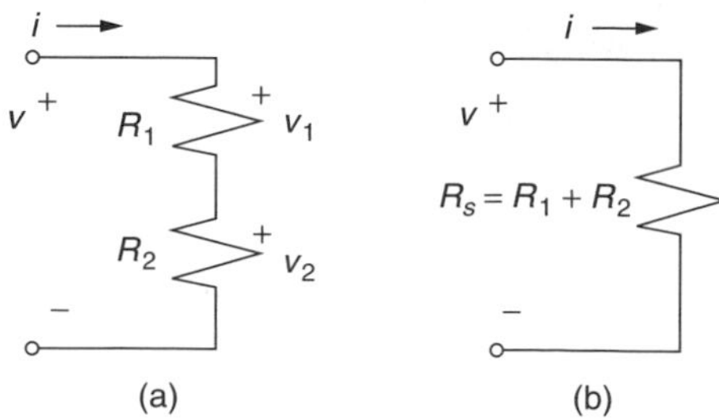

Figure 1.11 R_s is equivalent to resistors R_1 and R_2 in series.

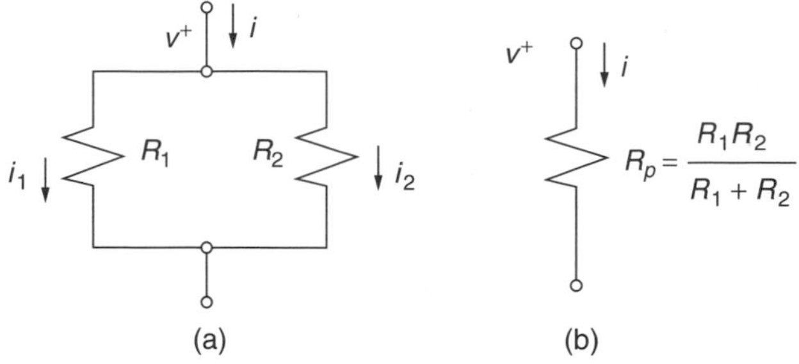

Figure 1.12 Equivalent resistance of resistors wired in parallel.

1.4.3 Resistors in Parallel

When circuit elements are wired together as in Fig. 1.12, so that the same voltage appears across each of them, they are said to be in *parallel*.

To find the equivalent resistance of two resistors in parallel, we can first incorporate Kirchhoff's current law followed by Ohm's law:

$$i = i_1 + i_2 = \frac{v}{R_1} + \frac{v}{R_2} = \frac{v}{R_p} \tag{1.13}$$

so that

$$\frac{1}{R_1} + \frac{1}{R_2} = \frac{1}{R_p} \quad \text{or} \quad G_1 + G_2 = G_p \tag{1.14}$$

Notice that one reason for introducing the concept of conductance is that the conductance of a parallel combination of n resistors is just the sum of the individual conductances.

For two resistors in parallel, the equivalent resistance can be found from Eq. (1.14) to be

$$R_p = \frac{R_1 R_2}{R_1 + R_2} \tag{1.15}$$

Notice that when R_1 and R_2 are of equal value, the resistance of the parallel combination is just one-half that of either one. Also, you might notice that the parallel combination of two resistors always has a lower resistance than either one of those resistors.

Example 1.3 Analyzing a Resistive Circuit. Find the equivalent resistance of the following network.

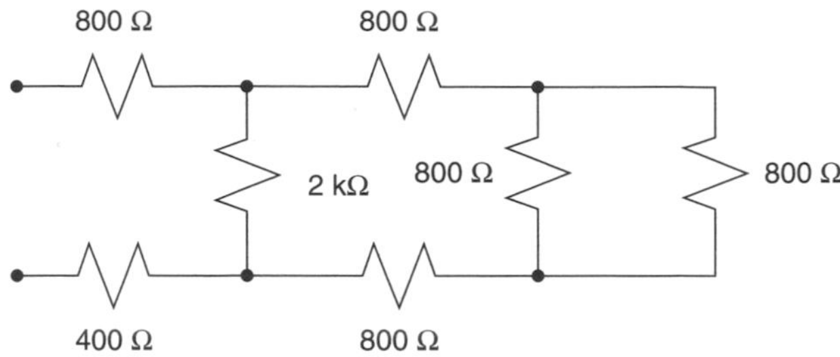

Solution. While this circuit may look complicated, you can actually work it out in your head. The parallel combination of the two 800-Ω resistors on the right end is 400 Ω, leaving the following equivalent:

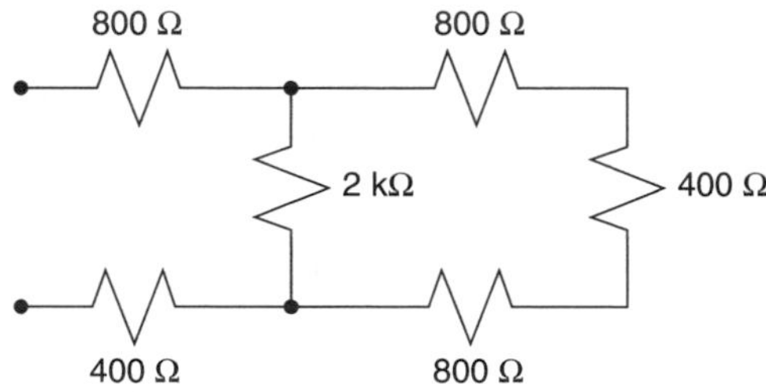

The three resistors on the right end are in series so they are equivalent to a single resistor of 2 kΩ (=800 Ω + 400 Ω + 800 Ω). The network now looks like the following:

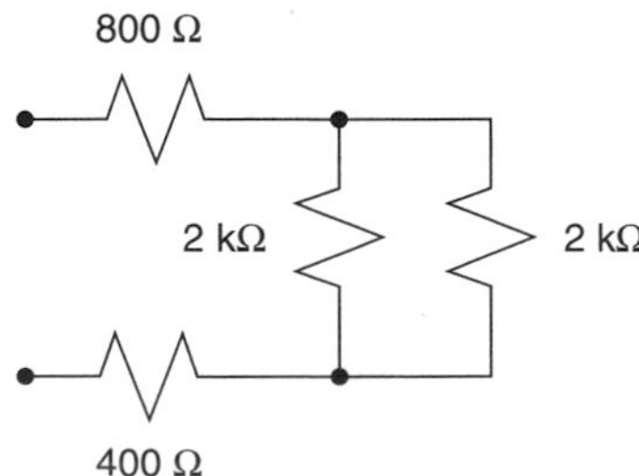

The two 2-kW resistors combine to 1 kΩ, which is in series with the 800-Ω and 400-Ω resistors. The total resistance of the network is thus 800 Ω + 1 kΩ + 400 Ω = 2.2 kΩ.

1.4.4 The Voltage Divider

A voltage divider is a deceptively simple, but surprisingly useful and important circuit. It is our first example of a two-port network. Two-port networks have a pair of input wires and a pair of output wires, as shown in Fig. 1.13.

The analysis of a voltage divider is a straightforward extension of Ohm's law and what we have learned about resistors in series.

As shown in Fig. 1.14, when a voltage source is connected to the voltage divider, an amount of current flows equal to

$$i = \frac{v_{\text{in}}}{R_1 + R_2} \tag{1.16}$$

Since $v_{\text{out}} = iR_2$, we can write the following voltage-divider equation:

$$v_{\text{out}} = v_{\text{in}} \left(\frac{R_2}{R_1 + R_2} \right) \tag{1.17}$$

Equation (1.17) is so useful that it is well worth committing to memory.

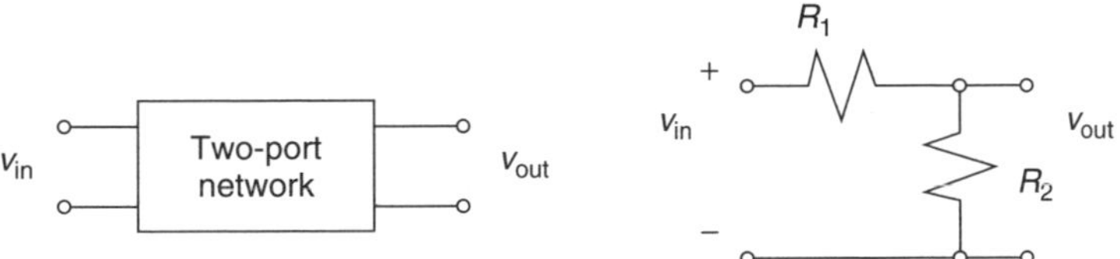

Figure 1.13 A voltage divider is an example of a two-port network.

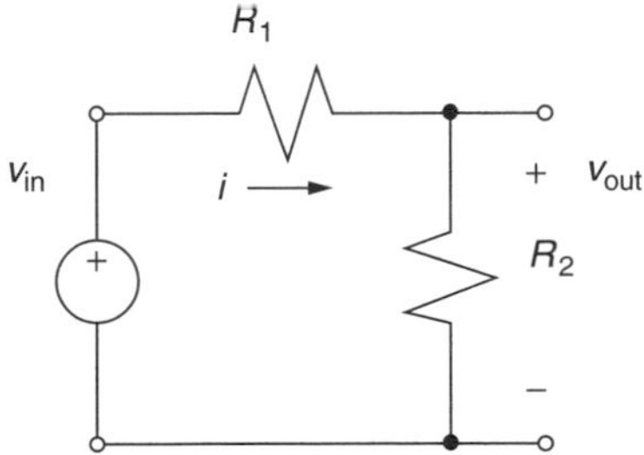

Figure 1.14 A voltage divider connected to an ideal voltage source.

Example 1.4 Analyzing a Battery as a Voltage Divider. Suppose an automobile battery is modeled as an ideal 12-V source in series with a 0.1-Ω internal resistance.

a. What would the battery output voltage drop to when 10 A is delivered?
b. What would be the output voltage when the battery is connected to a 1-Ω load?

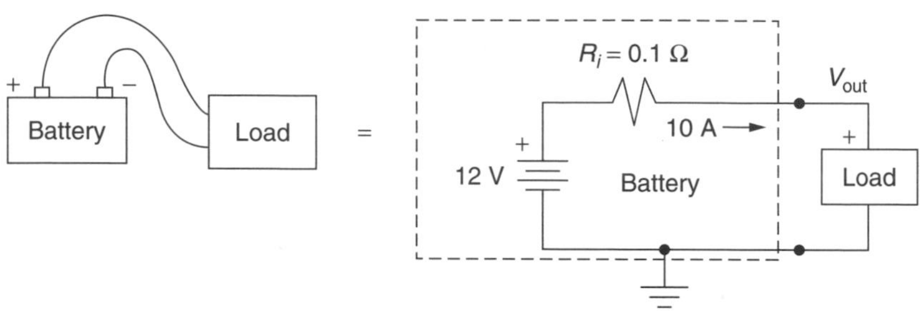

Solution

a. With the battery delivering 10 A, the output voltage drops to

$$V_{\text{out}} = \text{V}_B - IR_i = 12 - 10 \times 0.1 = 11\ \text{V}$$

b. Connected to a 1-Ω load, the circuit can be modeled as shown below:

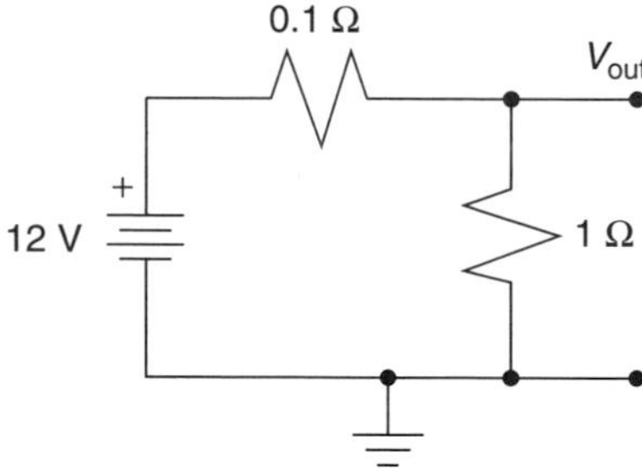

We can find V_{out} from the voltage divider relationship, (1.17):

$$v_{\text{out}} = v_{\text{in}}\left(\frac{R_2}{R_1 + R_2}\right) = 12\left(\frac{1.0}{0.1 + 1.0}\right) = 10.91\ \text{V}$$

1.4.5 Wire Resistance

In many circumstances connecting wire is treated as if it were perfect—that is, as if it had no resistance—so there is no voltage drop in those wires. In circuits delivering a fair amount of power, however, that assumption may lead to serious errors. Stated another way, an important part of the design of power circuits is choosing heavy enough wire to transmit that power without excessive losses. If connecting wire is too small, power is wasted and, in extreme cases, conductors can get hot enough to cause a fire hazard.

The resistance of wire depends primarily on its length, diameter, and the material of which it is made. Equation (1.18) describes the fundamental relationship for resistance (Ω):

$$R = \rho \frac{l}{A} \tag{1.18}$$

where ρ is the resistivity of the material, l is the wire length, and A is the wire cross-sectional area.

With l in meters (m) and A in m^2, units for resistivity ρ in the SI system are Ω-m (in these units copper has $\rho = 1.724 \times 10^{-8}$ Ω-m). The units often used in the United States, however, are tricky (as usual) and are based on areas expressed *in circular mils*. One circular mil is the area of a circle with diameter 0.001 in. (1 mil = 0.001 in.). So how can we determine the cross-sectional area of a wire (in circular mils) with diameter d (mils)? That is the same as asking how many 1-mil-diameter circles can fit into a circle of diameter d mils.

$$A = \frac{\dfrac{\pi}{4} d^2 \text{ sq mil}}{\dfrac{\pi}{4} \cdot 1^2 \text{ sq mil/cmil}} = d^2 \text{ cmil} \tag{1.19}$$

Example 1.5 From mils to Ohms. The resistivity of annealed copper at 20°C is 10.37 ohm-circular-mils/foot. What is the resistance of 100 ft of wire with diameter 80.8 mils (0.0808 in.)?

Solution

$$R = \rho \frac{l}{A} = 10.37\ \Omega - \text{cmil/ft} \cdot \frac{100 \text{ ft}}{(80.8)^2 \text{cmil}} = 0.1588\ \Omega$$

Electrical resistance of wire also depends somewhat on temperature (as temperature increases, greater molecular activity interferes with the smooth flow of electrons, thereby increasing resistance). There is also a phenomenon, called the *skin effect*, which causes wire resistance to increase with frequency. At higher frequencies, the inherent inductance at the core of the conductor causes current to flow less easily in the center of the wire than at the outer edge of conductor, thereby increasing the average resistance of the entire conductor. At 60 Hz, for modest loads (not utility power), the skin effect is insignificant. As to materials, copper is preferred, but aluminum, being cheaper, is sometimes used by professionals, but never in home wiring systems. Aluminum under pressure slowly

TABLE 1.3 Characteristics of Copper Wire

Wire Gage (AWG No.)	Diameter (inches)	Area cmils	Ohms per 100 ft[a]	Max Current (amps)
000	0.4096	168,000	0.0062	195
00	0.3648	133,000	0.0078	165
0	0.3249	106,000	0.0098	125
2	0.2576	66,400	0.0156	95
4	0.2043	41,700	0.0249	70
6	0.1620	26,300	0.0395	55
8	0.1285	16,500	0.0628	40
10	0.1019	10,400	0.0999	30
12	0.0808	6,530	0.1588	20
14	0.0641	4,110	0.2525	15

[a] dc, at 68°F.

deforms, which eventually loosens connections. That, coupled with the high-resistivity oxide that forms over exposed aluminum, can cause high enough I^2R losses to pose a fire hazard.

Wire size in the United States with diameter less than about 0.5 in. is specified by its American Wire Gage (AWG) number. The AWG numbers are based on wire resistance, which means that larger AWG numbers have higher resistance and hence smaller diameter. Conversely, smaller gage wire has larger diameter and, consequently, lower resistance. Ordinary house wiring is usually No. 12 AWG, which is roughly the diameter of the lead in an ordinary pencil. The largest wire designated with an AWG number is 0000, which is usually written 4/0, with a diameter of 0.460 in. For heavier wire, which is usually stranded (made up of many individual wires bundled together), the size is specified in the United States in thousands of circular mills (kcmil). For example, 1000-kcmil stranded copper wire for utility transmission lines is 1.15 in. in diameter and has a resistance of 0.076 ohms per mile. In countries using the metric system, wire size is simply specified by its diameter in millimeters. Table 1.3 gives some values of wire resistance, in ohms per 100 feet, for various gages of copper wire at 68°F. Also given is the maximum allowable current for copper wire clad in the most common insulation.

Example 1.6 Wire Losses. Suppose an ideal 12-V battery is delivering current to a 12-V, 100-W incandescent lightbulb. The battery is 50 ft from the bulb, and No. 14 copper wire is used. Find the power lost in the wires and the power delivered to the bulb.

Solution. The resistance, R_b, of a bulb designed to use 100 W when it is supplied with 12 V can be found from (1.7):

$$P = \frac{v^2}{R} \quad \text{so} \quad R_b = \frac{v^2}{P} = \frac{12^2}{100} = 1.44\ \Omega$$

From Table 1.3, 50 ft of 14 ga. wire has 0.2525 Ω/100 ft, so since we have 50 ft of wire to the bulb and 50 ft back again, the wire resistance is $R_w = 0.2525\ \Omega$. The circuit is as follows:

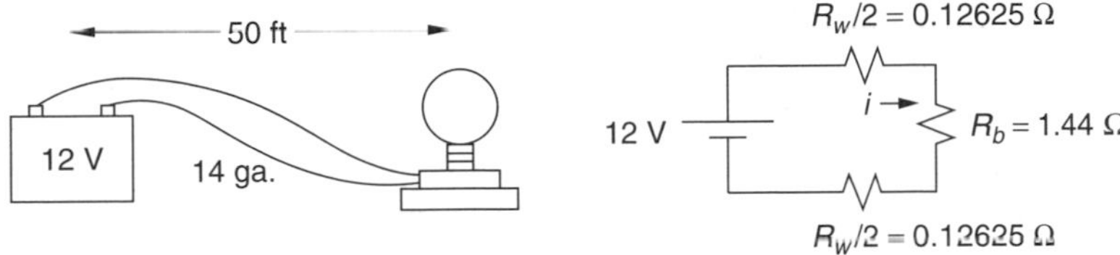

From Ohm's law, the current flowing in the circuit is

$$i = \frac{v}{R_{\text{tot}}} = \frac{12\ \text{V}}{(0.12625 + 0.12625 + 1.44)\ \Omega} = 7.09\ \text{A}$$

So, the power delivered to the lightbulb is

$$P_b = i^2 R_b = (7.09)^2 \cdot 1.44 = 72.4\ \text{W}$$

and the power lost in the wires is

$$P_w = i^2 R_w = (7.09)^2 \cdot 0.2525 = 12.7\ \text{W}$$

Notice that our bulb is receiving only 72.4 W instead of 100 W, so it will not be nearly as bright. Also note that the battery is delivering

$$\text{P}_{\text{battery}} = 72.4 + 12.7 = 85.1\ \text{W}$$

of which, quite a bit, about 15%, is lost in the wires (12.7/85.1 = 0.15).

Alternate Solution: Let us apply the concept of a voltage divider to solve this problem. We can combine the wire resistance going to the load with the wire resistance coming back, resulting in the simplified circuit model shown below:

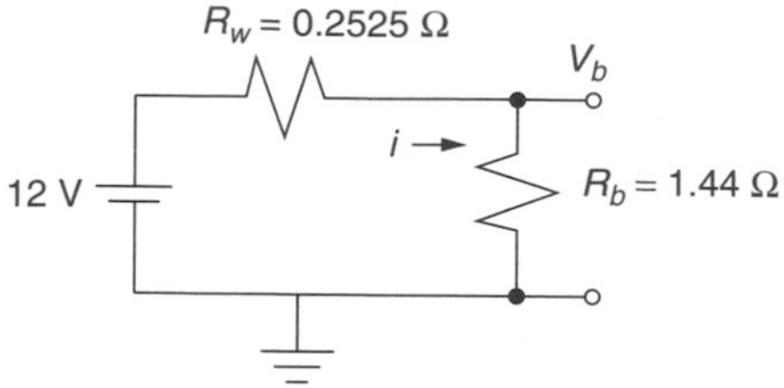

Using (1.17), the voltage delivered to the load (the lightbulb) is

$$v_{\text{out}} = v_{\text{in}}\left(\frac{R_2}{R_1 + R_2}\right) = 12\left(\frac{1.44}{0.2525 + 1.44}\right) = 10.21 \text{ V}$$

The 1.79-V difference between the 12 V supplied by the battery and the 10.21 V that actually appears across the load is referred to as the *voltage sag*.

Power lost in the wires is thus

$$P_w = \frac{V_w^2}{R_w} = \frac{(1.79)^2}{0.2525} = 12.7 \text{ W}$$

Example 1.6 illustrates the importance of the resistance of the connecting wires. We would probably consider 15% wire loss to be unacceptable, in which case we might want to increase the wire size (but larger wire is more expensive and harder to work with). If feasible, we could take the alternative approach to wire losses, which is to increase the supply voltage. Higher voltages require less current to deliver a given amount of power. Less current means less i^2R power losses in the wires as the following example demonstrates.

Example 1.7 Raising Voltage to Reduce Wire Losses Suppose a load that requires 120 W of power is located 50 ft from a generator. The load can be designed to operate at 12 V or 120 V. Using No. 14 wire, find the voltage sag and power losses in the connecting wire for each voltage.

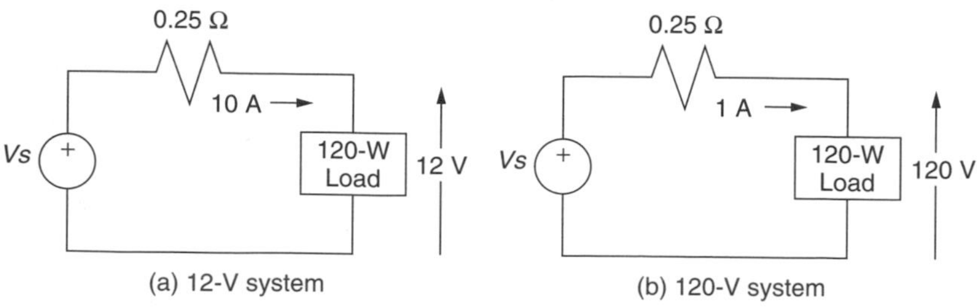

Solution. There are 100 ft of No. 14 wire (to the load and back) with total resistance of 0.2525 Ω (Table 1.3).

At 12 V: To deliver 120 W at 12 V requires a current of 10 A, so the voltage sag in the 0.2525-Ω wire carrying 10 A is

$$V_{\text{sag}} = iR = 10\ \text{A} \times 0.2525\ \Omega = 2.525\ \text{V}$$

The power loss in the wire is

$$P = i^2R = (10)^2 \times 0.2525 = 25.25\ \text{W}$$

That means the generator must provide $25.25 + 120 = 145.25$ W at a voltage of $12 + 2.525 = 14.525$ V. Wire losses are $25.25/145.25 = 0.174 = 17.4\%$ of the power generated. Such high losses are generally unacceptable.

At 120 V: The current required to deliver 120 W is only 1 A, which means the voltage drop in the connecting wire is only

$$\text{Voltage sag} = iR = 1\ \text{A} \times 0.2525\ \Omega = 0.2525\ \text{V}$$

The power loss in the wire is

$$P_w = i^2R = (1)^2 \times 0.2525 = 0.2525\ \text{W (1/100th that of the 12-V system)}$$

The source must provide 120 W + 0.2525 W = 120.2525 W, of which the wires will lose only 0.21%.

Notice that i^2R power losses in the wires are 100 times larger in the 12-V circuit, which carries 10 A, than they are in the 120-V circuit carrying only 1 A. That is, increasing the voltage by a factor of 10 causes line losses to decrease by a factor of 100, which is why electric power companies transmit their power at such high voltages.

1.5 CAPACITANCE

Capacitance is a parameter in electrical circuits that describes the ability of a circuit component to store energy in an electrical field. Capacitors are discrete components that can be purchased at the local electronics store, but the capacitance effect can occur whenever conductors are in the vicinity of each other. A capacitor can be as simple as two parallel conducting plates (Fig. 1.15), separated by a nonconducting dielectric such as air or even a thin sheet of paper.

If the surface area of the plates is large compared to their separation, the capacitance is given by

$$C = \varepsilon \frac{A}{d} \quad \text{farads} \tag{1.20}$$

where C is capacitance (farads, F), ε is permittivity (F/m), A is area of one plate (m^2), and d is separation distance (m).

Example 1.8 Capacitance of Two Parallel Plates. Find the capacitance of two 0.5-m^2 parallel conducting plates separated by 0.001 m of air with permittivity 8.8×10^{-12} F/m.

Solution

$$C = 8.8 \times 10^{-12}\ \text{F/m} \cdot \frac{0.5\ \text{m}^2}{0.001\ \text{m}} = 4.4 \times 10^{-9}\ \text{F} = 0.0044\ \mu\text{F} = 4400\ \text{pF}$$

Notice even with the quite large plate area in the example, the capacitance is a very small number. In practice, to achieve large surface area in a small volume, many capacitors are assembled using two flexible sheets of conductor, separated by a dielectric, rolled into a cylindrical shape with connecting leads attached to each plate.

Capacitance values in electronic circuits are typically in the microfarad (10^{-6} F $= \mu$F) to picofarad ($10^{-12} =$ pF) range. Capacitors used in utility power systems are much larger, and are typically in the millifarad range. Later, we will see how a different unit of measure, the kVAR, will be used to characterize the size of large, power-system capacitors.

While Eq. (1.20) can be used to determine the capacitance from physical characteristics, of greater importance is the relationship between voltage, current, and capacitance. As suggested in Fig. 1.15, when charge q builds up on the

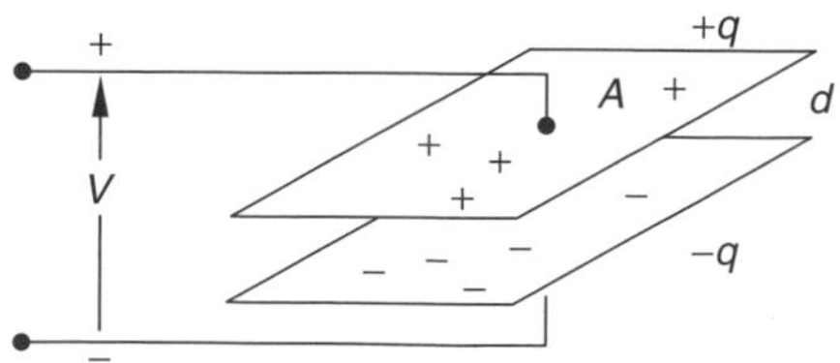

Figure 1.15 A capacitor can consist of two parallel, charged plates separated by a dielectric.

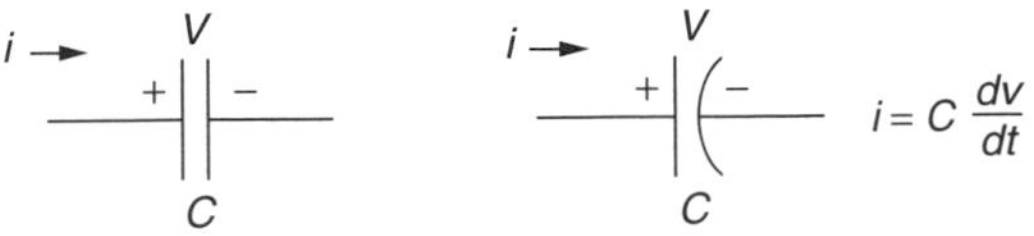

Figure 1.16 Two symbols for capacitors.

plates of a capacitor, a voltage v is created across the capacitor. This leads to the fundamental definition of capacitance, which is that capacitance is equal to the amount of charge required to create a 1-V potential difference between the plates:

$$C(\text{farads}) = \frac{q(\text{coulombs})}{v(\text{volts})} \tag{1.21}$$

Since current is the rate at which charge is added to the plates, we can rearrange (1.21) and then take the derivative to get

$$i = \frac{dq}{dt} = C\frac{dv}{dt} \tag{1.22}$$

The circuit symbol for a capacitor is usually drawn as two parallel lines, as shown in Fig. 1.16a, but you may also encounter the symbol shown in Fig. 1.16b. Sometimes, the term *condenser* is used for capacitors, as is the case in automobile ignition systems.

From the defining relationship between current and voltage (1.22), it can be seen that if voltage is not changing, then current into the capacitor has to be zero. That is, under dc conditions, the capacitor appears to be an open circuit, through which no current flows.

$$\text{dc:} \qquad \frac{dv}{dt} = 0, \quad i = 0, \tag{1.23}$$

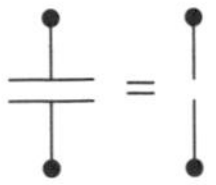

Kirchhoff's current and voltage laws can be used to determine that the capacitance of two capacitors in parallel is the sum of their capacitances and that the capacitance of two capacitors in series is equal to the product of the two over the sum, as shown in Fig. 1.17.

Another important characteristic of capacitors is their ability to store energy in the form of an electric field created between the plates. Since power is the rate of change of energy, we can write that energy is the integral of power:

$$W_c = \int P\,dt = \int vi\,dt = \int vC\frac{dv}{dt}\,dt = C\int v\,dv$$

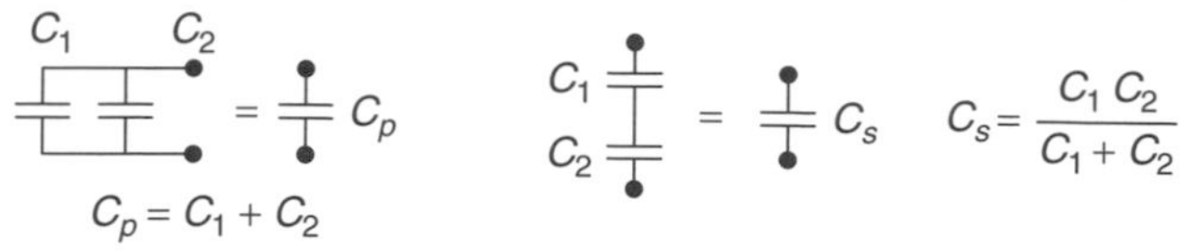

Figure 1.17 Capacitors in series and capacitors in parallel.

So, we can write that the energy stored in the electric field of a capacitor is

$$W_c = \tfrac{1}{2}Cv^2 \tag{1.24}$$

One final property of capacitors is that the *voltage across a capacitor cannot be changed instantaneously*. To change voltage instantaneously, charge would have to move from one plate, around the circuit, and back to the other plate in zero time. To see this conclusion mathematically, write power as the rate of change of energy,

$$P = \frac{dW}{dt} = \frac{d}{dt}\left(\frac{1}{2}Cv^2\right) = Cv\frac{dv}{dt} \tag{1.25}$$

and then note that if voltage could change instantaneously, dv/dt would be infinite, and it would therefore take infinite power to cause that change, which is impossible—hence, the conclusion that voltage cannot change instantaneously. An important practical application of this property will be seen when we look at rectifiers that convert ac to dc. Capacitors resist rapid changes in voltages and are used to smooth the dc voltage produced from such dc power supplies. In power systems, capacitors have a number of other uses that will be explored in the next chapter.

1.6 MAGNETIC CIRCUITS

Before we can introduce inductors and transformers, we need to understand the basic concept of electromagnetism. The simple notions introduced here will be expanded in later chapters when electric power quality (especially harmonic distortion), motors and generators, and fluorescent ballasts are covered.

1.6.1 Electromagnetism

Electromagnetic phenomena were first observed and quantified in the early nineteenth century—most notably, by three European scientists: Hans Christian Oersted, André Marie Ampère, and Michael Faraday. Oersted observed that a wire carrying current could cause a magnet suspended nearby to move. Ampère, in

1825, demonstrated that a wire carrying current could exert a force on another wire carrying current in the opposite direction. And Faraday, in 1831, discovered that current could be made to flow in a coil of wire by passing a magnet close to the circuit. These experiments provided the fundamental basis for the development of all electromechanical devices, including, most importantly, motors and generators.

What those early experiments established was that electrical current flowing along a wire creates a magnetic field around the wire, as shown in Fig. 1.18a. That magnetic field can be visualized by showing lines of magnetic flux, which are represented with the symbol ϕ. The direction of that field that can be determined using the "right hand rule" in which you imagine wrapping your right hand around a wire, with your thumb pointing in the direction of current flow. Your fingers then show the direction of the magnetic field. The field created by a coil of wire is suggested in Fig. 1.18b.

Consider an iron core wrapped with N turns of wire carrying current i as shown in Fig. 1.19. The magnetic field formed by the coil will take the path of least resistance—which is through the iron—in much the same way that electric current stays within a copper conductor. In essence, the iron is to a magnetic field what a wire is to current.

What Faraday discovered is that current flowing through the coil not only creates a magnetic field in the iron, it also creates a voltage across the coil that is proportional to the rate of change of magnetic flux ϕ in the iron. That voltage is called an electromotive force, or emf, and is designated by the symbol e.

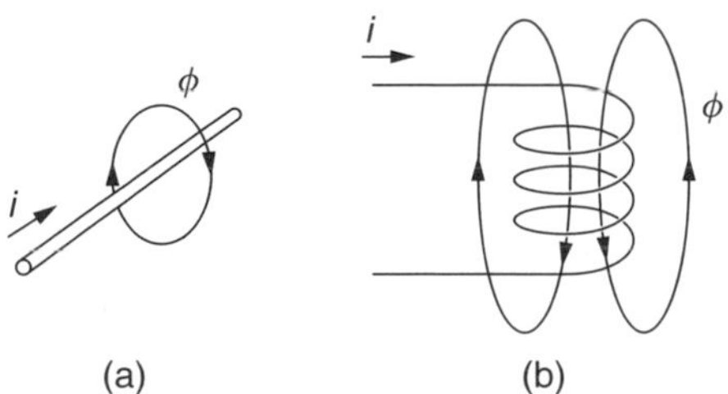

Figure 1.18 A magnetic field is formed around a conductor carrying current.

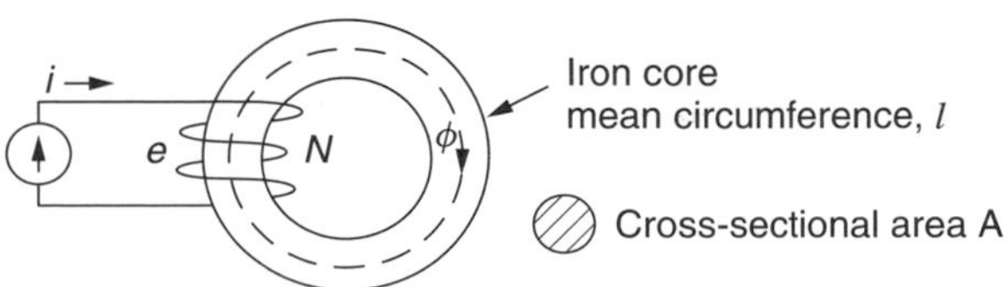

Figure 1.19 Current in the N-turn winding around an iron core creates a magnetic flux ϕ. An electromotive force (voltage) e is induced in the coil proportional to the rate of change of flux.

Assuming that all of the magnetic flux ϕ links all of the turns of the coil, we can write the following important relationship, which is known as *Faraday's law of electromagnetic induction*:

$$e = N\frac{d\phi}{dt} \tag{1.26}$$

The sign of the induced emf is always in a direction that opposes the current that created it, a phenomenon referred to as *Lenz's law*.

1.6.2 Magnetic Circuits

Magnetic phenomena are described using a fairly large number of terms that are often, at first, somewhat difficult to keep track of. One approach that may help is to describe analogies between electrical circuits, which are usually more familiar, and corresponding magnetic circuits. Consider the electrical circuit shown in Fig. 1.20a and the analogous magnetic circuit shown in Fig 1.20b. The electrical circuit consists of a voltage source, v, sending current i through an electrical load with resistance R. The electrical load consists of a long wire of length l, cross-sectional area A, and conductance ρ.

The resistance of the electrical load is given by (1.18):

$$R = \rho\frac{l}{A} \tag{1.18}$$

The current flowing in the electrical circuit is given by Ohm's law:

$$i = \frac{v}{R} \tag{1.5}$$

In the magnetic circuit of Fig. 1.20b, the driving force, analogous to voltage, is called the *magnetomotive force* (mmf), designated by $\mathfrak{F}$. The magnetomotive force is created by wrapping N turns of wire, carrying current i, around a toroidal

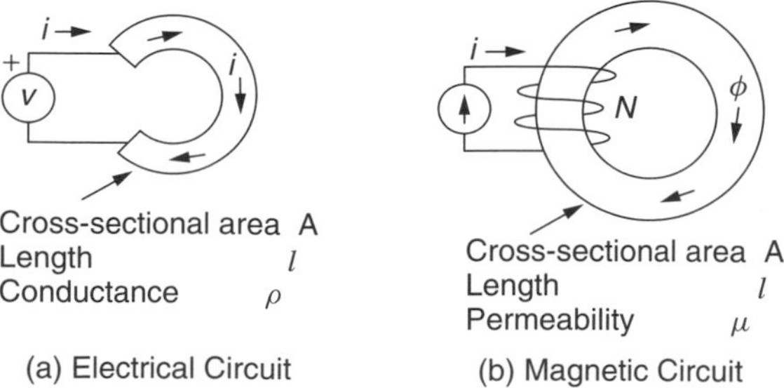

Figure 1.20 Analogous electrical and magnetic circuits.

core. By definition, the magnetomotive force is the product of *current* × *turns*, and has units of *ampere-turns*.

$$\textit{Magnetomotive force (mmf)}\mathfrak{F} = Ni \ \ (ampere - turns) \tag{1.27}$$

The response to that mmf (analogous to current in the electrical circuit) is creation of magnetic flux ϕ, which has SI units of *webers* (Wb). The magnetic flux is proportional to the mmf driving force and inversely proportional to a quantity called *reluctance* $\mathfrak{R}$, which is analogous to electrical resistance, resulting in the "Ohm's law" of magnetic circuits given by

$$\mathfrak{F} = \mathfrak{R}\,\phi \tag{1.28}$$

From (1.28), we can ascribe units for reluctance $\mathfrak{R}$ as amp-turns per weber (A-t/Wb).

Reluctance depends on the dimensions of the core as well as its materials:

$$reluctance = \mathfrak{R} = \frac{l}{\mu A} \qquad (A\text{-}t/Wb) \tag{1.29}$$

Notice the similarity between (1.29) and the equation for resistance given in (1.18).

The parameter in (1.29) that indicates how readily the core material accepts magnetic flux is the material's *permeability* μ. There are three categories of magnetic materials: *diamagnetic*, in which the material tends to exclude magnetic fields; *paramagnetic*, in which the material is slightly magnetized by a magnetic field; and *ferromagnetic*, which are materials that very easily become magnetized. The vast majority of materials do not respond to magnetic fields, and their permeability is very close to that of free space. The materials that readily accept magnetic flux—that is, ferromagnetic materials—are principally iron, cobalt, and nickel and various alloys that include these elements. The units of permeability are webers per amp-turn-meter (*Wb/A-t-m*).

The permeability of free space is given by

$$\text{Permeability of free space } \mu_0 = 4\pi \times 10^{-7} \text{ Wb/A-t-m} \tag{1.30}$$

Oftentimes, materials are characterized by their *relative permeability*, μ_r, which for ferromagnetic materials may be in the range of hundreds to hundreds of thousands. As will be noted later, however, the relative permeability is not a constant for a given material: It varies with the magnetic field intensity. In this regard, the magnetic analogy deviates from its electrical counterpart and so must be used with some caution.

$$\text{Relative permeability} = \mu_r = \frac{\mu}{\mu_0} \tag{1.31}$$

Another important quantity of interest in magnetic circuits is *the magnetic flux density*, B. As the name suggests, it is simply the "density" of flux given by the following:

$$\text{Magnetic flux density } B = \frac{\phi}{\text{A}} \text{ webers/m}^2 \text{ or teslas (T)} \tag{1.32}$$

When flux is given in webers (Wb) and area A is given in m^2, units for B are teslas (T). The analogous quantity in an electrical circuit would be the current density, given by

$$\text{Electric current density } J = \frac{i}{\text{A}} \tag{1.33}$$

The final magnetic quantity that we need to introduce is the *magnetic field intensity*, H. Referring back to the simple magnetic circuit shown in Fig. 1.20b, the magnetic field intensity is defined as the magnetomotive force (mmf) per unit of length around the magnetic loop. With N turns of wire carrying current i, the mmf created in the circuit is Ni ampere-turns. With l representing the mean path length for the magnetic flux, the magnetic field intensity is therefore

$$\text{Magnetic field intensity } H = \frac{Ni}{l} \text{ ampere-turns/meter} \tag{1.34}$$

An analogous concept in electric circuits is the electric field strength, which is voltage drop per unit of length. In a capacitor, for example, the intensity of the electric field formed between the plates is equal to the voltage across the plates divided by the spacing between the plates.

Finally, if we combine (1.27), (1.28), (1.29), (1.32), and (1.34), we arrive at the following relationship between magnetic flux density B and magnetic field intensity H:

$$B = \mu H \tag{1.35}$$

Returning to the analogies between the simple electrical circuit and magnetic circuit shown in Fig. 1.20, we can now identify equivalent circuits, as shown in Fig. 1.21, along with the analogs shown in Table 1.4.

TABLE 1.4 Analogous Electrical and Magnetic Circuit Quantities

Electrical	Magnetic	Magnetic Units
Voltage v	Magnetomotive force $\mathcal{F} = Ni$	Amp-turns
Current i	Magnetic flux ϕ	Webers Wb
Resistance R	Reluctance $\mathcal{R}$	Amp-turns/Wb
Conductivity $1/\rho$	Permeability μ	Wb/A-t-m
Current density J	Magnetic flux density B	$\text{Wb/m}^2 = \text{teslas T}$
Electric field E	Magnetic field intensity H	Amp-turn/m

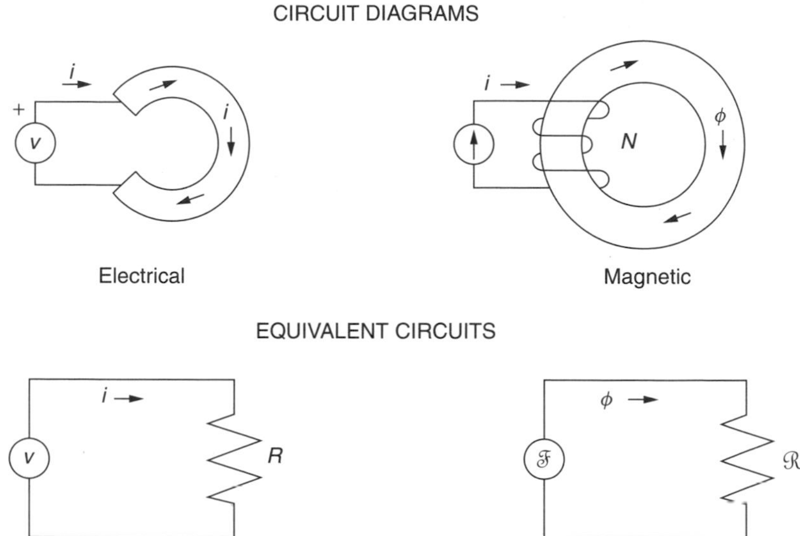

Figure 1.21 Equivalent circuits for the electrical and magnetic circuits shown.

1.7 INDUCTANCE

Having introduced the necessary electromagnetic background, we can now address inductance. Inductance is, in some sense, a mirror image of capacitance. While capacitors store energy in an electric field, inductors store energy in a magnetic field. While capacitors prevent voltage from changing instantaneously, inductors, as we shall see, prevent current from changing instantaneously.

1.7.1 Physics of Inductors

Consider a coil of wire carrying some current creating a magnetic field within the coil. As shown in Fig 1.22, if the coil has an air core, the flux can pretty much go where it wants to, which leads to the possibility that much of the flux will not link all of the turns of the coil. To help guide the flux through the coil, so that flux leakage is minimized, the coil might be wrapped around a ferromagnetic bar or ferromagnetic core as shown in Fig. 1.23. The lower reluctance path provided by the ferromagnetic material also greatly increases the flux ϕ.

We can easily analyze the magnetic circuit in which the coil is wrapped around the ferromagnetic core in Fig. 1.23a. Assume that all of the flux stays within the low-reluctance pathway provided by the core, and apply (1.28):

$$\phi = \frac{\mathcal{F}}{\mathcal{R}} = \frac{Ni}{\mathcal{R}} \tag{1.36}$$

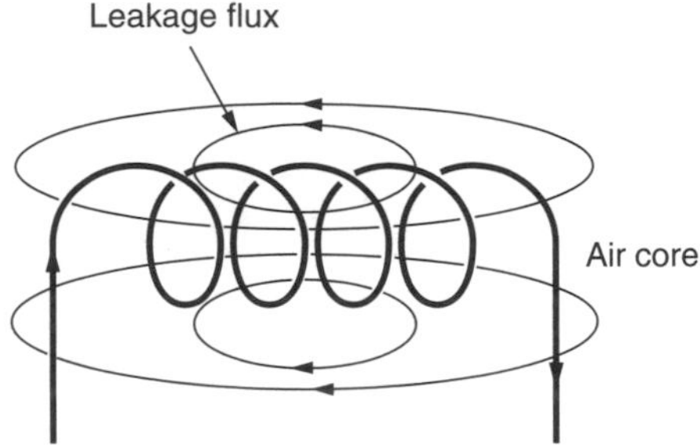

Figure 1.22 A coil with an air core will have considerable leakage flux.

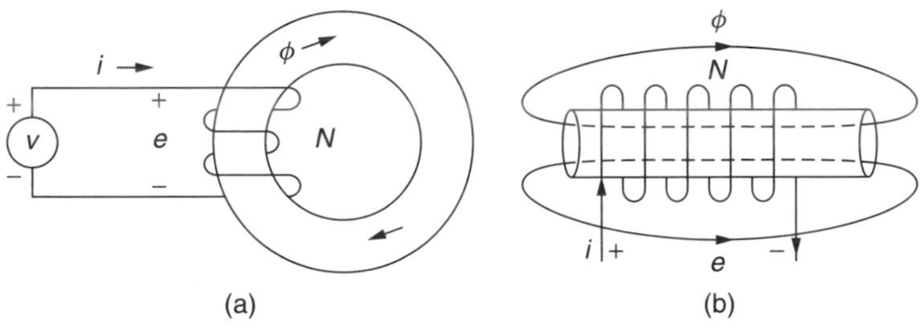

Figure 1.23 Flux can be increased and leakage reduced by wrapping the coils around a ferromagnetic material that provides a lower reluctance path. The flux will be much higher using the core (a) rather than the rod (b).

From Faraday's law (1.26), changes in magnetic flux create a voltage e, called the electromotive force (emf), across the coil equal to

$$e = N\frac{d\phi}{dt} \tag{1.26}$$

Substituting (1.36) into (1.26) gives

$$e = N\frac{d}{dt}\left(\frac{Ni}{\Re}\right) = \frac{N^2}{\Re}\frac{di}{dt} = L\frac{di}{dt} \tag{1.37}$$

where inductance L has been introduced and defined as

$$\text{Inductance } L = \frac{N^2}{\Re} \text{ henries} \tag{1.38}$$

Notice in Fig. 1.23a that a distinction has been made between e, the emf voltage induced across the coil, and v, a voltage that may have been applied to the circuit to cause the flux in the first place. If there are no losses in the connecting wires between the source voltage and the coil, then $e = v$ and we have the final defining relationship for an inductor:

$$v = L\frac{di}{dt} \tag{1.39}$$

As given in (1.38), inductance is inversely proportional to reluctance $\mathcal{R}$. Recall that the reluctance of a flux path through air is much greater than the reluctance if it passes through a ferromagnetic material. That tells us if we want a large inductance, the flux needs to pass through materials with high permeability (not air).

Example 1.9 Inductance of a Core-and-Coil. Find the inductance of a core with effective length $l = 0.1$ m, cross-sectional area $A = 0.001\ \text{m}^2$, and relative permeability μ_r somewhere between 15,000 and 25,000. It is wrapped with $N = 10$ turns of wire. What is the range of inductance for the core?

Solution. When the core's permeability is 15,000 times that of free space, it is

$$\mu_{\text{core}} = \mu_r\mu_0 = 15{,}000 \times 4\pi \times 10^{-7} = 0.01885\ \text{Wb/A-t-m}$$

so its reluctance is

$$\mathcal{R}_{\text{core}} = \frac{l}{\mu_{\text{core}}A} = \frac{0.1\ \text{m}}{0.01885\ (\text{Wb/A-t-m}) \times 0.001\ \text{m}^2} = 5305\ \text{A-t/Wb}$$

and its inductance is

$$L = \frac{N^2}{\mathcal{R}} = \frac{10^2}{5305} = 0.0188\ \text{henries} = 18.8\ \text{mH}$$

Similarly, when the relative permeability is 25,000 the inductance is

$$L = \frac{N^2}{\mathcal{R}} = \frac{N^2\mu_r\mu_0A}{l} = \frac{10^2 \times 25{,}000 \times 4\pi \times 10^{-7} \times 0.001}{0.1}$$
$$= 0.0314\ \text{H} = 31.4\ \text{mH}$$

The point of Example 1.9 is that the inductance of a coil of wire wrapped around a solid core can be quite variable given the imprecise value of the core's permeability. Its permeability depends on how hard the coil is driven by mmf so you can't just pick up an off-the-shelf inductor like this and know what its inductance is likely to be. The trick to getting a more precise value of inductance given the uncertainty in permeability is to sacrifice some amount of inductance by building into the core a small air gap. Another approach is to get the equivalent of an air gap by using a powdered ferromagnetic material in which the spaces between particles of material act as the air gap. The air gap reluctance, which is determined strictly by geometry, is large compared to the core reluctance so the impact of core permeability changes is minimized.

The following example illustrates the advantage of using an air gap to minimize the uncertainty in inductance. It also demonstrates something called *Ampère's circuital law*, which is the magnetic analogy to Kirchhoff's voltage law. That is, the rise in magnetomotive force (mmf) provided by N turns of wire carrying current i is equal to the sum of the mmf drops $\Re\,\phi$ around the magnetic loop.

Example 1.10 An Air Gap to Minimize Inductance Uncertainty. Suppose the core of Example 1.9 is built with a 0.001-m air gap. Find the range of inductances when the core's relative permeability varies between 15,000 and 25,000.

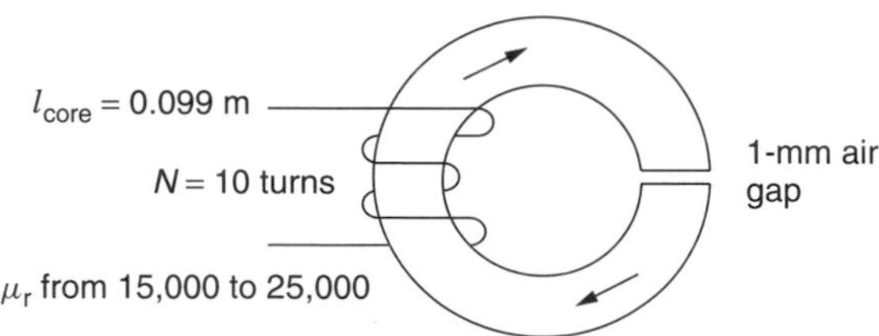

Solution. The reluctance of the ferromagnetic portion of the core when its relative permeability is 15,000 is

$$\Re_{core} = \frac{l_{core}}{\mu_{core}A} = \frac{0.099}{15{,}000 \times 4\pi \times 10^{-7} \times 0.001} = 5252 \text{ A-t/Wb}$$

And the air gap reluctance is

$$\Re_{air\ gap} = \frac{l_{air\ gap}}{\mu_0 A} = \frac{0.001}{4\pi \times 10^{-7} \times 0.001} = 795{,}775 \text{ A-t/Wb}$$

So the total reluctance of the series path consisting or core and air gap is

$$\Re_{Total} = 5252 + 795{,}775 = 801{,}027 \text{ A-t/Wb}$$

And the inductance is

$$L = \frac{N^2}{\Re} = \frac{10^2}{801{,}027} = 0.0001248 \text{ H} = 0.1248 \text{ mH}$$

When the core's relative permeability is 25,000, its reluctance is

$$\mathcal{R}_{\text{core}} = \frac{l_{\text{core}}}{\mu_{\text{core}} A} = \frac{0.099}{25{,}000 \times 4\pi \times 10^{-7} \times 0.001} = 3151 \text{ A-t/Wb}$$

And the new total inductance is

$$L = \frac{N^2}{\mathcal{R}} = \frac{10^2}{3151 + 795{,}775} = 0.0001251 \text{ H} = 0.1251 \text{ mH}$$

This is an insignificant change in inductance. A very precise inductance has been achieved at the expense of a sizable decrease in inductance compared to the core without an air gap.

1.7.2 Circuit Relationships for Inductors

From the defining relationship between voltage and current for an inductor (1.39), we can note that when current is not changing with time, the voltage across the inductor is zero. That is, for dc conditions an inductor looks the same as a short-circuit, zero-resistance wire:

$$\text{dc:}\ v = L\frac{di}{dt} = L \cdot 0 = 0 \tag{1.40}$$

When inductors are wired in series, the same current flows through each one so the voltage drop across the pair is simply:

$$v_{\text{series}} = L_1\frac{di}{dt} + L_2\frac{di}{dt} = (L_1 + L_2)\frac{di}{dt} = L_{\text{series}}\frac{di}{dt} \tag{1.41}$$

where L_{series} is the equivalent inductance of the two series inductors. That is,

$$L_{\text{series}} = L_1 + L_2 \tag{1.42}$$

Consider Fig. 1.24 for two inductors in parallel.

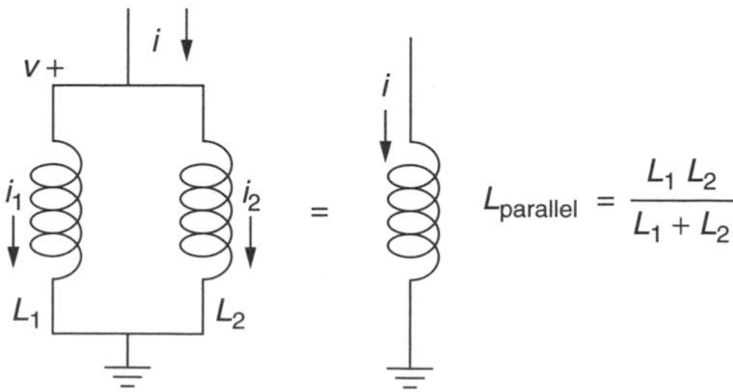

Figure 1.24 Two inductors in parallel.

The total current flowing is the sum of the currents:

$$i_{\text{parallel}} = i_1 + i_2 \tag{1.43}$$

The voltages are the same across each inductor, so we can use the integral form of (1.39) to get

$$\frac{1}{L_{\text{parallel}}} \int v\,dt = \frac{1}{L_1} \int v\,dt + \frac{1}{L_2} \int v\,dt \tag{1.44}$$

Dividing out the integral gives us the equation for inductors in parallel:

$$L_{\text{parallel}} = \frac{L_1\,L_2}{L_1 + L_2} \tag{1.45}$$

Just as capacitors store energy in their electric fields, inductors also store energy, but this time it is in their magnetic fields. Since energy W is the integral of power P, we can easily set up the equation for energy stored:

$$W_L = \int P\,dt = \int vi\,dt = \int \left(L\frac{di}{dt}\right) i\,dt = L \int i\,di \tag{1.46}$$

This leads to the following equation for energy stored in an inductor's magnetic field:

$$W_L = \tfrac{1}{2} L\,i^2 \tag{1.47}$$

If we use (1.47) to learn something about the power dissipated in an inductor, we get

$$P = \frac{dW}{dt} = \frac{d}{dt}\left(\frac{1}{2}Li^2\right) = Li\frac{di}{dt} \tag{1.48}$$

From (1.48) we can deduce another important property of inductors: *The current through an inductor cannot be changed instantaneously.* For current to change instantaneously, di/dt would be infinite, which (1.48) tells us would require infinite power, which is impossible. It takes time for the magnetic field, which is storing energy, to collapse. *Inductors, in other words, make current act like it has inertia.*

Now wait a minute. If current is flowing in the simple circuit containing an inductor, resistor, and switch shown in Fig 1.25, why can't you just open the switch and cause the current to stop instantaneously? Surely, it doesn't take infinite power to open a switch. The answer is that the current has to keep going for at least a short interval just after opening the switch. To do so, current momentarily must jump the gap between the contact points as the switch is

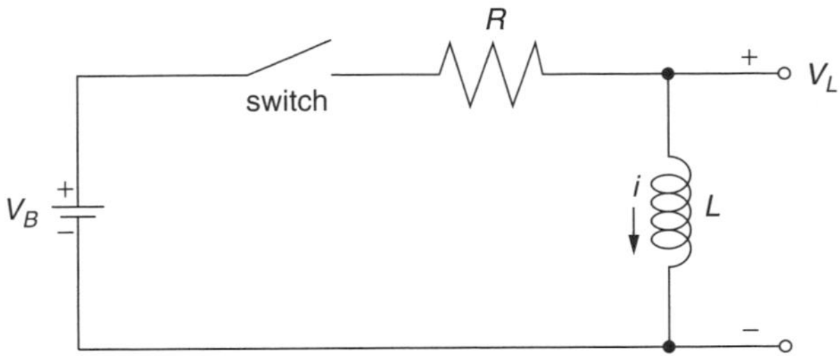

Figure 1.25 A simple R–L circuit with a switch.

opened. That is, the switch "arcs" and you get a little spark. Too much arc and the switch can be burned out.

We can develop an equation that describes what happens when an open switch in the R–L circuit of Fig. 1.25 is suddenly closed. Doing so gives us a little practice with Kirchhoff's voltage law. With the switch closed, the voltage rise due to the battery must equal the voltage drop across the resistance plus inductance:

$$V_B = iR + L\frac{di}{dt} \tag{1.49}$$

Without going through the details the solution to (1.49), subject to the initial condition that $i = 0$ at $t = 0$, is

$$i = \frac{V_B}{R}\left(1 - e^{-\frac{R}{L}t}\right) \tag{1.50}$$

Does this solution look right? At $t = 0$, $i = 0$, so that's OK. At $t = \infty$, $i = V_B/R$. That seems alright too since eventually the current reaches a steady-state, dc value, which means the voltage drop across the inductor is zero ($v_L = L\,di/dt - 0$). At that point, all of the voltage drop is across the resistor, so current is $i = V_B/R$. The quantity L/R in the exponent of (1.50) is called the *time constant*, τ.

We can sketch out the current flowing in the circuit of Fig. 1.25 along with the voltage across the inductor as we go about opening and closing the switch (Fig 1.26). If we start with the switch open at $t = 0^-$ (where the minus suggests just before $t = 0$), the current will be zero and the voltage across the inductor, V_L will be 0 (since $V_L = L\,di/dt$ and $di/dt = 0$).

At $t = 0$, the switch is closed. At $t = 0^+$ (just after the switch closes) the current is still zero since it cannot change instantaneously. With zero current, there is no voltage drop across the resistor ($v_R = i\ R$), which means the entire battery voltage appears across the inductor ($v_L = V_B$). Notice that there is no restriction on how rapidly voltage can change across an inductor, so an instantaneous jump is allowed. Current climbs after the switch is closed until dc conditions are reached, at which point $di/dt = 0$ so $v_L = 0$ and the entire battery voltage is dropped across the resistor. Current i asymptotically approaches V_B/R.

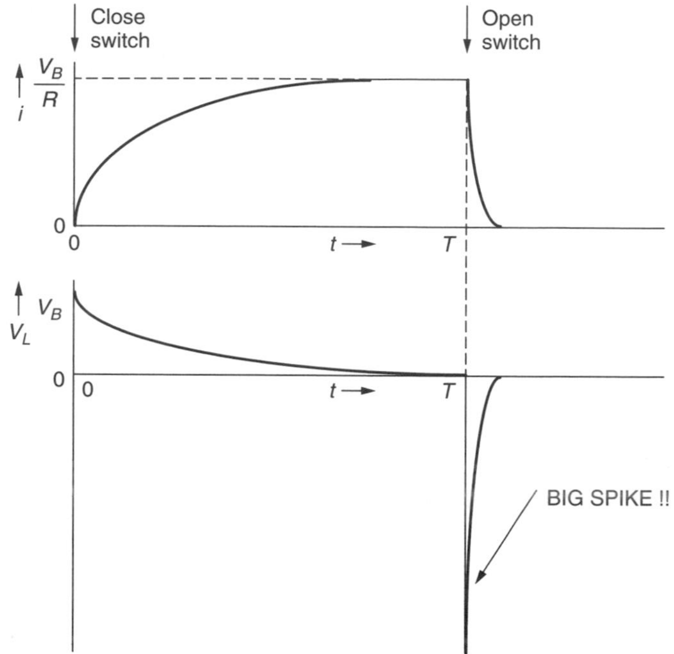

Figure 1.26 Opening a switch at $t = T$ produces a large spike of voltage across the inductor.

Now, at time $t = T$, open the switch. Current quickly, but not instantaneously, drops to zero (by arcing). Since the voltage across the inductor is $v_L = L\,di/dt$, and di/dt (the slope of current) is a very large negative quantity, v_L shows a precipitous, downward spike as shown in Fig. 1.26. This large spike of voltage can be much, much higher than the little voltage provided by the battery. In other words, with just an inductor, a battery, and a switch, we can create a very large voltage spike as we open the switch. This peculiar property of inductors is used to advantage in automobile ignition systems to cause spark plugs to ignite the gasoline in the cylinders of your engine. In your ignition system a switch opens (it used to be the points inside your distributor, now it is a transistorized switch), thereby creating a spike of voltage that is further amplified by a transformer coil to create a voltage of tens of thousands of volts—enough to cause an arc across the gap in your car's spark plugs. Another important application of this voltage spike is to use it to start the arc between electrodes of a fluorescent lamp.

1.8 TRANSFORMERS

When Thomas Edison created the first electric utility in 1882, he used dc to transmit power from generator to load. Unfortunately, at the time it was not possible to change dc voltages easily from one level to another, which meant transmission was at the relatively low voltages of the dc generators. As we have seen, transmitting significant amounts of power at low voltage means that high currents must flow, resulting in large i^2R power losses in the wires as well as high voltage drops between power plant and loads. The result was that power plants had to be located very close to loads. In those early days, it was not uncommon for power plants in cities to be located only a few blocks apart.

In a famous battle between two giants of the time, George Westinghouse solved the transmission problem by introducing ac generation using a transformer to boost the voltage entering transmission lines and other transformers to reduce the voltage back down to safe levels at the customer's site. Edison lost the battle but never abandoned dc—a decision that soon led to the collapse of his electric utility company.

It would be hard to overstate the importance of transformers in modern electric power systems. Transmission line power losses are proportional to the square of current and are inversely proportional to the square of voltage. Raising voltages by a factor of 10, for example, lowers line losses by a factor of 100. Modern systems generate voltages in the range of 12 to 25 kV. Transformers boost that voltage to hundreds of thousands of volts for long-distance transmission. At the receiving end, transformers drop the transmission line voltage to perhaps 4 to 25 kV at electrical substations for local distribution. Other transformers then drop the voltage to safe levels for home, office and factory use.

1.8.1 Ideal Transformers

A simple transformer configuration is shown in Fig. 1.27. Two coils of wire are wound around a magnetic core. As shown, the primary side of the transformer has N_1 turns of wire carrying current i_1, while the secondary side has N_2 turns carrying i_2.

If we assume an ideal core with no flux leakage, then the magnetic flux ϕ linking the primary windings is the same as the flux linking the secondary. From Faraday's law we can write

$$e_1 = N_1 \frac{d\phi}{dt} \tag{1.51}$$

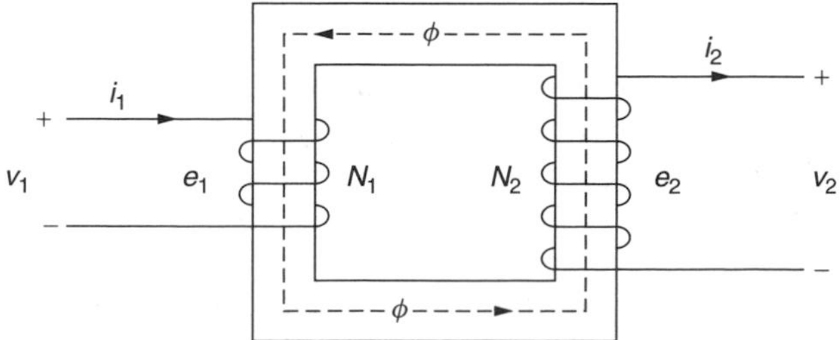

Figure 1.27 An idealized two-winding transformer.

and

$$e_2 = N_2 \frac{d\phi}{dt} \tag{1.52}$$

Continuing the idealization of the transformer, if there are no wire losses, then the voltage on the incoming wires, v_1, is equal to the emf e_1, and the voltage on the output wires, v_2, equals e_2. Dividing (1.52) by (1.51) gives

$$\frac{v_2}{v_1} = \frac{e_2}{e_1} = \frac{N_2(d\phi/dt)}{N_1(d\phi/dt)} \tag{1.53}$$

Before canceling out the $d\phi/dt$, note that we can only do so if $d\phi/dt$ is not equal to zero. That is, *the following fundamental relationship for transformers (1.53) is not valid for dc conditions*:

$$v_2 = \left(\frac{N_2}{N_1}\right) v_1 = (turns\ ratio) \cdot v_1 \tag{1.54}$$

The quantity in the parentheses is called *the turns ratio*. If voltages are to be raised, then the turns ratio needs to be greater than 1; to lower voltages it needs to be less than 1.

Does (1.54), which says that we can easily increase the voltage from primary to secondary, suggest that we are getting something for nothing? The answer is, as might be expected, no. While (1.54) suggests an easy way to raise ac voltages, energy still must be conserved. If we assume that our transformer is perfect; that is, it has no energy losses of its own, then power going into the transformer on the primary side, must equal power delivered to the load on the secondary side. That is,

$$v_1\, i_1 = v_2\, i_2 \tag{1.55}$$

Substituting (1.54) into (1.55) gives

$$i_2 = \left(\frac{v_1}{v_2}\right) i_1 = \left(\frac{N_1}{N_2}\right) i_1 \tag{1.56}$$

What (1.56) shows is that if we increase the voltage on the secondary side of the transformer (to the load), we correspondingly reduce the current to the load. For example, bumping the voltage up by a factor of 10 reduces the current delivered by a factor of 10. On the other hand, decreasing the voltage by a factor of 10 increases the current 10-fold on the secondary side.

Another important consideration in transformer analysis is what a voltage source "sees" when it sends current into a transformer that is driving a load. For example, in Fig. 1.28 a voltage source, transformer, and resistive load are shown. The symbol for a transformer shows a couple of parallel bars between the windings, which is meant to signify that the coil is wound around a metal (steel) core (not an air core). The dots above the windings indicate the polarity

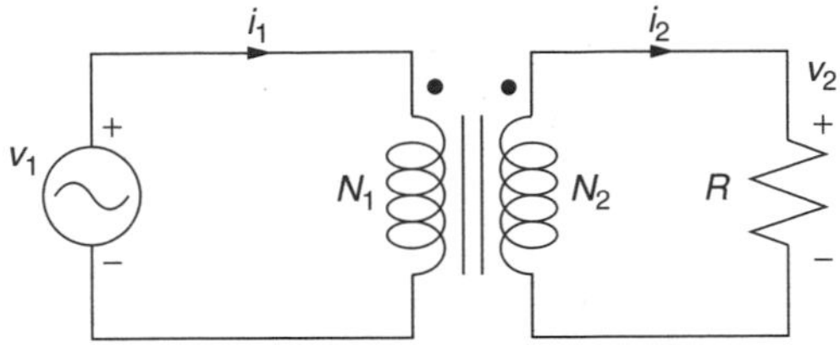

Figure 1.28 A resistance load being driven by a voltage source through a transformer.

of the windings. When both dots are on the same side (as in Fig. 1.28) a positive voltage on the primary produces a positive voltage on the secondary.

Back to the question of the equivalent load seen by the input voltage source for the circuit of Fig. 1.28. If we call that load R_{in}, then we have

$$v_1 = R_{\text{in}} i_1 \tag{1.57}$$

Rearranging (1.57) and substituting in (1.55) and (1.56) gives

$$R_{\text{in}} = \left(\frac{v_1}{i_1}\right) = \frac{(N_1/N_2)v_2}{(N_2/N_1)i_2} = \left(\frac{N_1}{N_2}\right)^2 \cdot \frac{v_2}{i_2} = \left(\frac{N_1}{N_2}\right)^2 R \tag{1.58}$$

where $v_2/i_2 = R$ is the resistance of the transformer load.

As far as the input voltage source is concerned, the load it sees is the resistance on the secondary side of the transformer divided by the square of the turns ratio. This is referred to as a resistance transformation (or more generally an *impedance* transformation).

Example 1.11 Some Transformer Calculations. A 120- to 240-V step-up transformer is connected to a 100-Ω load.

a. What is the turns ratio?
b. What resistance does the 120-V source see?
c. What is the current on the primary side and on the secondary side?

Solution

a. The turns ratio is the ratio of the secondary voltage to the primary voltage,

$$\text{Turns ratio} = \frac{N_2}{N_1} = \frac{v_2}{v_1} = \frac{240 \text{ V}}{120 \text{ V}} = 2$$

b. The resistance seen by the 120 V source is given by (1.58):

$$R_{\text{in}} = \left(\frac{N_1}{N_2}\right)^2 R = \left(\frac{1}{2}\right)^2 100 = 25\ \Omega$$

c. The primary side current will be

$$i_{\text{primary}} = \frac{v_1}{R_{\text{in}}} = \frac{120\ \text{V}}{25\ \Omega} = 4.8\ \text{A}$$

On the secondary side, current will be

$$i_{\text{secondary}} = \frac{v_2}{R_{\text{load}}} = \frac{240\ \text{V}}{100\ \Omega} = 2.4\ \text{A}$$

Notice that power is conserved:

$$v_1 \cdot i_1 = 120\ \text{V} \cdot 4.8\ \text{A} = 576\ \text{W}$$
$$v_2 \cdot i_2 = 240\ \text{V} \cdot 2.4\ \text{A} = 576\ \text{W}$$

1.8.2 Magnetization Losses

Up to this point, we have considered a transformer to have no losses of any sort associated with its performance. We know, however, that real windings have inherent resistance so that when current flows there will be voltage and power losses there. There are also losses associated with the magnetization of the core, which will be explored now.

The orientation of atoms in ferromagnetic materials (principally iron, nickel, and cobalt as well as some rare earth elements) are affected by magnetic fields. This phenomenon is described in terms of unbalanced spins of electrons, which causes the atoms to experience a torque, called a *magnetic moment*, when exposed to a magnetic field.

Ferromagnetic metals exist in a crystalline structure with all of the atoms within a particular portion of the material arranged in a well-organized lattice. The regions in which the atoms are all perfectly arranged is called a subcrystalline *domain*. Within each magnetic domain, all of the atoms have their spin axes aligned with each other. Adjacent domains, however, may have their spin axes aligned differently. The net effect of the random orientation of domains in an unmagnetized ferromagnetic material is that all of the magnetic moments cancel each other and there is no net magnetization. This is illustrated in Fig. 1.29a.

When a strong magnetic field H is imposed on the domains, their spin axes begin to align with the imposed field, eventually reaching saturation as shown in Fig. 1.29b. After saturation is reached, increasing the magnetizing force causes no increase in flux density, B. This suggests that the relationship between magnetic

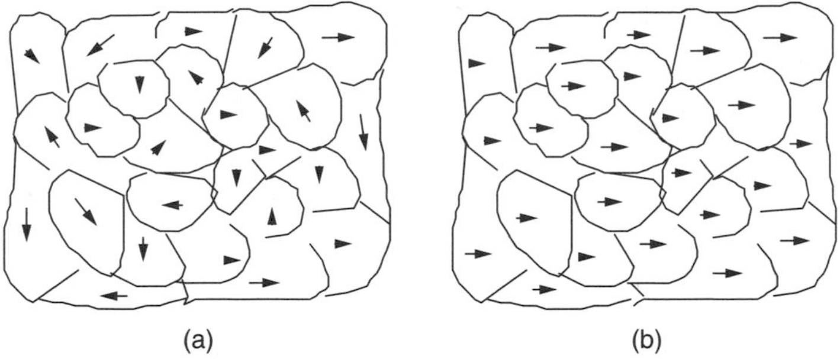

Figure 1.29 Representation of the domains in (a) an unmagnetized ferromagnetic material and (b) one that is fully magnetized.

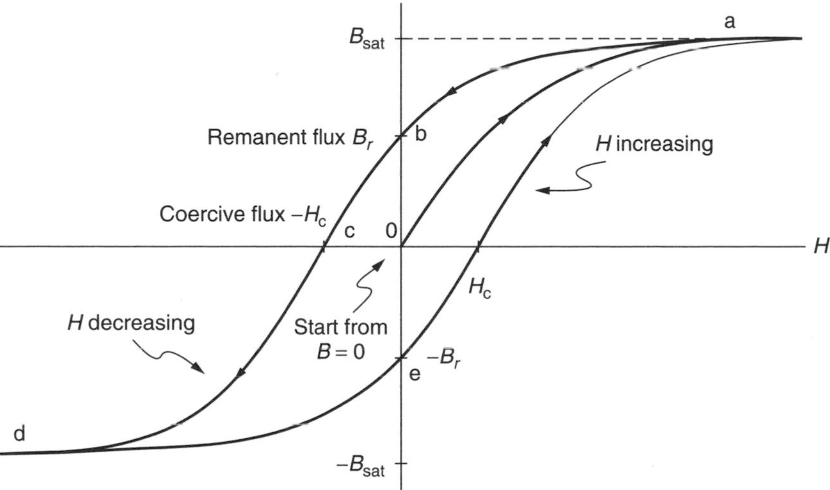

Figure 1.30 Cycling an imposed mmf on a ferromagnetic material produces a hysteresis loop.

field H and flux density B will not be linear, as was implied in (1.35), and in fact will exhibit some sort of s-shaped behavior. That is, permeability μ is not constant.

Figure 1.30 illustrates the impact that the imposition of a magnetic field H on a ferromagnetic material has on the resulting magnetic flux density B. The field causes the magnetic moments in each of the domains to begin to align. When the magnetizing force H is eliminated, the domains relax, but don't return to their original random orientation, leaving a remanent flux B_r; that is, the material becomes a "permanent magnet." One way to demagnetize the material is to heat it to a high enough temperature (called the *Curie temperature*) that the domains once again take on their random orientation. For iron, the Curie temperature is 770°C.

Consider what happens to the $B-H$ curve as the magnetic domains are cycled back and forth by an imposed ac magnetomagnetic force. On the $B-H$ curve of Fig 1.30, the cycling is represented by the path o–a followed by the path a–b. If the field is driven somewhat negative, the flux density can be brought back to zero (point c) by imposing a *coercive force*, H_c; forcing the applied mmf even more negative brings us to point d. Driving the mmf back in the positive direction takes us along path d–e–a.

The phenomenon illustrated in the $B-H$ curve is called hysteresis. Cycling a magnetic material causes the material to heat up; in other words, energy is being wasted. It can be shown that the energy dissipated as heat in each cycle is proportional to the area contained within the hysteresis loop. Each cycle through the loop creates an energy loss; therefore the rate at which energy is lost, which is power, is proportional to the frequency of cycling and the area within the hysteresis loop. That is, we can write an equation of the sort

$$\text{Power loss due to hysteresis} = k_1 f \tag{1.59}$$

where k_1 is just a constant of proportionality and f is the frequency.

Another source of core losses is caused by small currents, called *eddy currents*, that are formed within the ferromagnetic material as it is cycled. Consider a cross section of core with magnetic flux ϕ aligned along its axis as shown in Fig. 1.31a. We know from Faraday's law that anytime a loop of electrical conductor has varying magnetic flux passing through it, there will be a voltage (emf) created in that loop proportional to the rate of change of ϕ. That emf can create its own current in the loop. In the case of our core, the ferromagnetic material is the conductor, which we can think of as forming loops of conductor wrapped around flux creating the eddy currents shown in the figure.

To analyze the losses associated with eddy currents, imagine the flux as a sinusoidal, time-varying function

$$\phi = \sin(\omega t) \tag{1.60}$$

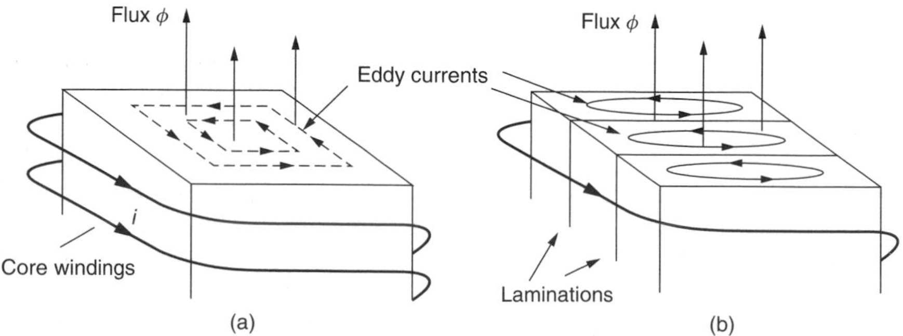

Figure 1.31 Eddy currents in a ferromagnetic core result from changes in flux linkages: (a) A solid core produces large eddy current losses. (b) Laminating the core yields smaller losses.

The emf created by changing flux is proportional to $d\phi/dt$

$$e = k_2 \frac{d\phi}{dt} = k_2 \omega \cos(\omega t) \tag{1.61}$$

where k_2 is just a constant of proportionality. The power loss in a conducting "loop" around this changing flux is proportional to voltage squared over loop resistance:

$$\text{Eddy current power loss} = \frac{e^2}{R} = \frac{1}{R}[k_2 \omega \cos(\omega t)]^2 \tag{1.62}$$

Equation (1.62) suggests that power loss due to eddy currents is inversely proportional to the resistance of the "loop" through which the current is flowing. To control power losses, therefore, there are two approaches: (1) Increase the electrical resistance of the core material, and (2) make the loops smaller and tighter. Tighter loops have more resistance (since resistance is inversely proportional to cross-sectional area through which current flows) and they contain less flux ϕ (emf is proportional to the rate of change of flux, not flux density).

Real transformer cores are designed to control both causes of eddy current losses. Steel cores, for example, are alloyed with silicon to increase resistance; otherwise, high-resistance magnetic ceramics, called ferrites, are used instead of conventional alloys. To make the loops smaller, cores are usually made up of many thin, insulated, lamination layers as shown in Fig. 1.31b.

The second, very important conclusion from Eq. (1.62) is that eddy current losses are proportional to frequency squared:

$$\text{Power loss due to eddy currents} = k_3 f^2 \tag{1.63}$$

Later, when we consider harmonics in power circuits, we will see that some loads cause currents consisting of multiples of the fundamental 60-Hz frequency. The higher-frequency harmonics can lead to transformer core burnouts due to the eddy current dependence on frequency squared.

Transformer hysteresis losses are controlled by using materials with minimal B–H hysteresis loop area. Eddy current losses are controlled by picking core materials that have high resistivity and then laminating the core with thin, insulated sheets of material. Leakage flux losses are minimized not only by picking materials with high permeability but also by winding the primary and secondary windings right on top of each other. A common core configuration designed for overlapping windings is shown in Fig. 1.32. The two windings are wrapped around the center section of core while the outer two sections carry the flux in closed loops. The top of a laminated slice of this core is a separate piece in order to facilitate wrapping the windings around core material. With the top off, a mechanical winder can easily wrap the core, after which the top bar is attached.

A real transformer can be modeled using a circuit consisting of an idealized transformer with added idealized resistances and inductors as shown in Fig. 1.33.

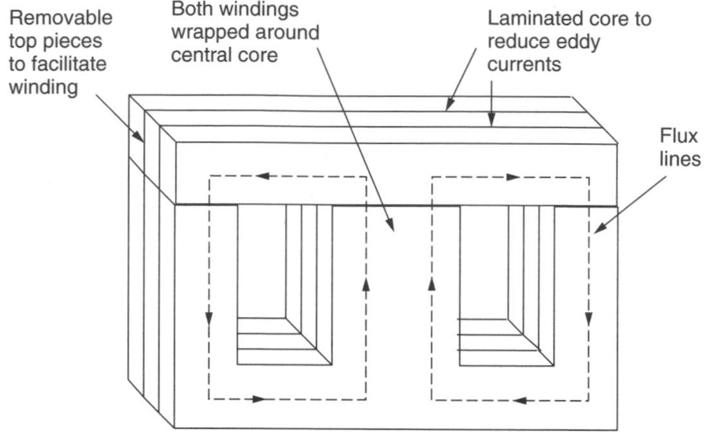

Figure 1.32 A type "E-1" laminated core for a transformer showing the laminations and the removable top pieces to enable machine winding. Windings are wound on top of each other on the central portion of the core.

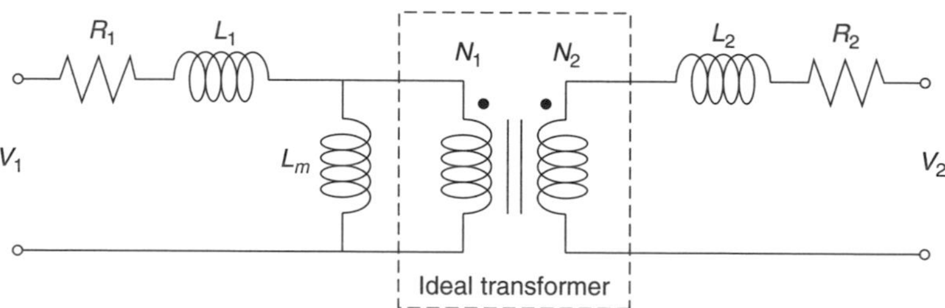

Figure 1.33 A model of a real transformer accounts for winding resistances, leakage fluxes, and magnetizing inductance.

Resistors R_1 and R_2 represent the resistances of the primary and secondary windings. L_1 and L_2 represent the inductances associated with primary and secondary leakage fluxes that pass through air instead of core material. Inductance L_m, *the magnetizing inductance*, allows the model to show current in the primary windings even if the secondary is an open circuit with no current flowing.

PROBLEMS

1.1. Either a resistor, capacitor or inductor is connected through a switch to a current source. At $t = 0$, the switch is closed and the following applied current results in the voltage shown. What is the circuit element and what is its magnitude?

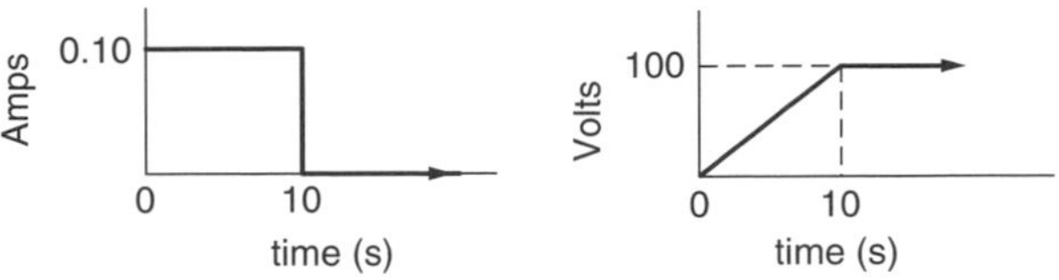

Figure P1.1

1.2. A voltage source produces the square wave shown below. The load, which is either an ideal resistor, capacitor or inductor, draws current current as shown below.

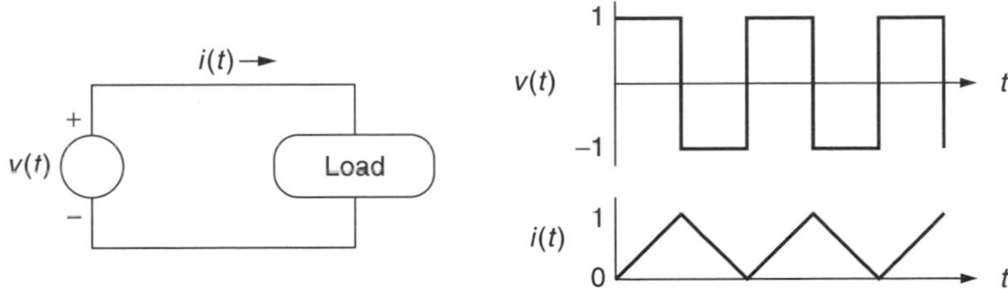

Figure P1.2

a. Is the "Load" a resistor, capacitor or inductor?

b. Sketch the power delivered to the load versus time.

c. What is the average power delivered to the load?

1.3. A single conductor in a transmission line dissipates 6,000 kWh of energy over a 24-hour period during which time the current in the conductor was 100 amps. What is the resistance of the conductor?

1.4. A core-and-coil inductor has a mean cross-sectional area of 0.004 m^2 and a mean circumference of 0.24 m. The iron core has a relative permeability of 20,000. It is wrapped with 100 turns carrying 1 amp of current.

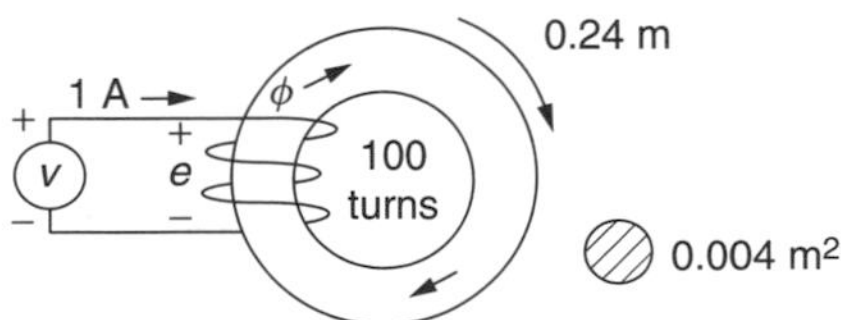

Figure P1.4

a. What is the reluctance of the core $\mathcal{R}$ (A-t/Wb)?

b. What is the inductance of the core and coil L (henries)?

c. What is the magnetic field intensity H (A-t/m)?

d. What is the magnetic flux density B (Wb/m^2)

1.5. The resistance of copper wire increases with temperature in an approximately linear manner that can be expressed as

$$R_{T2} = R_{T1}[1 + \alpha(T_2 - T_1)]$$

where $\alpha = 0.00393/^\circ\text{C}$. Assuming the temperature of a copper transmission line is the same as the ambient temperature, how hot does the weather have to get to cause the resistance of a transmission line to increase by 10% over its value at 20°C?

1.6. A 52-gallon electric water heater is designed to deliver 4800 W to an electric-resistance heating element in the tank when it is supplied with 240 V (it doesn't matter if this is ac or dc).

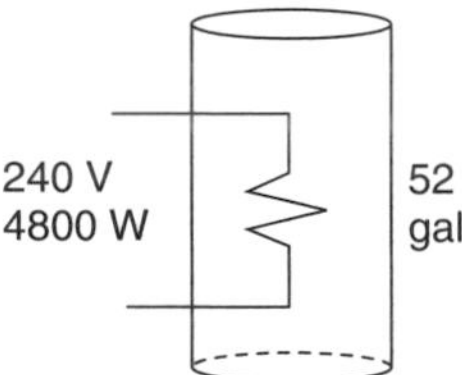

Figure P1.5

a. What is the resistance of the heating element?

b. How many watts would be delivered if the element is supplied with 208 V instead of 240 V?

c. Neglecting any losses from the tank, how long would it take for 4800 W to heat the 52 gallons of water from 60°F to 120°F? The conversion between kilowatts of electricity and Btu/hr of heat is given by 3412 Btu/hr = 1 kW. Also, one Btu heats 1 lb of water by 1°F and 1 gallon of water weighs 8.34 lbs.

1.7. Suppose an automobile battery is modeled as an ideal 12-V battery in series with an internal resistance of 0.01 Ω as shown in (a) below.

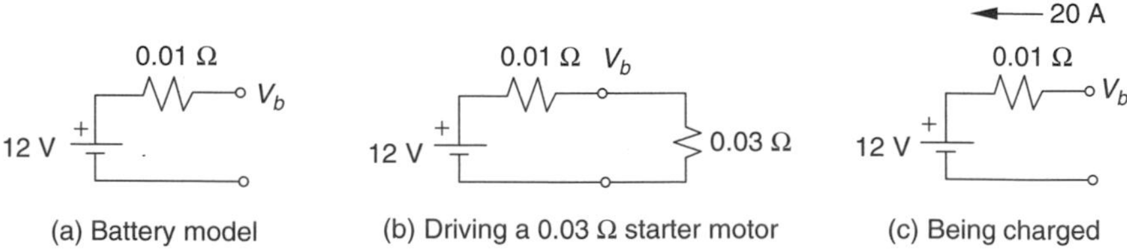

Figure P1.7

a. What current will be delivered when the battery powers a 0.03 Ω starter motor, as in (b)? What will the battery output voltage be?

b. What voltage must be applied to the battery in order to deliver a 20-A charging current as in (c)?

1.8. Consider the problem of using a low-voltage system to power a small cabin. Suppose a 12-V system powers a pair of 100-W lightbulbs (wired in parallel).

a. What would be the (filament) resistance of a bulb designed to use 100 W when it receives 12 V?

b. What would be the current drawn by two such bulbs if each receives a full 12 V?

c. What gage wire should be used if it is the minimum size that will carry the current.

d. Suppose a 12-V battery located 80-ft away supplies current to the pair of bulbs through the wire you picked in (c). Find:

1. The equivalent resistance of the two bulbs plus the wire resistance to and from the battery.
2. Current delivered by the battery
3. The actual voltage across the bulbs
4. The power lost in the wires
5. The power delivered to the bulbs
6. The fraction of the power delivered by the battery that is lost in the wires.

1.9. Repeat Problem 1.8 using a 60-V system using the same 12 gage wire.

1.10. Suppose the lighting system in a building draws 20 A and the lamps are, on the average, 100 ft from the electrical panel. Table 1.3 suggests that 12 ga wire meets code, but you want to consider the financial merits of wiring the circuit with bigger 10 ga wire. Suppose the lights are on 2500 hours per year and electricity costs $0.10 per kWh.

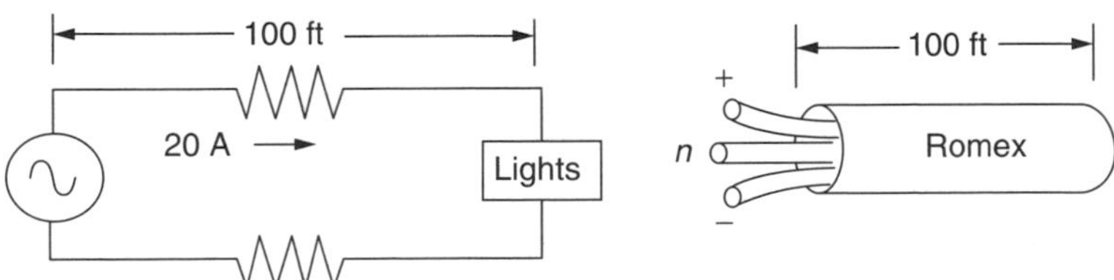

Figure P1.10

a. Find the energy savings per year (kWhr/yr) that would result from using 10 ga instead of 12 ga wire.

b. Suppose 12 ga wire costs $25 per 100 ft of "Romex" (2 conductors, each 100-ft long, plus a ground wire in a tough insulating sheath) and 10 ga costs $35 per 100 ft. What would be the "simple payback" period (simple payback = extra 1st cost/annual $ savings) when utility electricity costs $0.10/kWh?

c. An effective way to evaluate energy efficiency projects is by calculating the annual cost associated with conservation and dividing it by the annual energy saved. This is the *cost of conserved energy* (CCE) and is described more carefully in Section 5.4. CCE is defined as follows

$$\text{CCE} = \frac{\text{annual cost of saved electricity(\$/yr)}}{\text{annual electricity saved (kWhr/yr)}} = \frac{\Delta P \cdot \text{CRF}(i, n)}{\text{kWhr/yr}}$$

where ΔP is the extra cost of the conservation feature (heavier duty wire in this case), and CRF is the capital recovery factor (which means your annual loan payment on \$1 borrowed for n years at interest rate i.

What would be the "cost of conserved energy" CCE (cents/kWhr) if the building (and wiring) is being paid for with a 7-%, 20-yr loan with CRF = 0.0944/yr. How does that compare with the cost of electricity that you don't have to purchase from the utility at 10¢ /kWhr?

1.11. Suppose a photovoltaic (PV) module consists of 40 individual cells wired in series, (a). In some circumstances, when all cells are exposed to the sun it can be modeled as a series combination of forty 0.5-V ideal batteries, (b). The resulting graph of current versus voltage would be a straight, vertical 20-V line as shown in (c).

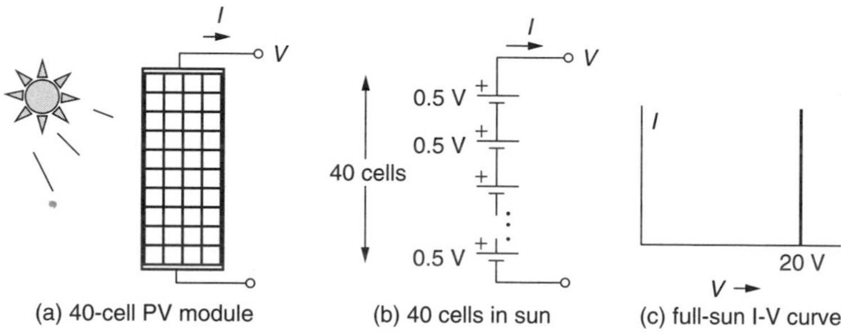

Figure P1.11

a. When an individual cell is shaded, it looks like a 5-Ω resistor instead of a 0.5-V battery, as shown in (d). Draw the I-V curve for the PV module with one cell shaded.

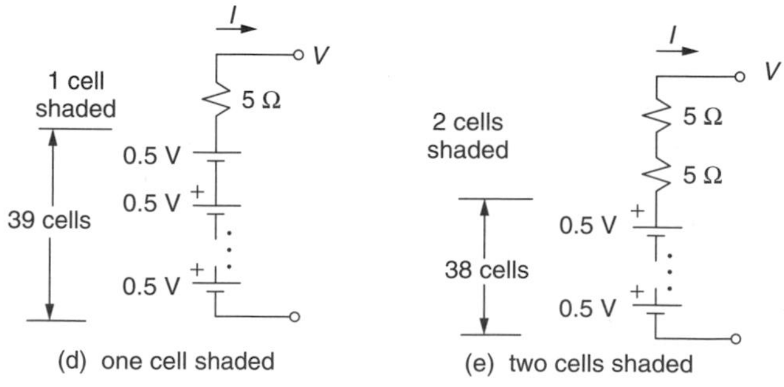

Figure P1.11

b. With two cells shaded, as in (e), draw the I-V curve for the PV module on the same axes as you have drawn the full-sun and 1-cell shaded I-V lines.

1.12. If the photovoltaic (PV) module in Problem 1.11 is connected to a 5-Ω load, find the current, voltage, and power that will be delivered to the load under the following conditions:

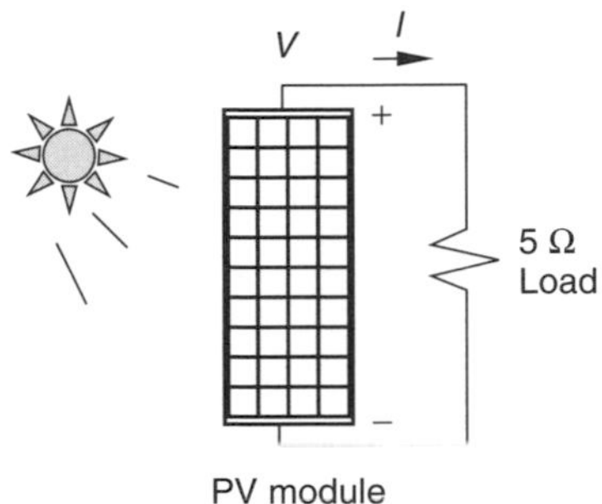

Figure P1.12

a. Every cell in the PV module is in the sun.

b. One cell is shaded.

c. Two cells are shaded.

Use the fact that the same current and voltage flows through both the PV module and the load so solve for I, V, and P.

1.13. When circuits involve a source and a load, the same current flows through each one and the same voltage appears across both. A graphical solution can therefore be obtained by simply plotting the current-voltage (I-V) relationship for the source onto the same axes that the I-V relationship for the load is plotted, and then finding the crossover point where both are satisfied simultaneously. This is an especially powerful technique when the relationships are nonlinear.

For the photovoltaic module supplying power to the 5-W resistive load in Problem 1.12, solve for the resulting current and voltage using the graphical approach:

a. For all cells in the sun.

b. For one cell shaded

c. For two cells shaded

CHAPTER 2

FUNDAMENTALS OF ELECTRIC POWER

2.1 EFFECTIVE VALUES OF VOLTAGE AND CURRENT

When voltages are nice, steady, dc, it is intuitively obvious what is meant when someone says, for example, "this is a 9-V battery." But what does it mean to say the voltage at the wall outlet is 120-V ac? Since it is ac, the voltage is constantly changing, so just what is it that the "120-V" refers to?

First, let us describe a simple sinusoidal current:

$$i = I_m \cos(\omega t + \theta) \tag{2.1}$$

where i is the current, a function of time; I_m is the magnitude, or amplitude, of the current; ω is the angular frequency (radians/s); and θ is the phase angle (radians). Notice that conventional notation uses lowercase letters for time-varying voltages or currents (e.g., i and v), while capitals are used for quantities that are constants (or parameters), (e.g., I_{m} or V_{rms}). Also note that we just as easily could have described the current with a sine function instead of cosine. A plot of (2.1) is shown in Fig. 2.1.

The frequency ω in (2.1) is expressed in radians per second. Equally common is to express the frequency f in hertz (Hz), which are "cycles per second." Since there are 2π radians per cycle, we can write

$$\omega = 2\pi \text{ (radians/cycle)} \cdot f \text{ (cycles/}s\text{)} = 2\pi f \tag{2.2}$$

Renewable and Efficient Electric Power Systems. By Gilbert M. Masters
ISBN 0-471-28060-7

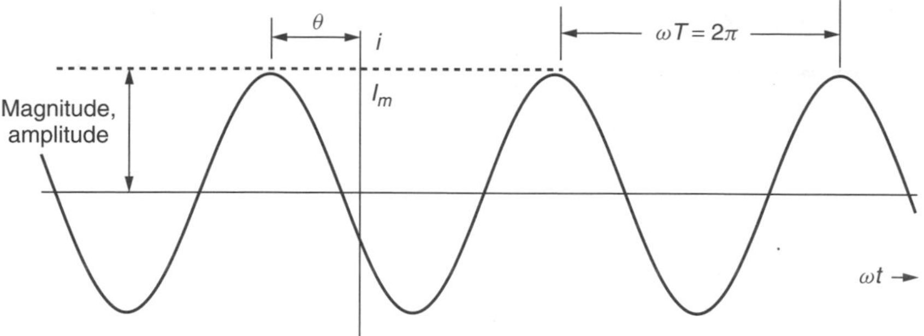

Figure 2.1 Illustrating the nomenclature for a sinusoidal function.

The sinusoidal function is periodic—that is, it repeats itself—so we can also describe it using its period, T:

$$T = 1/f \tag{2.3}$$

Thus, the sinusoidal current can have the following equivalent representations:

$$i = I_m \cos(\omega t + \theta) = I_m \cos(2\pi f\ t + \theta) = I_m \cos\left(\frac{2\pi}{T} t + \theta\right) \tag{2.4}$$

Suppose we have a portion of a circuit, consisting of a current i passing through a resistance R as shown in Fig. 2.2.

The instantaneous power dissipated by the resistor is

$$p = i^2 R \tag{2.5}$$

In (2.5), power is given a lowercase symbol to indicate that it is a time-varying quantity. The average value of power dissipated in the resistance is given by

$$P_{\text{avg}} = (i^2)_{\text{avg}} R = I_{\text{eff}}{}^2 R \tag{2.6}$$

In (2.6) an effective value of current, I_{eff}, has been introduced. The advantage of defining the effective value this way is that the resulting equation for average

Figure 2.2 A time-varying current i through a resistance R.

power dissipated looks very similar to the instantaneous power described by (2.5). This leads to a definition of the effective value of current given below:

$$I_{\text{eff}} = \sqrt{(i^2)_{\text{avg}}} = I_{\text{rms}} \tag{2.7}$$

The effective value of current is the square root of the mean value of current squared. That is, it is the root-mean-squared, or rms, value of current. The definition given in (2.7) applies to any current function, be it sinusoidal or otherwise.

To find the rms value of a function, we can always work it out formally using the following:

$$I_{\text{rms}} = \sqrt{(i^2)_{\text{avg}}} = \sqrt{\frac{1}{T}\int_0^T i^2(t)dt} \tag{2.8}$$

Oftentimes, however, it is easier simply to graph the square of the function and determine the average by inspection, as the following example illustrates.

Example 2.1 RMS value of a Square Wave. Find the rms value of a square-wave current that jumps back and forth from 0 to 2 A as shown below:

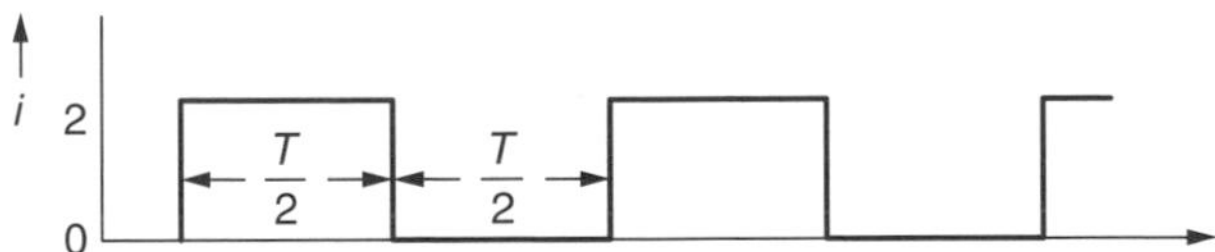

Solution We need to find the square root of the average value of the square of the current. The waveform for current squared is

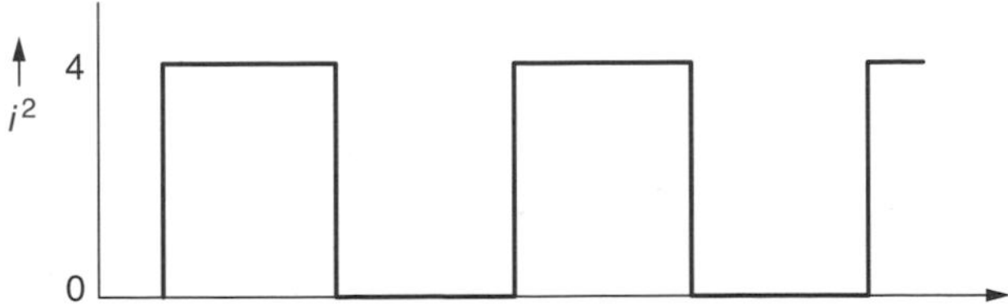

The average value of current squared is 2 by inspection (half the time it is zero, half the time it is 4):

$$I_{\text{rms}} = \sqrt{(i^2)_{\text{avg}}} = \sqrt{2}A$$

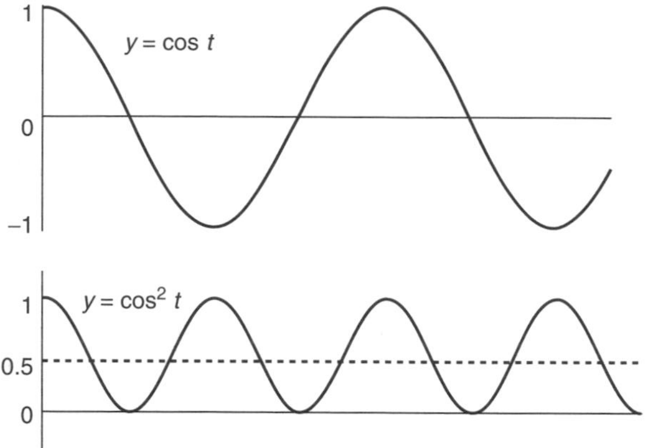

Figure 2.3 The average value of the square of a sinusoid is 1/2.

Let us derive the rms value for a sinusoid by using the simple graphical procedure. If we start with a sinusoidal voltage

$$v = V_m \cos \omega t \tag{2.9}$$

The rms value of voltage is

$$V_{\text{rms}} = \sqrt{(v^2)_{\text{avg}}} = \sqrt{(V_m{}^2 \cos^2 \omega t)_{\text{avg}}} = V_m \sqrt{(\cos^2 \omega t)_{\text{avg}}} \tag{2.10}$$

Since we need to find the average value of the square of a sine wave, let us graph $y = \cos^2 \omega t$ as has been done in Fig. 2.3.

By inspection of Fig. 2.3, the mean value of $\cos^2 \omega t$ is 1/2. Therefore, using (2.10), the rms value of a sinusoidal voltage is

$$\boxed{V_{\text{rms}} = V_m \sqrt{\frac{1}{2}} = \frac{V_m}{\sqrt{2}}} \tag{2.11}$$

This is a very important result:

The rms value of a sinusoid is the amplitude divided by the square root of 2. Notice this conclusion applies only to sinusoids! When an ac current or voltage is described (e.g., 120 V, 10 A), the values specified are always the rms values.

Example 2.2 Wall outlet voltage. Find an equation like (2.1) for the 120-V, 60-Hz voltage delivered to your home.

Solution From (2.11), the amplitude (magnitude, peak value) of the voltage is

$$V_m = \sqrt{2}V_{\text{rms}} = 120\sqrt{2} = 169.7 \text{ V}$$

The angular frequency ω is

$$\omega = 2\pi f = 2\pi 60 = 377 \text{ rad/s}$$

The waveform is thus

$$v = 169.7 \cos 377t$$

It is conventional practice to treat the incoming voltage as having zero phase angle, so that all currents will have phase angles measured relative to that reference voltage.

2.2 IDEALIZED COMPONENTS SUBJECTED TO SINUSOIDAL VOLTAGES

2.2.1 Ideal Resistors

Consider the response of an ideal resistor to excitation by a sinusoidal voltage as shown in Fig. 2.4.

The voltage across the resistance is the same as the voltage supplied by the source:

$$v = V_m \cos \omega t = \sqrt{2}V_{\text{rms}} \cos \omega t = \sqrt{2}V \cos \omega t \tag{2.12}$$

Notice the three ways that the voltage has been described: using the amplitude of the voltage V_m, the rms value of voltage V_{rms}, and the symbol V which, in this context, means the rms value. *We will consistently use current I or voltage V* (capital letters, without subscripts) to mean the rms values of that current or voltage.

The current that will pass through a resistor with the above voltage imposed will be

$$i = \frac{v}{R} = \frac{V_m}{R} \cos \omega t = \frac{\sqrt{2}V_{\text{rms}}}{R} \cos \omega t = \frac{\sqrt{2}V}{R} \cos \omega t \tag{2.13}$$

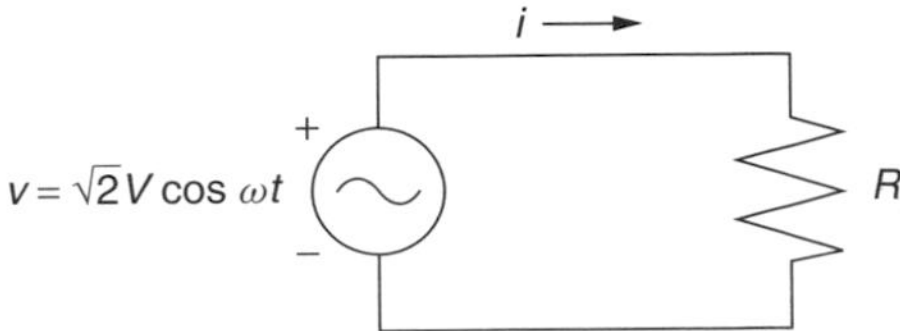

Figure 2.4 A sinusoidal voltage imposed on an ideal resistance.

Since the phase angle of the resulting current is the same as the phase angle of the voltage (zero), they are said to be *in phase* with each other. The rms value of current is therefore

$$I_{\text{rms}} = I = \frac{I_m}{\sqrt{2}} = \frac{\sqrt{2}V/R}{\sqrt{2}} = \frac{V}{R} \tag{2.14}$$

Notice how simple the result is: The rms current I is equal to the rms voltage V divided by the resistance R. We have, in other words, a very simple ac version of Ohm's law:

$$V = RI \tag{2.15}$$

where V and I are rms quantities.

Now let's look at the average power dissipated in the resistor.

$$P_{\text{avg}} = (vi)_{\text{avg}} = [\sqrt{2}\ V\cos\omega t \cdot \sqrt{2}\ I\cos\omega t]_{\text{avg}} = 2VI(\cos^2\omega t)_{\text{avg}} \tag{2.16}$$

The average value of $\cos^2\omega t$ is 1/2. Therefore,

$$P_{\text{avg}} = 2VI \cdot \frac{1}{2} = VI \tag{2.17}$$

In a similar way, it is easy to show the expected alternative formulas for average power are also true:

$$P_{\text{avg}} = VI = I^2R = \frac{V^2}{R} \tag{2.18}$$

Notice how the ac problem has been greatly simplified by using rms values of current and voltage. You should also note that the power given in (2.18) is the *average* power and not some kind of rms value. Since ac power is always interpreted to be average power, the subscript in P_{avg} is not usually needed.

Example 2.3 ac Power for a Lightbulb. Suppose that a conventional incandescent lightbulb uses 60 W of power when it supplied with a voltage of 120 V. Modeling the bulb as a simple resistance, find that resistance as well as the current that flows. How much power would be dissipated if the voltage drops to 110 V?

Solution Using (2.18), we have

$$R = \frac{V^2}{P} = \frac{(120)^2}{60} = 240\ \Omega$$

and

$$I = \frac{P}{V} = \frac{60}{120} = 0.5 \text{ A}$$

When the voltage sags to 110 V, the power dissipated will be

$$P = \frac{V^2}{R} = \frac{(110)^2}{240} = 50.4 \text{ W}$$

2.2.2 Idealized Capacitors

Recall the defining equation for a capacitor, which says that current is proportional to the rate of change of voltage across the capacitor. Suppose we apply an ac voltage of V volts (rms) across a capacitor, as shown in Fig. 2.5.

The resulting current through the capacitor will be

$$i = C\frac{dv}{dt} = C\frac{d}{dt}(\sqrt{2}V\cos\omega t) = -\omega C\sqrt{2}V\sin\omega t \tag{2.19}$$

If we apply the trigonometric identity that $\sin x = -\cos(x + \pi/2)$, we get

$$i = \sqrt{2}\omega CV\cos\left(\omega t + \frac{\pi}{2}\right) \tag{2.20}$$

There are several things to note about (2.20). For one, the current waveform is a sinusoid of the same frequency as the voltage waveform. Also note that there is a 90° phase shift ($\pi/2$ radians) between the voltage and current. The current is said to be *leading* the voltage by 90°. That the current leads the voltage should make some intuitive sense since charge must be delivered to the capacitor before it shows a voltage. The graph in Fig. 2.6 also suggests the idea that the current peaks 90° before the voltage peaks.

Finally, writing (2.20) in terms of (2.1) gives

$$i = \sqrt{2}\omega CV\cos\left(\omega t + \frac{\pi}{2}\right) = I_m\cos(\omega t + \theta) = \sqrt{2}I\cos(\omega t + \theta) \tag{2.21}$$

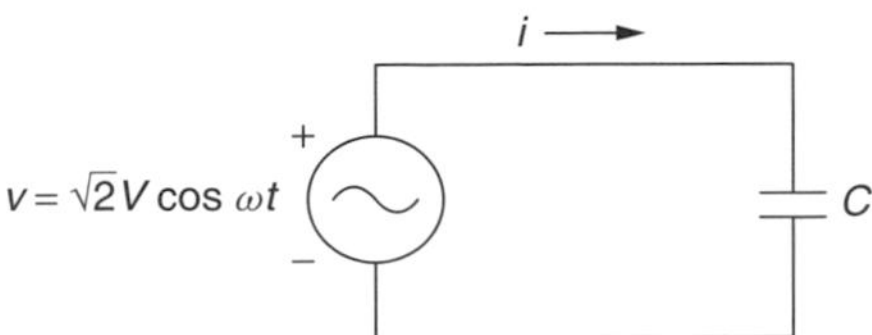

Figure 2.5 An ac voltage V, applied across a capacitor.

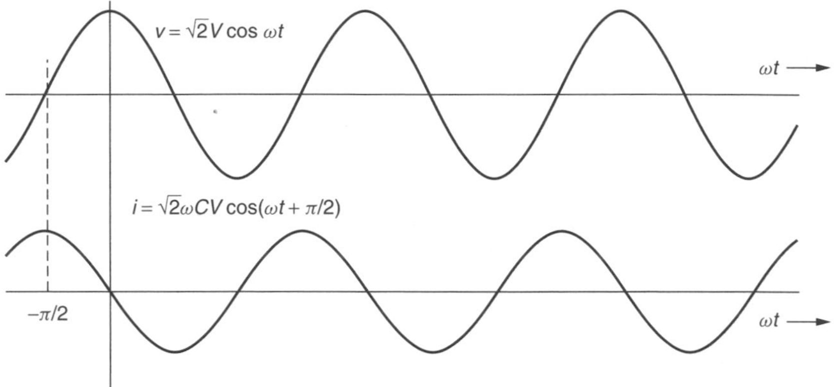

Figure 2.6 Current through a capacitor leads the voltage applied to it.

which says that the rms current I is given by

$$I = \omega CV \tag{2.22}$$

and the phase angle between current and voltage is

$$\theta = \pi/2 \tag{2.23}$$

Rearranging (2.22) gives

$$V = \left(\frac{1}{\omega C}\right) I \tag{2.24}$$

Equation (2.24) is beginning to look like an ac version of Ohm's law for capacitors. It should be used with caution, however, since it does not capture the notion that current and voltage are 90° out of phase with each other.

Also of interest is the average power dissipated by a capacitor subjected to a sinusoidal voltage. Since instantaneous power is the product of voltage and current, we can write

$$p = vi = \sqrt{2}V \cos \omega t \cdot \sqrt{2}I \cos\left(\omega t + \frac{\pi}{2}\right) \tag{2.25}$$

Using the trigonometric identity $\cos A \cdot \cos B = \frac{1}{2}[\cos(A + B) + \cos(A - B)]$ gives

$$p = 2VI \cdot \frac{1}{2}\left\{\cos\left(\omega t + \omega t + \frac{\pi}{2}\right) + \cos\left[\omega t - \left(\omega t + \frac{\pi}{2}\right)\right]\right\} \tag{2.26}$$

Since $\cos(-\pi/2) = 0$, this simplifies to

$$p = VI \cos\left(2\omega t + \frac{\pi}{2}\right) \tag{2.27}$$

Since the average value of a sinusoid is zero, (2.27) tells us that *the average power dissipated by a capacitor is zero.*

$$P_{\text{avg capacitor}} = 0 \tag{2.28}$$

Some of the time the capacitor is absorbing power (charging) and some of the time it is delivering power (discharging), but the average power is zero.

Example 2.4 Current in a Capacitor. A 120-V, 60-Hz ac source sends current to a 10-microfarad capacitor. Find the rms current flowing and write an equation for the current as a function of time.

Solution. The rms value of current is given by (2.22) as

$$I = \omega CV = 2\pi 60 \cdot 10 \times 10^{-6} \cdot 120 = 0.452 \text{ A}$$

The phase angle is $\theta = \pi/2$, so from (2.21) the complete expression for current is

$$i = \sqrt{2}I\cos(\omega t + \theta) = \sqrt{2} \cdot 0.452\cos\left(2\pi 60t + \frac{\pi}{2}\right) = 0.639\cos\left(377t + \frac{\pi}{2}\right)$$

2.2.3 Idealized Inductors

A sinusoidal voltage applied across an inductor is shown in Fig. 2.7.

We want to find the current through the inductor. Starting with the fundamental relationship for inductors,

$$v = L\frac{di}{dt} \tag{2.29}$$

and then solving for current:

$$i = \int di = \int \frac{v}{L}\,dt = \frac{1}{L}\int v\,dt \tag{2.30}$$

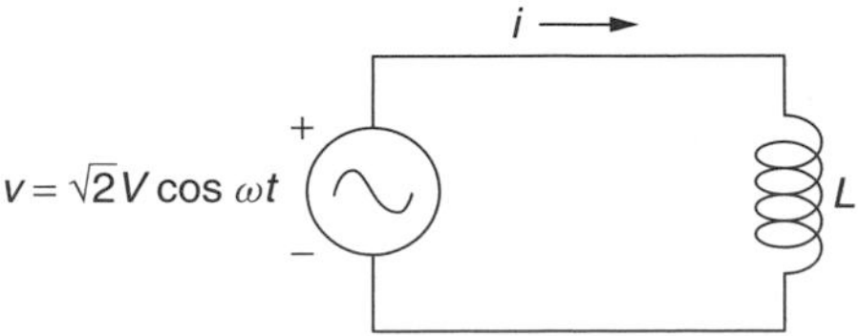

Figure 2.7 A sinusoidal voltage across an ideal inductor.

and inserting the equation for applied voltage

$$i = \frac{1}{L}\int \sqrt{2}V \cos\omega t \, dt = \frac{\sqrt{2}V}{L}\int \cos\omega t \, dt = \frac{\sqrt{2}V}{\omega L}\sin\omega t \tag{2.31}$$

Applying the trigonometric relationship sin ωt $= \cos(\omega t - \pi/2)$ gives

$$i = \left(\frac{1}{\omega L}\right)\sqrt{2}V \cos\left(\omega t - \frac{\pi}{2}\right) = \sqrt{2}I\cos(\omega t + \theta) \tag{2.32}$$

Equation (2.32) tells us that (1) the current through the inductor has the same frequency ω as the applied voltage, (2) the current *lags* behind the voltage by an angle $\theta = -\pi/2$, and (3) the rms value of current is

$$I = \left(\frac{1}{\omega L}\right)V \tag{2.33}$$

Rearranging (2.33) gives us something that looks like an ac version of Ohm's law for inductors:

$$V = (\omega L)I \tag{2.34}$$

So, for an inductor, you have to supply some voltage before current flows; for a capacitor, you need to supply current before voltage builds up. One way to remember which is which, is with the memory aid

"ELI the ICE man"

That is, for an inductor (L), voltage (E, as in emf) comes before current (I), while for a capacitor C, current (I) comes before voltage (E).

Finally, let us take a look at the power dissipated by an inductor:

$$p = vi = \sqrt{2}V\cos\omega t \cdot \sqrt{2}I\cos\left(\omega t - \frac{\pi}{2}\right) \tag{2.35}$$

Using the trigonometric identity for the product of two cosines, gives instantaneous power through the inductor

$$\begin{aligned} p &= 2VI \cdot \frac{1}{2}\left\{\cos\left(\omega t + \omega t - \frac{\pi}{2}\right) + \cos\left[\omega t - \left(\omega t - \frac{\pi}{2}\right)\right]\right\} \\ &= VI\cos\left(2\omega t - \frac{\pi}{2}\right) \end{aligned} \tag{2.36}$$

The average value of (2.36) is zero.

$$P_{\text{avg inductor}} = 0 \tag{2.37}$$

That is, an inductor is analogous to a capacitor in that it absorbs energy while current is increasing, storing that energy in its magnetic field, then it returns that energy when the current drops and the magnetic field collapses. *The net power dissipated when an inductor is subjected to an ac voltage is zero.*

2.3 POWER FACTOR

Those rather tedious derivations for the impact of ac voltages applied to idealized resistors, capacitors, and inductors has led to three simple but important conclusions. One is that the currents flowing through any of these components will have the same ac frequency as the source of the voltage that drives the current. Another is that there can be a phase shift between current and voltage. And finally, resistive elements are the only components that dissipate any net energy. Let us put these ideas together to analyze the generalized black box of Fig. 2.8.

The black box contains any number of idealized resistors, capacitors, and inductors, wired up any which way. The voltage source driving this box of components has rms voltage V, and we will arbitrarily assign it a phase angle of $\theta = 0$.

$$v = \sqrt{2}V \cos \omega t \tag{2.38}$$

Since the current delivered to the black box has the same frequency as the voltage source that drives it, we can write the following generalized current response as

$$i = \sqrt{2}I \cos(\omega t + \theta) \tag{2.39}$$

The instantaneous power supplied by the voltage source, and dissipated by the circuit in the box, is

$$p = vi = \sqrt{2}V \cos \omega t \cdot \sqrt{2}I \cos(\omega t + \theta) = 2VI[\cos \omega t \cdot \cos(\omega t + \theta)] \tag{2.40}$$

Once again, applying the identity $\cos A \cdot \cos B = \frac{1}{2}[\cos(A + B) + \cos(A - B)]$ gives

$$p = 2VI \left\{\tfrac{1}{2}[\cos(\omega t + \omega t + \theta) + \cos(\omega t - \omega t - \theta)]\right\} \tag{2.41}$$

so

$$p = VI \cos(2\omega t + \theta) + VI \cos(-\theta) \tag{2.42}$$

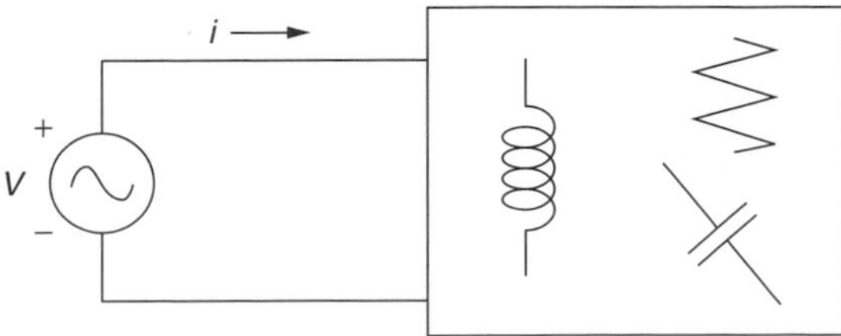

Figure 2.8 A black box of ideal resistors, capacitors, and inductors.

The average value of the first term in (2.42) is zero, and using $\cos x = \cos(-x)$ lets us write that the average power dissipated in the black box is given by

$$P_{\text{avg}} = VI\cos\theta = VI \times PF \tag{2.43}$$

Equation (2.43) is an important result. It says that the average power dissipated in the box is the product of the rms voltage supplied times the rms current delivered times the cosine of the angle between the voltage and current. The quantity $\cos\theta$ is called the *power factor (PF)*:

$$\text{Power factor} = \text{PF} = \cos\theta \tag{2.44}$$

The power expressed by (2.43) tells us the rate at which real work can be done in the black box. That black box, for example, might be a motor, in which case (2.43) gives us power to the motor in watts.

Why is power factor important? With an "ordinary" watt-hour meter on the premises, a utility customer pays only for watts of real power used within their factory, business, or home. The utility, on the other hand, has to cover the i^2R resistive power losses in the transmission and distribution wires that bring that power to the customer. When a customer has voltage and current way out of phase—that is, the power factor is "poor"—the utility loses more i^2R power on its side of the meter than occurs when a customer has a "good" power factor (PF $\approx$ 1.0).

Example 2.5 Good Versus Poor Power Factor. A utility supplies 12,000 V (12 kV) to a customer who needs 600 kW of real power. Compare the line losses for the utility when the customer's load has a power factor of 0.5 versus a power factor of 1.0.

Solution. To find the current drawn when the power factor is 0.5, we can start with (2.43):

$$P = VI \cdot PF$$

$$600 \text{ kW} = 12 \text{ kV} \cdot I(\text{A}) \cdot 0.5$$

so

$$I = \frac{600}{12 \times 0.5} = 100 \text{ A}$$

When the power factor is improved to 1.0, (2.43) now looks like

$$600 \text{ kW} = 12 \text{ kV} \cdot I(\text{A}) \cdot 1.0$$

so the current needed will be $I = \dfrac{600}{12} = 50$ A

When the power factor in the plant is improved from 0.5 to 1.0, the amount of current needed to do the same work in the factory is cut in half. The utility line losses are proportional to current squared, so line losses for this customer have been cut to one-fourth of their original value.

2.4 THE POWER TRIANGLE AND POWER FACTOR CORRECTION

Equation (2.43) sets up an important concept, called the *power triangle*. The actual power consumed by a circuit is the rate at which real work can be done (in watts). Bccausc voltagc V and current I may not be in phase, their product does not, in general, equal real power. Figure 2.9 sets up a power triangle in which the hypotenuse is the product of rms volts times rms amps. This leg is called the *apparent power, S*, and it has units of volt-amps (VA). Those volt-amps are resolved into the horizontal component $P = VI\cos\theta$, which is *real power* in watts. The vertical side of the triangle, $Q = VI\sin\theta$, is called *reactive power* and has units of VAR (which stands for volt-amps-reactive). Reactive VAR power is incapable of doing any work: It corresponds to voltage 90° out of phase with current so any work absorbed in one half of the cycle is returned, unchanged, in the other half.

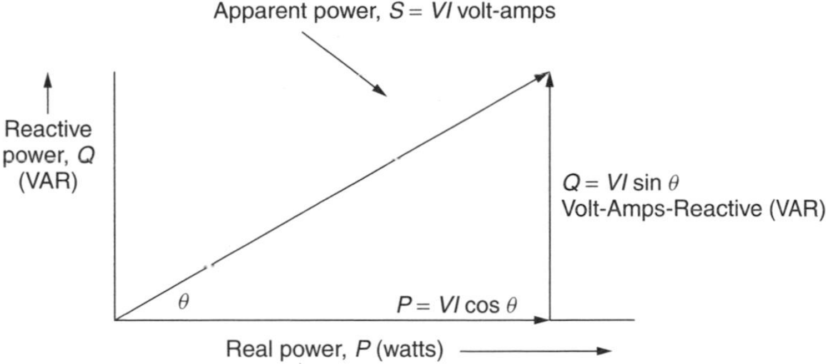

Figure 2.9 Showing apparent power S (volt-amps) resolved into reactive power (VAR) and real power P (watts).

Example 2.6 Power Triangle for a Motor. A 230-V induction motor draws 25 A of current while delivering 3700 W of power to its shaft. Draw its power triangle.

Solution

$$\text{Real power } P = 3700\text{ W} = 3.70\text{ kW}$$

$$\text{Apparent power } S = 25\text{ A} \times 230\text{ V} = 5750\text{ volt-amps} = 5.75\text{ kVA}$$

$$\text{Power factor PF} = \frac{\text{Real power}}{\text{Apparent power}} = \frac{3700 \text{ W}}{5750 \text{ VA}} = 0.6435$$

$$\text{Phase angle } \theta = \cos^{-1}(0.6435) = 50^\circ$$

$$\text{Reactive power } Q = S \sin\theta = 5750 \sin 50^\circ = 4400 \text{ VAR} = 4.40 \text{ kVAR}$$

The power triangle is therefore

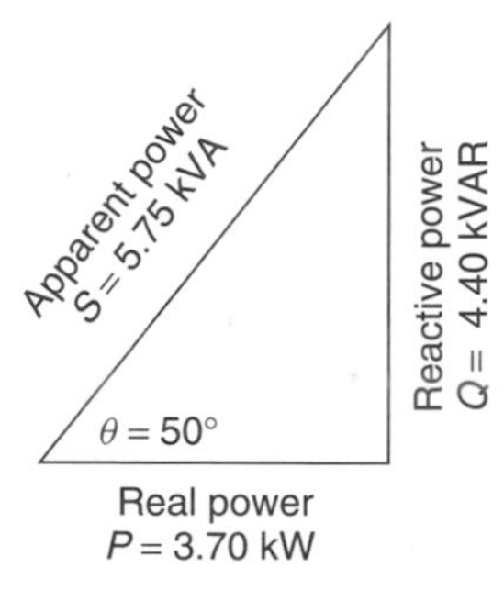

Utilities are very concerned about customers who draw a lot of reactive power—that is, customers with poor power factors. As suggested in Example 2.5, reactive power increases line losses for the utility, but doesn't result in any more kilowatt-hours (kWh) of energy sales to the customer. To discourage large customers from having a poor power factor, utilities will charge a penalty based on how low the power factor is, or they will charge not only for kWh of energy but also for kVAR of reactive power.

Many large customers have loads that are dominated by electric motors, which are highly inductive. It has been estimated that lagging power factor, mostly caused by induction motors, is responsible for as much as one-fifth of all grid losses in the United States, equivalent to about 1.5% of total national power generation and costs on the order of $2 billion per year. Another reason for concern about power factor is that transformers (on both sides of the meter) are rated in kVA, not watts, since it is heating caused by current flow that causes them to fail. By correcting power factor, a transformer can deliver more real power to the loads. This can be especially important if loads have increased to the point where the existing transformers can no longer handle the load without overheating and potentially burning out. Power factor correction can sometimes avoid the need for additional transformer capacity.

The question is, How can the power factor be brought closer to a perfect 1.0? The typical approach is fairly intuitive; that is, if the load is highly inductive, which most are, then try to offset that by adding capacitors as is suggested in Fig. 2.10. The idea is for the capacitor to provide the current that the inductance needs rather than having that come from the transformer. The capacitor, in turn,

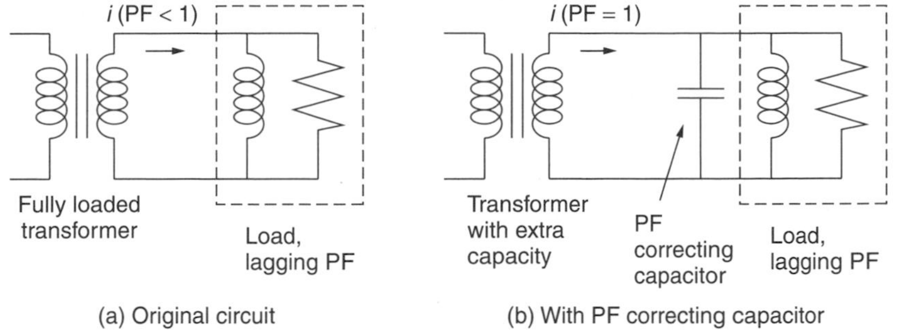

Figure 2.10 Correcting power factor for an inductive load by adding a parallel capacitor.

gets its current from the inductance. That is, the two reactive elements, capacitor and inductance, oscillate, sending current back and forth to each other.

Capacitors used for power factor compensation are rated by the volt-amps-reactive (VAR) that they supply at the system's voltage. When rated in these units, sizing a power-factor correcting capacitor is quite straightforward and is based on the kVAR of a capacitor offsetting some or all of the kVAR in the power triangle.

Example 2.7 Avoiding a New Transformer by Improving the Power Factor. A factory with a nearly fully loaded transformer delivers 600 kVA at a power factor of 0.75. Anticipated growth in power demand in the near future is 20%. How many kVAR of capacitance should be added to accommodate this growth so they don't have to purchase a larger transformer?

Solution. At PF = 0.75, the real power delivered at present is 0.75×600 kVA = 450 kW. And the phase angle is $\theta = \cos^{-1}(0.75) = 41.4^\circ$. If demand grows by 20%, then an additional 90 kW of real power will need to be supplied. At that point, if nothing is done, the new power triangle would show

$$\text{Real power } P = 450 + 90 = 540 \text{ kW}$$

$$\text{Apparent power } S = 540 \text{ kW}/0.75 = 720 \text{ kVA (too big for this transformer)}$$

$$\text{Reactive power } Q = VI \sin\theta = 720 \text{ kVA} \sin(41.4^\circ) = 476 \text{ kVAR}$$

For this transformer to still supply only 600 kVA, the power factor will have to be improved to at least

$$\text{PF} = 540 \text{ kW}/600 \text{ kVA} = 0.90$$

The phase angle now will be $\theta = \cos^{-1}(0.90) = 25.8°$. The reactive power will need to be reduced to

$$Q = 600\ \text{kVA}\sin 25.8° = 261\ \text{kVAR}$$

The difference in reactive power between the 476 kVAR needed without power factor correction and the desired 261 kVAR must be provided by the capacitor. Hence

$$\text{PF correcting capacitor} = 476 - 261 = 215\ \text{kVAR}$$

The power triangles before and after PF correction are shown below:

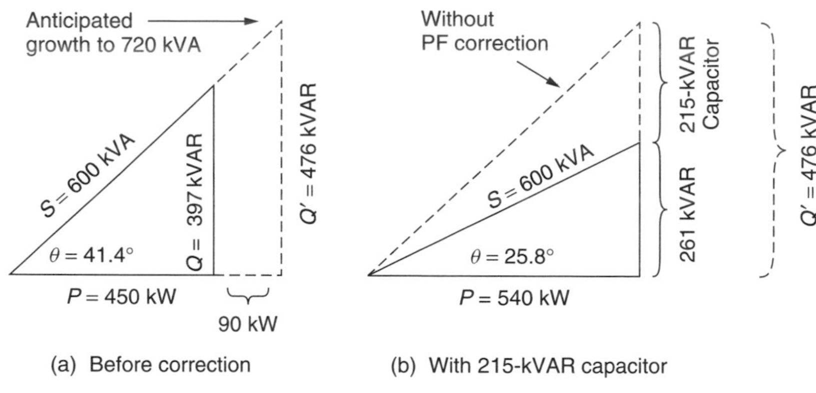

(a) Before correction (b) With 215-kVAR capacitor

While rating capacitors in terms of the VAR they provide is common in power systems, there are times when the actual value of capacitance is needed. From (2.24), we have a relationship between current through a capacitor and voltage across it:

$$V = \left(\frac{1}{\omega C}\right) I \tag{2.24}$$

The power through a capacitor is all reactive, so that

$$\text{VAR} = VI = V(\omega CV) = \omega CV^2 \tag{2.45}$$

That is, the conversion from VARs to farads for a capacitor is given by

$$C(\text{farads}) = \frac{\text{VARs}}{\omega V^2} \tag{2.46}$$

Notice, by the way, that the VAR rating of a capacitor depends on the square of the voltage. For example, a 100-VAR capacitor at 120 V would be a 400-VAR reactance at 240 V. That is, the VAR rating itself is meaningless without knowing the voltage at which the capacitor will be used.

Example 2.8 Power-Factor-Correcting Capacitor for a Motor What size capacitor would be needed to correct the power factor of the 230-V, 60-Hz, 5-hp motor in Example 2.6?

Solution. The capacitor must provide 4.40 kVAR of capacitive reactance to correct for the motor's 4.40 kVAR of inductive reactance. Since this is a 230-V motor operating at 60 Hz, (2.46) indicates that the capacitor should be

$$C = \frac{\text{VARs}}{\omega V^2} = \frac{4400}{2\pi \times 60 \times (230)^2} = 0.000221\ \text{F} = 221\ \mu\text{F}$$

2.5 THREE-WIRE, SINGLE-PHASE RESIDENTIAL WIRING

The wall receptacle at home provides single-phase, 60-Hz power at a nominal voltage of about 120 V (actual voltages are usually in the range of 110–120 V). Such voltages are sufficient for typical, low-power applications such as lighting, electronic equipment, toasters, and refrigerators. For appliances that requires higher power, such as an electric clothes dryer or an electric space heater, special outlets in your home provide power at a nominal 240 V. Running high-power equipment on 240 V rather than 120 V cuts current in half, which cuts i^2R heating of wires to one-fourth. That allows easy-to-work-with, 12-ga. wire to be used in a household, for both 120-V and 240-V applications. So, how is that 240 V provided?

Somewhere nearby, usually on a power pole or in a pad-mounted rectangular box, there is a transformer that steps down the voltage from the utility distribution system at typically 4.16 kV (though sometimes as high as 34.5 kV) to the 120 V/240 V household voltage. Figure 2.11 shows the basic three-wire, single-phase service drop to a home, including the transformer, electric meter, and circuit breaker panel box.

As shown in Fig. 2.11, by grounding the center tap of the secondary side of the transformer (the neutral, white wire), the top and bottom ends of the windings are at the equivalent of + 120 V and − 120 V. The voltage difference between the two "hot" sides of the circuit (the red and black wires) is 240 V. Notice the inherent safety advantages of this configuration: At no point in the home's wiring system is the voltage more than 120 V higher than ground.

The ± 120-V lines are 120 V (rms) with a 180° phase angle between them. In fact, it would be reasonable to say this is a two-phase system (but nobody does).

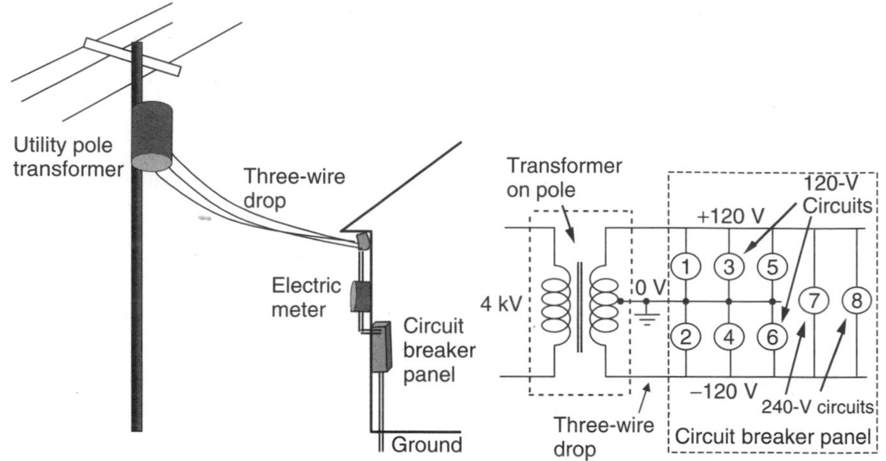

Figure 2.11 Three-wire, single-phase power drop, including the wiring in the breaker box to feed 120-V and 240-V circuits in the house.

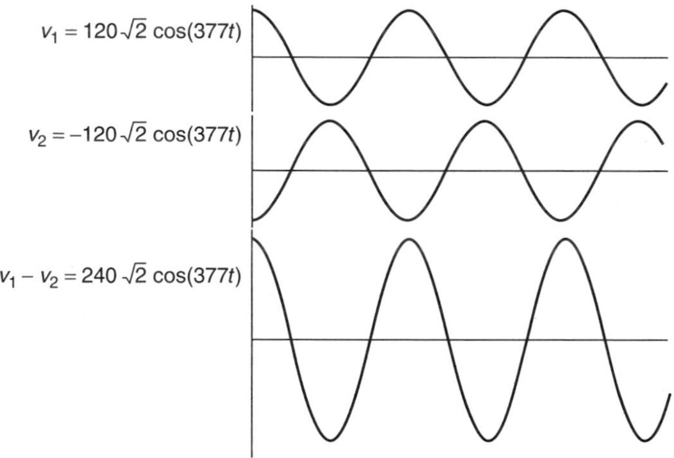

Figure 2.12 Waveforms for ± 120 V and the difference between them creating 240 V.

There are a number of ways to demonstrate the creation of 240 V across the two hot leads coming into a circuit. One is using algebra, which is modestly messy:

$$v_1 = 120\sqrt{2}\cos(2\pi \cdot 60t) = 120\sqrt{2}\cos 377t \tag{2.47}$$

$$v_2 = 120\sqrt{2}\cos(377t + \pi) = -120\sqrt{2}\cos 377t \tag{2.48}$$

$$v_1 - v_2 = 240\sqrt{2}\cos 377t \tag{2.49}$$

A second approach is to actually draw the waveforms, as has been done in Fig. 2.12.

Example 2.9 Currents in a Single-Phase, Three-Wire System. A three-wire, 120/240-V system supplies a residential load of 1200 W at 120 V on phase A, 2400 W at 120 V on phase B, and 4800 W at 240 V. The power factor for each load is 1.0. Find the currents in each of the three legs.

Solution. The 1200-W load at 120 V draws 10 A; the 2400-W load draws 20 A; and the 4800-W, 240-V load draws 20 A. A simple application of Kirchhoff's current law results in the following diagram. Notice that the sum of the currents at each node equals zero; also note that the currents are rms values that we can add directly since they each have current and voltage in phase.

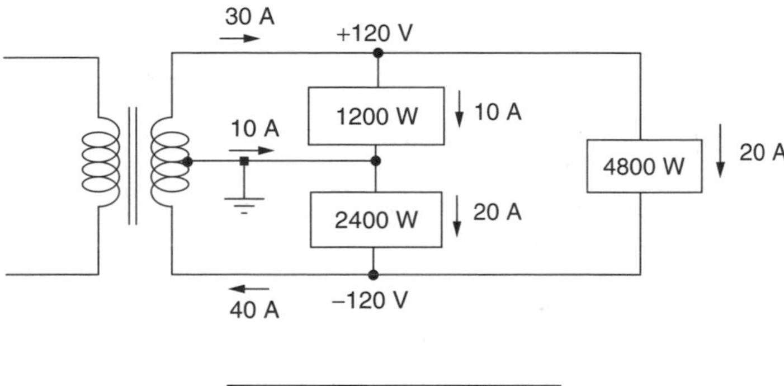

Example 2.9 illustrates the currents for a circuit with unequal power demand in each of the 120-V legs. Because the load is *unbalanced*, the current in the neutral wire is not equal to zero (it is 10 A). A *balanced* circuit has equal currents in each hot leg, and the current in the neutral is zero.

2.6 THREE-PHASE SYSTEMS

Commercial electricity is almost always produced with three-phase synchronous generators, and it is also almost always sent on its way along three-phase transmission lines. There are at least two good reasons why three-phase circuits are so common. For one, three-phase generators are much more efficient in terms of power per unit of mass and they operate much smoother, with less vibration, than single-phase generators. The second advantage is that three-phase transmission and distribution systems use their wires much more efficiently.

2.6.1 Balanced, Wye-Connected Systems

To understand the advantages of three-phase transmission lines, begin by comparing the three independent, single-phase circuits in Fig. 2.13a with the circuit shown in Fig. 2.13b. The three generators are the same in each case, so the total power delivered hasn't changed, but in Fig. 2.13b they are all sharing the same wire to return current to the generators. That is, by sharing the "neutral" return wire, only four wires are needed to transmit the same power as the six wires needed in the three single-phase circuits. That would seem to sound like a nice savings in transmission wire costs.

The potential problem with combining the neutral return wire for the three circuits in 2.13a is that we now have to size the return wire to handle the sum of the individual currents. So, maybe we haven't gained much after all in terms of saving money on the transmission cables. The key to making that return wire oversizing problem disappear is to be more clever in our choice of generators. Suppose that each generator develops the same voltage, but does so 120° out of phase with the other two generators, so that

$$v_a = V\sqrt{2}\cos(\omega t) \qquad \overline{V}_a = V\angle 0^\circ \tag{2.50}$$

$$v_b = V\sqrt{2}\cos(\omega t + 120^\circ) \qquad \overline{V}_b = V\angle 120^\circ \tag{2.51}$$

$$v_c = V\sqrt{2}\cos(\omega t + 240^\circ) = V\sqrt{2}\cos(\omega t - 120^\circ) \quad \overline{V}_c = V\angle 240^\circ = V\angle -120^\circ \tag{2.52}$$

Notice the simple vector notation introduced in (2.50) to (2.52). Voltages are described in terms of their rms values (e.g., V_b) and phase angle (e.g., 120°). This notation is commonly used in electrical engineering to represent ac voltages and currents as long as the frequency is not a concern.

Sizing the neutral return wire in Fig. 2.13b means that we need to look at currents flowing in each phase of the circuit so that we can add them up. The simplest situation to analyze occurs when each of the three loads are exactly the same so that the currents are all the same except for their phase angles. When

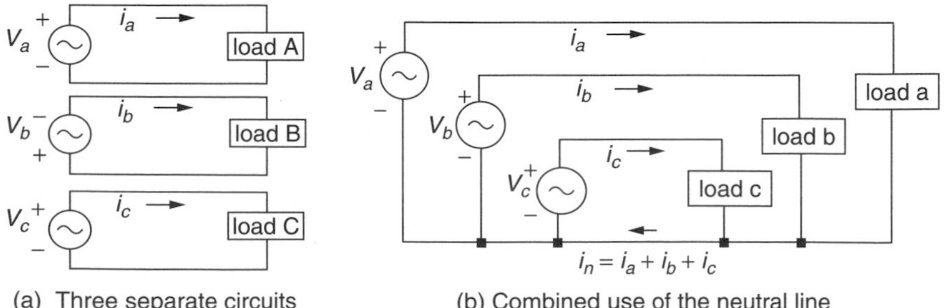

Figure 2.13 By combining the return wires for the circuits in (a), the same power can be sent using four wires instead of six. But it would appear the return wire could carry much more current than the supply lines.

that is the case, the three-phase circuit is said to be *balanced*. With balanced loads, the currents in each phase can be expressed as

$$i_a = I\sqrt{2}\cos(\omega t), \qquad \overline{I}_a = I\angle 0^\circ \tag{2.53}$$

$$i_b = I\sqrt{2}\cos(\omega t + 120^\circ), \qquad \overline{I}_b = I\angle 120^\circ \tag{2.54}$$

$$i_c = I\sqrt{2}\cos(\omega t + 240^\circ) = I\sqrt{2}\cos(\omega \mathrm{t} - 120^\circ), \qquad \overline{I}_c = I\angle 240^\circ = I\angle -120^\circ \tag{2.55}$$

The current flowing in the neutral wire is therefore

$$i_n = i_a + i_b + i_c = I\sqrt{2}[\cos(\omega t) + \cos(\omega t + 120^\circ) + \cos(\omega t - 120^\circ)] \tag{2.56}$$

This looks messy, but something great happens when you apply some trigonometry. Recall the identity

$$\cos A \cdot \cos B = \tfrac{1}{2}[\cos(A + B) + \cos(A - B)] \tag{2.57}$$

so that

$$\cos \omega t \cdot \cos(120^\circ) = \tfrac{1}{2}[\cos(\omega t + 120) + \cos(\omega t - 120^\circ)] \tag{2.58}$$

Substituting (2.58) into (2.56) gives

$$i_n = I\sqrt{2}[\cos \omega t + 2\cos \omega t \cdot \cos(120^o)] \tag{2.59}$$

But $\cos(120^\circ) = -1/2$, so that

$$i_n = I\sqrt{2}[\cos \omega t + 2\cos \omega t \cdot (-1/2)] = 0 \tag{2.60}$$

Now, we can see the startling conclusion: *For a balanced three-phase circuit, there is no current in the neutral wire .... in fact, for a balanced three-phase circuit, we don't even need the neutral wire!* Referring back to Fig. 2.13, what we have done is to go from six transmission cables for three separate, single-phase circuits to three transmission cables (of the same size) for a balanced three-phase circuit. In three-phase transmission lines, the neutral conductor is quite often eliminated or, if it is included at all, it will be a much smaller conductor designed to handle only modest amounts of current when loads are unbalanced.

While the algebra suggests transmission lines can do without their neutral cable, the story is different for three-phase loads. For three-phase *loads* (as opposed to three-phase transmission lines), the practice of undersizing neutral lines in wiring systems in buildings has had unexpected, dangerous consequences. As we will see later in this chapter, when loads include more and more computers, copy machines, and other electronic equipment, harmonics of the fundamental

60 Hz current are created, and those harmonics do not cancel out the way the fundamental frequency did in (2.60). The result is that undersized neutral lines in buildings can end up carrying much more current than expected, which can cause dangerous overheating and fires. Those same harmonics also play havoc on transformers in buildings as we shall see.

Figure 2.13b has been redrawn in its more conventional format in Fig. 2.14. As drawn, the configuration is referred to as a three-phase, four-wire, *wye connected* circuit. Later we will briefly look at another wiring system that creates circuits in which the connections form a *delta* rather than a wye. Figure 2.14 shows the specification of various voltages within the three-phase wye-connected system. The voltages measured with respect to the neutral wire—that is, V_a, V_b, and V_c—are called *phase voltages*. Voltages measured between the phases themselves are designated as follows: For example, the voltage at "a" with respect to the voltage at "b" is labeled V_{ab}. These voltages, V_{ab}, V_{ac}, and V_{bc}, are called *line voltages*. When the voltage on a transmission line or transformer is specified, it is always the line voltages that are being referred to.

Let us derive the relationship between phase voltages and line voltages. Letting the subscript "0" refer to the neutral line, we can write, for example, the line-to-line voltage between line a and line b:

$$v_{ab} = v_{a0} + v_{0b} = v_{a0} - v_{b0} \tag{2.61}$$

so

$$v_{ab} = V_a\sqrt{2}\cos\omega t - V_b\sqrt{2}\cos(\omega t - 120^\circ) \tag{2.62}$$

We will specify that the rms values of all of the phase voltages are the same (V_{ph}), so by symmetry the rms voltages of the line voltages are all the same (V_{line}). Now we can write

$$v_{\text{line}} = V_{\text{ph}}\sqrt{2}\cdot[\cos\omega t - \cos(\omega t - 120^\circ)] \tag{2.63}$$

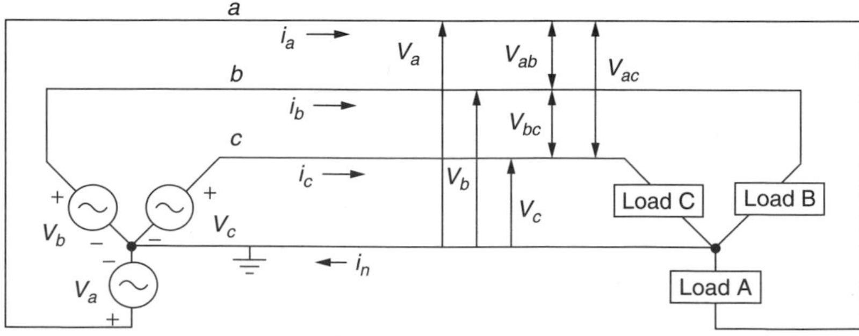

Figure 2.14 A four-wire, wye-connected, three-phase circuit, showing source and load.

Using the trigonometric identity $\cos x - \cos y = -2\sin\left[\frac{1}{2}(x+y)\right]\cdot\sin\left[\frac{1}{2}(x-y)\right]$ gives

$$v_{\text{line}} = V_{\text{ph}}\sqrt{2}\cdot(-2)\sin\left[\tfrac{1}{2}(\omega t + \omega t - 120^\circ)\right]\cdot\sin\left[\tfrac{1}{2}(\omega t - \omega t + 120^\circ)\right] \quad (2.64)$$

Assembling terms gives us

$$v_{\text{line}} = V_{\text{ph}}\sqrt{2}\cdot(-2)\sin(\omega t - 60^\circ)\cdot\sin(60^\circ) = V_{\text{ph}}\sqrt{2}\cdot(-2)\left(\frac{\sqrt{3}}{2}\right)\cdot\sin(\omega t - 60^\circ) \quad (2.65)$$

Finally, since $\sin x = -\cos(x + 90^\circ)$, we have

$$v_{\text{line}} = V_{\text{phase}}\sqrt{2}\cdot\sqrt{3}\cos(\omega t + 30^\circ) = V_{\text{line}}\sqrt{2}\cos(\omega t + \theta) \quad (2.66)$$

While (2.66) shows a phase shift, that is not the important result. What is important is the rms value of line voltage as a function of the rms value of phase voltage:

$$V_{\text{line}} = \sqrt{3}\ V_{\text{phase}} \quad (2.67)$$

To illustrate (2.67), the most widely used four-wire, three-phase service to buildings provides power at a line voltage of 208 V. With the neutral wire serving as the reference voltage, that means that the phase voltages are

$$V_{\text{phase}} = \frac{V_{\text{line}}}{\sqrt{3}} = \frac{208\ \text{V}}{\sqrt{3}} = 120\ \text{V} \quad (2.68)$$

For relatively high power demands, such as large motors, a line voltage of 480 V is often provided, which means that the phase voltage is $V_{\text{phase}} = 480/\sqrt{3} = 277$ V. The 277-V phase voltages are often used in large commercial buildings to power fluorescent lighting systems. A wiring diagram for a 480/277-V system, which includes a single-phase transformer to convert the 480-V line voltage into 120V/240V power, is shown in Fig. 2.15.

To find the power delivered in a balanced, three-phase system with a given line voltage, we need to consider all three kinds of power: apparent power S (VA), real power P (watts), and reactive power Q (VAR). The total apparent power is three times the apparent power in each phase:

$$S_{3\phi} = 3V_{\text{phase}}I_{\text{line}} \quad \text{(volt-amps)} \quad (2.69)$$

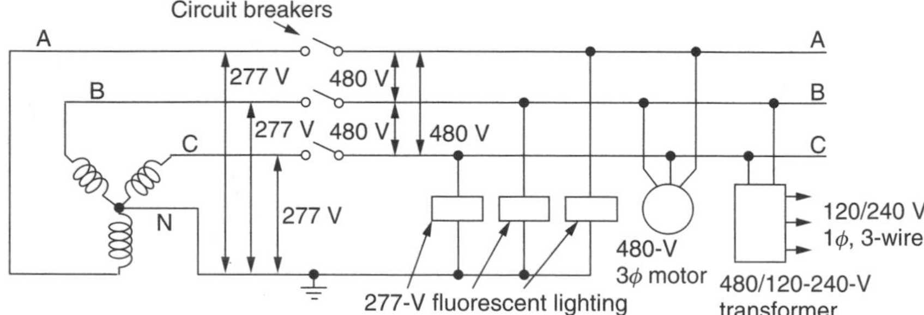

Figure 2.15 Example of a three-phase, 480-V, large-building wiring system that provides 480-V, 277-V, 240-V, and 120-V service. The voltage supply is represented by three coils, which are the three windings on the secondary side of the three-phase transformer serving the building. Source: Based on Stein and Reynolds (1992).

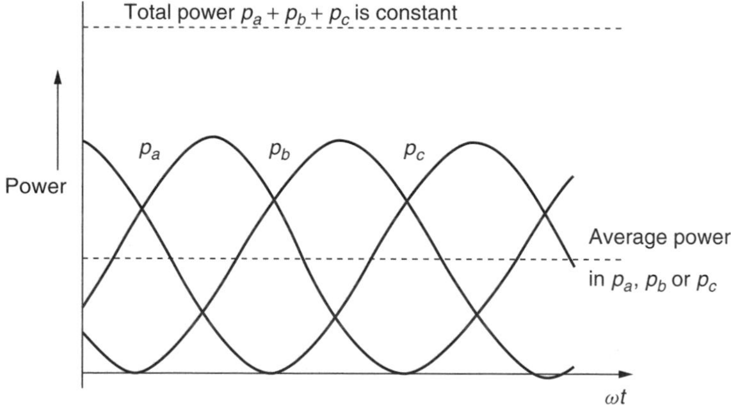

Figure 2.16 The sum of the three phases of power in balanced delta and wye loads is a constant, not a function of time.

where V_{phase} is the rms voltage in each phase, and I_{line} is the rms current in each phase (assumed to be the same in each phase). Applying (2.67) gives

$$S_{3\phi} = 3V_{\text{phase}}I_{\text{line}} = 3 \cdot \frac{V_{\text{line}}}{\sqrt{3}} \cdot I_{\text{line}} = \sqrt{3}V_{\text{line}}I_{\text{line}} \tag{2.70}$$

Similarly, the reactive power is

$$Q_{3\phi} = \sqrt{3}V_{\text{line}}I_{\text{line}}\sin\theta \quad \text{(VAR)} \tag{2.71}$$

where θ is the phase angle between line current and voltage, which is assumed to be the same for all three phases since this is a balanced load. Finally, the real power in a balanced three-phase circuit is given by

$$P_{3\phi} = \sqrt{3} V_{\text{line}} I_{\text{line}} \cos\theta \quad \text{(watts)} \tag{2.72}$$

While (2.72) gives an average value of real power, it can be shown algebraically that in fact the real power delivered is a constant that doesn't vary with time. The sketch shown in Fig. 2.16 shows how the summation of the power in each of the three legs leads to a constant total. That constant level of power is responsible for one of the advantages of three-phase power—that is, the smoother performance of motors and generators. For single-phase systems, instantaneous power varies sinusoidally leading to rougher motor operation.

Example 2.10 Correcting the Power Factor in a Three-Phase Circuit. Suppose that a shop is served with a three-phase, 208-V transformer. The real power demand of 80 kW is mostly single-phase motors, which cause the power factor to be a rather poor 0.5. Find the total apparent power, the individual line currents, and real power before and after the power factor is corrected to 0.9. If power losses before PF correction is 5% (4 kW), what will they be after power factor improvement?

Solution. Before correction, with 80 kW of real power being drawn, the apparent power before PF correction can be found from

$$P_{3\phi} = 80\text{ kW} = \sqrt{3} V_{\text{line}} I_{\text{line}} \cos\theta = S_{3\phi}\cos\theta = S_{3\phi} \cdot 0.5$$

so, the total apparent power is

$$S_{3\phi} = \frac{80\text{ kW}}{0.5} = 160\text{ kVA}$$

From (2.70), the current flowing in each leg of the three-phase system is

$$I_{\text{line}} = \frac{S_{3\phi}}{\sqrt{3} \cdot V_{\text{line}}} = \frac{160\text{ kVA}}{\sqrt{3} \cdot 208\text{ V}} = 0.444\text{ kA} = 444\text{ A}$$

After correcting the power factor to 0.9 the resulting apparent power is now

$$S_{3\phi} = \frac{P_{3\phi}}{\cos\theta} = \frac{80\text{ kW}}{0.9} = 88.9\text{ kVA}$$

The current in each line is now,

$$I_{\text{line}} = \frac{S_{3\phi}}{\sqrt{3} \cdot V_{\text{line}}} = \frac{88.9\text{ kVA}}{\sqrt{3} \cdot 208\text{ V}} = 0.247\text{ kA} = 247\text{ A}$$

Before correction, line losses are 5% of 80 kW, which is 4 kW. Since line losses are proportional to the square of current, the losses after power factor correction will be reduced to

$$\text{Losses after correction, } P_{\text{line}} = 4\text{ kW} \cdot \frac{(247)^2}{(444)^2} = 1.24\text{ kW}$$

So, line losses in the factory have been cut from 5 kW to 1.24 kW—a decrease of almost 70%.

Power factor adjustment not only reduces line losses (as the above example illustrates) but also makes a smaller, less expensive transformer possible. In a transformer, it is the heat given off as current runs through the windings that determines its rating. Transformers are rated by their voltage and their kVA limits, and not by the kW of real power delivered to their loads. In the above example, the kVA needed drops from 160 to 88.9 kVA—a reduction of 44%. That reduction could be used to accommodate future growth in factory demand without needing to buy a new, bigger transformer; or perhaps, when the existing transformer needs replacement, a smaller one could be purchased.

2.6.2 Delta-Connected, Three-Phase Systems

So far we have dealt only with three-phase circuits that are wired in the wye configuration, but there is another way to connect three-phase generators, transformers, transmission lines, and loads. The *delta* connection uses three wires

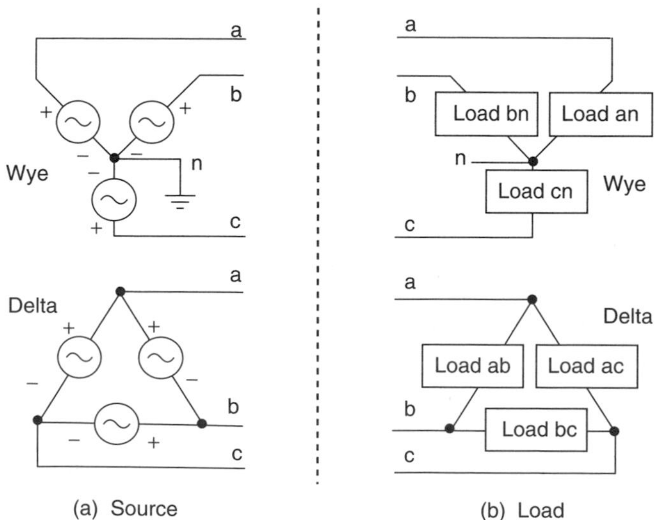

Figure 2.17 Wye and Delta sources and loads.

TABLE 2.1 Summary of Wye and Delta-Connected Current, Voltage, and Power Relationships[a]

Quantity	Wye-Connected	Delta-Connected
Current (rms)	$I_{\text{line}} = I_{\text{phase}}$	$I_{\text{line}} = \sqrt{3}\ I_{\text{phase}}$
Voltage (rms)	$V_{\text{line}} = \sqrt{3}\ V_{\text{phase}}$	$V_{\text{line}} = V_{\text{phase}}$
Power	$P_{3\phi} = \sqrt{3}\ V_{\text{line}} I_{\text{line}} \cos\theta$	$P_{3\phi} = \sqrt{3}\ V_{\text{line}} I_{\text{line}} \cos\theta$

[a]The angle θ for delta system is the phase angle of the loads

and has no inherent ground or neutral line (though oftentimes, one of the lines is grounded). The wiring diagrams for wye and delta connections are shown in Fig. 2.17. A summary of the key relationships between currents and voltages for wye-connected and delta-connected three-phase systems is presented in Table 2.1.

2.7 POWER SUPPLIES

Almost all electronic equipment these days requires a power supply to convert 120-V ac from the power lines into the low-voltage, 3- to 15-V dc needed whenever digital technologies are incorporated into the device. Everything from televisions to computers, copy machines, portable phones, electric-motor speed controls, and so forth—virtually anything that has an integrated circuit, digital display, or electronic control function—uses them. Even portable electronic products that operate with batteries will usually have an external power supply to recharge their batteries. It has been estimated that roughly 6% of the total electricity sold in the United States—some 210 billion kWh/yr—passes through these power supplies, with roughly one-third of that simply ending up as waste heat. Of the 2.5 billion power supplies in use in the United States, approximately 40% are ac adapters external to the device, while the remainder are mounted inside the appliance itself.

Power supplies can be categorized into somewhat traditional *linear* supplies, which are those that operate with transformers to drop the incoming ac voltage to an appropriate level, and *switching*, or *switch-mode*, power supplies that skip the transformer and do their voltage conversion using a technique based on the rapid on-and-off switching of a transistorized circuit. Linear power supplies typically operate in the 50–60% range of energy efficiency, while switching power supplies are closer to 70–80% energy efficient. Increasing the efficiency of both types of power supplies, along with replacing linear with switch-mode supplies, could save roughly 1% of U.S. electricity and $2.5 billion annually (Calwell and Reeder, 2002).

Figure 2.18 provides an example comparing the efficiencies of a 9-V linear power supply for a cordless phone versus one incorporating a switch-mode supply. The switching supply is far more efficient throughout the range of currents

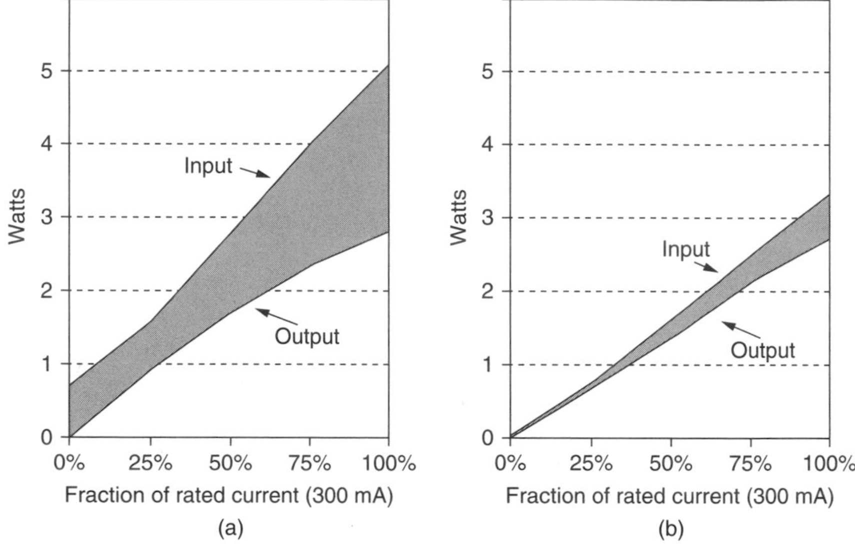

Figure 2.18 Power consumed by 9-V linear (a) and switch-mode power supplies (b) for a cordless phone. Source: Calwell and Reeder (2002).

drawn. Note, too, that the linear supply continues to consume power even when the device is delivering no power at all to the load. The wasted energy that occurs when appliances are apparently turned off, but continue to consume power, or when the appliance isn't performing its primary function, is referred to as standby power. A typical U.S. household has approximately 20 appliances that together consume about 500 kWh/yr in standby mode. That 5–8% of all residential electricity costs about $4 billion per year (Meier, 2002).

2.7.1 Linear Power Supplies

A device that converts ac into dc is called a *rectifier*. When a rectifier is equipped with a *filter* to help smooth the output, the combination of rectifier and filter is usually referred to as a dc *power supply*. In the opposite direction, a device that converts dc into ac is called an *inverter*.

The key component in rectifying an ac voltage into dc is a *diode*. A diode is basically a one-way street for current: It allows current to flow unimpeded in one direction, but it blocks current flow in the opposite direction. In the forward direction, an ideal diode looks just like a zero-resistance short circuit so that the voltage at both diode terminals is the same. In the reverse direction, no current flows and the ideal diode acts like an open circuit. Figure 2.19 summarizes the characteristics of an ideal diode, including its current versus voltage relationship.

Of course an idealized diode is only an approximation to the real thing. Real diodes are semiconductor devices, usually made of silicon, with the relationship

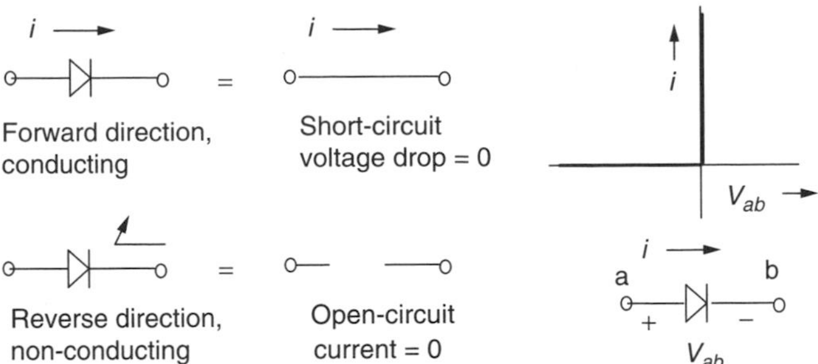

Figure 2.19 Characteristics of an ideal diode. In the forward direction it acts like a short circuit; in the reverse direction it appears to be an open circuit.

between current and voltage expressed as follows:

$$I = I_0(e^{V/E_T} - 1) \tag{2.73}$$

where I_0 is the reverse saturation current, usually much less than 1 μA; and E_T is the energy equivalent of temperature = 0.026 V at room temperature. The exponential in (2.73) indicates a very rapid rise in current for very modest changes in voltage. In the forward direction, the voltage drop for a silicon diode is often approximated to be on the order of 0.7 V, which leads to the sometimes useful diode model shown in Fig. 2.20. When that 0.7-V drop across a diode is unacceptably high, other, more expensive Shottkey diodes with voltage drops closer to 0.2 V are often used.

The simplest rectifier is comprised of just a single diode placed between the ac source voltage and the load as shown in Fig. 2.21. On the positive stroke of the input voltage, the diode is forward-biased, current flows, and the full voltage appears across the load. When the input voltage goes negative, however, current wants to go in the opposite direction, but it is prevented from doing so by the

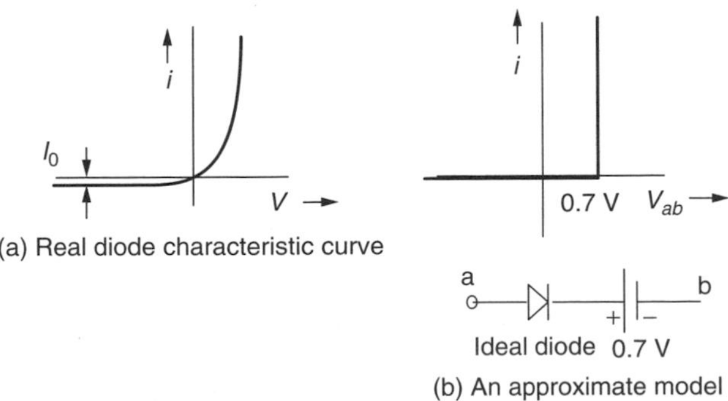

Figure 2.20 Real diode $I-V$ curve, and an often-used approximation.

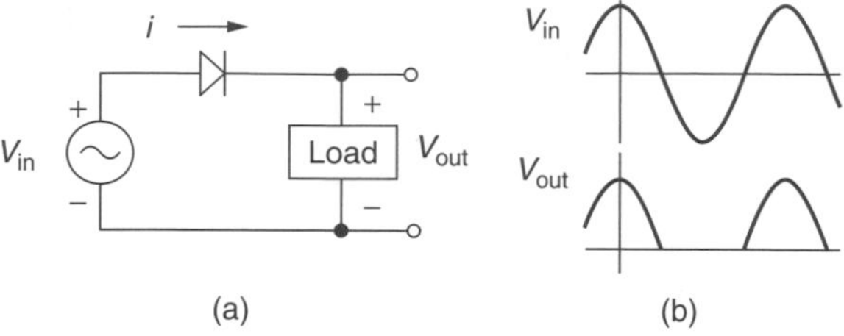

Figure 2.21 A half-wave rectifier: (a) The circuit. (b) The input and output voltages.

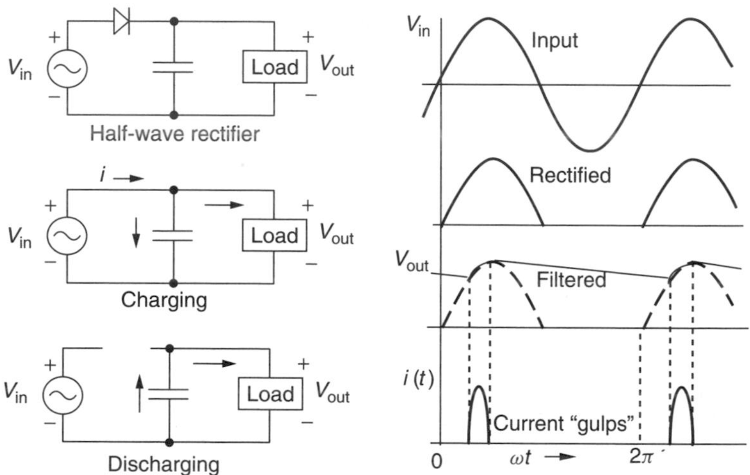

Figure 2.22 A half-wave rectifier with capacitor filter showing the gulps of current that occur during the brief periods when the capacitor is charging.

diode. No current flows, so there is no voltage drop across the load, leading to the output voltage waveform shown in Fig. 2.21b.

While the output voltage waveform in Fig. 2.21b doesn't look very much like dc, it does have an average value that isn't zero. The dc value of a waveform is defined to be the average value, so the waveform does have a dc component, but it also has a bunch of wiggles, called *ripple*, in addition to its dc level. The purpose of a filter is to smooth out those ripples. The simplest filter is just a big capacitor attached to the output, as shown in Fig. 2.22. During the last portion of the upswing of input voltage, current flows through the diode to the load and capacitor, and it charges the capacitor. Once the input voltage starts to drop, the diode cuts off and the capacitor then supplies current to the load. The resulting output voltage is greatly smoothed compared to the rectifier without the capacitor. Notice how current flows from the input source only for a short while in each cycle and does so very close to the times when the input voltage peaks. The circuit "gulps" current in a highly nonlinear way.

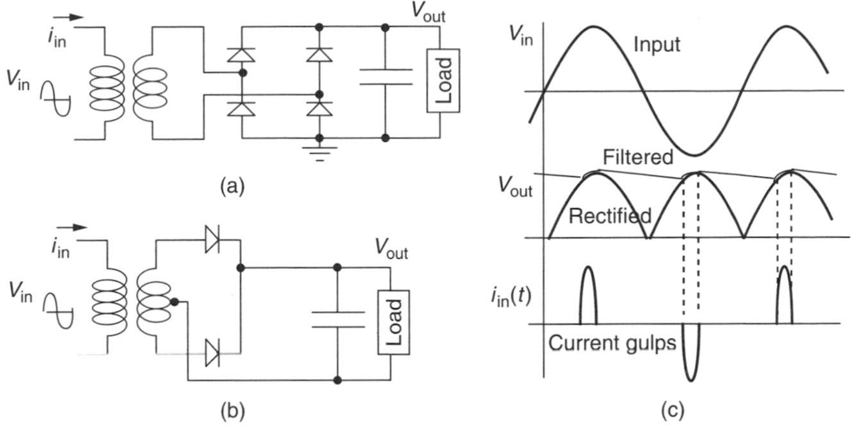

Figure 2.23 Full-wave rectifiers with capacitor filter showing gulps of current drawn from the supply. (a) A four-diode, bridge rectifier. (b) A two-diode, center-tapped transformer rectifier.

The ripple on the output voltage can be further reduced by using a full-wave rectifier instead of the half-wave version described above. Two versions of power supplies incorporating full-wave rectifiers are shown in Fig. 2.23: One is shown using a center-tapped transformer with just two diodes; the other uses a simpler transformer with a four-diode bridge rectifier. The transformers drop the voltage to an appropriate level. The capacitors smooth the full-wave-rectified output for the load. Numerous small, home electronic devices have transformer-based, battery-charging power supplies such as these, which plug directly into the wall outlet. A disadvantage of the transformer is that it is somewhat heavy and bulky and it continues to soak up power even if the electronic device it is powering is turned off. Either of these full-wave rectifier approaches results in a voltage waveform that has two positive humps per cycle as shown. This means that the capacitor filter is recharged twice per cycle instead of once, which smoothes the output voltage considerably. Notice that there are now two gulps of current from the source: one in the positive direction, one in the negative.

Three-phase circuits also use the basic diode rectification idea to produce a dc output. Figure 2.24a shows a three-phase, half-wave rectifier. At any instant the phase with the highest voltage will forward bias its diode, transferring the input voltage to the output. The result is an output voltage that is considerably smoother than a single-phase, full-wave rectifier.

The three-phase, full-wave rectifier shown in Fig. 2.24b is better still. The voltage at any instant that reaches the output is the difference between the highest of the three input voltages and the lowest of the three phase voltages. It therefore has a higher average voltage than the three-phase, half-wave rectifier, and it reaches its peak values with twice the frequency, resulting in relatively low ripple even without a filter. For very smooth dc outputs, an inductor (sometimes called a "choke") is put in series with the load to act as a smoothing filter.

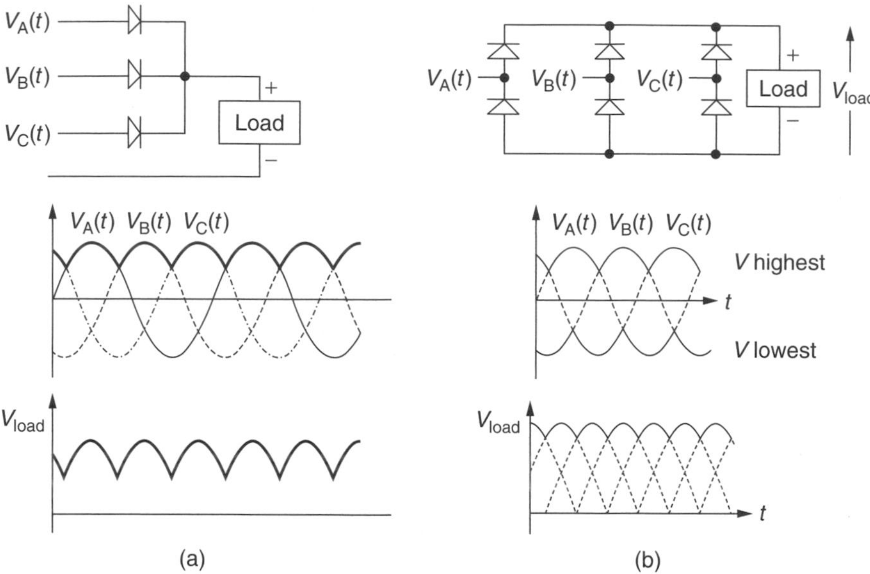

Figure 2.24 (a) Three-phase, half-wave rectifier. (b) Three-phase, full-wave rectifier. After Chapman (1999).

2.7.2 Switching Power Supplies

While most dc power supplies in the past were based on circuitry that dropped the incoming ac voltage with a transformer, followed by rectification and filtering, that is no longer the case. More often, modern power supplies skip the transformer altogether and rectify the ac at its incoming relatively high voltage (e.g., $120\sqrt{2} = 170$ V peak) and then use a *dc-to-dc converter*, to adjust the dc output voltage.

Actual dc-to-dc voltage conversion circuits are quite complex (see, for example, *Elements of Power Electronics*, P. T. Krein, 1998), but we can get a basic understanding of their operation by studying briefly the simple *buck converter* depicted in Fig. 2.25.

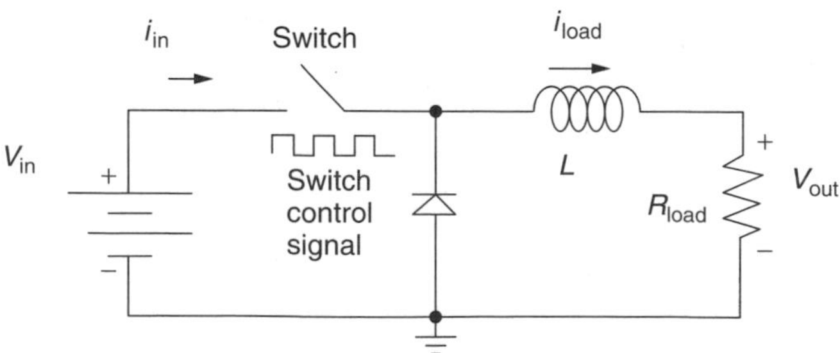

Figure 2.25 A dc-to-dc, step-down, voltage converter—sometimes called a *buck converter*.

In Fig. 2.25, the rectified, incoming (high) voltage is represented by an idealized dc source of voltage V_{in}. There is also a switch that allows this dc input voltage to be either (a) connected across the diode so that it can deliver current to the inductor and load or (b) disconnected from the circuit entirely. The switch is not something that is opened or closed manually, but is a semiconductor device that can be rapidly switched on and off with a electrical control signal. The switch itself is usually an insulated-gate bipolar transistor (IGBT), a silicon-controlled rectifier (SCR), or a gate-turnoff thyristor (GTO)—devices that you may have encountered in electronics courses, but which we don't need to understand here beyond the fact that they are voltage-controlled, on/off switches. The voltage control is a signal that we can think of as having a value of either 0 or 1. When the control signal is 1, the switch is closed (short circuit); when it is 0, the switch is open. This on/off control can be accomplished with associated digital (or sometimes analog) circuitry not shown here.

The basic idea behind the buck converter is that by rapidly opening and closing the switch, short bursts of current are allowed to flow through the inductor to the load (here represented by a resistor). With the switch closed, the diode is reverse-biased (open circuit) and current flows directly from the source to the load (Fig. 2.26a). When the switch is opened, as in Fig. 2.26b, the inductor keeps the current flowing through the load resistor and diode (remember, inductors act as "current momentum" devices—we can't instantaneously start or stop current through one). If the switch is flipped on and off with high enough frequency, the current to the load doesn't have much of a chance to build up or decay; that is, it is fairly constant, producing a dc voltage on the output.

The remaining feature to describe in the buck converter is the relationship between the dc input-voltage V_{in}, and the dc output voltage, V_{out}. It turns out that the relationship is a simple function of the *duty cycle* of the switch. The duty cycle, D, is defined to be the fraction of the time that the control voltage is a "1" and the switch is closed. Figure 2.27 illustrates the duty cycle concept.

We can now sketch the current through the load and the current from the source, as has been done in Fig. 2.28. When the switch is closed, the two currents are equal and rising. When the switch is open, the input current immediately drops to zero and the load current begins to sag. If the switching rate is fast enough (and they are designed that way), the rising and falling of these currents is essentially

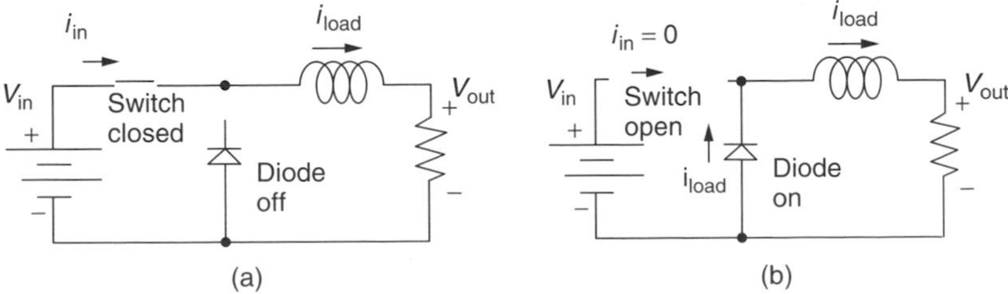

Figure 2.26 The buck converter with (a) switch closed and (b) switch open.

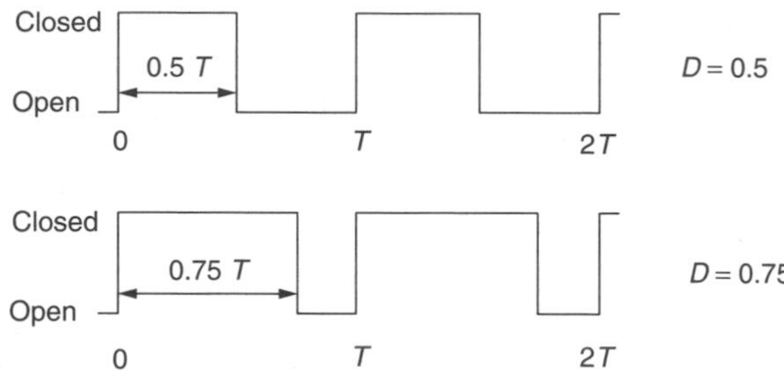

Figure 2.27 The fraction of the time that the switch in a dc-to-dc buck converter is closed is called the duty cycle, D. Two examples are shown.

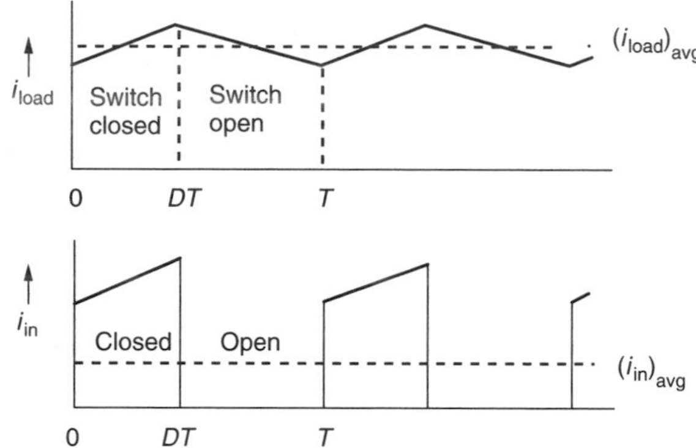

Figure 2.28 When the switch is closed, the input and load currents are equal and rising; when it is open, the input drops to zero and the load current sags.

linear so they appear as straight lines in the figure. With T representing the period of the switching circuit, then within each cycle the switch is closed DT seconds.

We are now ready to determine the relationship between input voltage and output voltage in the switching circuit. Start by using Fig. 2.28 to determine the relationship between average input current and average output current. While the switch is closed, the areas under the input-current and load-current curves are equal:

$$(i_{\text{in}})_{\text{avg}} \cdot T = (i_{\text{load}})_{\text{avg}} \cdot DT \qquad \text{so} \qquad (i_{\text{in}})_{\text{avg}} = D \cdot (i_{\text{load}})_{\text{avg}} \tag{2.74}$$

We will use an energy argument to determine the voltage relationship. Begin by writing the average power delivered to the circuit by the input voltage source:

$$(P_{\text{in}})_{\text{avg}} = (V_{\text{in}} \cdot i_{\text{in}})_{\text{avg}} = V_{\text{in}}(i_{\text{in}})_{\text{avg}} = V_{\text{in}} \cdot D \cdot (i_{\text{load}})_{\text{avg}} \tag{2.75}$$

The average power into the circuit equals the average power dissipated in the switch, diode, inductor, and load. If the diode and switch are ideal components, they dissipate no energy at all. Also, we know the average power dissipated by an ideal inductor as it passes through its operating cycle is also zero. That means the average input power equals the average power dissipated by the load. The power dissipated by the load is given by

$$(P_{\text{load}})_{\text{avg}} = (V_{\text{out}} \cdot i_{\text{load}})_{\text{avg}} = (V_{\text{out}})_{\text{avg}} \cdot (i_{\text{load}})_{\text{avg}} \tag{2.76}$$

Equating (2.75) and (2.76) gives

$$V_{\text{in}} \cdot D \cdot (i_{\text{load}})_{\text{avg}} = (V_{\text{out}})_{\text{avg}} \cdot (i_{\text{load}})_{\text{avg}} \tag{2.77}$$

which results in the relationship we have been looking for

$$(V_{\text{out}})_{\text{avg}} = D \cdot V_{\text{in}} \tag{2.78}$$

So, the only parameter that determines the dc-to-dc buck converter step-down voltage is the duty cycle of the switch.

As long as the switching cycle is fast enough, the load will be supplied with a very precisely controlled dc voltage. The buck converter is essentially, then, a dc transformer. Since the output voltage is controlled by the width of the control pulses, this approach is sometimes referred to as *pulse-width-modulation* (PWM). When the buck converter is fed by a rectified ac signal, perhaps with capacitor smoothing, the whole unit becomes a switch-mode power supply.

Buck converters are used in a great many electronic circuit applications besides switch-mode power supplies, including high-performance dc motor controllers, as the circuit of Fig. 2.29 suggests.

The dc motor is modeled as an inductor (the motor windings) in series with a back-emf voltage source. The back emf produced by the motor is proportional to the speed of the motor, $E = k\omega$. The voltage across the back emf of the motor is the same as that given in (2.78), which lets us solve for the motor speed:

$$E = D \cdot V_{\text{in}} = k\omega \qquad \text{so} \qquad \omega = \frac{D \cdot V_{\text{in}}}{k} \tag{2.79}$$

That is, the motor speed is directly proportional to the duty cycle of the switch. Pretty simple.

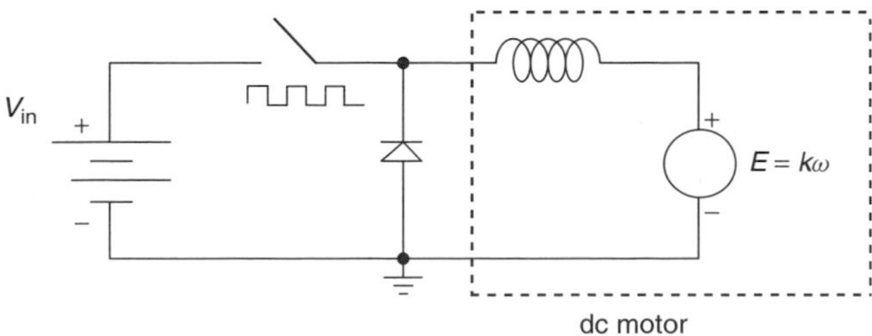

Figure 2.29 A buck converter used as a dc-motor speed controller.

Later, in Chapter 9, we will see how a somewhat similar circuit, which can both raise and lower dc voltages from one level to another, is used to enhance the performance of photovoltaic arrays.

2.8 POWER QUALITY

Utilities have long been concerned with a set of current and voltage irregularities, which are lumped together and referred to as *power quality* issues. Figure 2.30 illustrates some of these irregularities. Voltages that rise above, or fall below, acceptable levels, and do so for more than a few seconds, are referred to as undervoltages and overvoltages. When those abnormally high or low voltages are momentary occurrences lasting less than a few seconds, such as might be caused by a lightning strike or a car ramming into a power pole, they are called sag and swell incidents. Transient surges or spikes lasting from a few microseconds to milliseconds are often caused by lightning strikes, but can also be caused by the utility switching power on or off somewhere else in the system. Downed power lines can blow fuses or trip breakers, resulting in power interruptions or outages.

Power interruptions of even very short duration, sometimes as short as a few cycles, or voltage sags of 30% or so, can bring the assembly line of a factory to a standstill when programmable logic controllers reset themselves and adjustable speed drives on motors malfunction. Restarting such lines can cause delays and wastage of damaged product, with the potential to cost hundreds of thousands of dollars per incident. Outages in digital-economy businesses can be even more devastating.

While most of the power quality problems shown in Fig. 2.30 are caused by disturbances on the utility side of the meter, two of the problems are caused by

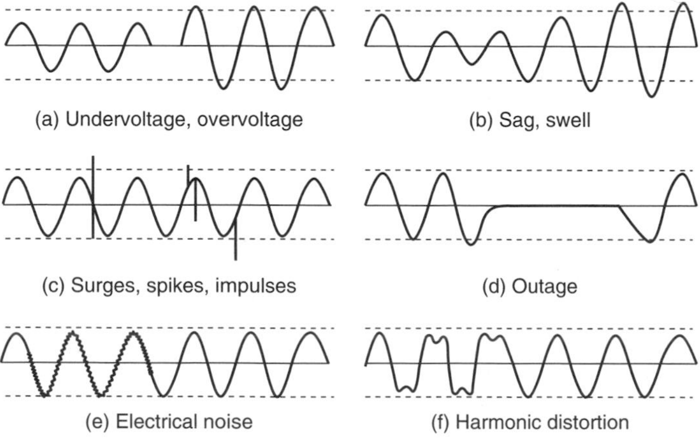

Figure 2.30 Power quality problems (Lamarre, 1991).

the customers themselves. As shown in Fig. 2.30e, when circuits are not well grounded, a continuous, jittery voltage "noise" appears on top of the sinusoidal signal. The last problem illustrated in Fig. 2.36f is harmonic distortion, which shows up as a continuous distortion of the normal sine wave. Solutions to power quality problems lie on both sides of the meter. Utilities have a number of technologies including filters, high-energy surge arrestors, fault-current limiters, and dynamic voltage restorers that can be deployed. Customers can invest in uninterruptible power supplies (UPS), voltage regulators, surge suppressors, filters, and various line conditioners. Products can be designed to be more tolerant of irregular power, and they can be designed to produce fewer irregularities themselves. And, as will be seen in Chapter 4, one of the motivations for distributed generation technologies, in which customers produce their own electricity, is increasing the reliability and quality of their electric power.

2.8.1 Introduction to Harmonics

Loads that are modeled using our basic components of resistance, inductance, and capacitance, when driven by sinusoidal voltage and current sources, respond with smooth sinusoidal currents and voltages of the same frequency throughout the circuit. As we have seen in the discussion on power supplies, however, electronic loads tend to draw currents in large pulses. Those nonlinear "gulps" of current can cause a surprising number of very serious problems ranging from blown circuit breakers to computer malfunctions, transformer failures, and even fires caused by overloaded neutral lines in three-phase wiring systems in buildings.

The harmonic distortion associated with gulps of current is especially important in the context of energy efficiency since some of the most commonly used efficiency technologies, including electronic ballasts for lighting systems and adjustable speed drives for motors, are significant contributors to the problem. Ironically, everything digital contributes to the problem, and at the same time it is those digital devices that are often the most sensitive to the distortion they create.

To understand harmonic distortion and its effects, we need to review the somewhat messy mathematics of periodic functions. Any periodic function can be represented by a Fourier series made up of an infinite sum of sines and cosines with frequencies that are multiples of the fundamental (e.g., 60 Hz) frequency. Frequencies that are multiples of the fundamental are called harmonics; for example, the third harmonic for a 60-Hz fundamental is 180 Hz.

The definition of a periodic function is that $f(t) = f(t+T)$, where T is the period. The Fourier series, or harmonic analysis, of any periodic function can be represented by

$$\begin{aligned} f(t) = \left(\frac{a_0}{2}\right) &+ a_1 \cos \omega t + a_2 \cos 2\omega t + a_3 \cos 3\omega t + \cdots \\ &+ b_1 \sin \omega t + b_2 \sin 2\omega t + b_3 \sin 3\omega t + \cdots \end{aligned} \tag{2.80}$$

where $\omega = 2\pi f = 2\pi/T$. The coefficients can be found from

$$a_n = \frac{2}{T}\int_0^T f(t)\cos n\omega t\, dt, \qquad n = 0, 1, 2\ldots \tag{2.81}$$

and

$$b_n = \frac{2}{T}\int_0^T f(t)\sin n\omega t\, dt, \qquad n = 1, 2, 3\ldots \tag{2.82}$$

Under special circumstances, the series in (2.80) simplifies. For example, when there is no dc component to the waveform (average value = 0), the first term, a_0, drops out:

$$a_0 = 0: \qquad \text{when average value, dc} = 0 \tag{2.83}$$

For functions with symmetry about the y-axis, the series contains only cosine terms. That is,

$$\text{cosines only:} \qquad \text{when } f(t) = f(-t) \tag{2.84}$$

For the series to contain only sine terms, it must satisfy the relation

$$\text{sines only:} \qquad \text{when } f(t) = -f(-t) \tag{2.85}$$

Finally, when a function has what is *called half-wave-symmetry* it contains no even harmonics. That is,

$$\text{no even harmonics:} \qquad \text{when } f\left(t + \frac{T}{2}\right) = -f(t) \tag{2.86}$$

Examples of these properties are illustrated in Fig. 2.31. Notice the waveform in Fig. 2.31c is very much like the current waveform for a full-wave rectifier, which says that its Fourier spectrum will have only cosine terms with only odd harmonics.

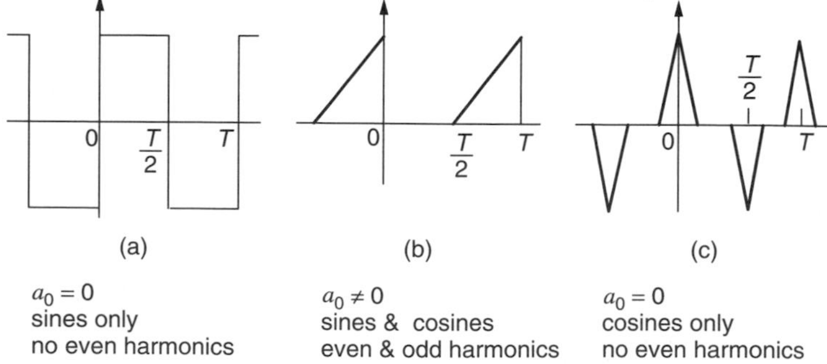

Figure 2.31 Examples of periodic functions, with indications of special properties of their Fourier series representations.

Example 2.11 Harmonic Analysis of a Square Wave. Find the Fourier series equivalent of the square wave in Fig. 2.31a, assuming that it has a peak value of 1 V.

Solution We know by inspection, using (2.83) to (2.86), that the series will have only sines, with no even harmonics. Therefore, all we need are the even coefficients b_n from (2.82):

$$b_n = \frac{2}{T}\int_0^T f(t)\sin n\omega t\,dt = \frac{2}{T}\left[\int_0^{T/2} 1\cdot\sin n\omega t\,dt + \int_{T/2}^{T}(-1)\cdot\sin n\omega t\,dt\right]$$

Recall that the integral of a sine is the cosine with a sign change, so

$$\begin{aligned} b_n &= \frac{2}{T}\left[\frac{-1}{n\omega}\cos n\omega t\Big|_{t=0}^{t=T/2} + \frac{1}{n\omega}\cos n\omega t\Big|_{t=T/2}^{t=T}\right] \\ &= \frac{2}{n\omega T}\left\{(-1)\left[\cos n\omega\frac{T}{2} - \cos n\omega\cdot 0\right] + \cos n\omega T - \cos n\omega\frac{T}{2}\right\} \end{aligned}$$

Substituting $\omega = 2\pi f = 2\pi/T$ gives

$$\begin{aligned} &= \frac{T}{2\pi}\cdot\frac{2}{nT}\left\{-\cos\left(\frac{2\pi}{T}\cdot\frac{nT}{2}\right) + 1 + \cos\left(\frac{2\pi}{T}\cdot nT\right) - \cos\left(\frac{2\pi}{T}\cdot\frac{nT}{2}\right)\right\} \\ &= \frac{1}{n\pi}[-2\cos n\pi + 1 + \cos 2n\pi] \end{aligned}$$

Since this is half-wave symmetric, there are no even harmonics; that is, n is an odd number. For odd values of n,

$$\cos n\pi = \cos\pi = -1 \quad \text{and} \quad \cos 2n\pi = \cos 0 = 1$$

That makes for a nice, simple solution:

$$b_n = \frac{4}{n\pi}$$

That is,

$$b_1 = \frac{4}{\pi} = 1.273, \qquad b_3 = \frac{4}{3\pi} = 0.424, \qquad b_5 = \frac{4}{5\pi} = 0.255\ldots$$

So the series is

$$\text{(Square wave with amplitude of 1)} = \frac{4}{\pi}\left[\sin\omega t + \frac{1}{3}\sin 3\omega t + \frac{1}{5}\sin 5\omega t + \cdots\right]$$

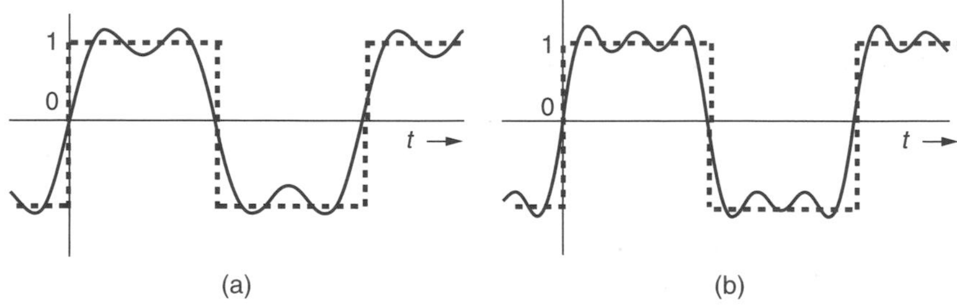

Figure 2.32 Showing the sum of the first two terms (a) and first three terms (b) of the Fourier series for a square wave along with the square wave that it is approximating.

To show how quickly the Fourier series for the square wave begins to approximate reality, Fig. 2.32 shows the sum of the first two terms of the series, and the sum of the first three terms, along with the square wave that it is approximating. Adding more terms, of course, will make the approximation more and more accurate.

The first few terms in the harmonic analysis of the square wave derived in Example 2.11 are presented in tabular form in Table 2.2. Two columns are shown for the harmonics: one that lists the amplitudes of each harmonic, and the other listing the harmonics as a percentage of the fundamental amplitude. Both presentations are commonly used.

Another convenient way to describe the harmonic analysis that the Fourier series provides is with bar graphs such as those shown in Fig. 2.33 for a square wave.

An example harmonic spectrum for the current drawn by an electronically ballasted compact fluorescent lamp (CFL) is shown in Fig. 2.34. Notice only odd harmonics are present, which is the case for all periodic half-wave symmetric

TABLE 2.2 Harmonic Analysis of a Square Wave with Amplitude 1[a]

Harmonic	Amplitude	Percentage of fundamental
1	1.273	100.0%
3	0.424	33.3%
5	0.255	20.0%
7	0.182	14.3%
9	0.141	11.1%
11	0.116	9.1%
13	0.098	7.7%
15	0.085	6.7%

[a]Only harmonics through the 15th are shown.

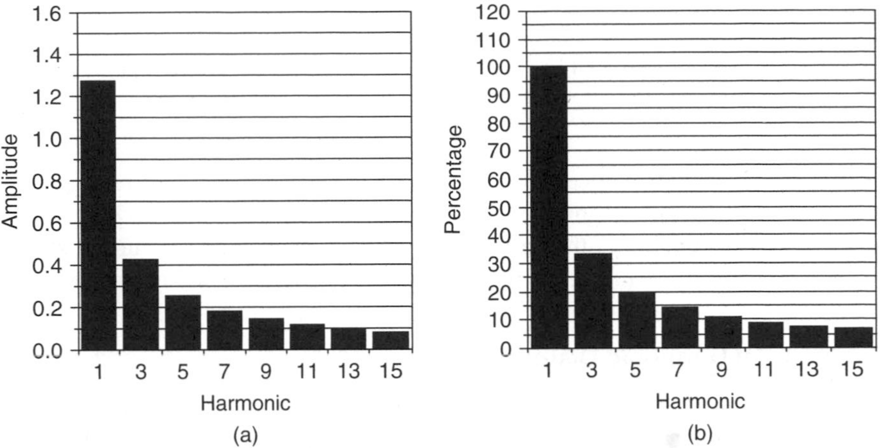

Figure 2.33 First few harmonics for a square wave with amplitude 1: (a) By amplitude of harmonics. (b) By percentage of fundamental.

waveforms. Also notice that the harmonics are still sizable as far out as they are shown, which is all the way out to the 49th harmonic!

We can reconstruct the waveform for the CFL whose current is shown in Fig. 2.34. The bulb is run on a 60-Hz ($2\pi \cdot 60 = 377$ rad/s) voltage source so, taking a few values from Fig. 2.34, we can write the first few terms of the

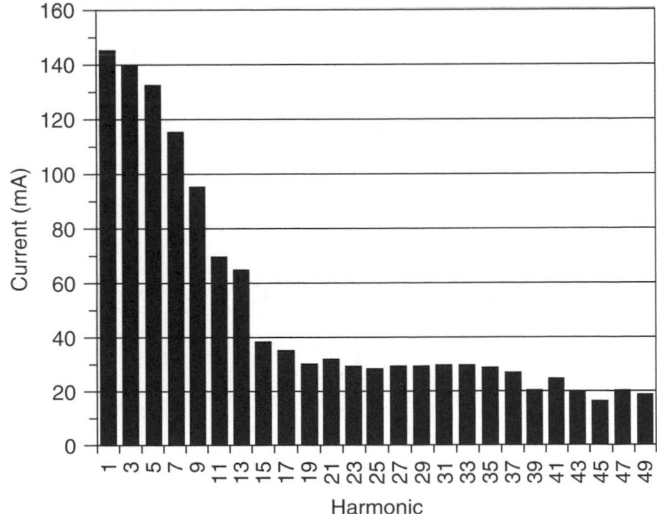

Figure 2.34 Harmonic current spectrum for an 18-W, electronic-ballast compact fluorescent lamp. Currents are rms values.

current as

$$i\ (A) = \sqrt{2}(0.145\cos\omega t + 0.140\cos 3\omega t + 0.132\cos 5\omega t + 0.115\cos 7\omega t + \cdots) \quad (2.87)$$

where $\omega = 377$ rad/s. Notice the $\sqrt{2}$ converts the rms values of current given in the figure to their full magnitude. A plot of (2.87) is shown in Fig. 2.35.

Even with just a few of the harmonics included, the "gulps" of current similar to those shown in the full-wave rectifier circuit of Fig. 2.23 are clearly evident.

2.8.2 Total Harmonic Distortion

While the Fourier series description of the harmonics in a periodic waveform contains all of the original information in the waveform, it is usually awkward to work with. There are several simpler quantitative measures that can be developed and more easily used. For example, suppose we start with a series representation of a current waveform that is symmetric about the y axis:

$$i = \sqrt{2}(I_1\cos\omega t + I_2\cos 2\omega t + I_3\cos 3\omega t + \cdots) \quad (2.88)$$

where I_n is the rms value of the current in the nth harmonic. The rms value of current is therefore

$$I_{\text{rms}} = \sqrt{(i^2)_{\text{avg}}} = \sqrt{\left[\sqrt{2}(I_1\cos\omega t + I_2\cos 2\omega t + I_3\cos 3\omega t + \cdots)\right]^2_{\text{avg}}} \quad (2.89)$$

While algebraically squaring that infinite series in the parentheses looks menacing, it falls apart when we fool around a bit with the algebra. Starting with the

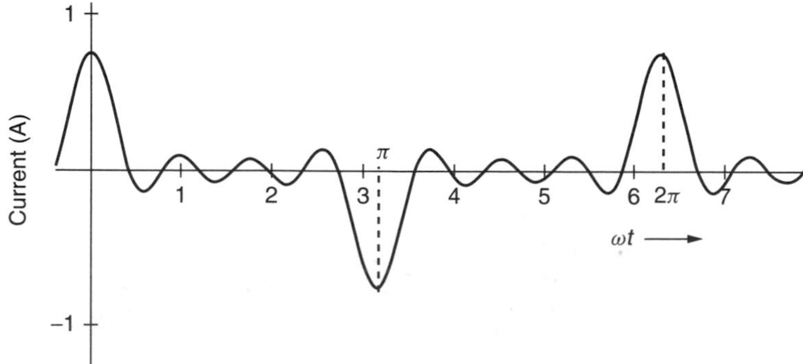

Figure 2.35 The sum of the first through seventh harmonics of current for the CFL of Fig. 2.34.

square of a sum of terms

$$(a+b+c+\cdots)^2 = a^2+b^2+c^2\cdots+2ab+2ac+2bc+\cdots \tag{2.90}$$

we see that each term is squared, and then in addition there are combinations of all possible cross products between terms. Therefore,

$$I_{\text{rms}} = \sqrt{\begin{aligned}&2[(I_1\cos\omega t)^2+(I_2\cos 2\omega t)^2+\cdots+(2I_1I_2\cos\omega t\cdot\cos 2\omega t)\\&\quad+(2I_1I_3\cos\omega t\cdot\cos 3\omega t)+\cdots]_{\text{avg}}\end{aligned}} \tag{2.91}$$

Fortunately, the average value of the product of two sinusoids of differing frequency is zero; that is,

$$[\cos n\omega t\cdot\cos m\omega t]_{\text{avg}} = 0 \qquad \text{for } n \neq m \tag{2.92}$$

Also, we note that

$$({I_n}^2\cos^2 n\omega t)_{\text{avg}} = \frac{{I_n}^2}{2} \tag{2.93}$$

That leaves

$$I_{\text{rms}} = \sqrt{2\left[\frac{{I_1}^2}{2}+\frac{{I_2}^2}{2}+\frac{{I_3}^2}{2}+\cdots\right]} = \sqrt{{I_1}^2+{I_2}^2+{I_3}^2\cdots} \tag{2.94}$$

So, the rms value of current when there are harmonics is just the square root of the sum of the squares of the individual rms values for each frequency. While this was derived for a Fourier series involving just a sum of cosines, it holds for the general case in which the sum involves cosines and sines.

In the United States, the most commonly used measure of distortion is called the *total harmonic distortion* (THD), which is defined as

$$\text{THD} = \frac{\sqrt{{I_2}^2+{I_3}^2+{I_4}^2+\cdots}}{I_1} \tag{2.95}$$

Notice that since THD is a ratio, it doesn't matter whether the currents in (2.95) are expressed as peak values or rms values. When they are rms values, we can recognize (2.95) to be the ratio of the rms current in all frequencies except the fundamental divided by the rms current in the fundamental. When no harmonics are present, the THD is zero. When there are harmonics, there is no particular limit to THD and it is often above 100%.

Example 2.12 Harmonic Distortion for a CFL. A harmonic analysis of the current drawn by a CFL yields the following data. Find the rms value of current and the total harmonic distortion (THD).

Harmonics	rms Current (A)
1	0.15
3	0.12
5	0.08
7	0.03
9	0.02

Solution From (2.94) the rms value of current is

$$I_{\text{rms}} = [(0.15)^2 + (0.12)^2 + (0.08)^2 + (0.03)^2 + (0.02)^2]^{1/2} = 0.211 \text{ A}$$

From (2.95) the total harmonic distortion is

$$\text{THD} = \frac{\sqrt{(0.12)^2 + (0.08)^2 + (0.03)^2 + (0.02)^2}}{0.15} = 0.99 = 99\%$$

so the total rms current in the harmonics is almost exactly the same as the rms current in the fundamental.

2.8.3 Harmonics and Voltage Notching

As we have seen, diode-rectifiers draw currents in bursts and those bursts can be described in terms of their THD. One effect of these current surges is that there will be a voltage drop in the connecting lines during those surges, which results in momentary drops in the voltage that reaches the load. To illustrate this problem, consider the circuit shown in Fig. 2.36 in which a sinusoidal voltage source sends current through a line to a load. To keep everything as simple as possible, the line impedance is modeled as just having resistance with no reactive component.

Figure 2.37 shows what we might expect if current surges are modeled as simple, rectangular pulses. During current pulses, there will be voltage drop in the line equal to line current times line resistance. The voltage reaching the load

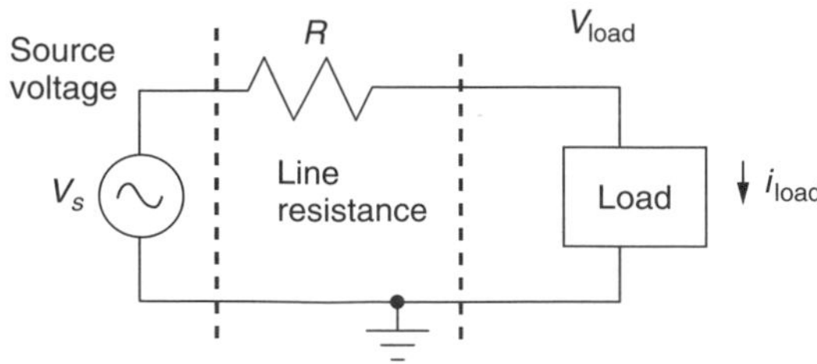

Figure 2.36 A simple circuit to illustrate voltage notching.

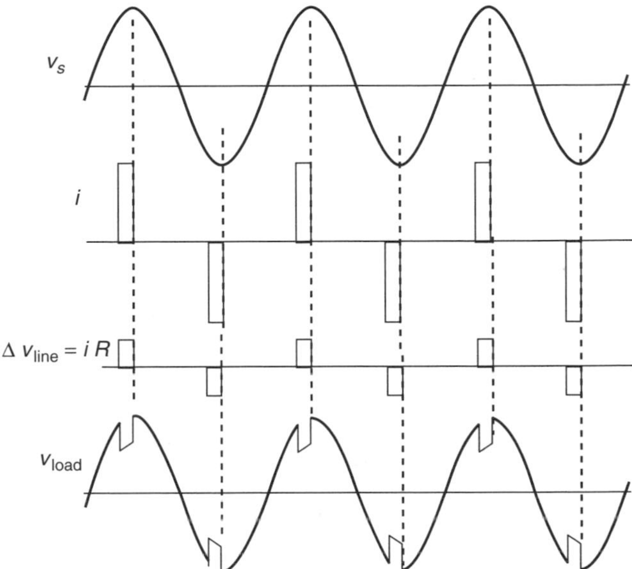

Figure 2.37 Source voltage, line current, voltage drop across the line, and resulting notched voltage to the load.

will correspondingly be decreased by that line voltage drop. In between pulses, the full source voltage reaches the load. The result is a "notch" in the load voltage as shown. Also, there will be harmonic distortion, this time on the voltage signal rather than the current, that can be analyzed and measured in the same ways we just dealt with current harmonic distortion.

It is important to note that the nonlinear load causing the notch may also be affected by the notch. Moreover, other loads that may be on the same circuit will also be affected. Those current surges could even cause notches in the voltage delivered to neighboring facilities. When notching is a potential problem, capacitors may be added to help smooth the waveform.

2.8.4 Harmonics and Overloaded Neutrals

One of the most important areas of concern associated with harmonics is the potential for the neutral wire in four-wire, three-phase, wye-connected systems to overheat, potentially to a degree sufficient to initiate combustion (Fig. 2.38).

The current in the neutral wire is always given by

$$i_n = i_A + i_B + i_C \tag{2.96}$$

For balanced three-phase currents, with no harmonics, we know from (2.56) and (2.60) that the current in the neutral wire is zero:

$$i_n = \sqrt{2}I_{\text{phase}}[\cos\omega t + \cos(\omega t + 120^\circ) + \cos(\omega t - 120^\circ)] = 0 \tag{2.97}$$

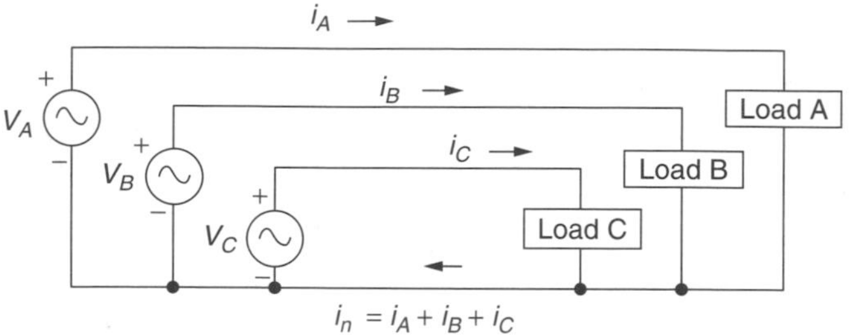

Figure 2.38 Showing the neutral line current in a four-wire, three-phase, wye-connected circuit.

In setting up a building's electrical system, every effort is made to distribute the loads so that approximate balance is maintained, with the intention of minimizing current flow in the neutral. Under the assumption that the neutral carries little or no current, early building codes allowed smaller neutral wires than the phase conductors. It has only been since the mid-1980s that the building code has required the neutral wire to be a full-size conductor.

The question now is, What happens when there are harmonics in the phase currents? Even if the loads are balanced, harmonic currents cause special problems for the neutral line. Suppose that each phase carries exactly the same current (shifted by 120° of course), but now there are third harmonics involved. That is,

$$\begin{aligned} i_A &= \sqrt{2}[I_1 \cos \omega t + I_3 \cos 3\omega t] \\ i_B &= \sqrt{2}\{I_1 \cos(\omega t + 120^\circ) + I_3 \cos[3(\omega t + 120^\circ)]\} \\ i_C &= \sqrt{2}\{I_1 \cos(\omega t - 120^\circ) + I_3 \cos[3(\omega t - 120^\circ)]\} \end{aligned} \tag{2.98}$$

Now, notice what happens to the third harmonics in phases B and C due to the following:

$$\cos[3(\omega t \pm 120^\circ)] = \cos(3\omega t \pm 360^\circ) = \cos 3\omega t \tag{2.99}$$

This means that the current in the neutral wire, which is the sum of each of the three phase currents, is

$$\begin{aligned} i_n = \sqrt{2}[&I_1 \cos \omega t + I_1 \cos(\omega t + 120^\circ) + I_1 \cos(\omega t - 120^\circ) + I_3 \cos 3\omega t \\ &+ I_3 \cos 3\omega t + I_3 \cos 3\omega t] \end{aligned} \tag{2.100}$$

Since the sum of the three cosines of equal magnitude—but each out of phase with the other by 120°—is zero, the fundamental currents drop out and we are left with

$$i_n = 3\sqrt{2} I_3 \cos 3\omega t \tag{2.101}$$

In terms of rms values,

$$I_n = 3I_3 \tag{2.102}$$

That is, the rms current in the neutral line is three times the rms current in each line's third harmonic. There is considerable likelihood, therefore, that harmonic-generating loads will cause the neutral to carry even more current than the phase conductors, not less!

The same argument about harmonics adding in the neutral applies to all of the harmonic numbers that are multiples of 3 (since $3 \times n \times 120^\circ = n360^\circ = 0^\circ$). That is, the third, sixth, ninth, twelfth, ... harmonics all add to the neutral current in an amount equal to three times their phase-current harmonics. Notice, by the way, that harmonics not divisible by 3 cancel out in the same way that the fundamental cancels—for example, the second harmonic:

$$i_{n-\text{2nd harmonic}} = \sqrt{2}I_2\{\cos(2\omega t) + \cos[2(\omega t + 120^\circ)] + \cos[2(\omega t - 120^\circ)]\} \tag{2.103}$$

The terms in the bracket are

$$\cos(2\omega t) + \cos(2\omega t + 240^\circ) + \cos(2\omega t - 240^\circ) = \cos 2\omega t + \cos(2\omega t - 120^\circ) + \cos(2\omega t + 120^\circ) = 0$$

so the second harmonic currents cancel just as the fundamental did. This will be the case for all harmonics *not* divisible by 3.

For currents that show half-wave symmetry, there are no even harmonics, so the only harmonics that appear on the neutral line for balanced loads of this sort will be third, 9th, 15th, 21st, ... etc. These odd harmonics, divisible by three, are called *triplen* harmonics.

Example 2.13 Neutral-Line Current. A four-wire, wye-connected balanced load has phase currents described by the following harmonics:

Harmonic	rms Current (A)
1	100
3	50
5	20
7	10
9	8
11	4
13	2

Find the rms current flowing in the neutral wire and compare it to the rms phase current.

Solution Only the harmonics divisible by three will contribute to neutral line current, so this means that all we need to consider are the third and ninth harmonics.

Assuming the fundamental is 60 Hz, the harmonics contribute

Third harmonic: $3 \times 50 = 150$ A at 180 Hz
Ninth harmonic: $3 \times 8 = 24$ A at 540 Hz

The rms current is the square root of the sum of the squares of the harmonic currents, so the neutral wire will carry

$$I_n = \sqrt{150^2 + 24^2} = 152 \text{ A}$$

The rms current in each phase will be

$$I_n = \sqrt{100^2 + 50^2 + 20^2 + 10^2 + 8^2 + 4^2 + 2^2} = 114 \text{ A}$$

The neutral current, rather than being smaller than the phase currents, is actually 33% higher!

2.8.5 Harmonics in Transformers

Recall from Chapter 1 that cyclic magnetization of ferromagnetic materials causes magnetic domains to flip back and forth. With each cycle, there are hysteresis losses in the magnetic material, which heat the core at a rate that is proportional to frequency.

$$\text{Power loss due to hysteresis} = k_1 f \tag{2.104}$$

Also recall that sinusoidal variations in flux within a magnetic core induces circulating currents within the core material itself. To help minimize these currents, silicon-alloyed steel cores or powdered ceramics, called ferrites, are used to increase the resistance to current. Also, by laminating the core, currents have to flow in smaller spaces, which also increases the path resistance. The important point is that those currents are proportional to the rate of change of flux and therefore the heating caused by those i^2R losses is proportional to frequency squared:

$$\text{Power losses due to eddy currents} = k_2 f^2 \tag{2.105}$$

Since harmonic currents in the windings of a transformer can have rather high frequencies, and since core losses depend on frequency—especially eddy-current losses, which are dependent on the square of frequency—harmonics can cause transformers to overheat. Even if the overheating does not immediately burn out the transformer, the durability of transformer-winding insulation is very dependent on temperature so harmonics can shorten transformer lifetime.

Concerns over harmonic distortion—especially when voltage distortion from one facility (e.g. a building) can affect loads for other customers on the same feeder—has led to the establishment of a set of THD limits set forth by the Institute for Electrical and Electronic Engineers (IEEE) known as the IEEE Standard 519–1992. The goal is to control voltage THD at the point where the utility feeder connects to the step-down transformer of a facility to limit the potential for one customer to affect another.

REFERENCES

Bosela, T. R. (1997). *Introduction to Electrical Power System Technology*, Prentice-Hall, Englewood Cliffs, NJ.

Calwell, C., and T. Reeder (2002). *Power Supplies: A Hidden Opportunity for Energy Savings*, Ecos Consulting for NRDC, May.

Chapman, S. J. (1999). *Electric Machinery Fundamentals*, 3rd ed., McGraw-Hill, New York.

Douglas, J. (1993). Solving Problems of Power Quality, *EPRI Journal*, December.

Krein, P. T. (1998). *Elements of Power Electronics*, Oxford University Press, New York.

Lamarre, L. (1991). Problems with Power Quality, *EPRI Journal*, July/August.

Meier, A. (2002). *Reducing Standby Power: A Research Report*, Lawrence Berkeley National Labs, April.

Stein, B., and J. S. Reynolds (1992). *Mechanical and Electrical Systems for Buildings*, 8th ed., John Wiley & Sons, New York.

PROBLEMS

2.1. In some parts of the world, the standard voltage is 50 Hz at 220 V. Write the voltage in the form

$$v = V_m \cos(\omega t)$$

2.2. A 120-V, 60-Hz source supplies current to a 1 μF capacitor, a 7.036-*H* inductor and a 120-Ω resistor, all wired in parallel.

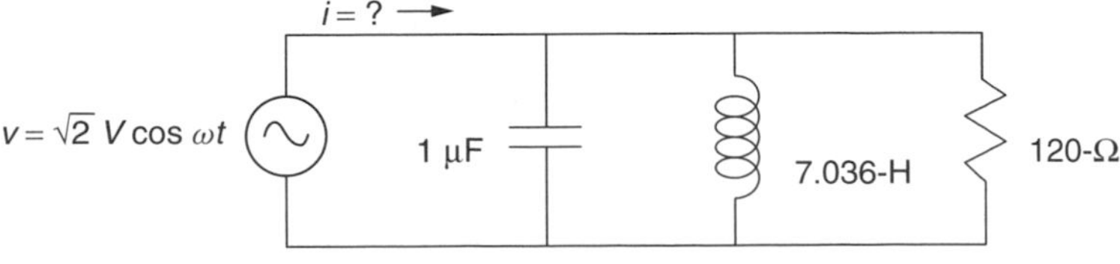

Figure P2.2

a. Find the rms value of current through the capacitor, the inductor, and the resistor.

b. Write the current through each component as a function of time

$$i = I\sqrt{2}\cos(\omega t + \theta)$$

c. What is the total current delivered by the 120-V source (both rms value and as a function of time)?

2.3. Inexpensive inverters often use a "modified" square wave approximation to a sinusoid. For the following voltage waveform, what would be the rms value of voltage?

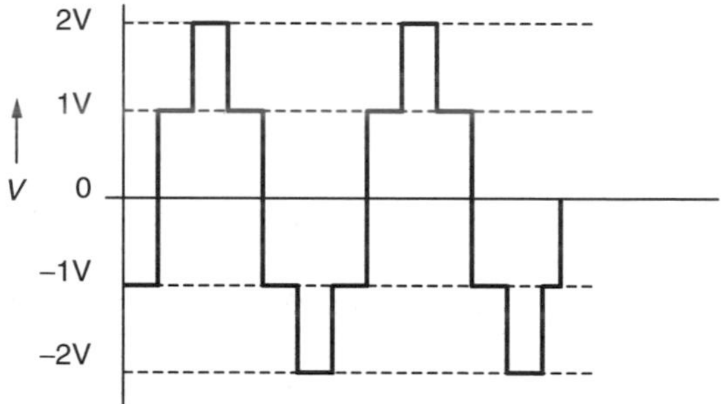

Figure P2.3

2.4. Find the rms value of voltage for the sawtooth waveform shown below. Recall from calculus how you find the average value of a periodic function

$$\overline{f(t)} = \frac{1}{T}\int_0^T f(t)\,dt.$$

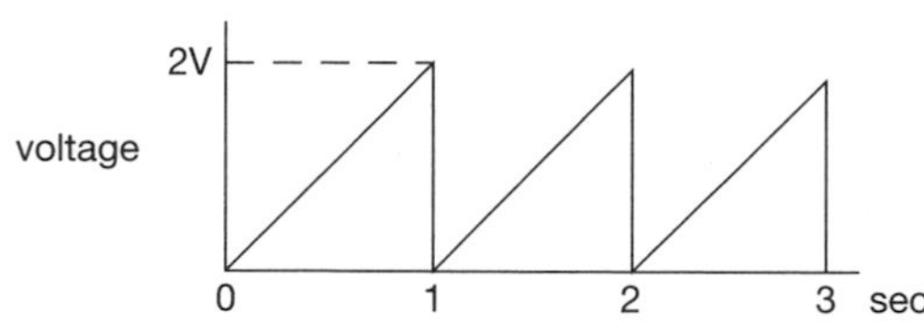

Figure P2.4

2.5 A load connected to a 120-V ac source draws "gulps" of current approximated by rectangular pulses of amplitude 10 A and duration 0.2 radians as shown below:

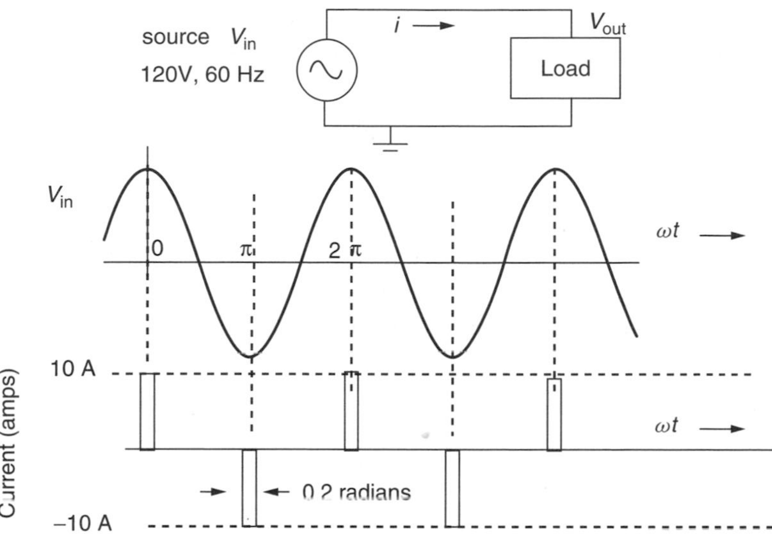

Figure P2.4a

a. What is the rms value of current?

b. Sketch the power delivered by the 120-V source as a function of time $p(t) = v(t)\ i(t)$. Assume the currents are drawn at a voltage equal to the peak voltage of the source.

c. From (b), find the average power $\overline{P}$ delivered by the source.

d. Using the relationship $\overline{P} = V_{\text{rms}} \cdot I_{\text{rms}} \cdot \text{PF}$, find the power factor.

e. Suppose the wires connecting the source to the load have resistance of 1 ohm.

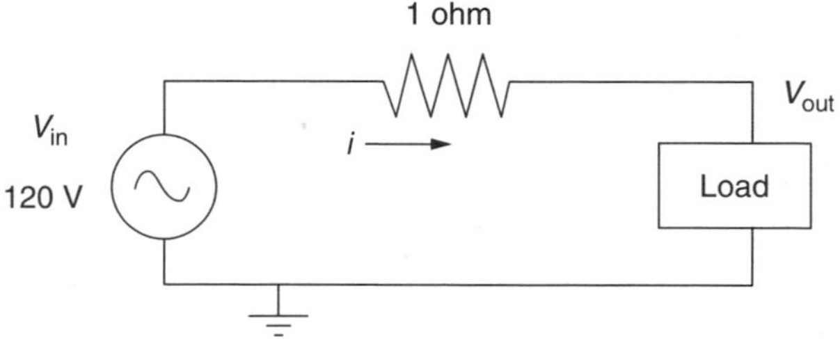

Figure P2.4b

When a gulp of current is drawn there will be a voltage drop (iR) in the wires causing a notch in the voltage reaching the load. Sketch the resulting voltage waveform across the load, labeling any significant values.

2.6. If a 480-V supply delivers 50 A with a power factor of 0.8, find the real power (kW), the reactive power (kVAR), and the apparent power (kVA). Draw the power triangle.

2.7. Suppose a utility charges its large industrial customers \$0.04/kWh for kWh of energy plus \$7/month per peak kVA (demand charge). Peak kVA means the highest level drawn by the load during the month. If a customer uses an average of 750 kVA during a 720-hr month, with a 1000-kVA peak, what would be their monthly bill if their power factor (PF) is 0.8? How much money could be saved each month if their real power remains the same but their PF is corrected to 1.0?

2.8. Suppose a motor with power factor 0.5 draws 3600 watts of real power at 240 V. It is connected to a transformer located 100 ft away.

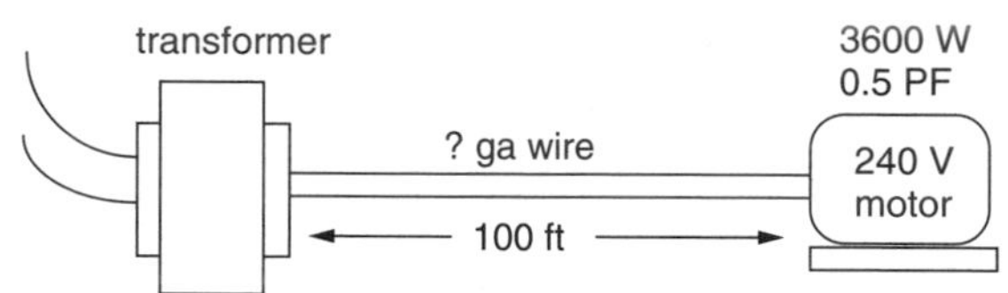

Figure P2.8

a. What minimum gage wire could be used to connect transformer to motor?

b. What power loss will there be in those wires?

c. Draw a power triangle for the wire and motor combination using the real power of motor plus wires and reactive power of the motor itself.

2.9. A motor with power factor 0.6 draws 4200 W of real power at 240 V from a 10 kVA transformer. Suppose a second motor is needed and it too draws 4200 W. With both motors on line, the transformer will be overloaded unless a power factor correction can be made.

a. What power factor would be needed to be able to continue to use the same 10-kVA transformer?

b. How many kVAR would need to be added to provide the needed power factor?

c. How much capacitance would be needed to provide the kVAR correction?

2.10. A transformer rated at 1000 kVA is operating near capacity as it supplies a load that draws 900-kVA with a power factor of 0.70.

a. How many kW of real power is being delivered to the load?

b. How much additional load (in kW of real power) can be added before the transformer reaches its full rated kVA (assume the power factor remains 0.70).

c. How much additional power (above the amount in a) can the load draw from this transformer without exceeding its 1000 kVA rating if the power factor is corrected to 1.0?

2.11. The current waveform for a half-wave rectifier with a capacitor filter looks something like the following:

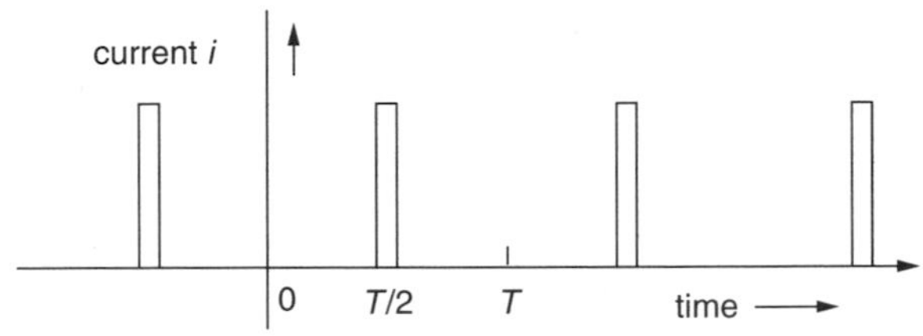

Figure P2.11

Since this is a periodic function, it can be represented by a Fourier series:

$$i(t) = \left(\frac{a_0}{2}\right) + \sum_{n=1}^{\infty} a_n \cos(n\omega t) + \sum_{m=1}^{\infty} b_m \sin(m\omega t)$$

Which of the following assertions are true and which are false?

a. The value of a_0 is zero.

b. There are no sine terms in the series.

c. There are no cosine terms in the series.

d. There are no even harmonics in the series.

2.12. The power system for a home is usually described as three-wire, single-phase power. It can however, also be thought of as "two-phase power" (analogous to three-phase systems).

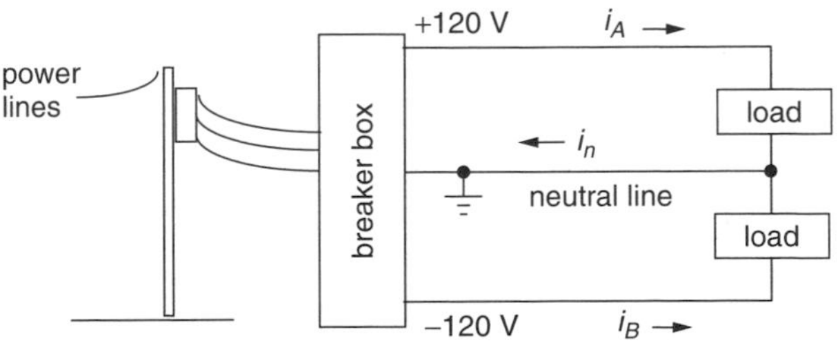

Figure P2.12

a. For balanced loads resulting in the currents

$$i_A = 5\sqrt{2}\cos(377\ t) \quad \text{and} \quad i_B = 5\sqrt{2}\cos(377\ t + \pi) \text{ Amps}$$

What is the rms current in the neutral line?

b. For the following currents with harmonics:

$$i_A = 5\sqrt{2}\cos(377\ t) + 4\sqrt{2}\cos[2(377\ t)] + 3\sqrt{2}\cos[3(377\ t)]$$
$$i_B = 5\sqrt{2}\cos(377\ t + \pi) + 4\sqrt{2}\cos[2(377\ t + \pi)] + 3\sqrt{2}\cos[3(377\ t + \pi)]$$

c. What is the harmonic distortion in each of those currents?

d. What is the rms current in the neutral line?

2.13. Consider a balanced three-phase system with phase currents shown below:

$$i_a = 5\sqrt{2}\cos\omega t \qquad\qquad + 4\sqrt{2}\cos 3\omega t$$
$$i_b = 5\sqrt{2}\cos(\omega t + 120^\circ) + 4\sqrt{2}\cos 3(\omega t + 120^\circ)$$
$$i_c = 5\sqrt{2}\cos(\omega t - 120^\circ) + 4\sqrt{2}\cos 3(\omega t - 120^\circ)$$

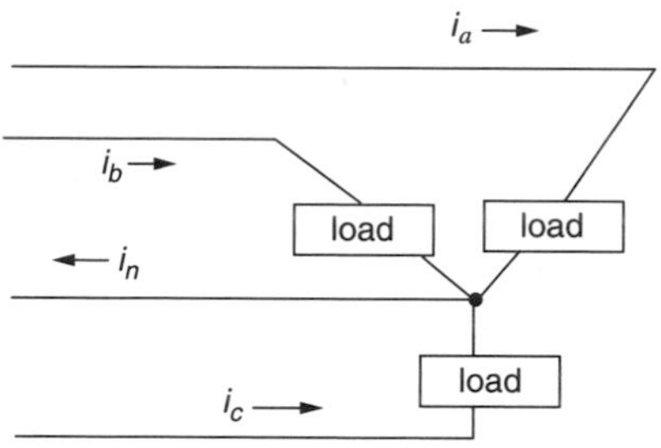

Figure P2.13

a. What is the rms value of the current in each phase?

b. What is the total harmonic distortion (THD) in each of the phase currents?

c. What is the current in the neutral as a function of time $i_n(t)$?

d. What is the rms value of the current in the neutral line? Compare it to the individual phase currents.

2.14. Consider a balanced three-phase system with phase currents shown below:

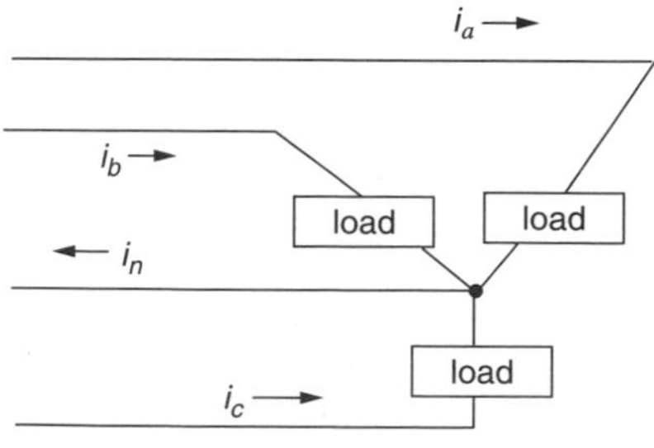

Figure P2.14

$$i_a = 5\sqrt{2}\cos \omega t \qquad + 4\sqrt{2}\cos 3\omega t \qquad + 3\sqrt{2}\cos 9\omega t$$

$$i_b = 5\sqrt{2}\cos(\omega t + 120^\circ) + 4\sqrt{2}\cos 3(\omega t + 120^\circ) + 3\sqrt{2}\cos 9(\omega t + 120^\circ)$$

$$i_c = 5\sqrt{2}\cos(\omega t - 120^\circ) + 4\sqrt{2}\cos 3(\omega t - 120^\circ) + 3\sqrt{2}\cos 9(\omega t - 120^\circ)$$

a. What is the rms current in each of the phases?
b. What is the THD in each of the phase currents?
c. What is the current in the neutral as a function of time $i_n(t)$?
d. What is the rms value of the current in the neutral line?

CHAPTER 3

THE ELECTRIC POWER INDUSTRY

Little more than a century ago there were no lightbulbs, refrigerators, air conditioners, or any of the other electrical marvels that we think of as being so essential today. Indeed, nearly 2 billion people around the globe still live without the benefits of such basic energy services. The electric power industry has since grown to be one of the largest enterprises on the planet, with annual sales of over $300 billion in the United States alone. It is also one of the most polluting of all industries, responsible for three-fourths of U.S. sulfur oxides (SO_X) emissions, one-third of our carbon dioxide (CO_2) and nitrogen oxides (NO_X) emissions, and one-fourth of particulate matter and toxic heavy metals emissions.

Twenty years ago, the industry consisted of regulated utilities with monopoly franchises quietly going about the business of generating power, sending it out over transmission and distribution lines to their own customers who dutifully paid their bills. That all began to change in the 1980s when it became apparent that new, small plants could generate electric power at lower cost than the price at which electricity was being sold by utilities. When large industrial customers realized they could save money by generating some, or all, of their own power on-site, or perhaps buy it directly from nonutility providers, the whole concept of electricity being a natural monopoly came into question. Just as many states began to move toward a new, competitive power industry, California's disastrous deregulation experience of 2000-2001 has put the question back on the table, unanswered.

Technology has provided the impetus for change, but to move an industry as large and important as this one requires an historic change in the regulatory

Renewable and Efficient Electric Power Systems. By Gilbert M. Masters
ISBN 0-471-28060-7

arena as well. The technological and regulatory systems that underlie the current electric power industry are the subject of this chapter.

3.1 THE EARLY PIONEERS: EDISON, WESTINGHOUSE, AND INSULL

The path that leads to today's enormous electric utility industry began in earnest in the nineteenth century with the scientific descriptions of electricity and magnetism by giants such as Hans Christian Oersted, André Marie Ampère, and James Clerk Maxwell. The development of electromechanical conversion technologies such as the direct-current dynamos developed in 1831 separately by Maxwell and H. Pixil of France, followed by Nicola Tesla's work on alternating-current generation and distribution along with polyphase, brushless ac induction motors in the 1880s, made it possible to imagine using electricity as a viable and versatile new source of power.

The first major electric power market developed around the need for illumination. Although many others had worked on the concept of electrically heating a filament to create light, it was Thomas Alva Edison who, in 1879, created the first workable incandescent lamp. Simultaneously, he launched the Edison Electric Light Company with the purpose of providing illumination, not just kilowatt-hours. To do so he provided not only the generation and transmission of electricity, but also the lamps themselves. In 1882, his company began distributing power primarily for lights, but also for electric motors, from his Pearl Street Station in Manhattan. This was to become the first investor-owned utility in the nation.

Edison's system was based on direct current, which he preferred in part because it provided flicker-free light but also because it enables easier speed control of dc motors. The downside of dc, however, was that in those days it was very difficult to change voltage from one level to another—something that became simple to do in ac after the invention of the transformer in 1883. That flaw in dc was critical since dc was less able to take advantage of the ability to reduce line losses by increasing the voltage of electricity as it goes onto transmission lines and then decreasing it back to safe levels at the customer's facility. Recall that line losses are proportional to the square of current, while the power delivered is the product of current and voltage. By doubling the voltage, for example, the same power can be delivered using half the current, which cuts I^2R line losses by a factor of four. Given dc's low-voltage transmission constraint, Edison's customers had to be located within just a mile or two of a generating station, which meant that power stations were beginning to be located every few blocks around the city.

Meanwhile, George Westinghouse recognized the advantages of ac for transmitting power over greater distances and, utilizing ac technologies developed by Tesla, launched the Westinghouse Electric Company in 1886. Within just a few years, Westinghouse was making significant inroads into Edison's electricity market and a bizarre feud developed between these two industry giants. Rather than hedge his losses by developing a competing ac technology, Edison stuck with dc and launched a campaign to discredit ac by condemning its high voltages as a safety hazard. To make the point, Edison and his assistant, Samuel Insull, began demonstrating its lethality by coaxing animals, including dogs, cats, calves and eventually even a horse, onto a metal plate wired to a 1000-volt ac generator, and then electrocuting them in front of the local press (Penrose, 1994). Edison and other proponents of dc continued the campaign by promoting the idea that capital punishment by hanging was horrific and could be replaced by a new, more humane approach based on electrocution. The result was the development of the electric chair, which claimed its first victim in 1890 in Buffalo, NY (also home of the nation's first commercially successful ac transmission system).

The advantages of high-voltage transmission, however, were overwhelming and Edison's insistence on dc eventually led to the disintegration of his electric utility enterprise. Through buyouts and mergers, Edison's various electricity interests were incorporated in 1892 into the General Electric Company, which shifted the focus from being a utility to manufacturing electrical equipment and end-use devices for utilities and their customers.

One of the first demonstrations of the ability to use ac to deliver power over large distances occurred in 1891 when a 106-mile, 30,000-V transmission line began to carry 75 kW of power between Lauffen and Frankfurt, Germany. The first transmission line in the United States went into operation in 1890 using 3.3-kV lines to connect a hydroelectric station on the Willamette River in Oregon to the city of Portland, 13 miles away. Meanwhile, the flicker problem for incandescent lamps with ac was resolved by trial and error with various frequencies until it was no longer noticeable. Surprisingly, it wasn't until the 1930s that 60 Hz finally became the standard in the United States. Some countries had by then settled on 50 Hz, and even today, some countries, such as Japan, use both.

Another important player in the evolution of electric utilities was Samuel Insull. Insull is credited with having developed the business side of utilities. It was his realization that the key to making money was to find ways to spread the high fixed costs of facilities over as many customers as possible. One way to do that was to aggressively market the advantages of electric power, especially for use during the daytime to complement what was then the dominant nighttime lighting load. In previous practice, separate generators were used for industrial facilities, street lighting, street cars, and residential loads, but Insull's idea was to integrate the loads so that he could use the same expensive generation and transmission equipment on a more continuous basis to satisfy them all. Since operating costs were minimal, amortizing high fixed costs over more kilowatt-hour sales results in lower prices, which creates more demand. With controllable transmission line losses and attention to financing, Insull promoted rural electrification, further extending his customer base.

With more customers, more evenly balanced loads, and modest transmission losses, it made sense to build bigger power stations to take advantage of economies of scale, which also contributed to decreasing electricity prices and increasing profits. Large, centralized facilities with long transmission lines

TABLE 3.1 Chronology of Major Electricity Milestones

Year	Event
1800	First electric battery (A. Volta)
1820	Relationship between electricity and magnetism confirmed (H. C. Oersted)
1821	First electric motor (M. Faraday)
1826	Ohm's law (G. S. Ohm)
1831	Principles of electromagnetism and induction (M. Faraday)
1832	First dynamo (H. Pixil)
1839	First fuel cell (W. Grove)
1872	Gas turbine patent (F. Stulze)
1879	First practical incandescent lamp (T. A. Edison and J. Swan, independently)
1882	Edison's Pearl Street Station opens
1883	Transformer invented (L. Gaulard and J. Gibbs)
1884	Steam turbine invented (C. Parsons)
1886	Westinghouse Electric formed
1888	Induction motor and polyphase AC systems (N. Tesla)
1889	Impulse turbine patent (L. Pelton)
1890	First single-phase ac transmission line (Oregon City to Portland)
1891	First three-phase ac transmission line (Germany)
1903	First successful gas turbine (France)
1907	Electric vacuum cleaner and washing machines
1911	Air conditioning (W. Carrier)
1913	Electric refrigerator (A. Goss)
1935	Public Utility Holding Company Act (PUHCA)
1936	Boulder dam completed
1962	First nuclear power station (Canada)
1973	Arab oil embargo, price of oil quadruples
1978	Public Utilities Regulatory Policies Act (PURPA)
1979	Iranian revolution, oil price triples; Three Mile Island nuclear accident
1983	Washington Public Power Supply System (WPPS) \$2.25 billion nuclear reactor bond default
1986	Chernobyl nuclear accident (USSR)
1990	Clean Air Act amendments introduce tradeable SO_2 allowances
1992	National Energy Policy Act (EPAct)
1998	California begins restructuring
2001	Restructuring collapses in California; Enron and Pacific Gas and Electric bankruptcy

required tremendous capital investments; to raise such large sums, Insull introduced the idea of selling utility common stock to the public.

Insull also recognized the inefficiencies associated with multiple power companies competing for the same customers, with each building its own power plants and stringing its own wires up and down the streets. The risk of the monopoly alternative, of course, was that without customer choice, utilities could charge whatever they could get away with. To counter that criticism, he helped establish the concept of regulated monopolies with established franchise territories and prices controlled by *public utility commissions* (PUCs). The era of regulation had begun.

The story of the evolution of the electric power industry continues in the latter part of this chapter in which the regulatory side of the industry is presented. Table 3.1 provides a quick chronological history of this evolution.

3.2 THE ELECTRIC UTILITY INDUSTRY TODAY

Electric utilities, monopoly franchises, large central power stations, and long transmission lines have been the principal components of the prevailing electric power paradigm since the days of Insull. Electricity generated at central power stations is almost always three-phase, ac power at voltages that typically range from about 14 kV to 24 kV. At the site of generation, transformers step up the voltage to long-distance transmission-line levels, typically in the range of 138 kV to 765 kV. Those voltages may be reduced for regional distribution using subtransmission lines that carry voltages in the range of 34.5 kV to 138 kV.

When electric power reaches major load centers, transformers located in distribution-system substations step down the voltage to levels typically between 4.16 kV and 34.5 kV, with 12.47 kV being the most common. Feeder lines carry power from distribution substations to the final customers. On power poles or in concrete-pad-mounted boxes, transformers again drop voltage to levels suitable for residential, commercial, and industrial uses. A sense of the overall utility generation, transmission, and distribution system is shown in Fig. 3.1.

3.2.1 Utilities and Nonutilities

Entities that provide electric power can be categorized as utilities or nonutilities depending on now their business is organized and regulated

Electric utilities traditionally have been given a monopoly franchise over a fixed geographical area. In exchange for that franchise, they have been subject to regulation by State and Federal agencies. A few large utilities are vertically integrated; that is, they own generation, transmission, and distribution infrastructure. Most, however, are just distribution utilities that purchase wholesale power,

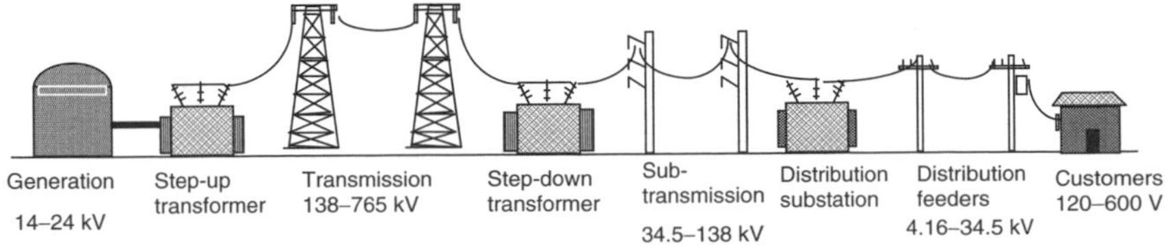

Figure 3.1 Conventional power generation, transmission, and distribution system.

which they sell to their retail customers using their monopoly distribution system. The roughly 3200 utilities in the United States can be subdivided into one of four categories of ownership: investor-owned, Federally owned, other publicly owned, and cooperatively owned.

Investor owned utilities (IOUs) are privately owned with stock that is publicly traded. They are regulated and authorized to receive an allowed rate of return on their investments. IOUs may sell power at wholesale rates to other utilities or they may sell directly to retail customers. While they comprise less than 5% of all U.S. utilities in number, they generate over two-thirds of U.S. electricity.

Federally owned utilities produce power at facilities run by entities such as the Tennessee Valley Authority (TVA), the U.S. Army Corps of Engineers, and the Bureau of Reclamation. The Bonneville Power Administration, the Western, Southeastern, and Southwestern Area Power Administrations, and the TVA market and sell power on a nonprofit basis mostly to Federal facilities, publicly owned utilities and cooperatives, and certain large industrial customers.

Other publicly owned utilities are State and Local government agencies that may generate some power, but which are usually just distribution utilities. They generally sell power at lower cost than IOUs because they are nonprofit and are often exempt from certain taxes. While two-thirds of U.S. utilities fall into this category, they sell only about 9% of total electricity.

Rural electric cooperatives were originally established and financed by the Rural Electric Administration in areas not served by other utilities. They are owned by groups of residents in rural areas and provide services primarily to their own members.

Nonutility generators (NUGs) are privately owned entities that generate power for their own use and/or for sale to utilities and others. They are distinct in that they do not operate transmission or distribution systems and are subject to different regulatory constraints than traditional utilities. Traditionally, NUGs have been industrial facilities generating on-site power for their own use. In the 1920s to 1940s, that amounted to roughly 20% of all U.S. generation, but by the 1970s they were generating an insignificant fraction. NUGs reemerged in the 1980s and 1990s when regulators began to open the grid to independent power producers. Then in the late 1990s, as part of the restructuring intended to create a more competitive industry, some utilities were required to sell off much of their own generation to private entities, transferring those generators to the NUG category. By 2001, NUGs were delivering over one-fourth of total U.S. generation (Fig. 3.2).

3.2.2 Industry Statistics

About 70% of U.S. electricity is generated in power stations that use energy derived from fossil fuels—coal, natural gas, and some oil. Nuclear and hydroelectric-power account for almost all of the remaining 30%, though there is a tiny fraction supplied by nonhydroelectric renewables. The energy going

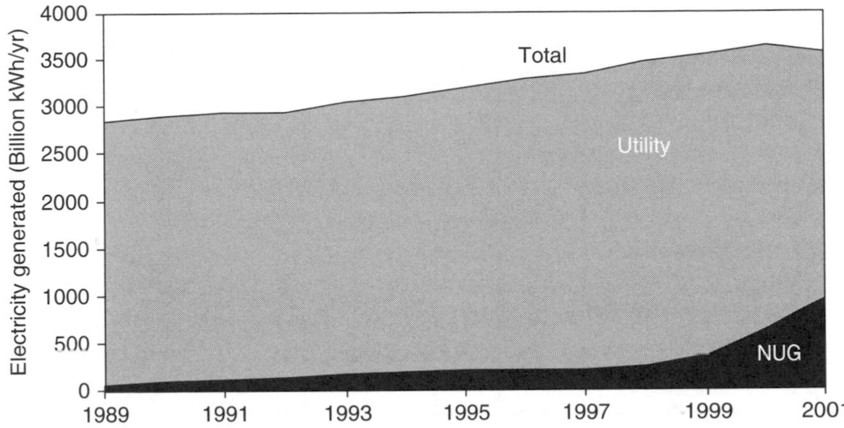

Figure 3.2 Nonutility generators have become a significant portion of total electricity generated in the United States. From *EIA Annual Energy Review 2001* (EIA, 2003).

into power plants is referred to as *primary energy* to distinguish it from *end-use energy*, which is the energy content of electricity that is actually delivered to customers. The numerical difference between primary and end-use energy is made up of losses during the conversion of fuel to electricity, losses in the transmission and distribution system (T&D), and energy used at the power plant itself for its own needs. Less than one-third of primary energy actually ends up in the form of electricity delivered to customers. For rough approximations, it is reasonable to estimate that for every 3 units of fuel into power plants, 2 units are wasted and 1 unit is delivered to end-users.

As shown in Fig. 3.3, coal is the dominant fuel, accounting for 52% of all power plant input energy, nuclear is 21%, natural gas 15%, and renewables (especially hydro and geothermal) 9%. Notice that petroleum is a very minor fuel in the electricity sector, about 3%, and that is almost all residual fuel oil—literally the bottom of the barrel—that has little value for anything else. That is, petroleum and electricity have very little to do with each other.

The distribution of power plant types around the country is very uneven as Fig. 3.4 suggests. The Pacific Northwest generates most of its power at large hydroelectric facilities owned by the Federal government. Coal is predominant in the midwestern and southern states, especially Ohio, West Virginia, Kentucky, and Tennessee. Texas, Louisiana, Oklahoma, and California derive significant fractions from natural gas, while what little oil-fired generation there is tends to be in Florida and New York.

The electricity equation has two sides, the supply side and the demand side, and both are critically important to understanding the industry. The end-use side of electricity is usually divided into three sectors, residential, commercial and industrial. In terms of kilowatt-hour sales (as opposed to electricity that

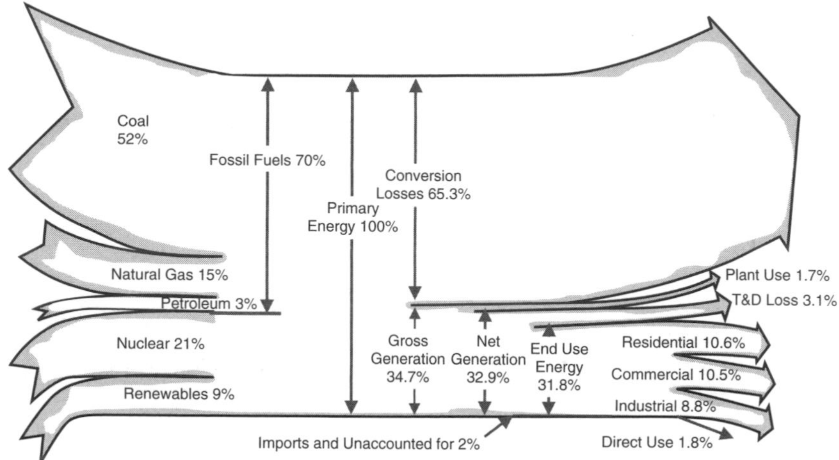

Figure 3.3 Electricity flows as a percentage of primary energy. Based on *EIA Annual Energy Review 2001* (EIA, 2003).

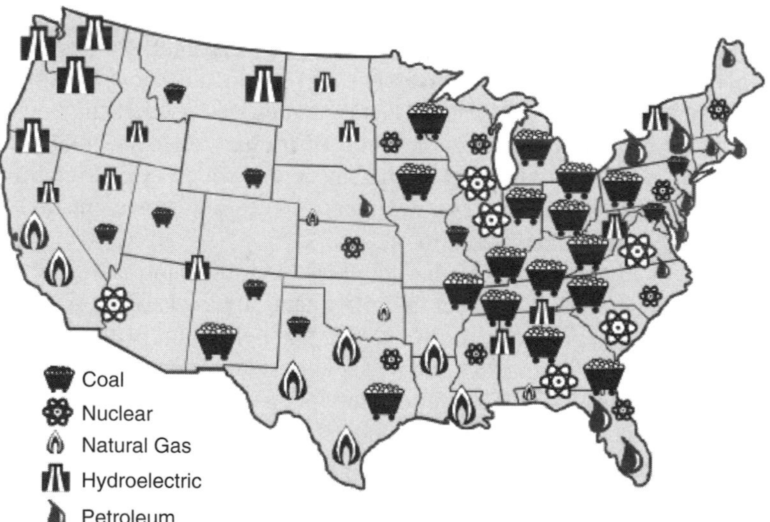

Figure 3.4 Energy sources for electricity generation by region. Each large icon represents about 10 GW of capacity, small ones about 5 GW. From *The Changing Structure of the Electric Power Industry 2000: An Update* (EIA, 2000).

is self-generated on site, which never enters the market), the buildings sector accounts for over two-thirds (36% is residential and 32% commercial) while industrial users purchase 29%. As shown in Fig. 3.5, in the residential sector, usage is spread over many miscellaneous loads, but the single largest ones are air conditioners and refrigerators. In the commercial sector, lighting is the dominant

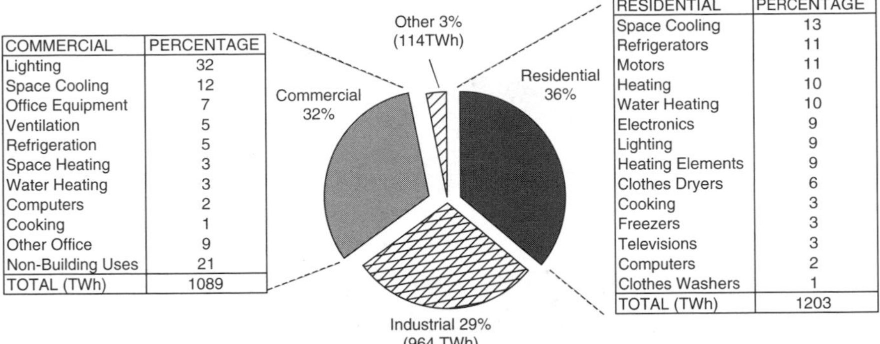

COMMERCIAL	PERCENTAGE
Lighting	32
Space Cooling	12
Office Equipment	7
Ventilation	5
Refrigeration	5
Space Heating	3
Water Heating	3
Computers	2
Cooking	1
Other Office	9
Non-Building Uses	21
TOTAL (TWh)	1089

RESIDENTIAL	PERCENTAGE
Space Cooling	13
Refrigerators	11
Motors	11
Heating	10
Water Heating	10
Electronics	9
Lighting	9
Heating Elements	9
Clothes Dryers	6
Cooking	3
Freezers	3
Televisions	3
Computers	2
Clothes Washers	1
TOTAL (TWh)	1203

Figure 3.5 Distribution of retail sales of electricity by end use. Residential and commercial buildings account for over two-thirds of sales. Total amounts in billions of kWh (TWh) are 2001 data. From EIA (2003).

single load with cooling also being of major importance. If the lighting and air-conditioning loads of residential and commercial buildings are combined, they account for almost one-fourth of all electricity sold. More importantly, those loads are the principal drivers of the peak demand for power, which for many utilities occurs in the mid-afternoon on hot, sunny days. It is the peak load that dictates the total generation capacity that must be built and operated.

As an example of the impact of lighting and air conditioning on the peak demand for power, Fig. 3.6 shows the California power demand on a hot, summer day. As can be seen, the industrial and agricultural loads are relatively constant, and the diurnal rise and fall of demand is almost entirely driven by the building sector. In terms of peak demand, buildings account for almost three-fourths of the load, with air conditioning and lighting alone driving over 40% of the total peak demand. Better buildings with greater use of natural daylighting, more efficient lamps, increased attention to reducing afternoon solar gains, greater use of natural-gas-fired absorption air-conditioning systems, load shifting by using ice made at night to cool during the day, and so forth, could make a significant difference in the number and type of power plants needed to meet demand.

Not only does the predominant fuel vary widely across the country, so does the resulting price of electricity. Figure 3.7 shows the statewide average price of electricity in 2001 ranged from a low of 4.24¢/kWh in Kentucky to a high of 14.05¢/kWh in Hawaii, with an overall average price of 7.32¢/kWh. Among the coterminous states, California's electricity was the most expensive in 2001 at 11.78¢/kWh, but that was when it was in the midst of its disastrous deregulation crisis (described later in the chapter); in 1998 it was 25% less, at 9.0 ¢/kWh.

While Fig. 3.7 shows the state-by-state average retail price of electricity, it greatly masks the large differences in price from one customer class to another. Industrial facilities are typically charged on the order of 40% less for their electricity than either residential or commercial customers. Industrial customers

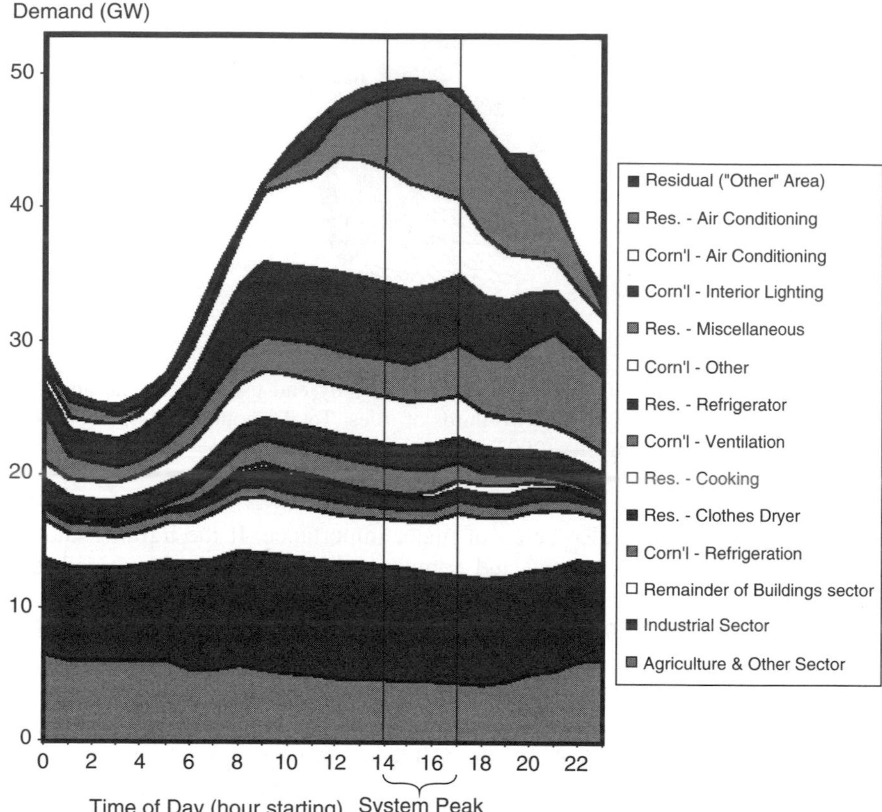

Figure 3.6 The load profile for the a peak summer day in California (1999) shows maximum demand occurs between 2 P.M. and 4 P.M. Lighting and air conditioning accounts for over 40% of the peak. End uses are ordered the same in the graph and legend. From Brown and Koomey (2002).

with more uniform energy demand can be served in large part by less expensive, base-load plants that run more or less continuously. The distribution systems serving utilities are more uniformly loaded, reducing costs, and certainly administrative costs to deal with customer billing and so forth are less. They also have more political influence.

Figure 3.8 shows the average prices charged for residential, commercial, and industrial customers, over time. It is interesting to note the sharp increases in prices that occurred in the 1970s and early 1980s, which can be attributed to increasing fuel costs associated with the spike in OPEC oil prices in 1973 and 1979, as well as the huge increase in spending for nuclear power plant construction during that era.

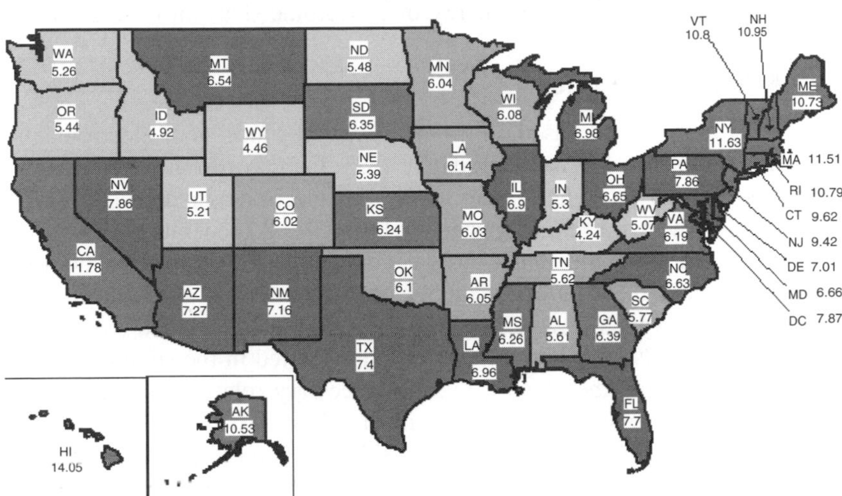

Figure 3.7 Average revenue per kilowatt-hour for all sectors by state, 2001. California in 1998 before restructuring was 9.03 ¢/kWh. *Source*: Energy Information Administration.

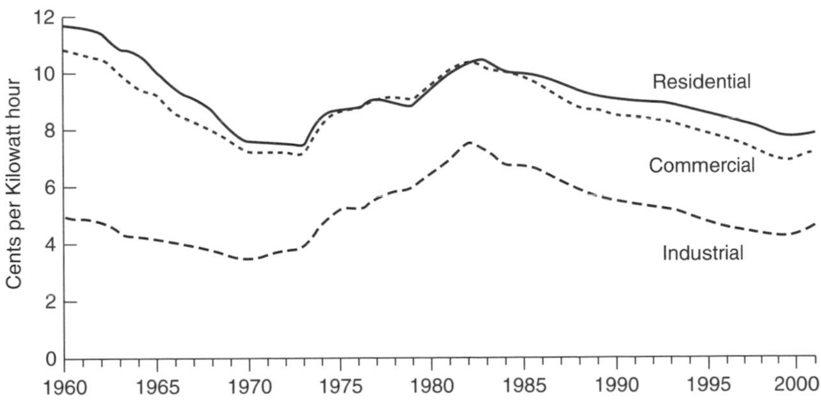

Figure 3.8 Average retail prices of electricity, by sector (constant $1996). From *EIA Annual Energy Review 2001* (EIA, 2003).

3.3 POLYPHASE SYNCHRONOUS GENERATORS

With the exception of minor amounts of electricity generated using internal combustion engines, fuel cells, or photovoltaics, the electric power industry is based on some energy source forcing a fluid (steam, combustion gases, water or air) to pass through turbine blades, causing a shaft to spin. The function of the generator, then, is to convert the rotational energy of the turbine shaft into electricity.

3.3.1 A Simple Generator

Electric generators are all based on the fundamental concepts of electromagnetic induction developed by Michael Faraday in 1831. Faraday discovered that moving a conductor through a magnetic field induces an electromagnetic force (emf), or voltage, across the wire, as suggested in Fig. 3.9a. A generator, very simply, is an arrangement of components designed to cause relative motion between a magnetic field and the conductors in which the emf is to be induced. Those conductors, out of which flows electric power, form what is called the *armature*. Most large generators have the armature windings fixed in the stationary portion of the machine (called the *stator*), and the necessary relative motion is caused by rotating the magnetic field, as shown in Fig. 3.9b.

To help illustrate generation of ac, consider the simple generator shown in Fig. 3.10. The rotor in this case is just a 2-pole magnet (1 north pole and 1 south pole), which for now we can consider to be just an ordinary permanent magnet. The stator consists of iron, shaped somewhat like a C (backwards, in this case), with some copper wire (the armature) wrapped around the iron. The purpose of the iron in the stator is to provide a low reluctance path for the magnetic flux lines, channeling as much flux as possible through the copper armature windings. You may recall from Chapter 1 that the low reluctance of ferromagnetic materials (e.g., iron) causes flux to prefer to stay in the iron, which is exactly analogous to current-carrying electrons wanting to stay in copper wires. And just as low-resistance copper wire allows more current to flow, low reluctance ferromagnetic materials allow more magnetic flux (flux doesn't "flow," however).

As the rotor turns, magnetic flux passes through the stator and the armature windings, in one direction, then diminishes to zero, then increases in the

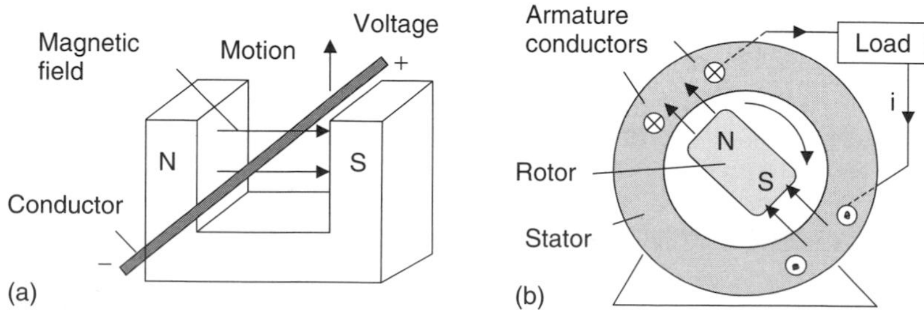

Figure 3.9 Voltage and current can be created by (a) moving a conductor through a magnetic field, or (b) moving the magnetic field past the conductors. The armature windings indicate current flow into the page with an "x" and current out of the page with a dot (the x is meant to resemble the feathers of an arrow moving away from you; the dot is the point of the arrow coming toward you).

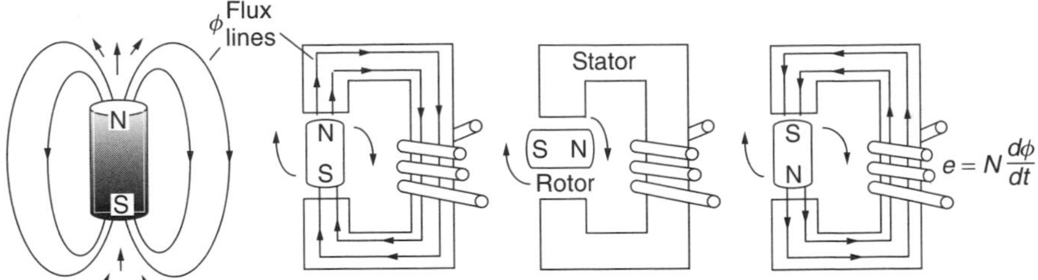

Figure 3.10 As the permanent-magnet rotor turns, it causes magnetic flux within the iron stator to vary (approximately) sinusoidally. The windings around the stator therefore see a time-varying flux, which creates a voltage across their terminals.

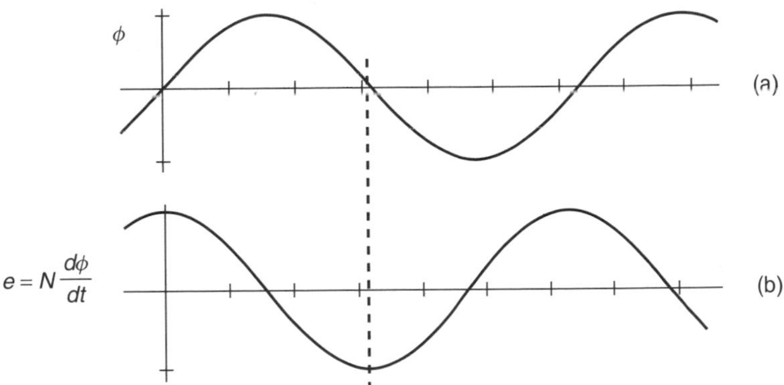

Figure 3.11 Changing flux in the stator creates an emf voltage across the windings.

other direction. Ideally, the flux ϕ would vary sinusoidally as suggested in Fig. 3.11a. From Faraday's law, whenever a winding links a time-varying amount of magnetic flux ϕ, there will be a voltage e (electromotive force) created across the winding:

$$e = N\frac{d\phi}{dt} \tag{3.1}$$

3.3.2 Single-Phase Synchronous Generators

Suppose we want to generate voltage at a frequency of 60 Hz so that it will match the frequency of conventional (U.S.) power. How fast would the rotor of the simple generator in Fig. 3.10 have to turn? Each revolution of the rotor gives one voltage cycle, so

$$N_s = \text{shaft rotation rate} = \frac{1 \text{ revolution}}{\text{cycle}} \times \frac{60 \text{ cycles}}{\text{sec}} \times \frac{60 \text{ sec}}{\text{min}} = 3600 \text{ rpm} \tag{3.2}$$

To generate 60 Hz using this 2-pole generator would therefore require that the rotor turn at a fixed rate of exactly 3600 rpm. Such a fixed-speed machine is called *a synchronous generator* since it is synchronized with the utility grid. Most conventional electric power is generated using synchronous generators (this is not the case for wind turbines, however).

While we could imagine the magnetic field in the rotor of a generator to be created using a permanent magnet, as suggested in Fig. 3.10, that would greatly limit the amount of power that could be generated. Instead, the magnetic field is created by sending dc through brushes and slip rings into conductors affixed to the rotor. Field windings may be imbedded into slots that run along the rotor as shown in Fig. 3.12a, or they may be wound around what are called *salient poles*, as shown in Fig. 3.12b. Salient pole rotors are less expensive to fabricate and are often used in slower-spinning hydroelectric generators. High-speed turbines and generators use round rotors, which are better able to handle the centrifugal forces and resulting stresses. Figure 3.13a shows a complete 2-pole generator with a round rotor, while Fig. 3.13b is a 4-pole generator with a salient rotor. Notice the 4-pole generator has four poles in both the rotor and stator.

Adding more poles allows the generator to spin more slowly while still producing a desired frequency for its output power. For example, when the rotor in a 4-pole machine makes one revolution, it generates two cycles on the output lines. This means that it only needs to rotate half as fast as the 2-pole machine, namely 1800 rpm, in order to generate 60-Hz ac. In general, rotor speed N_s as a function of number of poles p and output frequency required f is given by

$$N_s = \text{shaft rotation rate (rpm)} = \frac{1 \text{ revolution}}{(p/2) \text{ cycles}} \times \frac{f \text{ cycles}}{\text{s}} \times \frac{60 \text{ s}}{\text{min}} \tag{3.3}$$

$$N_s = \frac{120f}{p} \tag{3.4}$$

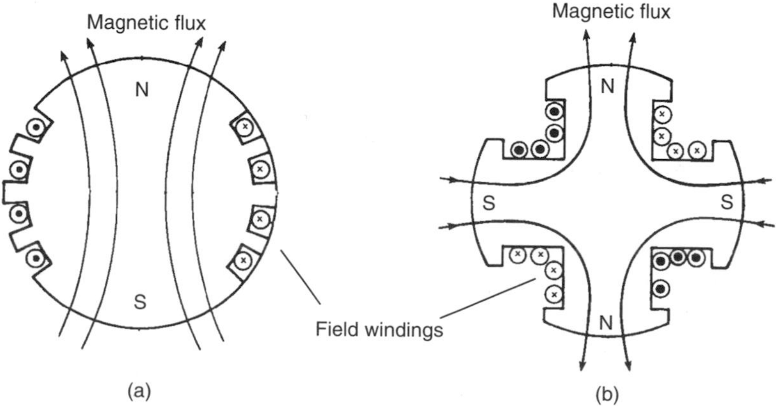

Figure 3.12 Field windings on (a) 2-pole, round rotor and (b) 4-pole, salient rotor.

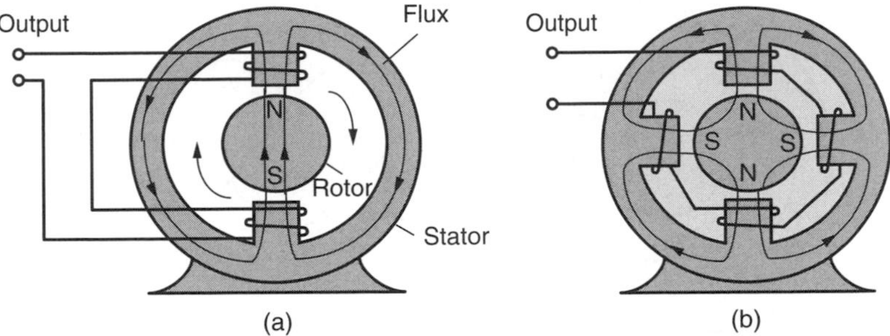

Figure 3.13 (a) A 2-pole machine has one N and one S pole on the rotor and on the stator. (b) A 4-pole machine has 4 poles on the rotor and 4 on the stator.

TABLE 3.2 Shaft Rotation (rpm) as a Function of Desired Output Frequency and Number of Poles

Poles p	50 Hz rpm	60 Hz rpm
2	3000	3600
4	1500	1800
6	1000	1200
8	750	900
10	600	720
12	500	600

While the United States uses 60 Hz exclusively for power, Europe and parts of Japan use 50 Hz. Table 3.2 provides a convenient summary of rotor speeds required for a synchronous generator to deliver power at 50 Hz and at 60 Hz.

3.3.3 Three-Phase Synchronous Generators

The machines shown thus far have been single-phase generators. In Chapter 2, we saw the value of 3-phase power, especially when large amounts of power are needed. To provide 3-phase power, we can keep the simple rotor with a single pair of north and south poles, but we need to add another winding to the stator as has been done in Fig. 3.14. Now during each revolution of the shaft, the rotor sweeps by each of the three stator windings, thereby inducing a voltage in each stator that is 120° out of phase with the adjacent windings.

Figure 3.15 shows a 4-pole, 3-phase generator with a salient-pole rotor, which means it spins at only half the rotor speed compared to a 2-pole machine.

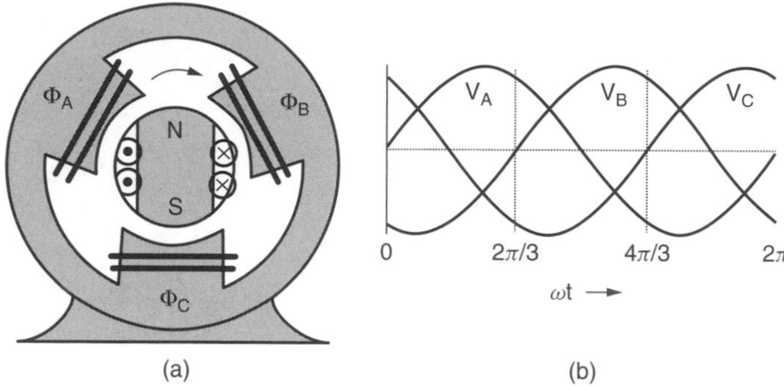

Figure 3.14 (a) A 2-pole, 3-phase synchronous generator. (b) Three-phase stator output voltage.

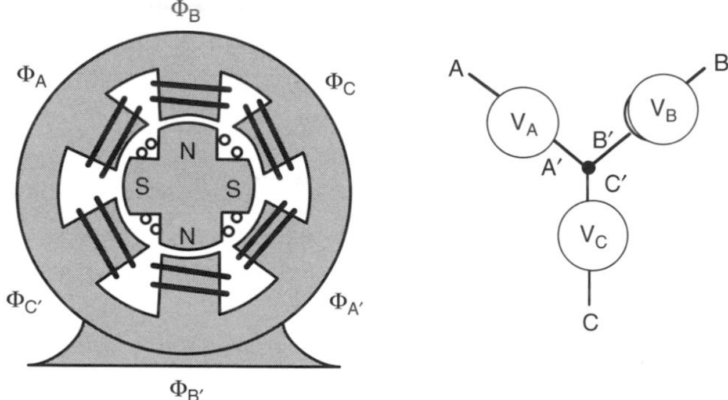

Figure 3.15 A 4-pole, 3-phase, wye-connected, synchronous generator with a 4-pole rotor. The dc rotor current needs to be delivered to the rotor through brushes and slip rings.

3.4 CARNOT EFFICIENCY FOR HEAT ENGINES

Over 90% of U.S. electricity is generated in power plants that convert heat into mechanical work. The heat may be the result of nuclear reactions, fossil-fuel combustion, or even concentrated sunlight focused onto a boiler. Almost all of this 90% is based on a heat source boiling water to make steam that spins a turbine and generator, but there is a rapidly growing fraction that is generated using gas turbines. The best new fossil-fuel power plants use a combination of both steam turbines and gas turbines to generate electricity with very high efficiency.

Steam engines, gas turbines, and internal-combustion engines are examples of machines that convert heat into useful work. What we are interested in here is, How efficiently can they do so? This same question will be asked when we describe fuel cells, photovoltaics, and wind turbines in future chapters, and in each case we will encounter quite interesting, fundamental limits to their maximum possible energy-conversion efficiencies.

3.4.1 Heat Engines

Very simply, a heat engine extracts heat Q_H from a high-temperature source, such as a boiler, converts part of that heat into work W, usually in the form of a rotating shaft, and rejects the remaining heat Q_C into a low-temperature sink such as the atmosphere or a local body of water. Figure 3.16 provides a general model describing such engines.

The thermal efficiency of a heat engine is the ratio of work done to input energy provided by the high-temperature source:

$$\text{Thermal efficiency} = \frac{\text{Net work output}}{\text{Total heat input}} = \frac{W}{Q_H} \tag{3.5}$$

Since energy is conserved,

$$Q_H = W + Q_C \tag{3.6}$$

which leads to another expression for efficiency:

$$\text{Thermal efficiency} = \frac{Q_H - Q_C}{Q_H} = 1 - \frac{Q_C}{Q_H} \tag{3.7}$$

3.4.2 Entropy and the Carnot Heat Engine

The most efficient heat engine that could possibly operate between a hot and cold thermal reservoir was first described back in the 1820s by the French engineer Sadi Carnot. To sketch out the basis for his famous equation, which links the

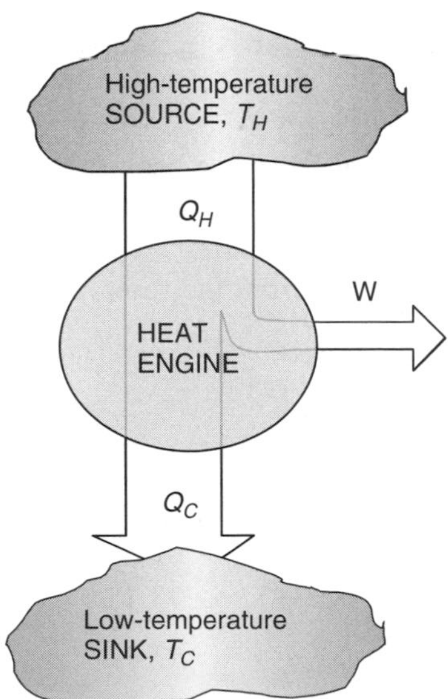

Figure 3.16 A heat engine converts some of the heat extracted from a high-temperature reservoir into work, rejecting the rest into a low-temperature sink.

maximum possible efficiency of a heat engine to the temperatures of the hot and cold reservoirs, we need to introduce the concept of *entropy*.

As is often the case in thermodynamics, the definition of this extremely important quantity is not very intuitive. It can be described as a measure of molecular disorder, or molecular randomness. At one end of the entropy scale is a pure crystalline substance at absolute zero temperature. Since every atom is locked into a predictable place, in perfect order, its entropy is defined to be zero. In general, substances in the solid phase have more ordered molecules and hence lower entropy than liquid or gaseous substances. When we burn some coal, there is more entropy in the gaseous end products than in the solid lumps we burned. That is, unlike energy, entropy is not conserved in a process. In fact, for every real process that occurs, disorder increases and the total entropy of the universe increases.

The concept of ever-increasing entropy is enormously important. It tells us that in any isolated system (e.g., the universe) in which the total energy cannot change, the only processes that can occur spontaneously are ones that result in an increase in the entropy of the system. One implication is that heat flows naturally from warm objects to cold ones, and not the other way around. It also dictates the direction of certain chemical reactions, as we'll see in Chapter 4 where entropy will be used to determine the maximum possible efficiency of fuel cells.

When we started the analysis of the heat engine in Fig. 3.16, we began by tabulating energy flows. The first law of thermodynamics treats energy in the form of heat transfer on an equal basis with energy that shows up as work done by the engine. For an entropy analysis, that is not the case. Work is considered to be an idealized process in which no increase in disorder occurs, and hence it has no accompanying entropy transfer. This is a key distinction. Processes involve heat transfer and work. Heat transfer is accompanied by entropy transfer, but work is entropy-free.

Obviously, to be a useful analysis tool, entropy must be described with equations as well as mental images. Going back to the heat engine, if an amount of heat Q is removed from a "large" thermal reservoir at temperature T (large enough that the temperature of the reservoir doesn't change as a result of this heat loss), the loss of entropy ΔS from the reservoir is defined as

$$\Delta S = \frac{Q}{T} \tag{3.8}$$

where T is an absolute temperature measured using either the Kelvin or Rankine scale. Conversions from Celsius to Kelvin and from Fahrenheit to Rankine are

$$\mathrm{K} = {}^\circ\mathrm{C} + 273.15 \tag{3.9}$$

$$\mathrm{R} = {}^\circ\mathrm{F} + 459.67 \tag{3.10}$$

Equation (3.8) suggests that entropy goes down as temperature goes up. We know that high-temperature heat is more useful than the same amount at lower temperature, which reminds us that entropy is not such a good thing. Less is better!

If we apply (3.8) to a heat engine, along with the requirement that entropy must increase during its operation, we can easily determine the maximum possible efficiency of such a machine. Since there is no entropy change associated with the work done, the requirement that entropy must increase (or, at best break even) tells us that the entropy added to the low-temperature sink must exceed the entropy removed from the high-temperature reservoir:

$$\frac{Q_C}{T_C} \geq \frac{Q_H}{T_H} \tag{3.11}$$

Rearranging (3.11)

$$\frac{Q_C}{Q_H} \geq \frac{T_C}{T_H} \tag{3.12}$$

and substituting into (3.7) gives us the following constraint on the efficiency of a heat engine:

$$\text{Thermal efficiency} = 1 - \frac{Q_C}{Q_H} \leq 1 - \frac{T_C}{T_H} \tag{3.13}$$

That is, the maximum possible efficiency of a heat engine is given by

$$\eta_{\text{max}} = 1 - \frac{T_C}{T_H} \tag{3.14}$$

This is the classical result described by Carnot. One immediate observation that can be made from (3.14) is that the maximum possible heat engine efficiency increases as the temperature of the hot reservoir increases or the temperature of the cold reservoir decreases. In fact, since neither infinitely high temperatures nor absolute zero temperatures are possible, we must conclude that no real engine can convert thermal energy into mechanical energy with 100% efficiency—there will always be waste heat rejected to the environment.

The following examples illustrate how (3.8) and (3.14) can be used in the entropy analysis of heat engines.

Example 3.1 Entropy Analysis of a Heat Engine. Consider a 40% efficient heat engine operating between a large, high-temperature reservoir at 1000 K (727°C) and a large, cold reservoir at 300 K (27°C).

a. If it withdraws 10^6 J/s from the high-temperature reservoir, what would be the rate of loss of entropy from that reservoir and what would be the rate of gain by the low-temperature reservoir?
b. Express the work done by the engine in watts.
c. What would be the total entropy gain of the system?

Solution

a. The loss of entropy from the high-temperature source would be

$$\Delta S_{\text{loss}} = \frac{Q_H}{T_H} = \frac{10^6 \text{ J/s}}{1000 \text{ K}} = 1000 \text{ J/s} \cdot \text{K}$$

Since 40% of the heat removed from the source is converted into work, the remaining 60% is heat transfer into the cold-temperature sink. The rate of entropy gain by the sink would be

$$\Delta S_{\text{gain}} = \frac{Q_C}{T_C} = \frac{0.60 \times 10^6 \text{ J/s}}{300 \text{ K}} = 2000 \text{ J/s} \cdot \text{K}$$

c. The heat engine converts 40% of its input energy into work, which is

$$\text{Work} = 0.40 \times 10^6 \text{ J / s} \times \frac{1 \text{ W}}{\text{J / s}} = 400 \text{ kW}$$

d. Since there is no entropy associated with the work done by the heat engine, the total change in entropy of the entire system is the loss from the source plus the gain to the sink:

$$\Delta S_{\text{total}} = -1000 + 2000 + 0 = +1000 \text{ J / s} \cdot \text{K}$$

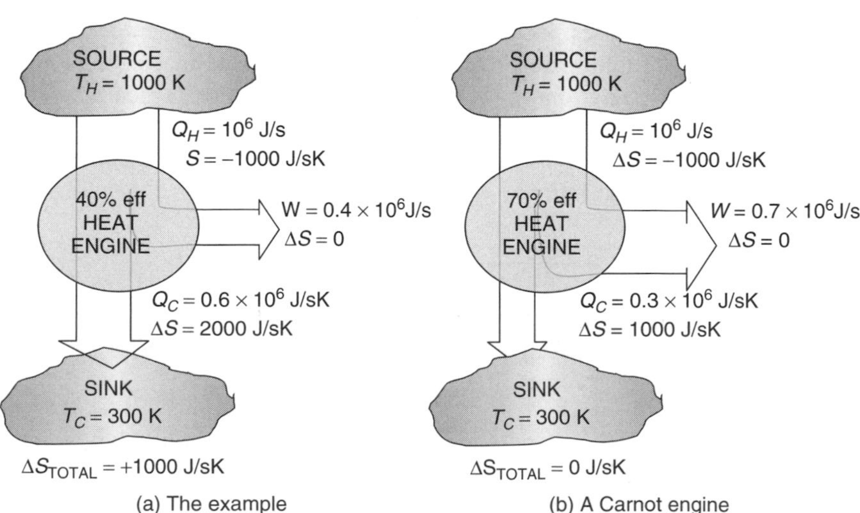

Figure 3.17 Energy and entropy analysis of two heat engines. The example heat engine (a) shows a net increase in entropy, while the Carnot engine (b) does not.

The fact that there was a net increase in entropy in Example 3.1 tells us the engine hasn't violated the Carnot efficiency limit, which from (3.14) we know would be 70% for this 1000 K source and 300 K sink. Figure 3.17 summarizes the energy and entropy analysis for the example heat engine as well as for a perfect Carnot engine.

3.5 STEAM-CYCLE POWER PLANTS

Conventional thermal power plants can be categorized by the thermodynamic cycles they utilize when converting heat into work. Utility-scale thermal power plants are based on either (a) the *Rankine cycle*, in which a working fluid is alternately vaporized and condensed, or (b) the *Brayton cycle*, in which the working fluid remains a gas throughout the cycle. Most *baseload* thermal power plants, which operate more or less continuously, are Rankine cycle plants in which steam is the working fluid. Most *peaking* plants, which are brought on line as needed to cover the daily rise and fall of demand, are gas turbines based on the Brayton cycle. The newest generation of thermal power plants use both cycles and are called combined-cycle plants.

3.5.1 Basic Steam Power Plants

The basic steam cycle can be used with any source of heat, including combustion of fossil fuels, nuclear fission reactions, or concentrated sunlight onto a boiler. The essence of a fossil-fuel-fired steam power plant is diagrammed in Fig. 3.18. In the steam generator, fuel is burned in a firing chamber surrounded by a boiler that transfers heat through metal tubing to the working fluid. Water circulating through the boiler is converted to high-pressure, high-temperature steam. During this conversion of chemical to thermal energy, losses on the order of 10% occur due to incomplete combustion and loss of heat up the stack.

High-pressure steam is allowed to expand through a set of turbine wheels that spin the turbine and generator shaft. For simplicity, the turbine in Fig. 3.18 is shown as a single unit, but for increased efficiency it may actually consist of two or sometimes three turbines in which the exhaust steam from a higher-pressure turbine is reheated and sent to a lower-pressure turbine, and so forth. The generator and turbine share the same shaft allowing the generator to convert the rotational energy of the shaft into electrical power that goes out onto transmission lines for distribution. A well-designed turbine may have an efficiency approaching 90%, while the generator may have a conversion efficiency even higher than that.

The spent steam is drawn out of the last turbine stage by the partial vacuum created in the condenser as the cooled steam undergoes a phase change back to the liquid state. The condensed steam is then pumped back to the boiler to be reheated, completing the cycle.

The heat released when the steam condenses is transferred to cooling water, which circulates through the condenser. Usually, cooling water is drawn from a

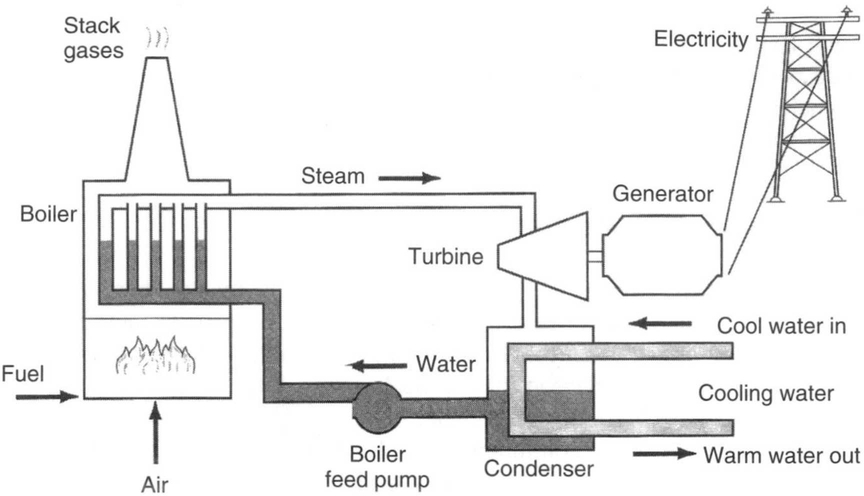

Figure 3.18 A fuel-fired, steam-electric power plant.

river, lake, or sea, heated in the condenser, and returned to that body of water, in which case the process is called *once-through cooling*. A more expensive approach, which has the dual advantages of requiring less water and avoiding the thermal pollution associated with warming up the receiving body of water, involves use of cooling towers that transfer the heat directly into the atmosphere. In either case, if we think of the power plant as a heat engine, it is the environment that acts as the heat sink so its temperature helps determine the overall efficiency of the power cycle.

Let us use the Carnot limit to estimate the maximum efficiency that a power plant such as that shown in Fig. 3.18 can possibly have. A reasonable estimate of T_H, the source temperature, might be the temperature of the steam from the boiler, which is typically around 600°C. For T_C we might use a typical condenser operating temperature of about 30°C. Using these values in (3.14), and remembering to convert temperatures to the absolute scale, gives

$$\text{Carnot efficiency} = 1 - \frac{T_C}{T_H} = 1 - \left(\frac{30 + 273}{600 + 273}\right) = 0.65 = 65\% \tag{3.15}$$

We know from Fig. 3.3 that the average efficiency of U.S. power plants is only about half this amount.

3.5.2 Coal-Fired Steam Power Plants

Coal-fired, Rankine cycle, steam power plants provide more than half of U.S. electricity and are responsible for three-fourths of the sulfur oxide (SO_x) emissions as well as a significant fraction of the country's carbon dioxide, particulate matter (PM), mercury, and nitrogen oxides (NO_x). Up until the 1960s, coal-fired power plants were notoriously dirty and not much was being done to control those emissions. That picture has changed dramatically, and now close to 40% of the cost of building a new coal plant is money spent to pay for pollution controls. Not only have existing levels of control been expensive in monetary terms, they also use up about 5% of the power generated, reducing the plant's overall efficiency.

Figure 3.19 shows some of the complexity that emission controls add to coal-fired steam power plants. In this plant, coal that has been crushed in a pulverizer is burned to make steam in a boiler for the turbine-generator system. The steam is then condensed, in this case using a cooling tower to dissipate waste heat to the atmosphere rather than a local body of water, and the condensate is then pumped back into the boiler. The flue gas from the boiler is sent to an electrostatic precipitator, which adds a charge to the particulates in the gas stream so that they can be attracted to electrodes, which collect them. Next, a wet scrubber sprays a limestone slurry over the flue gas, precipitating the sulfur and removing it in a sludge of calcium sulfide or calcium sulfate, which then must be treated and disposed of.

The thermal efficiency of power plants is often expressed as a *heat rate*, which is the thermal input (Btu or kJ) required to deliver 1 kWh of electrical output (1 Btu/kWh = 1.055 kJ/kWh). The smaller the heat rate, the higher the efficiency. In the United States, heat rates are usually expressed in Btu/kWh, which results in the following relationship between it and thermal efficiency, η:

$$\text{Heat rate (Btu/kWh)} = \frac{3412 \text{ Btu/kWh}}{\eta} \tag{3.16}$$

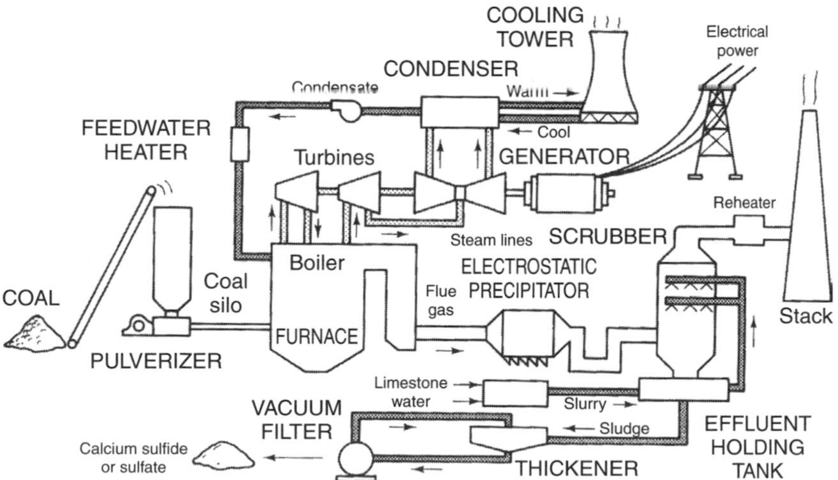

Figure 3.19 Typical modern coal-fired power plant using an electrostatic precipitator for particulate control and a limestone-based SO_2 scrubber. A cooling tower is shown for thermal pollution control. From Masters (1998).

Or, in SI units,

$$\text{Heat rate (kJ/kWh)} = \frac{1\ \text{(kJ/s)/kW} \times 3600\ \text{s/h}}{\eta} = \frac{3600\ \text{kJ/kWh}}{\eta} \qquad (3.17)$$

While Edison's first power plants in the 1880s had heat rates of about 70,000 Btu/kWh (≈5% efficient), the average new steam plant is about 34% efficient and has a heat rate of approximately 10,000 Btu/kWh. The best steam plants have efficiencies near 40%, but due to cost and technical complexity they are not widely used in the United States.

Example 3.2 Materials Balance for a Coal-Fired Steam Power Plant. Consider a power plant with a heat rate of 10,800 kJ/kWh burning bituminous coal with 75 percent carbon and a heating value (energy released when it is burned) of 27,300 kJ/kg. About 15% of thermal losses are up the stack, and the remaining 85% are taken away by cooling water.

a. Find the efficiency of the plant.
b. Find the mass of coal that must be provided per kWh delivered.
c. Find the rate of carbon and CO_2 emissions from the plant in kg/kWh.
d. Find the minimum flow of cooling water per kWh if its temperature is only allowed to increase by 10°C.

Solution

a. From (3.17), the efficiency of the plant is

$$\eta = \frac{3600\ \text{kJ/kWh}}{10,800\ \text{kJ/kWh}} = 0.333 = 33.3\%$$

that is, for each 3 units of input heat, the plant delivers 1 unit of electricity and 2 units of waste heat.

b. The rate at which coal needs to be burned is

$$\text{Coal rate} = \frac{10,800\ \text{kJ/kWh}}{27,300\ \text{kJ/kg}} = 0.396\ \text{kg coal/kWh}$$

c. With 75% of the coal being carbon, the carbon emission rate will be

$$\text{Carbon emissions} = 0.75 \times 0.396\ \text{kg/kWh} = 0.297\ \text{kgC/kWh}$$

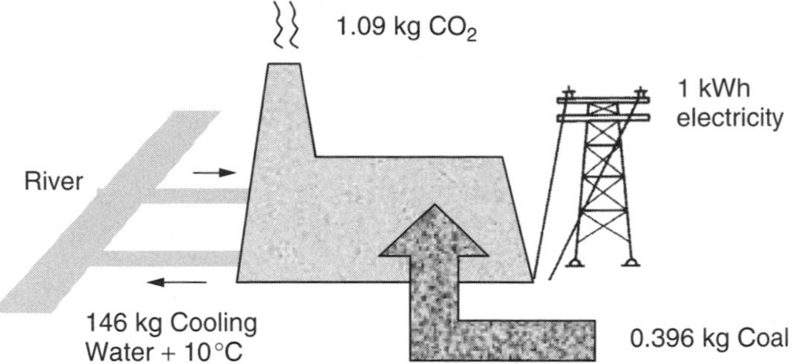

Figure 3.20 Mass flows to generate 1 kWh of electricity in a 33.3% efficient, coal-fired power plant burning bituminous coal.

Since the molecular weight of CO_2 is $12 + 2 \times 16 = 44$, there are 12 kg of C in 44 kg of CO_2. That translates into

$$CO_2 \text{ emissions} = 0.297 \text{ kgC/kWh} \times \left(\frac{44 \text{ kg}\, CO_2}{12 \text{ kg}\, C}\right) = 1.09 \text{ kg}\, CO_2/\text{kWh}$$

d. Two-thirds of the input energy is wasted, and 85% of that is removed by the cooling water. It takes 4.184 kJ of energy to raise 1 kg of water by 1°C (the specific heat), so the minimum flow rate for cooling water to increase by less than 10°C will be

$$\text{Cooling water} = \frac{0.85 \times \left(\frac{2}{3}\right) \times 10{,}800 \text{ kJ/kWh}}{4.184 \text{ kJ/kg}^\circ\text{C} \times 10^\circ\text{C}} = 146.3 \text{ kg/kWh}$$

A summary of Example 3.2 is shown in Fig. 3.20. In American units, it takes about 0.9 pounds of coal and 40 gallons of cooling water to generate 1 kWh of electricity while releasing about 2.4 pounds of CO_2 emissions.

3.6 COMBUSTION GAS TURBINES

The characteristics of combustion gas turbines for electricity generation are somewhat complementary to those of the steam turbine-generators just discussed. Steam power plants tend to be large, coal-fired units that operate best with fairly fixed loads. They tend to have high capital costs, largely driven by required emission controls, and low operating costs since they so often use low-cost boiler fuels such as coal. Once they have been purchased, they are cheap to operate so they usually are run more or less continuously. In contrast, gas turbines tend to be natural-gas-fired smaller units, which adjust quickly and easily to changing loads. They have low capital costs and relatively high fuel costs, which means they are most cost-effective as peaking power plants that run only intermittently. Historically, both steam and gas-turbine plants have had similar efficiencies, typically in the low 30% range.

3.6.1 Basic Gas Turbine

A basic gas turbine driving a generator is shown in Fig. 3.21. In it, fresh air is drawn into a compressor where spinning rotor blades compress the air, elevating its temperature and pressure. This hot, compressed air is mixed with fuel, usually natural gas, though LPG, kerosene, landfill gas, or oil are sometimes used, and subsequently burned in the combustion chamber. The hot exhaust gases are expanded in a turbine and released to the atmosphere. The compressor and turbine share a connecting shaft, so that a portion, typically more than half, of the rotational energy created by the spinning turbine is used to power the compressor.

Gas turbines have long been used in industrial applications and as such were designed strictly for stationary power systems. These *industrial gas turbines* tend to be large machines made with heavy, thick materials whose high thermal capacitance and moment of inertia reduces their ability to adjust quickly to changing loads. They are available in a range of sizes from hundreds of kilowatts to hundreds of megawatts. For the smallest units they are only about 20% efficient, but for turbines over about 10 MW they tend to have efficiencies of around 30%.

Another style of gas turbine takes advantage of the billions of dollars of development work that went into designing lightweight, compact engines for jet aircraft. The thin, light, super-alloy materials used in these *aeroderivative turbines* enable fast starts and quick acceleration, so they easily adjust to rapid load changes and numerous start-up/shut-down events. Their small size makes it easy to fabricate the complete unit in the factory and ship it to a site, thereby

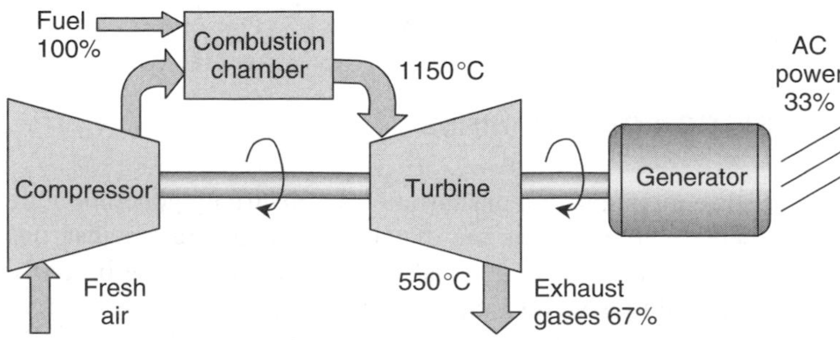

Figure 3.21 Basic gas turbine and generator. Temperatures and efficiencies are typical.

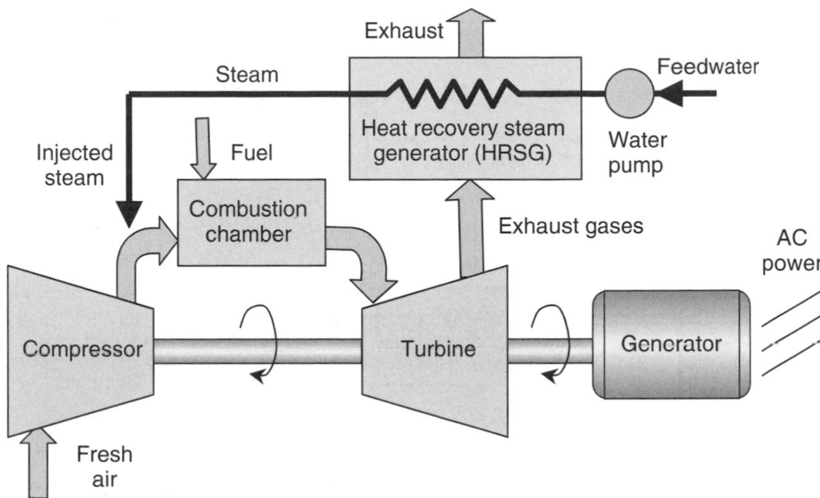

Figure 3.22 Steam-injected gas turbine (STIG) for increased efficiency and reduced NO_x emissions. Efficiencies may approach 45%.

reducing field installation time and cost. Aeroderivative turbines are available in sizes ranging from a few kilowatts up to about 50 MW. In their larger sizes, they achieve efficiencies exceeding 40%.

3.6.2 Steam-Injected Gas Turbines (STIG)

One way to increase the efficiency of gas turbines is to add a heat exchanger, called a heat recovery steam generator (HRSG), to capture some of the waste heat from the turbine. As shown in Fig. 3.22, water pumped through the HRSG turns to steam, which is injected back into the airstream coming from the compressor. The injected steam displaces a portion of the fuel heat that would otherwise be needed in the combustion chamber. These units, called steam-injected gas turbines (STIG), can have efficiencies approaching 45%. Moreover, the injected steam reduces combustion temperatures, which helps control NO_x emissions. They are considerably more expensive than simple gas turbines due to the extra cost of the HRSG, and the care that must be taken to purify incoming feedwater.

3.7 COMBINED-CYCLE POWER PLANTS

From our analysis of the Carnot cycle, we know that the maximum possible efficiency of a heat engine is limited by a low-temperature sink as well as by a high-temperature source. With exhaust gases leaving a gas turbine at temperatures frequently above 500°C, it should come at no surprise that overall

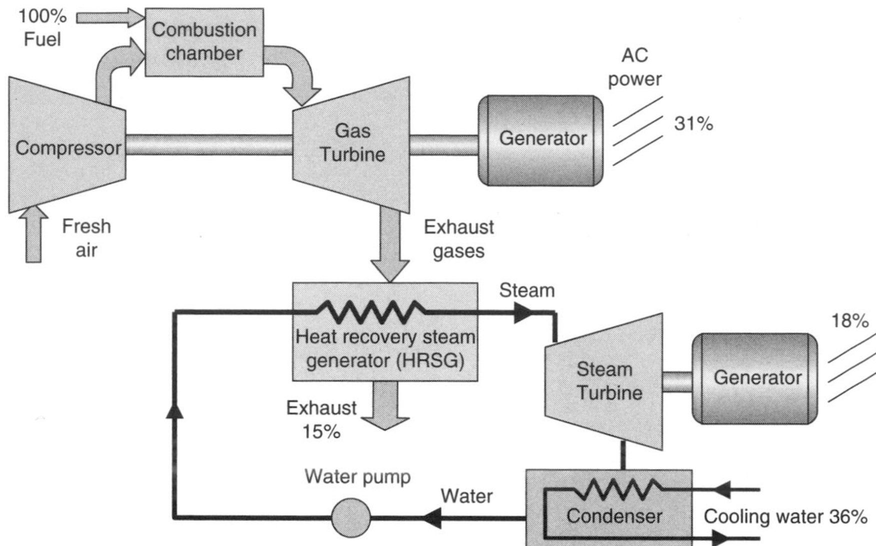

Figure 3.23 Combined-cycle power system with representative energy flows providing a total efficiency of 49%.

efficiencies are usually modest, in the 30% range, unless some use is made of that high-quality waste heat. A heat recovery steam generator (HRSG) can capture some of that waste heat for process steam or other thermal purposes, but if the goal is to maximize electricity output while treating any thermal benefits as secondary, the gas turbine waste heat can be used to power a second-stage steam turbine. When a gas turbine and steam turbine are coupled this way, the result is called a *combined-cycle* power plant. Working together, such combined-cycle plants have achieved fuel-to-electricity efficiencies in excess of 50%.

Figure 3.23 shows how the two turbines are linked. Exhaust gases from the gas turbine pass through a heat recovery steam generator (HRSG) before being allowed to vent to the atmosphere. The HRSG boils water, creating high-temperature, high-pressure steam that expands in the steam turbine. The rest of the steam cycle is the normal one: A partial vacuum is created in the condenser, drawing steam from the turbine, and the resulting condensate is pumped back through the HSRG to complete the cycle.

3.8 GAS TURBINES AND COMBINED-CYCLE COGENERATION

Exhaust temperatures from gas turbines frequently are above 500°C, which is high enough to be very useful for a number of applications including industrial process heating, absorption cooling, space heating, or, as we have seen, generation of additional electricity with a steam turbine. When a power plant produces useful

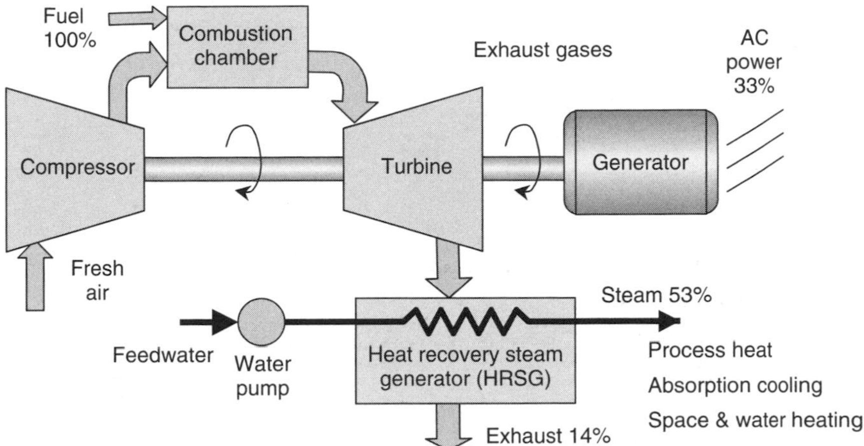

Figure 3.24 Simple-cycle gas turbine with a steam generator for cogeneration showing typical conversion efficiencies.

thermal energy as well as electrical power in a sequential process using a single fuel, it is called a *cogeneration* plant. The usual way to capture the waste heat in a gas-turbine cogeneration facility is with a heat recovery steam generator (HRSG), as shown in Fig. 3.24. Example conversion efficiencies of 33% for electricity and 53% for thermal output are shown.

The steam generated in the HRSG in Fig. 3.24 could, of course, be sent to a steam turbine to squeeze more electrical output from the fuel. In such a combined-cycle unit, the electrical efficiency goes up but there is less waste heat for cogeneration. Moreover, if it is a conventional combined-cycle plant, the steam turbine has been designed for maximum electrical efficiency, which means that the condenser operating temperature is as cold as it can be. While there would still be a lot of heat that could be transferred from condenser to a thermal load, it would be at such a low temperature that it wouldn't be of much use.

To use a combined-cycle plant for cogeneration often means substituting a noncondensing heat exchanger for the usual condenser. Without the low pressure normally created in the condenser, these back-pressure turbines generate more useful thermal energy at the expense of reduced electrical efficiency. Figure 3.25 shows a schematic of such a combined-cycle, cogeneration plant, including representative energy flows.

3.9 BASELOAD, INTERMEDIATE AND PEAKING POWER PLANTS

We saw from the load profile for a hot, summer day in California (Fig. 3.6) that the demand for electricity can vary considerably from day to night. It also exhibits weekly patterns, with diminished demands on weekends, as well as seasonally,

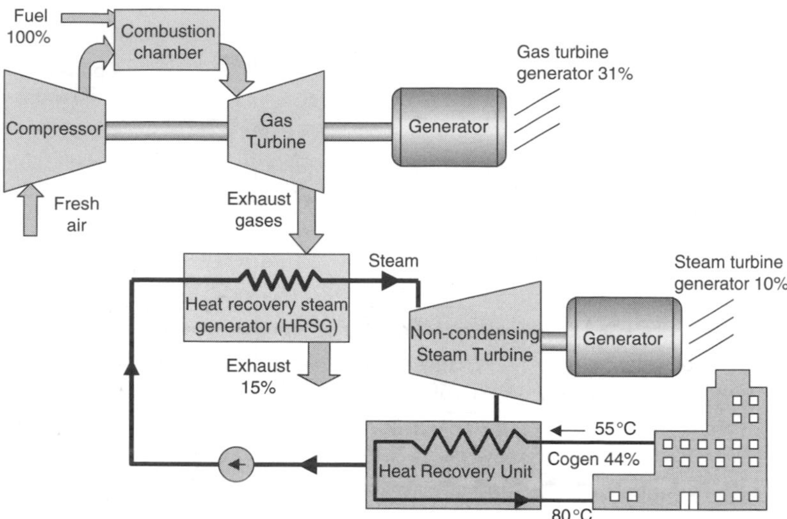

Figure 3.25 Representative energy flows for a combined-cycle, cogeneration plant with back-pressure steam turbine, delivering thermal energy to a district heating system.

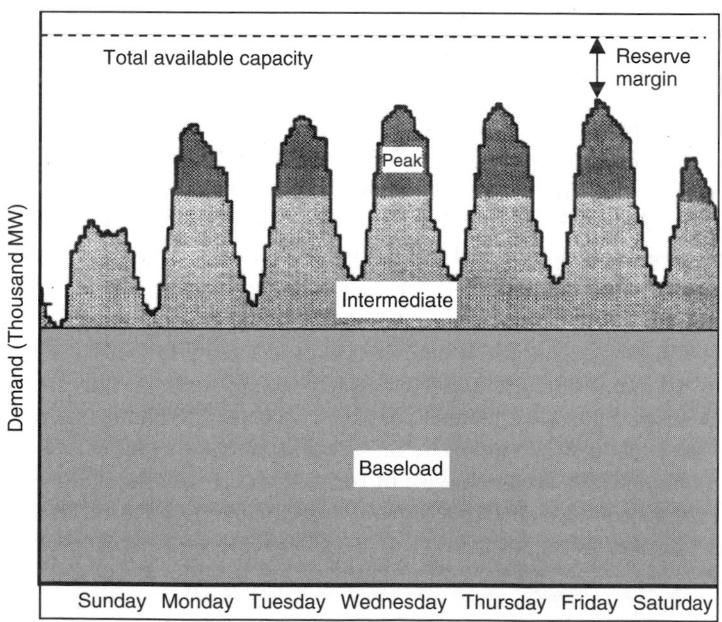

Figure 3.26 Example of weekly load fluctuations and roughly how power plants can be categorized as baseload, intermediate, or peaking plants.

with some utilities seeing their annual highest loads on hot summer days and others on cold winter mornings.

These fluctuations in demand suggest that during the peak demand, most of a utility's power plants will be operating, while in the valleys, many are likely to be idling or shut off entirely. In other words, many power plants don't operate with a schedule anything like full output all of the time. It has also been mentioned that some power plants, especially large coal-fired plants as well as hydroelectric plants, are expensive to build but relatively cheap to operate, so they should be run more or less continuously as *baseload* plants; others, such as simple-cycle gas turbines, are relatively inexpensive to build but expensive to operate. They are most appropriately used as *peaking* power plants, turned on only during periods of highest demand. Other plants have characteristics that are somewhere in between; these *intermediate* load plants are often run for most of the daytime and then cycled as necessary to follow the evening load. Figure 3.26 suggests these designations of baseload, intermediate, and peaking power plants applied to a weeklong demand curve.

An important question for utility planners is what combination of power plants will most economically meet the hour-by-hour power demands of their customers. While the details of such generation planning is beyond the scope of this book, we can get a good feel for the fundamentals with a few simple notions involving the economic characteristics of different types of power plants and how they relate to the loads they must serve.

3.9.1 Screening Curves

A very simple model of the economics of a given power plant takes all of the costs and puts them into two categories: fixed costs and variable costs. Fixed costs are monies that must be spent even if the power plant is never turned on, and they include such things as capital costs, taxes, insurance, and any fixed operations and maintenance costs that will be incurred even when the plant isn't operated. Variable costs are the added costs associated with actually running the plant. These are mostly fuel plus operations and maintenance costs. The first step in finding the optimum mix of power plants is to develop *screening curves* that show annual revenues required to pay fixed and variable costs as a function of hours per year that the plant is operated.

The capital costs of a power plant can be annualized by multiplying it by a quantity called the *fixed charge rate (FCR)*. The fixed charge rate accounts for interest on loans, acceptable returns for investors, fixed operation and maintenance (O&M) charges, taxes, and so forth. The FCR depends primarily on the cost of capital, so it may vary as interest rates change, but it is a number usually between 11% and 18% per year. On a per-kilowatt of rated power basis, the annualized fixed costs are computed from

$$\text{Fixed (\$/yr-kW)} = \text{Capital cost (\$/kW)} \times \text{Fixed charge rate (yr}^{-1}\text{)} \tag{3.18}$$

The variable costs, which are also annualized, depend on the unit cost of fuel, the O&M rate for actual operation of the plant, and the number of hours per year the plant is operated.

$$\text{Variable (\$/yr-kW)} = [\text{Fuel (\$/Btu)} \times \text{Heat rate (Btu/kWh)} + \text{O\&M (\$/kWh)}] \times \text{h/yr} \tag{3.19}$$

In (3.19) it is assumed that the plant runs at full rated power while it is operated, but no power at other times. Adjusting for less than full power is an easy modification that will be introduced later. Also, (3.19) assumes that the fuel cost is fixed, but it too is easily adjusted to account for fuel escalation and inflation. For our purposes here, these modifications are not important. They will, however, be included in the economic analysis of power plants presented in Chapter 5. Table 3.3 provides some representative costs for some of the most commonly used power plants.

Example 3.3 Cost of Electricity from a Coal-Fired Steam Plant. Find the annual revenue required for a pulverized-coal steam plant using parameters given in Table 3.3. Assume a fixed charge rate of 0.16/yr and assume that the plant operates at the equivalent rate of full power for 8000 hours per year. What should be the price of electricity from this plant?

Solution From (3.18) the annual fixed revenue required would be

$$\text{Fixed costs} = \$1400/\text{kW} \times 0.16/\text{yr} = \$224/\text{kW-yr}$$

TABLE 3.3 Example Cost Parameters for Power Plants

Technology	Fuel	Capital Cost (\$/kW)	Heat Rate (Btu/kWh)	Fuel Cost (\$/million Btu)	Variable O&M (¢/kWh)
Pulverized coal steam	Coal	1400	9,700	1.50	0.43
Advanced coal steam	Coal	1600	8,800	1.50	0.43
Oil/gas steam	Oil/Gas	900	9,500	4.60	0.52
Combined cycle	Natural gas	600	7,700	4.50	0.37
Combustion turbine	Natural gas	400	11,400	4.50	0.62
STIG gas turbine	Natural gas	600	9,100	4.50	0.50
New hydroelectric	Water	1900	—	0.00	0.30

Source: Based on data from Petchers (2002) and UCS (1992).

The variable cost for fuel and O&M, operating 8000 hours at full power, would be

$$\begin{aligned}\text{Variable} &= (\$1.50/10^6\ \text{Btu} \times 9700\ \text{Btu/kWh} + 0.0043\$/\text{kWh}) \times 8000\ \text{hr/yr} \\ &= \$150.80/\text{kW-yr}\end{aligned}$$

For a 1-kW plant,

$$\text{Electricity generated} = 1\ \text{kW} \times 8000\ \text{hr/yr} = 8000\ \text{kWh/yr}$$

$$\text{Price} = 1\ \text{kW} \times \frac{(224 + 150.80)\$/\text{yr-kW}}{8000\ \text{kWh/yr}} = \$0.0469/\text{kWh} = 4.69¢/\text{kWh}$$

In the above example, it was assumed that in a year with 8760 hours, the plant would operate at full power for 8000 hours and no power for 760 hours. The same 8000 kWh/yr could, of course, be the result of operating all 8760 hours, but not always at the full rated output. The resulting price of electricity would be the same in either case. One way to capture this subtlety is to introduce the notion of a *capacity factor* (CF):

$$\text{Annual output (kWh/yr)} = \text{Rated power (kW)} \times 8760\ \text{h/yr} \times \text{CF} \tag{3.20}$$

Solving (3.20) for CF gives another way to interpret capacity factor as the ratio of average power to rated power:

$$\text{CF} = \frac{\text{Average power (kW)} \times 8760\ \text{h/yr}}{\text{Rated power (kW)} \times 8760\ \text{h/yr}} = \frac{\text{Average power}}{\text{Rated power}} \tag{3.21}$$

Figure 3.27 shows how total revenues required for the coal plant in Example 3.3 vary as a function of its capacity factor (or as a function of hours per year at full power). Under the circumstances in the example, the average cost of electricity ($0.0469/kWh) is the slope of a line drawn to the point on the curve corresponding to the 8000 hours of operation (CF = 0.9132). Clearly, the average cost increases as CF decreases, which helps explain why peaking power plants that operate only a few hours each day have such high average cost of electricity.

When plots like that shown in Fig. 3.27 are drawn on the same axes for different power plants, the resulting screening curves provide the first step in determining the optimum mix of different power plant types. The screening curve for the pulverized coal plant in Fig. 3.27, along with analogous curves for the combined-cycle plant and the combustion turbine described in Table 3.3, are shown in Fig. 3.28. What these screening curves show is that the combustion turbine, which is cheap to build but expensive to operate, is the least-cost option as long as it doesn't operate more than 1675 h/yr ($\text{CF} \leq 0.19$), making it the best choice for peaking power plants. The coal-steam plant, with its high capital cost and low fuel cost, is the least expensive as long as it runs at least 6565 h/yr

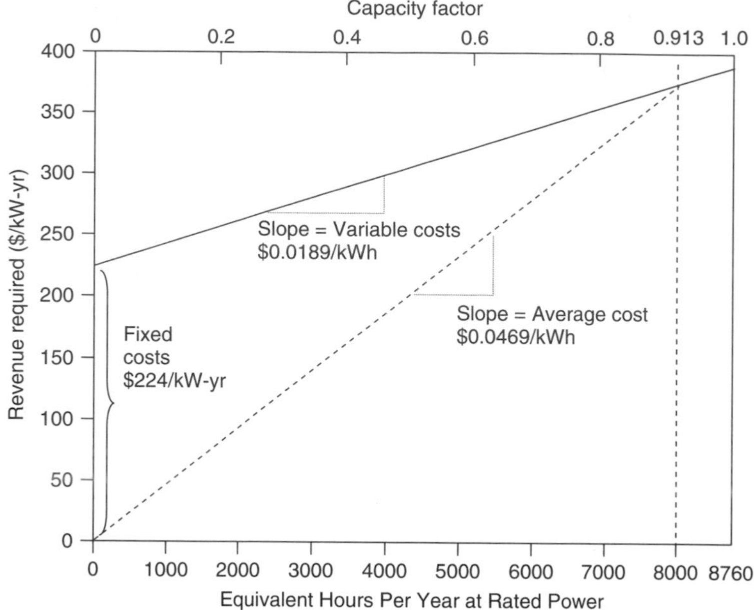

Figure 3.27 The average cost of electricity is the slope of the line drawn from the origin to point on the revenue curve that corresponds to the capacity factor. The data shown are for the coal plant in Example 3.3.

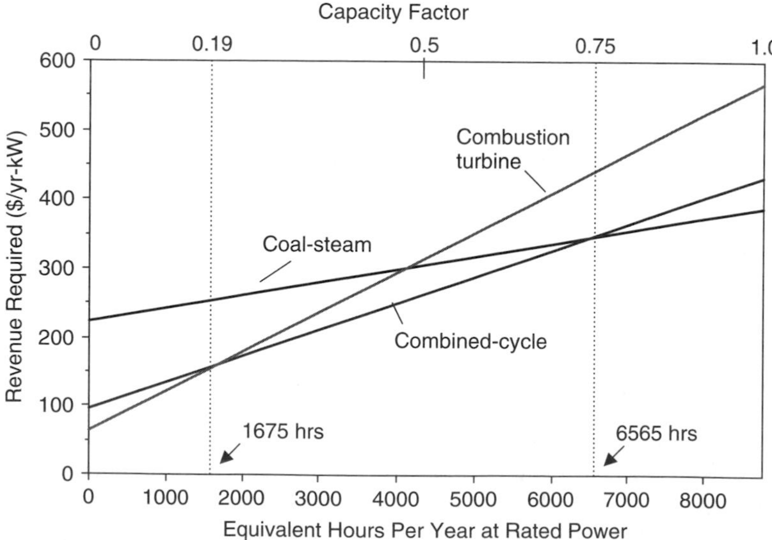

Figure 3.28 Screening curves for coal-steam, combustion turbine, and combined-cycle plants based on data in Table 3.3. For plants operated less than 1675 h/yr, combustion turbines are least expensive; for plants operated more than 6565 h/yr, a coal-steam plant is cheapest; otherwise, a combined-cycle plant is least expensive.

(CF $\geq$ 0.75), making it an ideal baseload plant. The combined cycle plant is the cheapest option if it runs somewhere between 1675 and 6565 h/yr (0.19 $\leq$ CF $\leq$ 0.75), which makes it well suited as an intermediate load plant.

3.9.2 Load–Duration Curves

We can imagine a load–time curve, such as that shown in Fig. 3.26, as being a series of one-hour power demands arranged in chronological order. Each slice of the load curve has a height equal to power (kW) and width equal to time (1 h), so its area is kWh of energy used in that hour. As suggested in Fig. 3.29, if we rearrange those vertical slices, ordering them from highest kW demand to lowest through an entire year of 8760 h, we get something called a *load–duration curve*. The area under the load–duration curve is the total kWh of electricity used per year.

A smooth version of a load-duration curve is shown in Fig. 3.30. Notice that the x axis is still measured in hours, but now a different way to interpret the curve presents itself. The graph tells how many hours per year the load is equal to or above a particular value. For example, in Fig. 3.30, the load is above 3000 MW

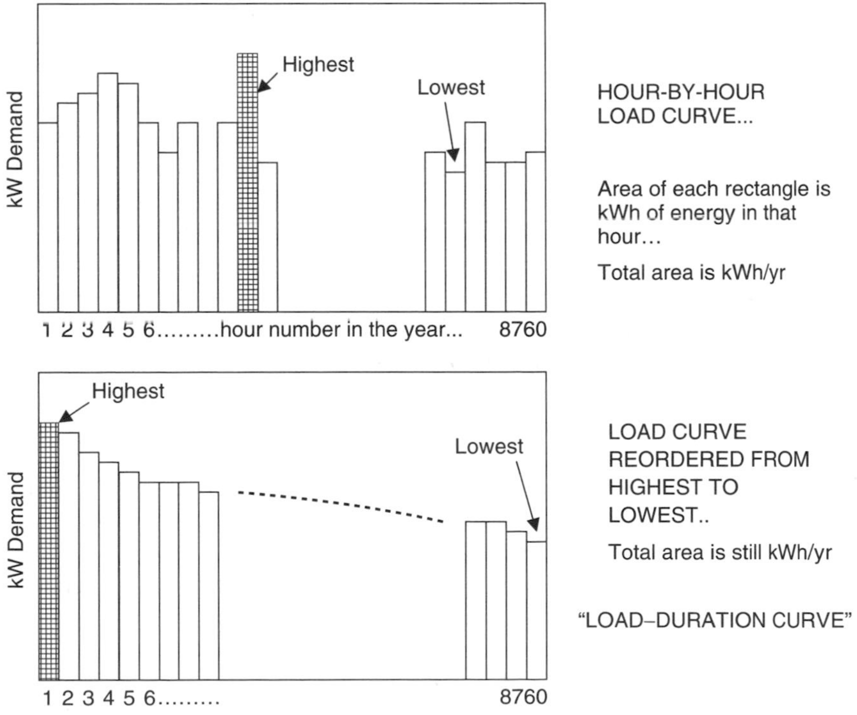

Figure 3.29 A load–duration curve is simply the hour-by-hour load curve rearranged from chronological order into an order based on magnitude. The area under the curve is the total kWh/yr.

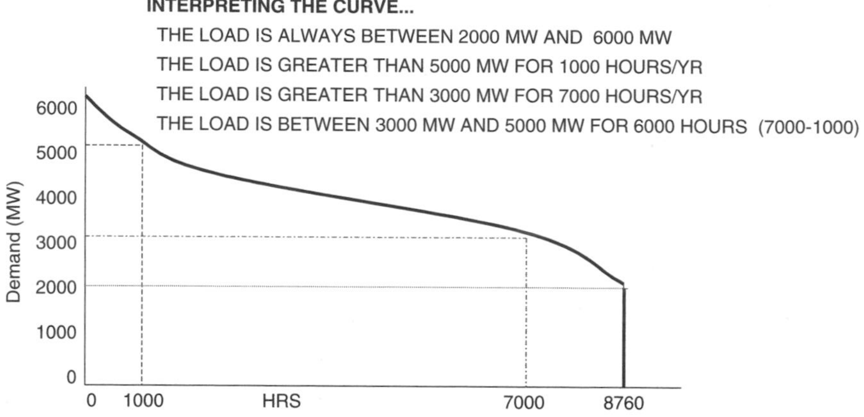

Figure 3.30 Interpreting a load–duration curve.

for 7000 h each year, and it is above 5000 MW for only 1000 h/yr. It is always above 2000 MW and never above 6000 MW.

By entering the crossover points in the resource screening curves into the load-duration curve, it is easy to come up with a first-cut estimate of the optimal mix of power plants. For example, the crossover between gas turbines and combined-cycle plants in Fig. 3.28 occurs at 1675 hours of operation, while the crossover between combined-cycle and coal-steam plants is at 6565 hours. Those are entered into the above load–duration curve and presented in Fig. 3.31. The screening curve tells us that coal plants are the best option as long as they operate for more than 6565 h/yr, and the load–duration curve indicates that the demand is at least 3500 MW for 6565 h. Therefore we should have 3500 MW of baseload, coal-steam plants in the mix. Combined-cycle plants need to operate at least 1675 h/yr and less than 6565 h to be most cost-effective. The screening curve tells us that 1200 MW of these intermediate plants would operate within that range. Since combustion turbines are the most cost effective as long as they don't operate more than 1675 h/yr, and the load is between 4700 MW and 6000 MW for 1675 h, the mix should contain 1300 MW of peaking gas turbines.

The generation mix shown on a load-duration curve allows us to find the average capacity factor for each type of generating plant in the mix, which will determine the average cost of electricity for each type. Figure 3.32 shows rectangular horizontal slices corresponding to energy that would be generated by each plant type if it operated continuously. The shaded portion of each slice is the energy actually generated. The ratio of shaded area to total rectangle area is the capacity factor for each. The baseload coal plants operate with a CF of about 91%, the intermediate-load combined-cycle plants operate with a CF of about 47%, and the peaking gas turbines have a CF of about 10%. Those capacity factors, combined with cost parameters from Table 3.3, allow us to determine the cost of electricity from each type of plant.

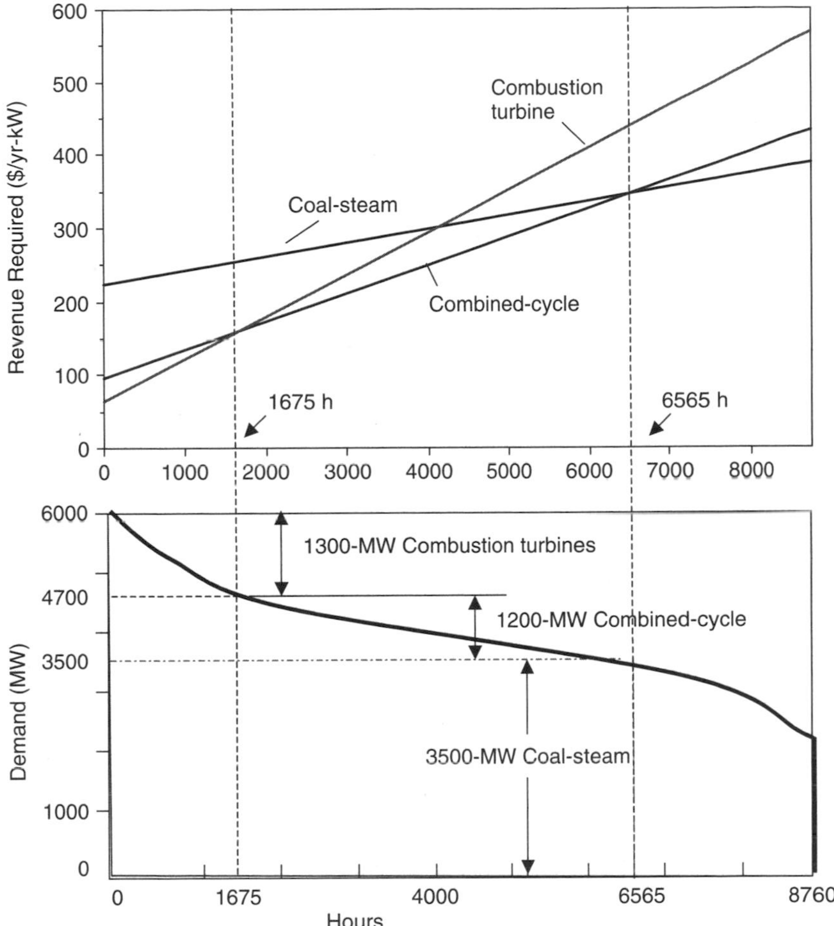

Figure 3.31 Plotting the crossover points from screening curves (Fig. 3.28) onto the load–duration curve (Fig. 3.30) to determine an optimum mix of power plants.

The process and results for our example utility are summarized in Table 3.4. The baseload plants deliver energy at 4.69¢/kWh, the intermediate plants for 6.23¢/kWh, and the combustion-turbine peakers for 12.87¢/kWh. The peaker plant electricity is so much more expensive in part because they have lower efficiency while burning the more expensive natural gas, but mostly because their capital cost is spread over so few kilowatt-hours of output since they are used so little.

Using screening curves for generation planning is merely a first cut at determining what a utility should build to keep up with changing loads and aging existing plants. Unless the load–duration curve already accounts for a cushion of excess capacity, called the *reserve margin*, the generation mix just estimated

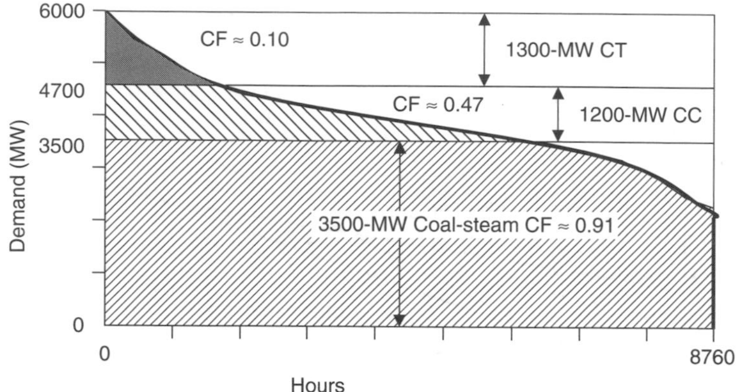

Figure 3.32 The fraction of each horizontal rectangle that is shaded is the capacity factor for that portion of the generation facilities.

TABLE 3.4 Unit Cost of Electricity from the Three Types of Generation for the Example Utility[a]

Generation Type	Rated Power (MW)	CF	Fixed Cost (million $/yr)	Variable ($/kWh)	Output (billion kWh/yr)	Total Cost (billion $/yr)	Unit Cost (¢/kWh)
Coal-steam	3500	0.91	784.0	0.0189	27.99	1.312	4.69
Combined-cycle	1200	0.47	115.2	0.0390	4.94	0.308	6.23
Combustion turbine	1300	0.10	83.2	0.0556	1.14	0.147	12.87

[a] Electricity from the peakers is expensive because peakers are used so little.

would have to be augmented to allow for plant outages, sudden peaks in demand, and other complicating factors.

The process of selecting which plants to operate at any given time is called *dispatching*. Since costs already incurred to build power plants (sunk costs) must be paid no matter what, it makes sense to dispatch plants in order of their operating costs, from lowest to highest. Renewables, with their intermittent operation but very low operating costs, should be dispatched first whenever they are available; so even though their capacity factors may be low, they are part of the baseload. A special case is hydroelectric plants, which must be operated with multiple constraints including the need for proper flows for downstream ecosystems, water supply, and irrigation while maintaining sufficient reserves to cover dry seasons. Hydro is especially useful as a dispatchable source that may supplement baseload, intermittent, or peak loads, especially when existing facilities are down for maintenance or other reasons.

3.10 TRANSMISSION AND DISTRIBUTION

While the generation side of electric power systems usually receives the most attention, the shift toward utility restructuring, along with the emergence of distributed generation systems, is causing renewed interest in the transmission and distribution (T&D) side of the business.

Figure 3.33 shows the relative capital expenditures on T&D over time compared with generation by U.S. investor-owned utilities. The most striking feature of the graph is the extraordinary period of power plant construction that lasted from the early 1970s through the mid-1980s, driven largely by huge spending for nuclear power stations. Except for that anomalous period, T&D construction has generally cost utilities more than they have spent on generation. In the latter half of the 1990s, T&D expenditures were roughly double that of generation, with most of that being spent on the distribution portion of T&D.

The utility grid system starts with transmission lines that carry large blocks of power, at voltages ranging from 161 kV to 765 kV, over relatively long distances from central generating stations toward major load centers. Lower-voltage subtransmission lines may carry it to distribution substations located closer to the loads. At substations, the voltage is lowered once again, to typically 4.16 to 24.94 kV and sent out over distribution feeders to customers. An example of a simple distribution substation is diagrammed in Fig. 3.34. Notice the combination of switches, circuit breakers, and fuses that protect key components and which allow different segments of the system to be isolated for maintenance or during emergency *faults* (short circuits) that may occur in the system.

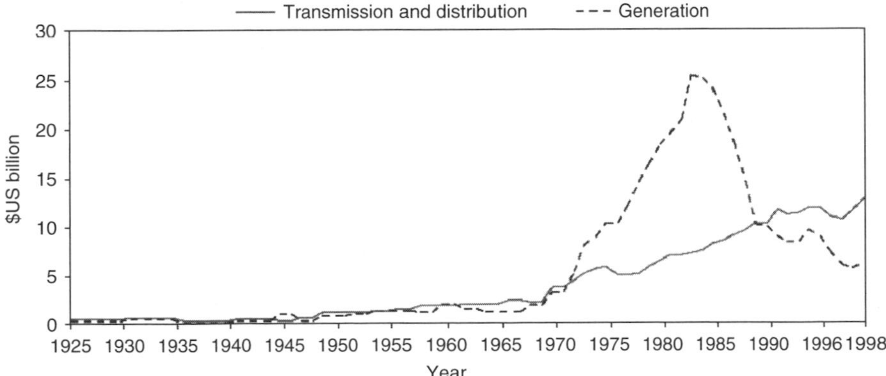

Figure 3.33 Transmission and distribution (T&D) construction expenditures at U.S. investor-owned utilities compared with generation. Except for the anomalous spurt in power plant construction during the 1970s and early 1980s, T&D costs have generally exceeded generation. From Lovins et al. (2002), using Edison Electric Institute data.

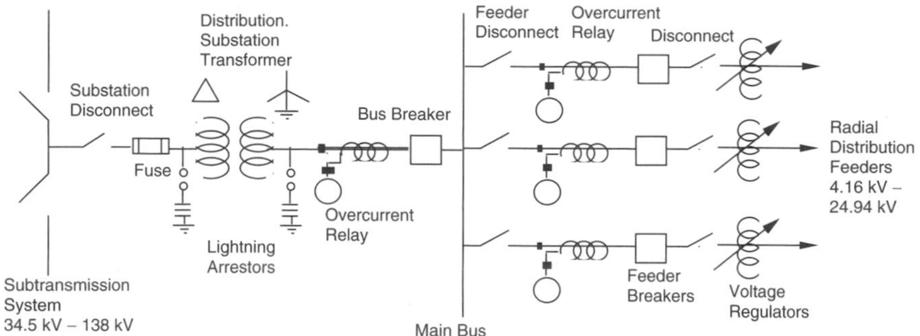

Figure 3.34 A simple distribution station. For simplication, this is drawn as a *one-line diagram*, which means that a single conductor on the diagram corresponds to the three lines in a three-phase system.

3.10.1 The National Transmission Grid

The United States has close to 275,000 miles of transmission lines, most of which carry high-voltage, three-phase ac power. Investor owned utilities (IOUs) own three-fourths of those lines (200,000 miles), with the remaining 75,000 miles owned by federal, public, and cooperative utilities. Independent power producers do not own transmission lines so their ability to wheel power to customers depends entirely on their ability to have access to that grid. As will be described in the regulatory section of this chapter, Federal Energy Regulatory Commission (FERC) Order 2000 is attempting to dramatically change the utility-ownership of the grid as part of its efforts to promote a fully competitive wholesale power market. Order 2000 encourages the establishment of independent regional transmission organizations (RTOs), which could shift transmission line ownership to a handful of separate transmission companies (TRANSCOs), or it could allow continued utility ownership but with control turned over to independent system operators (ISOs).

As shown in Fig. 3.35, the transmission network in the United States is organized around three major power grids: the Eastern Interconnect, the Western Interconnect, and the Texas Interconnect. Texas is unique in that its power does not cross state lines so it is not subject to control by the Federal Energy Regulatory Commission (FERC). Within each of these three interconnection zones, utilities buy and sell power among themselves. There are very limited interconnections between the three major power grids. After a major blackout in the Northeastern United States in 1965, the North American Electric Reliability Council (NERC) was formed to help coordinate bulk power policies that affect the reliability and adequacy of service within 10 designated regions of the U.S. grid.

While almost all power in the United States is transmitted over three-phase ac transmission lines, there are circumstances in which high-voltage

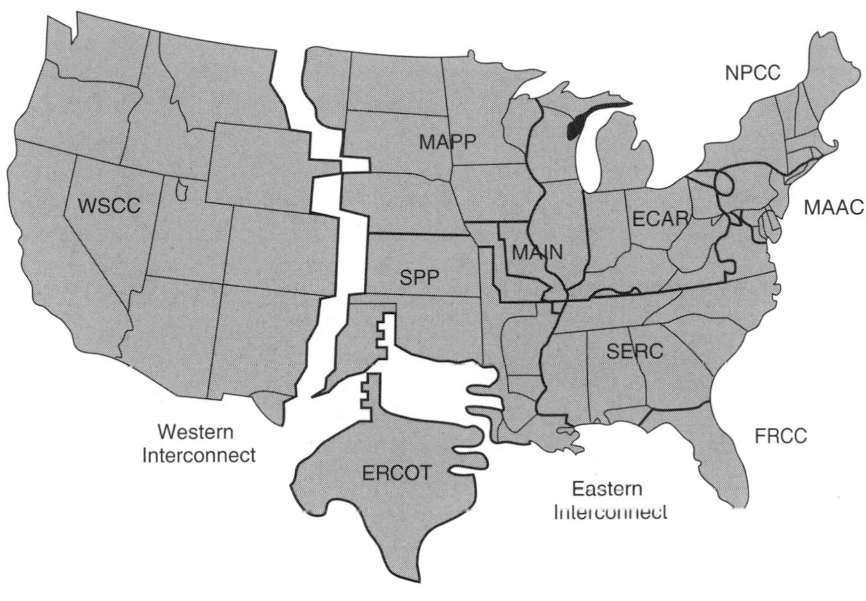

Figure 3.35 Transmission of U.S. electric power is divided into three quite separate power grids, which are further subdivided into 10 North American Electric Reliability Council Regions. ECAR, East Central Area Reliability Coordination Agreement; ERCOT, Electric Reliability Council of Texas; FRCC, Florida Reliability Coordinating Council; MAAC, Mid-Atlantic Area Council; MAPP, Mid-Continent Area Power Pool; MAIN, Mid-America Interconnected Network; NPCC, Northeast Power Coordinating Council; SERC, Southeastern Electric Reliability Council; SPP, Southwest Power Pool; WSCC, Western Systems Coordinating Council. (EIA 2001).

dc (HVDC) lines have certain benefits. They are especially useful for interconnecting the power grid in one part of the country to a grid in another area since problems associated with exactly matching frequency, phase, and voltages are eliminated in dc. An example of such a system is the 600-kV, 6000-MW Pacific Intertie between the Pacific Northwest and Southern California. Similar situations occur between countries, and indeed many of the HVDC links around the world are used to link the grid of one country to another—examples include: Norway–Denmark, Finland–Sweden, Sweden–Denmark, Canada–United States, Germany–Czechoslovakia, Austria–Hungary, Argentina–Brazil, France–England, and Mozambique–South Africa. The control and interfacing simplicity of dc makes HVDC links particularly well suited for connecting ac grids that operate with different frequencies, as is the case, for example, in Japan, with its 50-Hz and 60-Hz regions.

HVDC links require converters at both ends of the dc transmission line to rectify ac to dc and then to invert dc back to ac. The converters at each end can operate either as a rectifier or as an inverter, which allows power flow

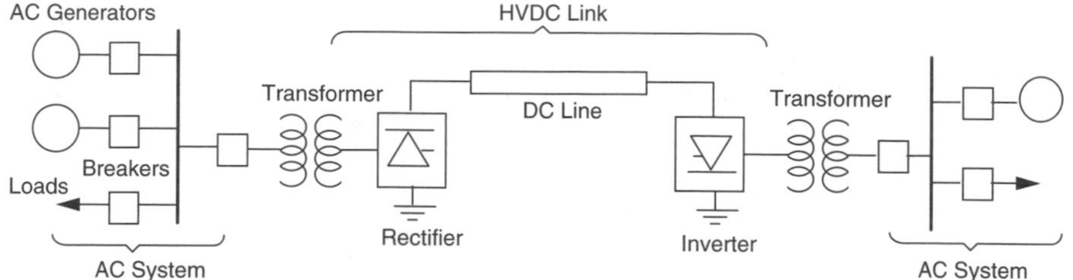

Figure 3.36 A one-line diagram of a dc link between ac systems. The inverter and rectifier can switch roles to allow bidirectional power flow.

in either direction. A simple one-line drawing of an HVDC link is shown in Fig. 3.36. HVDC lines offer the most economic form of transmission over very long distances—that is, distances beyond about 500 miles or so. For these longer distances, the extra costs of converters at each end can be more than offset by the reduction in transmission line and tower costs.

3.10.2 Transmission Lines

The physical characteristics of transmission lines depend very much on the voltages that they carry. Cables carrying higher voltages must be spaced further apart from each other and from the ground to prevent arcing from line to line, and higher current levels require thicker conductors. Table 3.5 lists the most common voltages in use in the United States along with their usual designation as being transmission, subtransmission, distribution, or utilization voltages.

Figure 3.37 shows examples of towers used for various representative transmission and subtransmission voltages. Notice that the 500-kV tower has three suspended connections for the three-phase current, but it also shows a fourth connection, namely, a ground wire above the entire structure. This ground wire not only serves as a return path in case the phases are not balanced, but also provides a certain amount of lightning protection.

TABLE 3.5 Nominal Standard T&D System Voltages

Transmission (kV)	Subtransmission (kV)	Distribution (kV)	Utilization (V)
765	138	24.94	600
500	115	22.86	480
345	69	13.8	240
230	46	13.2	208
161	34.5	12.47	120
		8.32	
		4.16	

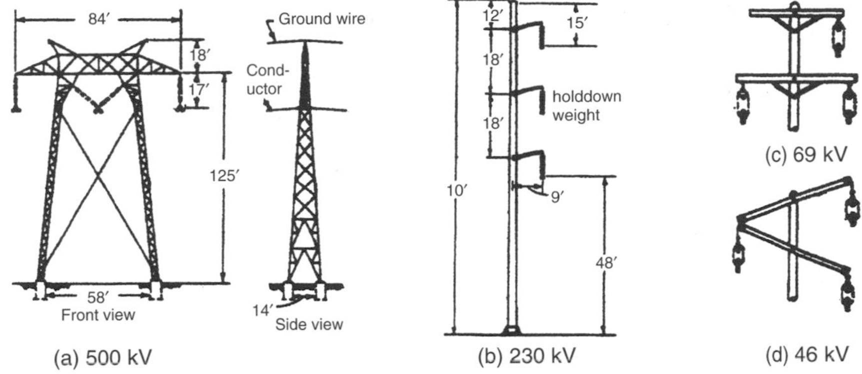

Figure 3.37 Examples of transmission towers: (a) 500 kV; (b) 230-kV steel pole; (c) 69-kV wood tower; (d) 46-kV wood tower.

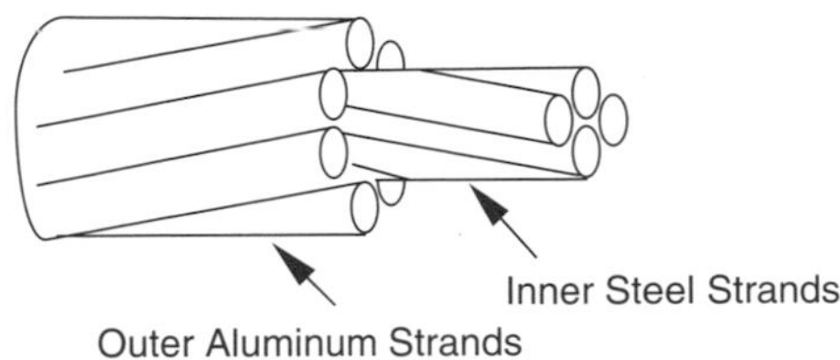

Figure 3.38 Aluminum conductor with steel reinforcing (ACSR).

Overhead transmission lines are usually uninsulated, stranded aluminum or copper wire that is often wrapped around a steel core to add strength (Fig. 3.38). The resistance of such cable is of obvious importance due to the i^2R power losses in the wires as they may carry hundreds of amps of current. Examples of cable resistances, diameters, and current-carrying capacity are shown in Table 3.6.

Example 3.4 Transmission Line Losses. Consider a 40-mile-long, three-phase, 230-kV (line-to-line) transmission system using 0.502-in.-diameter ACSR cable. The line supplies a three-phase, wye-connected, 100-MW load with a 0.90 power factor. Find the power losses in the transmission line and its efficiency. What savings would be achieved if the power factor could be corrected to 1.0?

Solution From Table 3.6, the cable has 0.7603-Ω/mile resistance, so each line has resistance

$$R = 40 \text{ mi} \times 0.7603\ \Omega/\text{mi} = 30.41\ \Omega$$

TABLE 3.6 Conductor Characteristics[a]

Conductor Material	Outer Diameter (in.)	Resistance (Ω/mile)	Ampacity (A)
ACSR	0.502	0.7603	315
ACSR	0.642	0.4113	475
ACSR	0.858	0.2302	659
ACSR	1.092	0.1436	889
ACSR	1.382	0.0913	1187
Copper	0.629	0.2402	590
Copper	0.813	0.1455	810
Copper	1.152	0.0762	1240
Aluminum	0.666	0.3326	513
Aluminum	0.918	0.1874	765
Aluminum	1.124	0.1193	982

[a] Resistances at 75°C conductor temperature and 60 Hz, ampacity at 25°C ambient, and 2-ft/s wind velocity.

Source: Data from Bosela (1997).

The phase voltage from line to neutral is given by (2.67)

$$V_{\text{phase}} = \frac{V_{\text{line}}}{\sqrt{3}} = \frac{230 \text{ kV}}{\sqrt{3}} = 132.79 \text{ kV}$$

The 100 MW of real power delivered is three times the power delivered in each phase. From (2.72)

$$P = 3\ V_{\text{phase}} I_{\text{line}} \times \text{PF} = 100 \times 10^6 \text{ W}$$

Solving for the line current gives

$$I_{\text{line}} = \frac{100 \times 10^6}{3 \times 132{,}790 \times 0.90} = 278.9 \text{ A}$$

Checking Table 3.6, this is less than the 315 A the cable is rated for (at 25°C).

The total line losses in the three phases is therefore

$$P = 3I^2R = 3 \times (278.9)^2 \times 30.41 = 7.097 \times 10^6\ W = 7.097 \text{ MW}$$

The overall efficiency of the transmission line is therefore

$$\text{Efficiency} = \frac{\text{Power delivered}}{\text{Input power}} = \frac{100}{100 + 7.097} = 0.9337 = 93.37\%$$

That is, there are 6.63% losses in the transmission line. The figure below summarizes the calculations.

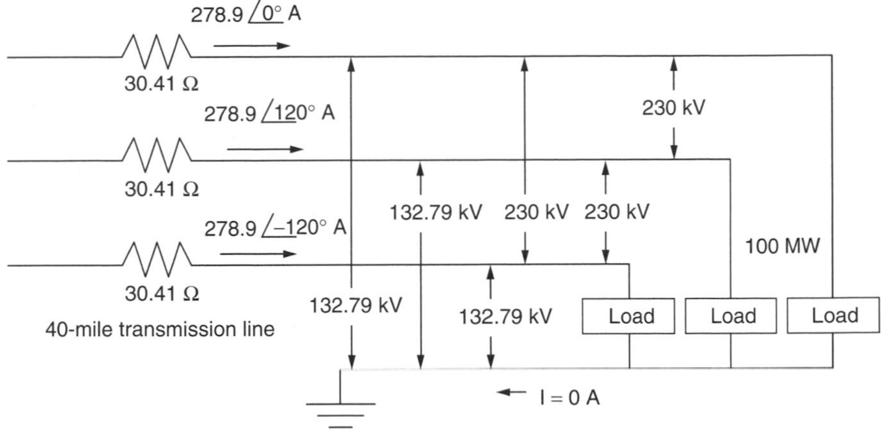

If the power factor could be corrected to 1.0, the line losses would be reduced to

$$P = 3 \times \left(\frac{100 \times 10^6}{3 \times 132,790 \times 1.0} \right)^2 \times 30.41 = 5.75 \text{ MW}$$

which is a 19% reduction in line losses.

As wires heat up due to high ambient temperatures and the self-heating associated with internal losses, they not only expand and sag, which increases the likelihood of arcing or shorting out, but their resistance also increases. For example, the resistance of both copper and aluminum conductors increases by about 4% for each 10°C of heating. Resistance also depends on the ac frequency of the current flowing. Due to a phenomenon called the *skin effect*, the resistance at 60 Hz for a transmission line is about 5–10% higher than its dc resistance. The skin effect is the result of induced emfs that oppose current flow by a greater amount at the center of a conductor than at its edge. The result is that current flows more easily through the outer edge of a conductor than at its center. Recent interest in dc transmission lines is partly based on the lower resistance posed by the wires when carrying dc.

3.11 THE REGULATORY SIDE OF ELECTRIC POWER

Edison and Westinghouse launched the electric power industry in the United States, but it was Insull who shaped what became the modern electric utility by bringing the concepts of regulated utilities with monopoly franchises into being. The economies of scale that went with increasingly large steam power plants led to an industry based on centralized generation coupled with a complex infrastructure of transmission lines and distribution facilities.

In the last two decades of the twentieth century, however, cheaper, smaller turbines, the discovery of the customer's side of the meter, and growing interest in renewable energy systems led to increasing pressure for change in the regulatory structure that guides and controls the industry.

A basic background in both electric power technology and the electricity industry regulatory structure are essential to understanding the future of this most important industry. What follows is a brief historical sketch of the evolution of the regulatory side of the business.

3.11.1 The Public Utility Holding Company Act of 1935 (PUHCA)

In the early part of the twentieth century, as enormous amounts of money were being made, utility companies began to merge and grow into larger conglomerates. A popular corporate form emerged, called a *utility holding company*. A holding company is a financial shell that exercises management control of one or more companies through ownership of their stock. Holding companies began to purchase each other and by 1929, 16 holding companies controlled 80% of the U.S. electricity market, with just three of them owning 45% of the total.

With so few entities having so much control, it should have come as no surprise that financial abuses would emerge. Holding companies formed pyramids with other holding companies, each owning stocks in subsequent layers of holding companies. An actual operating utility at the bottom found itself directed by layers of holding companies above it, with each layer demanding its own profits. At one point, these pyramids were sometimes ten layers thick. When the stock market crashed in 1929, the resulting Depression drove many holding companies into bankruptcy, causing investors to lose fortunes. Insull became somewhat of a scapegoat for the whole financial fiasco associated with holding companies, and he fled the country amidst charges of mail fraud, embezzlement, and bankruptcy violations, charges for which he was later cleared.

In response to these abuses, Congress created the *Public Utility Holding Company Act of 1935* (PUHCA) to regulate the gas and electric industries and prevent holding company excesses from reoccurring. Many holding companies were dissolved, their geographic size was limited, and the remaining ones came under control of the newly created Securities and Exchange Commission (SEC).

While PUHCA has been an effective deterrent to previous holding company financial abuses, recent changes in utility regulatory structures, with their goal of increasing competition, have led many to say that it has outlived its usefulness and should be modified or abolished. The main issue is a provision of PUHCA that restricts holding companies to business within a single integrated utility, which is a major deterrent to the modern pressure to allow wholesale wheeling of power from one region in the country to another.

3.11.2 The Public Utility Regulatory Policies Act of 1978 (PURPA)

With the country in shock from the oil crisis of 1973 and with the economies of scale associated with ever-larger power plants having pretty much played out, the country was drawn toward energy efficiency, renewable energy systems, and new, small, inexpensive gas turbines. To encourage these systems, President Carter signed the Public Utility Regulatory Policies Act of 1978 (PURPA).

There are two key provisions of PURPA; both of these relate to allowing independent power producers, under certain restricted conditions, to connect their facilities to the utility-owned grid. For one, PURPA allows certain industrial facilities and other customers to build and operate their own, small, on-site generators while remaining connected to the utility grid. Prior to PURPA, utilities could refuse service to such customers, which meant that self-generators had to provide all of their own power, all of the time, including their own redundant, back-up power systems. That had virtually eliminated the possibility of using efficient, economical on-site power production to provide just a portion of a customer's needs.

PURPA not only allowed grid interconnection, but also required utilities to purchase electricity from certain *qualifying facilities* (QFs) at a "just and reasonable price." The purchase price of QF electricity was to be based on what it would have cost the utility to generate the power itself or to purchase it on the open market (referred to as the *avoided cost*). This provision stimulated the construction of numerous renewable energy facilities, especially in California, since PURPA guaranteed a market, at a good price, for any electricity generated.

PURPA, as implemented by the Federal Regulatory Commission (FERC), allows interconnection to the grid by *Qualifying Small Power Producers* or *Qualifying Cogeneration Facilities*; both are referred to as QFs. Small power producers are less than 80 MW in size and use at least 75% wind, solar, geothermal, hydroelectric, or municipal waste as energy sources. Cogenerators are defined as facilities that produce both electricity and useful thermal energy in a sequential process from a single source of fuel, which may be entirely oil or natural gas. To encourage competition, ownership of QFs by *investor-owned electric utilities* (IOUs) was limited to 50%.

PURPA not only gave birth to the electric side of the renewable energy industry, but also enabled clear evidence to accrue which demonstrated that small, on-site generation could deliver power at considerably lower cost than the retail rates charged by utilities. Competition had begun.

3.11.3 The Energy Policy Act of 1992 (EPAct)

The Energy Policy Act of 1992 (EPAct) created even more competition in the electricity generation market by opening the grid to more than just the QFs identified in PURPA. A new category of access was granted to *exempt wholesale generators* (EWGs), which can be of any size, using any fuel and any generation technology, without the restrictions and ownership constraints that PURPA and PUHCA impose. EPAct allows EWGs to generate electricity in one location and sell it anywhere else in the country using someone else's transmission system to wheel their power from one location to another. The key restriction of an EWG is that it deals exclusively with the *wholesale wheeling* of power from the generator to a buyer who is not the final retail customer who uses that power.

Retail wheeling, in which generators wheel power over utility power lines and then sell it directly to retail customers, is the next step, still underway, in the process of creating effective competition for traditional utility generation. PURPA allows retail wheeling for QFs. Other *independent power producers* (IPPs) can also take advantage of retail wheeling if states allow it, but they continue to be regulated under the rules of FERC and PUHCA.

3.11.4 FERC Order 888 and Order 2000

While the 1992 EPAct allowed IPPs to gain access to the transmission grid, problems arose during periods when the transmission lines were being used to near capacity. In these and other circumstances, the investor-owned utilities (IOUs) that owned the lines favored their own generators, and IPPs were often denied access. In addition, the regulatory process administered by the Federal Energy Regulatory Commission (FERC) was initially cumbersome and inefficient. To eliminate such deterrents, the FERC issued Order 888 in 1996, which had as a principal goal the elimination of anti-competitive practices in transmission services by requiring IOUs to publish nondiscriminatory tariffs that applied to all generators.

Order 888 also encouraged formation of *independent system operators* (ISOs), which are nonprofit entities established to control operation of transmission facilities owned by traditional IOUs. Later, in December 1999, FERC issued Order 2000, which broadens its efforts to break up vertically integrated utilities by calling for the creation of *regional transmission organizations* (RTOs). RTOs can follow the ISO model in which ownership of the transmission system remains with the utilities, with the ISO being there to provide control of the system's operation, or they would be separate transmission companies (TRANSCOs), which would actually own the transmission facilities and operate them for a profit.

3.11.5 Utilities and Nonutility Generators

With PURPA and EPAct regulating certain categories of generators that are not owned and operated by traditional utilities, a whole new set of definitions has crept into the discussion of the electric power industry. Before addressing the emergence of competition, it is worth clarifying the players. The industry is now divided into utilities and nonutility generators (NUGs), and each has its own subcategories.

Recall that utilities are categorized as being investor-owned (IOUs), federally owned, such as the Bonneville Power Administration, publicly owned municipal and state utilities, and local rural cooperatives owned by the members they serve.

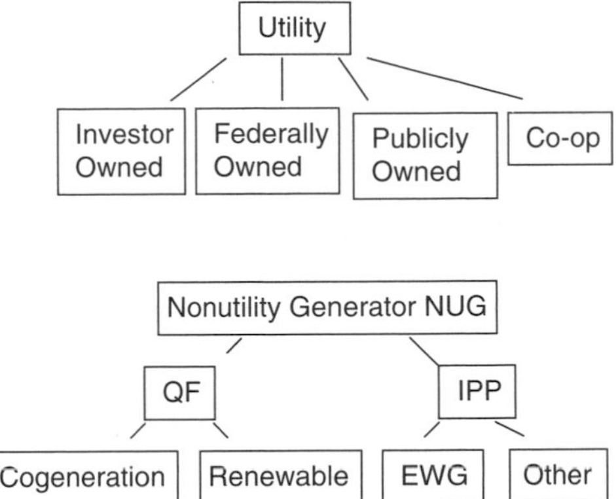

Figure 3.39 The electric power industry is now divided into utilities and nonutilities, with each having its own subset of categories and distinctive features.

Nonutilities, or NUGs, have their own somewhat confusing nomenclature.

Qualifying facilities (QFs) meet certain ownership, operating, and efficiency requirements as defined by PURPA. They may be cogenerators that sequentially produce thermal and electrical energy, or they may be small power producers that rely mostly on renewable energy sources. Utilities must purchase QF power at a price based on the utility's avoided cost.

Independent power producers (IPPs) are non-PURPA-regulated NUGs. A major category of IPP is defined by EPAct as exempt wholesale generators (EWGs). EWGs are exempt from PUHCA's corporate and geographic restrictions, but they are not allowed to wheel power directly to retail customers. Non-EWG IPPs are regulated by PUHCA, but they are allowed to sell to retail customers.

Figure 3.39 summarizes these various categories of utilities and nonutilities.

3.12 THE EMERGENCE OF COMPETITIVE MARKETS

Prior to PURPA, the accepted method of regulation was based on monopoly franchises, vertically integrated utilities that owned some or all of their own generation, transmission, and distribution facilities, and consumer protections based on strict control of rates and utility profits. In the final decades of the twentieth century, however, the successful deregulation of other traditional monopolies such as telecommunications, airlines, and the natural gas industry provided evidence that introducing competition in the electric power industry might also work there. While the disadvantages of multiple systems of wires to transmit and distribute power continue to suggest they be administered as regulated monopolies, there is no inherent reason why there shouldn't be competition between generators who want to put power onto those wires. The whole thrust of both PURPA and

EPAct was to begin the opening up of that grid to allow generators to compete for customers, thereby hopefully driving down costs and prices.

3.12.1 Technology Motivating Restructuring

Restructuring of the electric power industry has been motivated by the emergence of new, small power plants, especially gas turbines and combined-cycle plants, that offer both reduced first cost and operating costs compared with almost all of the generation facilities already on line. The economies of scale that motivated ever-larger power plants in the past seem to have played out, as illustrated in Fig. 3.40. By 2000 the largest plants being built were only a few hundred megawatts, whereas in the previous decade they were closer to 1400 MW. Table 3.7 shows the dramatic shift anticipated in power plant

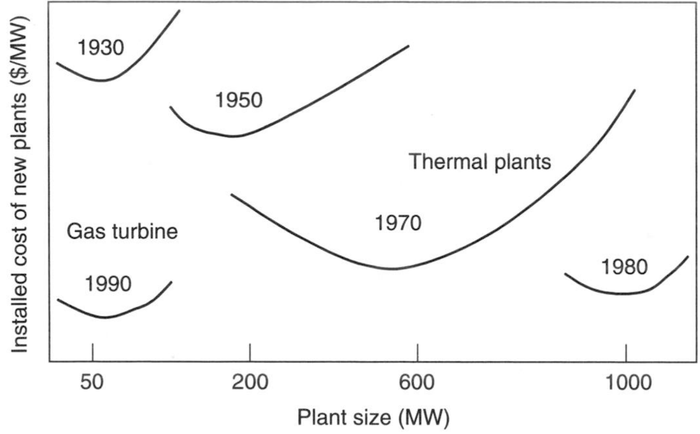

Figure 3.40 The era of bigger-is-better that dominated power plant construction for most of the twentieth century shifted rather abruptly around 1990. From Bayliss (1994).

TABLE 3.7 Total Projected Additions of Electricity Generating Capability, in units of Thousand Megawatts (GW), 1999–2020

Technology	Capacity Additions
Coal steam	21.1
Combined cycle	135.2
Combustion turbine/Diesel	133.8
Fuel cells	0.1
Renewable sources	9.7
TOTAL	300

Source: EIA (2000).

construction away from the workhorse, coal-fired steam turbines that now supply more than half of all electricity generated in the United States. In the projection shown, new coal-fired plants make up only 7% of the 300 GW of new power plant generation anticipated by 2020.

The emergence of small, less capital-intensive power plants helped independent power producers (IPPs) get into the power generation business. By the early 1990s the cost of electricity generated by IPPs was far less than the average price of power charged by regulated utilities. With EPAct opening the grid, large customers began to imagine how much better off they would be if they could just bypass the regulated utility monopolies and purchase power directly from those small, less expensive units. Large customers, with the wherewithal, threatened to leave the grid entirely and generate their own electricity, while others, when allowed, began to take advantage of retail wheeling to purchase power directly from IPPs.

Not only did small power plants become more cost effective, independent power producers found themselves with a considerable advantage over traditional regulated utilities. Even though utilities and IPPs had equal access to new, less expensive generation, the utilities had huge investments in their existing facilities, so the addition of a few low-cost new turbines had almost no impact on their overall average cost of generation.

To help utilities successfully compete with IPPs in the emerging competitive marketplace, FERC included in Order 888 a provision to allow utilities to speed up the recovery of costs on power plants that were no longer cost-effective. The argument was based on the idea that when utilities built those expensive power plants, they did so under a regulatory regime that allowed cost recovery of all prudent investments. To be fair and to help assure utility support for a new competitive market, FERC believed it was appropriate to allow utilities to recover those stranded asset costs even if that might delay the emergence of a competitive market. FERC then endorsed the concept of charging customers who chose to leave the utility system a departure fee to help pay for those stranded costs left behind.

3.12.2 California Begins to Restructure

With the grid opening up, a number of states began to develop restructuring programs in the 1990s. It is no surprise that the states that took the initiative tended to be those with higher utility costs.

While Fig. 3.7 showed *average* electricity prices, of greater importance to understanding the driving force behind restructuring is the price of electricity for large industrial customers. It is those customers who are most likely to threaten to purchase power from an IPP, generate their own electricity, or move the location of their operations to a state with lower electricity prices. Figure 3.41 shows the state-by-state average revenue per kWh for the industrial sector. States with the highest prices include California and much of New England. Both are areas where restructuring began.

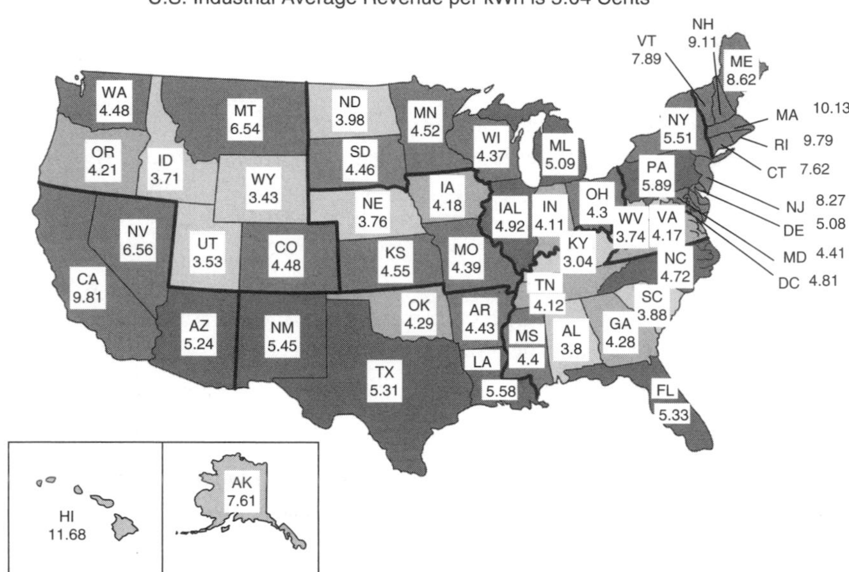

Figure 3.41 Average industrial sector revenue per kWh, 2001. California was 6.6¢/kWh in 1998, before restructuring. From EIA (2003).

In the mid-1990s, California's retail price of electricity for the industrial sector was among the highest in the nation, yet the on-peak, wholesale market price was less than one-fourth of that amount. Excess capacity in nearby states, along with plenty of hydroelectric power, kept the wholesale marginal cost of electricity low. The gap between wholesale and retail prices suggested that if utilities could just rid themselves of their QF obligations and quickly pay down their expensive existing generators, they would be able to dramatically lower prices by simply purchasing their power in the wholesale market. Worried about losing large industrial customers to other states with cheaper electric rates, the California public utility commission (CPUC) in 1994 issued a proposed rule to restructure the power industry with the goal of reducing the state's high price of electricity. The restructuring was unfortunately misnamed "deregulation," and that became the term that was used throughout the ensuing years of turmoil.

California's AB 1890 Barely two years after the CPUC proposed to restructure the electric power industry, in 1996 the California Legislature responded by enacting Assembly Bill 1890 (AB 1890). AB 1890 included the following major provisions:

- The wholesale market was opened to competition, and all customers would have a choice of electricity suppliers.
- Residential and small commercial customers would be given an immediate 4-year, 10% rate reduction to be funded by issuing bonds that would be

repaid after the 4-year period ended. Larger customers who stayed with their IOUs would have rates frozen at 1996 levels.

- Utilities would be given the opportunity to recover their stranded costs (especially nuclear and QF obligations) during that 4-year rate-freeze period. The assumption was that utilities would collect more revenue at the frozen, retail rates than would be needed to purchase power in the wholesale market, cover transmission and distribution costs, and make payments on their rate-reduction bonds. The difference between retail price and those various costs, called the competition transition charge (CTC), would be used to pay down stranded assets over the 4-year rate-freeze period.
- A public goods charge would be levied to continue support for programs focused on renewable energy, customer energy efficiency, and rate subsidies for low-income consumers.
- Utilities were to sell off much of their generating capacity to create new players in the market who would then compete to sell their power, thereby lowering prices.
- After the initial 4 years—that is, by 2002—rates would be unfrozen and the benefits associated with the free market would be realized. The rate freeze would end sooner for any utility that paid off its stranded assets in less than 4 years.

AB 1890 applied to the state's three major IOUs, Pacific Gas and Electric (PG&E), Southern California Edison (SCE), and San Diego Gas and Electric (SDG&E), which deliver roughly three-fourths of the electricity in California. The remaining one-fourth is mostly made up of municipal utilities, which were encouraged, but not required, to participate in the restructuring.

Following FERC's Order 888, California established a not-for-profit independent system operator (ISO) to operate and manage the state's transmission grid. The ISO was meant to assure access to the transmission system so that buyers could purchase electricity from any sellers that they may choose. In addition, the California Power Exchange (CalPX) was set up to match supply to demand using daily auctions. It accepted price and quantity bids from participants in both "Day-Ahead" and "Day-Of" markets from which the market clearing price at which energy was bought and sold was determined.

One of the benefits of the restructuring was that customers could choose their supplier of electricity. Competition was facilitated by energy service providers, or power marketers, who offered a range of generation combinations, including options for various fractions of renewables. To help make comparisons, utilities and power marketers provided power content labels specifying the fraction of their generation that was from renewables, coal, large hydroelectric, natural gas, and nuclear.

The basic structure of California's intended, restructured electricity system is diagrammed in Fig. 3.42.

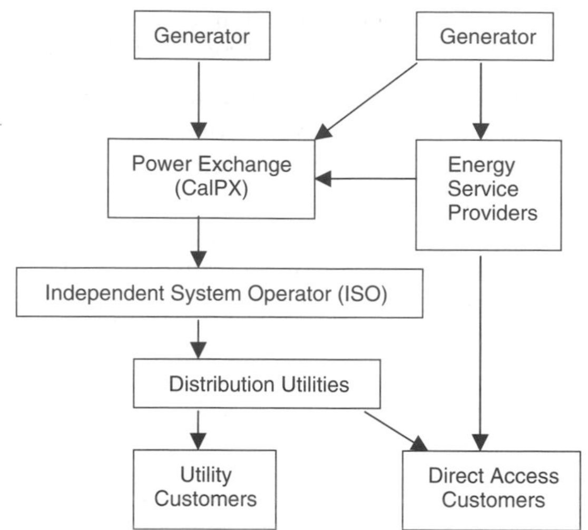

Figure 3.42 California's initial electric industry restructuring.

3.12.3 Collapse of "Deregulation" in California

California officially opened its wholesale electricity market to competition in March 1998. For the first two years, it worked quite well, with wholesale prices hovering around $35/MWh (3.5¢/kWh) which was far less than the frozen retail rates. Even after adding on the cost of transmission, distribution and rate-reduction bonds, utilities were still netting around $35/MWh to pay down stranded assets (Wolak, 2003). PG&E, SCE, and SDG&E had sold off approximately 18,000 MW of generation capacity, or roughly 40% of California's generating capacity, to five new entrants into the California market: Duke, Dynergy, Reliant, AES, and Mirant. Marketers of green power were beginning to make inroads into the residential market.

Then in the summer of 2000, it all began to unravel (Fig. 3.43). The cost of power on the wholesale market rose to unheard of levels. In August 2000, at an average price of $170/MWh (17¢/kWh) it was five times higher than it had been in the same month in 1999, and on one day it reached $800/MWh (80¢/kWh). Under the fixed retail-price constraints imposed by AB 1890, PG&E and SCE were forced to purchase wholesale power at higher prices than they could sell it for and began to rapidly acquire unsustainable levels of debt. SDG&E, however, had already paid off its stranded assets and therefore was able to pass these higher costs on to its customers, who, as a result, saw rates increase by as much as 300%.

By the end of 2000, Californians had paid $33.5 billion for electricity, nearly five times the $7.5 billion spent in 1999. Factors that contributed to the crisis included higher-than-normal natural gas prices, a drought that reduced the availability of imported electricity, especially from the northwest, reduced efforts by

California utilities to pursue customer efficiency programs in a deregulated environment, and insufficient new plant construction. Arguably, this construction was not necessitated by growth in California's electricity demand, which had been modest, but rather it was the lack of construction in adjacent states that had traditionally exported power to California. Those states were not keeping up with their own demands, which eroded the amounts usually available for California to import. But the principal cause was the discovery by new generation companies that they could make a lot more money manipulating the market than meeting the needs of California with adequate supplies. In November 2000, even FERC concluded that *unjust and unreasonable* summer wholesale prices resulted from the exercise of market power by generators.

In January 2001, California experienced almost daily Stage 3 emergencies and had several days with rolling blackouts. To have rolling blackouts in January when peak power demands are one third-lower than they are in the summer provided compelling evidence that generators had gamed the system by purposely removing capacity from the grid in order to raise prices and increase profits on the facilities that they continued to operate. In fact, wholesale rates in January 2001 far exceeded those of the previous summer and at one point reached $1500/MWh ($1.50/kWh). California paid as much for electricity in the first month and a half of 2001 as it had paid in all of 1999.

As Fig. 3.43 shows, wholesale prices remained extraordinarily high through the spring of 2001. By May, PG&E had declared bankruptcy and SCE was on the edge, the CalPX had been shut down, and CalISO began to purchase power as well as operate the grid. The crisis finally began to ease by the summer of 2001 after FERC instituted price caps on wholesale power, the Governor began to negotiate long-term power contracts, and the state's aggressive energy-conservation efforts began to pay off.

The true success story in relieving the anticipated summer 2001 meltdown was California's intensive and effective energy conservation program. Additional funds were devoted to energy efficiency programs including a "20/20" program that provided a 20% percent rate reduction to customers who cut energy use by

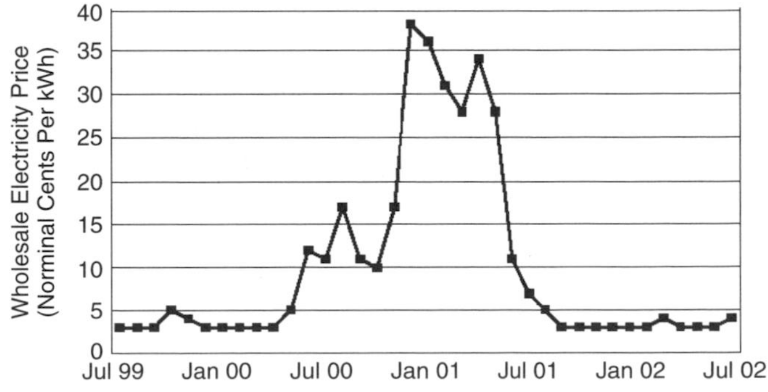

Figure 3.43 California wholesale electricity prices during the crisis of 2000–2001. From: Bachrach et al. (2003).

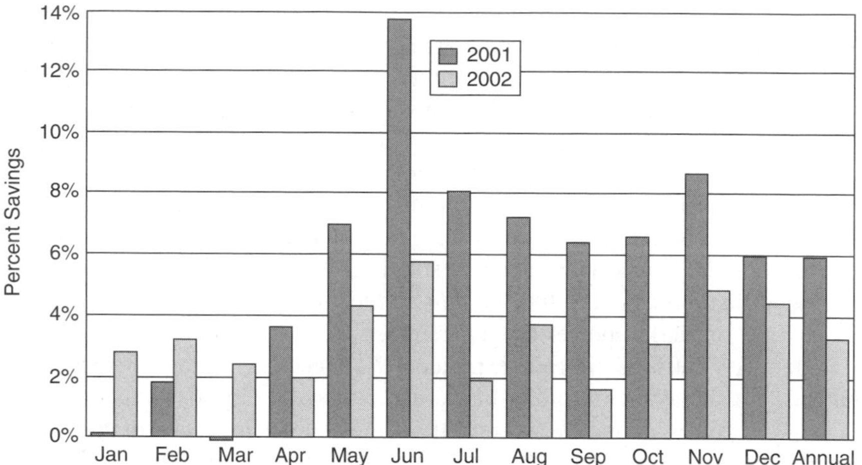

Figure 3.44 California weather-adjusted monthly electricity reductions relative to 2000 (Bachrach, 2003). Efficiency programs saved the state from what could have been a catastrophic 2001 summer.

20%—a target that over one-third of residential customers achieved. A number of specialized programs were created to focus on reducing peak demand. Relative to June 2000, energy use in June 2001, adjusted for weather, was down almost 14% and for the whole summer it was almost 8% lower. Those were gains during a year in which the economy grew by 2.3%. The estimated cost of the state's conservation program was less than 3 cents per kWh saved, an order of magnitude below the average wholesale price the previous winter and half the cost of the long-term contracts the state signed in March 2001 (Bachrach, 2003). Moreover, as Fig. 3.44 shows, even after the crisis had passed, the conservation gains showed every sign that they would persist.

In March 2003, the FERC issued a statement concluding that California electricity and natural gas prices were driven higher because of widespread manipulation and misconduct by Enron and more than 30 other energy companies during the 2000–2001 energy crisis.

While the momentum of the 1990s toward restructuring was shaken by the California experience, the basic arguments in favor of a more competitive electric power industry remain attractive. Analysis of the failure there has guided the restructuring that continues to proceed in a number of other states. It is hoped that they'll do better.

REFERENCES

Bachrach, D., M. Ardema, and A. Leupp (2003). *Energy Efficiency Leadership in California: Preventing the Next Crisis*, Natural Resources Defense Council, Silicon Valley Manufacturing Group, April.

Bayliss, C. (1994). Less is More: Why Gas Turbines Will Transform Electric Utilities, *Public Utility Fortnightly*, December 1, pp. 21–25.

Bosela, T. R. (1997). *Introduction to Electrical Power System Technology*, Prentice-Hall, Englewood Cliffs, NJ.

Brown, R. E., and J. Koomey (2002). *Electricity Use in California: Past Trends and Present Usage Patterns*, Lawrence Berkeley National Labs, LBL-47992, Berkeley, CA.

California Collaborative (1990). *An Energy Efficiency Blueprint for California*, January.

Culp, A. W., Jr. (1979). *Principles of Energy Conversion*, McGraw-Hill, New York.

Energy Information Administration (EIA, 2001). *The Changing Structure of the Electric Power Industry 2000: An Update*, DOE/EIA-0562(00), Washington, DC.

Energy Information Administration (EIA, 2003). *EIA Annual Energy Review 2001*, DOE/EIA-0348(2001), Washington, DC.

IPCC (1996). *Climate Change 1995: Impacts, Adaptations and Mitigation of Climate Change: Scientific-Technical Analyses*, Intergovernmental Panel on Climate Change, Cambridge University Press, New York.

Lovins, A. B., E. Datta, T. Feiler, K. Rabago, J. Swisher, A. Lehmann, and K. Wicker (2002). *Small is Profitable: The Hidden Economic Benefits of Making Electrical Resources the Right Size*, Rocky Mountain Institute, Snowmass, CO.

Masters, G. M. (1998). *Introduction to Environmental Engineering and Science*, 2nd ed., Prentice-Hall, Englewood Cliffs, NJ.

Penrose, J. E., Inventing Electrocution, *Invention and Technology*, Spring 1994, pp. 35–44.

Petchers, N. (2002). *Combined Heating, Cooling & Power: Technologies & Applications*, The Fairmont Press, Lilburn, GA.

Sant, R. (1993). Competitive Generation is Here, *The Electricity Journal*, August/September.

Union of Concerned Scientists (UCS, 1992). *America's Energy Choices: Investing in a Strong Economy and a Clean Environment*, Cambridge, MA.

Wolak, F. A. (2003). *Lessons from the California Electricity Crisis*, University of California Energy Institute, Center for the Study of Energy Markets, CSEMWP110.

PROBLEMS

3.1 A Carnot heat engine receives 1000 kJ/s of heat from a high temperature source at 600°C and rejects heat to a cold temperature sink at 20°C.

a. What is the thermal efficiency of this engine?

b. What is the power delivered by the engine in watts?

c. At what rate is heat rejected to the cold temperature sink?

d. What is the entropy change of the sink?

3.2 A heat engine supposedly receives 500 kJ/s of heat from an 1100-K source and rejects 300 kJ/s to a low-temperature sink at 300 K.

a. Is this possible or impossible?

b. What would be the net rate of change of entropy for this system?

3.3 A solar pond consists of a thin layer of fresh water floating on top of a denser layer of salt water. When the salty layer absorbs sunlight it warms up and much of that heat is held there by the insulating effect of the fresh water above it (without the fresh water, the warm salt water would rise to the surface and dissipate its heat to the atmosphere). A solar pond can easily be 100°C above the ambient temperature.

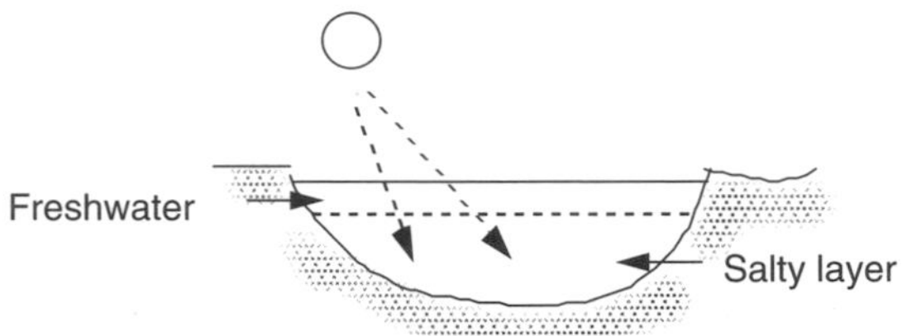

Figure P3.3

a. What is the maximum efficiency of a heat engine operating off of a 120°C solar pond on a 20°C day?

b. If a real engine is able to achieve half the efficiency of a Carnot engine, how many kilowatt hours of electricity could be generated per day from a 100 m × 100 m pond that captures and stores 50% of the 7 kWh/m^2 solar radiation striking the surface?

c. For a house that requires 500 kWh per (30-day) month, what area of pond would be needed?

3.4 A combined-cycle, natural-gas, power plant has an efficiency of 52%. Natural gas has an energy density of 55,340 kJ/kg and about 77% of the fuel is carbon.

a. What is the heat rate of this plant expressed as kJ/kWh and Btu/kWh?

b. Find the emission rate of carbon (kgC/kWh) and carbon dioxide (kgCO_2/kWh). Compare those with the average coal plant emission rates found in Example 3.2.

3.5 A new coal-fired power plant with a heat rate of 9000 Btu/kWh burns coal with an energy content of 24,000 kJ/kg. The coal content includes 62-% carbon, 2-% sulfur and 10-% unburnable minerals called *ash*.

a. What will be the carbon emission rate (g C/kWh)?

b. What will be the uncontrolled sulfur emission rate (g S/kWh)?

c. If 70% of the ash is released as particular matter from the stack (called *fly ash*), what would be the uncontrolled particulate emission rate (g/kWh)?

d. Since the Clean Air Act restricts SO_2 emissions to 130 g of sulfur per 10^6 kJ of heat into the plant, what removal efficiency does the scrubber need to have for this plant?

e. What efficiency does the particulate removal equipment need to have to meet the Clean Air Act standard of no more than 13 g of fly ash per 10^6 kJ of heat input?

3.6 Using the representative capital costs of power plant and fuels given in Table 3.3, compute the cost of electricity from the following power plants. For each, assume a fixed charge rate of 0.14/yr.

a. Pulverized coal steam plant with capacity factor CF = 0.7.

b. Advanced coal steam plant with CF = 0.8.

c. Oil/gas steam plant with CF = 0.5.

d. Combined cycle gas plant with CF = 0.5.

e. Gas-fired combustion turbine with CF = 0.2.

f. SIGT gas turbine with CF = 0.4.

g. New hydroelectric plant with CF = 0.6.

h. Wind turbine costing $800/kW, CF = 0.37, O&M − 0.60¢/kWh

3.7 Consider the following very simplified load duration curve for a small utility:

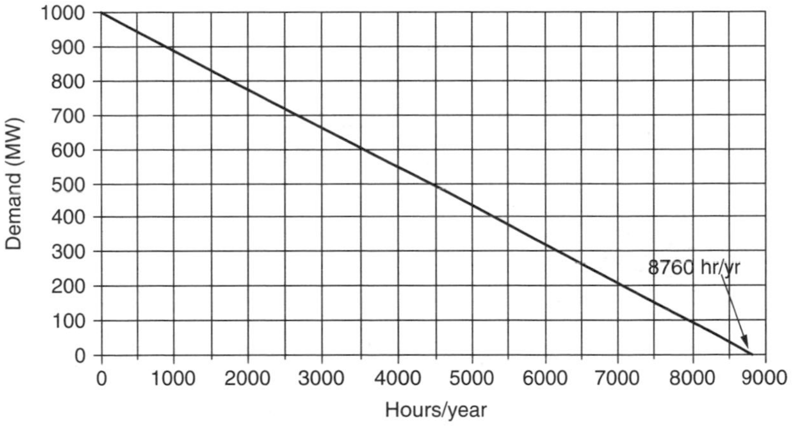

Figure P3.7

a. How many hours per year is the load less than 200 MW?

b. How many hours per year is the load between 300 MW and 600 MW?

c. If the utility has 500 MW of base-load coal plants, what would their average capacity factor be?

d. How many kWh would those coal plants deliver per year?

3.8 If the utility in Problem 3.7 has 400 MW of peaking power plants with the following "revenue required" curve, what would be the cost of electricity (¢/kWh) from these plants when operated as peakers?

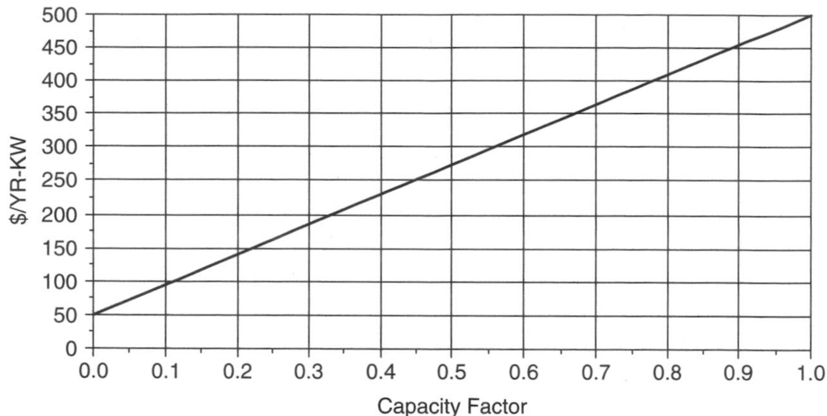

Figure P3.8

3.9 Consider screening curves for gas turbines and coal-fired power plants given below along with a load duration curve for a hypothetical utility:

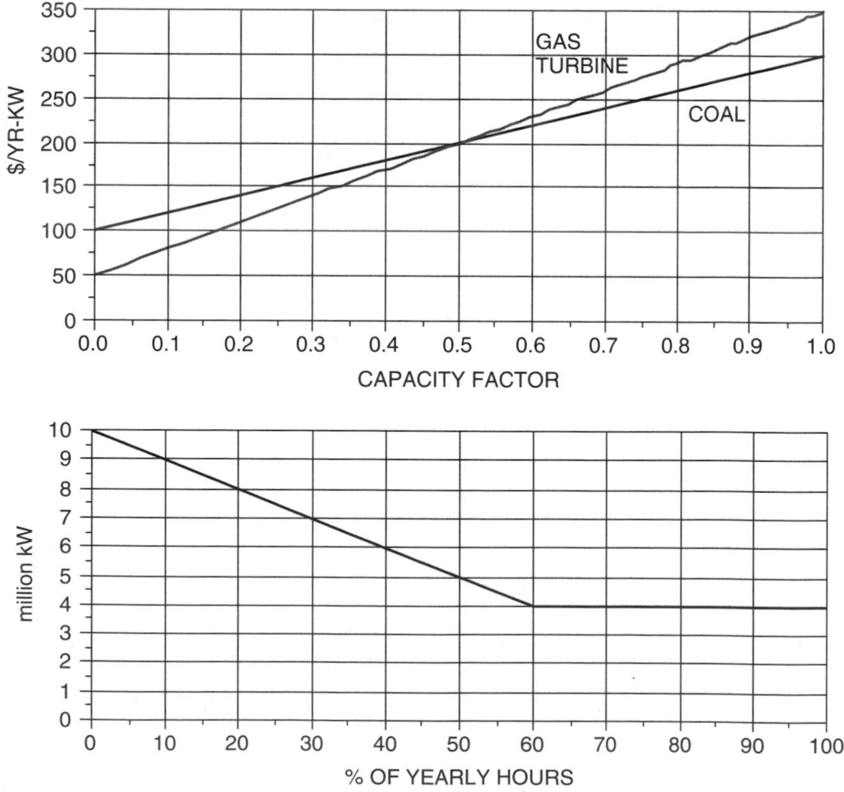

Figure P3.9

a. For a least-cost combination of power plants, how many MW of each kind of power plant should the utility have?

b. What would the cost of electricity be ($/kWh) for the gas turbines sized in (a)?

3.10 The following table gives capital costs and variable costs for a coal plant, a natural gas combined-cycle plant, and a natural-gas-fired gas turbine:

	COAL	NG CC	NG GT
Capital Cost ($/kW)	$ 1,500.00	$ 1,000.00	$ 500.00
Variable Cost (¢/kWh)	2.50	4.00	8.00

The utility uses a fixed charge rate of 0.10/yr for capital costs. Its load duration curve is shown as follows.

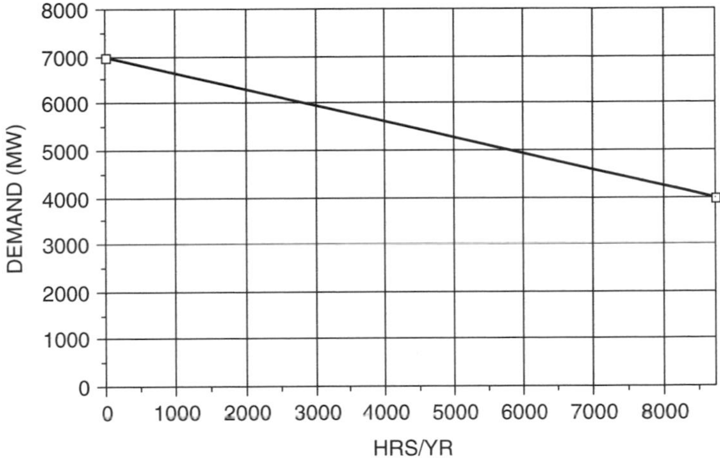

Figure P3.10

a. Draw the screening curves (Revenue Required $/yr-kW vs hr/yr) for each of the three types of power plants.

b. For a least-cost combination of power plants, how many MW of each kind of power plant should the utility have? If you plot this carefully, you can do it graphically. Otherwise you may need to solve algebraically.

c. Estimate the average capacity factor for each type of power plant.

d. How many MWhr of electricity would each type of power plant generate each year?

e. What annual revenue would the utility need to receive from each type of power plant?

f. What would be the cost of electricity (¢/kWh) from each type of power plant?

3.11 A 345 kV, three-phase transmission system uses 0.642-in. diameter ACSR cable to deliver 200 MW to a wye-connected load 100 miles away. Compute the line losses if the power factor is 0.90.

CHAPTER 4

DISTRIBUTED GENERATION

4.1 ELECTRICITY GENERATION IN TRANSITION

The traditional, vertically integrated utility incorporating generation, transmission, distribution, and customer energy services is in the beginning stages of what could prove to be quite revolutionary changes. The era of ever-larger central power stations seems to have ended. The opening of the transmission and distribution grid to independent power producers who offer cheaper, more efficient, smaller-scale plants is well underway. Attempts to restructure the regulatory side of utilities to help create competition among generators and allow customers to choose their source of power have been initiated in a number of states, but with mixed success. And, partly due to California's deregulation crisis of 2000–2001, the customer's side of the meter is being rediscovered and energy efficiency is enjoying a resurgence of attention.

On the customer side of the meter, the power business is beginning to look more like it did in the early part of the twentieth century when more than half of U.S. electricity was self-generated with small, isolated systems for direct use by industrial firms. Many of those systems were located in the basements of buildings, which were heated by the waste heat from the power plants. Those old steam-powered, engine generators used for heat and power have modern equivalents in the form of microturbines, fuel cells, internal-combustion engines, and small gas turbines. Using these technologies, customers are rediscovering the economic advantages of on-site *cogeneration* of heat and power, or *trigeneration* for heating, electric power, and cooling.

Renewable and Efficient Electric Power Systems. By Gilbert M. Masters
ISBN 0-471-28060-7

TABLE 4.1 Typical Power Plant Power Output and End-Use Power Demands (kW)

Generation	kW	End Use	kW
Large hydro dam or power-plant cluster	10,000,000	Portable computer	0.02
Nuclear plant	1,100,000	Desktop computer	0.1
Coal plant	600,000	Household average power (U.S.)	1
Combined-cycle turbines	250,000	Commercial customer average power	10
Simple-cycle gas turbine	150,000	Supermarket	100
Aeroderivative gas turbine cogeneration	50,000	Medium-sized office building	1,000
Molten carbonate fuel cell	4,000	Large factory	10,000
Wind turbine	1,500	Peak use of largest buildings	100,000
Fuel-cell powered automobile	60		
Microturbine	30		
Residential PEM fuel cell	5		
Residential photovoltaic system	3		

In addition to economic benefits, other motivations helping to drive the transition toward small-scale, decentralized energy systems include increased concern for environmental impacts of generation, most especially those related to climate change, increased concern for the vulnerability of our centralized energy systems to terrorist attacks, and increased demands for electricity reliability in the digital economy.

A sense of the dramatic decrease in scale that is underway is provided in Table 4.1, in which a number of generation technologies are listed along with typical power outputs. For comparison, some examples of power demands of typical end uses are also shown. While the power ratings of some of the distributed generation options may look trivially small, it is the potentially large numbers of replicated small units that will make their contribution significant. For example, the U.S. auto industry builds around 6 million cars each year. If half of those were 60-kW fuel-cell vehicles, the combined generation capacity of 5-year's worth of automobile production would be greater than the total installed capacity of all U.S. power plants.

4.2 DISTRIBUTED GENERATION WITH FOSSIL FUELS

Distributed generation (DG) is the term used to describe small-scale power generation, usually in sizes up to around 50 MW, located on the distribution system close to the point of consumption. Such generators may be owned by a utility or, more likely, owned by a customer who may use all of the power on site or who may sell a portion, or perhaps all of it, to the local utility. When there is waste heat available from the generator, the customer may be able to use it for such applications as process heating, space heating, and air conditioning, thereby increasing the overall efficiency from fuel to electricity and useful thermal energy.

The process of capturing and using waste heat while generating electricity is sometimes called *cogeneration* and sometimes *combined heat and power* (CHP). There are subtle distinctions between the terms. Some say cogeneration applies only to qualifying facilities (QFs) as defined by the Public Utilities Regulatory Policy Act (PURPA), and some say that CHP applies only to low-temperature heat used for space heating and cooling), but in this text the terms will be used interchangeably.

4.2.1 HHV and LHV

Before describing some of the emerging technologies for distributed generation, we need to sort out a subtle distinction having to do with the way power plant efficiencies are often presented. When a fuel is burned, some of the energy released ends up as latent heat in the water vapor produced (about 1060 Btu per pound of vapor, or 2465 kJ/kg). Usually that water vapor, along with the latent heat it contains, exits the stack along with all the other combustion gases, and its heating value is, in essence, lost. In some cases, however, that is not the case. For example, the most fuel-efficient, modern furnaces used for space-heating homes achieve their high efficiencies (over 90%) by causing the combustion gases to cool enough to condense the water vapor before it leaves the stack. Whether or not the latent heat in water vapor is included leads to two different values of what is called the *heat of combustion* for a fuel. The *higher heating value* (HHV), also known as the *gross heat of combustion*, includes the latent heat, while the *lower heating value* (LHV), or *net heat of combustion*, does not.

Examples of HHV and LHV for various fuels, along with the LHV/HHV ratios, are presented in Table 4.2. Since natural gas is a combination of methane,

TABLE 4.2 Higher Heating Value (HHV) and Lower Heating Value (LHV) for Various Fuels[a]

	Higher Heating Value HHV		Lower Heating Value LHV		
Fuel	Btu/lbm	kJ/kg	Btu/lbm	kJ/kg	LHV/HHV
Methane	23,875	55,533	21,495	49,997	0.9003
Propane	21,669	50,402	19,937	46,373	0.9201
Natural gas	22,500	52,335	20,273	47,153	0.9010
Gasoline	19,657	45,722	18,434	42,877	0.9378
No. 4 oil	18,890	43,938	17,804	41,412	0.9425

[a]The gases are based on dry, 60°F, 30-in. Hg conditions. Natural gas is a representative value.

Source: Based on Babcock and Wilcox (1992) and Petchers (2002).

ethane, and other gases, which varies depending on the source, the values given for HHV and LHV in the table are meant to be just representative, or typical, values. For natural gas, the difference between HHV and LHV is about 10%.

Since the efficiency of a power plant is output power divided by fuel-energy input, the question becomes, Which fuel value should be used, HHV or LHV? Unfortunately, both will be encountered. For large power stations, efficiency is almost always based on HHV, but for the most common distributed generation technologies such as microturbines and reciprocating engines, it is usually based on LHV. When it is important to reconcile the two, the following relationship may be used:

$$\text{Thermal efficiency (HHV)} = \text{Thermal efficiency (LHV)} \times \left(\frac{\text{LHV}}{\text{HHV}}\right) \tag{4.1}$$

where the LHV/HHV ratio can be found from Table 4.2.

Example 4.1 Microturbine Efficiency. A microturbine has a natural gas input of 13,700 Btu (LHV) per kWh of electricity generated. Find its LHV efficiency and its HHV efficiency.

Solution. In Section 3.5.2 the relationship between efficiency and heat rates (in American units) is given by (3.16)

$$\text{Efficiency} = \frac{3412 \text{ Btu/kWh}}{\text{Heat rate (Btu input/kWh output)}}$$

Using the LHV for fuel gives the LHV efficiency:

$$\text{Efficiency (LHV)} = \frac{3412 \text{ Btu/kWh}}{13{,}700 \text{ Btu/kWh}} = 0.2491 = 24.91\%$$

Using the LHV/HHV ratio of 0.9010 for natural gas (Table 4.2) in Eq. (4.1) gives the HHV efficiency for this turbine:

$$\text{Efficiency (HHV)} = 24.91\% \times 0.901 = 22.44\%$$

We will revisit the concept of LHV and HHV in the context of fuel cells later in the chapter.

4.2.2 Microcombustion Turbines

Gas turbines, used either in the simple-cycle mode or in combined-cycle power plants, were introduced in Chapter 3. Simple-cycle turbines used by utilities for peaking power plants typically generate anywhere from a few megawatts to hundreds of megawatts. Industrial facilities often use these same relatively large turbines to generate some of their own electricity—especially when they can take advantage of the waste heat.

More recently, a new generation of very small gas turbines has entered the marketplace. Often referred to as *microturbines*, these units generate anywhere from about 500 watts to several hundred kilowatts. Figure 4.1 illustrates the basic configuration including compressor, turbine, and permanent-magnet generator, in this case all mounted on a single shaft. Incoming air is compressed to three or four atmospheres of pressure and sent through a heat exchanger called a *recuperator*, where its temperature is elevated by the hot exhaust gases. By preheating the compressed incoming air, the recuperator helps boost the efficiency of the unit. The hot, compressed air is mixed with fuel in the combustion chamber and is burned. The expansion of hot gases through the turbine spins the compressor and generator. The exhaust is released to the atmosphere after transferring much of its heat to the incoming compressed air in the recuperator.

Example specifications of several microturbines are given in Table 4.3. For example, the Capstone Turbine Corporation manufactures several refrigerator-size microturbines that generate up to 30 kW and 60 kW. These turbines have only one moving part—the common shaft with compressor, turbine, and generator, which spins at up to 96,000 rpm on air bearings that require no lubrication. There are no gearboxes, lubricants, coolants, or pumps that could require maintenance. The generator creates variable frequency ac that is rectified to dc and

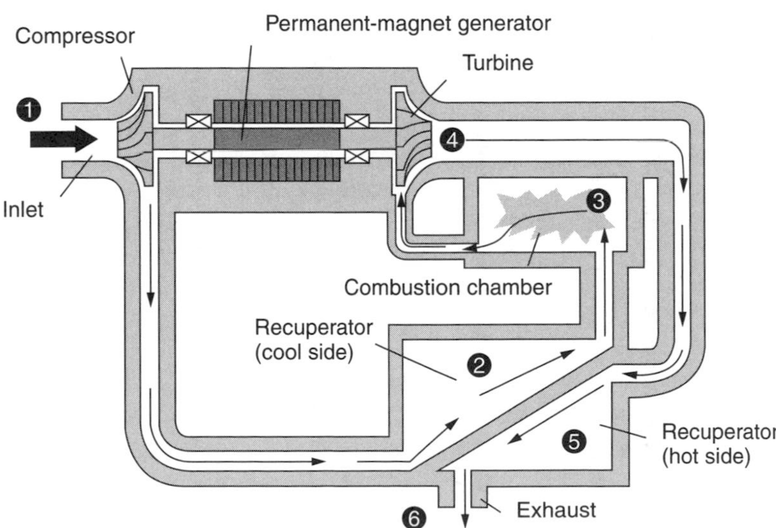

Figure 4.1 Microturbine power plant. Air is compressed (1), preheated in the recuperator (2), combusted with natural gas (3), expanded through the turbine (4), cooled in the recuperator (5), and exhausted (6). From Cler and Shepard (1996).

TABLE 4.3 Specifications for Example Microturbines[a]

Manufacturer, Model	Capstone C30	Capstone C60	Elliott TA 100R
Rated power	30 kW	60 kW	105 kW
Fuel input	390,130 Btu/hr (LHV)	724,000 Btu/h (LHV)	1,235,506 Btu/h (LHV)
Heat rate (LHV)	12,800 kJ (13,100 Btu)/kWh	12,900 kJ (12,200 Btu)/kWh	12,415 kJ (11,770 Btu)/kWh
Efficiency (LHV)	26%	28%	29%
Exhaust gas temp.	275°C (530°F)	305°C (580°F)	279°C (535°F)
NO_x emissions	< 9 ppmV (0.49 lb/MWh)	< 9 ppmV (0.49 lb/MWh)	< 24 ppmV (1.59 lb/MWh)
CO emissions	< 40 ppmV @ 15% O2	< 40 ppmV @ 15% O2	< 41 ppm @ 15% O2
Turbine rotation	96,000 rpm	96,000 rpm	45,000 rpm
Dimensions H, W, D	1.90 × 0.71 × 1.34 m (75 × 28 × 53″)	2.11 × 0.76 × 1.96 m (83 × 30 × 77″)	2.11 × 0.85 × 3.05 m (83 × 34 × 120″)
Weight	478 kg (1052 lb)	758 kg (1671 lb)	1845 kg (4000 lb)
Noise	65 dBA @ 10 m (33 ft)	70 dBA @ 10 m (33 ft)	70 dBA

[a]Emissions are for natural gas fuel.

Source: Capstone www.microturbine.com and Elliott www.elliottmicroturbines.com websites.

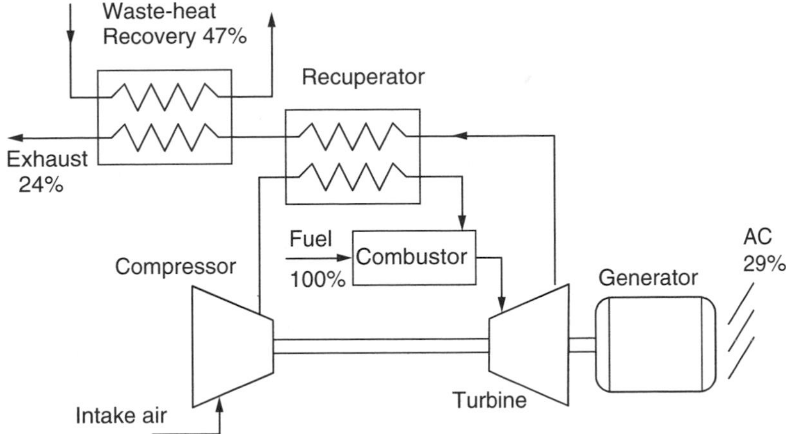

Figure 4.2 Example energy flows for a microturbine with recuperator and waste-heat recovery. The electric efficiency is based on LHV.

inverted back to 50- or 60-Hz ac electricity ready for use. These are designed so that 2 to 20 units can easily be stacked in parallel to generate multiples of 30 kW or 60 kW.

The Elliott 100-kW turbine described in Table 4.3 is an oil-cooled, recuperated unit designed for combined heat and power. It can deliver 105 kW of electrical power plus 172 kW of thermal energy for water heating with an overall thermal efficiency from fuel to electricity and useful heat of over 75%. An example of its energy flows in a combined heat and power mode is shown in Fig. 4.2.

Example 4.2 Combined Heat and Power. The Elliott TA 100A at its full 105-kW output burns 1.24×10^6 Btu/hr of natural gas. Its waste heat is used to supplement a boiler used for water and space heating in an apartment house. The design calls for water from the boiler at 120°F to be heated to 140°F and returned to the boiler. The system operates in this mode for 8000 hours per year.

a. If 47% of the fuel energy is transferred to the boiler water, what should the water flow rate be?
b. If the boiler is 75% efficient, and it is fueled with natural gas costing $6 per million Btu, how much money will the microturbine save in displaced boiler fuel?
c. If utility electricity costs $0.08/kWh, how much will the microturbine save in avoided utility electricity?

d. If O&M is \$1500/yr, what is the net annual savings for the microturbine?
e. If the microturbine costs \$220,000, what is the ratio of annual savings to initial investment (called the initial rate of return)?

Solution

a. The heat Q required to raise a substance with specific heat c and mass flow rate $\dot{m}$ by a temperature difference of ΔT is

$$Q = \dot{m} c \Delta T$$

Since it takes 1 Btu to raise 1 lb of water by 1°F, and one gallon of water weighs 8.34 lb, we can write

$$\begin{aligned}\text{Water flow rate } \dot{m} &= \frac{0.47 \times 1.24 \times 10^6 \text{ Btu/h}}{1 \text{ Btu/lb}^\circ\text{F} \times 20^\circ\text{F} \times 8.34 \text{ lb/gal} \times 60 \text{ min/h}} \\ &= 58 \text{ gpm}\end{aligned}$$

b. The fuel displaced by not using the 75 percent efficient boiler is worth

$$\begin{aligned}\text{Fuel savings} &= \frac{0.47 \times 1.24 \times 10^6 \text{ Btu/h}}{0.75} \times \frac{\$6.00}{10^6 \text{ Btu}} \times 8000 \text{ hr/yr} \\ &= \$37{,}300/\text{yr}\end{aligned}$$

c. The utility electricity savings is

$$\begin{aligned}\text{Electric utility savings} &= 105 \text{ kW} \times 8000 \text{ hr/yr} \times \$0.08/\text{kWh} \\ &= \$67{,}200/\text{yr}\end{aligned}$$

d. The cost of fuel for the microturbine is

$$\begin{aligned}\text{Microturbine fuel cost} &= 1.24 \times 10^6 \text{ Btu/h} \times \frac{\$6.00}{10^6 \text{ Btu}} \times 8000 \text{ h/yr} \\ &= \$59{,}520/\text{yr}\end{aligned}$$

So the net annual savings of the microturbine, including \$1500/yr in O&M, is

$$\begin{aligned}\text{Microturbine savings} &= (\$37{,}300 + \$67{,}200) - \$59{,}520 - \$1{,}500 \\ &= \$43{,}480/\text{yr}\end{aligned}$$

e. The initial rate of return on this investment would be

$$\text{Initial rate of return} = \frac{\text{Annual savings}}{\text{Initial investment}} = \frac{\$43{,}480/\text{yr}}{\$220{,}000} = 0.198$$
$$= 19.8\%/\text{yr}$$

(Chapter 5 presents more sophisticated investment analysis techniques.)

4.2.3 Reciprocating Internal Combustion Engines

Distributed generation today is dominated by installations that utilize reciprocating—that is, piston-driven—internal combustion engines (ICEs) connected to constant-speed ac generators. They are readily available in sizes that range from 0.5 kW to 6.5 MW, with electrical efficiencies in the vicinity of 37–40% (LHV basis). They can be designed to run on a number of fuels, including gasoline, natural gas, kerosene, propane, fuel oil, alcohol, waste-treatment plant digester gas, and hydrogen. They are the least expensive of the currently available DG technologies, and when burning natural gas they are relatively clean. Reciprocating engines make up a large fraction of the current market for combined heat and power.

Most of these engines are conventional, four-stroke reciprocating engines very much like the ones found in automobiles and trucks. The four-stroke cycle, as illustrated in Fig. 4.3, consists of an intake stroke, a compression stroke, a power stroke, and an exhaust stroke. During the intake stroke, the piston moves downward. The partial vacuum created draws in air, or a mixture of air and vaporized fuel, through the open intake valve. During the compression stroke, the piston moves upward with both the intake and exhaust valves closed. The gases are compressed and heated, and near the top of the stroke combustion is initiated. The hot, expanding gases drive the piston downward in the power stroke forcing the crankshaft to rotate. In the final stroke, the rising piston forces the hot exhaust gases to exit via the now-open exhaust valve, completing the cycle.

An alternative to the four-stroke engine is one in which every other stroke is a power stroke. In these two-stroke engines, as the piston approaches its lowest point in the power stroke, an exhaust port opens, thereby allowing exhaust gases to escape while the intake port opens to allow fresh fuel and air to enter. Both close at the start of the compression stroke. Two-stroke engines have greater power for their size, but because neither intake nor exhaust functions are particularly effective, their efficiency is not as good as four-stroke engines and they produce much higher emissions. Since air quality is so often a critical constraint in siting distributed generation, two-stroke engines have limited potential.

There are two important variations on the four-stroke engine: *spark-ignited* (Otto-cycle) and *compression-ignition* (Diesel-cycle). Spark-ignited engines burn

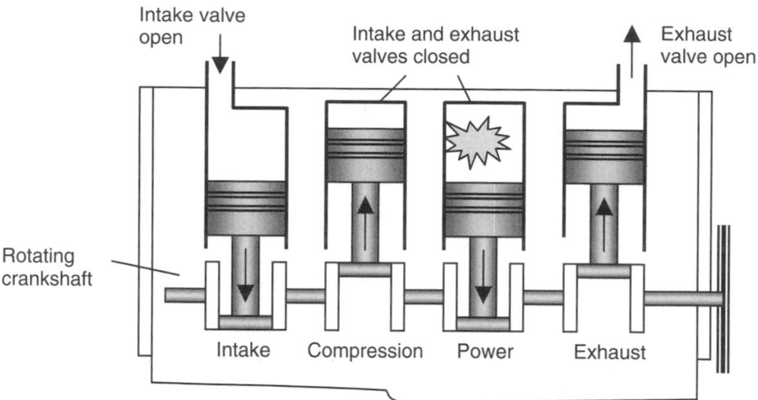

Figure 4.3 Basic four-stroke, internal combustion engine.

gasoline or other easily ignitable fuels such as natural gas or propane. Combustion is initiated by an externally timed spark, which ignites the air–fuel mixture that entered the cylinder during the intake stroke and which has been compressed during the subsequent compression stroke.

In contrast, compression-ignition engines burn heavier petroleum distillates such as diesel or fuel oil. These fuels are not premixed with air, but instead are injected under high pressure directly into the cylinder toward the end of the compression cycle. Self-ignited diesel engines have much higher compression ratios than do spark-ignited engines, which means that during the compression stroke the air is heated to much higher temperatures. As the pressure increases, so does the temperature until a point is reached at which spontaneous combustion causes the fuel to explode, initiating the power stroke. In addition to having higher compression ratios, diesel engines also experience much more sudden fuel ignition—more of an explosive "bang" than occurs in the slower-burning air–fuel mixture of a spark-ignited engine. Both of these factors mean that diesel engines must be built to be stronger and heavier than their spark-ignited counterparts. In spite of these more severe materials requirements, diesel engines tend to have higher efficiencies, which means for the same output they can be physically smaller in size and somewhat less expensive. Generally speaking, they require more maintenance than spark-ignition engines and their emissions are more problematic, making them more difficult to site in areas where air quality is an issue.

The efficiency of internal combustion engines can be increased by pressurizing the air, or air–fuel mixture, before it is enters the cylinder (called *charged aspiration*). This can be done with a turbocharger, which is a small turbine driven by the exhaust gases, or a supercharger, which is mechanically driven by an auxiliary shaft from the engine. Thermal efficiencies (from fuel to horsepower) of as high as 41% LHV (38% HHV) for spark-ignited engines and 46% LHV (44% HHV) for Diesel-cycle engines have been achieved. An added benefit of charged aspiration is that it allows the engine to run with a leaner air–fuel mixture, which helps lower combustion temperature and hence lowers NO_x emissions.

Since emissions are a major constraint, most of the interest in reciprocating engines for distributed generation is focused on engines that burn the cleanest fuel—natural gas. The U.S. Department of Energy (DOE), along with Caterpillar, Cummins, and Waukesha Engines and a number of universities, has initiated an Advanced Reciprocating Engines Systems (ARES) program to help improve natural-gas engine efficiency while lowering emissions, with a goal of reaching 50% electrical efficiency and 80+% CHP efficiency, with NO_x emissions below 0.1 g NO_x per brake-horsepower-hour, at a capital cost of \$400 to \$450 per kW. Figure 4.4 shows a typical heat balance for today's 39.1% efficient engine along

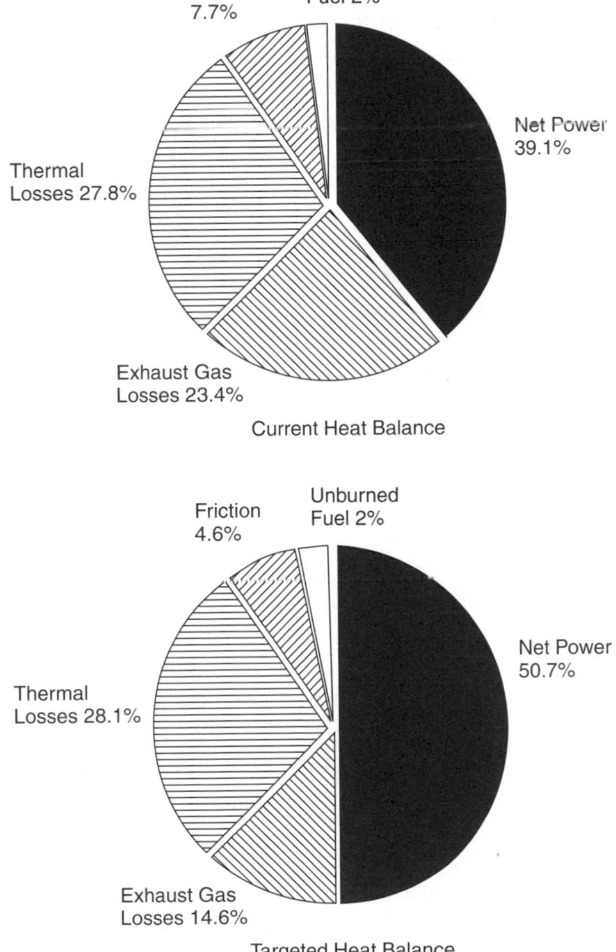

Figure 4.4 Heat balance for conventional high-efficiency reciprocating engines, along with the goals of the ARES program for next-generation engines. Based on McNeely (2003).

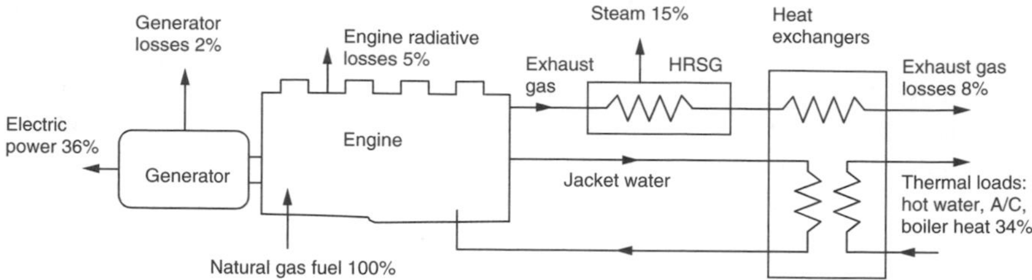

Figure 4.5 Heat balance for an example reciprocating-engine cogeneration system with overall efficiency of 85%. Based on data from Petchers.

with the targeted heat balance for the next generation, 50% efficient, natural-gas fueled, reciprocating engines.

For combined heat and power applications, waste heat from reciprocating engines can be tapped mainly from exhaust gases and cooling water that circulates around cylinders in the engine jacket, with additional potential from oil and turbo coolers. While engine exhaust and cooling water each provide about half of the useful thermal energy in a reciprocating-engine cogeneration system, the exhaust is at much higher temperature (around 450°C versus 100°C) and hence is more versatile. Figure 4.5 shows an example heat balance for a reciprocating-engine cogeneration system utilizing a heat-recovery steam generator (HRSG) for steam along with other heat exchangers to provide lower temperature energy for such loads as absorption air-conditioning, hot water, and boiler heat. The conversion efficiency from fuel to electricity (36%) and usable heat (49%) gives an overall thermal efficiency of 85%.

4.2.4 Stirling Engines

The spark-ignition and Diesel-cycle reciprocating engines just described are examples of internal combustion engines. That is, combustion takes place inside the engine itself. An alternative approach is *external combustion*, in which energy is supplied to the working fluid from a source outside of the engine. A steam-cycle power plant is one example of an external combustion engine. A Stirling-cycle engine is an example of a piston-driven reciprocating engine that relies on external rather than internal combustion. As such, it can run on any virtually any fuel or other source of high temperatures such as concentrated sunlight shining onto a black absorber plate.

Stirling-cycle engines were patented in 1816 by a minister in the Church of Scotland, Robert Stirling. He was apparently motivated, in part, by concern for the safety of his parishioners who were at some risk from poor-quality steam engines that had a tendency, in those days, to violently and unexpectedly explode. His engine had no such problem since they were designed to operate at relatively low pressures. The first known application of a Stirling engine was in 1872 by John Ericsson, a British/American inventor. Thereafter, Stirling engines were used quite extensively until the early 1900s, at which point advances in steam and spark-ignition engines, with higher efficiencies and greater versatility, pretty much eliminated them from the marketplace. They are now enjoying somewhat of a comeback, however, especially as an efficient technology for converting concentrated sunlight into electricity.

The basic operation of a Stirling engine is explained in Fig. 4.6. In this case, the engine consists of two pistons in the same cylinder—one on the hot side of the engine, the other on the cold side—separated by a "short-term" thermal energy storage device called a *regenerator*. Unlike an internal-combustion engine, the gas, which may be just air, but is more likely to be nitrogen, helium, or hydrogen, is permanently contained in the cylinder. The regenerator may be just a wire or ceramic mesh or some other kind of porous plug with sufficient mass to allow it to maintain a good thermal gradient from one face to the other. It also needs to be porous enough to allow gas in the cylinder to be pushed through it first in one direction, then the other. As the gas passes through the regenerator, it either picks up heat or drops it off, depending on which way the gas is moving.

As shown, the space on the left-hand side of the regenerator is kept hot with some source of heat, which might be a continuously burning flame or perhaps concentrated sunlight. On the right-hand side the space is kept cold by radiative cooling or active cooling with a circulating heat-exchange fluid. If it is actively cooled, that becomes a source of heat for cogeneration. In other words, this is a heat engine operating between a hot source and a cold sink. As such, its efficiency is constrained by the Carnot-cycle limit (interestingly, Carnot did not develop his famous relationship until well after Stirling's patent).

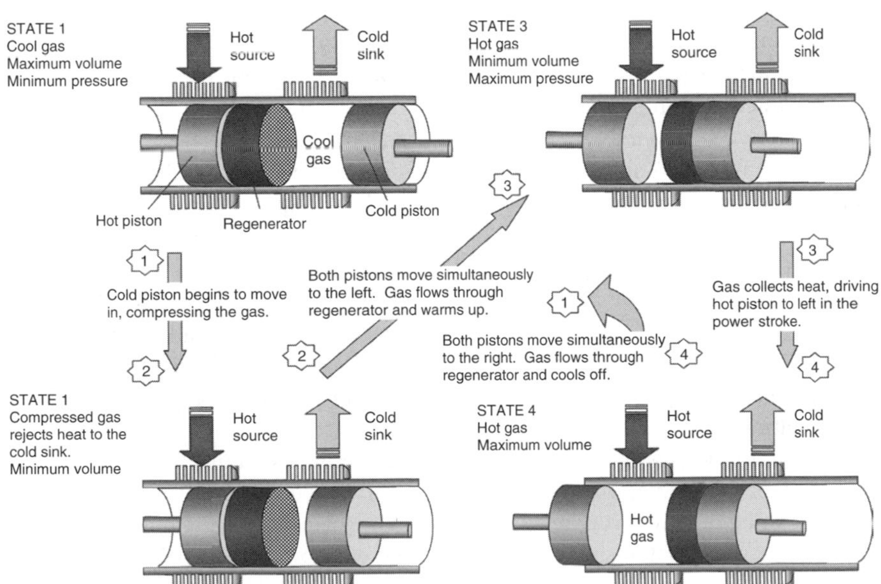

Figure 4.6 The four states and transitions of the Stirling cycle.

The Stirling cycle consists of four states and four transitions between those states. As shown in Fig. 4.6, the cycle starts with the hot piston up against the hot face of the regenerator, while the cold piston is fully extended away from the cold face of the regenerator. In State 1, essentially all of the gas is cool (there is an insignificant amount within the pores of the regenerator), the pressure is at its lowest, and the gas volume is the highest it will get. The following describes the transitions from one state to another:

① to ②. The hot piston remains stationary while the cold one moves to the left, compressing the gas while transferring heat to the cold sink. Ideally, this is an isothermal process; that is, the temperature of the gas remains constant at T_C.

② to ③. Both pistons now move simultaneously to the left at the same rate and by the same amount. The gas passing through the regenerator into the hot space picks up heat in the regenerator and increases in temperature and pressure while the volume remains constant.

③ to ④. The gas in the hot space absorbs energy from the hot source and isothermally expands, pushing the hot piston to the left. This is the power stroke.

④ to ①. Both pistons now move simultaneously to the right while maintaining a constant volume. The gas passing through the regenerator into the cold space transfers heat to the regenerator and drops in temperature and pressure.

Piston movement in the above machine can be controlled by a rotating crankshaft to which the pistons are connected. And, of course, a generator could be attached to that rotating shaft to produce electricity.

The idealized pressure–volume diagram shown in Fig. 4.7 may make it easier to understand the states and transitions in the Stirling cycle. In thermodynamics, $P-v$ diagrams, such as this one, greatly enhance intuition, especially since the area contained within the process curves represents the net work produced during the cycle.

Stirling engines in sizes ranging from less than 1 kW up to about 25 kW are beginning to be made commercially available. While their efficiency is still relatively low, typically less than 30%, rapid progress is being made toward boosting that into the range in which they would be competitive with internal-combustion engines. Since fuel is burned slowly and constantly, with no explosions, these engines are inherently quiet, which could make them especially attractive for automobiles, boats, recreational vehicles, and even small aircraft. In fact, that quietness has been used to advantage in Stirling engine propulsion systems for submarines.

Potential markets include small-scale portable power systems for battery charging and other off-grid applications. They could be especially useful in developing countries since they will run on any fuel, including biomass; or when incorporated in parabolic dish systems, they can use solar energy. Cogeneration is possible using the cooling water that maintains the cold sink, so heat-and-power systems for homes are likely. With higher efficiencies, their quiet, vibration-free operation,

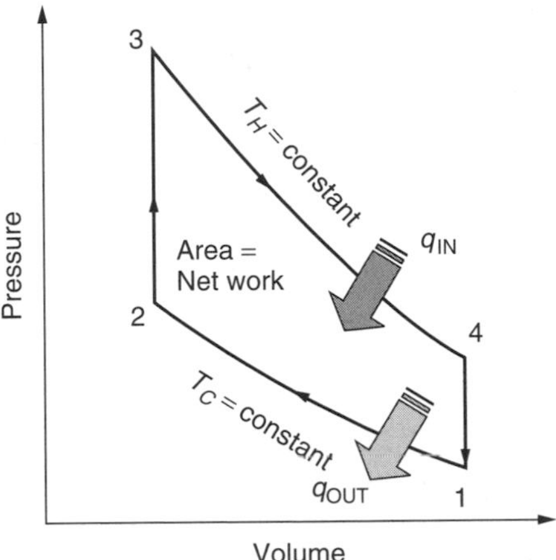

Figure 4.7 Pressure–volume diagram for an ideal Stirling cycle.

very low emissions when burning natural gas, simplicity, and potentially high reliability could make them an attractive alternative in the near future.

4.3 CONCENTRATING SOLAR POWER (CSP) TECHNOLOGIES

While the already mentioned combustion turbines, reciprocating engines, and conventional Stirling-cycle engines have many desirable attributes, including the ability to take advantage of cogeneration, they do still rely on fossil fuels, which means they pollute, making them hard to site, and they are vulnerable to the economic uncertainties associated with fuel-price volatility. An alternative approach is based on capturing sunlight. That can be done with photovoltaics, which are covered extensively later in this book, or with concentrating solar power (CSP) technologies.

CSP technologies convert sunlight into thermal energy to run a heat engine to power a generator. With the environment providing the cold temperature sink, the maximum efficiency of a heat engine is directly related to how hot the high-temperature source can be made. Without concentration, sunlight can't provide high enough temperatures to make the thermodynamic efficiency of a heat engine worth pursuing. But with concentration, it is an entirely different story. There are three successfully demonstrated approaches to concentrating sunlight that have been pursued with some vigor: parabolic dish systems with Stirling engines, linear solar-trough systems, and heliostats (mirrors) reflecting sunlight onto a power tower.

4.3.1 Solar Dish/Stirling Power Systems

Dish/Stirling systems use a concentrator made up of multiple mirrors that approximate a parabolic dish. The dish tracks the sun and focuses it onto a thermal receiver. The thermal receiver absorbs the solar energy and converts it to heat that is delivered to a Stirling engine. The receiver can be made up of a bank of tubes containing a heat transfer medium, usually helium or hydrogen, which also serves as the working fluid for the engine. Another approach is based on heat pipes in which the boiling and condensing of an intermediate fluid is used to transfer heat from the receiver to the Stirling engine. The cold side of the Stirling engine is maintained using a water-cooled, fan-augmented radiator system similar to that in an ordinary automobile. Being a closed system, very little make-up water is required, which can be a major advantage over other CSP technologies. Some of the existing units have been designed to operate in a hybrid mode in which fuel is burned to heat the engine when solar is inadequate. As a hybrid, the output becomes a reliable source of power with no backup needed for inclement weather or nighttime loads.

With average efficiencies of over 20% and the record measured peak efficiency of nearly 30%, dish/Stirling systems currently exceed the efficiency of any other solar conversion technology.

Two competing dish/Stirling system technologies have been successfully demonstrated. In one, the dish is built by Science Applications International Corporation (SAIC) and the engine by Sterling Thermal Motors (STM). The other is a Boeing/Stirling Energy Systems (SES) power plant. Both provide on the order of 25 kW per system with conversion efficiencies from direct-beam solar radiation to electrical power of over 20%.

The SAIC dish/Stirling system is illustrated in Fig. 4.8. The dish itself is made up of an array of 16 stretched-membrane, mirrored facets. Each facet consists of a steel ring approximately 3.2 m in diameter, with thin stainless steel membranes stretched over both sides of the ring to form a structure that resembles a drum. The top membrane is made highly reflective by laminating either a thin glass mirror or a silverized polymer reflective film onto the membrane. By partially evacuating the space between the membranes, the shape of the mirrored surface can be made slightly concave, allowing each facet to be focused appropriately onto the receiver.

Sunlight, concentrated by the SAIC dish, is absorbed in the receiver to provide 725°C heat to the Stirling engine. The STM engine is made up of four cylinders, each with a double-acting piston, arranged in a square pattern. The connecting rods for the pistons cause a swashplate to convert their motion to the rotary motion needed by the generator. The efficiency of these engines from heat to mechanical power is over 36%.

Table 4.4 shows the efficiencies and power outputs for the SAIC/STM system from sunlight to net power delivered. Even though the Stirling engine itself is 36.1% efficient, by the time losses at the reflector, receiver, gear box, and generator are added to parasitic power needed to operate the system, the overall efficiency is just under 21%. In good locations, with these efficiencies, the land area required is about four acres per megawatt of power.

Dish/Stirling systems can be stand-alone power plants that don't need access to fuel lines or sources of cooling water. Not needing water except to wash the

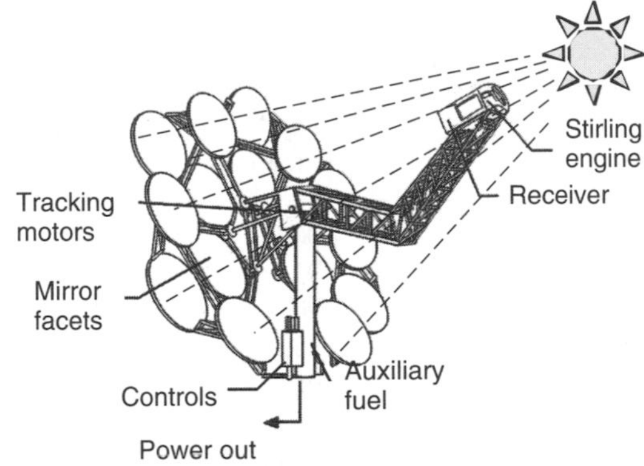

(a) The complete SAIC/STM system

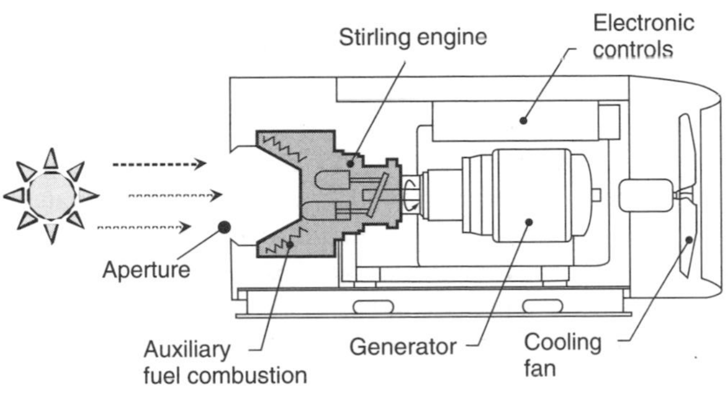

(b) The STM Stirling engine.

Figure 4.8 The SAIC/STM dish/Stirling system.

mirrors once in a while, they are ideal for power generation in sunny desert areas where lack of cooling water is a major constraint. Many load centers, including those in southern California, Arizona, and Nevada, are located close to such sunny deserts so transmission distances could be short. Unless they are hybridized with fuel augmentation, there are no air emissions, which means that the lengthy process of obtaining air permits can be eliminated. With short construction times, easy permitting, and relatively small 25-kW modular sizes, which simplifies financing, the time from project design to delivered power can be very short—on the order of just a year or so. Short lead times means that capacity can be added incrementally as needed to follow load growth, avoiding the periods of oversupply that characterize large-scale generation facilities (see Section 5.7.1).

4.3.2 Parabolic Troughs

As of 2003, the world's largest solar power plant is a 354-MW parabolic-trough facility located in the Mojave Desert near Barstow, California called the Solar

TABLE 4.4 Second-Generation SAIC/STM Dish/Stirling System Efficiencies[a]

Step Description	Step Efficiency (%)	Cumulative Efficiency (%)	Power (kW)
Solar insolation	100.0	100	113.5
Reflected by mirrors	93.1	93.1	105.7
Intercepted at aperture	90.3	84.1	95.4
Absorbed by receiver	85.0	71.5	81.1
Receiver heat loss	98.0	70.0	79.5
Engine mechanical efficiency	36.1	25.3	28.7
Gear box efficiency	98.0	24.8	28.1
Gross generator output	92.0	22.8	25.9
Electrical parasites	−2.2 kW	20.9	23.7

[a]Insolation is direct normal solar radiation at 1000 W/m^2.
Source: Davenport et al. (2002).

Electric Generation System (SEGS). SEGS consists of nine large arrays made up of rows of parabolic-shaped mirrors that reflect and concentrate sunlight onto linear receivers located along the foci of the parabolas. The receivers, or heat collection elements (HCE), consist of a stainless steel absorber tube surrounded by a glass envelope with the vacuum drawn between the two to reduce heat losses. A heat transfer fluid circulates through the receivers, delivering the collected solar energy to a somewhat conventional steam turbine/generator to produce electricity. The SEGS collectors, with over 2 million m^2 of surface area, run along a north–south axis, and they rotate from east to west to track the sun throughout the day. Figure 4.9 illustrates the parabolic trough concept.

The first plant, SEGS I, is a 13.4-MW facility built in 1985, while the final plant, SEGS IX, produces 80 MW and was completed in 1991. SEGS I

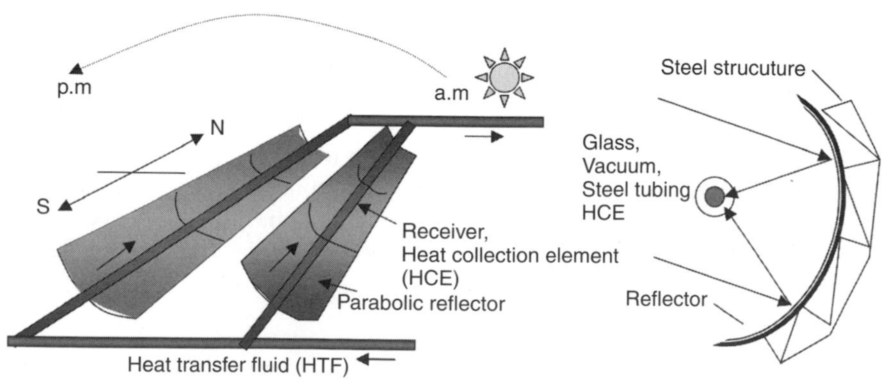

Figure 4.9 Parabolic trough solar collector system.

was designed with thermal storage to enable it to operate several extra hours each day after the sun had gone down. Storage was based on elevating the temperature of a highly flammable mineral oil, called Caloria. Unfortunately, an accident in 1999 set the storage unit on fire and it was destroyed. Later SEGS plants do not include storage, but have been designed to operate as hybrids with back-up energy supplied by fossil fuels to run the steam/turbine when solar is not available. Future designs may once again include heat storage, but the medium will likely be based on molten salts rather than mineral oil.

Figure 4.10 presents an overall system diagram for a typical parabolic trough power plant. The heat transfer fluid (HTF) is heated to approximately 400°C in the receiver tubes along the parabola foci. The HTF passes through a series of heat exchangers to generate high-pressure superheated steam for the turbine. The system shown includes the possibility for thermal storage as well as fossil-fuel-based auxiliary heat to run the plant when solar is insufficient. Two options for hybrid operation are indicated. One is based on a natural-gas-fired HTF heater in parallel with the solar array. The other is an optional gas boiler to generate steam for the turbine, which makes it virtually the same as a complete, conventional steam-cycle power plant with an auxiliary solar unit.

Solar trough systems have thus far been coupled with conventional steam-cycle power plants, which means that cooling water is needed for their condensers. Once-through cooling requires enormous amounts of water; and even when

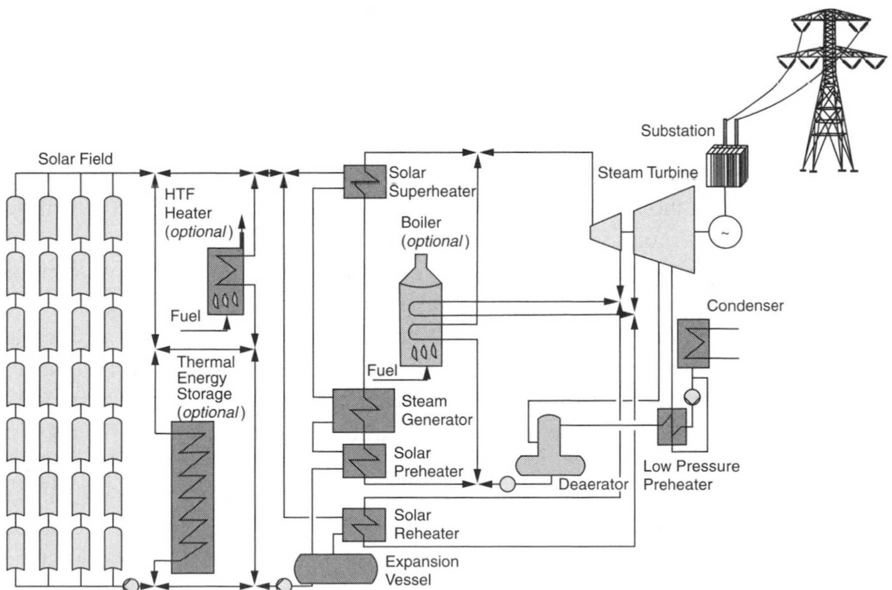

Figure 4.10 Solar trough system coupled with a steam turbine/generator, including optional heat storage and two approaches to fossil-fuel hybridization. From NREL website.

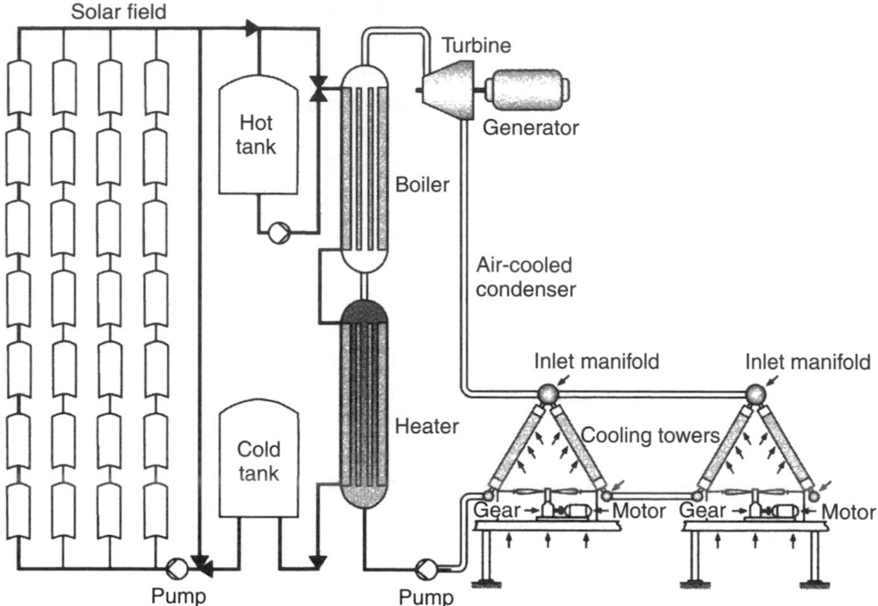

Figure 4.11 An organic Rankine-cycle, solar-trough system that eliminates the need for cooling water. From NREL, in Price et al. (2002).

cooling towers are used, the water demand is still high. Since these plants are likely to be located in desert areas, where water is precious, it would be highly advantageous to develop a trough technology that might eliminate this constraint. One approach, shown in Fig. 4.11, is based on a modified, organic fluid, Rankine-cycle technology currently used in geothermal power plants. The key is the use of an organic fluid that can be condensed at above-atmospheric pressures using air-cooled, fan-driven, cooling towers. This technology is being evaluated for relatively small plants, on the order of 100 kW to 10 MW, which could make them attractive for remote or distributed generation applications.

The SEGS experience has demonstrated that solar trough systems can be reliable and reasonably cost effective. Operating experience with the later 80-MW SEGS plants show overall annual solar-to-electricity efficiencies of around 10% with generation costs of about 12¢/kWh (in 2001 dollars), which makes them the least expensive source of solar electricity today. Advanced, 200-MW plants with more efficient thermal storage units and refinements in receivers coupled with general cost reductions associated with gained experience suggest that electricity from future solar-trough systems could be close to 5¢/kWh (Price et al., 2002). While this technology was first demonstrated on a large scale in the United States, much of the current development is taking place in other countries, including Spain, Greece, Italy, India, and Mexico.

4.3.3 Solar Central Receiver Systems

Another approach to achieving the concentrated sunlight needed for solar thermal power plants is based on a system of computer-controlled mirrors, called *heliostats*, which bounce sunlight onto a receiver mounted on top of a tower (Fig. 4.12).

The evolution of power towers began in 1976 with the establishment of the National Solar Thermal Test Facility at Sandia National Laboratories in Albuquerque, New Mexico. That soon led to the construction of number of test facilities around the world, the largest of which was a 90-m-tall, 10-MW power tower, called *Solar One*, built near Barstow, California. In Solar One, water was pumped up to the receiver where it was turned into steam that was brought back down to a steam turbine/generator. Steam could also be diverted to a large, thermal storage tank filled with oil, rock, and sand to test the potential for continued power generation during marginal solar conditions or after the sun had set. While thermal storage as a concept was successfully demonstrated, there was a sizable mismatch between the storage tank temperature and the temperature needed by the turbine for maximum efficiency.

Solar One operated from 1982 to 1988, after which time it was dismantled; parts of it, including the 1818 heliostats and the tower itself, were reused in another 10-MW test facility called *Solar Two*. While Solar One used water as the heat exchange medium, Solar Two used molten nitrate salts (60% sodium nitrate and 40% potassium nitrate). A two-tank, molten-salt thermal storage system replaced the original oil/rock storage tank in the configuration shown in Fig. 4.13. The molten-salt system has proven to be a great success. Its temperature of 565°C well matches the needs of the steam turbine, and its round-trip efficiency (the ratio of thermal energy out to thermal energy in) was greater than 97%. It was designed to provide enough storage to deliver the full 10-MW output of the plant for an extra three hours past sunset. With reduced output, it could deliver power for much longer periods from stored solar energy.

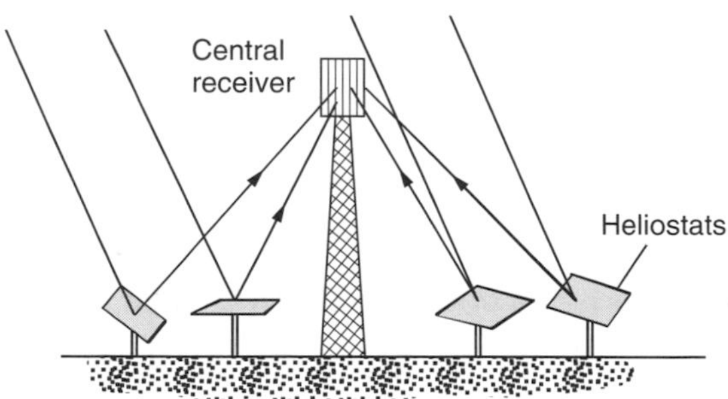

Figure 4.12 Central receiver system (CRS) with heliostats to reflect sunlight onto a central receiver.

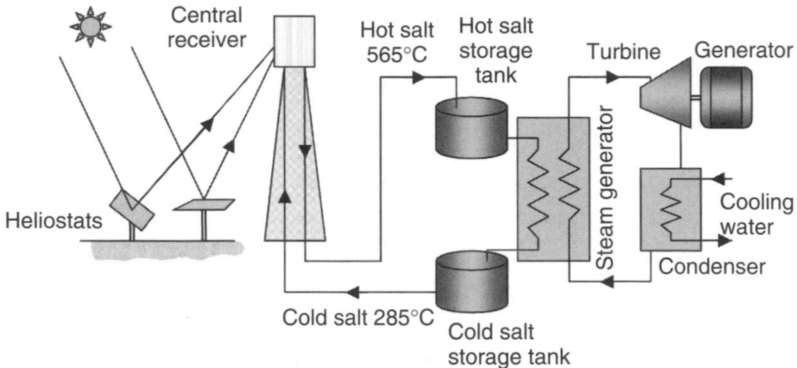

Figure 4.13 Schematic of a molten-salt central receiver system.

After operating for three years, Solar Two was decommissioned in 1999. Since then, a design team from the United States and Spain has proposed construction of a 15-MW *Solar Tres*, to be located somewhere in Spain. This system will have a much higher ratio of heliostat area to rated generator power and a much larger molten-salt thermal storage tank, which will enable it to operate at full capacity for over 12 h on stored energy alone. Systems with more collector area than is needed to run the turbines while the sun is out, augmented with more thermal storage to save that extra solar energy until it is needed later, could become reliable sources of power, ready when needed to meet typical afternoon and evening peak demands. This dispatchability advantage of power towers with thermal storage would add considerably to their value.

The future of central receiver systems is promising, but somewhat uncertain. Molten-salt power towers are the most well understood technology, but to be cost effective they need to be made much larger—probably on the order of 100 MW, which means that the first ones will cost hundreds of millions of dollars. The financial risk for the first plants will likely be a significant deterrent. There are also emerging power tower technologies that use air as the working fluid rather than molten salt. Solar-heated air could be used directly in a steam generator for a conventional Rankine-cycle power plant, or it might be used to preheat air leaving the compressor on its way to the combustor of a gas turbine in a combined-cycle hybrid power plant.

4.3.4 Some Comparisons of Concentrating Solar Power Systems

The three approaches to concentrating solar thermal power systems—dish Stirling, parabolic trough, and central receiver—can be compared from a variety of perspectives. Although they share the same fundamental approach of using mirrored surfaces to reflect and concentrate sunlight onto a receiver creating high enough temperatures to run a heat engine with reasonable efficiency, they are in many ways quite different from one another.

With regard to efficiency, all three of these technologies incorporate heat engines, which means the higher the temperature of the heat source, the greater the potential efficiency. The key to high temperatures is the intensity of solar radiation focused onto the receiver, which is usually expressed in dimensionless "suns" of concentration where the reference point of 1 sun means no concentration. Dish Stirling systems achieve concentration ratios of about 3000 suns, power towers about 1000 suns, and parabolic trough systems about 100 suns. Not surprisingly, the corresponding efficiencies of these technologies follow the same ranking, with dish Stirling the highest and parabolic trough the lowest.

There are a number of measures of efficiency, including peak efficiency under design conditions, annual efficiency as measured in the field, and land area required per unit of electrical output. The current annual efficiency from sunlight hitting collectors to electrical energy delivered to the grid for these systems is approximately: Dish Stirling 21%, power towers 16%, and parabolic troughs 14%. In terms of land area required, however, power towers suffer because of the empty space between tower and mirrors, so the rankings shift some. Dish Stirling requires about 4 acres per MW, parabolic troughs about 5 acres/MW, and power towers about 8 acres/MW.

Another important concern for solar systems in general is whether they can deliver electricity whenever it is needed. All three of these CSP technologies can be hybridized using fossil fuel auxiliary heat sources, so they are the somewhat the same in that regard. Another way to achieve reliability, however, is with thermal storage; in that regard, parabolic troughs and power towers have an advantage over dish Stirling engines. When thermal storage is the backup rather than fuel combustion, systems are easier to permit since they can be 100% solar.

Since all CSP technologies need to be able to focus the suns rays, they will most likely be used in areas with very clear skies. If those are desert areas, minimizing the need for cooling water can be a significant concern. In that regard, Dish Stirling engines, which need no cooling water have the advantage over current designs for troughs and towers. They also make very little noise and have a relatively low profile so they may be easier to site close to residential loads.

During the early stages of development, and as technologies begin to be deployed, economic risks are incurred and the scale of the investments required can be an important determinant of the speed with which markets expand. Small-scale systems cost less per modular unit so the financial risks associated with the first few units are similarly small. Dish Stirling systems appear to be appropriately sized at about 25 kW each, but economies of scale play a bigger role for troughs and towers and they may be most economical in unit sizes of about 100 MW. It seems likely to be easier to find investors willing to help develop $100,000 dish systems, working out the bugs and improving the technology as they go along, than to assemble the hundreds of millions of dollars needed for a single trough or tower system. Wind turbines, with their explosive growth, have certainly benefited from the fact that they too are small in scale.

4.4 BIOMASS FOR ELECTRICITY

Biomass energy systems utilize solar energy that has been captured and stored in plant material during photosynthesis. While the overall efficiency of conversion of sunlight to stored chemical energy is low, plants have already solved the two key problems associated with all solar energy technologies—that is, how to collect the energy when it is available, and how to store it for use when the sun isn't shining. Plants have also very nicely dealt with the greenhouse problem since the carbon released when they use that stored energy for respiration is the same carbon they extracted in the first place during photosynthesis. That is, they get energy with no net carbon emissions.

While there is already a sizable agricultural industry devoted to growing crops specifically for their energy content, it is almost entirely devoted to converting plant material into alcohol fuels for motor vehicles. On the other hand, biomass for electricity production is essentially all waste residues from agricultural and forestry industries and, to some extent, municipal solid wastes. Since it is based on wastes that must be disposed of anyway, biomass feedstocks for electricity production may have low-cost, no-cost, or even negative-cost advantages.

Currently there are about 14 GW of installed generation capacity powered by biomass in the world, with about half of that being in the United States. About two-thirds of the biomass power plants in the United States cogenerate both electricity and useful heat. Virtually all biomass power plants operate on a conventional steam–Rankine cycle. Since transporting their rather disbursed fuel sources over any great distances could be prohibitively expensive, biomass power plants tend to be small and located near their fuel source, so they aren't able to take advantage of the economies of scale that go with large steam plants. To offset the higher cost of smaller plants, lower-grade steel and other materials are often used, which requires lower operating temperatures and pressures and hence lower efficiencies. Moreover, biomass fuels tend to have high water content and are often wet when burned, which means that wasted energy goes up the stack as water vapor. The net result is that existing biomass plants tend to have rather low efficiencies—typically less than 20%. Even though the fuel may be very inexpensive, those low efficiencies translate to reasonably expensive electricity, which is currently around 9¢/kWh.

An alternative approach to building small, inefficient plants dedicated to biomass power production is to burn biomass along with coal in slightly modified, conventional steam-cycle power plants. Called co-firing, this method is an economical way to utilize biomass fuels in relatively efficient plants. And, since biomass burns cleaner than coal, overall emissions are correspondingly reduced in co-fired facilities.

New combined-cycle power plants don't need to be large to be efficient, so it is interesting to contemplate a new generation of biomass power plants based on gas turbines rather than steam. The problem is, however, that gas turbines cannot run directly on biomass fuels since the resulting combustion products would damage the turbine blades, so an intermediate step would have to be introduced. By gasifying the fuel first and then cleaning the gas before combustion, it would be possible to use biomass with gas turbines. Indeed, considerable work has already been done *on coal-integrated gasifier/gas turbine* CIG/GT systems. CIG/GT systems have not been commercialized, however, in part due to gas-cleaning problems caused by the high sulfur content of coal. Biomass, however, tends to have very little sulfur, so this problem is minimized. Moreover, biomass tends to gasify easier than coal, so it may well be that it will be easier to develop biomass-integrated gasifier/gas turbine (BIG/GT) systems than to develop CIG/GTs. Biomass fuels do, however, have rather high nitrogen content, and control of NO_X would have to be addressed. It is projected that BIG/GT plants could generate electricity at about 5¢/kWh.

A two-step process for gasifying biomass fuels is illustrated in Fig. 4.14. In the first step, the raw biomass fuel is heated, causing it to undergo a process called *pyrolysis* in which the volatile components of the biomass are vaporized. As the fuel is heated, moisture is first driven off; then at a temperature of about 400°C the biomass begins to break down, yielding a product gas, or *syngas*, consisting mostly of hydrogen (H_2), carbon monoxide (CO), methane (CH_4), carbon dioxide (CO_2), and nitrogen (N_2) as well as tar. The solid byproducts of pyrolysis are char (fixed carbon) and ash. In the second step, char heated to about 700°C reacts with oxygen, steam, and hydrogen to provide additional syngas. The heat needed to drive both steps comes from the combustion of some of the char.

The relative percentages of each gas vary considerably, depending on the technology used in the gasifier, and so does the heating value of the mix. High heating values (HHV) of syngas ranges from about 5.4 to 17.4 MJ/m^3. For comparison, natural gas typically has an HHV of anywhere from 36 to 42 MJ/m^3.

Pyrolysis reactions can also be used to convert biomass to liquid, as diagrammed in Fig. 4.15. Different temperatures and reaction rates produce energy-rich vapors that can be condensed into a liquid "biocrude" oil with a heating value about 60% that of diesel fuel.

Another technology for converting biomass to energy is based on the anaerobic (without oxygen) decomposition of organic materials by microorganisms to produce a biogas consisting primarily of methane and carbon dioxide. The biochemical reactions taking place are extremely complex, but in simplified terms

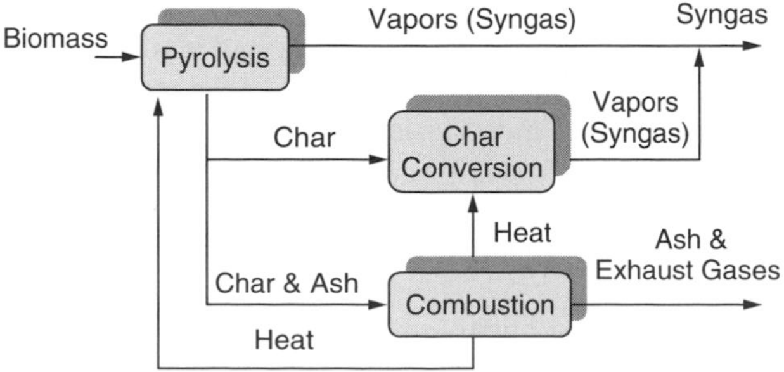

Figure 4.14 Biomass gasification process.

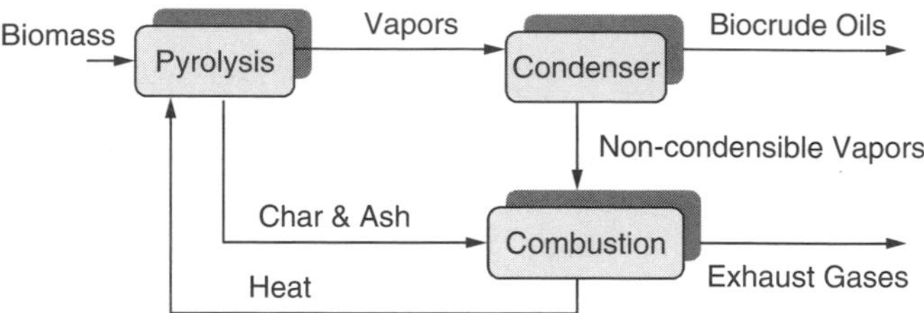

Figure 4.15 Biomass pyrolysis used to create liquid biocrude oil.

they can be summarized as

$$C_nH_aO_bN_cS_d + H_2O + \text{bacteria} \rightarrow CH_4 + CO_2 + NH_3 + H_2S + \text{new cells} \tag{4.2}$$

where $C_nH_aO_bN_cS_d$ is meant to represent organic matter, and the equation is obviously not chemically balanced. The technology is conceptually simple, consisting primarily of a closed tank, called a digester, in which the reactions can take place in the absence of oxygen. The desired end product is methane, CH_4 and the relative amount of methane in the digester gas is typically between 55% and 75%.

Anaerobic digesters are fairly common in municipal wastewater treatment plants where their main purpose is to transform sewage sludge into innocuous, stabilized end products that can be easily disposed of in landfills or, sometimes, recycled as soil conditioners. Anaerobic digesters can be used with other biomass feedstocks including food processing wastes, various agricultural wastes, municipal solid wastes, bagasse, and aquatic plants such as kelp and water hyacinth. When the biogas is treated to remove its sulfur, the resulting gas can be burned in reciprocating engines to produce electricity and usable waste heat.

4.5 MICRO-HYDROPOWER SYSTEMS

Hydroelectric power is a very significant source, accounting for 19% of the global production of electricity. In a number of countries in Africa and South America, it is the source of more than 90% of their electric power. Hydropower generates 9% of U.S. electricity, which may sound modest, but it is still almost an order of magnitude more than the combined output of all of the other renewables. Almost all of that hydropower is generated in large-scale projects, which are sometimes defined to be those larger than 30 MW in capacity. Small-scale hydropower systems are considered to be those that generate between 100 kW and 30 MW, while micro-hydro plants are smaller than 100 kW. Our interest here is in micro-hydropower.

The simplest micro-hydro plants *are run-of-the-river* systems, which means that they don't include a dam. As such, they don't cause nearly the ecosystem disruption of their dams and reservoir counterparts. An example of a run-of-the-river

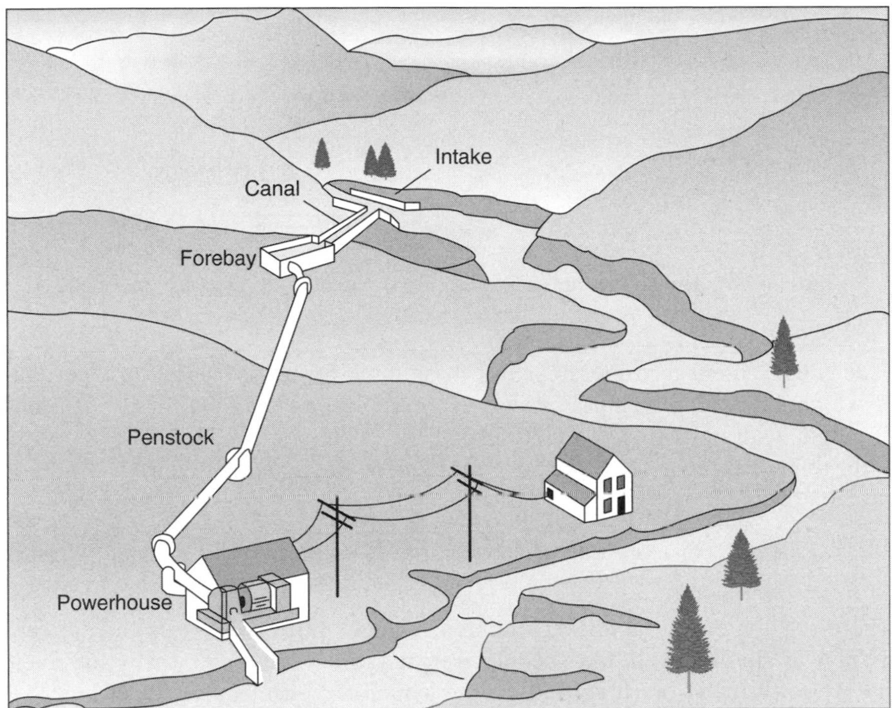

Figure 4.16 A micro-hydropower system in which water is diverted into a penstock and delivered to a powerhouse below. *Source*: Small Hydropower Systems, DOE/GO-102001-1173, July 2001.

system is shown in Fig. 4.16. A portion of the river is diverted into a pipeline, called a *penstock*, that delivers water under pressure to a hydraulic turbine/generator located in a powerhouse located at some elevation below the intake. Depending on how the system has been designed, the powerhouse may also contain a battery bank to help provide for peak demands that exceed the average generator output.

4.5.1 Power From a Micro-Hydro Plant

The energy associated with water manifests itself in three ways: as potential energy, pressure energy, and kinetic energy. The energy in a hydroelectric system starts out as potential energy by virtue of its height above some reference level—in this case, the height above the powerhouse. Water under pressure in the penstock is able to do work when released, so there is energy associated with that pressure as well. Finally, as water flows there is the kinetic energy that is associated with any mass that is moving. Figure 4.17 suggests the transformations between these forms of energy as water flows from the forebay, through the penstock, and out of a nozzle.

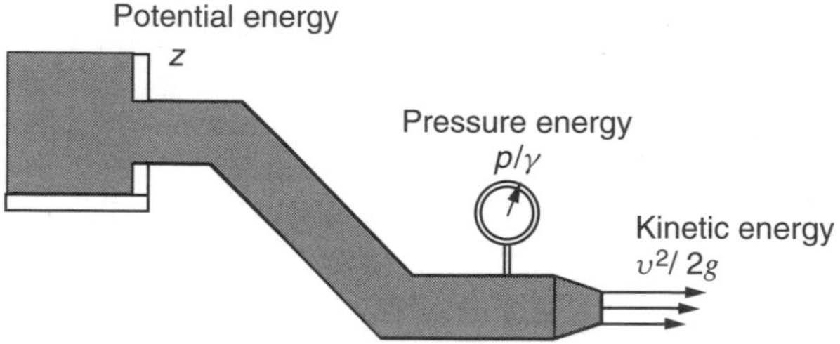

Figure 4.17 Transformations of energy from potential, to pressure, to kinetic.

It is convenient to express each of these three forms of energy on a per unit of weight basis, in which case energy is referred to as *head* and has dimensions of length, with units such as "feet of head" or "meters of head." The total energy is the sum of the potential, pressure, and kinetic head and is given by

$$\text{Energy head} = z + \frac{p}{\gamma} + \frac{v^2}{2g} \tag{4.3}$$

where z is the elevation above a reference height (m) or (ft), p is the pressure (N/m^2) or (lb/ft^2), γ is the specific weight (N/m^3) or (lb/ft^3), v is the average velocity (m/s) or (ft/s), and g is gravitational acceleration (9.81 m/s^2) or (32.2 ft/s^2).

In working with micro-hydropower systems, especially in the United States, mixed units are likely to be incurred, so Table 4.5 is presented to help sort them out.

TABLE 4.5 Useful Conversions for Water

	American	SI
1 cubic foot =	7.4805 gal	0.02832 m^3
1 foot per second =	0.6818 mph	0.3048 m/s
1 cubic foot per second =	448.8 gal/min (gpm)	0.02832 m^3/s
Water density =	62.428 lb/ft^3	1000 kg/m^3
1 pound per square inch =	2.307 ft of water	6896 N/m^2
1 kW =	737.56 ft-lb/s	1000 N-m/s

Example 4.3 Mixed Units in the American System. Suppose a 4-in.-diameter penstock delivers 150 gpm of water through an elevation change of 100 feet. The pressure in the pipe is 27 psi when it reaches the powerhouse. What fraction of the available head is lost in the pipe? What power is available for the turbine?

Solution. From (4.3)

$$\text{Pressure head} = \frac{p}{\gamma} = \frac{27\ \text{lb/in.}^2 \times 144\ \text{in.}^2/\text{ft}^2}{62.428\ \text{lb/ft}^3} = 62.28\ \text{ft}$$

To find velocity head, we need to use $Q = vA$, where Q is flow rate, v is velocity, and A is cross-sectional area:

$$v = \frac{Q}{A} = \frac{150\ \text{gal/min}}{(\pi/4)(4/12\ \text{ft})^2 \times 60\ \text{s/min} \times 7.4805\ \text{gal/ft}^3} = 3.830\ \text{ft/s}$$

So, from (4.3) the velocity head is

$$\text{Velocity head} = \frac{v^2}{2g} = \frac{(3.830\ \text{ft/s})^2}{2 \times 32.2\ \text{ft/s}^2} = 0.228\ \text{ft}$$

The total head left at the bottom of the penstock is the sum of the pressure and velocity head, or 62.28 ft + 0.228 ft = 62.51 ft. Notice the velocity head is negligible. Since we started with 100 ft of head, pipe losses are 100 – 62.51 = 37.49 ft, or 37.49%.

Using conversions from Table 4.5, while carefully checking to see that units properly cancel, the power represented by a flow of 150 gpm at a head of 62.51 ft is

$$P = \frac{150\ \text{gal/min} \times 62.428\ \text{lb/ft}^3 \times 62.51\ \text{ft}}{60\ \text{s/min} \times 7.4805\ \text{gal/ft}^3 \times 737.56\ \text{ft-lb/s-kW}} = 1.77\ \text{kW}$$

The power theoretically available from a site is proportional to the difference in elevation between the source and the turbine, called the *head* H, times the rate at which water flows from one to the other, Q. Using a simple dimensional analysis, we can write that

$$\text{Power} = \frac{\text{Energy}}{\text{Time}} = \frac{\text{Weight}}{\text{Volume}} \times \frac{\text{Volume}}{\text{Time}} \times \frac{\text{Energy}}{\text{Weight}} = \gamma \text{QH} \tag{4.4}$$

Substituting appropriate units in both the SI and American systems results in the following key relationships for water:

$$P(W) = 9810\ Q(\text{m}^3/\text{s})\ H(\text{m}) = \frac{Q(\text{gpm})\ H(\text{ft})}{5.3} \tag{4.5}$$

While (4.5) makes no distinction between a site with high head and low flow versus one with the opposite characteristics, the differences in physical facilities are considerable. With a high-head site, lower flow rates translate into smaller-diameter piping, which is more readily available and a lot easier to work with, as well as smaller, less expensive turbines. Home-scale projects with modest flows and decent heads can lead to quick, simple, cost-effective systems.

The head H given in (4.5) is called the *gross head*, call it H_G, because it does not include pipe losses that decrease the power available for the turbine. The *net head*, H_N, will be the gross head (the actual elevation difference) minus the head loss in the piping. Those losses are a function of the pipe diameter, the flow rate, the length of the pipe, how smooth the pipe is, and how many bends, valves, and elbows the water has to pass through on its way to the turbine. Figure 4.18 illustrates the difference between gross and net head.

4.5.2 Pipe Losses

Figure 4.19 shows the friction loss, expressed as feet of head per 100 feet of pipe, for PVC and for polyethylene (poly) pipe of various diameters. PVC pipe has lower friction losses and it is also less expensive than poly pipe, but small-diameter poly may be easier to install since it is somewhat flexible and can be purchased in rolls from 100 to 300 feet long. Larger-diameter poly comes in shorter lengths that can be butt-welded on site. Both need to be protected from sunlight, since ultraviolet exposure makes these materials brittle and easier to crack.

From (4.5), we can write that the energy delivered by a micro-hydro system is given by

$$P(W)_{\text{delivered}} = \frac{eQ(\text{gpm})\ H_N(\text{ft})}{5.30}, \qquad P(\text{kW}) = 9.81eQ(\text{m}^3/\text{s})\ H_N(\text{m}) \quad (4.6)$$

where e is the efficiency of the turbine/generator. From (4.6) comes the following very handy *rules-of-thumb* for ≈50%-efficient turbine/generator systems (which is in the right ballpark for a micro-hydro plant):

$$P(W) \approx \frac{Q(\text{gpm})\ H_N(\text{ft})}{10}, \qquad P(\text{kW}) \approx 5Q(\text{m}^3/\text{s})\ H_N(\text{m}) \quad (4.7)$$

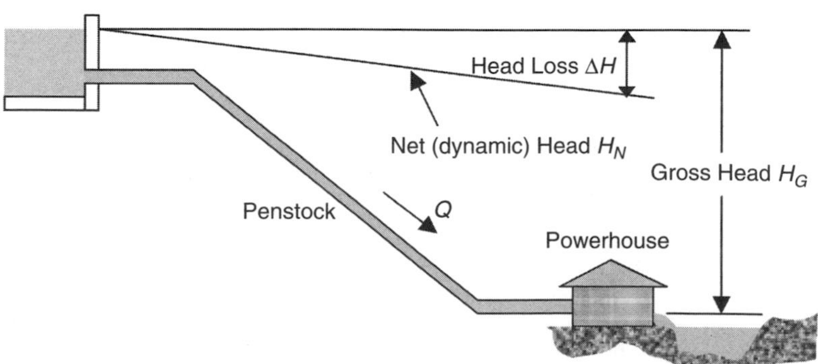

Figure 4.18 Net head is what remains after pipe losses are accounted for.

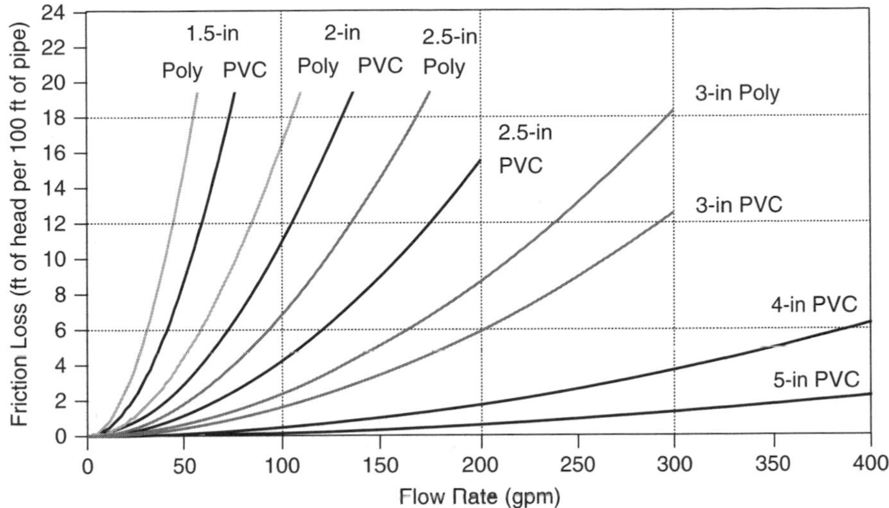

Figure 4.19 Friction head loss, in feet of head per 100 ft of pipe, for 160 psi PVC piping and for polyethylene, SDR pressure-rated pipe.

Example 4.4 Power from a Small Source. Suppose 150 gpm of water is taken from a creek and delivered through 1000 ft of 3-in.-diameter polyethylene pipe to a turbine 100 ft lower than the source. Use the rule-of-thumb (4.7) to estimate the power delivered by the turbine/generator. In a 30-day month, how much electric energy would be generated?

Solution. From Fig. 4.19, at 150 gpm, 3-in. poly loses about 5 ft of head for every 100 ft of length. Since we have 1000 ft of pipe, the friction loss is

$$1000 \text{ ft} \times 5 \text{ ft}/100 \text{ ft} = 50 \text{ ft of head loss}$$

The net head available to the turbine is

$$\text{Net head} = \text{Gross head} - \text{Friction head} = 100 \text{ ft} - 50 \text{ ft} = 50 \text{ ft}$$

The rule-of-thumb estimate for electrical power delivered is

$$P(W) \approx \frac{Q(\text{gpm}) \times H_N \text{ (ft)}}{10} = \frac{150 \text{ gpm} \times 50 \text{ ft}}{10} = 750 \text{ W}$$

The monthly electricity supplied would be about 24 hr $\times$ 30 d/mo $\times$ 0.75 kW $=$ 540 kWh. This is roughly the average electrical energy used by a typical U.S. household that does not use electricity for heating, cooling, or water heating.

Example 4.4 would be a poor design since half of the power potentially available is lost in the piping. Larger-diameter pipe will always reduce losses and increase power delivered, so bigger is definitely better from an energy perspective. But, large pipe costs more, especially when the cost of larger valves and other fittings is included, and it is more difficult to work with. Since the cost of the piping is often a significant fraction of the total cost of a micro-hydro project, an economic analysis should be used to decide upon the optimum pipe diameter.

Suppose, however, that the pipe size has already been determined for one reason or another. Perhaps it is already the largest diameter conveniently available, perhaps it is as much as can be afforded, or perhaps it has already been purchased. The question arises as to whether there is an optimum flow rate through the pipe. If flow is too high, friction eats up too much of the power and output drops. By slowing down the flow rate, friction losses are reduced but so is the power delivered. So, there must be some ideal flow rate that balances these two competing phenomena and produces the maximum power for a given pipe size.

A derivation of the theoretical maximum power delivered by a pipeline is quite straightforward. The curves given in Fig. 4.19 can be approximated quite accurately by the following relationship:

$$\Delta H = kQ^2 \tag{4.8}$$

where k is just an arbitrary constant. The power delivered by the pipeline will be the net head times the flow, with an appropriate constant of proportionality c that depends on the units used:

$$P = cH_N Q = c(H_G - \Delta H)Q = c(H_G - kQ^2)Q \tag{4.9}$$

For maximum power, we can take the derivative with respect to Q and set it equal to zero:

$$\frac{dP}{dQ} = 0 = H_G - 3kQ^2 = H_G - 3\Delta H \tag{4.10}$$

That leads to the conclusion that the theoretical maximum power delivered by a pipeline occurs when pipe losses are one-third of the gross head:

$$\Delta H = \tfrac{1}{3} H_G \tag{4.11}$$

Example 4.5 Optimum Flow Rate. Find the optimum flow rate for the 1000 ft of 3-in. poly pipe in Example 4.4. Gross head is 100 ft.

Solution. Applying (4.11), for friction to be one-third of 100 ft of gross head means 33.3 ft of pipe losses in 1000 ft of pipe. That translates to 3.33 ft of loss per 100 ft of pipe. For 3-in. poly, Fig. 4.19 indicates that a flow rate of about 120 gpm will be optimum.

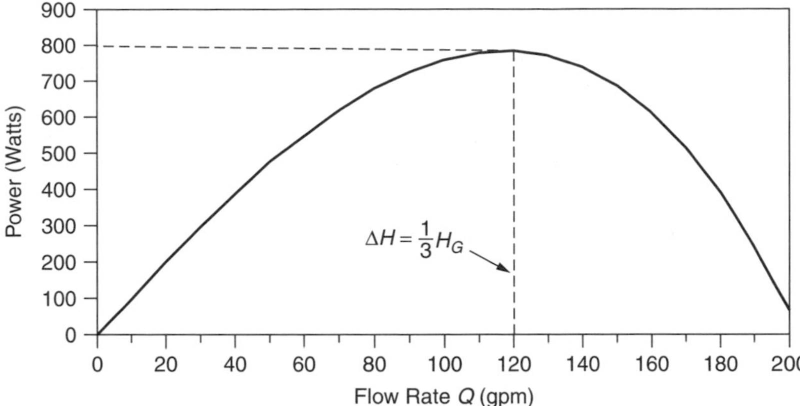

Figure 4.20 Showing the influence of flow rate on power delivered from a penstock (values correspond to Example 4.4).

At 120 gpm and with net head equal to two-thirds of the gross head, power delivered will be

$$P(W) \approx \frac{Q(\text{gpm}) \times H_N(\text{ft})}{10} = \frac{120 \text{ gpm} \times \frac{2}{3} \times 100 \text{ ft}}{10} = 800 \text{ W}$$

For comparison, in Example 4.3, with 150 gpm of flow only 750 W were generated. The relationship between flow and power for this pipeline is shown in Fig. 4.20.

While the above illustrates how slowing the flow may increase power delivered by the pipeline, that doesn't mean it would help to simply put a valve in line that is then partially shut off. That would waste more power than would have been gained. Instead, it is the function of a properly designed nozzle to control that flow without much power loss. And, once again, it should be pointed out that the first approach to increasing power delivered is always to consider a larger pipe. Keeping flow rates less than about 5 ft per second and friction losses less than 20% seem to be good design guidelines.

4.5.3 Measuring Flow

Obviously, a determination of the available water flow is essential to planning and designing a system. In some circumstances, the source may be so abundant and the demand so low that a very crude assessment may be all that is necessary. If, however, the source is a modest creek, or perhaps just a spring, especially if it is one whose flow is seasonal, much more careful observations and measurements may be required before making a hydropower investment. In those circumstances, regular measurements taken over at least a full year should be made.

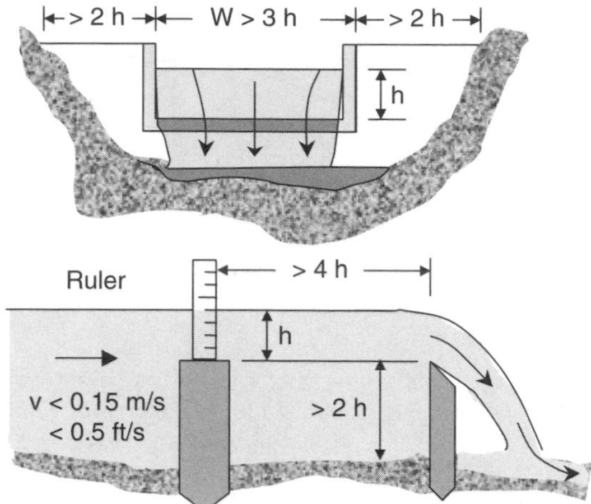

Figure 4.21 A rectangular weir used to measure flows. Based on Inversin (1986).

Methods of estimating stream flow vary from the very simplest bucket-and-stopwatch approach to more sophisticated approaches, including one in which stream velocity measurements are made across the entire cross section of the creek using a propeller-or-cup-driven current meter. For micro-hydro systems, oftentimes the best approach involves building a temporary plywood, concrete, or metal wall, called a weir, across the creek. The height of the water as it flows through a notch in the weir can be used to determine flow.

The notch in the weir may have a number of different shapes, including rectangular, triangular, and trapezoidal. It needs to have a sharp edge to it so that the water drops off immediately as it crosses the weir. For accuracy, it must create a very slow moving pool of water behind it so that the surface is completely smooth as it approaches the weir, and a ruler of some sort needs to be set up so that the height of the water upstream can be measured. When the geometric relationships for the rectangular weir shown in Fig. 4.21 are followed, and height h is more than about 5 cm, or 2 in., the flow can be estimated from the following:

$$Q = 1.8(W - 0.2\ h)h^{3/2} \qquad \text{with } Q(m^3/\text{s}), h(\text{m}), W(\text{m}) \tag{4.12}$$

$$Q = 2.9(W - 0.2\ h)h^{3/2} \qquad \text{with } Q(\text{gpm}), h(\text{in.}), W(\text{in.}) \tag{4.13}$$

Example 4.6 Designing a Weir. Design a weir to be able to measure flows expected to be at least 100 gpm following the constraints given in Fig. 4.21.

Solution. At the 100-gpm low-flow rate, we'll use the suggested minimum water surface height above the notch of 2 in. From (4.13),

$$W \leq \frac{Q}{2.9\ h^{3/2}} + 0.2\ h = \frac{100}{2.9 \times 2^{3/2}} + 0.2 \times 2 = 12.6 \text{ in}$$

To give the weir the greatest measurement range under the constraint that $h \leq W/3$, suggests making $W = 12.6$ in. The maximum water height above the weir can then be $h = 12.6/3 = 4.2$ in.

From (4.13), the greatest flow that could be measured with the 12.6 in. × 4.2 in. notch would be

$$Q = 2.9(W - 0.2\ h)h^{3/2} = 2.9(12.6 - 0.2 \times 4.2)4.2^{3/2} = 294 \text{ gpm}$$

4.5.4 Turbines

Just as energy in water manifests itself in three forms—potential, pressure, and kinetic head—there are also three different approaches to transforming that waterpower into the mechanical energy needed to rotate the shaft of an electrical generator. *Impulse turbines* capture the kinetic energy of high-speed jets of water squirting onto buckets along the circumference of a wheel. In contrast, water velocity in a *reaction turbine* plays only a modest role, and instead it is mostly the pressure difference across the runners, or blades, of these turbines that creates the desired torque. Generally speaking, impulse turbines are most appropriate in high-head, low-flow circumstances, while the opposite is the case for reaction turbines. And, finally, the slow-moving, but powerful, traditional overshot *waterwheel* converts potential energy into mechanical energy. The slow rotational rates of waterwheels are a poor match to the high speeds needed by generators, so they are not used for electric power.

Impulse turbines are the most commonly used turbines in micro-hydro systems. The original impulse turbine was developed and patented by Lester Pelton in 1880, and its modern counterparts continue to carry his name. In a *Pelton wheel*, water squirts out of nozzles onto sets of twin buckets attached to the rotating wheel. The buckets are carefully designed to extract as much of the water's kinetic energy as possible while leaving enough energy in the water to enable it to leave the buckets without interfering with the incoming water. A diagram of a four-nozzle Pelton wheel is shown in Fig. 4.22. These turbines have efficiencies typically in the range of 70–90%.

The flow rate to a Pelton wheel is controlled by nozzles, which for simple residential systems are much like those found on ordinary irrigation sprinklers. When water exits a nozzle, its pressure head is converted to kinetic energy, which

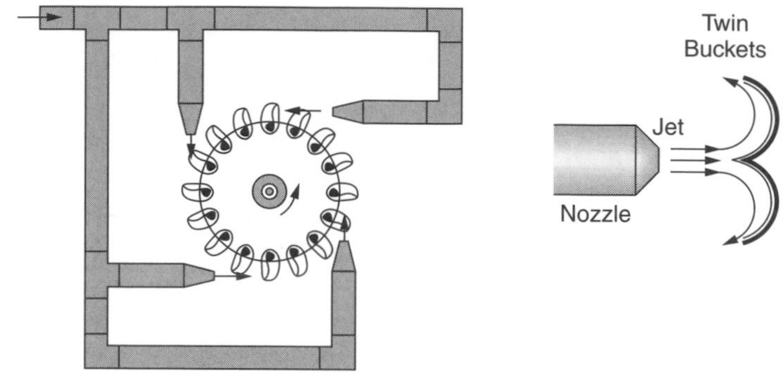

Figure 4.22 A four-nozzle Pelton turbine.

means that we can use (4.3) to determine the flow velocity.

$$H_N = \frac{v^2}{2g} \qquad \text{so that} \qquad v = \sqrt{2gH_N} \tag{4.14}$$

From flow velocity v and flow rate Q we can determine an appropriate diameter for the water jets. For nozzles with needle valves, the jet diameter is on the order of 10–20% smaller than the nozzle, but for simple, home-scale systems, the jet and nozzle diameters are nearly equal. Using $Q = vA$ along with (4.14), we can determine the jet diameter d for a turbine with n nozzles:

$$Q = vA = \sqrt{2gH_N}\left(\frac{\pi}{4}\right)nd^2 \tag{4.15}$$

Solving for jet diameter gives

$$d = \frac{0.949}{(gH_N)^{1/4}}\sqrt{\frac{Q}{n}} \tag{4.16}$$

Example 4.7 Nozzles for a Pelton Turbine. A penstock provides 150 gpm (0.334 cfs) with 50 ft of head to a Pelton turbine with 4 nozzles. Assuming jet and nozzle diameters are the same, pick a nozzle diameter.

Solution. From (4.16)

$$d = \frac{0.949}{(32.2\ \text{ft/s}^2 \times 50\ \text{ft})^{1/4}}\sqrt{\frac{0.334\ \text{ft}^3/\text{s}}{4}} = 0.0433\ \text{ft} = 0.52\ \text{in.}$$

The efficiency of the original Pelton design suffers somewhat at higher flow rates because water trying to leave the buckets tends to interfere with the incoming jet. Another impulse turbine, called a *Turgo wheel*, is similar to a Pelton, but the runner has a different shape and the incoming jet of water hits the blades somewhat from one side, allowing exiting water to leave from the other, which greatly reduces the interference problem. The Turgo design also allows the jet to spray several buckets at once; this spins the turbine at a higher speed than a Pelton, which makes it somewhat more compatible with generator speeds.

There is another impulse turbine, called a *cross-flow* turbine, which is especially useful in low-to-medium head situations (5–20 m). This turbine is also known as a Banki, Mitchell, or Ossberger turbine—names that reflect its inventor, principal developer, and current manufacturer. These turbines are especially simple to fabricate, which makes them popular in developing countries where they can be built locally.

For low head installations with large flow rates, reaction turbines are the ones most commonly used. Rather than having the runner be shot at by a stream of water, as is the case with impulse turbines, reaction turbine runners are completely immersed in water and derive their power from the mass of water moving through them rather than the velocity. Most reaction turbines used in micro-hydro installations have runners that look like an outboard motor propeller. The propeller may have anywhere from three to six blades, which for small systems are usually fixed pitch. Larger units that include variable-pitch blades and other adjustable features are referred to as Kaplan turbines. An example of a right-angle-drive system in which a bulb contains the gearing between the turbine and an externally mounted generator is shown in Fig. 4.23.

4.5.5 Electrical Aspects of Micro-Hydro

Larger micro-hydro systems may be used as a source of ac power that is fed directly into utility lines using conventional synchronous generators and grid interfaces. Since the output frequency is determined by the rpm of the generator, very precise speed controls are necessary. While mechanical governors and hand-operated control valves have been traditionally used, modern systems have electronic governors controlled by microprocessors.

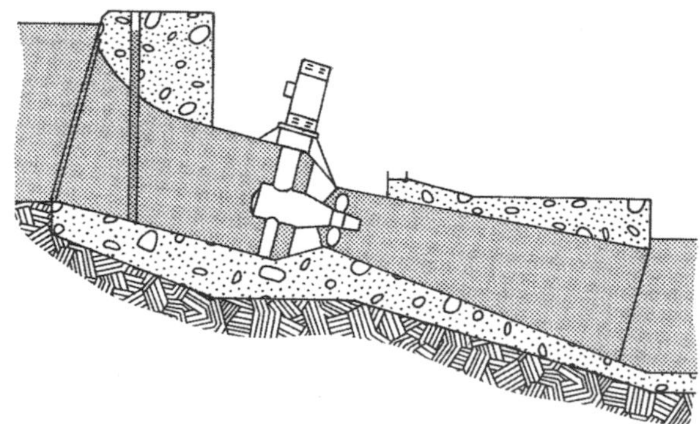

Figure 4.23 A right-angle-drive propeller turbine. Inversin (1986).

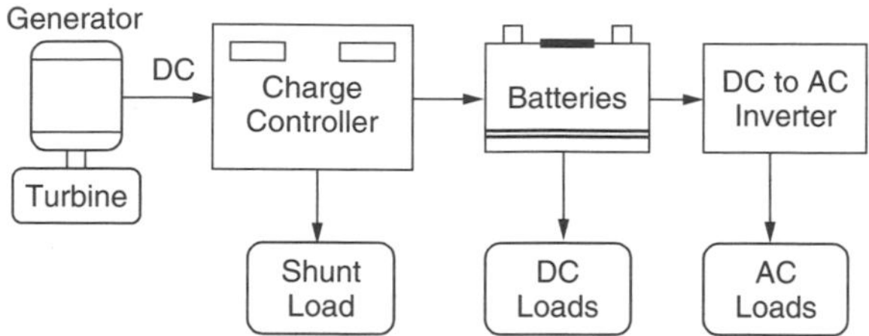

Figure 4.24 Electrical block diagram of a battery-based micro-hydro system.

On the other end of the scale, home-size micro-hydro systems usually generate dc, which is used to charge batteries. An exception would be the case in which utility power is conveniently available, in which case a grid-connected system in which the meter spins in one direction when demand is less than the hydro system provides, and the other way when it doesn't, would be simpler and cheaper than the battery-storage approach. The electrical details of grid-connected systems as well as stand-alone systems with battery storage are covered in some detail in the *Photovoltaic Systems* chapter of this text (Chapter 9).

The battery bank in a stand-alone micro-hydro system allows the hydro system, including pipes, valves, turbine, and generator, to be designed to meet just the average daily power demand, rather than the peak, which means that everything can be smaller and cheaper. Loads vary throughout the day, of course, as appliances are turned on and off, but the real peaks in demand are associated with the surges of current needed to start the motors in major appliances and power tools. Batteries handle that with ease. Since daily variations in water flow are modest, micro-hydro battery storage systems can be sized to cover much shorter outages than weather-dependent PV systems must handle. Two days of storage is considered reasonable.

A diagram of the principal electrical components in a typical battery-based micro-hydro system is given in Fig. 4.24. To keep the batteries from being damaged by overcharging, the system shown includes a charge controller that diverts excess power from the generator to a shunt load, which could be, for example, the heating element in an electric water heater tank. Other control schemes are possible, including use of regulators that either (a) adjust the flow of water through the turbine or (b) modulate the generator output by adjusting the current to its field windings. As shown, batteries can provide dc power directly to some loads, while other loads receive ac from an inverter.

4.6 FUEL CELLS

> I believe that water will one day be employed as a fuel, that hydrogen and oxygen which constitute it, used singly or together, will furnish an inexhaustible source of heat and light.
>
> —Jules Verne, *Mysterious Island,* 1874

The portion of the above quote in which Jules Verne describes the joining of hydrogen and oxygen to provide a source of heat and light is a remarkably accurate description of one of the most promising new technologies now nearing commercial reality—the fuel cell. However, he didn't get it quite right since more energy is needed to dissociate water into hydrogen and oxygen than can be recovered so water itself cannot be considered a fuel.

Fuel cells convert chemical energy contained in a fuel (hydrogen, natural gas, methanol, gasoline, etc.) directly into electrical power. By avoiding the intermediate step of converting fuel energy first into heat, which is then used to create mechanical motion and finally electrical power, fuel cell efficiency is not constrained by the Carnot limits of heat engines (Fig. 4.25). Fuel-to-electric power efficiencies as high as 65% are likely, which gives fuel cells the potential to be roughly twice as efficient as the average central power station operating today.

Fuel cells have other properties besides high efficiency that make them especially appealing. The usual combustion products (SO_x, particulates, CO, and various unburned or partially burned hydrocarbons) are not emitted, although there may be some thermal NO_x when fuel cells operate at high temperatures. They are vibration-free and almost silent, which, when coupled with their lack of emissions, means they can be located very close to their loads—for example, in the basement of a building. Being close to their loads, they not only avoid transmission and distribution system losses, but their waste heat can be used to cogenerate electricity and useful heat for applications such as space heating, air-conditioning, and hot water. Fuel-cell cogeneration systems can have overall efficiencies from fuel to electricity and heat of over 80%. High overall efficiency not only saves fuel but also, if that fuel is a hydrocarbon such as natural gas, emissions of the principal greenhouse gas, CO_2, are reduced as well. In fact,

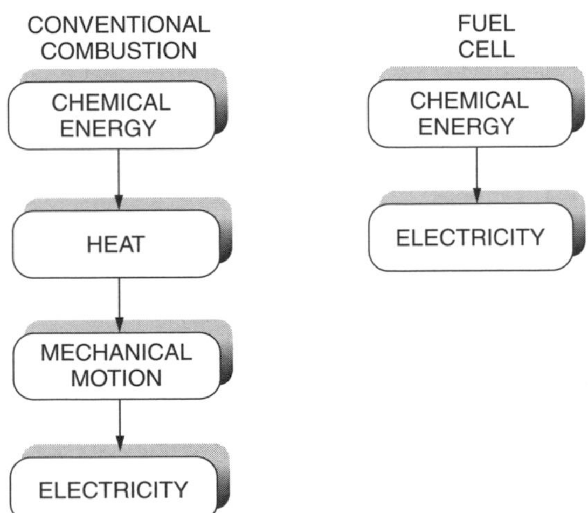

Figure 4.25 Conversion of chemical energy to electricity in a fuel cell is not limited by the Carnot efficiency constraints of heat engines.

if fuel cells are powered by hydrogen obtained by electrolysis of water using renewable energy sources such as wind, hydroelectric, or photovoltaics, they have no greenhouse gas emissions at all. Fuel cells are easily modulated to track short-term changes in electrical demand, and they do so with modest compromises in efficiency. Finally, they are inherently modular in nature, so that small amounts of generation capacity can be added as loads grow rather than the conventional approach of building large, central power stations in anticipation of load growth.

4.6.1 Historical Development

While fuel cells are now seen as a potentially dominant distributed generation technology for the twenty-first century, it is worth noting that they were first developed more than 160 years ago. Sir William Grove, the English scientist credited with the invention of the original galvanic cell battery, published his original experiments on what he called a "gaseous voltaic battery" in 1839 (Grove, 1839). He described the effects caused by his battery as follows: "A shock was given which could be felt by five persons joining hands, and which when taken by a single person was painful." Interestingly, this same phenomenon is responsible for the way that the organs and muscles of an electric eel supply their electric shock.

Grove's battery depended upon a continuous supply of rare and expensive gases, and corrosion problems were expected to result in a short cell lifetime, so the concept was not pursued. Fifty years later, Mond and Langer picked up on Grove's work and developed a 1.5-W cell with 50% efficiency, which they named a "fuel cell" (Mond and Langer, 1890). After another half century of little progress, Francis T. Bacon, a descendent of the famous seventeenth-century scientist, began work in 1932 that eventually resulted in what is usually thought to be the first practical fuel cell. By 1952, Bacon was able to demonstrate a 5-kW alkaline fuel cell (AFC) that powered, among other things, a 2-ton capacity fork-lift truck. In the same year, Allis Chalmers demonstrated a 20-horsepower fuel-cell-powered tractor.

Fuel cell development was greatly stimulated by NASA's need for on-board electrical power for spacecraft. The Gemini series of earth-orbiting missions used fuel cells that relied on permeable membrane technology, while the later Apollo manned lunar explorations and subsequent Space Shuttle flights have used advanced versions of the alkaline fuel cells originally developed by Bacon. Fuel cells not only provide electrical power, their byproduct is pure water, which is used by astronauts as a drinking water supply. For longer missions, however, photovoltaic arrays, which convert sunlight into electric power, have become the preferred technology.

Fuel cells for cars, buildings, central power stations, and spacecraft were the subject of intense development efforts in the last part of the twentieth century. Companies with major efforts in these applications include: Ballard Power Systems, Inc. (Canada), General Electric Company, the International Fuel Cells division of United Technologies Corp. and its ONSI subsidiary,

Plug Power, Analytic Power, General Motors, H-Power, Allison Chalmers, Siemens, ELENCO (Belgium), Union Carbide, Exxon/Ashthom, Toyota, Mazda, Honda, Toshiba, Hitachi Ltd., Ishikawajima-Harima Heavy Industries, Deutsche Aerospace, Fuji Electric, Mitsubishi Electric Corp. (MELCO), Daimler Chrysler, Ford, Energy Research Corporation, M-C Power Corp., Siemens-Westinghouse, CGE, DenNora, and Ansaldo. Clearly, there is an explosion of activity on the fuel cell front.

4.6.2 Basic Operation of Fuel Cells

There are many variations on the basic fuel cell concept, but a common configuration looks something like Fig. 4.26. As shown there, a single cell consists of two porous gas diffusion electrodes separated by an electrolyte. It is the choice of electrolyte that distinguishes one fuel cell type from another.

The electrolyte in Fig. 4.26 consists of a thin membrane that is capable of conducting positive ions but not electrons or neutral gases. Guided by the flow field plates, fuel (hydrogen) is introduced on one side of the cell while an oxidizer (oxygen) enters from the opposite side. The entering hydrogen gas has a slight tendency to dissociate into protons and electrons as follows:

$$H_2 \leftrightarrow 2H^+ + 2e^- \tag{4.17}$$

This dissociation can be encouraged by coating the electrodes or membrane with catalysts to help drive the reaction to the right. Since the hydrogen gas releases

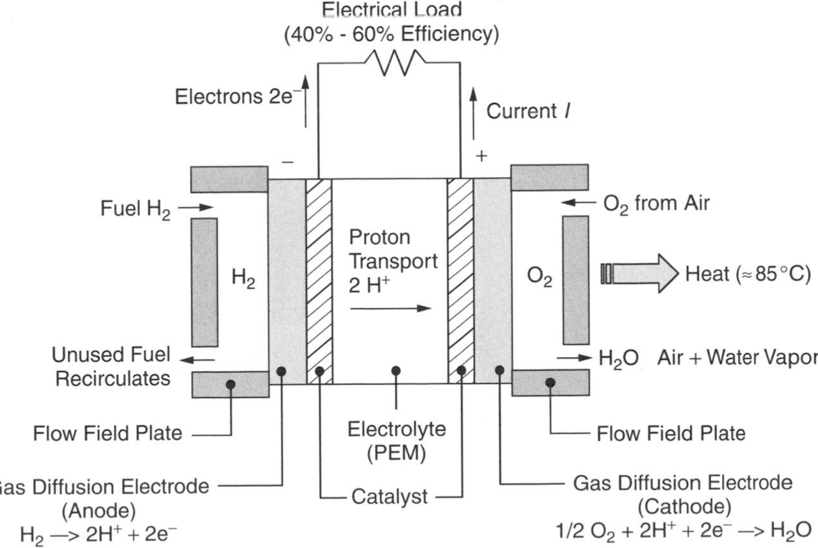

Figure 4.26 Basic configuration of a proton-exchange membrane (PEM) fuel cell.

protons in the vicinity of the electrode on the left (the anode), there will be a concentration gradient across the membrane between the two electrodes. This gradient will cause protons to diffuse through the membrane leaving electrons behind. As a result, the cathode takes on a positive charge with respect to the anode. Those electrons that had been left behind are drawn toward the positively charged cathode; but since they can't pass through the membrane, they must find some other route. If an external circuit is created between the electrodes, the electrons will take that path to get to the cathode. The resulting flow of electrons through the external circuit delivers energy to the load (remember that conventional current flow is opposite to the direction that electrons move, so current I "flows" from cathode to anode).

A single cell, such as that shown in Fig. 4.26, typically produces on the order of 1 V or less under open circuit conditions and produces about 0.5 V under normal operating conditions. To build up the voltage, cells can be stacked in series. To do so, the gas flow plates inside the stack are designed to be *bipolar*; that is, they carry both the oxygen and hydrogen used by adjacent cells as is suggested in Fig. 4.27.

4.6.3 Fuel Cell Thermodynamics: Enthalpy

The fuel cell shown in Fig. 4.26 is described by the following pair of half-cell reactions:

$$\text{Anode:} \quad H_2 \rightarrow 2H^+ + 2e^- \tag{4.18}$$

$$\text{Cathode:} \quad \tfrac{1}{2}O_2 + 2H^+ + 2e^- \rightarrow H_2O \tag{4.19}$$

When combined, (4.18) and (4.19) result in the same equation that we would write for ordinary combustion of hydrogen:

$$H_2 + \tfrac{1}{2}O_2 \rightarrow H_2O \tag{4.20}$$

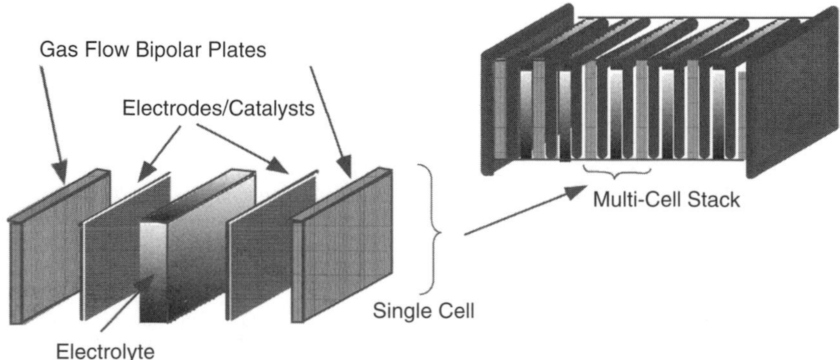

Figure 4.27 A multicell stack made up of multiple cells increases the voltage. After Srinivasan et al. (1999).

The reaction described in (4.20) is *exothermic*; that is, it liberates heat (as opposed to *endothermic* reactions, which need heat to be added to make them occur). Since (4.20) is exothermic, it will occur spontaneously—the hydrogen and oxygen want to combine to form water. Their eagerness to do so provides the energy that the fuel cell uses to deliver electrical energy to its load. The questions, of course, are how much energy is liberated in reaction (4.20) and how much of that can be converted to electrical energy. To answer those questions, we need to describe three quantities from thermodynamics: *enthalpy*, *free energy*, and *entropy*. Unfortunately, the precise definitions of each of these quantities tend not to lend themselves to easy, intuitive interpretation. Moreover, they have very subtle properties that are beyond our scope here, and there are risks in trying to present a simplified introduction.

The enthalpy of a substance is defined as the sum of its *internal energy* U and the product of its volume V and pressure P:

$$\text{Enthalpy } H = U + PV \tag{4.21}$$

The internal energy U of a substance refers to its microscopic properties, including the kinetic energies of molecules and the energies associated with the forces acting between molecules, between atoms within molecules, and within atoms. The total energy of that substance is the sum of its internal energy plus the observable, macroscopic forms such as its kinetic and potential energies. The units of enthalpy are usually kJ of energy per mole of substance.

Molecules in a system possess energy in various forms such as *sensible* and *latent* energy, which depends on temperature and state (solid, liquid, gaseous), *chemical energy* (associated with the molecular structure), and *nuclear energy* (associated with the atomic structure). For a discussion of fuel cells, it is changes in chemical energy that are of interest, and those changes are best described in terms of enthalpy changes. For example, we can talk about the potential energy of an object as being its weight times its height above some reference elevation. Our choice of reference elevation does not matter as long as we are only interested in the change in potential energy as an object is raised against gravity from one elevation to another. The same concept applies for enthalpy. We need to describe it with respect to some arbitrary reference conditions.

In the case of enthalpy a reference temperature of 25°C and a reference pressure of 1 atmosphere (standard temperature and pressure, STP) are assumed. It is also assumed that the reference condition for the chemically stable form of an element at 1 atmosphere and 25°C is zero. For example, the stable form of oxygen at STP is gaseous O_2, so the enthalpy for $O_2(g)$ is zero, where (g) just means it is in the gaseous state. On the other hand, since atomic oxygen (O) is not stable, its enthalpy is not zero (it is, in fact, 247.5 kJ/mol). Notice that the state of a substance at STP matters. Mercury, for example, at 1 atmosphere and 25°C is a liquid, so the standard enthalpy for Hg(l) is zero, where (l) means the liquid state.

One way to think about enthalpy is that it is a measure of the energy that it takes to form that substance out of its constituent elements. The difference between the enthalpy of the substance and the enthalpies of its elements is called the *enthalpy of formation*. It is in essence the energy stored in that substance due to its chemical composition. A short list of enthalpies of formation at STP conditions appears in Table 4.6. To remind us that a particular value of enthalpy (or other thermodynamic properties such as entropy and free energy) is at STP conditions, a superscript "o" is used (e.g. H°).

Table 4.6 also includes two other quantities, the absolute entropy S° and the Gibbs free energy G°, which will be useful when we try to determine the maximum possible fuel cell efficiency. Notice when a substance has negative enthalpy of formation, it means that the chemical energy in that substance is less than that of the constituents from which it was formed. That is, during its formation, some of the energy in the reactants didn't end up as chemical energy in the final substance.

In chemical reactions, the difference between the enthalpy of the products and the enthalpy of the reactants tells us how much energy is released or absorbed in the reaction. When there is less enthalpy in the final products than in the reactants, heat is liberated—that is, the reaction is exothermic. When it is the other way around, heat is absorbed and the reaction is endothermic.

If we analyze the reaction in (4.20), the enthalpies of H_2 and O_2 are zero so the enthalpy of formation is simply the enthalpy of the resulting H_2O. Notice in Table 4.6 that the enthalpy of H_2O depends on whether it is liquid water or gaseous water vapor. When the result is liquid water:

$$H_2 + \tfrac{1}{2}O_2 \rightarrow H_2O(l) \qquad \Delta H = -285.8 \text{ kJ} \tag{4.22}$$

When the resulting product is water vapor:

$$H_2 + \tfrac{1}{2}O_2 \rightarrow H_2O(g) \qquad \Delta H = -241.8 \text{ kJ} \tag{4.23}$$

TABLE 4.6 Enthalpy of Formation H°, Absolute Entropy S°, and Gibbs Free Energy G° at 1 atm, 25°C for Selected Substances

Substance	State	H° (kJ./mol)	S° (kJ/mol-K)	G° (kJ/mol)
H	Gas	217.9	0.114	203.3
H_2	Gas	0	0.130	0
O	Gas	247.5	0.161	231.8
O_2	Gas	0	0.205	0
H_2O	Liquid	−285.8	0.0699	−237.2
H_2O	Gas	−241.8	0.1888	−228.6
C	Solid	0	0.006	0
CH_4	Gas	−74.9	0.186	−50.8
CO	Gas	−110.5	0.197	−137.2
CO_2	Gas	−393.5	0.213	−394.4
CH_3OH	Liquid	−238.7	0.1268	−166.4

The negative signs for the enthalpy changes in (4.22) and (4.23) tell us these reactions are exothermic; that is, heat is released. The difference between the enthalpy of liquid water and gaseous water vapor is 44.0 kJ/mol. Therefore, that amount is the familiar latent heat of vaporization of water. Recall that latent heat is what distinguishes the higher heating valuez (HHV) of a hydrogen-containing fuel and the lower heating value LHV. The HHV includes the 44.0 kJ/mol of latent heat in the water vapor formed during combustion, while the LHV does not.

Example 4.8 The High Heating Value (HHV) for Methane. Find the HHV of methane CH_4 in kJ/mol and kJ/kg when it is oxidized to CO_2 and liquid H_2O.

Solution. The reaction is written below, and beneath it are enthalpies taken from Table 4.6. Notice that we must balance the equation so that we know how many moles of each constituent are involved.

$$\begin{array}{ccccccc} CH_4(g) & + & 2\,O_2(g) & \rightarrow & CO_2(g) & + & 2H_2O(l) \\ (-74.9) & & 2 \times (0) & & (-393.5) & & 2 \times (-285.8) \end{array}$$

Notice, too, that we have used the enthalpy of liquid water to find the HHV.

The difference between the total enthalpy of the reaction products and the reactants is

$$\begin{aligned} \Delta H &= [(-393.5) + 2 \times (-285.8)] - [(-74.9) + 2 \times (0)] \\ &= -890.2 \text{ kJ/mol of } CH_4 \end{aligned}$$

Since the result is negative, heat is released during combustion; that is, it is exothermic. The HHV is the absolute value of ΔH, which is 890.2 kJ/mol.

Since there are $12.011 + 4 \times 1.008 = 16.043$ g/mol of CH_4, the HHV can also be written as

$$\text{HHV} = \frac{890.2 \text{ kJ/mol}}{16.043 \text{ g/mol}} \times 1000 \text{ g/kg} = 55{,}490 \text{ kJ/kg}$$

4.6.4 Entropy and the Theoretical Efficiency of Fuel Cells

While the enthalpy change tells us how much energy is released in a fuel cell, it doesn't tell us how much of that energy can be converted directly into electricity. To figure that out, we need to review another thermodynamic quantity, *entropy*. Entropy has already been introduced in the context of heat engines in Section 3.4.2, where it was used to help develop the Carnot efficiency limit. In a similar fashion, the concept of entropy will help us develop the maximum efficiency of a fuel cell.

To begin, let us note that all energy is not created equal. That is, for example, 1 joule of energy in the form of electricity or mechanical work is much more useful than a joule of heat. We can convert that joule of electricity or work into heat with 100% conversion efficiency, but we cannot get back the joule of electricity or work from just a single joule of heat. What this suggests is that there is a hierarchy of energy forms, with some being "better" than others. Electricity and mechanical energy (doing work) are of the highest quality. In theory we could go back and forth between electricity and mechanical work with 100% conversion efficiency. Heat energy is of much lower quality, with low-temperature heat being of lower quality than high-temperature heat. So, where does chemical energy fit in this scheme? It is better than thermal, but worse than mechanical and electrical. Entropy will help us figure out just where it stands.

Recall that when an amount of heat Q is removed from a thermal reservoir large enough that its temperature T does not change during the process (i.e., the process is isothermal), the loss of entropy ΔS from the reservoir is defined to be

$$\Delta S = \frac{Q}{T} \tag{4.24}$$

With Q measured in kilojoules (kJ) and T in kelvins ($K = {}^\circ\mathrm{C} + 273.15$), the units of entropy are kJ/K. Recall, too, that entropy is only associated with heat transfer and that electrical or mechanical work is perfect so that these forms have zero entropy. And, finally, remember that in any real system, if we carefully add up all of the entropy changes, the second law of thermodynamics requires that there be an overall increase in entropy. Now let us apply these ideas to a fuel cell.

Consider Fig. 4.28, which shows a fuel cell converting chemical energy into electricity and waste heat. The fuel cell reactions (4.22) and (4.23) are exothermic, which means that their enthalpy changes ΔH are negative. Working with negative quantities leads to awkward nomenclature, which we can avoid by saying that those reactions act as a source of enthalpy H that can be converted to heat and work as Fig. 4.28 implies.

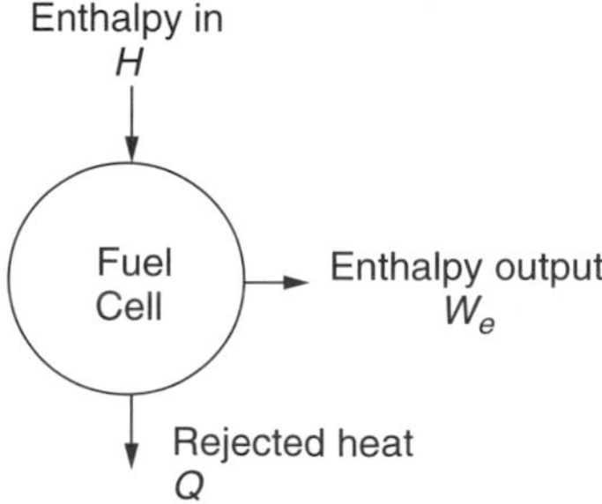

Figure 4.28 The energy balance for a fuel cell.

The cell generates an amount of electricity W_e and rejects an amount of thermal energy Q to its environment. Since there is heat transfer and it is a real system, there must be an increase in entropy. We can use that requirement to determine the minimum amount of rejected heat and therefore the maximum amount of electric power that the fuel cell will generate. To do so, we need to carefully tabulate the entropy changes occurring in the cell:

$$H_2 + \tfrac{1}{2}O_2 \rightarrow H_2O + Q \tag{4.25}$$

where we have included the fact that heat Q will be released. The entropy of the reactants H_2 and O_2 will disappear, but new entropy will appear in the H_2O that is formed plus the entropy that appears in the form of heat Q. As long as the process is isothermal, which is a reasonable assumption for a fuel cell, we can write the entropy appearing in the rejected heat as

$$\Delta S = \frac{Q}{T} \tag{4.26}$$

What about the entropy associated with the work done, W_e? Since there is no heat transfer in electrical (or mechanical) work, that entropy is zero.

To make the necessary tabulation, we need values of the entropy of the reactants and products. And, as usual, we need to define reference conditions. It is conventional practice to declare that the entropy of a pure crystalline substance at zero absolute temperature is zero (the "third law of thermodynamics"). The entropy of a substance under other conditions, referenced to the zero base conditions, is called the *absolute entropy* of that substance, and those values are tabulated in a number of places. Table 4.6 gives absolute entropy values, S°, for several substances under STP conditions (25°C, 1 atm).

The second law of thermodynamics requires that in a real fuel cell there be a net increase in entropy (an ideal cell will release just enough heat to make the increase in entropy be zero). Therefore, we can write that the entropy that shows up in the rejected heat and the product water (liquid water) must be greater than the entropy contained in the reactants (H_2 and O_2):

$$\text{Entropy gain} \geq \text{Entropy loss} \tag{4.27}$$

$$\frac{Q}{T} + \sum S_{\text{products}} \geq \sum S_{\text{reactants}} \tag{4.28}$$

which leads to

$$Q \geq T\left(\sum S_{\text{reactants}} - \sum S_{\text{products}}\right) \tag{4.29}$$

Equation (4.29) tells us the minimum amount of heat that must appear in the fuel cell. That is, we cannot convert all of the fuel's energy into electricity—we are stuck with some thermal losses. At least our thermal losses are going to be less than if we tried to generate electricity with a heat engine.

We can now easily determine the maximum efficiency of the fuel cell. From Fig. 4.28, the enthalpy supplied by the chemical reaction H equals the electricity produced W_e plus the heat rejected Q:

$$H = W_e + Q \tag{4.30}$$

Since it is the electrical output that we want, we can write the fuel cell's efficiency as

$$\eta = \frac{W_e}{H} = \frac{H - Q}{H} = 1 - \frac{Q}{H} \tag{4.31}$$

To find the maximum efficiency, all we need to do is plug in the theoretical minimum amount of heat Q from (4.29).

Example 4.9 Minimum Heat Released from a Fuel Cell. Suppose a fuel cell that operates at 25°C (298 K) and 1 atm forms liquid water (that is, we are working with the HHV of the hydrogen fuel):

$$H_2 + \tfrac{1}{2}O_2 \rightarrow H_2O(l) \qquad \Delta H = -285.8 \text{ kJ/mol of } H_2$$

a. Find the minimum amount of heat rejected per mole of H_2.
b. What is the maximum efficiency of the fuel cell?

Solution

a. From the reaction, 1 mole of H_2 reacts with 1/2 mole of O_2 to produce 1 mole of liquid H_2O. The loss of entropy by the reactants per mole of H_2 is found using values given in Table 4.6:

$$\begin{aligned}\sum S_{\text{reactants}} &= 0.130 \text{ kJ/mol-K} \times 1 \text{ mol } H_2 \\ &\quad + 0.205 \text{ kJ/mol-K} \times 0.5 \text{ mol } O_2 \\ &= 0.2325 \text{ kJ/K}\end{aligned}$$

The gain in entropy in the product water is

$$\sum S_{\text{product}} = 0.0699 \text{ kJ/mol-K} \times 1 \text{ mol } H_2O(l) = 0.0699 \text{ kJ/K}$$

From (4.29), the minimum amount of heat released during the reaction is therefore

$$\begin{aligned}Q_{\text{min}} &= T\left(\sum S_{\text{reactants}} - \sum S_{\text{products}}\right) \\ &= 298 \text{ K } (0.2325 - 0.0699) \text{ kJ/K} = 48.45 \text{ kJ per mole } H_2\end{aligned}$$

b. From (4.22), the enthalpy made available during the formation of liquid water from H_2 and O_2 is $H = 285.8$ kJ/mol of H_2.
The maximum efficiency possible occurs when Q is a minimum; thus from (4.31)

$$\eta_{max} = 1 - \frac{Q_{min}}{H} = 1 - \frac{48.45}{285.8} = 0.830 = 83.0\%$$

4.6.5 Gibbs Free Energy and Fuel Cell Efficiency

The chemical energy released in a reaction can be thought of as consisting of two parts: an entropy-free part, called *free energy* ΔG, that can be converted directly into electrical or mechanical work, plus a part that must appear as heat Q. The "G" in free energy is in honor of Josiah Willard Gibbs (1839–1903), who first described its usefulness, and the quantity is usually referred to as Gibbs free energy. The free energy G is the enthalpy H created by the chemical reaction, minus the heat that must be liberated, $Q = T\ \Delta S$, to satisfy the second law.

The Gibbs free energy ΔG corresponds to the maximum possible, entropy-free, electrical (or mechanical) output from a chemical reaction. It can be found at STP using Table 4.6 by taking the difference between the sum of the Gibbs energies of the reactants and the products:

$$\Delta G = \sum G_{\text{products}} - \sum G_{\text{reactants}} \tag{4.32}$$

This means that the maximum possible efficiency is just the ratio of the Gibbs free energy to the enthalpy change ΔH in the chemical reaction

$$\eta_{max} = \frac{\Delta G}{\Delta H} \tag{4.33}$$

Example 4.10 Maximum Fuel Cell Efficiency Using Gibbs Free Energy. What is the maximum efficiency at STP of a proton-exchange-membrane (PEM) fuel cell based on the higher heating value (HHV) of hydrogen?

Solution. The HHV corresponds to liberated water in the liquid state so that the appropriate reaction is

$$H_2 + \tfrac{1}{2}O_2 \rightarrow H_2O(l) \qquad \Delta H = -285.8 \text{ kJ/mol of } H_2$$

From Table 4.6 the Gibbs free energy of the reactants H_2 and of O_2 are both zero, and that of the product, liquid water, is -237.2 kJ. Therefore, from 4.32)

$$\Delta G = -237.2 - (0 + 0) = -237.2 \text{ kJ/mol}$$

So, from (4.33),

$$\eta_{\max} = \frac{\Delta G}{\Delta H} = \frac{-237.2 \text{ kJ/mol}}{-285.8 \text{ kJ/mol}} = 0.830 = 83.0\%$$

This is the same answer that we found in Example 4.9 using entropy.

4.6.6 Electrical Output of an Ideal Cell

The Gibbs free energy ΔG is the maximum possible amount of work or electricity that a fuel cell can deliver. Since work and electricity can be converted back and forth without loss, they are referred to as reversible forms of energy. For an ideal hydrogen fuel cell, the maximum possible electrical output is therefore equal to the magnitude of ΔG. For a fuel cell producing liquid water, this makes the maximum electrical output at STP equal to

$$W_e = |\Delta G| = 237.2 \text{ kJ per mol of } H_2 \tag{4.34}$$

To use (4.34) we just have to adjust the units so that the electrical output W_e will have the conventional electrical units of volts, amps, and watts. To do so, let us introduce the following nomenclature along with appropriate physical constants:

$$\begin{aligned}
q &= \text{charge on an electron} = 1.602 \times 10^{-19} \text{ coulombs} \\
N &= \text{Avogadro's number} = 6.022 \times 10^{23} \text{ molecules/mol} \\
v &= \text{volume of 1 mole of ideal gas at STP} = 22.4 \text{ liter/mol} \\
n &= \text{rate of flow of hydrogen into the cell (mol/s)} \\
I &= \text{current (A), where 1 A} = 1 \text{ coulomb/s} \\
V_R &= \text{ideal (reversible) voltage across the two electrodes (volts)} \\
P &= \text{electrical power delivered (W)}
\end{aligned}$$

For each mole of H_2 into an ideal fuel cell, two electrons will pass through the electrical load (see Fig. 4.26). We can therefore write that the current flowing through the load will be

$$\begin{aligned}
I(\text{A}) &= n\left(\frac{\text{mol}}{\text{s}}\right) \cdot 6.022 \times 10^{23}\left(\frac{\text{molecules } H_2}{\text{mol}}\right) \cdot \frac{2 \text{ electrons}}{\text{molecule } H_2} \cdot 1.602 \\
&\quad \times 10^{-19}\left(\frac{\text{coulombs}}{\text{electron}}\right) \\
I(\text{A}) &= 192,945n
\end{aligned} \tag{4.35}$$

Using (4.34), the ideal power (watts) delivered to the load will be the 237.2 kJ/mol of H_2 times the rate of hydrogen use:

$$P(\text{W}) = 237.2(\text{kJ/mol}) \times n(\text{mol/s}) \times 1000(\text{J/kJ}) \cdot \frac{1\ \text{W}}{\text{J/s}} = 237{,}200n \qquad (4.36)$$

And the reversible voltage produced across the terminals of this ideal fuel cell will be

$$V_R = \frac{P(\text{W})}{I(\text{A})} = \frac{237{,}200n}{192{,}945n} = 1.229\ V \qquad (4.37)$$

Notice the voltage does not depend on the input rate of hydrogen. It should also be noted that the ideal voltage drops with increasing temperature, so that at the more realistic operating temperature of a PEM cell of about 80°C, V_R is closer to 1.18 V.

We can now easily find the hydrogen that needs to be supplied to this ideal fuel cell per kWh of electricity generated.

$$\text{Hydrogen rate} = \frac{n(\text{mol/s}) \times 2(\text{g/mol}) \times 3600\ \text{s/h}}{237{,}200n(\text{W}) \times 10^{-3}(\text{kW/W})} = 30.35\ \text{g}\,\text{H}_2/\text{kWh} \qquad (4.38)$$

4.6.7 Electrical Characteristics of Real Fuel Cells

Just as real heat engines don't perform nearly as well as a perfect Carnot engine, real fuel cells don't deliver the full Gibbs free energy either. *Activation* losses result from the energy required by the catalysts to initiate the reactions. The relatively slow speed of reactions at the cathode, where oxygen combines with protons and electrons to form water, tends to limit fuel cell power. *Ohmic* losses result from current passing through the internal resistance posed by the electrolyte membrane, electrodes, and various interconnections in the cell. Another loss, referred to as *fuel crossover*, results from fuel passing through the electrolyte without releasing its electrons to the external circuit. And finally, *mass transport* losses result when hydrogen and oxygen gases have difficulty reaching the electrodes. This is especially true at the cathode if water is allowed to build up, clogging the catalyst. For these and other reasons, real fuel cells, in general, generate only about 60–70% of the theoretical maximum.

Figure 4.29 shows the relationship between current and voltage for a typical fuel cell (photovoltaic I–V curves bear a striking resemblance to those for a fuel cell). Notice that the voltage at zero current, called the *open-circuit voltage*, is a little less than 1 V, which is about 25% lower than the theoretical value of 1.229 V. Also shown is the product of voltage and current, which is power. Since power at zero current, or at zero voltage, is zero, there must be a point somewhere in between at which power is a maximum. As shown in the figure, that maximum corresponds to operation of the fuel cell at between 0.4

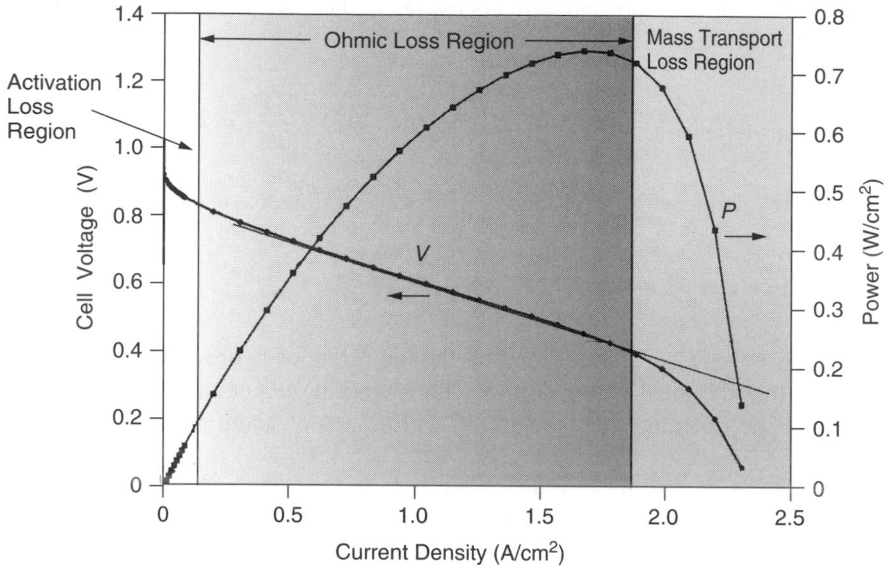

Figure 4.29 The voltage–current curve for a typical fuel cell. Also shown is the power delivered, which is the product of voltage and current.

and 0.5 V per cell. The three regions shown on the graph point out the ranges of currents in which activation, ohmic, and mass-transport losses are individually most important.

Over most of the length of the fuel cell $I-V$ graph, voltage drops linearly as current increases. This suggests a simple equivalent circuit consisting of a voltage source in series with some internal resistance. Fitting the $I-V$ curve in the ohmic region for the fuel cell shown in Fig. 4.29 yields the following approximate relationship:

$$V = 0.85 - 0.25J = 0.85 - \frac{0.25}{A}I \tag{4.39}$$

where A is cell area (cm^2), I is current (amps), and J is current density (A/cm^2).

Example 4.11 Rough Parameters of a Home-Scale Fuel Cell Stack. A 1-kW fuel cell operating on a continuous basis would provide all of the electrical needs of a typical U.S. house. If such a fuel cell stack generates 48 V dc with cells operating at 0.6 V each, how many cells of the type described by (4.39) would be needed and what should be the membrane area of each cell?

Solution. With 0.6-V cells all wired in series, $48/0.6 = 80$ cells would be needed to generate 48 V dc. The current that needs to flow through each cell is

$$I = \frac{P}{V} = \frac{1000 \text{ W/80 cells}}{0.6 \text{ V/cell}} = 20.83 \text{ A}$$

Using (4.39) to find the area of each cell yields

$$0.6 = 0.85 - \frac{0.25}{A} \times 20.83 \qquad A = 20.83 \text{ cm}^2$$

4.6.8 Types of Fuel Cells

To this point in this chapter, the fuel cell reactions and explanations have been based on the assumption that hydrogen H_2 is the fuel, Eq. (4.18) and (4.19) describe the reactions, and the electrolyte passes protons from anode to cathode through a membrane. While it is true that these are the most likely candidates for vehicles and small, stationary power systems, there are competing technologies that use other electrolytes and which have other distinctive characteristics that may make them more suitable in some applications.

Proton Exchange Membrane Fuel Cells (PEMFC) Originally known as *Solid Polymer Electrolyte* (SPE) fuel cells, and sometimes now called *polymer electrolyte membrane* fuel cells, these are the furthest along in their development, in part because of the early stimulus provided by the Gemini space program, and especially now since they are the leading candidates for use in hybrid electric vehicles (HEVs). Their efficiencies are the highest available at around 45% (HHV). Currently operating units range in size from 30 W to 250 kW.

PEM cells generate over 0.5 W/cm^2 of membrane at around 0.65 V per cell and a current density of 1 A/cm^2. To control water evaporation from the membranes, these cells require active cooling to keep temperatures in the desired operating range of 50°C to 80°C. With such low temperatures, waste heat cogeneration is restricted to simple water or space heating applications, which is fine for residential power systems. One limitation of PEM cells is their need for very pure hydrogen as their fuel source. Hydrogen reformed from hydrocarbon fuels such as methanol (CH_3OH) or methane CH_4 often contains carbon monoxide (CO), which can lead to CO poisoning of the catalyst. When CO adsorbs onto the surface of the anode catalyst, it diminishes the availability of sites where the hydrogen reactions need to take place. Minimizing CO poisoning, managing water and heat in the cell stack, and developing lower-cost materials and manufacturing techniques are current challenges for PEM cells.

Direct Methanol Fuel Cells (DMFC) These cells use the same polymer electrolytes as PEM cells do, but they offer the significant advantage of being able to utilize a liquid fuel, methanol (CH_3OH), instead of gaseous hydrogen. Liquid fuels are much more convenient for portable applications such as motor vehicles as well as small, portable power sources for everything from cell phones and lap-top computers to replacements for diesel-engine generators.

The chemical reactions taking place at the anode and cathode are as follows:

$$CH_3OH + H_2O \rightarrow CO_2 + 6H^+ + 6e^- \qquad \text{(Anode)} \tag{4.40}$$

$$\frac{1}{2}O_2 + 2H^+ + 2e^- \rightarrow H_2O \qquad \text{(Cathode)} \tag{4.41}$$

for an overall reaction of

$$CH_3OH + \frac{3}{2}O_2 \rightarrow CO_2 + 2H_2O \qquad \text{(Overall)} \tag{4.42}$$

Significant technical challenges remain, including control of excessive fuel crossover through the membrane concern for methanol toxicity, and reducing catalyst poisoning by CO and other methanol reaction products. The advantages of portability and simplified fuel handling, however, almost guarantee that these will be commercially available in the very near future.

Phosphoric Acid Fuel Cells (PAFC) These fuel cells were introduced into the marketplace in the 1990s, and there are hundreds of 200-kW units built by the ONSI division of IFC currently in operation. Their operating temperature is higher than that of PEMFCs (close to 200°C), which makes the waste heat more usable for absorption air conditioning as well as water and space heating in buildings.

The electrochemical reactions taking place in a PAFC are the same ones that occur in a PEM cell, but the electrolyte in this case is phosphoric acid rather than a proton exchange membrane. These cells tolerate CO better than PEM cells, but they are quite sensitive to H_2S. Although there are already a number of PAFCs in use, their future will be closely tied to whether higher production levels will be able to reduce manufacturing costs to the point where they will be competitive with other cogeneration technologies.

Alkaline Fuel Cells (AFC) These highly efficient and reliable fuel cells were developed for the Apollo and Space Shuttle programs. Their electrolyte is potassium hydroxide (KOH), and the charge carrier is OH^- rather than H^+ ions. The electrochemical reactions are as follows:

$$H_2 + 2\ OH^- \rightarrow 2H_2O + 2e^- \qquad \text{(Anode)} \tag{4.43}$$

$$\frac{1}{2}O_2 + H_2O + 2e^- \rightarrow 2OH^- \qquad \text{(Cathode)} \tag{4.44}$$

The major problem with alkaline fuel cells is their intolerance for exposure to CO_2, even at the low levels found in the atmosphere. Since air is the source of O_2 for the cathodic reactions, it is unlikely that these will be used in terrestrial applications.

Molten-Carbonate Fuel Cells (MCFC) These fuel cells operate at very high temperatures, on the order of 650°C, which means that the waste heat is of high enough quality that it can be used to generate additional power in accompanying steam or gas turbines. At this high temperature, there is the potential for the fuel cell waste heat to be used to directly convert, or *reform*, a hydrocarbon fuel, such as methane, into hydrogen by the fuel cell itself. Moreover, the usual accompanying CO in fuel reforming does not poison the catalyst and, in fact, becomes part of the fuel. Efficiencies of 50–55% are projected for internal-reforming MCFCs. With combined-cycle operation, electrical efficiencies of 65% are projected, and cogeneration efficiencies of over 80% are possible.

In an MCFC the conducting ion is carbonate $CO_3{}^{2-}$ rather than H^+, and the electrolyte is molten lithium, potassium, or sodium carbonate. At the cathode, CO_2 and O_2 combine to form carbonate ions, which are conducted through the electrolyte to the anode where they combine with hydrogen to form water and carbon dioxide as shown in the following electrochemical reactions:

$$H_2 + CO_3{}^{2-} \rightarrow H_2O + CO_2 + 2e^- \qquad \text{(Anode)} \tag{4.45}$$

$$\frac{1}{2}O_2 + CO_2 + 2e^- \rightarrow CO_3{}^{2-} \qquad \text{(Cathode)} \tag{4.46}$$

Notice the overall reaction is the same as that described earlier for a "generic" fuel cell

$$H_2 + \frac{1}{2}O_2 \rightarrow H_2O \qquad \text{(Overall)} \tag{4.47}$$

MCFCs operate in a very corrosive environment, and the challenges associated with devising appropriate materials that will operate with suitably long lifetimes are significant.

Solid Oxide Fuel Cells (SOFC) SOFCs and MCFCs are competing for the future large power station market. Both operate at such high temperatures (MCFC, 650°C; SOFC, 750–1000°C) that their waste heat can be used for combined-cycle steam or combined cycle gas turbines, and both can take advantage of those temperatures to do internal fuel reforming. The SOFC is physically smaller than an MCFC for the same power, and it may ultimately have greater longevity.

The electrolyte in an SOFC is a solid ceramic material made of zirconia and yttria, which is very unlike the liquids and solid polymers used in every other type of fuel cell. The charge carrier that is transported across the electrolyte is the oxide O^{2-} ion, which is formed at the cathode when oxygen combines with electrons from the anode. At the anode, the oxide ion combines with hydrogen to form water and electrons, as shown below:

$$H_2 + O^{2-} \rightarrow H_2O + 2e^- \qquad \text{(Anode)} \tag{4.48}$$

$$\frac{1}{2}O_2 + 2e^- \rightarrow O^{2-} \qquad \text{(Cathode)} \tag{4.49}$$

Efficiencies for SOFCs of 60% for electric power and greater than 80% for cogeneration are projected. When combined with a gas turbine, such as is suggested in Fig. 4.30, electrical efficiencies approaching 70% (LHV) may be achievable.

A summary of the main characteristics of these various categories of fuel cells is presented in Table 4.7.

4.6.9 Hydrogen Production

With the exception of DMFCs, fuel cells require a source of hydrogen H_2 for the anodic reactions. For those that operate at higher temperatures (MCFCs and SOFCs), methane may be reformed to yield hydrogen as part of the fuel cell system itself; but in general, obtaining a supply of hydrogen of sufficient purity and at a reasonable cost is a major hurdle that must be dealt with before large-scale commercialization of fuel cells will be achieved.

Hydrogen as a fuel has many desirable attributes. When burned, it yields only small amounts of NO_X created when combustion temperatures are high enough to cause the nitrogen and oxygen in air to combine, and when used in fuel cells, the only end product is water. Given its low density, it readily escapes from confined environments so that it is less likely to concentrate into dangerous pools the way that gasoline fumes, for example, do. It is, however, not an energy source. It is,

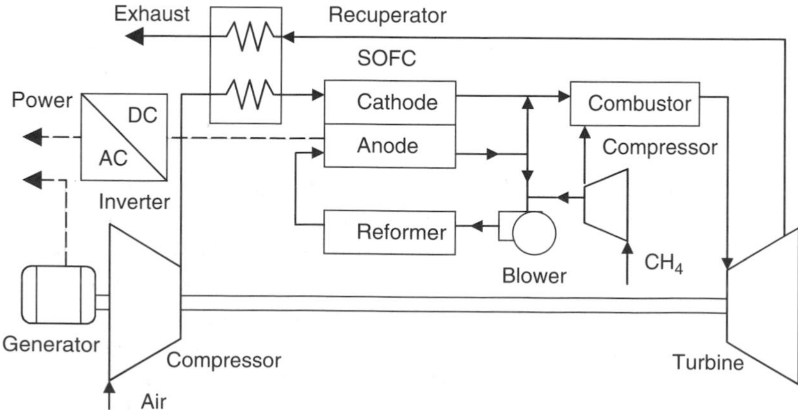

Figure 4.30 Gas turbines with pressurized solid-oxide fuel cells may be capable of nearly 70% LHV efficiency.

TABLE 4.7 Essential Characteristics of Different Types of Fuel Cells

	Type of Fuel Cell[a]					
Characteristic	PEMFC	DMFC	AFC	PAFC	MCFC	SOFC
Electrolyte	Proton exchange membrane	Proton exchange membrane	Potassium hydroxide (8–12N)	Phosphoric acid (85–100%)	Molten carbonates (Li, K, Na)	Solid oxide (ZrO_2–Y_2O_3)
Operating temperature	50–90°C	50–90°C (≤ 130°C)	50–250°C	180–200°C	650°C	750–1050°C
Charge carrier	H^+	H^+	OH^-	H^+	CO_3^{2-}	O^{2-}
Electrocatalyst	Pt	Pt	Pt, Ni/NiO_x	Pt	Ni/$LiNiO_x$	Ni/Perovskites
Fuel	H_2 (pure or reformed)	CH_3OH	H_2	H_2 (reformed)	H_2 and CO reformed and CH_4	H_2 and CO reformed and CH_4
Poison	CO > 10 ppm	Adsorbed intermediates	CO, CO_2	CO > 1% H_2S > 50 ppm	H_2S > 0.5 ppm	H_2S > 1 ppm
Applications	Portable, transportation	Portable, transportation	Space	Power generation, cogeneration, transportation	Power generation, cogeneration	Power generation, cogeneration

[a]PEMFC, proton exchange membrane fuel cell; DMFC, direct methanol fuel cell; AFC, alkaline fuel cell; PAFC, phosphoric acid fuel cell; MCFC, molten-carbonate fuel cell; SOFC, solid-oxide fuel cell.

Source: Srinivasan et al. (1999).

like electricity, a high-quality energy carrier that is not naturally available in the environment. It must be manufactured, which means an energy investment must be made to create the desired hydrogen fuel.

The main technologies currently in use for hydrogen production are steam reforming of methane (SMR), partial oxidation (POX), and electrolysis of water. More exotic methods of production in the future may include photocatalytic, photoelectrochemical, or biological production of hydrogen using sunlight as the energy source.

Methane Steam Reforming (MSR) About 5% of U.S. natural gas is already converted to hydrogen for use in ammonia production, oil refining, and a variety of other chemical processes. Almost all of that is done with steam methane reformers. After some gas cleanup, especially to remove sulfur, a mixture of natural gas and steam is passed through a catalyst at very high temperature (700–850°C), producing a synthesis gas, or syngas, consisting of CO and H_2:

$$CH_4 + H_2O \rightarrow CO + 3H_2 \tag{4.50}$$

The above reaction is endothermic; that is, heat must be added, which may be provided in part by burning some of the methane as fuel.

The hydrogen concentration in the syngas is then increased using a water-gas shift reaction:

$$CO + H_2O \rightarrow CO_2 + H_2 \tag{4.51}$$

This reaction is exothermic, which means some of the heat released can be used to drive (4.50). The resulting syngas in (4.51) is 70–80% H_2, with most of the remainder being CO_2 plus small quantities of CO, H_2O, and CH_4. Final processing includes removal of CO_2 and conversion of remaining CO into methane in a reverse of reaction (4.50). The overall energy efficiency of SMR hydrogen production is typically 75–80%, but higher levels are achievable.

Partial Oxidation (POX) These systems are based on methane (or other hydrocarbon fuels) being partially oxidized in the following exothermic reaction:

$$CH_4 + \frac{1}{2}O_2 \rightarrow CO + 2H_2 \tag{4.52}$$

Since (4.52) is exothermic, it produces its own heat, which makes it potentially simpler than the MSR process since it can eliminate the MSR heat exchanger required to transfer heat from (4.51) to (4.50). After the partial oxidation step, a conventional shift reaction can be used to concentrate the H_2 in the resulting syngas.

Gasification of Biomass, Coal, or Wastes As mentioned in Section 4.4, gasification of biomass or other solid fuels such as coal or municipal wastes by high-temperature pyrolysis can be used to produce hydrogen. In fact, that was the primary method of hydrogen production before natural gas become so widely available. With the likelihood of relatively inexpensive technology to remove CO_2 from the resulting syngas, there is growing interest in coal gasification for hydrogen production, followed by capture and sequestration of CO_2 in deep saline aquifers or depleted gas fields. Some researchers hope such carbon sequestration may provide a way to continue to exploit the earth's large coal resources with minimal carbon emissions.

Electrolysis of Water In reactions that are the reverse of conventional fuel cells, electrical current forced through an electrolyte can be used to break apart water molecules, releasing hydrogen and oxygen gases:

$$2H_2O \rightarrow 2H_2 + O_2 \tag{4.53}$$

In fact, the same membranes that are used in PEM cells can be used in low-temperature electrolyzers. Similarly, solid-oxide electrolytes can be used for high-temperature electrolysis.

A sketch of an electrolysis cell that uses a proton exchange membrane is shown in Fig. 4.31. De-ionized water introduced into the oxygen side of the cell dissociates into protons, electrons, and oxygen. The oxygen is liberated, the protons pass through the membrane, and the electrons take the external path through the power source to reach the cathode where they reunite with protons to form hydrogen gas. Overall efficiency can be as high as 85%.

Hydrogen produced by electrolysis has the advantage of being highly purified, so the problems of catalytic CO poisoning that some fuel cells are subject to is not a concern. When the electricity for electrolysis is generated using a renewable energy system, such as wind, hydro, or photovoltaic power, hydrogen is produced without emission of any greenhouse gases. And, as Fig. 4.32 suggests, when the resulting hydrogen is subsequently converted back to electricity using fuel cells,

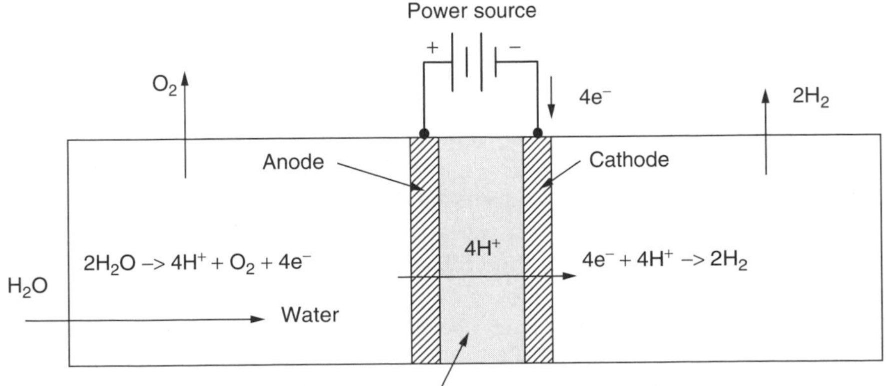

Figure 4.31 A proton exchange membrane used to electrolyze water.

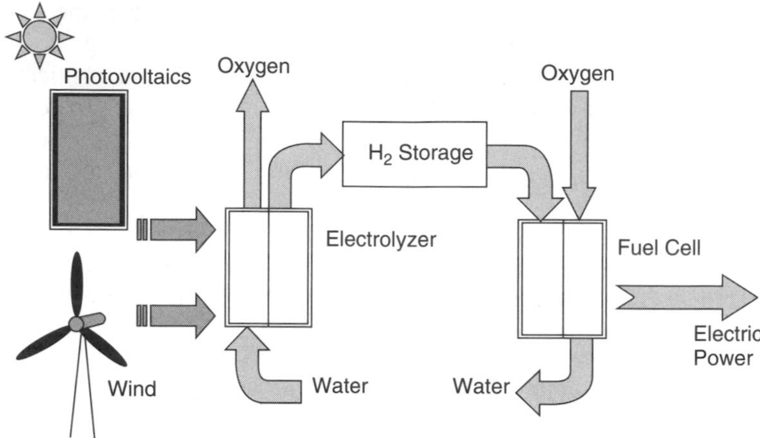

Figure 4.32 Renewable energy sources coupled with fuel cells can provide electric power where and when it is required, cleanly, and sustainably.

the ultimate goals of carbon-free electricity, available whenever it is needed, whether or not the sun is shining or the wind is blowing, without depleting scarce nonrenewable resources, can become an achievable reality.

REFERENCES

Babcock and Wilcox (1992). *Steam*, 40th ed., Babcock & Wilcox, Barberton, OH.

Cler, G. I., and M. Shepard (1996). *Distributed Generation: Good Things Are Coming in Small Packages*, Esource Tech Update, TU-96-12, Esource, Boulder, CO.

Davenport, R. L., B. Butler, R. Taylor, R. Forristall, S. Johansson, J. Ulrich, C. VanAusdal, and J. Mayette (2002). Operation of Second-Generation Dish/Stirling Power Systems, *Proceedings of the 31st ASES Annual Conference*, June 15–20.

Grove, W. R. (1839). On Voltaic Series in Combinations of Gases by Platinum, *Philosophical Magazine*, vol. 14, pp. 127–130.

Inversin, A. R. (1986). *Micro-Hydropower Sourcebook*, National Rural Electrification Cooperative Association, Arlington, VA.

McNeely, M. (2003). ARES—Gas Engines for Today and Beyond, *Diesel and Gas Turbine Worldwide*, May, pp. 1–6.

Mond, L. L., and C. Langer (1890). Proceedings of the Royal Society (London), Services, A, vol. 46, pp. 296–304.

Petchers, N. (2002). *Combined Heating, Cooling and Power*: Technologies and Applications, The Fairmont Press, Lilburn, GA.

Price, H., E. Lupfert, D. Kearney, E. Zarra, G. Cohen, R. Gee, and R. Mahoney (2002). Advances in Parabolic Trough Solar Power Technology, *Journal of Solar Engineering*, May, pp. 109–125.

Srinivasan, S., R. Mosdale, P. Stevens, and C. Yang (1999). Fuel Cells: Reaching the Era of Clean and Efficient Power Generation in the Twenty-First Century. *Annual Review of Energy and Environment*, pp. 281–328.

PROBLEMS

4.1 A natural-gas-fired microturbine has an overall efficiency of 26% when expressed on an LHV basis. Using data from Table 4.2, find the efficiency expressed on an HHV basis.

4.2 On an HHV basis, a 600-MW coal-fired power plant has a heat rate of 9700 Btu/kWh. The particular coal being burned has an LHV of 5957 Btu/lbm and an HHV of 6440 Btu/lbm.

a. What is its HHV efficiency?

b. What is its LHV efficiency?

c. At what rate will coal have to be supplied to the plant (tons/hr)?

4.3 A natural-gas fueled, 250-kW, solid-oxide fuel cell with a heat rate of 7260 Btu/kWh costs $1.5 million. In its cogeneration mode, 300,000 Btu/hr of exhaust heat is recovered, displacing the need for heat that would have been provided from a 75% efficient gas-fired boiler. Natural gas costs $5 per million Btu and electricity purchased from the utility costs $0.10/kWh. The system operates with a capacity factor of 80%.

a. What is the value of the fuel saved by the waste heat ($/yr)?

b. What is the savings associated with not having to purchase utility electricity ($/yr)?

c. What is the annual cost of natural gas for the CHP system?

d. With annual O&M costs equal to 2% of the capital cost, what is the net annual savings of the CHP system?

e. What is the simple payback (ratio of initial investment to annual savings)?

4.4 Suppose 200 gpm of water is taken from a creek and delivered through 800 ft of 3-inch diameter PVC pipe to a turbine100 ft lower than the source. If the turbine/generator has an efficiency of 40%, find the electrical power that would be delivered. In a 30-day month, how much energy would be provided?

4.5 The site in Problem 4.4 has a flow rate of 200 gpm, 100-ft elevation change, and 800-ft length of pipe, but there is excessive friction loss in the pipe.

a. What internal pipe diameter would keep flow to less than a recommended speed of 5 ft/sec.

b. Assuming locally available PVC pipe comes in 1-in diameter increments (2-in, 3-in, etc), pick a pipe size closest to the recommended diameter.

c. Find the kWh/month delivered by the 40% efficient turbine/generator.

d. With a 4-nozzle pelton wheel, what diameter jets would be appropriate?

4.6 The rectangular weir flow equation is based on water height h above the notch being at least 2 inches while the notch width must be at least 3 h. Design a notch (width and height) that will be able to measure the maximum flow when the minimum flow is estimated to be 200 gpm. What maximum flow rate could be accommodated?

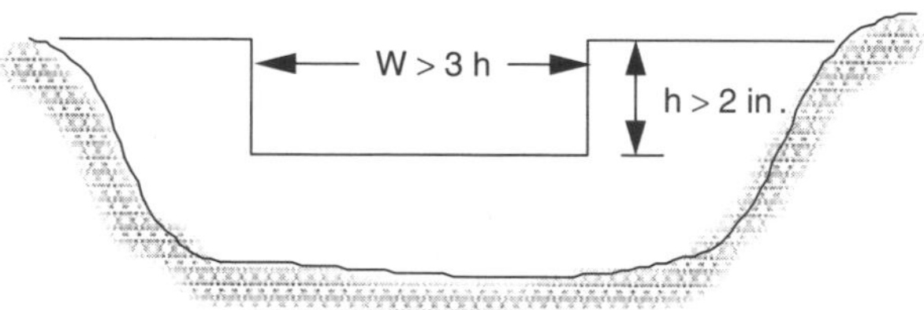

Figure P4.6

4.7 Use enthalpies from Table 4.6 to compute the following heating values (kJ/kg) when they are oxidized to CO_2 and H_2O:

a. LHV of H_2 (molecular weight (MW) = 2.016)

b. HHV of H_2

c. LHV of methane, CH_4 (MW = 16.043)

d. LHV of liquid methanol CH_3OH (MW = 32.043)

e. HHV of methanol, CH_3OH (l)

4.8 Suppose a methanol fuel cell forms liquid water during its operation. Assuming everything is at STP conditions:

a. What minimum amount of heat must be rejected in order to satisfy entropy constraints (kJ/mol)?

b. What maximum efficiency could the fuel cell have using the calculation in (b)?

c. What maximum efficiency could the fuel cell have using the Gibbs free energy approach?

4.9 Equation (4.38) indicates that an ideal PEM cell needs 30.35 g H_2 to produce one kWh of electrical output. Suppose an electric car can travel 10 miles per kWh. How much hydrogen would be needed to give the car a range of 300 miles if the source of electricity is an on-board fuel cell with an efficiency that is 50% of the ideal?

4.10 An ideal PEM fuel cell with an efficiency of 83% needs 30.35 g of hydrogen to generate 1 kWh. For a real cell with half that efficiency, how much hydrogen per day would be needed to power a house that uses 25 kWh per day?

CHAPTER 5

ECONOMICS OF DISTRIBUTED RESOURCES

In this chapter, the techniques needed to evaluate the economics of both sides of the electric meter—the demand side and the supply side—will be explored. Some of this is simply applied engineering economy, but the energy systems that we need to evaluate, especially those involving cogeneration, sometimes require special perspectives to adequately describe their economic advantages.

5.1 DISTRIBUTED RESOURCES (DR)

The traditional focus in electric power planning has been on generation resources—forecasting demand, perhaps by extrapolating existing trends, and then trying to select the most cost-effective combination of new power plants to meet that forecast. In the final decades of the twentieth century, however, there was an important shift in which it was recognized that the real need was for energy services—illumination, for example—not raw kilowatt-hours, and a least-cost approach to providing those energy services should include programs to help customers use energy more efficiently. Out of that recognition a process called *integrated resource planning* (IRP) emerged in which both supply-side and demand-side resources were evaluated, including environmental and social costs, to come up with a least-cost plan to meet the wants and needs of customers for energy services. Building new power plants had to compete with helping customers install more efficient lamps and motors.

Renewable and Efficient Electric Power Systems. By Gilbert M. Masters
ISBN 0-471-28060-7

More recently, with increased attention to the electricity grid and the emergence of efficient, cost-effective cogeneration, IRP now recognizes three kinds of electricity resources: *generation resources*, especially the distributed generation (DG) technologies explored in the last chapter; *grid resources*, which move electricity from generators to customers; and *demand-side resources*, which link electricity to energy services. These three *distributed resources* (DR) are all equally valid, comparable resources that need to be evaluated as part of a least-cost planning process. In this chapter, the focus is on distributed resources that are relatively small in scale and located somewhat near the end-user. Examples of such distributed resources are shown in Fig. 5.1. In an astonishing book by Amory Lovins et al. *Small is Profitable: The Hidden Economic Benefits of Making Electrical Resources the Right Size* (2002), over 200 benefits of distributed resources are explored and explained.

Improving efficiency is an important energy resource in every electric power system. On the generation side, efficiency in the context of renewable energy technologies is important because it helps reduce the size and cost of the systems. Other important new technologies, including combined-cycle turbines, microturbines, Stirling engines, and, to some extent, fuel cells, rely on fossil fuels, so efficiency has added importance because it helps extend the lifetime of nonrenewable resources while reducing the political and environmental consequences associated with their mining, processing, and use.

For many renewable and nonrenewable generation systems, their efficiency is greatly enhanced when the cogeneration of electricity and usable waste heat is incorporated into the energy conversion processes. The usability of waste heat, of course, depends on what task is to be accomplished. Low-temperature heat ($\approx$50–80°C) can be used for water heating and perhaps space heating; medium temperatures ($\approx$80–100°C) can be used for space heating and air conditioning; and with temperatures over 100°C, steam can be generated for process heating and other industrial purposes. In some circumstances, cogeneration not only reduces total energy consumption, but can also cut the demand for electric power itself

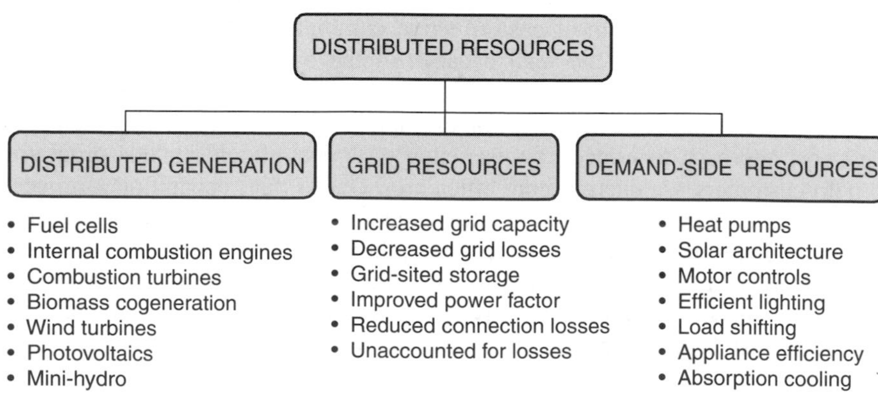

Figure 5.1 Examples of distributed resources. (Based on Lovins et al. (2002).

by allowing fuel rather than electricity to provide the desired energy service. An example is air conditioning that can be shifted from electrically powered, vapor-compression chillers to absorption chillers that run on waste heat.

On the demand side, technologies that use electricity more efficiently—better lamps and fixtures, more efficient motors and controls, heat pumps for water and space heating, and more efficient appliances—are the major energy resource, but finding ways to totally eliminate the need for power is also a resource. When we focus on what we want energy for, rather than how many kilowatt hours are usually needed, important perspectives emerge. Illumination, for example, can be provided by burning coal in a power plant, sending the generated electricity through a transmission and distribution system to the filament of a lightbulb where it is converted mostly into heat plus a little bit of light. In a big building, this heat may have to be removed from the conditioned space, which means burning more coal to generate electricity for the air conditioner. The process can be made more efficient by using fluorescent rather than incandescent lighting; or, better still, the need for artificial light, and the cooling that often accompanies it, can be minimized, or even eliminated entirely, by proper manipulation of natural daylight through well-designed fenestration (windows). Natural daylight, for example, may do more than just reduce the electricity demand for illumination, it may also increase worker productivity the value of which may far exceed the reduction in utility bills.

5.2 ELECTRIC UTILITY RATE STRUCTURES

An essential step in any economic calculation for a distributed resource (DR) project is a careful analysis of the cost of electricity and/or fuel that will be displaced by the proposed system. Our focus in this section is on electric utility rate structures, which are critical factors for customers evaluating a DR project intended to reduce electricity purchases.

Electric rates vary considerably, depending not only on the utility itself, but also on the electrical characteristics of the specific customer purchasing the power. The rate structure for a residential customer will typically include a basic fee to cover costs of billing, meters, and other equipment, plus an energy charge based on the number of kilowatt-hours of energy used. Commercial and industrial customers are usually billed not only for energy (kilowatt-hours) but also for the peak amount of power that they use (kilowatts). That *demand charge* for power ($/mo per kW) is the most important difference between the rate structures designed for small customers versus large ones. Large industrial customers may also pay additional fees if their power factor—that is, the phase angle between the voltage supplied and the current drawn—is outside of certain bounds.

5.2.1 Standard Residential Rates

Consider the example rate structure for a residential customer shown in Table 5.1. Notice that it includes three tiers based on monthly kWh consumed, and notice

TABLE 5.1 Example Standard Residential Electric Rate Schedule for a California Utility

Tier Level	Winter: November–April		Summer: May–October	
Tier I	First 620kWh	7.378¢/kWh	First 700kWh	8.058¢/kWh
Tier II	621–825	12.995¢/kWh	701–1000	13.965¢/kWh
Tier III	Over 825	14.231¢/kWh	Over 1000	15.688¢/kWh

that the rates increase with increasing demand. This is an example of what is called an *inverted block rate structure*, designed to discourage excessive consumption. Not that long ago, the most common structures were based on *declining block rates*, which made electricity cheaper as the customer's demand increased, which of course discouraged conservation. This particular rate structure is for a summer peaking utility (the Sacramento Municipal Utility District, 2003), so the rates increase somewhat in summer to encourage customers to conserve during the peak season.

Example 5.1 Calculating a Simple Residential Utility Bill. Suppose that a customer subject to the rate structure in Table 5.1 uses 1200 kWh/mo during the summer.

a. What would be the total cost of electricity ($/mo, ignoring the monthly service charge)?

b. What would be the value (¢/kWh) of an efficiency project that cuts the demand to 900 kWh/mo?

Solution:

a. The total monthly bill includes 700 kWh @ 8.058¢, 300 kWh @ 13.965¢, and 200 kWh @ 15.688¢, for a total of

$$700 \times \$0.08058 + 300 \times \$0.13965 + 200 \times \$0.15688 = \ \$129.68/\text{mo}$$

b. If the demand is reduced to 900 kWh/mo, the bill would be

$$700 \times \$0.08058 + 200 \times \$0.13965 = \$84.34/\text{mo}$$

The savings per kWh is ($129.68 – $84.34)/300 kWh = $0.1511/kWh

---

TABLE 5.2 Example Residential Time-of-Use (TOU) Rate Schedule

	November–April		May–October	
On-peak	7–10 A.M., 5–8 P.M.	8.335 ¢/kWh	2–8 P.M.	19.793 ¢/kWh
Off-peak	All other times	7.491 ¢/kWh	All other times	8.514 ¢/kWh

5.2.2 Residential Time-of-Use (TOU) Rates

In an effort to encourage customers to shift their loads away from the peak demand times, some utilities are beginning to offer residential time-of-use (TOU) rates. For many utilities the peak demand occurs on hot, summer afternoons when air conditioners are humming (see, for example, Fig. 3.6), so TOU rates in such circumstances charge more for electricity during summer afternoons. Conversely, at night when there is idle capacity, rates may be significantly lower.

Table 5.2 presents an example of a residential TOU rate schedule for a summer-peaking utility. While there is not much time/price differential in the winter, in the summer the TOU rate schedule is very costly during the on-peak times. In this case, however, the off-peak summer rate is still rather high so a careful calculation would need to be made to determine whether the TOU rate or the regular residential rate schedule in Table 5.1 would be most cost-effective for an individual homeowner.

An especially interesting calculation is one for a TOU customer with photovoltaics (PVs) on the roof displacing the need for expensive on-peak utility electricity. Some utilities allow true *net-metering* with TOU rates, in which case monthly net energy consumed or generated is billed or credited to the customer at the applicable TOU rate. The following example shows how this incentive works.

Example 5.2 PVs, TOU Rates, and Net Metering. During the summer a rooftop PV system generates 10 kWh/day during the off-peak hours and 10 kWh/day during the on-peak hours. Suppose too, that the customer uses 2 kWh/day on-peak and 18 kWh/day off-peak. That is, the PVs generate 20 kWh/day and the household consumes 20 kWh/day.

	PV supply	Demand
On-peak	10kWh	2kWh
Off-peak	10kWh	18kWh
Total	20kWh/day	20kWh/day

For a 30-day month in the summer, find the electric bill for this customer if the TOU rates of Table 5.2 apply.

Solution During the on-peak hours, the customer generates 10 kWh and uses 2 kWh, so there would be a credit of

$$\text{On-peak credits} = 8\ \text{kWh/day} \times \$0.19793/\text{kWh} \times 30\ \text{day/mo} = \$47.50$$

During the off-peak hours, the customer generates 10 kWh and uses 18 kWh, so the bill for those hours would be

$$\text{Off-peak bill} = 8\ \text{kWh/day} \times \$0.08514/\text{kWh} \times 30\ \text{day/mo} = \$20.43/\text{mo}$$

So the net bill for the month would be

$$\text{Net bill} = \$20.43 - \$47.50 = -\$27.07\text{mo}$$

That is, the utility would owe the customer $27.07 for this month. Most likely in other months there will be actual bills against which this amount would be credited.

Notice that the bill would have been zero, instead of the $27.07 credit, had this customer elected the standard rate schedule of Table 5.1 instead of the TOU rates.

5.2.3 Demand Charges

The rate structures that apply to commercial and industrial customers usually include a monthly demand charge based on the highest amount of power drawn by the facility. That demand charge may be especially severe if the customer's peak corresponds to the time during which the utility has its maximum demand since at those times the utility is running its most expensive peaking power plants.

In the simplest case, the demand charge is based on the peak demand in a given month, usually averaged over a 15-minute period, no matter what time of day it occurs. An example of a typical rate structure with such a demand charge is given in Table 5.3.

Example 5.3 Impact of Demand Charges. During the summer, a small commercial building that uses 20,000 kWh per month has a peak demand of 100 kW.

a. Compute the monthly bill (ignoring fixed customer charges).
b. How much does the electricity cost for a 100-W computer that is used 6 h a day for 22 days in the month? The computer is turned on during the period when the peak demand is reached for the building. How much is that in ¢/kWh?

TABLE 5.3 Electricity Rate Structure Including Monthly Demand Charges

	Winter Oct–May	Summer June–Sept
Energy charges	\$0.0625/kWh	\$0.0732/kWh
Demand charges	\$7/mo-kW	\$9/mo-kW

Solution

a. The monthly bill is made up of energy and demand charges:

$$\text{Energy charge} = 20{,}000\text{ kWh} \times \$0.0732/\text{kWh} = \$1464/\text{mo}$$

$$\text{Demand charge} = 100\text{ kW} \times \$9/\text{mo-kW} = \$900/\text{mo}$$

For a total of \$1464 + \$900 = \$2364/mo (38% of which is demand)

b. The computer uses 0.10 kW × 6 h/d × 22 day/mo = 13.2 kWh/mo

$$\text{Energy charge} = 13.2\text{ kWh/mo} \times \$0.0732/\text{kWh} = \$0.97/\text{mo}$$

$$\text{Demand charge} = 0.10\text{ kW} \times \$9/\text{mo-kW} = \$0.90/\text{mo}$$

$$\text{Total cost} = \$0.97 + \$0.90 = \$1.87/\text{mo}$$

On a per kilowatt-hour basis, the computer costs

$$\text{Electricity} = \frac{\$1.87/\text{mo}}{13.2\text{ kWh/mo}} = \$0.142/\text{kWh}$$

Notice how the demand charge makes the apparent cost of electricity for the computer (14.2¢/kWh) nearly double the 7.32 ¢/kWh price of electric energy.

5.2.4 Demand Charges with a Ratchet Adjustment

The demand charge in the rate schedule shown in Table 5.3 applies to the peak demand for each particular month in the year. The revenue derived from demand charges that may only be monetarily significant for just one month of the year, however, may not be sufficient for the utility to pay for the peaking power plant they had to build to supply that load. To address that problem, it is common to have a ratchet adjustment built into the demand charges. For example, the monthly demand charges may be ratcheted to a level of perhaps 80% of the annual peak demand. That is, if a customer reaches a highest annual peak demand of 1000 kW, then for every month of the year the demand charge will be based on consumption of at least 0.80×1000 kW = 800 kW. This can lead to some rather extraordinary penalties for customers who add a few kilowatts to their load right at the time of their annual peak; conversely, it provides considerable incentive to reduce their highest peak demand.

Example 5.4 Impact of Ratcheted Demand Charges on an Efficiency Project. A customer's highest demand for power comes in August when it reaches 100 kW. The peak in every other month is less than 70 kW. A proposal to dim the lights for 3 h during each of the 22 workdays in August will reduce the August peak by 10 kW. The utility's energy charge is 8¢/kWh and its demand charge is $9/kW-mo with an 80% ratchet on the demand charges.

a. What is the current annual cost due to demand charges?
b. What annual savings in demand and energy charges will result from dimming the lights?
c. What is the equivalent savings expressed in ¢/kWh?

Solution

a. At $9/kW-mo, the current demand charge in August will be

$$\text{August} = 100\text{ kW} \times \$9/\text{kW-mo} = \$900$$

For the other 11 months, the minimum demand charge will be based on 80 kW, which is higher than the actual demand:

$$\begin{aligned}\text{Sept–July demand charge} &= 0.8 \times 100\text{ kW} \times \$9/\text{kW-mo} \times 11\text{ mo}\\ &= \$7920\end{aligned}$$

So the total annual demand charge will be

$$\text{Annual} = \$900 + \$7920 = \$8820$$

b. By reducing the August demand by 10 kW, the annual demand charges will now be

$$\text{August} = 90\text{ kW} \times \$9/\text{kW-mo} = \$810$$

$$\text{Sept–July} = 0.8 \times 90\text{ kW} \times \$9/\text{kW-mo} \times 11\text{ mo} = \$7128$$

$$\text{Total annual demand charge} = \$810 + \$7128 = \$7938$$

$$\text{Annual demand savings} = \$8820 - \$7938 = \$882$$

$$\text{August energy savings} = 3\ \text{h/d} \times 10\ \text{kW} \times 22\ \text{days} \times \$0.08/\text{kWh}$$
$$= \$52.80$$
$$\text{Total Annual Savings} = \$882 + \$52.80 = \$934.80$$

Notice that the demand savings is 94.4% of the total savings!

c. Dimming the lights saved 3 h/d × 10 kW × 22 d = 660 kWh and $934.80, which on a per kWh basis is

$$\text{Savings} = \frac{\$934.80}{660\ \text{kWh}} = \$1.42/\text{kWh}$$

In other words, the business saves $1.42 for each kWh that it saves, which is about 18 times more than would be expected if just the $0.08/kWh cost of energy is considered.

5.2.5 Load Factor

The ratio of a customer's average power demand to its peak demand, called the *load factor*, is a useful way for utilities to characterize the cost of providing power to that customer:

$$\text{Load factor (\%)} = \frac{\text{Average power}}{\text{Peak power}} \times 100\% \tag{5.1}$$

For example, a customer with a peak demand of 100 kW that uses 876,000 kWh/yr (8760 h/yr × 100 kW) would have an annual load factor of 100%. Another customer using the same 876,000 kWh/yr with a peak demand of 200 kW would have a load factor of 50%. For this example, the utility would need twice as much generation capacity and twice as much transmission and distribution capacity to serve the customer with the lower load factor, which means that the rate structure must be designed to help recover those extra costs. Since they use the same amount of energy, the $/kWh charge won't differentiate between the two, but the demand charges $/kW-mo will. A stiff demand charge will encourage customers to shed some of their peak power, perhaps by shifting it to other times of day to even out their demand.

Example 5.5 Impact of Load Factor on Electricity Costs. Two customers each use 100,000 kWh/mo. One (customer A) has a load factor of 15% and the other (customer B) has a 60% load factor. Using a rate structure with energy charges of $0.06/kWh and demand charges of $10/kW-mo, compare their monthly utility bills.

Solution. They both have the same energy costs: 100,000 kWh/mo × \$0.06/kWh = \$6000/mo

Using (5.1), the peak demand for A is

$$\text{Peak(A)} = \frac{100{,}000\ \text{kWh/mo}}{15\% \times 24\ \text{h/day} \times 30\ \text{day/mo}} \times 100\% = 925.9\ \text{kW}$$

which, at \$10/kW-mo, will incur demand charges of \$9259/mo.

The peak demand for B is

$$\begin{aligned}\text{Peak(B)} &= \frac{100{,}000\ \text{kWh/mo}}{60\% \times 24\ \text{h/day} \times 30\ \text{day/mo}} \times 100\% \\ &= 231.5\ \text{kW} \qquad \text{costing \$2315/mo}\end{aligned}$$

The total monthly bill for A with the poor load factor is nearly twice as high as for B (\$15,259 for A and \$8315 for B).

5.2.6 Real-Time Pricing (RTP)

While time-of-use (TOU) rates attempt to capture the true cost of utility service, they are relatively crude since they only differentiate between relatively large blocks of time (peak, partial-peak, and off-peak, for example) and they typically only acknowledge two seasons: summer and non-summer. The ideal rate structure would be one based on real-time pricing (RTP) in which the true cost of energy is reflected in rates that change throughout the day, each and every day. With RTP, there would be no demand charges, just energy charges that might vary, for example, on an hourly basis.

Some utilities now offer one-day-ahead, hour-by-hour, real-time pricing. When a customer knows that tomorrow afternoon the price of electricity will be high, they can implement appropriate measures to respond to that high price. With the price of electricity more accurately reflecting the real, almost instantaneous, cost of power, it is hoped that market forces will encourage the most efficient management of demand.

5.3 ENERGY ECONOMICS

There are many ways to calculate the economic viability of distributed generation and energy efficiency projects. The capital cost of equipment, the operation and maintenance costs, and the fuel costs must be combined in some manner so that a comparison may be made with the costs of not doing the project. The treatment presented here, although somewhat superficial, is intended to provide a reasonable start to the financial evaluation—enough at least to know whether the project deserves further, more careful, analysis.

5.3.1 Simple Payback Period

One of the most common ways to evaluate the economic value of a project is with a simple payback analysis. This is just the ratio of the extra first cost ΔP to the annual savings, S:

$$\text{Simple payback} = \frac{\text{Extra first cost } \Delta P(\$)}{\text{Annual savings S}(\$/\text{yr})} \tag{5.2}$$

For example, an energy-efficient air conditioner that costs an extra $1000 and which saves $200/yr in electricity would have a simple payback of 5 years.

Simple payback has the advantage of being the easiest to understand of all economic measures, but it has the unfortunate problem of being one of the least convincing ways to present the economic advantages of a project. Surveys consistently show that individuals, and corporations alike, demand very short payback periods—on the order of only a few years—before they are willing to consider an energy investment. The 5-year payback in the above example would probably be too long for most decision makers; yet, for example, if the air conditioner lasts for 10 years, the extra cost is equivalent to an investment that earns a tax-free annual return of over 15%. Almost anyone with some money to invest would jump at the chance to earn 15%, yet most would not choose to put it into a more efficient air conditioner.

Simple payback is also one of the most misleading measures since it doesn't include anything about the longevity of the system. Two air conditioners may both have 5-year payback periods, but even though one lasts for 20 years and the other one falls apart after 5, the payback period makes absolutely no distinction between the two.

5.3.2 Initial (Simple) Rate-of-Return

The initial (or simple) rate of return is just the inverse of the simple payback period. That is, it is the ratio of the annual savings to the extra initial investment:

$$\text{Initial (simple) rate of return} = \frac{\text{Annual savings S }(\$/\text{yr})}{\text{Extra first cost } \Delta P(\$)} \tag{5.3}$$

Just as the simple payback period makes an investment look worse than it is, the initial rate of return does the opposite and makes it look too good. For example, if an efficiency investment with a 20% initial rate of return, which sounds very good, lasts only 5 years, then just as the device finally pays for itself, it dies and the investor has earned nothing. On the other hand, if the device has a long lifetime, the simple return on investment is a good indicator of the true value of the investment as will be shown in the next section.

Even though the initial rate of return may be misleading, it does often serve a useful function as a convenient "minimum threshold" indicator. If the investment has an initial rate of return below the threshold, there is no need to proceed any further.

5.3.3 Net Present Value

The simple payback period and rate of return are just that, too simple. Any more serious analysis involves taking into account the time value of money—that is, the fact that one dollar 10 years from now isn't as good as having one dollar in your pocket today. To account for these differences, a present worth analysis in which all future costs are converted into an equivalent *present value* or *present worth* (the terms are used interchangeably) is often required.

Begin by imagining making an investment today of P into an account earning interest i. After 1 year the account will have earned interest iP so it will then be worth $P + iP = P(1 + i)$; after 2 years it will have $P(1 + i)^2$, and so forth. This says that the future amount of money F in an account that starts with P, which earns annual interest i over a period of n years, will be

$$F = P\ (1 + i)^n \tag{5.4}$$

Rearranging (5.4) gives us a relationship between a future amount of money F and what it should be worth to us today P:

$$P = \frac{F}{(1 + i)^n} \tag{5.5}$$

When converting a future value F into a present worth P, the interest term i in (5.5) is usually referred to as a discount rate d. The discount rate can be thought of as the interest rate that could have been earned if the money had been put into the best alternative investment. For example, if an efficiency investment is projected to save \$1000 in fuel in the fifth year, and the best alternative investment is one that would have earned 10%/yr interest, the present worth of that \$1000 would be

$$P = \frac{F}{(1 + d)^n} = \frac{\$1000}{(1 + 0.10)^5} = \$620.92 \tag{5.6}$$

That is, a person with a discount rate of 10% should be neutral about the choice between having \$620.92 in his or her pocket today, or having \$1000 in 5 years. Or stated differently, that person should be willing to spend as much as \$620.92 today in order to save \$1000 worth of energy 5 years from now. Notice that the higher the discount rate, the less valuable a future payoff becomes. For instance, with a 20% discount rate, that \$1000 in 5 years is equivalent to only \$401.87 today.

When viewed from the perspective of a neutral decision—that is, would someone be just as happy with P now or F later—the discount rate takes on added meaning and may not refer to just the best alternative investment. Ask people without significant financial resources—for instance, college students—about the choice between \$621 today or \$1000 in 5 years, and chances are high that they would much prefer the \$621 today. In 5 years, college students anticipate having a lot more money, so \$1000 then wouldn't mean nearly as much as \$621 today when they are so strapped for cash. That is, their personal discount rate is probably much higher than 10%. Similarly, if there is significant risk in the proposition, then factoring in the probability that there will be no F in the future means that the discount rate would need to be much higher than would be suggested by the interest that a conventional alternative investment might earn. Deciding just what is an appropriate discount rate for an investment in energy efficiency or distributed generation is often the most difficult, and critical, step in a present value analysis.

Frequently, a distributed generation or efficiency investment will deliver financial benefits year after year. To find the present value P of a stream of annual cash flows A, for n years into the future, with a discount rate d, we can introduce a conversion factor called the *present value function* (PVF):

$$P = A \cdot \text{PVF}(d, n) \tag{5.7}$$

For a series of n annual \$1 amounts that start 1 year from the present, PVF is the summation of the present values:

$$\text{PVF}(d, n) = \frac{1}{1+d} + \frac{1}{(1+d)^2} + \cdots + \frac{1}{(1+d)^n} \tag{5.8}$$

A series analysis of (5.8) yields the following:

$$\text{PVF}(d, n) = \text{Present value function} = \frac{(1+d)^n - 1}{d(1+d)^n} \tag{5.9}$$

With all of the variables expressed in annual terms, the units of PVF will be years.

The present value of all costs, present and future, for a project is called the *life-cycle cost* of the system under consideration. When a choice is to be made between two investments, the present value, or life-cycle cost, for each, is computed and compared. The difference between the two is called the *net present value* (NPV) of the lower-cost alternative.

Example 5.6 Net Present Value of an Energy-Efficient Motor. Two 100-hp electric motors are being considered—call them "good" and "premium." The good motor draws 79 kW and costs \$2400; the premium motor draws 77.5 kW and costs \$2900. The motors run 1600 hours per year with electricity costing \$0.08/kWh. Over a 20-year life, find the net present value of the cheaper alternative when a discount rate of 10% is assumed.

Solution. The annual electricity cost for the two motors is

$$A(\text{good}) = 79 \text{ kW} \times 1600 \text{ h/yr} \times \$0.08/\text{kWh} = \$10{,}112/\text{yr}$$
$$A(\text{premium}) = 77.5 \text{ kW} \times 1600 \text{ h/yr} \times \$0.08/\text{kWh} = \$9920/\text{yr}$$

Notice how the annual energy cost of a motor is far more than the initial cost.

The present value factor for these 20-year cash flows with a 10% discount rate is

$$\text{PVF}(d, n) = \frac{(1+d)^n - 1}{d(1+d)^n} = \frac{(1+0.10)^{20} - 1}{0.10(1+0.10)^{20}} = 8.5136 \text{ yr}$$

The present value of the two motors, including first cost and annual costs, is therefore

$$P(\text{good}) = \$2400 + 8.5136 \text{ yr} \times \$10{,}112/\text{yr} = \$88{,}489$$
$$P(\text{premium}) = \$2900 + 8.5136 \text{ yr} \times \$9920/\text{yr} = \$87{,}354$$

The premium motor is the better investment with a net present value of

$$\text{NPV} = \$88{,}489 - \$87{,}354 = \$1{,}135$$

The net present value calculation can be simplified by comparing the present value of all of those future fuel savings ΔA with the extra first cost of the more efficient product ΔP.

$$\text{NPV} = \text{present value of annual savings} - \text{added first cost of better product}$$
$$\text{NPV} = \Delta A \times \text{PVF}(d, n) - \Delta P \tag{5.10}$$

Using (5.10) with the data in Example 5.6 gives

$$\text{NPV} = (\$10{,}112 - \$9920)/\text{yr} \times 8.5136 \text{ yr} - (\$2900 - \$2400) = \$1135$$

which agrees with the example.

5.3.4 Internal Rate of Return (IRR)

The *internal rate of return* (IRR) is perhaps the most persuasive measure of the value of an energy-efficiency or distributed-generation project. It is also the trickiest to compute. The IRR allows the energy investment to be directly compared with the return that might be obtained for any other competing investment. IRR is the discount rate that makes the net present value of the energy investment equal to zero. In the simple case of a first-cost premium ΔP for the more efficient product, which results in an annual fuel savings ΔA, it is the discount rate that makes the net present value in (5.10) be zero:

$$\text{NPV} = \Delta A \times \text{PVF(IRR}, n) - \Delta P \quad = 0 \tag{5.11}$$

Rearranging (5.11), and realizing that $\Delta P/\Delta A$ is just the simple payback period introduced earlier, gives the following convenient relationship for finding the internal rate of return:

$$\text{PVF(IRR}, n) = \frac{\Delta P}{\Delta A} = \text{Simple payback period} \tag{5.12}$$

Solving (5.12) is not straightforward and may require some trial-and-error estimates of the discount rate until the equation balances. Many spreadsheet programs, and some of the more powerful pocket calculators, will also do the calculation automatically. If the calculation is to be done by hand, it helps to have precalculated values for the present value function such as those presented in Table 5.4. To use Table 5.4, enter the table in the row corresponding to the project lifetime and move across until the simple payback period $\Delta P/\Delta A$ is reached. The IRR is the interest rate in that column of the table. Of course, some interpolation may be called for.

For example, the premium air conditioner in Example 5.6 costs an extra \$1000 and saves \$200 per year in electricity, so $\Delta P/\Delta A = \$1000/(\$200/\text{yr}) = 5.0$ years. With a lifetime of 10 years, Table 5.4 suggests that it would have an IRR of just over 15%. If it lasts 20 years, its IRR would be between 19% and 21%. To find the exact value would require iteration on (5.12) or interpolation in Table 5.4.

Table 5.4 can also be used to determine the IRR when the decision maker wants an energy payback, with interest, within a certain number of years. For

TABLE 5.4 Present Value Function to Help Estimate the Internal Rate of Return[a]

Life (years)	9%	11%	13%	15%	17%	19%	21%	23%	25%	27%	29%	31%	33%	35%	37%	39%
1	0.92	0.90	0.88	0.87	0.85	0.84	0.83	0.81	0.80	0.79	0.78	0.76	0.75	0.74	0.73	0.72
2	1.76	1.71	1.67	1.63	1.59	1.55	1.51	1.47	1.44	1.41	1.38	1.35	1.32	1.29	1.26	1.24
3	2.53	2.44	2.36	2.28	2.21	2.14	2.07	2.01	1.95	1.90	1.84	1.79	1.74	1.70	1.65	1.61
4	3.24	3.10	2.97	2.85	2.74	2.64	2.54	2.45	2.36	2.28	2.20	2.13	2.06	2.00	1.94	1.88
5	3.89	3.70	3.52	3.35	3.20	3.06	2.93	2.80	2.69	2.58	2.48	2.39	2.30	2.22	2.14	2.07
6	4.49	4.23	4.00	3.78	3.59	3.41	3.24	3.09	2.95	2.82	2.70	2.59	2.48	2.39	2.29	2.21
7	5.03	4.71	4.42	4.16	3.92	3.71	3.51	3.33	3.16	3.01	2.87	2.74	2.62	2.51	2.40	2.31
8	5.53	5.15	4.80	4.49	4.21	3.95	3.73	3.52	3.33	3.16	3.00	2.85	2.72	2.60	2.48	2.38
9	6.00	5.54	5.13	4.77	4.45	4.16	3.91	3.67	3.46	3.27	3.10	2.94	2.80	2.67	2.54	2.43
10	6.42	5.89	5.43	5.02	4.66	4.34	4.05	3.80	3.57	3.36	3.18	3.01	2.86	2.72	2.59	2.47
15	8.06	7.19	6.46	5.85	5.32	4.88	4.49	4.15	3.86	3.60	3.37	3.17	2.99	2.83	2.68	2.55
20	9.13	7.96	7.02	6.26	5.63	5.10	4.66	4.28	3.95	3.67	3.43	3.21	3.02	2.85	2.70	2.56
25	9.82	8.42	7.33	6.46	5.77	5.20	4.72	4.32	3.98	3.69	3.44	3.22	3.03	2.86	2.70	2.56
30	10.27	8.69	7.50	6.57	5.83	5.23	4.75	4.34	4.00	3.70	3.45	3.22	3.03	2.86	2.70	2.56

[a] Enter the row corresponding to project life, and move across until values close to the simple payback period, $\Delta P/\Delta A$, are reached. IRR is the interest rate in that column. For example, a 10-year project with a 5-year payback has an internal rate of return of just over 15%.

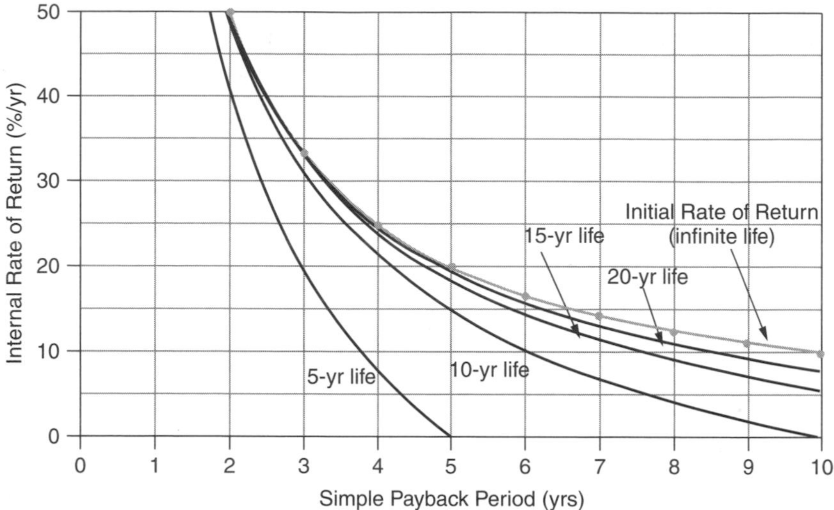

Figure 5.2 Internal rate of return as a function of the simple payback period, with project life as a parameter. When the lifetime greatly exceeds the simple payback, IRR approaches the initial rate of return.

example, the premium motor in Example 5.6 costs an extra \$500 and saves \$192/yr, giving it a simple payback period of $\Delta P/\Delta A = \$500/\$192 = 2.60$ years. Assuming a 20-year life and using Table 5.4, the internal rate of return is between 37% and 39% (it is actually 38.34%). Suppose management decides that it wants to earn all of its extra \$500 investment back sooner—in 5 years, for example—with interest. That is equivalent to saying the efficiency device has only a 5-year life. Table 5.4 indicates that the investment would still earn, almost 27% compounded annual interest. That's a pretty good investment!

Since it is so common for the simple payback period to be used to judge a project, and it is so often misleading, it may be handy to use Fig. 5.2 to translate payback into the more persuasive internal rate of return. Notice that for long-lived projects, the initial rate of return is a good approximation to IRR.

5.3.5 NPV and IRR with Fuel Escalation

The chances are that the cost of fuel in the future will be higher than it is today, which means that the annual amount of money saved by an efficiency measure could increase with time. To account for that potential it is handy to have a way to include a fuel price escalation factor in the present worth analysis.

Begin by rewriting (5.8) so that it is the sum of present values for an annual amount that is worth \$1 at time $t = 0$, but becomes $\$(1 + e)$ at $t = 1$ year, and escalates to $\$(1 + e)^n$ in the nth year

$$\text{PVF}(d, e, n) = \frac{1+e}{1+d} + \frac{(1+e)^2}{(1+d)^2} + \ldots + \left(\frac{1+e}{1+d}\right)^n \tag{5.13}$$

where d is the buyer's discount rate and e is the escalation rate of the annual savings. A comparison of (5.8) with (5.13) lets us write an equivalent discount rate d' when there is fuel savings escalation e

$$\frac{1+e}{1+d} = \frac{1}{1+d'} \tag{5.14}$$

Solving,

$$\text{Equivalent discount rate with fuel escalation} = d' = \frac{d-e}{1+e} \tag{5.15}$$

The equivalent discount rate given in (5.15) can be used in all of the present value relationships, including (5.6), (5.9), (5.10), and (5.12).

Example 5.7 Net Present Value of Premium Motor with Fuel Escalation. The premium motor in Example 5.6 costs an extra \$500 and saves \$192/yr at today's price of electricity. If electricity rises at an annual rate of 5%, find the net present value of the premium motor if the best alternative investment earns 10%.

Solution. Using (5.15), the equivalent discount rate with fuel escalation is

$$d' = \frac{d-e}{1+e} = \frac{0.10-0.05}{1+0.05} = 0.04762$$

From (5.9), the present value function for 20 years of escalating savings is

$$\text{PVF}(d', n) = \frac{(1+d')^n - 1}{d'(1+d')^n} = \frac{(1+04762)^{20} - 1}{0.04762(1+0.04762)^{20}} = 12.717 \text{ yr}$$

From (5.10), the net present value is

$$\text{NPV} = \Delta A \times \text{PVF}(d', n) - \Delta P \quad = \$192/\text{yr} \times 12.717 \text{ yr} - \$500 = \$1942$$

(Without fuel escalation, the net present value of the premium motor was only \$1135.)

To find the IRR when there is fuel escalation, the present value of the escalating series of annual savings must equal the extra initial principal,

$$\text{NPV} = \Delta A \times \text{PVF}(d', n) - \Delta P \quad = 0 \tag{5.16}$$

$$\text{PVF}(d', n) = \frac{\Delta P}{\Delta A} = \text{Simple payback period} \tag{5.17}$$

where ΔA is the annual savings at $t = 0$. The quantity d' can be found by trial-and-error or interpolation using Table 5.4 or Fig. 5.2 as a guide, or with a special calculator. Comparing (5.12) with (5.17), we can see that d' will be the same as the internal rate of return that is found without fuel escalation, which we will call IRR_0. Replacing d' with IRR_0 in (5.15) gives

$$\text{IRR}_0 = \frac{d - e}{1 + e} \tag{5.18}$$

Since d in (5.18) is the buyer's discount rate that results in a NPV of zero, it is the same as the internal rate of return with fuel escalation, which we will call IRR_e:

$$\text{IRR}_e = \text{IRR}_0(1 + e) + e \tag{5.19}$$

Example 5.8 IRR for an HVAC Retrofit Project with Fuel Escalation. Suppose the energy-efficiency retrofit of a large building reduces the annual electricity demand for heating and cooling from 2.3×10^6 kWh to 0.8×10^6 kWh and the peak demand for power from by 150 kW. Electricity costs \$0.06/kWh and demand charges are \$7/kW-mo, both of which are projected to rise at an annual rate of 5%. If the project costs \$500,000, what is the internal rate of return over a project lifetime of 15 years?

Solution. The initial annual savings will be

$$\text{Energy savings} = (2.3 - 0.8) \times 10^6 \text{ kWh/yr} \times \$0.06/\text{kWh} = \$90{,}000/\text{yr}$$

$$\text{Demand savings} = 150 \text{ kW} \times \$7/\text{kW-mo} \times 12 \text{ mo/yr} = \$12{,}600/\text{yr}$$

$$\text{Total annual savings} = \Delta A = \$90{,}000 + \$12{,}600 = \$102{,}600/\text{yr}$$

The simple payback will be

$$\text{Simple payback period} = \frac{\Delta P}{\Delta A} = \frac{\$500{,}000}{\$102{,}600/\text{yr}} = 4.87 \text{ yr}$$

From Table (5.4), the internal rate of return without fuel escalation IRR_0 is very close to 19%.

From (5.19) the internal rate of return with fuel escalation is

$$\text{IRR}_e = \text{IRR}_0(1 + e) + e = 0.19(1 + 0.05) + 0.05 = 0.2495 \approx 25\%/\text{yr}$$

5.3.6 Annualizing the Investment

In many circumstances the extra capital required for an energy investment will be borrowed from a lending company, obtained from investors who require a

TABLE 5.5 Capital Recovery Factors as a Function of Interest Rate and Loan Term

Years	3%	4%	5%	6%	7%	8%	9%	10%	11%	12%	13%
5	0.2184	0.2246	0.2310	0.2374	0.2439	0.2505	0.2571	0.2638	0.2706	0.2774	0.2843
10	0.1172	0.1233	0.1295	0.1359	0.1424	0.1490	0.1558	0.1627	0.1698	0.1770	0.1843
15	0.0838	0.0899	0.0963	0.1030	0.1098	0.1168	0.1241	0.1315	0.1391	0.1468	0.1547
20	0.0672	0.0736	0.0802	0.0872	0.0944	0.1019	0.1095	0.1175	0.1256	0.1339	0.1424
25	0.0574	0.0640	0.0710	0.0782	0.0858	0.0937	0.1018	0.1102	0.1187	0.1275	0.1364
30	0.0510	0.0578	0.0651	0.0726	0.0806	0.0888	0.0973	0.1061	0.1150	0.1241	0.1334

return on their investments, or taken from one's own accounts. In all of these circumstances, the economic analysis can be thought of as a loan that converts the extra capital cost into a series of equal annual payments that eventually pay off the loan with interest. Even if the money is not actually borrowed, the same approach can be used to annualize the cost of the energy investment, which has many useful applications. The key equation is

$$A = P \times \text{CRF}(i, n) \tag{5.20}$$

where A represents annual loan payments (\$/yr), P is the principal borrowed (\$), i is the interest rate (e.g. 10% corresponds to $i = 0.10$/yr), and n is the loan term (yrs), and

$$\text{CRF}(i, n) = \text{Capital recovery factor}(\text{yr}^{-1}) = \frac{i(1+i)^n}{(1+i)^n - 1} \tag{5.21}$$

Notice that the capital recovery factor (CRF) is just the inverse of the present value function (PVF). Since we are treating the first cost of the investment as a loan, we have gone back to using an interest rate i rather than a discount rate d.

A short table of values for the CRF is given in Table 5.5.

Equations (5.20) and (5.21) were written as if the loan payments are made only once each year. They are easily adjusted to find monthly payments by dividing the annual interest rate i by 12 and multiplying the loan term n by 12, leading to the following:

$$\text{CRF}(i, n) = \frac{(i/12)[1 + (i/12)]^{12n}}{[1 + (i/12)]^{12n} - 1} \text{ per month} \tag{5.22}$$

Example 5.9 Comparing Annual Costs to Annual Savings. An efficient air conditioner that costs an extra \$1000 and saves \$200 per year is to be paid for with a 7% interest, 10-year loan.

a. Find the annual monetary savings.
b. Find the ratio of annual benefits to annual costs.

Solution. From (5.21), the capital recovery factor will be

$$\text{CRF}(0.07, 10) = \frac{0.07(1+0.07)^{10}}{(1+0.07)^{10}-1} = 0.14238/\text{yr}$$

The annual payments will be $A = \$1000 \times 0.14238/yr = \$142.38/\text{yr}$.

a. The annual savings will be \$200 – \$142.38 = \$57.62/yr. Notice that by annualizing the costs the buyer makes money every year so the notion that a 5-year payback period might be considered unattractive becomes irrelevant.
b. The benefit/cost ratio would be

$$\text{Benefit/Cost} = \frac{\$200/\text{yr}}{\$142.38/\text{yr}} = 1.4$$

The annualized cost method is also applicable to investments in generation capacity, as the following example illustrates.

Example 5.10 Cost of Electricity from a Photovoltaic System. A 3-kW photovoltaic system, which operates with a capacity factor (CF) of 0.25, costs \$10,000 to install. There are no annual costs associated with the system other than the payments on a 6%, 20-year loan. Find the cost of electricity generated by the system (¢/kWh).

Solution. From either (5.21) or Table 5.5, the capital recovery factor is 0.0872/yr. From (5.20) the annual payments will be

$$A = P \times \text{CRF}(0.06, 20) = \$10{,}000 \times 0.0872/\text{yr} = \$872/\text{yr}$$

To find the annual electricity generated, recall the definition of capacity factor (CF) from (3.20):

$$\text{Annual energy (kWh/yr)} = \text{Rated power (kW)} \times 8760 \text{ hr/yr} \times \text{CF}$$

In this case

$$\text{kWh/yr} = 3 \text{ kW} \times 8760 \text{ h/yr} \times 0.25 = 6570 \text{ kWh/yr}$$

The cost of electricity from the PV system is therefore

$$\text{Cost of PV electricity} = \frac{\$872/\text{yr}}{6570 \text{ kWh/yr}} = \$0.133/\text{kWh} = 13.3¢/\text{kWh}$$

5.3.7 Levelized Bus-Bar Costs

To do an adequate comparison of cost per kilowatt-hour from a renewable energy system versus that for a fossil-fuel-fired power plant, the potential for escalating future fuel costs must be accounted for. To ignore that key factor is to ignore one of the key advantages of renewable energy systems; that is, their independence from the uncertainties associated with future fuel costs.

The cost of electricity per kilowatt-hour for a power plant has two key components—an up-front fixed cost to build the plant plus an assortment of costs that will be incurred in the future. In the usual approach to cost estimation, a present value calculation is first performed to find an equivalent initial cost, and then that amount is spread out into a uniform series of annual costs. The ratio of the equivalent annual cost ($/yr) to the annual electricity generated (kWh/year) is called the *levelized, bus-bar* cost of power (the "bus-bar" refers to the wires as they leave the plant boundaries).

In the first step, the present value of all future costs must be found, including the impacts of inflation. To keep things simple, we'll assume that the annual costs today are A_0, and that they escalate due to inflation (and other factors) at the rate e. Figure 5.3 illustrates the concept.

The present value of these escalating annual costs over a period of n years is given by

$$\text{PV(annual costs)} = A_0 \cdot \text{PVF}(d', n) \tag{5.23}$$

where d' is the equivalent discount rate including inflation introduced in (5.15).

Having found the present value of those future costs, we now want to find an equivalent annual cost using the capital recovery factor:

$$\text{Levelized annual costs} = A_0[\text{PVF}(d', n) \cdot \text{CRF}(d, n)] \tag{5.24}$$

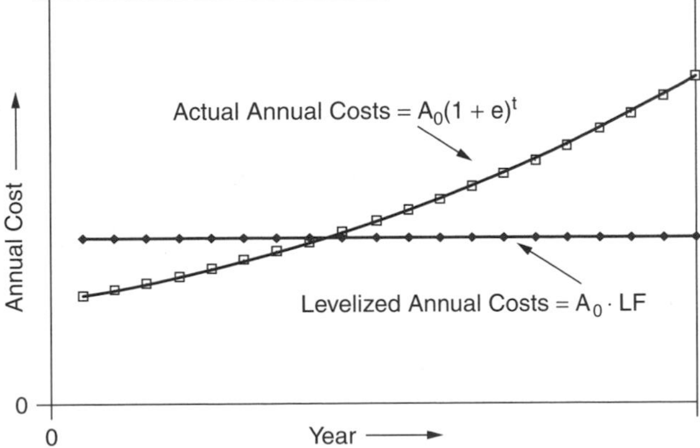

Figure 5.3 Levelizing annual costs when there is fuel escalation.

The product in the brackets, called the *levelizing factor*, is a multiplier that converts the escalating annual fuel and O&M costs into a series of equal annual amounts:

$$\text{Levelizing factor (LF)} = \left[\frac{(1+d')^n - 1}{d'(1+d')^n}\right] \cdot \left[\frac{d(1+d)^n}{(1+d)^n - 1}\right] \tag{5.25}$$

Notice that when there is no escalation ($e = 0$), then $d' = d$ and the levelizing factor is just unity.

The impact of the levelizing factor can be very high, as is illustrated in Fig. 5.4. For example, if fuel prices increase at 5%/yr for an owner with a 10% discount rate, the levelizing factor is 1.5. If they increase at 8.3%/yr, the impact is equivalent to an annualized cost of fuel that is double the initial cost.

Normalizing the levelized annual costs to a per kWh basis can be done using the heat rate of the plant (Btu/kWh), the initial fuel cost ($/Btu), the per kWh O&M costs and the levelizing factor:

$$\text{Levelized annual costs(\$/kWh)} = \left[\text{Heat rate}\left(\frac{\text{Btu}}{\text{kWh}}\right) \times \text{Fuel}\left(\frac{\$}{\text{Btu}}\right) + \text{O \& M}\left(\frac{\$}{\text{kWh}}\right)\right]_0 \times \text{LF} \tag{5.26}$$

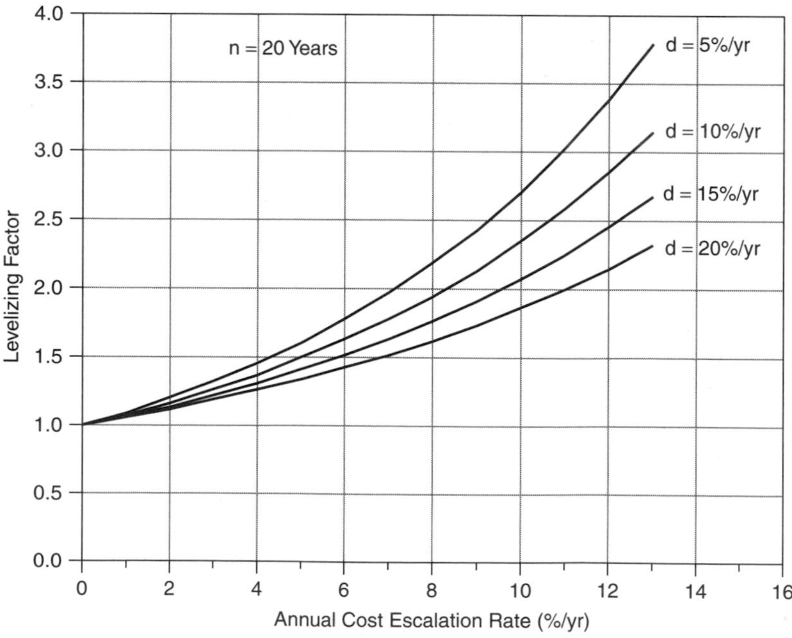

Figure 5.4 Levelizing Factor for a 20-year term as a function of the escalation rate of annual costs, with the owner's discount rate as a parameter. With a discount rate of 10%/yr and fuel escalation rate of 5%/yr, the levelized cost of fuel is 1.5 times the initial cost.

Just as the future cost of fuel and O&M needs to be levelized, so does the capital cost of the plant. To do so, it is handy to combine the CRF with other costs that depend on the capital cost of the plant into a quantity called the *fixed charge rate* (FCR). The fixed charge rate covers costs that are incurred even if the plant doesn't operate, including depreciation, return on investment, insurance, and taxes. Fixed charge rates vary depending on plant ownership and current costs of capital, but tend to be in the range of 10–18% per year. The governing equation that annualizes capital costs is then

$$\text{Levelized fixed cost(\$/kWh)} = \frac{\text{Capital cost(\$/kW)} \times \text{FCR(1/yr)}}{8760 \text{ h/yr} \times \text{CF}} \tag{5.27}$$

where CF is the capacity factor of the plant.

Table 3.3 in Chapter 3 provides estimates for some of the key variables in (5.26) and (5.27).

Example 5.11 Cost of Electricity from a Microturbine. A microturbine has the following characteristics:

Plant cost = \$850/kW
Heat rate = 12,500 Btu/kWh
Capacity factor = 0.70
Initial fuel cost = \$4.00/10^6 Btu
Variable O&M cost = \$0.002/kWh
Fixed charge rate = 0.12/yr
Owner discount rate = 0.10/yr
Annual cost escalation rate = 0.06/yr

Find its levelized (\$/kWh) cost of electricity over a 20-year lifetime.

Solution. From (5.27), the levelized fixed cost is

$$\text{Levelized fixed cost} = \frac{\$850/\text{kW} \times 0.12/\text{yr}}{8760 \text{ h/yr} \times 0.70} = \$0.0166/\text{kWh}$$

Using (5.26), the initial annual cost for fuel and O&M is

$$A_0 = 12{,}500 \text{ Btu/kWh} \times \$4.00/10^6 \text{ Btu} + \$0.002/\text{kWh} = \$0.052/\text{kWh}$$

This needs to be levelized to account for inflation. From (5.15), the inflation adjusted discount rate d' would be

$$d' = \frac{d - e}{1 + e} = \frac{0.10 - 0.06}{1 + 0.06} = 0.037736$$

From (5.25), the levelizing factor for annual costs is

$$\text{Levelizing factor (LF)} = \left[\frac{(1.037736)^{20} - 1}{0.037736(1.037736)^{20}}\right] \cdot \left[\frac{0.10(1.10)^{20}}{(1.10)^{20} - 1}\right] = 1.628$$

The levelized annual cost is therefore

$$\text{Levelized annual cost} = A_0 \text{ LF} = \$0.052/\text{kWh} \times 1.628 = \$0.0847/\text{kWh}$$

The levelized fixed plus annual cost is

$$\text{Levelized bus-bar cost} = \$0.0166/\text{kWh} + \$0.0847/\text{kWh} = \$0.1013/\text{kWh}$$

5.3.8 Cash-Flow Analysis

One of the most flexible and powerful ways to analyze an energy investment is with a cash-flow analysis. This technique easily accounts for complicating factors such as fuel escalation, tax-deductible interest, depreciation, periodic maintenance costs, and disposal or salvage value of the equipment at the end of its lifetime. In a cash-flow analysis, rather than using increasingly complex formulas to characterize these factors, the results are computed numerically using a spreadsheet. Each row of the resulting table corresponds to one year of operation, and each column accounts for a contributing factor. Simple formulas in each cell of the table enable detailed information to be computed for each year along with very useful summations.

Table 5.6 shows an example cash-flow analysis for a \$1000, 6%, 10-year loan used to pay for a conservation measure that, at the time of loan initiation, saves a homeowner \$150/yr in electricity. This savings in the electric bill is expected to increase 5% per year. The homeowner has a personal discount factor of 10%. Since this is a home loan, any interest paid on the loan will qualify as a tax deduction and the homeowner's federal (and perhaps state) income taxes will go down accordingly. Let's work our way through the spreadsheet.

Begin with the loan payments. From (5.21), the capital recovery factor CRF(0.06, 10) can easily be found to be 0.13587/yr. Since the loan is for \$1000, this means 10 annual payments of \$135.87 must be made. At time $t = 0$, the \$1000 loan begins and the borrower has use of that money for a full year before making the first payment. This means, with 6% interest, $0.06 \times \$1000 = \$60$ in interest is owed at the end of the first, which comes out of the first \$135.87 payment. The difference $\$135.87 - \$60 = \$75.87$ is applied to the loan balance, bringing it down from \$1000 to \$924.13 at time $t = 1$ year. In the next year, $0.06 \times \$924.13 = \$55.45$ pays the interest, and $\$135.87 - \$55.45 = \$80.42$ is applied to the principal. As expected, when the tenth payment is made, the loan is completely paid off.

TABLE 5.6 Cash-Flow Analysis for a \$1000, 6%, 10-yr Loan, Showing Fuel Escalation and Income Tax Savings on Loan Interest[a]

Loan principal(\$) =	1000.00	Energy savings (kWh/yr) =	1500
Interest =	0.06	Price at $t = 0$ (\$/kWh) =	0.10
Loan term (yrs) =	10	Savings at $t = 0$ (\$/yr)	150
CRF(i, n) per yr =	0.13587	Escalating at (%/yr) =	5
Payments (\$/yr) =	135.87		
Tax bracket =	0.305	Personal discount rate =	0.10

End of Year	Loan Payment	Interest	Delta Principal	Loan Balance	Tax Savings	Loan Cost	Electric Savings	Net Savings	PV Savings	Cum PV Savs
0	0.00	0.00	0.00	1000.00	0.00	0.00	0.00	0.00	0.00	0.00
1	135.87	60.00	75.87	924.13	18.30	117.57	157.50	39.93	36.30	36.30
2	135.87	55.45	80.42	843.71	16.91	118.96	165.38	46.42	38.36	74.66
	1000 × 0.13587	0.06 × 924.13	135.87 − 55.45	924.13 − 80.42	0.305 × 55.45	135.87 − 16.91	1.05 × 157.50	165.38 − 118.96	$\frac{46.42}{(1.10)^2}$	38.36 + 36.30
9	135.87	14.95	120.92	128.18	4.56	131.31	232.70	101.39	43.00	369.75
10	135.87	7.69	128.18	0.00	2.35	133.52	244.33	110.81	42.72	412.48

Loan paid off (0.00) — PV cumulative savings (412.48)

[a] The cash flow is always positive even though this energy saving investment has an uninspiring simple payback of 6.67 years.

In most circumstances, interest on home loans is tax deductible, which means it reduces the income that is taxed by the I.R.S. To determine the tax benefit associated with the interest portion of the loan payments, we need to learn something about how income taxes are calculated. Table 5.7 shows the 2002 federal personal income tax brackets for married couples filing jointly. Similar tables are available for individuals with different filing status. For a family earning between \$109,250 and \$166,500, for example, every additional dollar of income has 30.5¢ of taxes taken out of it. On the other hand, if the income that has to be reported to the I.R.S. can be reduced by one dollar, that will save 30.5¢ in taxes. The 30.5% number is called the *marginal tax bracket* (MTB). Most (but not all) states also have their own income taxes, which increases an individual's total MTB. For example, in California, a married couple in the 30.5% federal bracket will have a combined state and federal MTB of about 37%.

Having teased out the interest in each year's payments for the example loan in Table 5.6, we can now determine the income-tax advantages associated with that interest. When the first payment of \$135.87 was made on the loan, \$60 was tax-deductible interest. That means the taxpayer's net income (net income = gross income − tax deductions) will be reduced by that \$60 deduction. With the taxpayer in the 30.5% tax bracket, that buyer's income taxes in the first year will be reduced by \$60 × 0.305 = \$18.30. Therefore, the loan in that first year really only costs the homeowner \$135.87 − \$18.30 = \$117.57.

TABLE 5.7 Federal Income Tax Brackets for Married Couples Filing Jointly, 2002

Income Over...	But Not Over...	Federal Tax Is...	Of the Amount Over
\$0	\$45,200	15%	\$0
45,200	109,250	\$6,780 + 27.5%	45,200
109,250	166,500	24,394 + 30.5%	109,250
166,500	297,350	41,855 + 35.5%	166,500
297,350	—	88,307 + 39.1%	297,350

The spreadsheet shown in Table 5.6 also includes an electricity savings associated with the efficiency measure worth \$150 per year at $t = 0$. This rises by 5% each year, so by the time the first payment is made the fuel savings is $1.05 \times \$150 = \$157.50$. The total savings at the time of that first annual payment is therefore

$$\begin{aligned}\text{First-year savings} &= \$157.50(\text{electricity}) + \$18.30(\text{taxes}) - \$135.87(\text{loan}) \\ &= \$39.93\end{aligned}$$

This homeowner has a personal discount rate of 10%, so the first year's savings has a present value of $\$39.93/(1.10) = \$36.30$. Continuing through the 10 years and adding up the cumulative present-values of each year's savings results in a total present value of \$412.48. That is, if this family goes ahead with the energy efficiency project, it is financially the same as getting a free system, with the same annual bills as would have been received without the system, plus a check for \$412.48. If the system lasts longer than 10 years, the benefits would be even greater.

5.4 ENERGY CONSERVATION SUPPLY CURVES

By converting all of the costs of an energy efficiency measure into a uniform series of annual costs, and dividing that by the annual energy saved, a convenient and persuasive measure of the value of saved energy can be found. The resulting *cost of conserved energy* (CCE) has units of \$/kWh, which makes it directly comparable to the \$/kWh cost of generation.

$$\text{CCE} = \frac{\text{Annualized cost of conservation}(\$/\text{yr})}{\text{Annual energy saved (kWh/yr)}} \tag{5.28}$$

When the only cost of the conservation measure is the extra initial capital cost, the annualized cost of conservation in the numerator is easy to obtain using an appropriate capital recovery factor (CRF). In more complicated situations, it may be necessary to do a levelized cost analysis in which the present value of all future costs is obtained, and then that is annualized using CRF and a fuel-savings levelizing factor.

Example 5.12 CCE for a Lighting Retrofit Project. It typically costs about \$50 in parts and labor to put in new lamps and replace burned out ballasts in a conventional four-lamp fluorescent fixture. For \$65, more efficient ballasts and lamps can be used in the replacement, which will maintain the same illumination but will decrease the power needed by the fixture from 170 W to 120 W. For an office in which the lamps are on 3000 h/yr, what is the cost of conserved energy for the better system if it is financed with a 15-yr, 8% loan, assuming that the new components last at least that long? Electricity from the utility costs 8¢/kWh.

Solution. The extra cost is \$65 – \$50 = \$15. From Table 5.5, CRF(0.08, 15) is 0.1168/yr so the annualized cost of the improvement is

$$A = P \times CRF(i, n) = \$15 \times 0.1168/\text{yr} = \$1.75/\text{yr}$$

The annual energy saved is

$$\text{Saved energy} = (170 - 120)\text{W} \times 3000\ \text{h/yr} \div (1000\ \text{W/kW}) = 150\ \text{kWh/yr}$$

The cost of conserved energy is

$$\text{CCE} = \frac{\$1.75/\text{yr}}{150\ \text{kWh/yr}} = \$0.0117/\text{kWh} = 1.17¢/\text{kWh}$$

The choice is therefore to spend 8¢ to purchase 1 kWh for illumination or spend 1.17¢ to avoid the need for that kWh. In either case, the amount of illumination is the same. Notice, too, that there will also be a reduction in demand charges with the more efficient system.

While CCE provides another measure of the economic benefits of a single efficiency measure for an individual or corporation, it has greater application as a policy tool for energy forecasters. By analyzing a number of efficiency measures and then graphing their potential cumulative savings, policy makers can estimate the total energy reduction that might be achievable at a cost less than that of purchased electricity.

To illustrate the procedure, consider four hypothetical conservation measures A, B, C, and D. Suppose they have individual costs of conserved energy and individual annual energy savings values as shown in Table 5.8. If we do Measure A, 300 kWh/yr will be saved at a cost of 1¢/kWh. If we do A and B, another

TABLE 5.8 Hypothetical Example of Four Independent Conservation Measures

Conservation Measure	CCE (¢/kWh)	Saved Energy (kWh/yr)	Cumulative Energy Saved (kWh/yr)	Cumulative Cost (¢/yr)
A	1	300	300	300
B	2	200	500	700
C	3	500	1000	2200
D	10	200	1200	4200

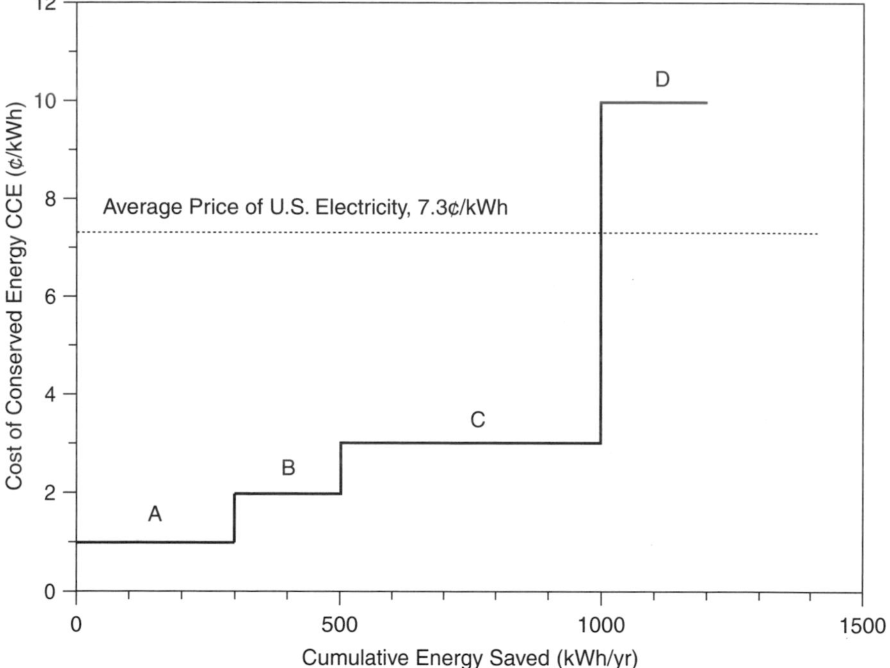

Figure 5.5 Energy conservation supply curve for the example in Table 5.8.

200 kWh/yr will be saved (assuming these are independent measures for which the energy savings of one doesn't affect the savings of the other), for a total of 500 kWh/yr saved. Similarly, total costs can be added up; so, for example, doing all four measures will save 1200 kWh/yr at a total cost of 4200¢/yr. A plot of the marginal cost of conserved energy (¢/kWh) versus the cumulative energy saved (kWh/yr) is called an *energy conservation supply curve*. Such a curve for this example is presented in Fig. 5.5.

Continuing the example, the average retail price of U.S. electricity at 7.3¢/kWh is also shown on Fig. 5.5. Measures A, B, and C each save energy at less than that price, and so they would be cost effective to implement, saving a total of 1000 kWh/yr. Measure D, which costs 10¢/kWh, is not cost effective since it would be cheaper to purchase utility electricity at 7.3¢. But, what if we did all four measures at the same time? A total of 1200 kWh/yr would be saved at a total cost of 4200¢/yr, for an *average* CCE of 2.85¢/kWh. An economist would argue that when the marginal cost of an efficiency measure is above the price of electricity, it shouldn't be implemented. An environmentalist, however, might argue that the full package of measures saves more energy while still saving society money and should be done. The point is that both arguments will be encountered, so it is important to be clear about whether assertions about cumulative energy efficiency potential are based on marginal CCE or average CCE.

An example of data derived for a real conservation supply curve by the National Academy of Sciences (1992) for U.S. buildings is given in Table 5.9. As shown there, 12 electricity efficiency measures were analyzed ranging from painting rooftops white and planting more trees to reduce air-conditioning loads, to increased efficiency in residential and commercial lighting, water heating, space heating, cooling, and ventilation systems. As is customary for such studies, the CCE is based on a *real* (net of inflation), rather than *nominal* (includes inflation), discount rate in order to exclude the uncertainties of inflation. In this case, a real discount rate of 6% has been used.

The data from Table 5.9 have been plotted in the conservation supply curve shown in Fig. 5.6. All 12 measures are cost effective when compared to the 7.3¢/kWh average price of electricity in the United States, yielding a total potential

TABLE 5.9 Data for an Energy Conservation Supply Curve for U.S. Residential Buildings Calculated Using a Discount Rate of 6% (Real)

	Conservation Measure	CCE $d = 0.06$ (¢/kWh)	Energy Savings (TWh/yr)	Cumulative Savings (TWh/yr)
1	White surfaces and urban trees	0.53	45	45
2	Residential lighting	0.88	56	101
3	Residential water heating	1.26	38	139
4	Commercial water heating	1.37	9	148
5	Commercial lighting	1.45	167	315
6	Commercial cooking	1.50	6	321
7	Commercial cooling	1.91	116	437
8	Commercial refrigeration	2.18	21	458
9	Residential appliances	3.34	103	561
10	Residential space heating	3.65	105	666
11	Commercial and Industrial space heating	3.96	22	688
12	Commercial ventilation	6.83	45	733

Source: National Academy of Sciences (1992).

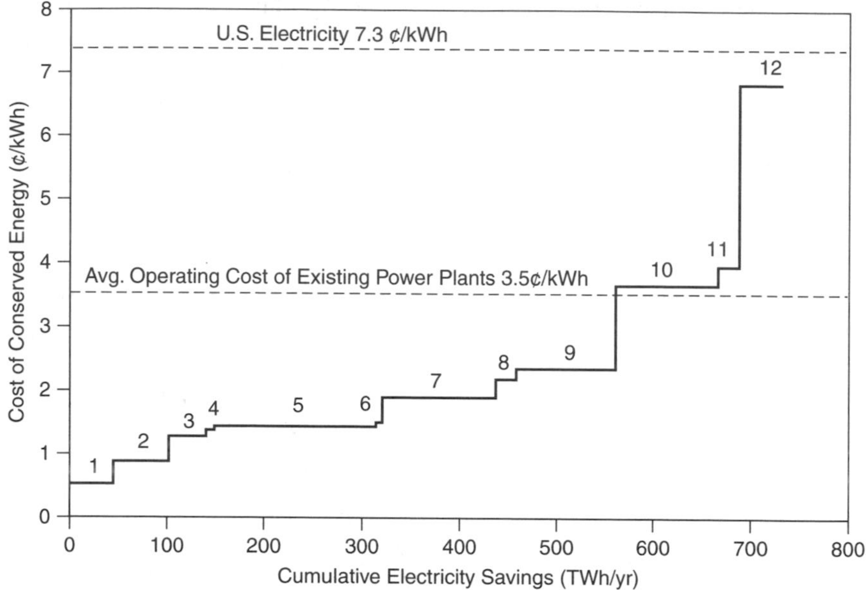

Figure 5.6 Electricity conservation supply curve for U.S. buildings (National Academy of Sciences, 1992).

savings of 733 billion kWh. If all 12 were implemented, they would reduce building electricity consumption by one-third at an average cost of just over 2.4¢/kWh, which is less than the average running cost of U.S. power plants.

5.5 COMBINED HEAT AND POWER (CHP)

Distributed generation (DG) technologies include some that produce usable waste heat as well as electricity (e.g. combustion turbines and fuel cells), and some that don't (e.g., photovoltaics and wind turbines). For those that do, the added monetary value of the waste heat can greatly enhance the economics of the DG technology. The value of waste heat is related to its temperature, distance from generation to where it will be used, the relative timing and magnitude of electricity demand and thermal demand, and cost of the fuel that the cogeneration displaces.

Higher-temperature heat is more versatile since it can be used in a variety of applications such as process steam, absorption cooling, and space heating, while simple water heating may be the only application of low-temperature heat. Distance from generator to load is important since transporting heat over long distances can be costly and wasteful. Finding applications where the timing and magnitude of electric and thermal demands are aligned can greatly affect the economics of cogeneration. For example, if the thermal application is for space heating alone, then the heat may have no value all summer long, whereas if it is used for space heating and cooling, it may be useful all year round. And finally, if the waste heat is displacing the need to purchase an expensive fuel, such as propane or electricity, the overall economics are obviously enhanced.

5.5.1 Energy-efficiency Measures of Combined Heat and Power (Cogeneration)

With combined heat and power (CHP), it is a little tricky to allocate the costs and benefits of the plant since the value of a unit of electricity is so much higher than a unit of thermal energy, yet both will be produced with the same fuel source. This dilemma requires some creative accounting.

The simplest approach to describing the efficiency of a cogeneration plant is to simply divide the total output energy (electrical plus thermal) by the total thermal input, remembering to use the same units for each quantity:

$$\text{Overall thermal efficiency} = \left(\frac{\text{Electrical} + \text{Thermal ouput}}{\text{Thermal input}}\right) \times 100\% \tag{5.29}$$

While this measure is often used, it doesn't distinguish between the value of recovered heat and electrical output. For example, a simple 75% efficient boiler that generates no electricity would have an overall thermal efficiency of 75%, while cogeneration that delivers 35% of its fuel energy as electricity and 40% of it as recovered heat would also have the same overall efficiency of 75%. Clearly, the true cogeneration plant is producing a much more valuable output that isn't recognized by the simple relationship given in (5.29).

A better way to evaluate CHP is by comparing the cogeneration of heat and power, in the same unit, to generation of electricity in one unit plus a separate boiler to provide the equivalent amount of heat. This method, however, requires an estimate of the efficiency of the separate boiler.

For example, consider Fig. 5.7 in which a CHP plant converts 30% of its fuel into electricity while capturing 48% of the input energy as thermally useful heat, for an overall thermal efficiency of 78%. Suppose we compare this plant with 33.3%-efficient grid electricity plus an 80% efficient boiler for heat, as shown in Fig. 5.8. To generate the 30 units of electricity from the grid, at 33.3% efficiency, requires 90 units of thermal energy. And, to produce 48 units of heat from the 80% efficient boiler, another 60 units of thermal energy are required. That is, with separate electrical and thermal generation a total of $60 + 90 = 150$ units of input energy are required, but with cogeneration only 100 units were needed—a savings of one-third.

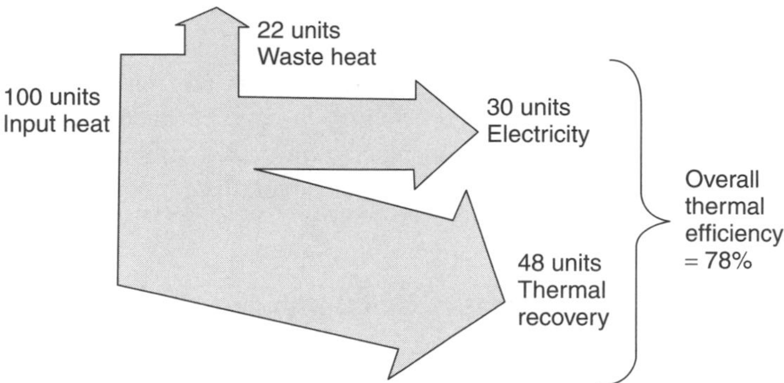

Figure 5.7 A cogeneration plant with an overall thermal efficiency of 78%.

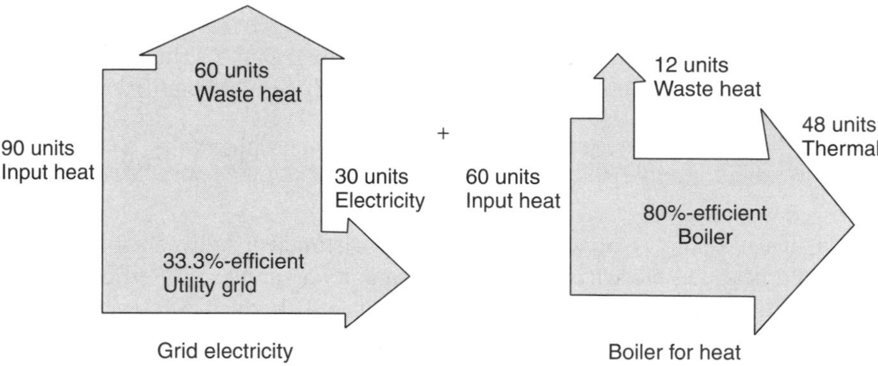

Figure 5.8 Separate generation of heat in an 80%-efficient boiler, and electricity from the grid at 33.3% efficiency, requires 150 units of thermal input compared to the 100 needed by the CHP plant of Fig. 5.7.

There are several ways to capture the essence of Figs. 5.7 and 5.8. One approach is to compare the overall thermal efficiency (5.29) with and without cogeneration. For example, using the data given in those figures,

$$\text{Overall thermal efficiency (with CHP)} = \left(\frac{30 + 48}{100}\right) \times 100\% = 78\% \tag{5.30}$$

$$\text{Overall thermal efficiency (without CHP)} = \left(\frac{30 + 48}{90 + 60}\right) \times 100\% = 52\% \tag{5.31}$$

By comparing the overall thermal efficiencies, the improvement caused by CHP is easy to determine. The improvement is called the *overall energy savings*, which is defined as

$$\text{Overall energy savings} = \left(1 - \frac{\text{Thermal input with CHP}}{\text{Thermal input without CHP}}\right) \times 100\% \quad (5.32)$$

where, of course, the two approaches (with and without CHP) must deliver the same electrical and thermal outputs.

For the data in Figs. 5.7 and 5.8,

$$\text{Overall energy savings} = \left(1 - \frac{100}{90 + 60}\right) \times 100\% = 33.3\% \quad (5.33)$$

While the previous example illustrates the overall energy (or fuel) savings for society gained by cogeneration, a slightly different perspective might be taken by an industrial facility when it considers CHP. Under the assumption that the facility needs heat anyway, say for process steam, the extra thermal input needed to generate electricity using cogeneration can be described using a quantity called the *Energy-Chargeable-to-Power* (ECP)

$$\text{ECP} = \frac{\text{Total thermal input} - \text{Displaced thermal input}}{\text{Electrical output}} \quad (5.34)$$

where the displaced thermal input is based on the efficiency of the boiler if one had been used. The units of ECP are the same as would be found for the heat rate of a conventional power plant, that is Btu/kWh or kJ/kWh.

Example 5.13 Cost of Electricity from a CHP Microturbine An industrial facility that needs a continuous supply of process heat is considering a 30 kW microturbine to help fill that demand. Waste heat recovery will offset fuel needed by its existing 75-percent efficient boiler. The microturbine has a 29% electrical efficiency and it recovers 47% of the fuel energy as usable heat. Find the Energy-Chargeable-to-Power (ECP)

Solution. Using the energy conversion factor of 3412 Btu/kWh, we can find the thermal input to the 29 percent efficient, 30 kW microturbine:

$$\text{Microturbine input} = \left(\frac{30\text{ kW}}{0.29}\right) \times 3412\text{ Btu/kWh} = 352{,}966\text{ Btu/hr}$$

Since 47 percent of that is delivered as usable heat,

$$\text{Microturbine usable heat} = 0.47 \times 352{,}966\text{ Btu/hr} = 165{,}894\text{ Btu/hr}$$

The displaced fuel for the 75 percent efficient boiler will be

$$\text{Displaced boiler fuel} = \frac{165{,}894\text{ Btu/hr}}{0.75} = 221{,}192\text{ Btu/hr}$$

The calculations thus far obtained are shown below:

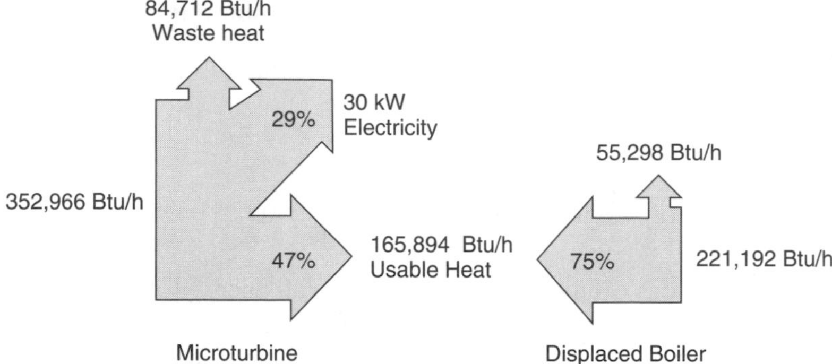

Substituting these values into (5.34) gives the Energy Chargeable to Power

$$\text{ECP} = \frac{\text{Extra CHP thermal input}}{\text{Electrical output}} = \frac{(352{,}966 - 221{,}192)\ \text{Btu/hr}}{30\ \text{kW}}$$
$$= 4393\ \text{Btu/kWh}$$

This 4393 Btu/kWh can be compared with the heat rate of conventional power plants, which is typically around 10,300 Btu/kWh. That is, given the need for process heat anyway, it can be argued that the bonus from CHP is the ability to generate electricity with less than half as much fuel as must be burned at an average thermal power plant.

5.5.2 Impact of Usable Thermal Energy on CHP Economics

The energy chargeable to power (ECP) depends on the amount of usable heat recovered as well as the efficiency of the boiler or furnace that would have generated the heat had it been provided separately. Heat recovery is important to the economics since, in a sense, it helps subsidize the cost of the more important output—electricity—so we should look more carefully at the role it plays.

When the *ECP* is modified to account for the cost of fuel, a simple measure of the added cost of electricity with cogeneration, called the *operating cost chargeable to power (CCP)*, can be found from

$$\text{Cost chargeable to power (CCP)} = \text{ECP} \times \text{Unit cost of energy} \tag{5.35}$$

CCP will have units of \$/kWh.

Example 5.14 Cost Chargeable to Power for a CHP Microturbine. Suppose the 30-kW microturbine in Example 5.13 costs \$50,000 and has an annual O&M cost of \$1200 per year. It operates 8000 hours per year and the owner uses a fixed charge rate of 12%/yr. Natural gas for the microturbine and existing boiler costs \$4 per million Btu.

a. Find the operating cost chargeable to power (CCP)
b. What is the cost of electricity from the microturbine?
c. If the facility currently pays 6.0¢/kWh for energy, plus demand charges of \$7/kW-mo, what would be the annual monetary savings of the microturbine?

Solution.

a. In Example 5.13, the energy cost chargeable to power for this microturbine was found to be 4393 Btu/kWh. From (5.35), the cost chargeable to power is

$$\text{CCP} = 4393\ \text{Btu/kWh} \times \$4/10^6\ \text{Btu} = \$0.0145/\text{kWh}$$

That is, choosing to generate on-site power will cost 1.45¢/kWh for fuel.

b. The amortized cost of the microturbine is $\$50,000 \times 0.12/\text{yr} = \$6,000/\text{yr}$.

$$\text{Annual operations and maintenance} = \$1200/\text{yr}$$

$$\text{Annual fuel cost for electricity} = 30\ \text{kW} \times 8000\ \text{hr/yr} \times \$0.0145/\text{kWh}$$
$$= \$3480/\text{yr}$$

$$\text{Electricity cost} = \frac{(\$6000 + \$1200 + \$3480)/\text{yr}}{30\ \text{kW} \times 8000\ \text{h/yr}} = \$0.0445/\text{kWh}$$

c. The value of the energy saved would be

$$\text{Energy savings} = 30\ \text{kW} \times 8000\ \text{hr/yr} \times (\$0.06 - \$0.0445)/\text{kWh}$$
$$= \$3720/\text{yr}$$

Assuming the peak demand is reduced by the full 30 kW saves

$$\text{Demand savings} = 30\ \text{kW} \times \$7/\text{mo-kW} \times 12\ \text{mo/yr} = \$2520/\text{yr}$$

Notice the value of the demand reduction is a significant fraction of the total value.

The total savings would be

$$\text{Total annual savings} = \$3720 + \$2520 = \$6240/\text{yr}$$

This is net profit since the capital cost of the turbine was included in the fixed charge rate.

The ECP defined in (5.34) can be expressed more generally in terms of electrical and boiler efficiency. Referring to Fig. 5.9 and using American units, we can write for 1 Btu of fuel input to the system:

$$\text{Electrical output (kWh)} = \frac{1\ (\text{Btu}) \times \eta_P}{3412\ (\text{Btu/kWh})} \tag{5.36}$$

$$\text{Displaced thermal input (Btu)} = \frac{1\ (\text{Btu}) \times \eta_H}{\eta_B} \tag{5.37}$$

where η_P is the efficiency with which 1 unit of CHP fuel is converted into electricity, η_H is the efficiency with which 1 unit of CHP fuel is converted into useful heat, and η_B is the efficiency of the boiler/heater that isn't used because of the CHP. The energy chargeable to power (ECP) is

$$\text{ECP} = \frac{\text{Total thermal input} - \text{Displaced thermal input}}{\text{Electrical output}}$$

$$= \frac{\left(1 - \dfrac{\eta_H}{\eta_B}\right)}{\eta_P} \times 3412\ \text{Btu/kWh} \tag{5.38}$$

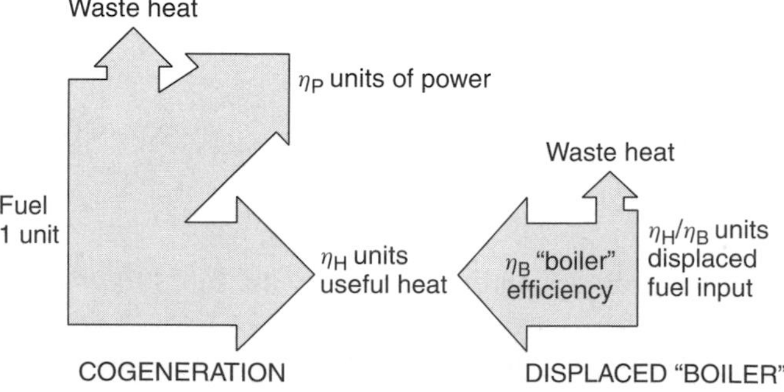

Figure 5.9 The economics of cogeneration are affected by the efficiency with which CHP fuel is converted into power and useful heat, as well as the efficiency of the boiler (or furnace or steam generator) that CHP heat displaces.

which simplifies to

$$\text{ECP} = \frac{3412}{\eta_P}\left(1 - \frac{\eta_H}{\eta_B}\right)\text{Btu/kWh} = \frac{3600}{\eta_P}\left(1 - \frac{\eta_H}{\eta_B}\right)\text{kJ/kWh} \tag{5.39}$$

Example 5.15 Cost of Electricity from a Fuel Cell. A fuel cell with an integral reformer generates heat and electricity for an apartment house from natural gas fuel. The heat is used for domestic water heating, displacing gas needed by the apartment house boiler. The following data describe the system:

Fuel cell rated output = 10 kW
Capacity factor CF = 0.90
Fuel cell/reformer electrical efficiency $\eta_P = 0.40$ (40%)
Fuel to useful heat efficiency $\eta_H = 0.42$
Boiler efficiency $\eta_B = 0.85$
Capital cost of the system = \$30,000
Paid for with an 8%, 20-year loan
Price of natural gas \$0.80/therm (\$8/10^6 Btu)

a. Find the energy, and cost of fuel, chargeable to power.
b. Find the cost of electricity (ignore inflation and tax benefits).
c. How many gallons of water per day could be heated from 60°F to 140°F?

Solution

a. From (5.39), the energy chargeable to power is

$$\text{ECP} = \frac{3412}{\eta_P}\left(1 - \frac{\eta_H}{\eta_B}\right)\text{Btu/kWh} = \frac{3412}{0.40}\left(1 - \frac{0.42}{0.85}\right) = 4315\ \text{Btu/kWh}$$

The fuel cost chargeable to power is

$$\text{CCP} = 4315\ \text{Btu/kWh} \times \$8/10^6\ \text{Btu} = \$0.0345/\text{kWh}$$

b. With a 90% capacity factor, the annual electricity delivered will be

$$\text{Electricity} = 10\ \text{kW} \times 8760\ \text{h/yr} \times 0.90 = 78{,}840\ \text{kWh/yr}$$

(sufficient for about 10 homes, or maybe 20 apartments)

From (5.21) or Table 5.5, the capital recovery factor is CRF(0.08, 20) = 0.1019/yr, so the annualized capital cost of the system is

$$A = P \times \text{CRF}(i, n) = \$30{,}000 \times 0.1019/\text{yr} = \$3057/\text{yr}$$

On a per kilowatt-hour basis, the annualized capital cost is

$$\text{Annualized capital cost} = \frac{\$3057/\text{yr}}{78{,}840\ \text{kWh/yr}} = \$0.0388/\text{kWh}$$

The fuel plus capital cost of electricity is therefore

$$\text{Electricity} = 3.45¢/\text{kWh} + 3.88¢/\text{kWh} = 7.33¢/\text{kWh}$$

This is about the same as the average U.S. price for electricity, and considerably less than the average residential price.
The gas cost for the fuel cell will increase over time due to inflation, but so will the competing price of electricity, so these factors tend to cancel each other.

c. Since the system is 40% efficient at converting fuel to 10 kW of electricity, the thermal input for the fuel cell is

$$\text{Fuel cell input} = \frac{10\ \text{kW}_\text{e}}{0.40} \times 3412\ \text{Btu/kWh} = 85{,}300\ \text{Btu/h}$$

Since 42% of the fuel input is converted to usable heat, the energy delivered to hot water while the unit is running at full power is

$$\text{Hot water} = \frac{85{,}300 \times 0.42\ \text{Btu/h}}{8.34\ \text{lb/gal} \times 1\ \text{Btu/lb°F} \times (140 - 60)\text{°F}} = 53.7\ \text{gal/h}$$

At a 90% capacity factor the daily hot water delivered would be

$$\text{Hot water} = 53.7\ \text{gal/h} \times 24\ \text{h/d} \times 0.90 = 1160\ \text{gal/day}$$

(Domestic consumption of hot water averages around 20 gal/person-day, so this is enough for almost 60 people.)

The calculation in Example 5.15 is based on reasonable near-term estimates for fuel cell costs and efficiencies, and the resulting cost of electricity, at 7.3 ¢/kWh, look quite attractive. A key reason for it looking so good is that the thermal output is helping to pay for the electricity from the fuel cell. To illustrate the importance of the cogenerated heat, Fig. 5.10 shows a sensitivity analysis in

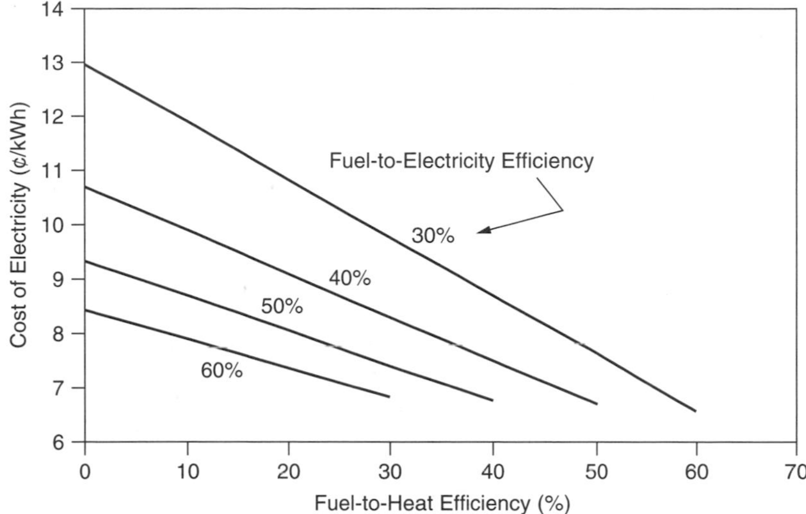

Figure 5.10 Cost of electricity from a fuel cell depends on the amount of waste heat utilized. Assumptions: CF = 0.9, $\eta_B = 0.85$, fuel = \$8/MMbtu, plant cost = \$3/W, loan 8%, 20 yr.

which electricity cost is plotted against the fraction of input energy that ends up as useful heat.

5.5.3 Design Strategies for CHP

The design of CHP systems is inherently complex since it involves a balancing act between the amount of power and usable heat that the cogenerator delivers and the amount of heat and power needed by the end use for that energy. While the power-to-heat ratio (P/H) of the cogeneration equipment may be reasonably constant, that is often not the case for the end-user for whom the demand for heat and for power often varies with the seasons. It may also vary throughout the day.

An industrial plant may have a relatively constant demand for heat and power, but that is not at all the case for a typical residential building. Imagine, for example, a residence with a relatively constant thermal demand for hot water, while its need for space heating is high in the winter and nonexistent in the summer. Moreover, its electrical demand may peak in the summer if it cools with a conventional, electricity-driven air conditioner, dip in the spring and fall when there is no heating or air conditioning, and rise somewhat again in the winter when more time is spent indoors. Figure 5.11 illustrates these not unusual circumstances.

Imagine trying to design a cogeneration system, say a fuel cell, with a P/H ratio of about 1/1, to deliver heat and electricity for the building in Fig. 5.11. The

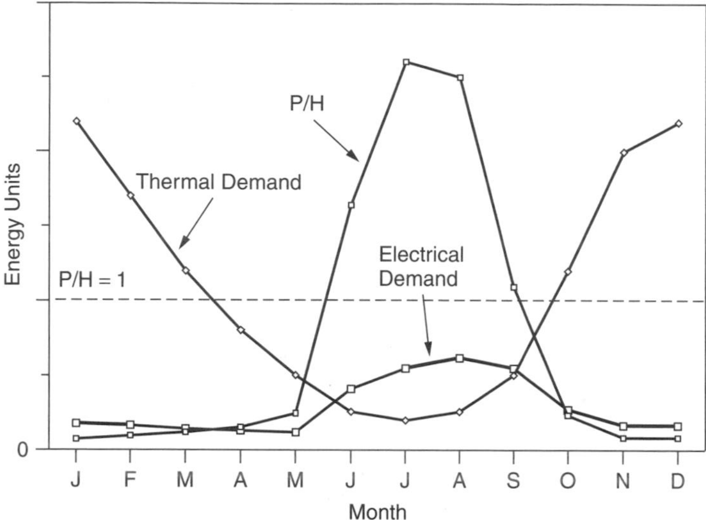

Figure 5.11 Thermal and electrical demand for a hypothetical residential building with summer air conditioning and winter space heating, expressed in equivalent units of energy. The ratio of power to heat for this load varies greatly from month to month.

building has an annual P/H ratio of about 1/3, but on a monthly basis it varies over a tremendous range from roughly 2.5 in the summer to 0.1 in the winter.

If, for example, the fuel cell is sized to deliver all of the electricity needed at the peak time of summer, and it continuously runs at that peak, then there will be extra power during most of the year. That extra power probably won't be wasted since it would likely be sold to the utility grid. With this sizing and operating protocol, almost all of the thermal output could be used with the exception of some excess heat from June to September and, assuming sales to the grid, all of the power would be of value.

If, on the other hand, the fuel cell is sized to deliver all of the thermal demand in the middle of winter, then it needs to have roughly four times the capacity of a system sized to meet the peak electrical load. If this much larger machine runs at full output all year long, then a very high fraction of the thermal output will be wasted and most of the electricity will have to be sold to the grid. As Fig. 5.10 demonstrates, the economics using this strategy will be severely compromised since there isn't much thermal output to subsidize the electrical output.

Another strategy would be to design for the thermal peak in winter, but modulate the fuel cell output so that it follows the thermal or electrical load. This isn't a very good idea either since the system will work with a very poor annual capacity factor, which increases the annualized capital cost. Both strategies based on sizing for the peak thermal load would not seem to be as cost effective as sizing for the peak electrical load.

A further complication results from the hour-to-hour variation in the heat and power demands for loads. For industrial facilities and office buildings, heat and power needs may vary somewhat in unison, being high together during the workday and lower at night. But for residential loads, especially in summer, the electrical and thermal demands may be completely unsynchronized with each other, in which case some sort of diurnal thermal storage system may be appropriate.

Even from this simple example, it should be apparent that designing the most cost-effective cogeneration system, one that takes maximum advantage of the waste heat, is a challenging exercise.

5.6 COOLING, HEATING, AND COGENERATION

The cost effectiveness of fuel-cell and other cogeneration technologies is quite dependent on the ability to utilize the available thermal energy as well as the electric power generated. However, as Fig. 5.11 suggested, the summer power demand for cooling and the winter thermal demand for heating cause the ratio of power-to-heat for many buildings to vary widely through the seasons, which complicates greatly the design of an economically viable CHP system. And since the buildings sector is almost three-fourths of the grid's total peak demand, and much of the rise in peak during the summer is associated with air conditioning, it is a load worth special attention.

There are several ways to smooth the power-to-heat ratio in buildings by using alternative methods for heating and cooling. Heat pumps, for example, can smooth the ratio by substituting electricity for heat in the winter, while absorption cooling systems can smooth if by substituting thermal for electrical power during the summer.

5.6.1 Compressive Refrigeration

A conventional vapor-compression chiller is shown in Fig. 5.12. It consists of a refrigerant that cycles through four major system components: compressor, condenser, expansion valve, and evaporator. In the compressor, the refrigerant, which enters as a cold, low-pressure vapor, is pumped up to a high pressure. When a gas is compressed, its temperature and pressure increase, so the refrigerant enters the condenser at a relatively high temperature. In the condenser, the refrigerant transfers heat to warm outside air (or perhaps to water from a cooling tower), causing the refrigerant to release its latent heat as it changes state from vapor to liquid. The high-pressure liquid refrigerant then passes through an expansion valve, where the pressure is suddenly released, causing the liquid to flash to vapor and it emerges as a very cold, low-pressure mixture of liquid and vapor. As it passes through the evaporator, the refrigerant continues to vaporize, absorbing heat from its environment as it changes state from liquid to a vapor.

The amount of heat the refrigerant absorbs and releases as it changes phase, and the temperature at which those transitions occur depend on the particular refrigerant being used. Traditionally, the refrigerants of choice have been

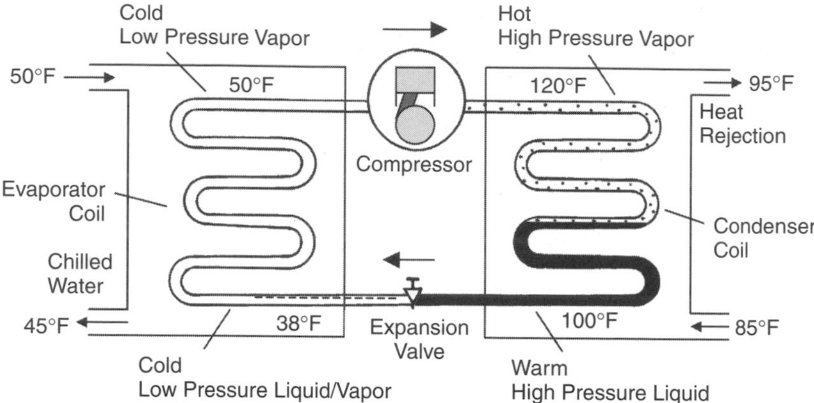

Figure 5.12 Conventional vapor compression chiller. Temperatures are representative for a building cooling system.

chlorofluorocarbons (CFCs), which are inert chemicals containing just chlorine, fluorine, and carbon. CFCs were developed to satisfy the need for a nontoxic, nonflammable, efficient refrigerant for home refrigerators. Before that time, the most common refrigerants were either toxic, noxious, or highly flammable or required high operating pressures, which necessitated heavy, bulky equipment. When they were first introduced in the 1930s, CFCs were hailed as wonder chemicals since they have none of those undesirable attributes. In 1974, however, it was discovered that CFCs were destroying the stratospheric ozone layer that shields us from lethal doses of ultraviolet radiation. That discovery, and subsequent international efforts to ban CFCs, has led to a scramble in the cooling industry to find suitable replacements.

CFCs were also used in the manufacture of various rigid and flexible plastic foams, including rigid foam insulation used in refrigerators and buildings. The simultaneous need to replace CFC refrigerants as well as CFC-containing foam insulation raised concerns that the electricity demand of refrigerators, which are the largest source of consumption in many households, would increase. In fact, the opposite occurred. With increasingly stringent California, and then Federal, refrigerator efficiency standards, along with a modest amount of government R&D stimulation, the refrigerator industry has accomplished a remarkable technological achievement. A modern refrigerator now uses one-third the electricity of the typical 1974 model, while its size and features have increased, and all of its ozone-depleting substances have been removed. Figure 5.13 shows this amazing improvement.

The efficiency of the basic refrigeration cycle can be expressed in dimensionless terms as a coefficient of performance (COP_R):

$$\text{COP}_\text{R} = \frac{\text{Desired output}}{\text{Work input}} = \frac{Q_L}{W} \tag{5.40}$$

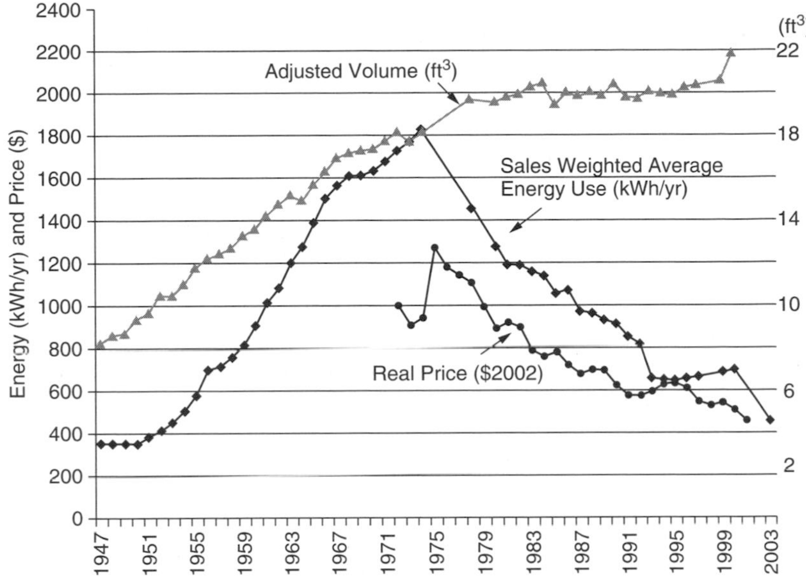

Figure 5.13 Energy use, size, and real price ($2002) of U.S. refrigerators updated from Goldstein and Geller (1999).

where W is the work put into the compressor and Q_L is the heat extracted from the refrigerated space. In the United States, the efficiency is often expressed in terms of an *energy efficiency rating* (*EER*), which is defined as

$$\text{Energy efficiency rating (EER)} = \frac{\text{Heat removal rate } Q_L \text{ (Btu/h)}}{\text{Input power } W \text{ (watts)}} \tag{5.41}$$

Typical air conditioners have values of EER that range from about 9 Btu/Wh to 17 Btu/Wh. Note that EER and COP_R are related by

$$\text{EER (Btu/Wh)} = \text{COP}_\text{R} \times 3.412 \text{ Btu/Wh} \tag{5.42}$$

Another measure of air conditioner efficiency is the *seasonal energy efficiency rating* (SEER), which has the same Btu/Wh units. While the EER is an instantaneous measure, the SEER is intended to provide an average rating accounting for varying outdoor temperatures and losses due to cycling effects.

The traditional unit of cooling capacity for U.S. air conditioners originated in the days when the only source of cooling was melting ice. By definition, one "ton" of cooling is the rate at which heat is absorbed when a one-ton block of ice melts in 24 hours. Since it takes 144 Btu to melt one pound of ice, and there are 2000 pounds in one (short) ton,

$$1 \text{ ton of cooling} = \frac{2000 \text{ lb} \times 144 \text{ Btu/lb}}{24 \text{ h}} = 12{,}000 \text{ Btu/h} \tag{5.43}$$

This leads to another measure of the efficiency of an air conditioning chiller, which is the tons of cooling it produces per kW of electrical power into the compressor:

$$\text{kW/ton} = \frac{12{,}000 \text{ Btu/h-ton}}{\text{EER (Btu/h-W)} \times 1000 \text{ W/kW}} = \frac{12}{\text{EER}} \tag{5.44}$$

An average, home-size, 3-ton air conditioner uses roughly 3 kW, but the best ones will only need 2 kW to do the same job. For comparison, a 100-ton rooftop chiller for commercial buildings draws on the order of 100 kW and, depending on climate, cools about 30,000 square feet of office space.

The heart of a cooling system is the chiller, which for a vapor compression system is the unit described in Fig. 5.12. How the coldness created in the evaporator heat exchanger reaches the desired destination is very dependent on the particular application. In a simple room air conditioner, a fan mixes warm return air from the conditioned space, with some fresh ambient air, and blows it directly across the cold evaporator coil for cooling. In larger buildings, chilled water lines may be used to carry cold water from the chiller to cooling coils located in forced air ducts around the building. For district heating and cooling systems there may be underground steam and chilled water lines running from a cogeneration plant to nearby buildings. Figure 5.14 shows the Stanford University campus energy facility, which includes a 50-MW combined cycle CHP plant, multiple vapor-compression and absorption chillers, and a 96,000 ton-hour underground water-and-ice storage vault. At night, when electricity is cheap, chillers make ice that can be melted the next day to supplement the cooling capacity of the plant.

5.6.2 Heat Pumps

Notice that the basic vapor-compression refrigeration cycle can be used to do heating as well as cooling by, in essence, turning the machine around as shown

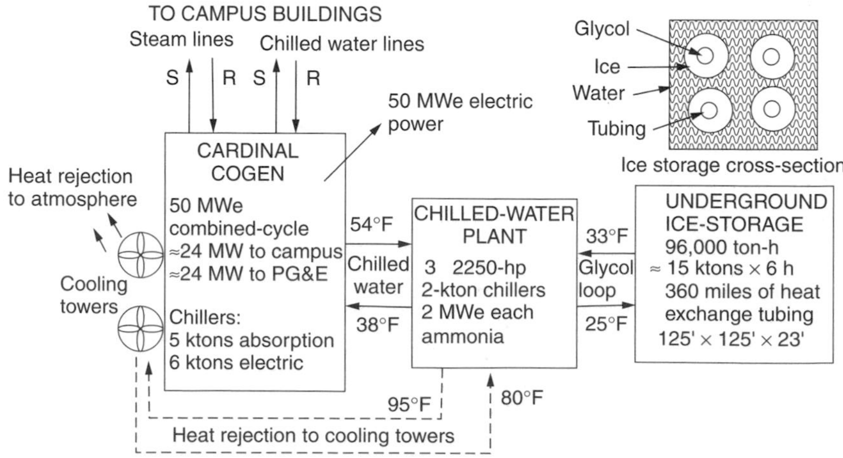

Figure 5.14 The Stanford University combined-cycle CHP plant with ice storage.

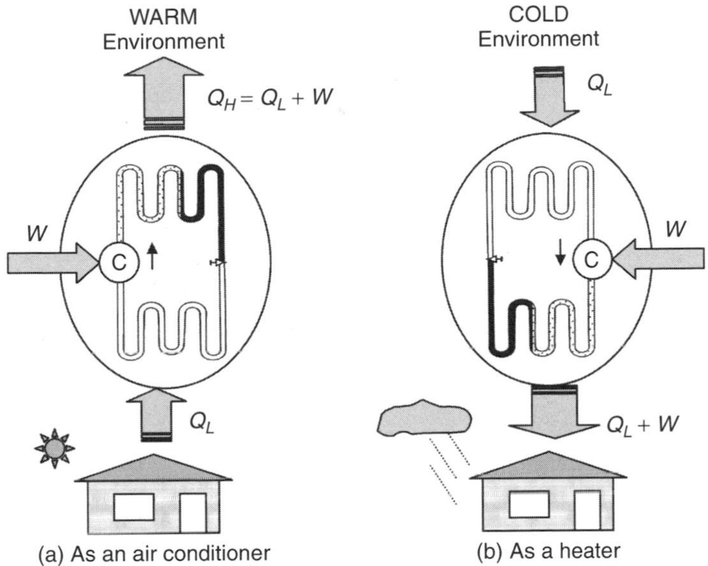

Figure 5.15 A heat pump acts as an air conditioner or a heater depending on which direction the evaporator and condenser coils face.

in Fig. 5.15. *Heat pumps* have this ability to be reversed, so they can serve as an air conditioner in the summer and a heater in the winter. Heat pumps may seem to defy some law of thermodynamics, since they take heat out of a cold space and pump it to a hot one, but of course they don't. When used as heaters, they are nothing more than a refrigerator with its door wide open to the ambient and its condenser coils facing into the kitchen, trying to cool the whole outside world just the way it tries to cool the warm beer on the shelf.

As a heater, a heat pump theoretically delivers the sum of the heat removed from the environment plus the compressor energy, which leads to the following expression for the COP for heat pumps:

$$\mathrm{COP_{HP}} = \frac{\text{Desired output}}{\text{Work input}} = \frac{Q_H}{W} = \frac{Q_L + W}{W} = 1 + \frac{Q_L}{W} > 1 \qquad (5.45)$$

Notice, by comparison with (5.40), there is a simple relationship between the coefficient of performance as a refrigerator or air conditioner ($\mathrm{COP_R}$) and as a heat pump ($\mathrm{COP_{HP}}$):

$$\mathrm{COP_{HP}} = \mathrm{COP_R} + 1 \qquad (5.46)$$

Equation (5.45) tells us that a heat pump, in its heating mode, has a $\mathrm{COP_{HP}}$ greater than unity, which means that more heat goes into the conditioned space than the energy content of the electricity running the compressor. The actual COP will depend on the ambient temperature, dropping as the outside temperature

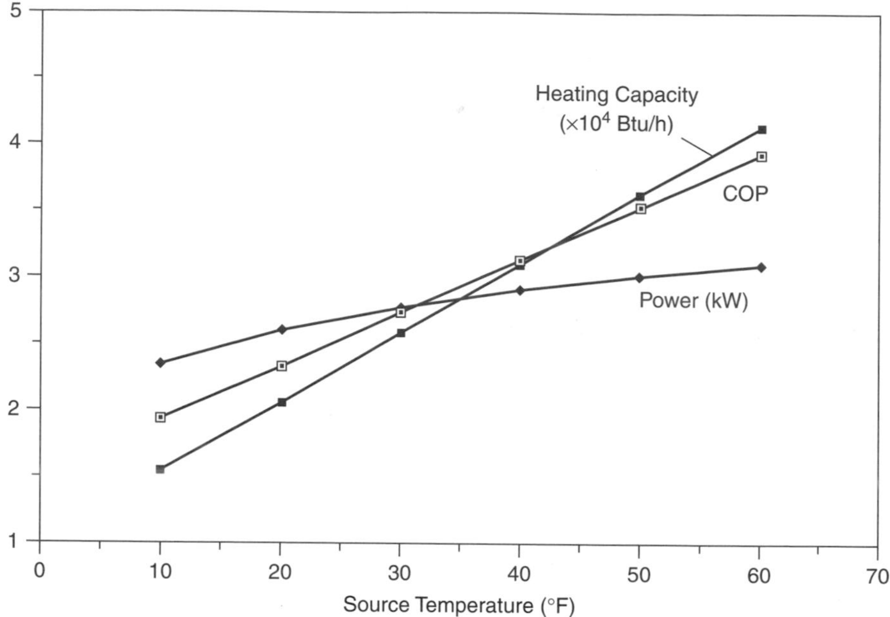

Figure 5.16 Temperature dependence of COP, power demand, and heating capacity for a home-size, 3-ton heat pump.

drops. An example of the temperature dependence of the COP, power demand, and heating capacity of a 3-ton heat pump is shown in Fig. 5.16. Notice that as it gets colder outside, the heater output drops at the same time that the demand for heat rises. What this means is that at some ambient temperature, called the balance point temperature, the heat pump can no longer deliver enough heat to satisfy the demand and supplementary heating must be provided. This supplement is typically built-in strip resistance heaters, which, when they are turned on, cuts the overall COP of the heat pump dramatically. It is not unusual for the overall COP to approach unity as ambient drops below freezing. One way, however, to imagine maintaining the high efficiency of a heat pump in cold weather is to couple it with a source of waste heat, such as a CHP fuel cell.

In mild climates where electricity is relatively inexpensive, many homes are heated with straight electric resistance heating. A simple air-source heat pump in those circumstances can usually reduce the electricity demand by at least half. In climates characterized by very cold winters and very hot summers, heat pumps that use a local body of water, or the ground itself, as the source and sink for heating and cooling can be even more efficient. Since the ground temperature ten-to-twenty feet below the surface often hovers around 50°F, a heat pump with its source at that temperature delivers heat with very high COP. Moreover, in the summer, rejecting heat into the cool ground rather than into hot ambient air also improves its efficiency as an air conditioner. Such ground-source heat pumps are

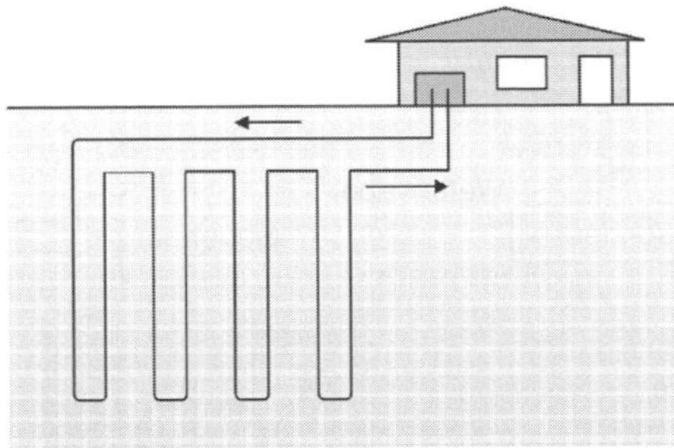

Figure 5.17 Ground source heat pump for high heating and cooling efficiency.

expensive, but can be the most efficient of all methods currently available to heat and cool a home. A simple schematic of such a system, with vertical closed heat exchange loops buried in the ground, is shown in Fig. 5.17.

5.6.3 Absorption Cooling

As was shown in Fig. 3.6, one of the biggest contributors to the peak demand for power in a utility system is space cooling. For residential and commercial customers, the cost of air conditioning is often a major fraction of their electricity bills—especially when the tariff includes demand charges. While most cooling systems in use today are based on electricity-powered, vapor-compression-cycle technologies, there are heat-driven alternatives that make especially good sense as part of a cogeneration system. The most developed of these systems is based on the absorption refrigeration cycle shown in Fig. 5.18.

In an absorption cooling system, the mechanical compressor in a conventional vapor compression system is replaced with components that perform the same function, but do so thermochemically. In Fig. 5.18, everything outside of the dotted box is conventional; that is, it has the same components as would be found in a compressive refrigeration system. A refrigerant, in this case water vapor, leaves the thermochemical compressor under pressure. As it passes through the condenser, it changes state, releasing heat to the environment, and emerges as pressurized liquid water. When its pressure is released in the expansion valve, it flashes back to the vapor state in the evaporator, drawing heat from the refrigerated space. The refrigerant then must be compressed to start another pass around the loop.

The key to the dotted box on the right of the figure is the method by which the vaporized refrigerant leaving the evaporator is re-pressurized using heat rather than a compressor. The system shown in Fig. 5.18 uses water vapor as the refrigerant and uses lithium bromide (LiBr) as the absorbent into which the water vapor dissolves. A very small pump shuttles the water–LiBr solution from the absorber to the generator. In the generator, heat is added, which pressurizes the dissolved water vapor and drives it out of solution. The pressurized water vapor leaving the

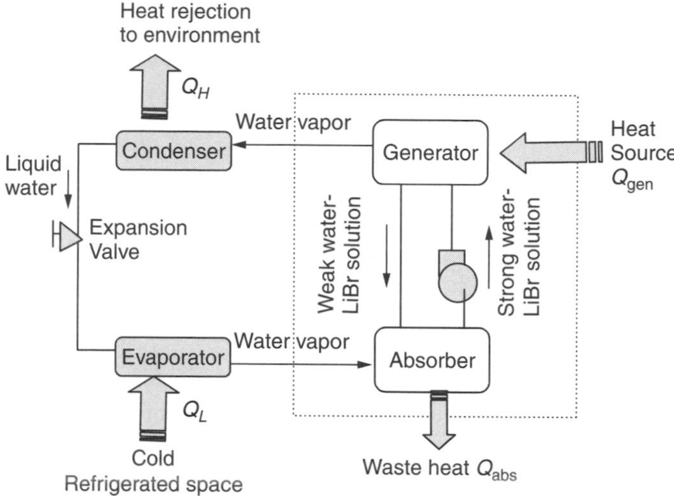

Figure 5.18 Lithium bromide absorption chiller schematic. The compressor in a standard compressive refrigeration system has been replaced with a generator and absorber that perform the same function but do so with heat instead of mechanical power.

generator is ready to pass through the condenser and the rest of the conventional refrigeration system. Meanwhile, the lithium bromide returns to the absorber, never having left the system. The absorption of refrigerant into the absorbent is an exothermic reaction, so waste heat is released from the absorber as well as the condenser.

The COP of an absorption refrigeration system is given by

$$\mathrm{COP_R} = \frac{\text{Desired output (cooling)}}{\text{Required input (heat + very little electricity)}} \approx \frac{Q_L}{Q_{\text{gen}}} \tag{5.47}$$

For a good LiBr system, the COP is currently about 1.0 to 1.1, but new "triple effect" absorption units promise COPs better than 1.5.

A competing absorption cycle chiller uses water as the absorbent and uses ammonia as the refrigerant. These NH_3–H_2O chillers date back to the mid-1850s and were originally popular as ice makers. While their COPs are not as good as those based on LiBr, they do create lower temperatures and, because they operate with higher pressures, the piping and equipment can be physically smaller.

As is suggested in Fig. 5.19, the advantage of absorption cooling coupled with cogeneration is the smoothing of the demand for thermal energy throughout the year, which significantly improves the economics of CHP electricity.

5.6.4 Desiccant Dehumidification

Another technology that can take advantage of thermal energy to provide cooling is based on the ability of some materials, called desiccants, to dehumidify air

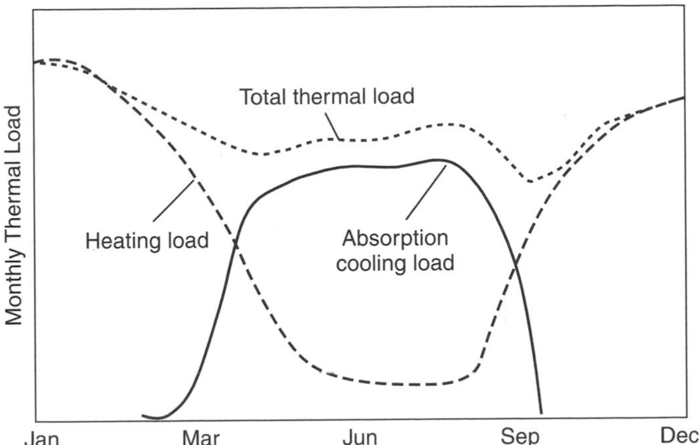

Figure 5.19 Absorption cooling helps smooth the annual thermal demand for a building.

by adsorbing water vapor onto their surfaces. The cooling load associated with cooling outside air, or recirculated indoor air, is a combination of (a) sensible cooling, in which the temperature of air is lowered, and (b) latent cooling, in which water vapor is removed. Especially in humid climates, the latent cooling load is by far the most important. In the usual compressive refrigeration system, air is dehumidified by cooling it well below its dew point to condense out some of the water vapor, and then it is reheated to a comfortable supply temperature. All of that takes considerable compressor power. When an inexpensive source of relatively low-temperature heat is available, however, desiccants offer an alternative, less electricity-intensive approach.

Desiccants, such as natural or synthetic zeolites, powdered silica gels, and certain solid polymers, have a strong affinity for moisture, which means that when contact is made, water vapor readily condenses, or adsorbs, onto their surfaces. The resulting release of latent heat raises the desiccant temperature, so the overall result of an air stream passing through a desiccant is dehumidification with an increase in air temperature. The accumulating moisture in the desiccant must of course be removed, or desorbed, and that is accomplished by driving a very hot air stream through the desiccant in a process called regeneration. The sorption and desorption of moisture is often done with a rotating wheel, as shown in Fig. 5.20.

The remaining components in a desiccant air-conditioning system are there to provide sensible cooling to the now hot, dry supply air and to provide the heat needed to regenerate the desiccant. It is this latter step that dominates the energy requirement of a desiccant system, and that is why the availability of an inexpensive source of heat becomes so important for these systems. Waste heat from an industrial process, thermal energy from a cogeneration facility, or heat from a solar collector system all could provide the necessary thermal input to these systems. An example of a desiccant cooling system is shown in Fig. 5.21.

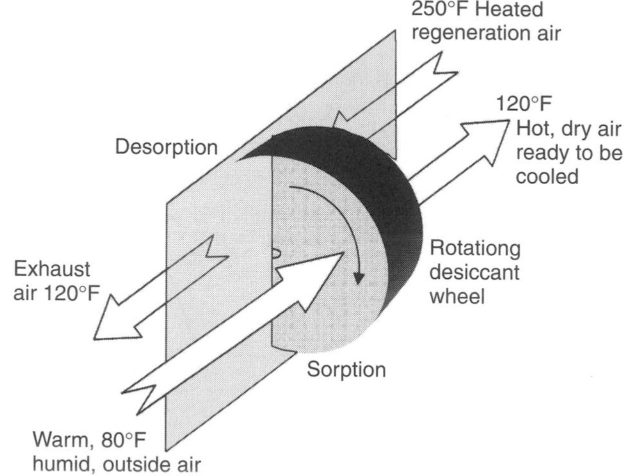

Figure 5.20 Desiccant wheel as part of a dehumidification and cooling system.

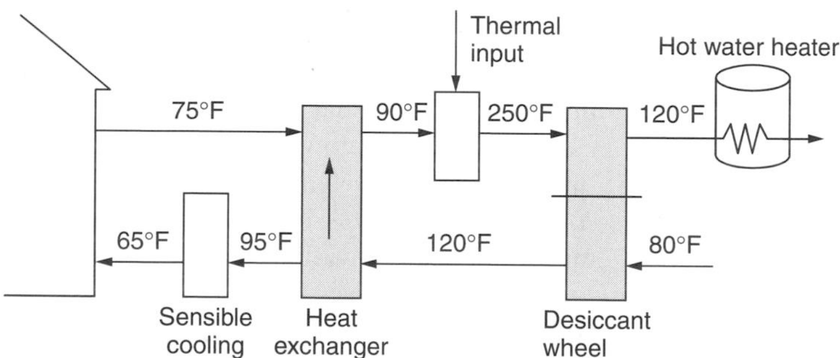

Figure 5.21 A desiccant refrigeration system for space cooling and domestic hot water. Temperatures are illustrative.

5.7 DISTRIBUTED BENEFITS

There are a number of subtle but important benefits associated with distributed generation, which have only recently begun to be appreciated and assessed (Swisher, 2002). In addition to direct energy savings, increased fuel efficiency with cogeneration, and reduced demand charges for larger customers, there are a number of other DG attributes that can significantly add value, including

- *Option Value:* Small increments in generation can track load growth more closely, reducing the costs of unused capacity.
- *Deferral Value:* Easing bottlenecks in distribution networks can save utilities costs.

- *Engineering Cost savings:* Voltage and power factor improvements and other ancillary benefits provide grid value.
- *Customer Reliability Value:* Reduced risk of power outages and better power quality can provide major benefits to some customers.
- *Environmental Value:* Reduced carbon emissions for CHP systems will have value if/when carbon taxes are imposed; for fuel cells, since they are emission-free, ease of permitting has value.

5.7.1 Option Values

In the context of distributed generation, the *option* part of *option value* refers to the choice of the incremental size of new generation capacity; that is, the buyer has the option to purchase a large power plant that will satisfy future growth for many years, or a series of small ones that will each, perhaps, only satisfy growth for a year or two. The *value* part refers to the economic advantages that go with small increments, which are better able to track load growth.

Figure 5.22 illustrates the comparison between adding fewer, larger increments of capacity versus smaller, but more frequent, increments. Smaller increments, such as distributed generation, better track the changing load and result in less idle capacity on line over time. The advantage is especially apparent when forecasted load growth doesn't materialize and a large incremental power plant may result in long-term overbuilt capacity that remains idle but still incurs costs.

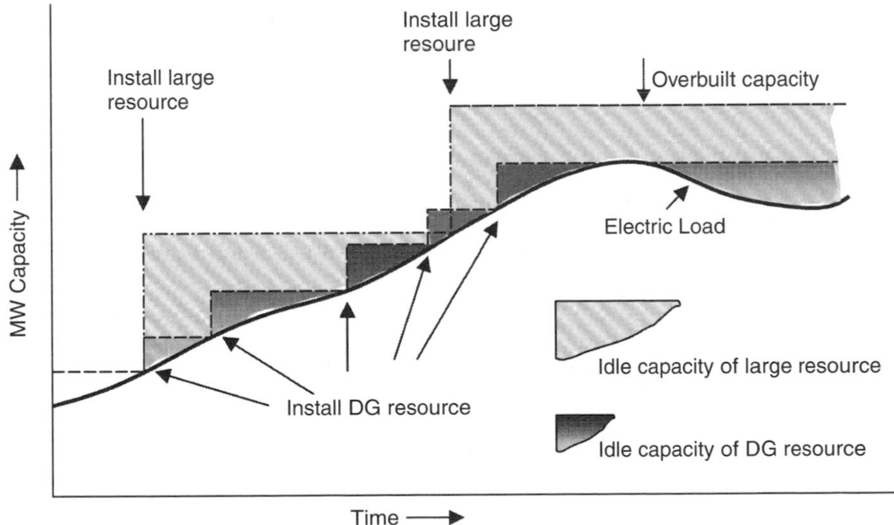

Figure 5.22 Smaller distributed generation increments better track the changing electric load with less idle capacity than fewer, larger plants. Less idle capacity translates into decreased costs. Based on Swisher (2002).

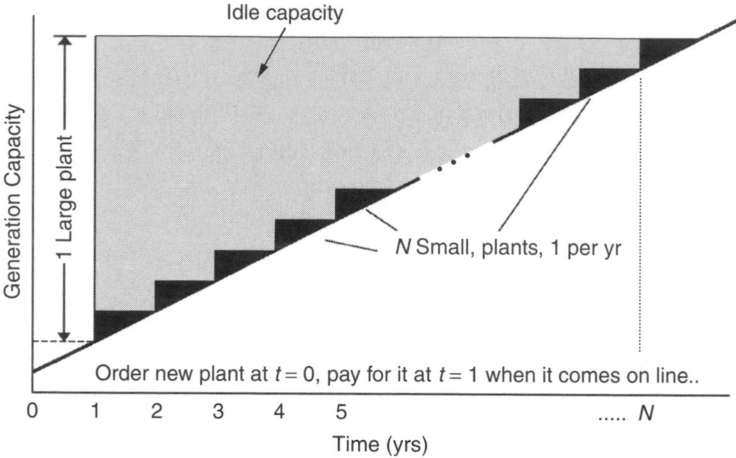

Figure 5.23 Comparing the present worth of a single large plant capable of supplying N years of growth with N plants, each satisfying 1 year of growth. For simplicity, both are assumed to have 1-year construction lead times.

One way to quantify the cost advantage of the option value is to do a present worth analysis of the capital cost of a single large plant compared with a series of small ones. Figure 5.23 shows a comparison of one large power plant, which can supply N years of load growth, with N small plants, each able to supply one year's worth of growth. Let us begin by assuming that it only takes 1 year between the time a plant is ordered and the time it comes on line, so at time $t = 0$, either a large plant or a small plant must be ordered. Finally, assume that full payment for a plant is due when it comes on line.

Let

$$\text{Annual growth in demand} = \Delta P \text{ (kW)}$$

$$\text{Cost of a small plant} = P_S \text{ (\$/kW)} \times \Delta P \text{ (kW)}$$

$$\text{Cost of a large plant} = P_L \text{ (\$/kW)} \times N\Delta P \text{ (kW)}$$

$$\text{Discount rate} = d \text{ (yr}^{-1}\text{)}$$

The present value of the single large plant (PV_L) is its initial cost discounted back 1 year to $t = 0$:

$$PV_L = \frac{P_L N \Delta P}{(1+d)} \tag{5.48}$$

For the sequence of N small plants, the present value of N payments of $P_S \Delta P$, with the first one at $t = 1$ year, is given by

$$PV_S = P_S \Delta P \,\text{PVF}(d, N) \tag{5.49}$$

where PVF(d, N) is the present value function used to find the present value of a series of N equal annual payments starting at $t = 1$. If we equate the present values of the large plant and the N small ones, we get

$$\frac{P_L N \Delta P}{(1+d)} = P_S \; \Delta P \; \text{PVF}(d, N) \tag{5.50}$$

and, using (5.9) for PVF(d,N) gives

$$\frac{P_S(\$/\text{kW})}{P_L(\$/\text{kW})} = \frac{N}{(1+d)\text{PVF}(d, N)} = \frac{N}{(1+d)}\left[\frac{d(1+d)^N}{(1+d)^N - 1}\right]$$
$$= N\left[\frac{d(1+d)^{N-1}}{(1+d)^N - 1}\right] \tag{5.51}$$

Equation (5.51) is the ratio of the small-plant to large plant capital costs (\$/kW) that makes them equivalent on a present worth basis.

Example 5.16 Present Value Benefit of Small Increments of Capacity. Use (5.51) to find the present value advantage of eight small plants, each supplying 1 year's worth of growth, over one large plant satisfying 8 years of growth. Use a discount rate of 10%/yr. Each plant takes 1 year to build. If the large one costs \$1000/kW, how much can the small ones cost to be equivalent?

Solution. Using (5.51), we obtain

$$\frac{P_S(\$/\text{kW})}{P_L(\$/\text{kW})} = 8\left[\frac{0.1(1+0.1)^7}{(1+0.1)^8 - 1}\right] = 1.363$$

This means the small DG plants can cost \$1363/kW and they still would be equivalent to the single large one at \$1000/kW.

Even though the calculation in Example 5.16 is impressive enough as it stands, it omits another important advantage of smaller plants. Large plants tie up capital for a longer period of time before the plant can be designed, permitted, built, and turned on. That longer lead time costs money, so let's modify our analysis to include it.

Imagine the initial cost of the large plant being spread over years 1 through L, where L is the lead time. Also, suppose payments on the N small plants begin in year L, as shown in Fig. 5.24. We continue to assume the small plants can

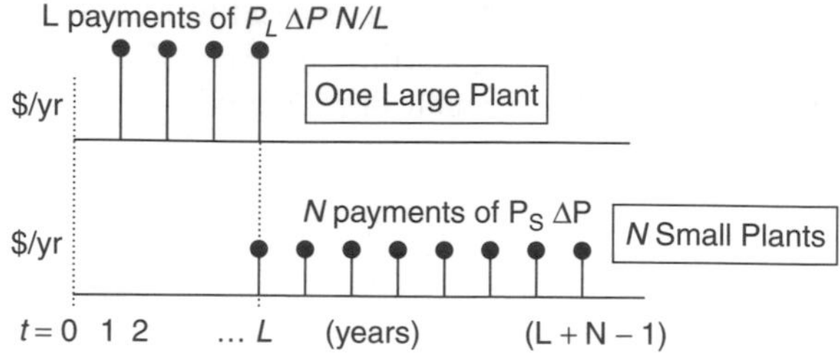

Figure 5.24 The payments made on one large plant spread over the L years of lead time that it takes to bring it on line, and annual payments made on N small plants. The first of each type comes on line in year L.

be built in one year. The present value of the payments made on the large plant costing $P_L N \Delta P$ spread out over L years is given by

$$\text{PV}_L = \frac{P_L N \Delta P}{L} \cdot \text{PVF}(d, L) \tag{5.52}$$

And, for the N small plants, each costing $P_S \Delta P$ over years L to $L + N - 1$, we have

$$\text{PV}_S = \frac{1}{(1+d)^{L-1}}[P_S \Delta P \cdot \text{PVF}(d, N)] \tag{5.53}$$

where the term in the brackets finds the present value of the series of payments, discounted back to year $L - 1$, and the initial term discounts that back to $t = 0$. Setting the present value of the large plant (5.52) equal to the present value of the series of small ones (5.53) results in the following ratio of capital costs for small and large plants, including the extra lead time for the large one:

$$\frac{P_S(\$/\text{kW})}{P_L(\$/\text{kW})} = \frac{N(1+d)^{L-1}}{L} \cdot \frac{\text{PVF}(d, L)}{\text{PVF}(d, N)} \tag{5.54}$$

where N is the number of small plants, each taking 1 year to build and satisfying 1 yr of growth; L represents the years of lead time for the single large plant; and d is the discount rate (yr^{-1}).

Example 5.17 Option Value of Small Plants Including Short Lead-Time Advantage. Find the capital cost that could be paid for eight small plants, each sized to supply one year of load growth and each taking 1 year to build, compared with one large plant that supplies 8 years of growth. The large plant costs \$1000/ kW and has a lead time of 4 years to build. Use a discount rate of 10%/yr.

Solution. To use(5.54) we need

$$\text{PVF}(d, L) = \frac{(1+d)^L - 1}{d(1+d)^L} = \frac{(1+0.1)^4 - 1}{0.1(1+0.1)^4} = 3.1698$$

and

$$\text{PVF}(d, N) = \frac{(1+0.1)^8 - 1}{0.1(1+0.1)^8} = 5.3349$$

so

$$\frac{P_S(\$/\text{kW})}{P_L(\$/\text{kW})} = \frac{N(1+d)^{L-1}}{L} \cdot \frac{\text{PVF}(d, L)}{\text{PVF}(d, N)} = \frac{8(1+0.1)^{4-1}}{4} \times \frac{3.1698}{5.3349} = 1.582$$

The capital cost of the small plants can be \$1582/kW, and they would still be equivalent to one large plant costing \$1000/kW.

Table 5.10 has been derived from (5.54) to show the cost premium that smaller plants can enjoy and still have the same present value as a single larger one. An additional advantage of the small plants is that they will have less unused capacity if load doesn't grow as fast as forecast.

Another advantage of meeting projected demand with incremental capacity additions is the reduced need for working capital to build the plants. Putting enormous amounts of capital into a huge, billion-dollar power plant and then

TABLE 5.10 Capital Cost Premium that *N* Smaller Plants Can Have Versus One Larger One and Still Have the Same Present Value[a]

N	$L = 1$ yr	2 yr	3 yr	4 yr	5 yr
1	0%	5%	10%	16%	22%
2	5%	10%	16%	22%	28%
3	10%	15%	21%	27%	34%
4	15%	20%	27%	33%	40%
5	20%	26%	32%	39%	46%
6	25%	32%	38%	45%	53%
7	31%	37%	44%	52%	60%
8	36%	43%	50%	58%	66%
9	42%	49%	57%	65%	73%
10	48%	55%	63%	72%	81%

[a] L is years of lead time to build the larger plant. Assumes discount rate of 10%/yr and a 1-year lead time for small plants.

Source: Swisher (2002).

waiting 5 or so years to start earning revenue could be perceived by the capital markets to be a riskier investment than would be the case for a series of short-lead-time, modular units. The reduced risk in the modular approach can translate into lower cost of capital, providing another financial advantage to distributed generation.

5.7.2 Distribution Cost Deferral

The utility distribution system, which sends power to every house and store in town, is a very expensive asset that is typically greatly underutilized except in certain critical locations during certain times of the day and year. As Fig. 5.25 attests, distribution systems constitute a significant fraction—on the order of half in recent years—of annual utility construction cost. Running power lines up and down every street to serve residential customers is expensive, and it is one reason why residential electric rates are so much higher than those for large, industrial facilities with dedicated feeders. In terms of the overall cost of running a utility, including construction costs, fuel, administrative costs, billing, and operations and maintenance, the distribution system accounts for about 20% of the total.

As Fig. 5.26 indicates, distribution system assets are not only expensive, but they are often utilized far less than is the case for generation. For example, only one-third of the time is this particular feeder operating above 50% of its capacity—or conversely, two-thirds of the time there is more than twice as much distribution capacity as is needed. Compare this with generation, where that level of underutilization never occurs.

Distribution systems are plagued by bottlenecks. When a new housing development or shopping center gets built, distribution feeders and the local substation may have to be upgraded even though their current capacity may be exceeded for only a few hours each day during certain months of the year. The effect of that localized growth can ripple from feeder to substation to transmission lines to power plants.

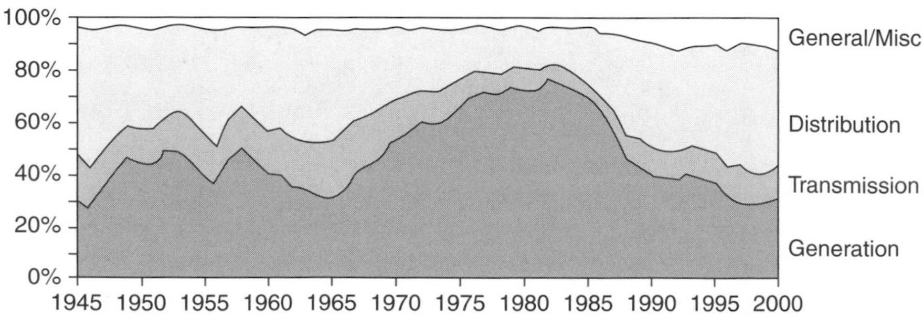

Figure 5.25 Allocation of U.S. investor-owned utilities' construction expenditures. From Lovins et al. (2002).

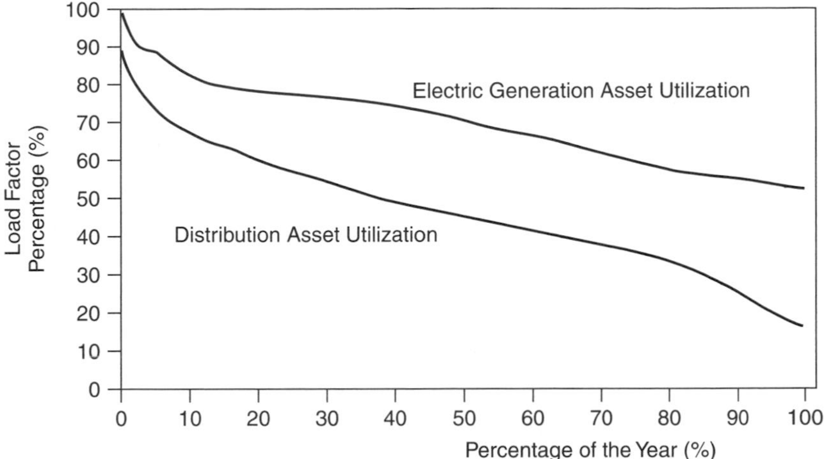

Figure 5.26 Comparing the load–duration curve for a typical feeder in the distribution system of a large utility with a curve for generation assets. The distribution system assets are utilized far less than the generation system. From Ianucci (1992).

Studies of area-and-time (ATS) costs can pinpoint those portions of a distribution network where the marginal cost of delivering another kilowatt makes distributed generation and demand-side management particularly cost effective. Figure 5.27, for example, shows an extreme case in which an ATS study identified a particular feeder in which the marginal cost of peak power rose to over $3.50/kWh during a few hours a day over just a few weeks each year. At such a site, almost any distributed generation or energy conservation system would be cost effective.

On-site generation by utility customers can help avoid or delay the need for distribution system upgrades, leading to more efficient use of existing facilities. Customers in portions of the distribution grid where capacity constraints are imminent could in the future be provided with incentives to generate some of their own power (or shed some of their load), especially at times of peak power demand. Area-and-time-based differential pricing of grid services could well become an important driver of distributed generation.

5.7.3 Electrical Engineering Cost Benefits

Besides capacity deferrals, distributed generation can also decrease costs in the distribution network by helping to improve the efficiency of the grid. Power injected into the grid by local distributed generation resources helps to reduce losses in several ways.

Without DG, the current supplied to a distribution network by the transmission system must be sufficient to satisfy all of the loads, from one end to the other. When current flows through conductors, there is voltage drop due to Ohm's law,

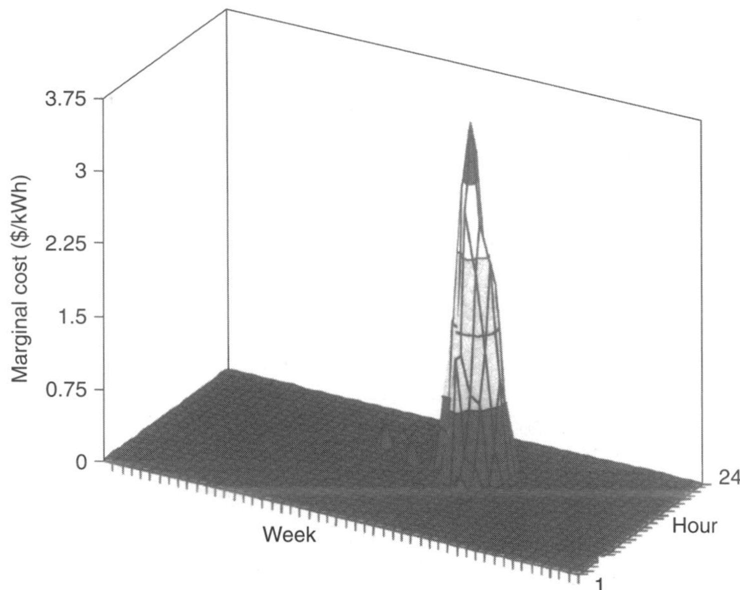

Figure 5.27 Area-and-time (ATS) studies can identify feeders where distributed generation and demand-side management programs can be particularly cost effective. This is a particular PG&E feeder in the early 1990s. From Swisher and Orans (1996).

$\Delta V = iR$, and there is power loss due to i^2R heating of the wires. The longer the distance and the greater the current, the more there will be voltage drop and power loss in the wires. If, however, a distributed generation source provides some of its own power, or if loads can be reduced through customer efficiency, the current and power that the grid needs to supply will drop and so will grid losses. Moreover, if the DG source actually delivers power to the grid, line losses will be reduced even more.

Injecting power onto the grid not only provides voltage support to offset iR drops and reduce i^2R losses, it can also raise the power factor of the lines. Recall that when voltage and current are out of phase with each other, more current must be provided to deliver the same amount of true power (watts) capable of doing work. Improving the power factor is usually accomplished by adding banks of capacitors to the line, but it can also be helped if DG systems are designed to inject appropriately phased reactive power. Improved power factor reduces line current, which reduces voltage sag and line losses. On top of that, a better power factor also helps distribution transformers waste less energy, supply more power, and extend their lifetime.

5.7.4 Reliability Benefits

Most power outages are caused by faults in the transmission and distribution system—wind and lightning, vehicular accidents, animals shorting out the wires—not generation failures. To the extent that a customer can provide some fraction of their own power during those outages, especially to critical loads such as computers and other digital equipment, the value of the added reliability can easily surpass the cost of generation by orders of magnitude. Such emergency standby power is now often provided with back-up generators, but such systems are usually not designed to be operated continuously, nor are they permitted to do so given their propensity to pollute. This means that they don't provide any energy payback under normal circumstances. Battery systems and other uninterruptible power supplies (UPS) can't cover sustained outages on their own without additional backup generators, so they have the same disadvantages of not being able to help pay for themselves by routinely delivering kilowatt-hours.

Fuel cells, on the other hand, can operate in parallel with the grid. With natural gas reformers, or sufficient stored hydrogen, they can cover extended power outages. With no emissions or noise they be housed within a building and can be permitted to run continuously so they are an investment with annual returns in addition to providing protection against utility outages.

5.7.5 Emissions Benefits

As concerns about climate change grow, there is increasing attention to the role of carbon emissions from power plants. The shift from large, coal-fired power plants to smaller, more efficient gas turbines and combined-cycle plants fueled by natural gas can greatly reduce those emissions. Reductions result from both the increased efficiency that many of these plants have and the lower carbon intensity (kgC/GJ) of natural gas. As Table 5.11 indicates, natural gas emits only about half the carbon per unit of energy when it is burned as does coal.

Calculating the reduction in emissions in conventional power plants that would be gained by fuel switching from coal to gas is straightforward. To do so for cogeneration is less well defined. One approach to calculating carbon emissions from CHP plants is to use the energy chargeable-to-power (ECP) measure. Recall that ECP subtracts the displaced boiler fuel no longer needed from the total input energy, which, in essence, attributes all of the fuel (and carbon) savings to the electric power output. The following example illustrates this approach.

TABLE 5.11 Carbon Intensity of Fossil Fuels Based on High-Heating Values (HHV)[a]

	Energy Density (kJ/kg)	Carbon Content (%)	Carbon Intensity (kgC/GJ)
Anthracite coal	34,900	92	26.4
Bituminous coal	27,330	75	27.4
Crude oil	42,100	80	19.0
Natural gas	55,240	77	13.9

[a]HHV includes latent heat in exhaust water vapor.

Source: IPCC (1996) and Culp (1979).

Example 5.18 Carbon Emission Reductions with CHP. Use the ECP method to determine the reduction in carbon emissions associated with a natural-gas-fired, combined-cycle CHP having 41% electrical efficiency and 44% thermal efficiency. Assume that the thermal output would have come from an 83% efficient boiler. Compare it to that of a 33.3% efficient conventional bituminous-coal-fired power plant.

Solution. The energy chargeable to power is given by (5.39)

$$\begin{aligned}\text{ECP} &= \frac{\text{Total thermal input} - \text{Displaced thermal input}}{\text{Electrical output}}\\ &= \frac{3600}{\eta_P}\left(1 - \frac{\eta_H}{\eta_B}\right) \text{kJ/kWh}\\ \text{ECP} &= \frac{3600}{0.41}\left(1 - \frac{0.44}{0.83}\right) = 4126 \text{ kJ/kWh}\end{aligned}$$

Using the 13.9-kgC/GJ carbon intensity of natural gas provided in Table 5.11 gives

$$\begin{aligned}\text{Carbon chargeable to power} &= \frac{4126 \text{ kJ/kWh} \times 13.9 \text{ kgC/GJ}}{10^6 \text{ kJ/GJ}}\\ &= 0.0573 \text{ kgC/kWh}\end{aligned}$$

The coal plant has no displaced thermal input, so its ECP is the full

$$\text{ECP} = \frac{3600}{0.333} = 10{,}811 \text{ kJ/kWh}$$

Using 27.4 kgC/GJ as its carbon intensity from Table 5.11, we obtain

$$\begin{aligned}\text{Carbon chargeable to power} &= \frac{10{,}811 \text{ kJ/kWh} \times 27.4 \text{ kgC/GJ}}{10^6 \text{ kJ/GJ}}\\ &= 0.296 \text{ kgC/kWh}\end{aligned}$$

The efficient CHP combined-cycle plant reduces carbon emissions by 81%.

Application of the above method for evaluating carbon emissions to a number of power plants results in the bar chart shown in Fig. 5.28. The bar heights represent emissions compared to those from the 33.3%-efficient coal plant just evaluated.

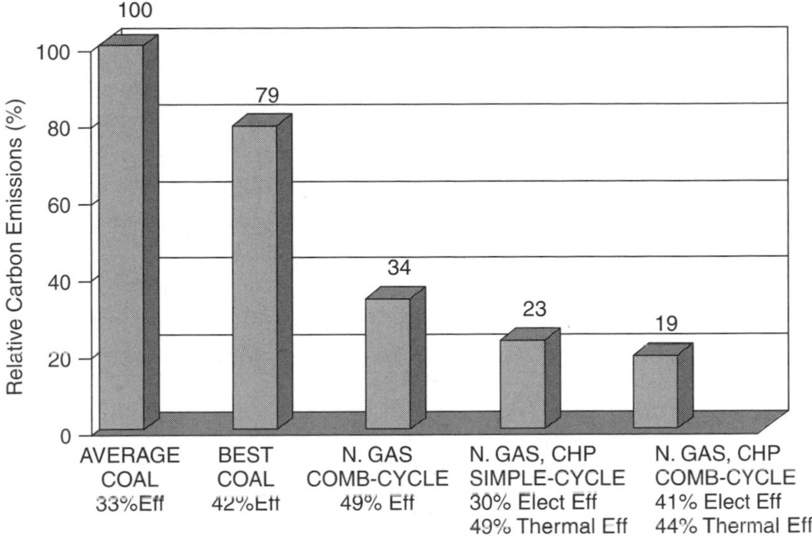

Figure 5.28 Relative carbon emissions for various power plants compared to the average coal plant. CHP units assume thermal output replaces an 83% efficient gas boiler.

5.8 INTEGRATED RESOURCE PLANNING (IRP) AND DEMAND-SIDE MANAGEMENT (DSM)

In the 1980s, regulators began to recognize that energy conservation could be treated as a "source" of energy that could be directly compared with traditional supply sources. If utilities could help customers be more efficient in their use of electricity, delivering the same energy service with fewer kilowatt-hours, and if they could do so at lower cost than supplying energy, then it would be in the public's interest to encourage that to occur. What emerged is a process called *integrated resource planning* (IRP) or, as it is sometimes referred to, *least-cost planning* (LCP).

The new and defining element of IRP was the incorporation of utility programs that were designed to control energy consumption on the customer's side of the electric meter. These are known as *demand-side management* (DSM) programs. While DSM most often refers to programs designed to save energy, it has been defined in a broader sense to refer to any program that attempts to modify customer energy use. As such it includes:

1. *Conservation/energy efficiency programs* that have the effect of reducing consumption during most or all hours of customer demand.
2. *Load management programs* that have the effect of reducing peak demand or shifting electric demand from the hours of peak demand to non-peak-time periods.

3. *Fuel substitution programs* that influence a customer's choice between electric or natural gas service from utilities. For example, the electricity needed for air conditioning can be virtually eliminated by replacing a compressive refrigeration system with one based on absorption cooling.

DSM programs have included a wide range of strategies, such as (1) energy information programs, including energy audits; (2) rebates on energy-efficient appliances and other devices; (3) incentives to help energy service companies (ESCOs) reduce commercial and industrial customer demand for their clients; (4) load control programs to remotely control customer appliances such as water heaters and air conditioners; (5) tariffs designed to shift or reduce loads (time-of-use rates, demand charges, real-time pricing, interruptible rates).

5.8.1 Disincentives Caused by Traditional Rate-Making

Electric utilities have traditionally made their money by selling kilowatts of power and kilowatt-hours of energy, so the question arises as to how they could possibly find it in their best interests to sell less, rather than more, electricity. To understand that challenge, we need to explore the role of state-run public utility commissions (PUCs) in determining the rates and profits that investor-owned utilities (IOUs) have traditionally been allowed to earn.

A fairly common rate-making process is based on utilities providing evidence for costs, and expected demand, to their PUC in what is usually referred to as a *general rate case* (GRC). General rate cases often focus only on the non-fuel component of utility costs (depreciation of equipment, taxes, non-fuel operation and maintenance costs, return on investment, and general administrative expenses). Dividing these revenue requirements by expected energy sales results in a base-case ratepayer cost per kWh. Fuel costs, which may change more rapidly than the usual schedule of general rate case hearings, may be treated in separate proceedings. In general, changes in fuel costs are simply passed on to the ratepayers as a separate charge in their bills.

To understand how this process encourages sales of kWh and discourages DSM, consider the simple graph of revenue requirements versus expected energy sales shown in Fig. 5.29. In this example, the utility has a fixed annual revenue requirement of $300 million needed to recover capital costs of equipment along with other fixed costs, plus an additional amount that depends on how many kilowatt-hours are generated. In this example, each additional kilowatt-hour of electricity generated costs an additional amount of one cent. This 1¢/kWh is called the *short-run marginal cost*, which means that it is based on operating existing capacity longer. There is also *long-run marginal cost*, which applies when the additional power needed triggers an expansion of the existing power plants, transmission lines, or distribution system.

If the utility in Fig. 5.29 projects that it will sell 10 billion kWh/yr, then annual revenues of $400 million would be needed. The ratio of the two is an average base rate of 4¢/kWh, which is what the utility would be allowed to charge until the next general rate case is heard.

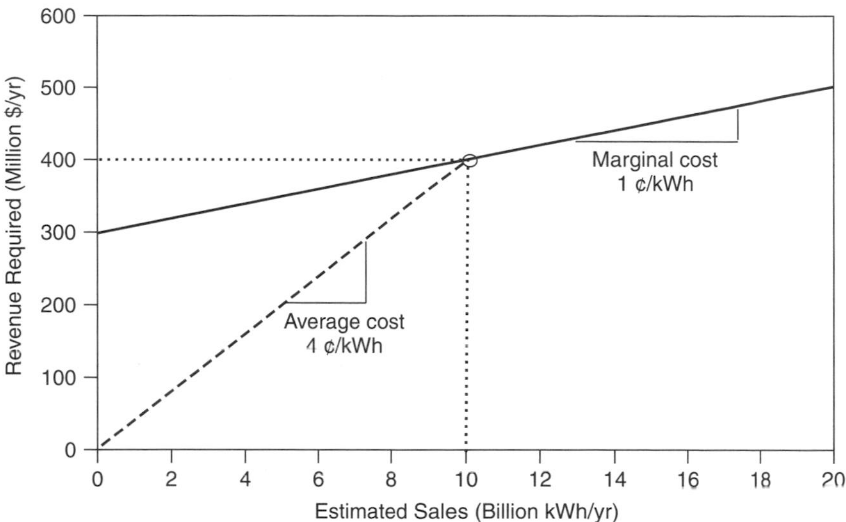

Figure 5.29 Estimated sales of 10 billion kWh would require $400 million in revenue, so the (non-fuel) base rate would be 4¢/kWh.

Consider now the perverse incentives that would encourage this utility to sell more kWh than the 10 billion estimated in Fig. 5.29. Having established a price of 4¢/kWh, any sales beyond the estimated 10 billion kWh/yr would yield revenues of 4¢ for each extra kWh sold. Since each extra kWh generated has a marginal cost of only 1¢, and it is being sold at 4¢, there will be a net 3¢ profit for each extra kWh that can be sold.

Notice, too, that the example utility loses money if it sells less than 10 billion kWh. Generating one fewer kWh reduces costs by 1¢, but reduces revenue by 4¢. In other words, this rather standard practice of rate-making not only encourages utilities to want to sell more energy, it also strongly discourages conservation that would reduce those sales.

5.8.2 Necessary Conditions for Successful DSM Programs

As the above example illustrates, traditional rate-making has tended to reward energy sales and penalize energy conservation, no matter what the relative costs of supply-side and demand-side options might be. During the 1980s and early 1990s, energy planners and utility commissions grappled with the problem of finding fair and equitable rate-making procedures that would reverse these tendencies, allowing cost-effective energy-efficiency to compete. Three conditions were found to be necessary for DSM to be successful:

1. *Decoupling* utility sales from utility profits.
2. *Recovery of DSM program costs* to allow utilities to earn profits on DSM.
3. *DSM incentives* to encourage utilities to prefer DSM over generation.

While there are several approaches that regulators have taken to decouple utility profits from sales, the *electric rate adjustment mechanism* (ERAM) is a good example of one that works quite effectively. ERAM simply incorporates any revenue collected, above or below 1 year's forecasted amount, into the following year's authorized base-rate revenue, thereby eliminating incentives to sell more kWh and removing disincentives to reduce kWh sales. Clearly, decoupling is a necessary condition for DSM to work, but it is not sufficient.

The second condition allows utilities to earn profits on demand-side programs just as they do on supply-side investments. Recovery of DSM program costs can be accomplished by including them in the rate base alongside capital costs of power plants and other utility infrastructure. In this way, supply-side and demand-side investments would both earn whatever rate of return the regulators would allow. Alternatively, DSM costs might be included as an expense that is simply passed along to the ratepayer.

Experience has shown that just decoupling sales and profits, along with rate-basing or expensing DSM costs, does not provide sufficient encouragement for utilities to actively pursue customer energy efficiency. Additional incentives for the utilities and their stockholders seems to be needed. Shared savings programs have emerged in which shareholders are allowed to keep some fraction of the net savings that DSM programs provide. The following example illustrates this idea.

Example 5.19 A Shared-Savings Program. Suppose that a utility offers a \$2 rebate on 18-W, 10,000-h, compact fluorescent lamps (CFLs) that produce as much light as the 75-W incandescents they are intended to replace. Suppose that it costs the utility an additional \$1 per CFL to administer the program. The utility has a marginal cost of electricity of 3¢/kWh, and 1 million customers are expected to take advantage of the rebate program.

If a shared savings program allocates 15% of the net utility savings to shareholders and 85% to ratepayers, what would be the DSM benefit to each?

Solution

$$\text{Energy savings} = (75 - 18)\ \text{W/CFL} \times 10\ \text{kh} \times \$0.03/\text{kWh} \times 10^6\ \text{CFL} = \$17.1\ \text{million}$$

$$\text{Program cost} = (\$2 + \$1)/\text{CFL} \times 10^6\ \text{CFL} = \$3\ \text{million}$$

$$\text{Net savings to utility} = \$17.1\ \text{million} - \$3\ \text{million} = \$14.1\ \text{million}$$

$$\text{Shareholder benefit} = 15\% \times \$14.1\ \text{million} = \$2.12\ \text{million}$$

$$\text{Ratepayer savings} = 85\% \times \$14.1\ \text{million} = \$11.99\ \text{million}$$

In the above example, it was quite easy to calculate the energy savings associated with replacing 75-W incandescents with 18-W CFLs, assuming that their rated 10,000-h lifetime is realized. For many other efficiency measures, however, the savings are less easy to calculate with such confidence. Exactly how much energy will be saved by upgrading the insulation in someone's home, for example, is not at all easy to predict.

Further complicating load reduction estimates are the problems of *free riders* and *take-backs*. Free riders are customers who would have purchased their energy efficiency devices even if the utility did not offer an incentive. In the above example, it was assumed that all 1 million rebates went to customers who would not have purchased CFLs without them. Take-backs refer to customers who change their behavior after installing energy-efficiency devices. For example, after insulating a home a user may be tempted to turn up the thermostat, thereby negating some fraction of the energy savings.

While free riders, take-backs, and uncertainties in estimating energy savings may result in less DSM value to the utility than anticipated, one can argue that these are more than offset by the environmental advantages of not generating unneeded electricity. Some utilities have, in fact, tried to incorporate environmental externalities into the avoided cost benefits of DSM.

5.8.3 Cost Effectiveness Measures of DSM

The goal of integrated resource planning is to minimize the cost to customers of reliable energy services by suitably incorporating supply-side and demand-side resources. A proper measure of cost-effectiveness on the demand side, unfortunately, is largely in the eyes of the beholder; that is, the most cost-effective DSM strategy for stockholders may be different from the best strategy as seen by utility customers. In fact, utility customers can also see things differently depending on whether the customer is someone who takes advantage of the conservation incentives, or whether it is a customer who either cannot, or will not, take advantage of the benefit (e.g., they already purchased the most efficient refrigerator on their own before an incentive was offered). To account for such differences of perspective, a number of standardized DSM cost-effectiveness measures have been devised.

The evaluation of DSM cost-effectiveness measures is complicated by such factors as what type of DSM it is (e.g. appliance efficiency, commercial peak-power reduction, fuel switching) and what assumptions are appropriate to determine future costs and savings. The following descriptions are vastly simplified for clarity; for greater detail, see, for example, CPUC (1987).

Ratepayer Impact Measure (RIM) Test. The RIM test is primarily a measure of what happens to utility rates as a result of the DSM program. For a DSM program to be cost effective using the RIM test, utility rates must not increase.

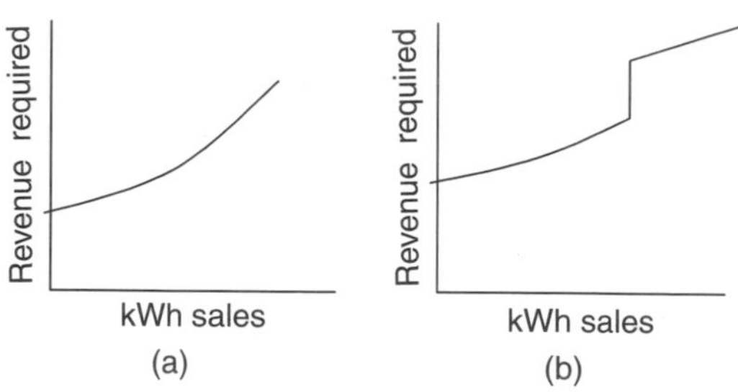

Figure 5.30 Marginal cost can be higher than average cost when the cost curve is nonlinear. (a) Smoothly increasing costs. (b) Jumps caused by capacity expansion.

That is, even nonparticipants must not see an increase in their electric bills (or at least a lower increase than they would have seen without DSM), which is why this is sometimes referred to as the *no-losers test*. Clearly, this is a difficult test to pass. For example, any reduction in demand for the utility modeled in Fig. 5.29 will result in an increase in the average cost of electricity. But, this example utility has a simple, linear relationship between demand and price, which may not always be the case. Increasing demand may mean that less efficient, more expensive power plants get used more often (Fig. 5.30a) or it may trigger the need for new generation, transmission, and distribution capacity (Fig. 5.30b). Avoiding or delaying new capacity expansions is often a major driver motivating DSM programs.

A necessary condition for a measure to pass the RIM test is that the marginal cost of electricity (including fuel costs) must be greater than the average cost. Moreover, the only way to keep rates from increasing is to have the DSM program cost less than the difference between the marginal cost of electricity and the average cost. For example, if the average cost of a kWh of electricity is 5¢ and the marginal cost is 6¢, then a DSM program has to cost less than 1¢ to save a kWh to pass the RIM test.

Example 5.20 The RIM Test for a CFL Program. Suppose that the utility offering the CFL rebate program in Example 5.18 has an average cost of electricity of 2¢/kWh. Will the program pass the RIM test?

Solution. Since the marginal cost of electricity (3¢/kWh) is more than the average cost (2¢/kWh), the DSM program might pass the RIM test. To check, however, we need to find the per kWh cost of the CFL program. With a \$2 rebate and a \$1 administrative cost, DSM costs \$3 per CFL so the cost of conserved energy is

$$\text{DSM cost} = \frac{\$3/\text{CFL}}{(75-18)W/\text{CFL} \times 10\text{ kh}} = \$0.0052/\text{kWh} = 0.52¢/\text{kWh}$$

To pass the RIM test, the DSM cost must be less than the difference between the marginal and average cost to the utility:

$$\begin{aligned}\text{Marginal cost} - \text{Average cost} &= 3\ \text{¢/kWh} - 2\ \text{¢/kWh}\\ &= 1\ \text{¢/kWh} > 0.52\ \text{¢/kWh}\end{aligned}$$

So the RIM test is satisfied and electric rates will decrease with this program.

Total Resource Cost (TRC) Test. The TRC test asks whether society as a whole is better off with DSM. It is again a cost–benefit calculation in which benefits result from reduced costs of fuel, operation, maintenance, and transmission losses, as well as potential reduction in needed power plant capacity, as before. Total costs are somewhat different since they not only include DSM administrative costs incurred by the utility, but also add any extra cost the customer pays when purchasing the more efficient product. Since the utility pays the rebate and the customer collects it, the net impact of the rebate itself on societal cost is zero, so it is simply not included in the TRC test on either side of the benefit–cost calculation. For measures passing the TRC test, utility *rates* may go up or down, but average utility *bills* will decrease. Nonparticipants may see an increase in their bills. The TRC test is the most commonly used measure of DSM cost effectiveness.

Example 5.21 The TRC Test for a CFL Program. Suppose 75-W incandescents costing \$0.50 each have an expected lifetime of 1000 h. The utility is offering a \$2 rebate to encourage customers to replace those with 10,000-h, 18-W CFLs that normally cost \$7. The utility has a marginal cost of \$0.03/kWh. It will cost the utility \$1 per CFL to administer the program. Does this measure satisfy the TRC test?

Solution. Since the TRC test excludes the cost of the rebate, the utility cost per CFL is just \$1 per CFL.

Ignoring the rebate, the customer spends \$7 for a 10,000-h CFL. To obtain the same light would require purchasing ten 1000-h incandescents at 50¢ each, for a total of \$5. The extra cost to the consumer is therefore \$7 − \$5 = \$2 for each CFL.

Total cost to society of the conservation measure (regardless of the amount of rebate) is the sum of the utility cost and consumer cost, or \$1 + \$2 = \$3 for each CFL.

The avoided cost benefit seen by the utility is

$$\begin{aligned}\text{Avoided cost benefit} &= (75 - 18)\ \text{W} \times 10\ \text{kh} \times \$0.03/\text{kWh}\\ &= \$17.10\ \text{per CFL}\end{aligned}$$

Since the cost to the participant and the utility ($3) is less than the benefit associated with not generating that extra energy ($17.10), the measure is cost effective using the TRC test.

It is interesting to note that a DSM measure can pass the RIM test yet still flunk the TRC test, and vice versa. In other words, neither is inherently more stringent, though it is more likely that a DSM program will pass the TRC test than RIM.

A variation on the TRC test is called the *societal test*, the difference being that the latter includes quantified impacts of environmental externalities in the cost and benefit calculations.

Utility Cost Test (UCT). The *utility cost test* (UCT) is similar to the total resource cost test in that it compares costs and benefits of DSM, but it does so only from the perspective of the utility itself. Since this is a utility perspective, the cost of DSM includes both the rebate and the administrative costs. If UCT is satisfied, total energy bills go down; rates may go up or down, so nonparticipants may see higher bills. Participants will see lower energy bills but, after paying for more expensive equipment, may end up paying more for the same energy service (e.g., illumination).

Applying UCT to Example 5.20 results in utility costs of $1 + $2 = $3 and avoided-cost benefits of $17.10, so from the perspective of the utility the measure is cost effective.

5.8.4 Achievements of DSM

Decoupling of utility profits from sales, adopting least-cost planning methods to allow energy efficiency to compete with energy supply, and encouraging cost-effective DSM with additional incentives to shareholders led to dramatic increases in the early 1990s in the use of energy-efficient technologies to help reduce demand growth. Many utilities embraced DSM and pursued it aggressively. For example, Pacific Gas and Electric (PG&E), one of the nation's largest utilities, proclaimed in its *1992 Annual Report* that its business strategy "...emphasizes meeting growth in electric demand through improved customer energy efficiency and renewables... efficiency is the most cost-effective way to meet our customers' growing electric needs... three-fourths of anticipated growth to 2000 will be met by customer energy efficiency."

Utility spending on DSM programs rose rapidly in the early 1990s, leveled off, then dropped for the rest of the decade in spite of the fact that the average cost of the energy saved was close to 3¢/kWh in those later years (Fig. 5.31). This is less than the average running cost of power plants, so it was cheaper to save energy than to turn on the generators. The slump in spending can be attributed in large part to the shift in focus in the mid-1990s to restructuring the

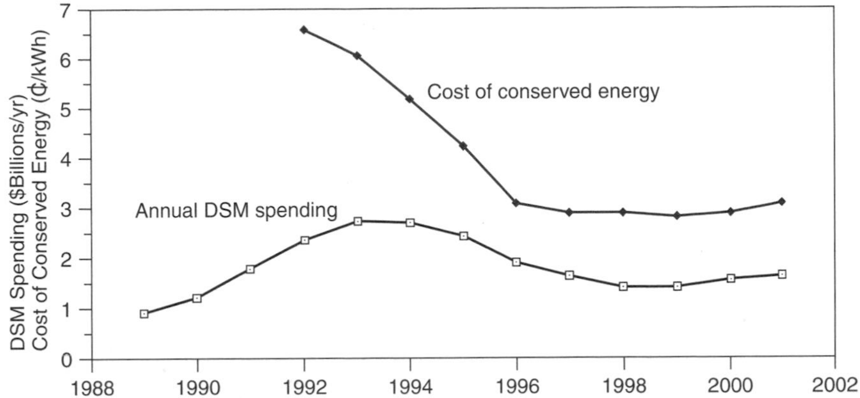

Figure 5.31 Annual U.S. DSM spending and cost of conserved energy. Based on USEIA Annual Electric Utility Data reports.

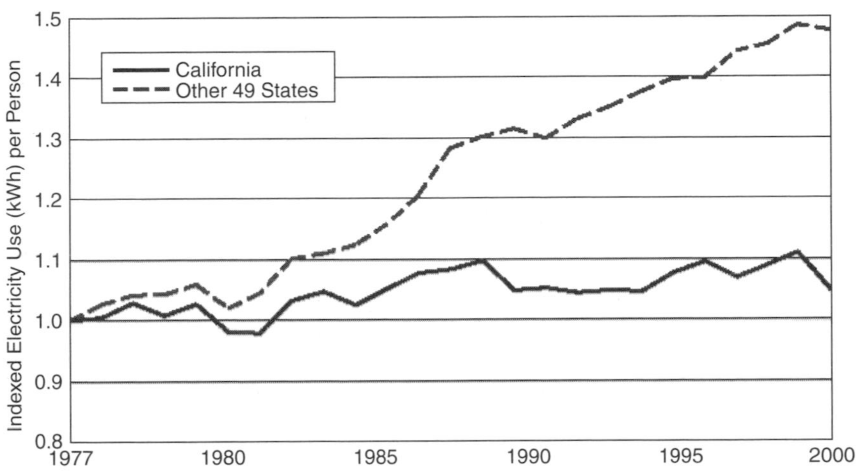

Figure 5.32 California's aggressive DSM programs were very effective in controlling growth in electricity demand. From Bachrach (2003).

electric power industry. For example, part of the 1996 legislation that set California's deregulation in motion abandoned the decoupling of sales from profits, recreating the original disincentives to DSM. After the state's 2000–2001 disastrous experience with deregulation, however, decoupling was reauthorized and through vigorous customer energy efficiency programs, the state began to return to near-normal conditions.

Throughout the 1980s and through much of the 1990s, California was a leader in DSM; as a result, in spite of the booming economic growth that characterized much of that period, per capita growth in electricity use there rose at a far slower rate than it did in the other 49 states (Fig. 5.32).

REFERENCES

Bachrach, D., M. Ardenna, and A. Leupp (2003). Energy Efficiency Leadership in California: Preventing the Next Crisis, Natural Resources Defense Council, Silicon Valley Manufacturing Group, April.

California Collaborative (1990). *An Energy Efficiency Blueprint for California*, January.

California Public Utilities Commission (CPUC) and California Energy Commission (1987). *Economic Analysis of Demand-Side Management Programs, Standard Practice Manual*, P400-87-006.

Cler, G. I., and M. Shepard (1996). *Distributed Generation: Good Things Are Coming in Small Packages*, Esource Tech Update, TU-96-12, Esource, Boulder, CO.

Culp, A. W. Jr. (1979). *Principles of Energy Conversion*, McGraw Hill, N.Y.

Eto, J. (1996). *The Past, Present, and Future of U.S. Utility Demand-Side Management Programs*, Lawrence Berkeley National Laboratory, LBNL-39931, December.

Goldstein, D. B., and H. S. Geller (1999). Equipment Efficiency Standards: Mitigating Global Climate Change at a Profit, *Physics & Society*, vol. 28, no. 2.

Ianucci, J. (1992). "The Distributed Utility: One view of the future," *Distributed Utility—Is This the Future?* EPRI, PG&E and NREL conference, December.

IPCC. (1996). Climate Change 1995: Impacts, Adaptations and Mitigation of Climate Change: Scientific-Technical Analyses, Intergovernmental Panel on Climate Change, Cambridge University Press.

Lovins, A. B, E. K. Datta, T. Feiler, K. R. Rabago, J. N. Swisher, A. Lehmann, and K. Wicker (2002). *Small Is Profitable: The Hidden Economic Benefits of Making Electrical Resources the Right Size*. Rocky Mountain Institute, Snowmass, CO.

National Academy of Sciences (1992). *Policy Implications of Greenhouse Warming: Mitigation, Adaptation, and the Science Base*. National Academic Press, Washington, DC.

Swisher, J. N. (2002). *Cleaner Energy, Greener Profits: Fuel Cells as Cost-Effective Distributed Energy Resources*, Rocky Mountain Institute, Snowmass, CO.

Swisher, J. N., and R. Orans (1996). The Use of Area-Specific Utility Costs to Target Intensive DSM Campaigns, *Utilities Policy*, vol. 5, pp 3–4.

PROBLEMS

5.1 An office building is billed based on the rate structure given in Table 5.3. As a money saving strategy, during the 21 days per month that the office is occupied, it has been decided to shed 10 kW of load for an hour during the period of peak demand in the summer.

a. How much energy will be saved (kWh/mo)?

b. By what amount will their bill be reduced ($/mo) during those months?

c. What is the value of the energy saved in ¢/kWh?

5.2 Suppose a small business can elect to use either the time-of-use (TOU) rate schedule shown below or the rate structure involving a demand charge. During the peak demand period they use 100 kW of power and 24,000 kWh/month, while off-peak they use 20 kW and 10,000 kWh/month.

A. TOU Rate Schedule		B. Demand Charge Schedule	
On-peak	12¢/kWh	Energy charge	6¢/kWh
Off-peak	7¢/kWh	Demand charge	\$9/mo-kW

a. Which rate schedule would give the lowest bills?

b. What would their load factor be (assume a 30-day month)?

5.3 An office building is cooled with an air conditioning (A/C) system having a COP = 3. Electricity costs \$0.10/kWh plus demand charges of \$10/mo-kW. Suppose a 100-W computer is left on 8 hs/day, 5 days/week, 50 weeks/yr, during which time the A/C is always on and the peak demand occurs.

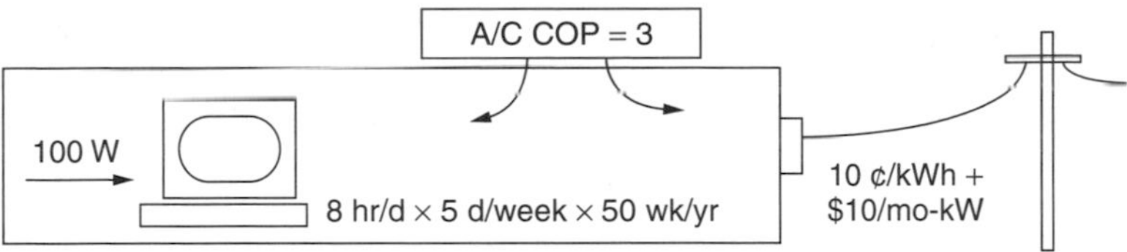

Figure P5.3

a. Find the annual cost of electricity for the computer and its associated air conditioning.

b. Find the total electricity cost per kWh used by the computer. Compare that with the 10¢/kWh cost of electric energy.

5.4 Better windows for a building adds \$3/ft^2 of window but saves \$0.55/ft^2 per year in reduced heating, cooling and lighting costs. With a discount rate of 12%:

a. What is the net present value (NPV) of the better windows over a 30-year period with no escalation in the value of the annual savings?

b. What is the internal rate of return (IRR) with no escalation rate?

c. What is the NPV if the savings escalates at 7%/yr due to fuel savings?

d. What is the IRR with that fuel escalation rate?

5.5 A 30 kW photovoltaic system on a building reduces the peak demand by 25 kW and reduces the annual electricity demand by 60,000 kWh/yr. The PV system costs \$135,000 to install, has no annual maintenance costs, and has an expected lifetime of 30 years. The utility rate structure charges \$0.07/kWh and \$9/kW per month demand charge.

a. What annual savings in utility bills will the PVs deliver?

b. What is the internal rate of return on the investment with no escalation in utility rates?

c. If the annual savings on utility bills increases 6% per year, what is the IRR?

5.6 A small, 10-kW wind turbine that costs $15,000 has a capacity factor of 0.25. If it is paid for with a 6-%, 20-year loan, what is the cost of electricity generated (¢/kWh)?

5.7 An office building needs a steady 200 tons of cooling from 6 A.M. to 6 P.M., 250 days/yr (3000 hrs/yr). The electricity rate schedule is given in the table below.

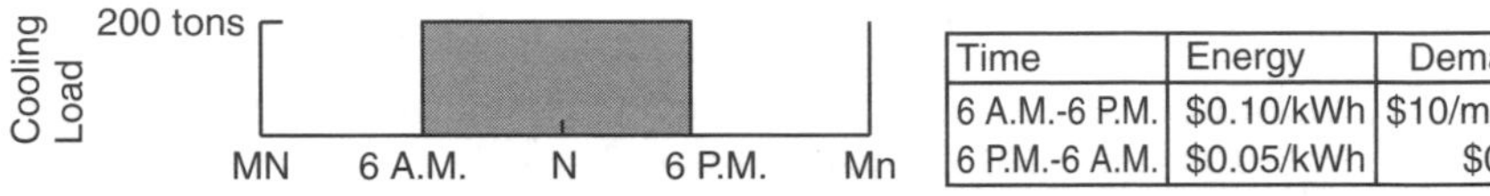

Time	Energy	Demand
6 A.M.-6 P.M.	$0.10/kWh	$10/mo-kW
6 P.M.-6 A.M.	$0.05/kWh	$0

Figure P5.7a

A thermal energy storage (TES) system is to be used to help shift the load to off-peak times. Option (a) uses a 200 kW, 200-ton chiller that runs 12 hours each night making ice that is melted during the day. Option (b) uses a 100 kW, 100-ton chiller that runs 24 hrs/day—the chiller delivers 100 tons during the day and melting ice provides the other 100 tons.

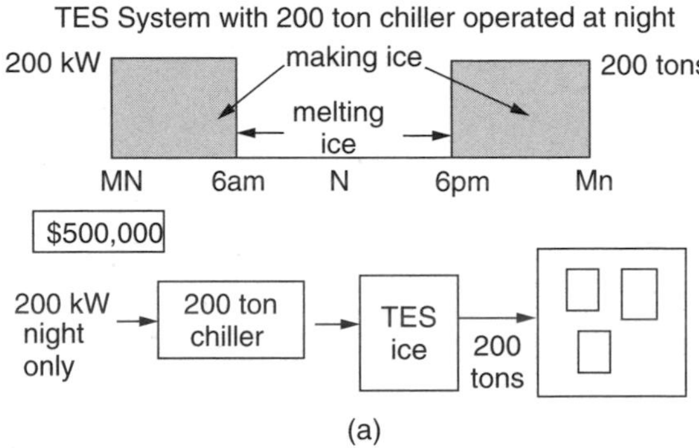

(a)

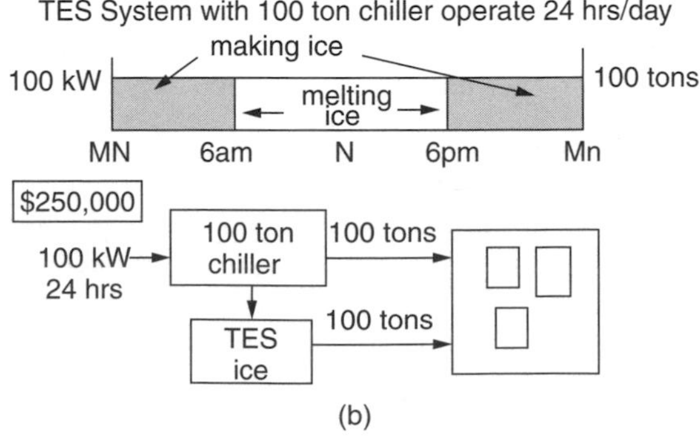

(b)

Figure P5.7b

Option (a) costs $500,000 to build, while (b) costs $250,000.

Using a CRF = 0.10/yr, compute the annual cost to own and operate each system. Which is the cheaper system?

5.8 A 2 kW photovoltaic (PV) system with capacity factor 0.20 costs $8000 after various incentives have been accounted for. It is to be paid for with a 5%, 20-year loan. Since the household has a net income of $130,000 per year, their marginal federal tax bracket is 30.5%.

a. Do a calculation by hand to find the cost of PV electricity in the first year.

b. Set up a spreadsheet to show the annual cash flow and annual ¢/kWh for this system. What is the cost of PV electricity in the 20th year?

5.9 The cost of fuel for a small power plant is currently $10,000 per year. The owner's discount rate is 12 percent and fuel is projected to increase at 6% per year over the 30-yr life of the plant. What is the levelized cost of fuel?

5.10 A photovoltaic system that generates 8000 kWh/yr costs $15,000. It is paid for with a 6%, 20-year loan.

a. Ignoring any tax implications, what is the cost of electricity from the PV system?

b. With local utility electricity costing 11¢/kWh, at what rate would that price have to escalate over the 20-year period in order for the levelized cost of utility electricity be the same as the cost of electricity from the PV system? Use Figure 5.4 and assume the buyer's discount rate is 15%.

5.11 A 50 MW combined-cycle plant with a heat rate of 7700 Btu/kWh operates with a 50-% capacity factor. Natural gas fuel now costs $5 per million Btu (MMBtu). Annual O&M is 0.4¢/kWh. The capital cost of the plant is $600/kW. The utility uses a fixed charge rate (FCR) of 0.12/yr.

a. What is the annualized capital cost of the plant ($/yr)?

b. What is the annual electricity production (kWh/yr)?

c. What is the annual fuel and O&M cost ($/kWh)?

d. What is the current cost of electricity from the plant ($/kWh)?

e. If fuel and O&M costs escalate at 6 percent per year and the utility uses a discount rate of 0.14/yr, what will be the levelized bus-bar cost of electricity from this plant over its 30-yr life?

5.12 Duplicate the spreadsheet shown in Table 5.6, then adjust it for a photovoltaic system that costs $12,000, generates 8000 kWh/yr (which avoids purchasing that much from the utility) and is paid for with a 6%, 20-year loan. Keep the tax bracket (30.5%), initial cost of utility electricity (10¢/kWh), utility electricity rate escalation (5%/yr), and personal discount factor the same. What is the first year net cash flow? What is the cumulative net present value of the system?

5.13 Consider the following energy conservation supply curve:

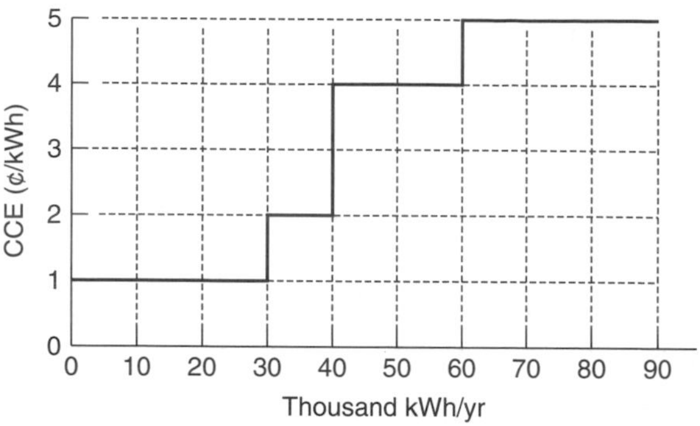

Figure P5.13

a. How much energy can be saved at a marginal cost of less than 3 ¢/kWh?

b. How much energy can be saved at an average cost of less than 3 ¢/kWh?

5.14 Four measures are being proposed for a small industrial facility to reduce its need to purchase utility electricity. The following table describes each measure and gives its capital cost and annual kWh savings. The owner uses a capital recovery factor of 0.06/yr.

TABLE P5.14

	Capital Cost ($)	Savings (kWh/yr)
A. Better lighting system	100,000	150,000
B. High efficiency motors	20,000	120,000
C. More efficient chiller	180,000	180,000
D. Photovoltaic system	150,000	60,000

a. Compute the cost of conserved energy (CCE) for each measure and plot an appropriate energy conservation supply curve.

b. How much energy can be saved (kWh/yr) if the criterion is that each measure must have a CCE of no more than 6¢/kWh?

c. How much energy can be saved (kWh/yr) if the criterion is that the average cost of energy saved is no more than 6¢/kWh?

5.15 Consider a 7-kW fuel cell used in a combined heat and power application (CHP) for a proposed green dorm. The fuel cell converts natural gas to electricity with 35% efficiency while 40% of the input fuel ends up as useful heat for the dorm's water, space heating and space cooling demand. Natural gas costs 90¢/therm (1 therm = 100,000 Btu).

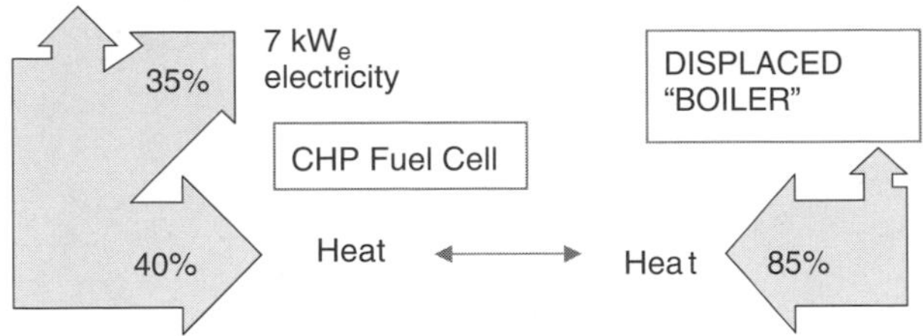

Figure P5.15

The fuel cell costs $25,000 installed, which is to be paid for with a 6%, 20-year loan. It operates with a capacity factor of 90%. Its (useful) waste heat displaces the need for heat from an 85% efficient, natural-gas-fired boiler.

a. Find the energy chargeable to power (ECP)

b. Find the fuel cost chargeable to power (CCP)

c. Find the cost of electricity (¢/kWh) from the fuel cell.

5.16 Consider a 40% efficient, 100-kW microturbine having a capacity factor of 0.8.

a. How many kWh/yr will the turbine generate?

b. Suppose the system is rigged for CHP such that half of the waste heat from the turbine is recovered and used to offset the need for fuel in an 85% efficient, natural-gas-fired boiler. If gas costs $5 per million Btu, what is the economic value of that waste heat recovery ($/yr)?

c. What economic value (¢/kWh) is associated with the waste heat recovery when used to offset the cost of electricity generated by the turbine?

5.17 Consider the option value associated with building ten 100-MW power plants to meet 10 years' worth of demand versus building one 1000-MW plant. The small plants can be built with a 1-year lead time while the large plant has a 5-year lead time. If the single large plant costs $800/kW, how much can each of the smaller plants cost to be equivalent on a present worth basis using a discount factor of 10%?

a. Solve by using Table 5.10

b. Solve by doing the calculations that produced that entry in the table.

5.18 Use the energy chargeable to power method to calculate the carbon emissions (gC/kWh) associated with a natural-gas-fueled microturbine with 30-% electrical efficiency and 40-% thermal efficiency. Assume the thermal output displaces the need for an 80-% efficient gas boiler. Compare that with the 296 gC/kWh of a 33.3-% efficient coal plant without CHP.

CHAPTER 6

WIND POWER SYSTEMS

6.1 HISTORICAL DEVELOPMENT OF WIND POWER

Wind has been utilized as a source of power for thousands of years for such tasks as propelling sailing ships, grinding grain, pumping water, and powering factory machinery. The world's first wind turbine used to generate electricity was built by a Dane, Poul la Cour, in 1891. It is especially interesting to note that La Cour used the electricity generated by his turbines to electrolyze water, producing hydrogen for gas lights in the local schoolhouse. In that regard we could say that he was 100 years ahead of his time since the vision that many have for the twenty-first century includes photovoltaic and wind power systems making hydrogen by electrolysis to generate electric power in fuel cells.

In the United States the first wind-electric systems were built in the late 1890s; by the 1930s and 1940s, hundreds of thousands of small-capacity, wind-electric systems were in use in rural areas not yet served by the electricity grid. In 1941 one of the largest wind-powered systems ever built went into operation at Grandpa's Knob in Vermont. Designed to produce 1250 kW from a 175-ft-diameter, two-bladed prop, the unit had withstood winds as high as 115 miles per hour before it catastrophically failed in 1945 in a modest 25-mph wind (one of its 8-ton blades broke loose and was hurled 750 feet away).

Renewable and Efficient Electric Power Systems. By Gilbert M. Masters
ISBN 0-471-28060-7

Subsequent interest in wind systems declined as the utility grid expanded and became more reliable and electricity prices declined. The oil shocks of the 1970s, which heightened awareness of our energy problems, coupled with substantial financial and regulatory incentives for alternative energy systems, stimulated a renewal of interest in windpower. Within a decade or so, dozens of manufacturers installed thousands of new wind turbines (mostly in California). While many of those machines performed below expectations, the tax credits and other incentives deserve credit for shortening the time required to sort out the best technologies. The wind boom in California was short-lived, and when the tax credits were terminated in the mid-1980s, installation of new machines in the United States stopped almost completely for a decade. Since most of the world's wind-power sales, up until about 1985, were in the United States, this sudden drop in the market practically wiped out the industry worldwide until the early 1990s.

Meanwhiley, wind turbine technology development continued—especially in Denmark, Germany, and Spain—and those countries were ready when sales began to boom in the mid-1990s. As shown in Fig. 6.1, the global installed capacity of wind turbines has been growing at over 25% per year.

Globally, the countries with the most installed wind capacity are shown in Fig. 6.2. As of 2003, the world leader is Germany, followed by Spain, the United States, Denmark, and India. In the United States, California continues to have the most installed capacity, but as shown in Fig. 6.3, Texas is rapidly closing the gap. Large numbers of turbines have been installed along the Columbia River Gorge in the Pacific Northwest, and the windy Great Plains states are experiencing major growth as well.

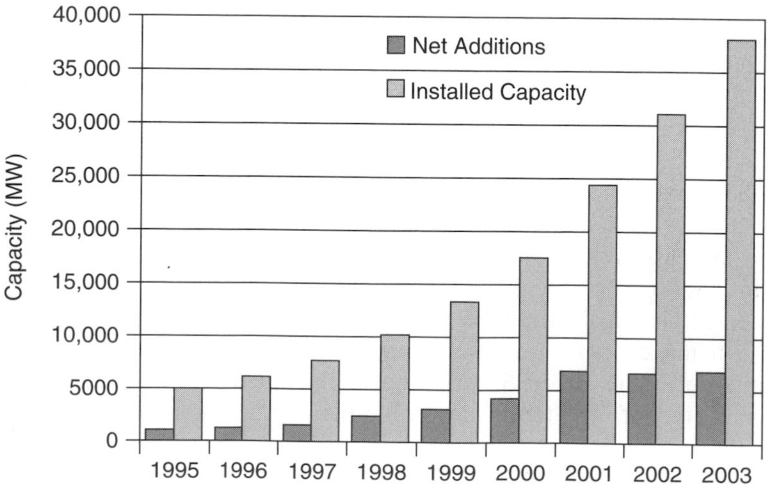

Figure 6.1 Worldwide installed wind-power capacity and net annual additions to capacity have grown by over 25% per year since the mid-1990s. Data from AWEA.

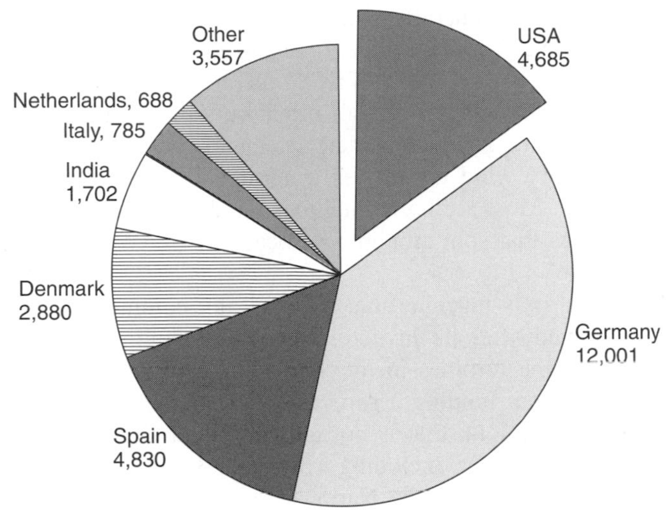

Figure 6.2 Total installed capacity in 2002, by country. AWEA data.

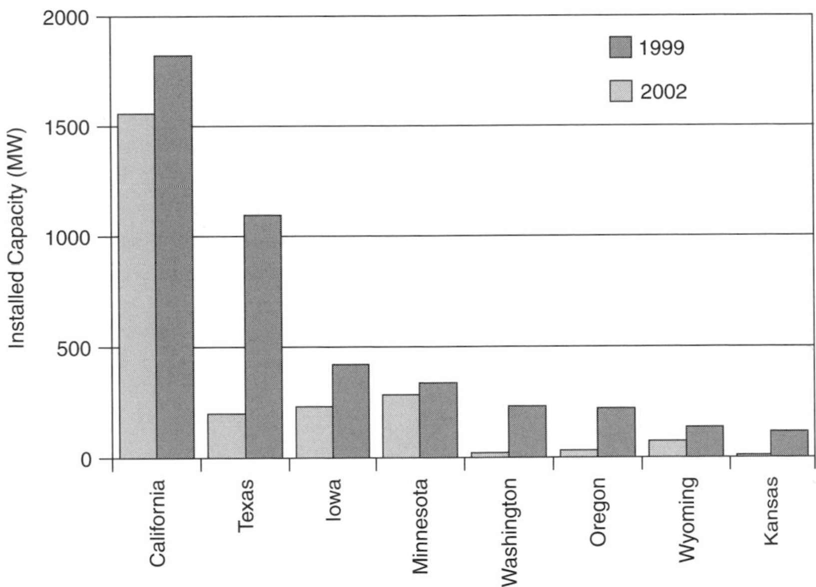

Figure 6.3 Installed wind capacity in the United States in 1999 and 2002.

6.2 TYPES OF WIND TURBINES

Most early wind turbines were used to grind grain into flour, hence the name "windmill." Strictly speaking, therefore, calling a machine that pumps water or generates electricity a windmill is somewhat of a misnomer. Instead, people are using more accurate, but generally clumsier, terminology: "Wind-driven generator," "wind generator," "wind turbine," "wind-turbine generator" (WTG), and "wind energy conversion system" (WECS) all are in use. For our purposes, "wind turbine" will suffice even though often we will be talking about system components (e.g., towers, generators, etc.) that clearly are not part of a "turbine."

One way to classify wind turbines is in terms of the axis around which the turbine blades rotate. Most are horizontal axis wind turbines (HAWT), but there are some with blades that spin around a vertical axis (VAWT). Examples of the two types are shown in Fig. 6.4.

The only vertical axis machine that has had any commercial success is the Darrieus rotor, named after its inventor the French engineer G. M. Darrieus, who first developed the turbines in the 1920s. The shape of the blades is that which would result from holding a rope at both ends and spinning it around a vertical axis, giving it a look that is not unlike a giant eggbeater. Considerable development of these turbines, including a 500-kW, 34-m diameter machine, was undertaken in the 1980s by Sandia National Laboratories in the United States. An American company, FloWind, manufactured and installed a number of these wind turbines before leaving the business in 1997.

The principal advantage of vertical axis machines, such as the Darrieus rotor, is that they don't need any kind of yaw control to keep them facing into the wind. A second advantage is that the heavy machinery contained in the *nacelle* (the housing around the generator, gear box, and other mechanical components) can be located down on the ground, where it can be serviced easily. Since the heavy equipment is not perched on top of a tower, the tower itself need not be structurally as strong as that for a HAWT. The tower can be lightened even further when guy wires are used, which is fine for towers located on land but not for offshore installations. The blades on a Darrieus rotor, as they spin around, are almost always in pure tension, which means that they can be relatively lightweight

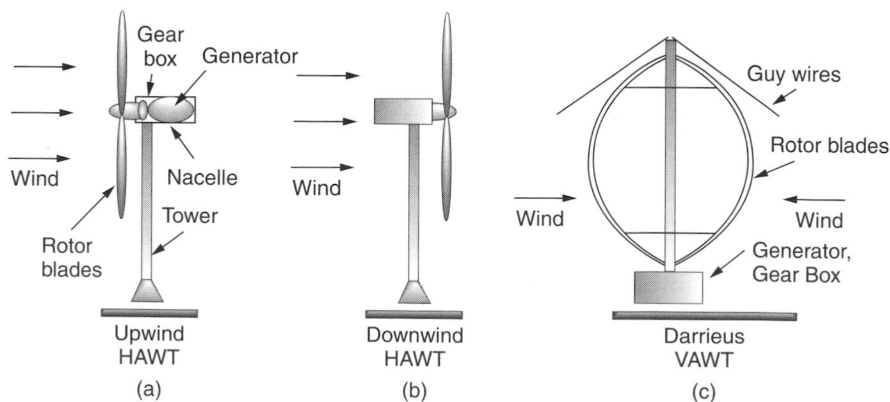

Figure 6.4 Horizontal axis wind turbines (HAWT) are either upwind machines (a) or downwind machines (b). Vertical axis wind turbines (VAWT) accept the wind from any direction (c).

and inexpensive since they don't have to handle the constant flexing associated with blades on horizontal axis machines.

There are several disadvantages of vertical axis turbines, the principal one being that the blades are relatively close to the ground where windspeeds are lower. As we will see later, power in the wind increases as the cube of velocity so there is considerable incentive to get the blades up into the faster windspeeds that exist higher up. Winds near the surface of the earth are not only slower but also more turbulent, which increases stresses on VAWTs. Finally, in low-speed winds, Darrieus rotors have very little starting torque; in higher winds, when output power must be controlled to protect the generator, they can't be made to spill the wind as easily as pitch-controlled blades on a HAWT.

While almost all wind turbines are of the horizontal axis type, there is still some controversy over whether an upwind machine or a downwind machine is best. A downwind machine has the advantage of letting the wind itself control the yaw (the left–right motion) so it naturally orients itself correctly with respect to wind direction. They do have a problem, however, with wind shadowing effects of the tower. Every time a blade swings behind the tower, it encounters a brief period of reduced wind, which causes the blade to flex. This flexing not only has the potential to lead to blade failure due to fatigue, but also increases blade noise and reduces power output.

Upwind turbines, on the other hand, require somewhat complex yaw control systems to keep the blades facing into the wind. In exchange for that added complexity, however, upwind machines operate more smoothly and deliver more power. Most modern wind turbines are of the upwind type.

Another fundamental design decision for wind turbines relates to the number of rotating blades. Perhaps the most familiar wind turbine for most people is the multibladed, water-pumping windmill so often seen on farms. These machines are radically different from those designed to generate electricity. For water pumping, the windmill must provide high starting torque to overcome the weight and friction of the pumping rod that moves up and down in the well. They must also operate in low windspeeds in order to provide nearly continuous water pumping throughout the year. Their multibladed design presents a large area of rotor facing into the wind, which enables both high-torque and low-speed operation.

Wind turbines with many blades operate with much lower rotational speed than those with fewer blades. As the rpm of the turbine increases, the turbulence caused by one blade affects the efficiency of the blade that follows. With fewer blades, the turbine can spin faster before this interference becomes excessive. And a faster spinning shaft means that generators can be physically smaller in size.

Most modern European wind turbines have three rotor blades, while American machines have tended to have just two. Three-bladed turbines show smoother operation since impacts of tower interference and variation of windspeed with height are more evenly transferred from rotors to drive shaft. They also tend to be quieter. The third blade, however, does add considerably to the weight and cost of the turbine. A three-bladed rotor also is somewhat more difficult to hoist up to the nacelle during construction or blade replacement. It is interesting to note that one-bladed turbines (with a counterweight) have been tried, but never deemed worth pursuing.

6.3 POWER IN THE WIND

Consider a "packet" of air with mass m moving at a speed v. Its kinetic energy K.E., is given by the familiar relationship:

$$\text{K.E.} = \frac{1}{2} m v^2 \tag{6.1}$$

Since power is energy per unit time, the power represented by a mass of air moving at velocity v through area A will be

$$\text{Power through area } A = \frac{\text{Energy}}{\text{Time}} = \frac{1}{2}\left(\frac{\text{Mass}}{\text{Time}}\right) v^2 \tag{6.2}$$

The mass flow rate $\dot{m}$, through area A, is the product of air density ρ, speed v, and cross-sectional area A:

$$\left(\frac{\text{Mass passing through A}}{\text{Time}}\right) = \dot{m} = \rho A v \tag{6.3}$$

Combining (6.3) with (6.2) gives us an important relationship:

$$P_w = \tfrac{1}{2} \rho A v^3 \tag{6.4}$$

In S.I. units; P_w is the power in the wind (watts); ρ is the air density (kg/m^3) (at 15°C and 1 atm, $\rho = 1.225$ kg/m^3); A is the cross-sectional area through which the wind passes (m^2); and v = windspeed normal to A (m/s) (a useful conversion: 1 m/s = 2.237 mph).

A plot of (6.4) and a table of values are shown in Fig. 6.5. Notice that the power shown there is per square meter of cross section, a quantity that is called the *specific power* or *power density*.

Notice that the power in the wind increases as the *cube* of windspeed. This means, for example, that doubling the windspeed increases the power by eightfold. Another way to look at it is that the energy contained in 1 hour of 20 mph winds is the same as that contained in 8 hours at 10 mph, which is the same as that contained in 64 hours (more than 2 $\frac{1}{2}$ days) of 5 mph wind. Later we will see that most wind turbines aren't even turned on in low-speed winds, and (6.4) reminds us that the lost energy can be negligible.

Equation (6.4) also indicates that wind power is proportional to the swept area of the turbine rotor. For a conventional horizontal axis turbine, the area

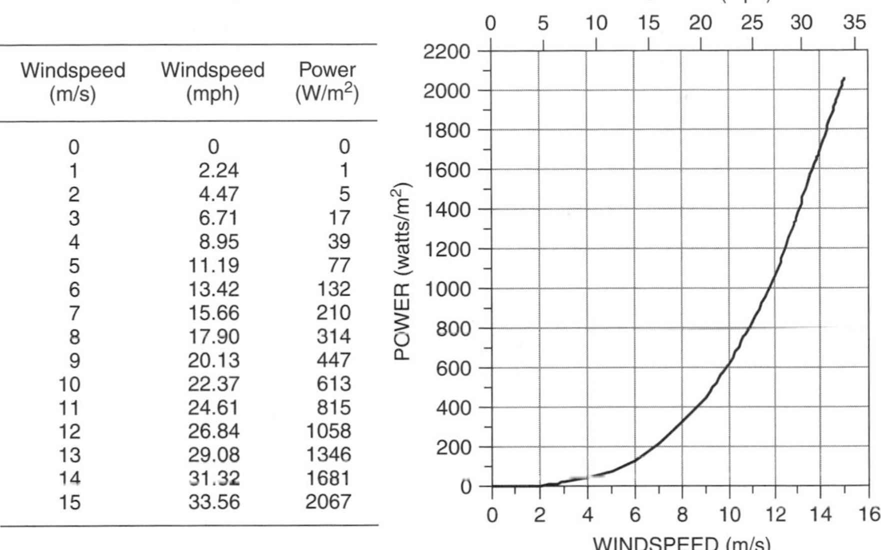

Windspeed (m/s)	Windspeed (mph)	Power (W/m²)
0	0	0
1	2.24	1
2	4.47	5
3	6.71	17
4	8.95	39
5	11.19	77
6	13.42	132
7	15.66	210
8	17.90	314
9	20.13	447
10	22.37	613
11	24.61	815
12	26.84	1058
13	29.08	1346
14	31.32	1681
15	33.56	2067

Figure 6.5 Power in the wind, per square meter of cross section, at 15°C and 1 atm.

A is obviously just $A = (\pi/4)D^2$, so windpower is proportional to the square of the blade diameter. Doubling the diameter increases the power available by a factor of four. That simple observation helps explain the economies of scale that go with larger wind turbines. The cost of a turbine increases roughly in proportion to blade diameter, but power is proportional to diameter squared, so bigger machines have proven to be more cost effective.

The swept area of a vertical axis Darrieus rotor is a bit more complicated to figure out. One approximation to the area is that it is about two-thirds the area of a rectangle with width equal to the maximum rotor width and height equal to the vertical extent of the blades, as shown in Fig. 6.6.

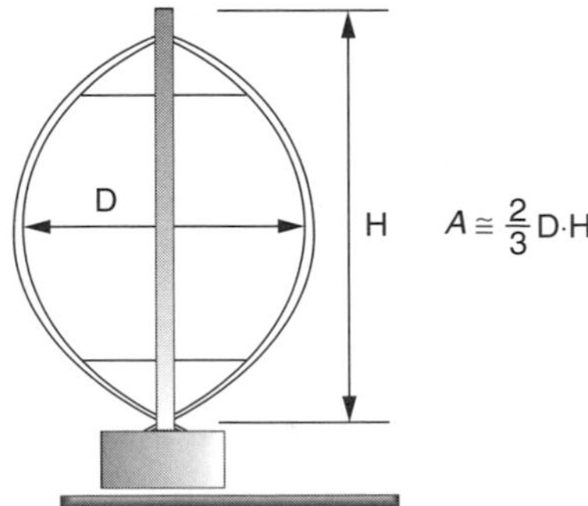

Figure 6.6 Showing the approximate area of a Darrieus rotor.

Of obvious interest is the energy in a combination of windspeeds. Given the nonlinear relationship between power and wind, we can't just use *average* windspeed in (6.4) to predict total energy available, as the following example illustrates.

Example 6.1 Don't Use Average Windspeed. Compare the energy at 15°C, 1 atm pressure, contained in 1 m^2 of the following wind regimes:

a. 100 hours of 6-m/s winds (13.4 mph),
b. 50 hours at 3 m/s plus 50 hours at 9 m/s (i.e., an average windspeed of 6 m/s)

Solution

a. With steady 6 m/s winds, all we have to do is multiply power given by (6.4) times hours:

$$\text{Energy (6 m/s)} = \tfrac{1}{2}\rho A v^3 \Delta t = \tfrac{1}{2} \cdot 1.225\ \text{kg/m}^3 \cdot 1\ \text{m}^2 \cdot (6\ \text{m/s})^3 \cdot 100\ \text{h}$$
$$= 13{,}230\ \text{Wh}$$

b. With 50 h at 3 m/s

$$\text{Energy (3 m/s)} = \tfrac{1}{2} \cdot 1.225\ \text{kg/m}^3 \cdot 1\ \text{m}^2 \cdot (3\ \text{m/s})^3 \cdot 50\ \text{h} = 827\ \text{Wh}$$

And 50 h at 9 m/s contain

$$\text{Energy (9 m/s)} = \tfrac{1}{2} \cdot 1.225\ \text{kg/m}^3 \cdot 1\ \text{m}^2 \cdot (9\ \text{m/s})^3 \cdot 50\ \text{h} = 22{,}326\ \text{Wh}$$

for a total of $827 + 22{,}326 = 23{,}152$ Wh

Example 6.1 dramatically illustrates the inaccuracy associated with using average windspeeds in (6.4). While both of the wind regimes had the same *average* windspeed, the combination of 9-m/s and 3-m/s winds (average 6 m/s) produces 75% more energy than winds blowing a steady 6 m/s. Later we will see that, under certain common assumptions about windspeed probability distributions, energy in the wind is typically almost twice the amount that would be found by using the average windspeed in (6.4).

6.3.1 Temperature Correction for Air Density

When wind power data are presented, it is often assumed that the air density is 1.225 kg/m^3; that is, it is assumed that air temperature is 15°C (59°F) and pressure is 1 atmosphere. Using the ideal gas law, we can easily determine the air density under other conditions.

$$PV = nRT \tag{6.5}$$

where P is the absolute pressure (atm), V is the volume (m^3), n is the mass (mol), R is the ideal gas constant $= 8.2056 \times 10^{-5}\ m^3 \cdot atm \cdot K^{-1} \cdot mol^{-1}$, and T is the absolute temperature (K), where $K = °C + 273.15$. One atmosphere of pressure equals 101.325 kPa (Pa is the abbreviation for pascals, where 1 Pa = 1 newton/m^2). One atmosphere is also equal to 14.7 pounds per square inch (psi), so 1 psi = 6.89 kPa. Finally, 100 kPa is called a *bar* and 100 Pa is a millibar, which is the unit of pressure often used in meteorology work.

If we let M.W. stand for the molecular weight of the gas (g/mol), we can write the following expression for air density, ρ:

$$\rho(\text{kg/m}^3) = \frac{\text{n(mol)} \cdot \text{M.W.(g/mol)} \cdot 10^{-3}(\text{kg/g})}{V(\text{m}^3)} \tag{6.6}$$

Combining (6.5) and (6.6) gives us the following expression:

$$\rho = \frac{P \times \text{M.W.} \times 10^{-3}}{RT} \tag{6.7}$$

All we need is the molecular weight of air. Air, of course, is a mix of molecules, mostly nitrogen (78.08%) and oxygen (20.95%), with a little bit of argon (0.93%), carbon dioxide (0.035%), neon (0.0018%), and so forth. Using the constituent molecular weights ($N_2 = 28.02$, $O_2 = 32.00$, $Ar = 39.95$, $CO_2 = 44.01$, $Ne = 20.18$), we find the equivalent molecular weight of air to be 28.97 ($0.7808 \times 28.02 + 0.2095 \times 32.00 + 0.0093 \times 39.95 + 0.00035 \times 44.01 + 0.000018 \times 20.18 = 28.97$).

Example 6.2 Density of Warmer Air. Find the density of air at 1 atm and 30°C (86°F)

Solution. From (6.7),

$$\rho = \frac{1\ \text{atm} \times 28.97\ \text{g/mol} \times 10^{-3}\ \text{kg/g}}{8.2056 \times 10^{-5}\text{m}^3 \cdot \text{atm}/(\text{K} \cdot \text{mol}) \times (273.15 + 30)\ \text{K}} = 1.165\ \text{kg/m}^3$$

which is a 5% decrease in density compared to the reference 1.225 kg/m^3; since power is proportional to density, it is also a 5% decrease in power in the wind.

TABLE 6.1 Density of Dry Air at a Pressure of 1 Atmosphere[a]

Temperature (°C)	Temperature (°F)	Density (kg/m^3)	Density Ratio (K_T)
−15	5.0	1.368	1.12
−10	14.0	1.342	1.10
−5	23.0	1.317	1.07
0	32.0	1.293	1.05
5	41.0	1.269	1.04
10	50.0	1.247	1.02
15	**59.0**	**1.225**	1.00
20	68.0	1.204	0.98
25	77.0	1.184	0.97
30	86.0	1.165	0.95
35	95.0	1.146	0.94
40	104.0	1.127	0.92

[a]The density ratio K_T is the ratio of density at T to the density at the standard (boldfaced) 15°C.

For convenience, Table 6.1 shows air density for a range of temperatures.

6.3.2 Altitude Correction for Air Density

Air density, and hence power in the wind, depends on atmospheric pressure as well as temperature. Since air pressure is a function of altitude, it is useful to have a correction factor to help estimate wind power at sites above sea level.

Consider a static column of air with cross section A, as shown in Fig. 6.7. A horizontal slice of air in that column of thickness dz and density ρ will have mass $\rho A\ dz$. If the pressure at the top of the slice due to the weight of the air above it *is* $P(z + dz)$, then the pressure at the bottom of the slice, $P(z)$, will be

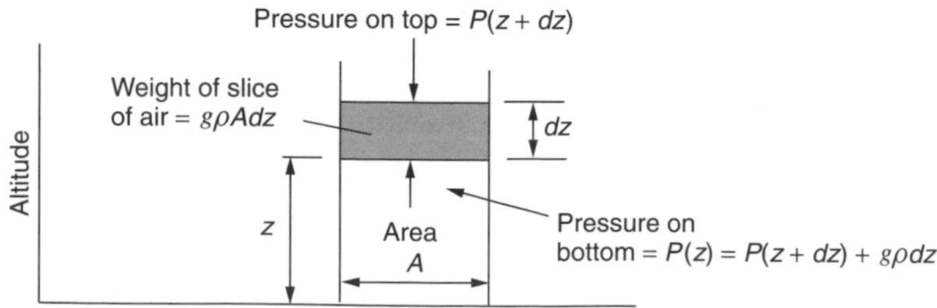

Figure 6.7 A column of air in static equilibrium used to determine the relationship between air pressure and altitude.

$P(z + dz)$ plus the added weight per unit area of the slice itself:

$$P(z) = P(z + dz) + \frac{g\rho A dz}{A} \tag{6.8}$$

where g is the gravitational constant, 9.806 m/s^2. Thus we can write the incremental pressure dP for an incremental change in elevation, dz as

$$dP = P(z + dz) - P(z) = -g\ \rho\, dz \tag{6.9}$$

That is,

$$\frac{dP}{dz} = -\rho g \tag{6.10}$$

The air density ρ given in (6.10) is itself a function of pressure as described in (6.7), so we can now write

$$\frac{dP}{dz} = -\left(\frac{g\ \text{M.W.} \times 10^{-3}}{R \cdot T}\right) \cdot P \tag{6.11}$$

To further complicate things, temperature throughout the air column is itself changing with altitude, typically at the rate of about 6.5°C drop per kilometer of increasing elevation. If, however, we make the simplifying assumption that T is a constant throughout the air column, we can easily solve (6.11) while introducing only a slight error. Plugging in the constants and conversion factors, while assuming 15°C, gives

$$\frac{dP}{dz} = -\left[\frac{9.806(\text{m/s}^2) \times 28.97(\text{g/mol}) \times 10^{-3}(\text{kg/g})}{8.2056 \times 10^{-5}(\text{m}^3 \cdot \text{atm} \cdot \text{K}^{-1} \cdot \text{mol}^{-1}) \times 288.15\ \text{K}}\right] \times \left(\frac{\text{atm}}{101,325\ \text{Pa}}\right) \cdot \left(\frac{1\ \text{Pa}}{\text{N/m}^2}\right)\left(\frac{1\ \text{N}}{\text{kg} \cdot \text{m/s}^2}\right) \cdot P$$

$$\frac{dP}{dz} = -1.185 \times 10^{-4} P \tag{6.12}$$

which has solution,

$$P = P_0 e^{-1.185 \times 10^{-4} \text{H}} = 1(\text{atm}) \cdot e^{-1.185 \times 10^{-4} \text{H}} \tag{6.13}$$

where P_0 is the reference pressure of 1 atm and H is in meters.

Example 6.3 Density at Higher Elevations. Find the air density (a), at 15°C (288.15 K), at an elevation of 2000 m (6562 ft). Then (b) find it assuming an air temperature of 5°C at 2000 m.

Solution

a. From (6.13), $P = 1 \text{ atm} \times e^{-1.185\times10^{-4}\times2000} = 0.789 \text{ atm}$
From (6.7),

$$\rho = \frac{P \cdot \text{M.W.} \cdot 10^{-3}}{R \cdot T}$$
$$= \frac{0.789(\text{atm}) \times 28.97(\text{g/mol}) \times 10^{-3}(\text{kg/g})}{8.2056 \times 10^{-5}(\text{m}^3 \cdot \text{atm} \cdot \text{K}^{-1} \cdot \text{mol}^{-1}) \times 288.15 \text{ K}}$$
$$= 0.967 \text{ kg/m}^3$$

b. At 5°C and 2000 m, the air density would be

$$\rho = \frac{0.789(\text{atm}) \times 28.97(\text{g/mol}) \times 10^{-3}(\text{kg/g})}{8.2056 \times 10^{-5}(\text{m}^3 \cdot \text{atm} \cdot \text{K}^{-1} \cdot \text{mol}^{-1}) \times (273.15 + 5) \text{ K}}$$
$$= 1.00 \text{ kg/m}^3$$

Table 6.2 summarizes some pressure correction factors based on (6.13). A simple way to combine the temperature and pressure corrections for density is as follows:

$$\rho = 1.225 K_T K_A \tag{6.14}$$

In (6.14), the correction factors K_T for temperature and K_A for altitude are tabulated in Tables 6.1 and 6.2.

TABLE 6.2 Air Pressure at 15°C as a Function of Altitude

Altitude (meters)	Altitude (feet)	Pressure (atm)	Pressure Ratio (K_A)
0	0	1	1
200	656	0.977	0.977
400	1312	0.954	0.954
600	1968	0.931	0.931
800	2625	0.910	0.910
1000	3281	0.888	0.888
1200	3937	0.868	0.868
1400	4593	0.847	0.847
1600	5249	0.827	0.827
1800	5905	0.808	0.808
2000	6562	0.789	0.789
2200	7218	0.771	0.771

Example 6.4 Combined Temperature and Altitude Corrections. Find the power density (W/m^2) in 10 m/s wind at an elevation of 2000 m and a temperature of 5°C.

Solution. Using K_T and K_A factors from Tables 6.1 and 6.2 along with (6.14) gives

$$\rho = 1.225 K_T K_A = 1.225 \times 1.04 \times 0.789 = 1.00\ \text{kg/m}^3$$

which agrees with the answer found in Example 6.3. The power density in 10 m/s winds is therefore

$$\frac{P}{A} = \frac{1}{2}\rho v^3 = \frac{1}{2} \cdot 1.00 \cdot 10^3 = 500\ \text{W/m}^2$$

6.4 IMPACT OF TOWER HEIGHT

Since power in the wind is proportional to the cube of the windspeed, the economic impact of even modest increases in windspeed can be significant. One way to get the turbine into higher winds is to mount it on a taller tower. In the first few hundred meters above the ground, wind speed is greatly affected by the friction that the air experiences as it moves across the earth's surface. Smooth surfaces, such as a calm sea, offer very little resistance, and the variation of speed with elevation is only modest. At the other extreme, surface winds are slowed considerably by high irregularities such as forests and buildings.

One expression that is often used to characterize the impact of the roughness of the earth's surface on windspeed is the following:

$$\left(\frac{v}{v_0}\right) = \left(\frac{H}{H_0}\right)^{\alpha} \tag{6.15}$$

where v is the windspeed at height H, v_0 is the windspeed at height H_0 (often a reference height of 10 m), and α is the friction coefficient.

The friction coefficient α is a function of the terrain over which the wind blows. Table 6.3 gives some representative values for rather loosely defined terrain types. Oftentimes, for rough approximations in somewhat open terrain a value of 1/7 (the "one-seventh" rule-of-thumb) is used for α.

While the power law given in (6.15) is very often used in the United States, there is another approach that is common in Europe. The alternative formulation is

$$\left(\frac{v}{v_0}\right) = \frac{\ln(H/z)}{\ln(H_0/z)} \tag{6.16}$$

TABLE 6.3 Friction Coefficient for Various Terrain Characteristics

Terrain Characteristics	Friction Coefficient α
Smooth hard ground, calm water	0.10
Tall grass on level ground	0.15
High crops, hedges and shrubs	0.20
Wooded countryside, many trees	0.25
Small town with trees and shrubs	0.30
Large city with tall buildings	0.40

TABLE 6.4 Roughness Classifications for Use in (6.16)

Roughness Class	Description	Roughness Length $z(m)$
0	Water surface	0.0002
1	Open areas with a few windbreaks	0.03
2	Farm land with some windbreaks more than 1 km apart	0.1
3	Urban districts and farm land with many windbreaks	0.4
4	Dense urban or forest	1.6

where z is called the roughness length. A table of roughness classifications and roughness lengths is given in Table 6.4. Equation (6.16) is preferred by some since it has a theoretical basis in aerodynamics while (6.15) does not.* In this chapter, we will stick with the exponential expression (6.15). Obviously, both the exponential formulation in (6.15) and the logarithmic version of (6.16) only provide a first approximation to the variation of windspeed with elevation. In reality, nothing is better than actual site measurements.

Figure 6.8a shows the impact of friction coefficient on windspeed assuming a reference height of 10 m, which is a commonly used standard elevation for an anemometer. As can be seen from the figure, for a smooth surface ($\alpha = 0.1$), the wind at 100 m is only about 25% higher than at 10 m, while for a site in a "small town" ($\alpha = 0.3$), the wind at 100 m is estimated to be twice that at 10 m. The impact of height on power is even more impressive as shown in Fig. 6.8b.

*When the atmosphere is thermally neutral—that is, it cools with a gradient of −9.8°C/km—the air flow within the boundary layer should theoretically vary logarithmically, starting with a windspeed of zero at a distance above ground equal to the roughness length.

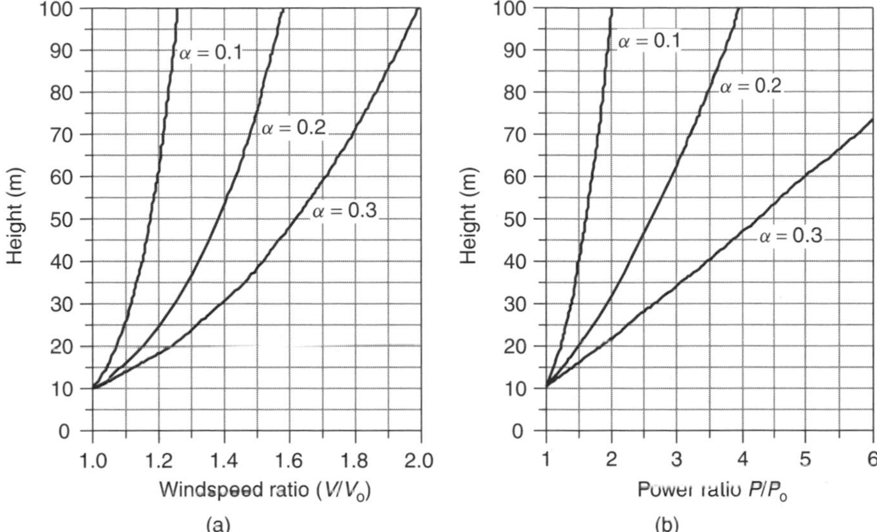

Figure 6.8 Increasing (a) windspeed and (b) power ratios with height for various friction coefficients α using a reference height of 10 m. For $\alpha = 0.2$ (hedges and crops) at 50 m, windspeed increases by a factor of almost 1.4 and wind power increases by about 2.6.

Example 6.5 Increased Windpower with a Taller Tower. An anemometer mounted at a height of 10 m above a surface with crops, hedges, and shrubs shows a windspeed of 5 m/s. Estimate the windspeed and the specific power in the wind at a height of 50 m. Assume 15°C and 1 atm of pressure.

Solution. From Table 6.3, the friction coefficient α for ground with hedges, and so on, is estimated to be 0.20. From the 15°C, 1-atm conditions, the air density is $\rho = 1.225$ kg/m^3. Using (6.15), the windspeed at 50 m will be

$$v_{50} = 5 \cdot \left(\frac{50}{10}\right)^{0.20} = 6.9 \text{ m/s}$$

Specific power will be

$$P_{50} = \tfrac{1}{2}\rho v^3 = 0.5 \times 1.225 \times 6.9^3 = 201 \text{ W/m}^2$$

That turns out to be more than two and one-half times as much power as the 76.5 W/m^2 available at 10 m.

Since power in the wind varies as the cube of windspeed, we can rewrite (6.15) to indicate the relative power of the wind at height H versus the power at the reference height of H_0:

$$\left(\frac{P}{P_0}\right) = \left(\frac{1/2\rho A v^3}{1/2\rho A v_0^3}\right) = \left(\frac{v}{v_0}\right)^3 = \left(\frac{H}{H_0}\right)^{3\alpha} \tag{6.17}$$

In Figure 6.8b, the ratio of wind power at other elevations to that at 10 m shows the dramatic impact of the cubic relationship between windspeed and power. Even for a smooth ground surface—for instance, for an offshore site—the power doubles when the height increases from 10 m to 100 m. For a rougher surface, with friction coefficient $\alpha = 0.3$, the power doubles when the height is raised to just 22 m, and it is quadrupled when the height is raised to 47 m.

Example 6.6 Rotor Stress. A wind turbine with a 30-m rotor diameter is mounted with its hub at 50 m above a ground surface that is characterized by shrubs and hedges. Estimate the ratio of specific power in the wind at the highest point that a rotor blade tip reaches to the lowest point that it falls to.

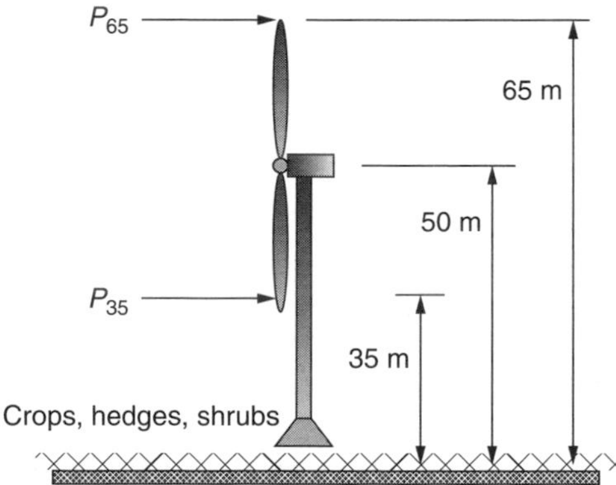

Solution. From Table 6.3, the friction coefficient α for ground with hedges and shrubs is estimated to be 0.20. Using (6.17), the ratio of power at the top of the blade swing (65 m) to that at the bottom of its swing (35 m) will be

$$\left(\frac{P}{P_0}\right) = \left(\frac{H}{H_0}\right)^{3\alpha} = \left(\frac{65}{35}\right)^{3\times 0.2} = 1.45$$

The power in the wind at the top tip of the rotor is 45% higher than it is when the tip reaches its lowest point.

Example 6.6 illustrates an important point about the variation in windspeed and power across the face of a spinning rotor. For large machines, when a blade is at its high point, it can be exposed to much higher wind forces than when it is at the bottom of its arc. This variation in stress as the blade moves through a complete revolution is compounded by the impact of the tower itself on wind-speed—especially for downwind machines, which have a significant amount of wind "shadowing" as the blades pass behind the tower. The resulting flexing of a blade can increase the noise generated by the wind turbine and may contribute to blade fatigue, which can ultimately cause blade failure.

6.5 MAXIMUM ROTOR EFFICIENCY

It is interesting to note that a number of energy technologies have certain fundamental constraints that restrict the maximum possible conversion efficiency from one form of energy to another. For heat engines, it is the Carnot efficiency that limits the maximum work that can be obtained from an engine working between a hot and a cold reservoir. For photovoltaics, we will see that it is the band gap of the material that limits the conversion efficiency from sunlight into electrical energy. For fuel cells, it is the Gibbs free energy that limits the energy conversion from chemical to electrical forms. And now, we will explore the constraint that limits the ability of a wind turbine to convert kinetic energy in the wind to mechanical power.

The original derivation for the maximum power that a turbine can extract from the wind is credited to a German physicist, Albert Betz, who first formulated the relationship in 1919. The analysis begins by imagining what must happen to the wind as it passes through a wind turbine. As shown in Fig. 6.9, wind approaching from the left is slowed down as a portion of its kinetic energy is extracted by the turbine. The wind leaving the turbine has a lower velocity and its pressure is reduced, causing the air to expand downwind of the machine. An envelope drawn around the air mass that passes through the turbine forms what is called a *stream tube*, as suggested in the figure.

So why can't the turbine extract all of the kinetic energy in the wind? If it did, the air would have to come to a complete stop behind the turbine, which, with nowhere to go, would prevent any more of the wind to pass through the rotor. The downwind velocity, therefore, cannot be zero. And, it makes no sense for the downwind velocity to be the same as the upwind speed since that would mean the turbine extracted no energy at all from the wind. That suggests that there must be some ideal slowing of the wind that will result in maximum power extracted by the turbine. What Betz showed was that an ideal wind turbine would slow the wind to one-third of its original speed.

In Fig. 6.9, the upwind velocity of the undisturbed wind is v, the velocity of the wind through the plane of the rotor blades is v_b, and the downwind velocity is v_d. The mass flow rate of air within the stream tube is everywhere the same, call it $\dot{m}$. The power extracted by the blades P_b is equal to the difference in kinetic energy between the upwind and downwind air flows:

$$P_b = \frac{1}{2}\dot{m}(v^2 - v_d^2) \tag{6.18}$$

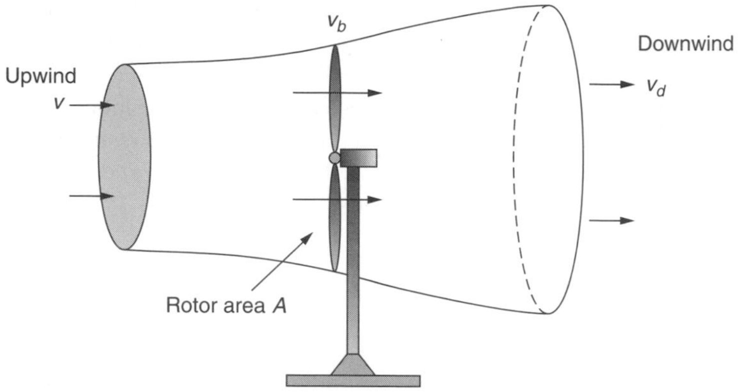

Figure 6.9 Approaching wind slows and expands as a portion of its kinetic energy is extracted by the wind turbine, forming the stream tube shown.

The easiest spot to determine mass flow rate $\dot{m}$ is at the plane of the rotor where we know the cross-sectional area is just the swept area of the rotor A. The mass flow rate is thus

$$\dot{m} = \rho A v_b \tag{6.19}$$

If we now make the assumption that the velocity of the wind through the plane of the rotor is just the average of the upwind and downwind speeds (Betz's derivation actually does not depend on this assumption), then we can write

$$P_b = \frac{1}{2}\rho A\left(\frac{v + v_d}{2}\right)(v^2 - v_d^2) \tag{6.20}$$

To help keep the algebra simple, let us define the ratio of downstream to upstream windspeed to be λ:

$$\lambda = \left(\frac{v_d}{v}\right) \tag{6.21}$$

Substituting (6.21) into (6.20) gives

$$P_b = \frac{1}{2}\rho A\left(\frac{v + \lambda v}{2}\right)(v^2 - \lambda^2 v^2) = \underbrace{\frac{1}{2}\rho A v^3}_{\text{Power in the wind}} \cdot \underbrace{\left[\frac{1}{2}(1+\lambda)(1-\lambda^2)\right]}_{\text{Fraction extracted}} \tag{6.22}$$

Equation (6.22) shows us that the power extracted from the wind is equal to the upstream power in the wind multiplied by the quantity in brackets. The quantity in the brackets is therefore the fraction of the wind's power that is extracted by the blades; that is, it is the efficiency of the rotor, usually designated as C_p.

$$\text{Rotor efficiency} = C_P = \tfrac{1}{2}(1+\lambda)(1-\lambda^2) \tag{6.23}$$

So our fundamental relationship for the power delivered by the rotor becomes

$$P_b = \tfrac{1}{2}\rho A v^3 \cdot C_p \tag{6.24}$$

To find the maximum possible rotor efficiency, we simply take the derivative of (6.23) with respect to λ and set it equal to zero:

$$\frac{dC_p}{d\lambda} = \frac{1}{2}[(1+\lambda)(-2\lambda) + (1-\lambda^2)] = 0$$

$$= \frac{1}{2}[(1+\lambda)(-2\lambda) + (1+\lambda)(1-\lambda)] = \frac{1}{2}(1+\lambda)(1-3\lambda) = 0$$

which has solution

$$\lambda = \frac{v_d}{v} = \frac{1}{3} \tag{6.25}$$

In other words, the blade efficiency will be a maximum if it slows the wind to one-third of its undisturbed, upstream velocity.

If we now substitute $\lambda = 1/3$ into the equation for rotor efficiency (6.23), we find that the theoretical maximum blade efficiency is

$$\text{Maximum rotor efficiency} = \frac{1}{2}\left(1+\frac{1}{3}\right)\left(1-\frac{1}{3^2}\right) = \frac{16}{27} = 0.593 = 59.3\% \tag{6.26}$$

This conclusion, that the maximum theoretical efficiency of a rotor is 59.3%, is called the *Betz efficiency* or, sometimes, *Betz' law*. A plot of (6.22), showing this maximum occurring when the wind is slowed to one-third its upstream rate, is shown in Fig. 6.10.

The obvious question is, how close to the Betz limit for rotor efficiency of 59.3 percent are modern wind turbine blades? Under the best operating conditions, they can approach 80 percent of that limit, which puts them in the range of about 45 to 50 percent efficiency in converting the power in the wind into the power of a rotating generator shaft.

For a given windspeed, rotor efficiency is a function of the rate at which the rotor turns. If the rotor turns too slowly, the efficiency drops off since the blades are letting too much wind pass by unaffected. If the rotor turns too fast, efficiency is reduced as the turbulence caused by one blade increasingly affects the blade that follows. The usual way to illustrate rotor efficiency is to present it as a function of its *tip-speed ratio* (TSR). The tip-speed-ratio is the speed at which the outer tip of the blade is moving divided by the windspeed:

$$\text{Tip-Speed-Ratio (TSR)} = \frac{\text{Rotor tip speed}}{\text{Wind speed}} = \frac{\text{rpm} \times \pi D}{60\ v} \tag{6.27}$$

where rpm is the rotor speed, revolutions per minute; D is the rotor diameter (m); and v is the wind speed (m/s) upwind of the turbine.

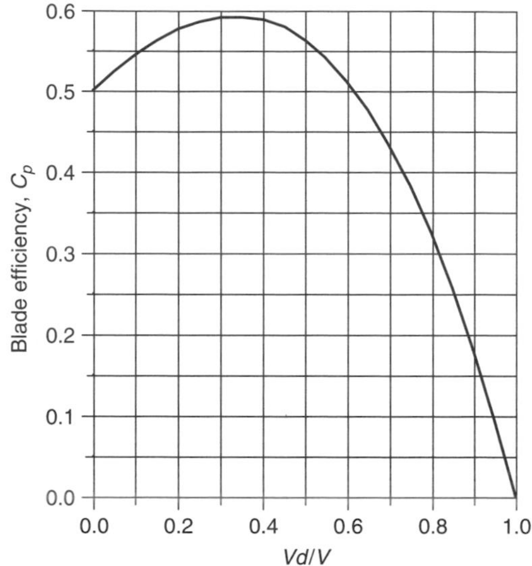

Figure 6.10 The blade efficiency reaches a maximum when the wind is slowed to one-third of its upstream value.

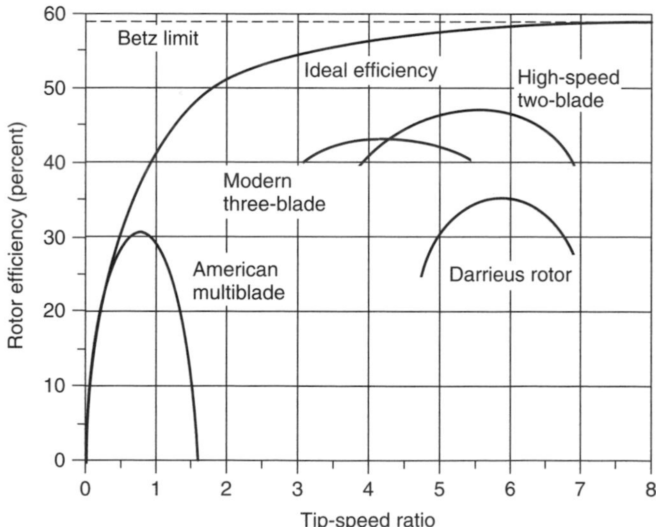

Figure 6.11 Rotors with fewer blades reach their optimum efficiency at higher rotational speeds.

A plot of typical efficiency for various rotor types versus TSR is given in Fig. 6.11. The American multiblade spins relatively slowly, with an optimal TSR of less than 1 and maximum efficiency just over 30%. The two- and three-blade rotors spin much faster, with optimum TSR in the 4–6 range and maximum efficiencies of roughly 40–50%. Also shown is a line corresponding to an "ideal efficiency," which approaches the Betz limit as the rotor speed increases. The curvature in the maximum efficiency line reflects the fact that a slowly turning rotor does not intercept all of the wind, which reduces the maximum possible efficiency to something below the Betz limit.

Example 6.7 How Fast Does a Big Wind Turbine Turn? A 40-m, three-bladed wind turbine produces 600 kW at a windspeed of 14 m/s. Air density is the standard 1.225 kg/m^3. Under these conditions,

a. At what rpm does the rotor turn when it operates with a TSR of 4.0?
b. What is the tip speed of the rotor?
c. If the generator needs to turn at 1800 rpm, what gear ratio is needed to match the rotor speed to the generator speed?
d. What is the efficiency of the complete wind turbine (blades, gear box, generator) under these conditions?

Solution

a. Using (6.27),

$$\text{rpm} = \frac{\text{TSR} \times 60\ v}{\pi D} = \frac{4 \times 60\ \text{s/min} \times 14\ \text{m/s}}{40\pi\ \text{m/rev}} = 26.7\ \text{rev/min}$$

That's about 2.2 seconds per revolution ... pretty slow!

b. The tip of each blade is moving at

$$\text{Tip speed} = \frac{26.7\ \text{rev/min} \times \pi 40\ \text{m/rev}}{60\ \text{s/ min}} = 55.9\ \text{m/s}$$

Notice that even though 2.2 s/rev sounds slow; the tip of the blade is moving at a rapid 55.9 m/s, or 125 mph.

c. If the generator needs to spin at 1800 rpm, then the gear box in the nacelle must increase the rotor shaft speed by a factor of

$$\text{Gear ratio} = \frac{\text{Generator rpm}}{\text{Rotor rpm}} = \frac{1800}{26.7} = 67.4$$

d. From (6.4) the power in the wind is

$$P_w = \frac{1}{2}\rho A v_w{}^3 = \frac{1}{2} \times 1.225 \times \frac{\pi}{4} \times 40^2 \times 14^3 = 2112\ \text{kW}$$

so the overall efficiency of the wind turbine, from wind to electricity, is

$$\text{Overall efficiency} = \frac{600\text{ kW}}{2112\text{ kW}} = 0.284 = 28.4\%$$

Notice that if the rotor itself is about 43% efficient, as Fig. 6.11 suggests, then the efficiency of the gear box times the efficiency of the generator would be about 66% (43% × 66% = 28.4%).

The answers derived in the above example are fairly typical for large wind turbines. That is, a large turbine will spin at about 20–30 rpm; the gear box will speed that up by roughly a factor of 50–70; and the overall efficiency of the machine is usually in the vicinity of 25–30%. In later sections of the chapter, we will explore these factors more carefully.

6.6 WIND TURBINE GENERATORS

The function of the blades is to convert kinetic energy in the wind into rotating shaft power to spin a generator that produces electric power. Generators consist of a rotor that spins inside of a stationary housing called a stator. Electricity is created when conductors move through a magnetic field, cutting lines of flux and generating voltage and current. While small, battery-charging wind turbines use dc generators, grid-connected machines use ac generators as described in the following sections.

6.6.1 Synchronous Generators

In Chapter 3, the operation of synchronous generators, which produce almost all of the electric power in the world, were described. Synchronous generators are forced to spin at a precise rotational speed determined by the number of poles and the frequency needed for the power lines. Their magnetic fields are created on their rotors. While very small synchronous generators can create the needed magnetic field with a permanent magnet rotor, almost all wind turbines that use synchronous generators create the field by running direct current through windings around the rotor core.

The fact that synchronous generator rotors needs dc current for their field windings creates two complications. First, dc has to be provided, which usually means that a rectifying circuit, called the *exciter*, is needed to convert ac from the grid into dc for the rotor. Second, this dc current needs to make it onto the spinning rotor, which means that slip rings on the rotor shaft are needed, along with brushes that press against them. Replacing brushes and cleaning up slip rings adds to the maintenance needed by these synchronous generators. Figure 6.12 shows the basic system for a wind turbine with a synchronous generator, including

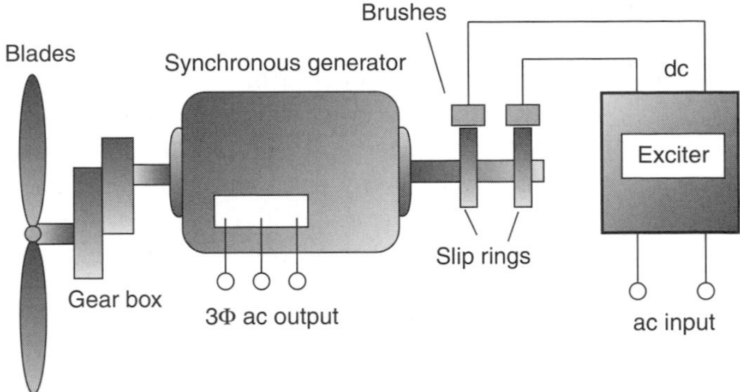

Figure 6.12 A three-phase synchronous generator needs dc for the rotor windings, which usually means that slip rings and brushes are needed to transfer that current to the rotor from the exciter.

a reminder that the generator and blades are connected through a gear box to match the speeds required of each.

6.6.2 The Asynchronous Induction Generator

Most of the world's wind turbines use induction generators rather than the synchronous machines just described. In contrast to a synchronous generator (or motor), induction machines do not turn at a fixed speed, so they are often described as *asynchronous* generators. While induction generators are uncommon in power systems other than wind turbines, their counterpart, induction motors, are the most prevalent motors around—using almost one-third of all the electricity generated worldwide. In fact, an induction machine can act as a motor or generator, depending on whether shaft power is being put into the machine (generator) or taken out (motor). Both modes of operation, as a motor during start-up and as a generator when the wind picks up, take place in wind turbines with induction generators. As a motor, the rotor spins a little slower than the synchronous speed established by its field windings, and in its attempts to "catch up" it delivers power to its rotating shaft. As a generator, the turbine blades spin the rotor a little faster than the synchronous speed and energy is delivered into its stationary field windings.

The key advantage of asynchronous induction generators is that their rotors do not require the exciter, brushes, and slip rings that are needed by most synchronous generators. They do this by creating the necessary magnetic field in the stator rather than the rotor. This means that they are less complicated and less expensive and require less maintenance. Induction generators are also a little more forgiving in terms of stresses to the mechanical components of the wind turbine during gusty wind conditions.

Rotating Magnetic Field. To understand how an induction generator or motor works, we need to introduce the concept of a rotating magnetic field. Begin by imagining coils imbedded in the stator of a three-phase generator as shown in Fig. 6.13. These coils consist of copper conductors running the length of the stator, looping around, and coming back up the other side. We will adopt the convention that positive current in any phase means that current flows from the unprimed letter to the primed one (e.g., positive i_A means that current flows from A to A'). When current in a phase is positive, the resulting magnetic field is drawn with a bold arrow; when it is negative, a dashed arrow is used. And remember the arrow symbolism: a "+" at the end of a wire means current flow into the page, while a dot means current flow out of the page.

Now, consider the magnetic fields that result from three-phase currents flowing in the stator. In Fig. 6.14a, the clock is stopped at $\omega t = 0$, at which point i_A reaches its maximum positive value, and i_B and i_C are both negative and equal in magnitude. The magnetic flux for each of the three currents is shown, the sum of which is a flux arrow that points vertically downward. A while later, let us stop the clock at $\omega t = \pi/3 = 60^\circ$. Now $i_A = i_B$ and both are positive, while i_C is now its maximum negative value, as shown in Fig. 6.14b. The resultant sum of the fluxes has now rotated 60° in the clockwise direction. We could continue this exercise for increasing values of ωt and we would see the resultant flux continuing to rotate around. This is an important concept for the inductance generator: With three-phase currents flowing in the stator, a rotating magnetic field is created inside the generator. The field rotates at the synchronous speed N_s determined by the frequency of the currents f and the number of poles p. That is, $N_s = 120f/p$, as was the case for a synchronous generator.

The Squirrel Cage Rotor. A three-phase induction generator must be supplied with three-phase ac currents, which flow through its stator, establishing the rotating magnetic field described above. The rotor of many induction generators (and motors) consists of a number of copper or aluminum bars shorted together at their ends, forming a cage not unlike one you might have to give your pet rodent some

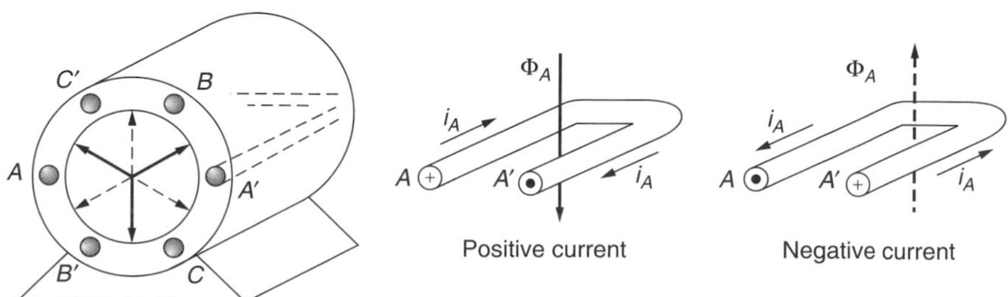

Figure 6.13 Nomenclature for the stator of an inductance generator. Positive current flow from A to A' results in magnetic flux Φ_A represented by a bold arrow pointing downward. Negative current (from A' to A) results in magnetic flux represented by a dotted arrow pointing up.

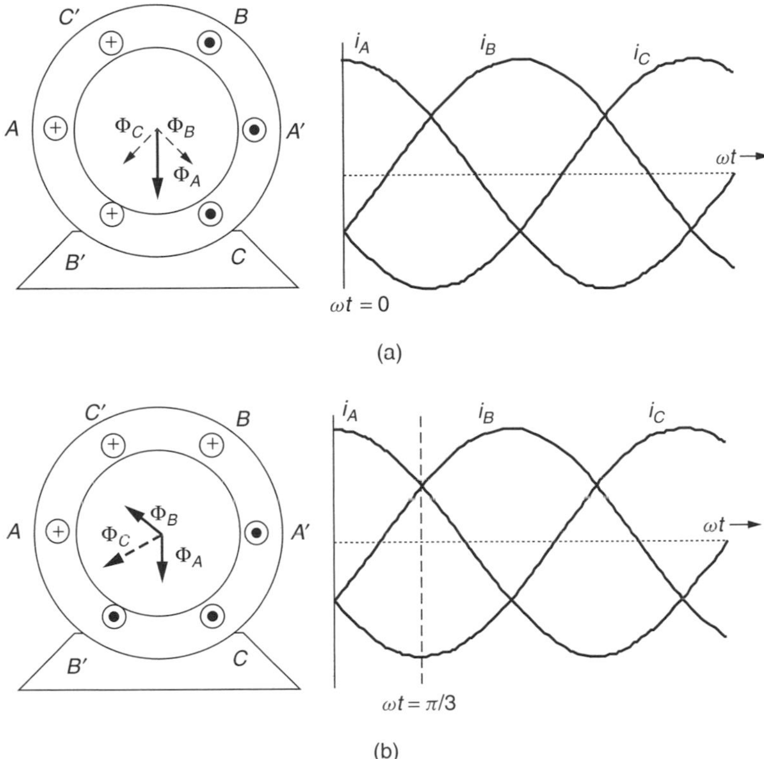

Figure 6.14 (a) At $\omega t = 0$, i_A is a positive maximum while i_B and i_C are both negative and equal to each other. The resulting sum of the magnetic fluxes points straight down; (b) at $\omega t = \pi/3$, the magnetic flux vectors appear to have rotated clockwise by 60°.

exercise. They used to be called "squirrel" cage rotors, but now they are just cage rotors. The cage is then imbedded in an iron core consisting of thin (0.5 mm) insulated steel laminations. The laminations help control eddy current losses (see Section 1.8.2). Figure 6.15 shows the basic relationship between stator and rotor, which can be thought of as a pair of magnets (in the stator) spinning around the cage (rotor).

To understand how the rotating stator field interacts with the cage rotor, consider Fig. 6.16a. The rotating stator field is shown moving toward the right, while the conductor in the cage rotor is stationary. Looked at another way, the stator field can be thought to be stationary and, relative to it, the conductor appears to be moving to the left, cutting lines of magnetic flux as shown in Fig. 6.16b. Faraday's law of electromagnetic induction (see Section 1.6.1) says that whenever a conductor cuts flux lines, an emf will develop along the conductor and, if allowed to, current will flow. In fact, the cage rotor has thick conductor bars with very little resistance, so lots of current can flow easily. That rotor current, labeled i_R in Fig. 6.16b, will create its own magnetic field, which wraps around

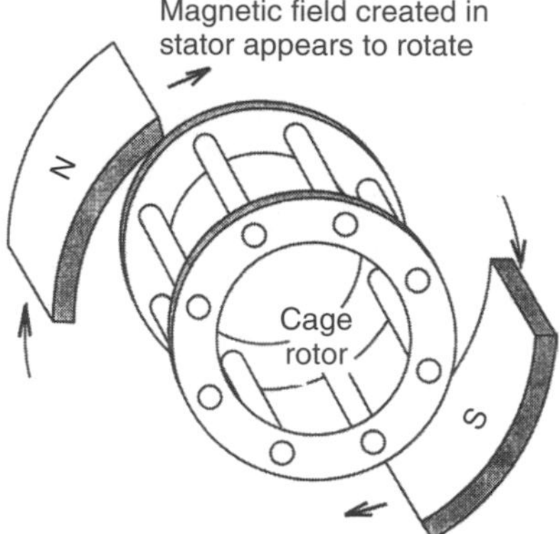

Figure 6.15 A cage rotor consisting of thick, conducting bars shorted at their ends, around which circulates a rotating magnetic field.

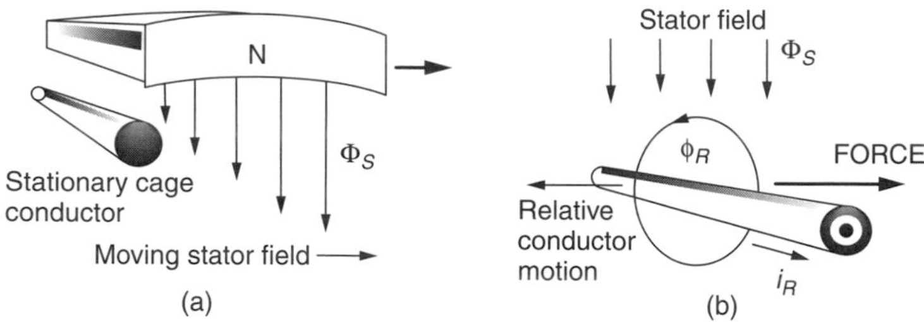

Figure 6.16 In (a) the stator field moves toward the right while the cage rotor conductor is stationary. As shown in (b), this is equivalent to the stator field being stationary while the conductor moves to the left, cutting lines of flux. The conductor then experiences a force that tries to make the rotor want to catch up to the stator's rotating magnetic field.

the conductor. The rotor's magnetic field then interacts with the stator's magnetic field, producing a force that attempts to drive the cage conductor to the right. In other words, the rotor wants to spin in the same direction that the rotating stator field is revolving—in this case, clockwise.

The Inductance Machine as a Motor. Since it is easier to understand an induction motor than an induction generator, we'll start with it. The rotating magnetic field in the stator of the inductance machine causes the rotor to spin in the same direction. That is, the machine is a motor—an induction motor. Notice that there are no electrical connections to the rotor; no slip rings or brushes are required. As the rotor approaches the synchronous speed of the rotating magnetic field, the relative motion between them gets smaller and smaller and less and less force is exerted on the rotor. If the rotor could move at the synchronous speed, there would be no relative motion, no current induced in the cage conductors, and no force developed to keep the rotor going. Since there will always be friction to

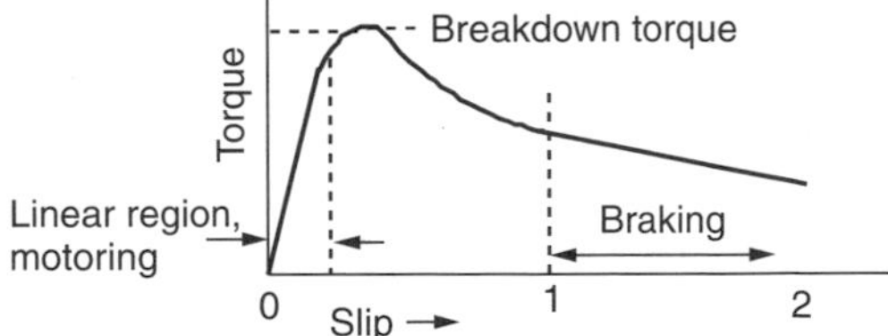

Figure 6.17 The torque–slip curve for an inductance motor.

overcome, the induction machine operating as a motor spins at a rate somewhat slower than the synchronous speed determined by the stator. This difference in speed is called slip, which is defined mathematically as

$$s = \frac{N_S - N_R}{N_S} = 1 - \frac{N_R}{N_S} \tag{6.28}$$

where s is the rotor slip, N_S is the no-load synchronous speed $= 120f/p$ rpm, where f is frequency and p is poles, and N_R is the rotor speed.

As the load on the motor increases, the rotor slows down, increasing the slip, until enough torque is generated to meet the demand. In fact, for most induction motors, slip increases quite linearly with torque within the usual range of allowable slip. There comes a point, however, when the load exceeds what is called the "breakdown torque" and increasing the slip no longer satisfies the load and the rotor will stop (Fig. 6.17). If the rotor is forced to rotate in the opposite direction to the stator field, the inductance machine operates as a brake.

Example 6.8 Slip for an induction motor A 60-Hz, four-pole induction motor reaches its rated power when the slip is 4%. What is the rotor speed at rated power?

The no-load synchronous speed of a 60-Hz, four-pole motor is

$$N_s = \frac{120f}{p} = \frac{120 \times 60}{4} = 1800 \text{ rpm}$$

From (6.28) at a slip of 4%, the rotor speed would be

$$N_R = (1 - s)N_S = (1 - 0.04) \cdot 1800 = 1728 \text{ rpm}$$

The Inductance Machine as a Generator. When the stator is provided with three-phase excitation current and the shaft is connected to a wind turbine and gearbox, the machine will start operation by motoring up toward its synchronous speed. When the windspeed is sufficient to force the generator shaft to exceed synchronous speed, the induction machine automatically becomes a three-phase generator delivering electrical power back to its stator windings. But where does the three-phase magnetization current come from that started this whole process? If it is grid-connected, the power lines provide that current. It is possible, however, to have an induction generator provide its own ac excitation current by incorporating external capacitors, which allows for power generation without the grid.

The basic concept for a self-excited generator is to create a resonance condition between the inherent inductance of the field windings in the stator and the external capacitors that have been added. A capacitor and an inductor connected in parallel form the basis for electronic oscillators; that is, they have a resonant frequency at which they will spontaneously oscillate if given just a nudge in that direction. That nudge is provided by a remnant magnetic field in the rotor. The oscillation frequency, and hence the rotor excitation frequency, depends on the size of the external capacitors, which provides one way to control wind turbine speed. In Fig. 6.18, a single-phase, self-excited, induction generator is diagrammed showing the external capacitance.

So how fast does an inductance generator spin? The same slip factor definition as was used for inductance motors applies [Eq. (6.28)], except that now the slip will be a negative number since the rotor spins faster than synchronous speed. For grid-connected inductance generators, the slip is normally no more than about 1%. This means, for example, that a two-pole, 60-Hz generator with synchronous speed 3600 rpm will turn at about

$$N_R = (1 - s)N_S = [1 - (-0.01)] \cdot 3600 = 3636 \text{ rpm}$$

An added bonus with induction generators is they can cushion the shocks caused by fast changes in wind speed. When the windspeed suddenly changes, the slip increases or decreases accordingly, which helps absorb the shock to the wind turbine mechanical equipment.

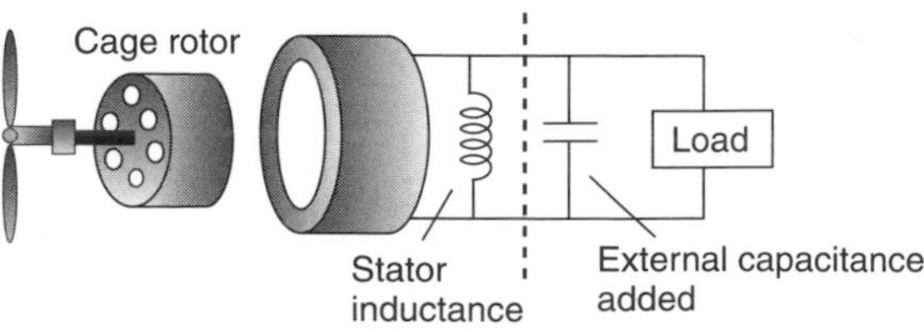

Figure 6.18 A self-excited inductance generator. External capacitors resonate with the stator inductance causing oscillation at a particular frequency. Only a single phase is shown.

6.7 SPEED CONTROL FOR MAXIMUM POWER

In this section we will explore the role that the gear box and generator have with regard to the rotational speed of the rotor and the energy delivered by the machine. Later, we will describe the need for speed control of rotor blades to be able to shed wind to prevent overloading the turbine's electrical components in highwinds.

6.7.1 Importance of Variable Rotor Speeds

There are other reasons besides shedding high-speed winds that rotor speed control is an important design task. Recall Fig. 6.11, in which rotor efficiency C_p was shown to depend on the tip-speed ratio, TSR. Modern wind turbines operate best when their TSR is in the range of around 4–6, meaning that the tip of a blade is moving 4–6-times the wind speed. Ideally, then, for maximum efficiency, turbine blades should change their speed as the windspeed changes. Figure 6.19 illustrates this point by showing an example of blade efficiency versus wind speed with three discrete steps in rotor rpm as a parameter. Unless the rotor speed can be adjusted, blade efficiency C_p changes as wind speed changes. It is interesting to note, however, that C_p is relatively flat near its peaks so that continuous adjustment of rpm is only modestly better than having just a few discrete rpm steps available.

While Fig. 6.19 shows the impact of rotor speed on blade efficiency, what is more important is electric power delivered by the wind turbine. Figure 6.20

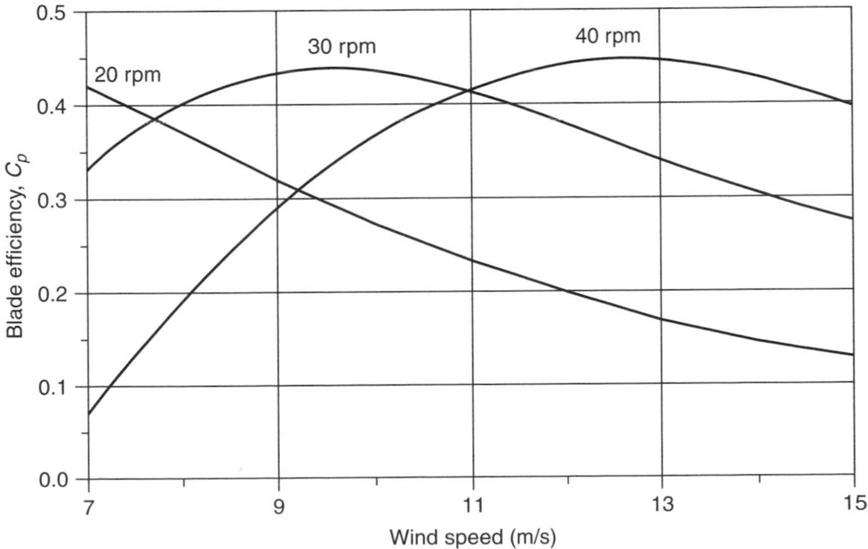

Figure 6.19 Blade efficiency is improved if its rotation speed changes with changing wind speed. In this figure, three discrete speeds are shown for a hypothetical rotor.

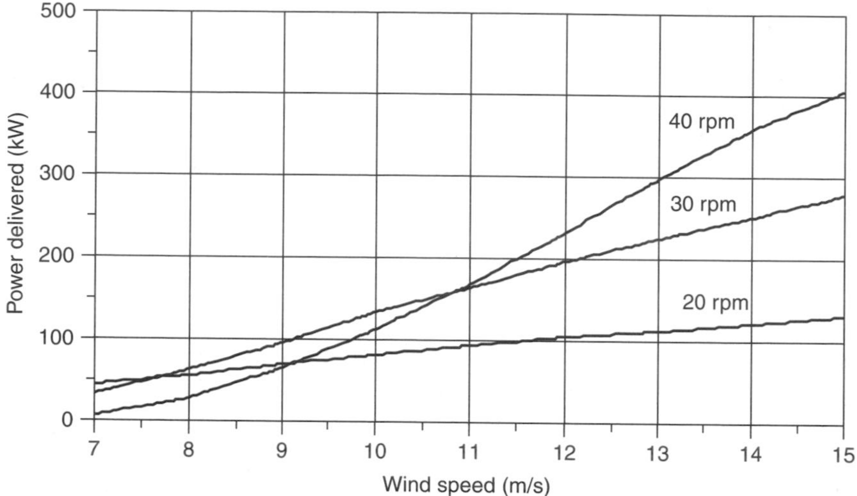

Figure 6.20 Example of the impact that a three-step rotational speed adjustment has on delivered power. For winds below 7.5 m/s, 20 rpm is best; between 7.5 and 11 m/s, 30 rpm is best; and above 11 m/s, 40 rpm is best.

shows the impact of varying rotor speed from 20 to 30 to 40 rpm for a 30-m rotor with efficiency given in Fig. 6.19, along with an assumed gear and generator efficiency of 70%.

While blade efficiency benefits from adjustments in speed as illustrated in Figs. 6.19 and 6.20, the generator may need to spin at a fixed rate in order to deliver current and voltage in phase with the grid that it is feeding. So, for grid-connected turbines, the challenge is to design machines that can somehow accommodate variable rotor speed and somewhat fixed generator speed—or at least attempt to do so. If the wind turbine is not grid-connected, the generator electrical output can be allowed to vary in frequency (usually it is converted to dc), so this dilemma isn't a problem.

6.7.2 Pole-Changing Induction Generators

Induction generators spin at a frequency that is largely controlled by the number of poles. A two-pole, 60-Hz generator rotates at very close to 3600 rpm; with four poles it rotates at close to 1800 rpm; and so on. If we could change the number of poles, we could allow the wind turbine to have several operating speeds, approximating the performance shown in Figs. 6.19 and 6.20. A key to this approach is that as far as the rotor is concerned, the number of poles in the stator of an induction generator is irrelevant. That is, the stator can have external connections that switch the number of poles from one value to another without needing any change in the rotor. This approach is common in household appliance motors such as those used in washing machines and exhaust fans to give two- or three-speed operation.

6.7.3 Multiple Gearboxes

Some wind turbines have two gearboxes with separate generators attached to each, giving a low-wind-speed gear ratio and generator plus a high-wind-speed gear ratio and generator.

6.7.4 Variable-Slip Induction Generators

A normal induction generator maintains its speed within about 1% of the synchronous speed. As it turns out, the slip in such generators is a function of the dc resistance in the rotor conductors. By purposely adding variable resistance to the rotor, the amount of slip can range up to around 10% or so, which would mean, for example, that a four-pole, 1800-rpm machine could operate anywhere from about 1800 to 2000 rpm. One way to provide this capability is to have adjustable resistors external to the generator, but the trade-off is that now an electrical connection is needed between the rotor and resistors. That can mean abandoning the elegant cage rotor concept and instead using a wound rotor with slip rings and brushes similar to what a synchronous generator has. And that means more maintenance will be required.

Another way to provide variable resistance for the rotor is to physically mount the resistors and the electronics that are needed to control them on the rotor itself. But then you need some way to send signals to the rotor telling it how much slip to provide. In one system, called Opti Slip®, an optical fiber link to the rotor is used for this communication.

6.7.5 Indirect Grid Connection Systems

In this approach, the wind turbine is allowed to spin at whatever speed that is needed to deliver the maximum amount of power. When attached to a synchronous or induction generator, the electrical output will have variable frequency depending on whatever speed the wind turbine happens to have at the moment. This means that the generator cannot be directly connected to the utility grid, which of course requires fixed 50- or 60-Hz current.

Figure 6.21 shows the basic concept of these indirect systems. Variable-frequency ac from the generator is rectified and converted into dc using high-power transistors. This dc is then sent to an inverter that converts it back to ac, but this time with a steady 50- or 60-Hz frequency. The raw output of an inverter is pretty choppy and needs to be filtered to smooth it. As described in Chapter 2, any time ac is converted to dc and back again, there is the potential for harmonics to be created, so one of the challenges associated with these variable-speed, indirect wind turbine systems is maintaining acceptable power quality.

In addition to higher annual energy production, variable-speed wind turbines have an advantage of greatly minimizing the wear and tear on the whole system caused by rapidly changing wind speeds. When gusts of wind hit the turbine, rather than having a burst of torque hit the blades, drive shaft, and gearbox,

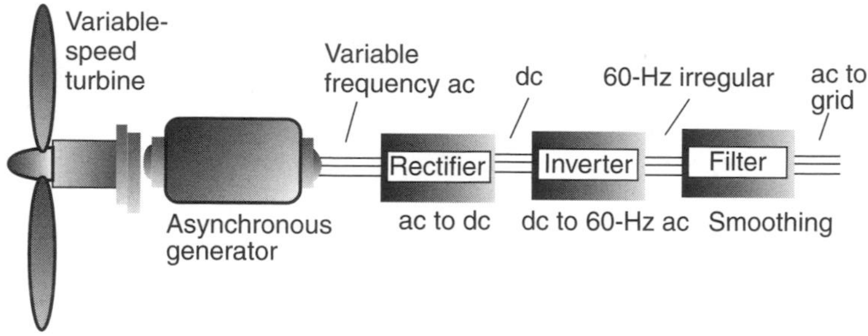

Figure 6.21 Variable-frequency output of the asynchronous generator is rectified, inverted, and filtered to produce acceptable 60-Hz power to the grid.

the blades merely speed up, thereby reducing those system stresses. In addition, some of that extra energy in those gusts can be captured and delivered.

6.8 AVERAGE POWER IN THE WIND

Having presented the equations for *power* in the wind and described the essential components of a wind turbine system, it is time to put the two together to determine how much *energy* might be expected from a wind turbine in various wind regimes,

The cubic relationship between power in the wind and wind velocity tells us that we cannot determine the average power in the wind by simply substituting average windspeed into (6.4). We saw this in Example 6.1. We can begin to explore this important nonlinear characteristic of wind by rewriting (6.4) in terms of average values:

$$P_{\text{avg}} = \left(\tfrac{1}{2}\rho A v^3\right)_{\text{avg}} = \tfrac{1}{2}\rho A (v^3)_{\text{avg}} \tag{6.29}$$

In other words, we need to find the average value of the cube of velocity. To do so will require that we introduce some statistics.

6.8.1 Discrete Wind Histogram

We are going to have to work with the mathematics of probability and statistics, which may be new territory for some. To help motivate our introduction to this material, we will begin with some simple concepts involving discrete functions involving windspeeds, and then we can move on to more generalized continuous functions.

What do we mean by the *average* of some quantity? Suppose, for example, we collect some wind data at a site and then want to know how to figure out the average windspeed during the measurement time. The average wind speed can be thought of as the total meters, kilometers, or miles of wind that have blown past the site, divided by the total time that it took to do so. Suppose, for example, that during a 10-h period, there were 3 h of no wind, 3 h at 5 mph, and 4 h at 10 mph. The average windspeed would be

$$v_{\text{avg}} = \frac{\text{Miles of wind}}{\text{Total hours}} = \frac{3\text{ h} \cdot 0\text{ mile/hr} + 3\text{ h} \cdot 5\text{ mile/h} + 4\text{ h} \cdot 10\text{ mile/h}}{3 + 3 + 4\text{ h}}$$
$$= \frac{55\text{ mile}}{10\text{ h}} = 5.5\text{ mph} \tag{6.30}$$

By regrouping some of the terms in (6.30), we could also think of this as having no wind 30% of the time, 5 mph for 30% of the time, and 10 mph 40% of the time:

$$v_{\text{avg}} = \left(\frac{3\text{ h}}{10\text{ h}}\right) \times 0\text{ mph} + \left(\frac{3\text{ h}}{10\text{ h}}\right) \times 5\text{ mph} + \left(\frac{4\text{ h}}{10\text{ h}}\right) \times 10\text{ mph} = 5.5\text{ mph} \tag{6.31}$$

We could write (6.30) and (6.31) in a more general way as

$$v_{\text{avg}} = \frac{\sum_i [v_i \cdot (\text{hours @ } v_i)]}{\sum \text{hours}} = \sum_i [v_i \cdot (\text{fraction of hours @ } v_i)] \tag{6.32}$$

Finally, if those winds were typical, we could say that the probability that there is no wind is 0.3, the probability that it is blowing 5 mph is 0.3, and the probability that it is 10 mph is 0.4. This lets us describe the average value in probabilistic terms:

$$v_{\text{avg}} = \sum_i [v_i \cdot \text{probability}(v = v_i)] \tag{6.33}$$

We know from (6.29) that the quantity of interest in determining average power in the wind is not the average value of v, but the average value of v^3. The averaging process is exactly the same as our simple example above, yielding the following:

$$(v^3)_{\text{avg}} = \frac{\sum_i [{v_i}^3 \cdot (\text{hours @ } v_i)]}{\sum \text{hours}} = \sum_i [{v_i}^3 \cdot (\text{fraction of hours @ } v_i)] \tag{6.34}$$

Or, in probabilistic terms,

$$(v^3)_{\text{avg}} = \sum_i [{v_i}^3 \cdot \text{probability}(v = v_i)] \tag{6.35}$$

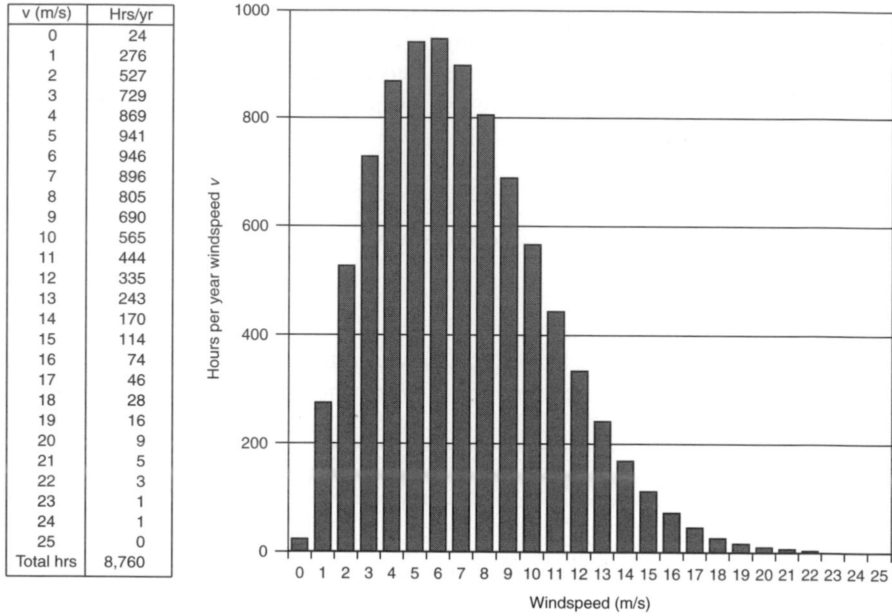

v (m/s)	Hrs/yr
0	24
1	276
2	527
3	729
4	869
5	941
6	946
7	896
8	805
9	690
10	565
11	444
12	335
13	243
14	170
15	114
16	74
17	46
18	28
19	16
20	9
21	5
22	3
23	1
24	1
25	0
Total hrs	8,760

Figure 6.22 An example of site data and the resulting wind histogram showing hours that the wind blows at each windspeed.

Begin by imagining that we have an anemometer that accumulates site data on hours per year of wind blowing at 1 m/s (0.5 to 1.5 m/s), at 2 m/s (1.5 to 2.5 m/s), and so on. An example table of such data, along with a histogram, is shown in Fig. 6.22.

Example 6.9 Average Power in the Wind. Using the data given in Fig. 6.22, find the average windspeed and the average power in the wind (W/m^2). Assume the standard air density of 1.225 kg/m^3. Compare the result with that which would be obtained if the average power were miscalculated using just the average windspeed.

Solution. We need to set up a spreadsheet to determine average wind speed v and the average value of v^3. Let's do a sample calculation of one line of a spreadsheet using the 805 h/yr at 8 m/s:

$$\text{Fraction of annual hours at 8 m/s} = \frac{805\ \text{h/yr}}{24\ \text{h/d} \times 365\ \text{d/yr}} = 0.0919$$

$$v_8 \cdot \text{Fraction of hours at 8 m/s} = 8\ \text{m/s} \times 0.0919 = 0.735$$

$$(v_8)^3 \cdot \text{Fraction of hours at 8 m/s} = 8^3 \times 0.0919 = 47.05$$

The rest of the spreadsheet to determine average wind power using (6.29) is as follows:

Wind Speed v_i (m/s)	Hours @ v_i per year	Fraction of Hours @ v_i	$v_i \times$ Fraction Hours @ v_i	$(v_i)^3$	$(v_i)^3 \times$ fraction Hours @ v_i
0	24	0.0027	0.000	0	0.00
1	276	0.0315	0.032	1	0.03
2	527	0.0602	0.120	8	0.48
3	729	0.0832	0.250	27	2.25
4	869	0.0992	0.397	64	6.35
5	941	0.1074	0.537	125	13.43
6	946	0.1080	0.648	216	23.33
7	896	0.1023	0.716	343	35.08
8	805	0.0919	0.735	512	47.05
9	690	0.0788	0.709	729	57.42
10	565	0.0645	0.645	1,000	64.50
11	444	0.0507	0.558	1,331	67.46
12	335	0.0382	0.459	1,728	66.08
13	243	0.0277	0.361	2,197	60.94
14	170	0.0194	0.272	2,744	53.25
15	114	0.0130	0.195	3,375	43.92
16	74	0.0084	0.135	4,096	34.60
17	46	0.0053	0.089	4,913	25.80
18	28	0.0032	0.058	5,832	18.64
19	16	0.0018	0.035	6,859	12.53
20	9	0.0010	0.021	8,000	8.22
21	5	0.0006	0.012	9,261	5.29
22	3	0.0003	0.008	10,648	3.65
23	1	0.0001	0.003	12,167	1.39
24	1	0.0001	0.003	13,824	1.58
25	0	0.0000	0.000	15,625	0.00
Totals:	8760	1.000	7.0		653.24

The average windspeed is

$$v_{\text{avg}} = \sum_i [v_i \cdot (\text{Fraction of hours @ } v_i)] = 7.0 \text{ m/s}$$

The average value of v^3 is

$$(v^3)_{\text{avg}} = \sum_i [v_i{}^3 \cdot (\text{Fraction of hours @ } v_i)] = 653.24$$

The average power in the wind is

$$P_{\text{avg}} = \tfrac{1}{2}\rho(v^3)_{\text{avg}} = 0.5 \times 1.225 \times 653.24 = 400\ \text{W/m}^2$$

If we had miscalculated average power in the wind using the 7 m/s *average* windspeed, we would have found:

$$P_{\text{average}}(\text{WRONG}) = \tfrac{1}{2}\rho(v_{\text{avg}})^3 = 0.5 \times 1.225 \times 7.0^3 = 210\ \text{W/m}^2$$

In the above example, the ratio of the average wind power calculated correctly using $(v^3)_{\text{avg}}$ to that found when the average velocity is (mis)used is $400/210 = 1.9$. That is, the correct answer is nearly twice as large as the power found when average windspeed is substituted into the fundamental wind power equation $P = \frac{1}{2}\rho A v^3$. In the next section we will see that this conclusion is always the case when certain probability characteristics for the wind are assumed.

6.8.2 Wind Power Probability Density Functions

The type of information displayed in the discrete windspeed histogram in Fig. 6.22 is very often presented as a continuous function, called a probability density function (p.d.f.). The defining features of a p.d.f., such as that shown in Fig. 6.23, are that the area under the curve is equal to unity, and the area under the curve

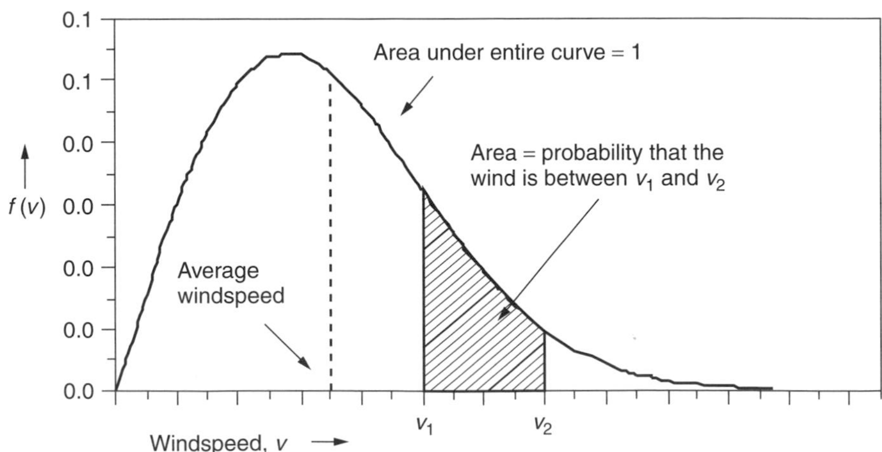

Figure 6.23 A windspeed probability density function (p.d.f).

between any two windspeeds equals the probability that the wind is between those two speeds.
Expressed mathematically,

$$f(v) = \text{windspeed probability density function}$$

$$\text{probability } (v_1 \le v \le v_2) = \int_{v_1}^{v_2} f(v)\, dv \tag{6.36}$$

$$\text{probability } (0 \le v \le \infty) = \int_{0}^{\infty} f(v)\, dv = 1 \tag{6.37}$$

If we want to know the number of hours per year that the wind blows between any two windspeeds, simply multiply (6.36) by 8760 hours per year:

$$\text{hours/yr } (v_1 \le v \le v_2) = 8760 \int_{v_1}^{v_2} f(v)\, dv \tag{6.38}$$

The average windspeed can be found using a p.d.f. in much the same manner as it was found for the discrete approach to wind analysis (6.33):

$$v_{\text{avg}} = \int_{0}^{\infty} v \cdot f(v)\, dv \tag{6.39}$$

The average value of the cube of velocity, also analogous to the discrete version in (6.35), is

$$(v^3)_{\text{avg}} = \int_{0}^{\infty} v^3 \cdot f(v)\, dv \tag{6.40}$$

6.8.3 Weibull and Rayleigh Statistics

A very general expression that is often used as the starting point for characterizing the statistics of windspeeds is called the *Weibull probability density function*:

$$f(v) = \frac{k}{c}\left(\frac{v}{c}\right)^{k-1} \exp\left[-\left(\frac{v}{c}\right)^{k}\right] \qquad \text{Weibull p.d.f.} \tag{6.41}$$

where k is called the *shape parameter*, and c is called the *scale parameter*.

As the name implies, the shape parameter k changes the look of the p.d.f. For example, the Weibull p.d.f. with a fixed scale parameter ($c = 8$) but varying shape parameters k is shown in Fig. 6.24. For $k = 1$, it looks like an exponential decay function; it would probably not be a good site for a wind turbine since most of the winds are at such low speeds. For $k = 2$, the wind blows fairly

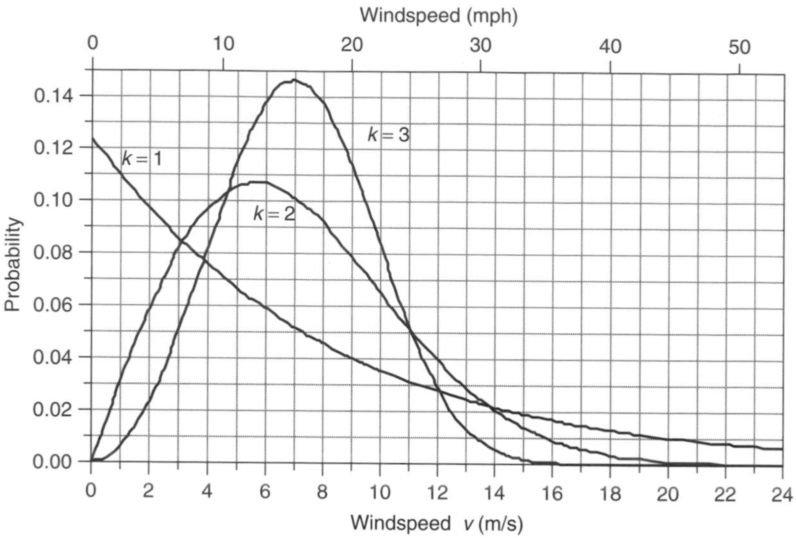

Figure 6.24 Weibull probability density function with shape parameter $k = 1$, 2, and 3 (with scale parameter $c = 8$).

consistently, but there are periods during which the winds blow much harder than the more typical speeds bunched near the peak of the p.d.f. For $k = 3$, the function resembles the familiar bell-shaped curve, and the site would be one where the winds are almost always blowing and doing so at a fairly constant speed, such as the trade winds do.

Of the three Weibull p.d.f.s in Fig. 6.24, intuition probably would lead us to think that the middle one, for which $k = 2$, is the most realistic for a likely wind turbine site; that is, it has winds that are mostly pretty strong, with periods of low wind and some really good high-speed winds as well. In fact, when little detail is known about the wind regime at a site, the usual starting point is to assume $k = 2$. When the shape parameter k is equal to 2, the p.d.f. is given its own name, the *Rayleigh* probability density function:

$$f(v) = \frac{2v}{c^2} \exp\left[-\left(\frac{v}{c}\right)^2\right] \qquad \text{Rayleigh p.d.f.} \qquad (6.42)$$

The impact of changing the scale parameter c for a Rayleigh p.d.f. is shown in Fig. 6.25. As can be seen, larger-scale factors shift the curve toward higher windspeeds. There is, in fact, a direct relationship between scaling factor c and average wind speed $\overline{v}$. Substituting the Rayleigh p.d.f. into (6.39) and referring to a table of standard integrals yield

$$\overline{v} = \int_0^\infty v \cdot f(v)\, dv = \int_0^\infty \frac{2v^2}{c^2} \exp\left[-\left(\frac{v}{c}\right)^2\right] = \frac{\sqrt{\pi}}{2} c \cong 0.886c \qquad (6.43)$$

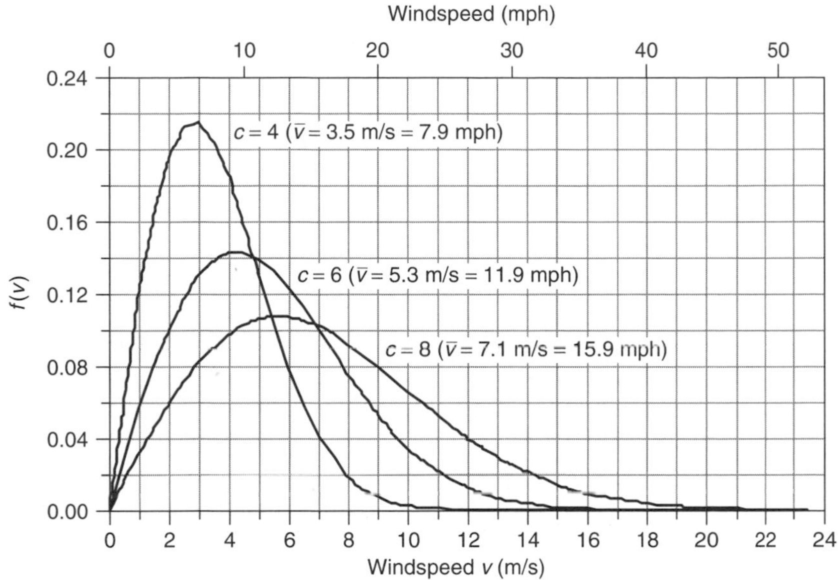

Figure 6.25 The Rayleigh probability density function with varying scale parameter c. Higher scaling parameters correspond to higher average windspeeds.

Or, the other way around:

$$c = \frac{2}{\sqrt{\pi}}\overline{v} \cong 1.128\ \overline{v} \tag{6.44}$$

Even though (6.44) was derived for Rayleigh statistics, it is quite accurate for a range of shape factors k from about 1.5 to 4 (Johnson, 1985). Substituting (6.44) into (6.42) gives us a more intuitive way to write the Rayleigh p.d.f. in terms of average windspeed $\overline{v}$:

$$f(v) = \frac{\pi v}{2\overline{v}^2} \exp\left[-\frac{\pi}{4}\left(\frac{v}{\overline{v}}\right)^2\right] \qquad \text{Rayleigh} \tag{6.45}$$

6.8.4 Average Power in the Wind with Rayleigh Statistics

The starting point for wind prospecting is to gather enough site data to at least be able to estimate average windspeed. That can most easily be done with an anemometer (which spins at a rate proportional to the wind speed) that has a revolution counter calibrated to indicate miles of wind that passes. Dividing miles of wind by elapsed time gives an average wind speed. These "wind odometers" are modestly priced (about $200 each) and simple to use. Coupling average windspeed with the assumption that the wind speed distribution follows Rayleigh statistics enables us to find the average power in the wind.

Substituting the Rayleigh p.d.f. (6.42) into (6.40) lets us find the average value of the cube of windspeed:

$$(v^3)_{\text{avg}} = \int_0^\infty v^3 \cdot f(v)dv = \int_0^\infty v^3 \cdot \frac{2v}{c^2} \exp\left[-\left(\frac{v}{c}\right)^2\right] dv = \frac{3}{4}c^3\sqrt{\pi} \quad (6.46)$$

Using (6.44) gives an alternative expression:

$$(v^3)_{\text{avg}} = \frac{3}{4}\sqrt{\pi}\left(\frac{2\overline{v}}{\sqrt{\pi}}\right)^3 = \frac{6}{\pi}\overline{v}^3 = 1.91\ \overline{v}^3 \quad (6.47)$$

Equation (6.47) is very interesting and very useful. It says that if we assume Rayleigh statistics then the average of the cube of windspeed is just 1.91 times the average wind speed cubed. Therefore, assuming Rayleigh statistics, we can rewrite the fundamental relationship for average power in the wind as

$$\overline{P} = \frac{6}{\pi} \cdot \frac{1}{2}\rho A\overline{v}^3 \qquad \text{(Rayleigh assumptions)} \quad (6.48)$$

That is, with Rayleigh statistics, *the average power in the wind is equal to the power found at the average windspeed multiplied by 6/π or 1.91.*

Example 6.10 Average Power in the Wind. Estimate the average power in the wind at a height of 50 m when the windspeed at 10 m averages 6 m/s. Assume Rayleigh statistics, a standard friction coefficient $\alpha = 1/7$, and standard air density $\rho = 1.225\ \text{kg/m}^3$.

Solution. We first adjust the winds at 10 m to those expected at 50 m using (6.15):

$$\overline{v}_{50} = \overline{v}_{10}\left(\frac{H_{50}}{H_{10}}\right)^\alpha = 6 \cdot \left(\frac{50}{10}\right)^{1/7} = 7.55\ \text{m/s}$$

So, using (6.48), the average wind power density would be

$$\overline{P}_{50} = \frac{6}{\pi} \cdot \frac{1}{2}\rho\overline{v}^3 = \frac{6}{\pi} \cdot \frac{1}{2} \cdot 1.225 \cdot (7.55)^3 = 504\ \text{W/m}^2$$

We also could have found average power at 10 m and then adjust it to 50 m using (6.17):

$$\overline{P}_{10} = \frac{6}{\pi} \cdot \frac{1}{2} \cdot 1.225 \cdot 6^3 = 252.67\ \text{W/m}^2$$

$$\overline{P}_{50} = \overline{P}_{10}\left(\frac{H_{50}}{H_{10}}\right)^{3\alpha} = 252.67\left(\frac{50}{10}\right)^{3\times 1/7} = 504\ \text{W/m}^2$$

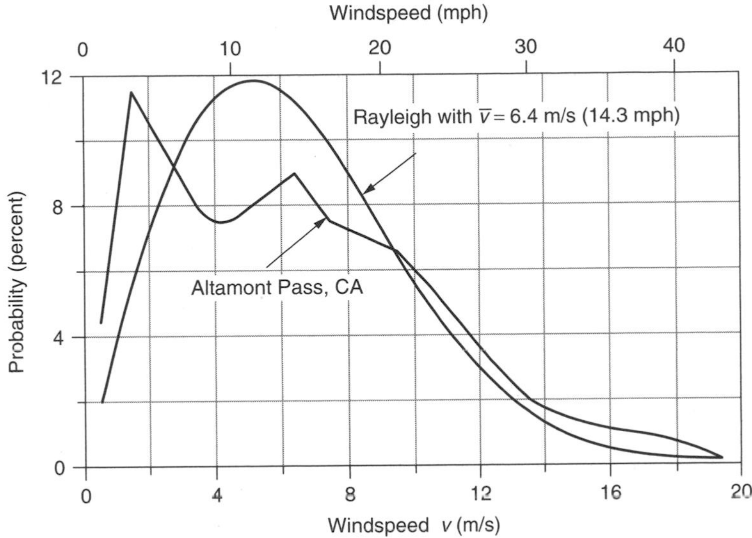

Figure 6.26 Probability density functions for winds at Altamont Pass, CA., and a Rayleigh p.d.f. with the same average wind speed of 6.4 m/s (14.3 mph). From Cavallo et al. (1993).

Lest we become too complacent about the importance of gathering real wind data rather than relying on Rayleigh assumptions, consider Fig. 6.26, which shows the probability density function for winds at one of California's biggest wind farms, Altamont Pass. Altamont Pass is located roughly midway between San Francisco (on the coast) and Sacramento (inland valley). In the summer months, rising hot air over Sacramento draws cool surface air through Altamont Pass, creating strong summer afternoon winds, but in the winter there isn't much of a temperature difference and the winds are generally very light unless a storm is passing through. The windspeed p.d.f. for Altamont clearly shows the two humps that correspond to not much wind for most of the year, along with very high winds on hot summer afternoons. For comparison, a Rayleigh p.d.f. with the same annual average wind speed as Altamont (6.4 m/s) has also been drawn in Fig. 6.26.

6.8.5 Wind Power Classifications and U.S. Potential

The procedure demonstrated in Example 6.10 is commonly used to estimate average wind power density (W/m^2) in a region. That is, measured values of average wind speed using an anemometer located 10 m above the ground are used to estimate average windspeed and power density at a height 50 m above the ground. Rayleigh statistics, a friction coefficient of 1/7, and sea-level air density at 0°C of 1.225 kg/m^3 are often assumed. A standard wind power classification scheme based on these assumptions is given in Table 6.5.

A map of the United States showing regions of equal wind power density based on the above assumptions is shown in Fig. 6.27. As can be seen, there is a broad

TABLE 6.5 Standard Wind Power Classifications[a]

Wind Power Class	Avg Windspeed at 10 m (m/s)	Avg Windspeed at 10 m (mph)	Wind Power Density at 10 m (W/m^2)	Wind Power Density at 50 m (W/m^2)
1	0–4.4	0–9.8	0–100	0–200
2	4.4–5.1	9.8–11.4	100–150	200–300
3	5.1–5.6	11.4–12.5	150–200	300–400
4	5.6–6.0	12.5–13.4	200–250	400–500
5	6.0–6.4	13.4–14.3	250–300	500–600
6	6.4–7.0	14.3–15.7	300–400	600–800
7	7.0–9.5	15.7–21.5	400–1000	800–2000

[a] Assumptions include Rayleigh statistics, ground friction coefficient $\alpha = 1/7$, sea-level 0°C air density 1.225 kg/m^3, 10-m anemometer height, 50-m hub height.

band of states stretching from Texas to North Dakota with especially high wind power potential, including large areas with Class 4 or better winds (over 400 W/m^2).

Translating available wind power from maps such as shown in Fig. 6.27 into estimates of electrical energy that can be developed is an especially important exercise for energy planners and policy makers. While the resource may be available, there are significant land use questions that could limit the acceptability of any given site. Flat grazing lands would be easy to develop, and the impacts on current usage of such lands would be minimal. On the other hand, developing sites in heavily forested areas or along mountain ridges, for example, would be

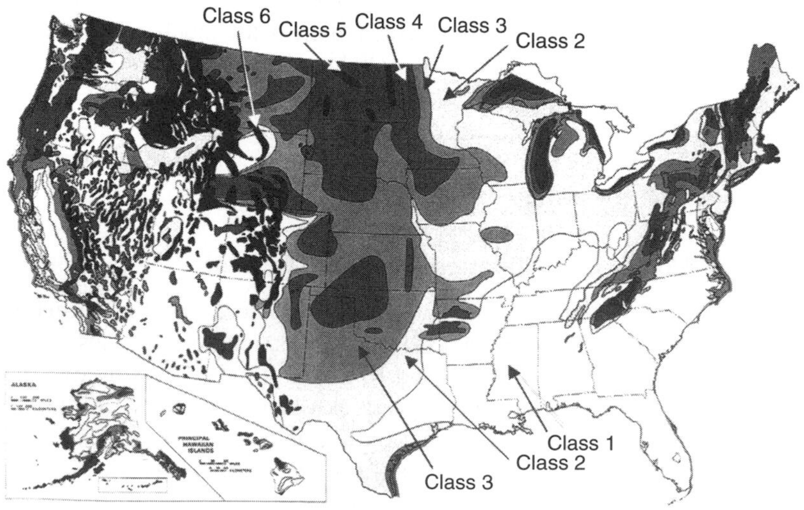

Figure 6.27 Average annual wind power density at 50-m elevation. From NREL Wind Energy Resource Atlas of the United States.

TABLE 6.6 Energy Potential for Class 3 or Higher Winds, in billion kWh/yr, Including Environmental and Land Use Constraints

Rank State	Potential	Percent of United States[a]	Rank State	Potential	Percent of United States[a]
1 North Dakota	1210	35%	11 Colorado	481	14%
2 Texas	1190	34%	12 New Mexico	435	12%
3 Kansas	1070	31%	13 Idaho	73	2%
4 South Dakota	1030	29%	14 Michigan	65	2%
5 Montana	1020	29%	15 New York	62	2%
6 Nebraska	868	25%	16 Illinois	61	2%
7 Wyoming	747	21%	17 California	59	2%
8 Oklahoma	725	21%	18 Wisconsin	58	2%
9 Minnesota	657	19%	19 Maine	56	2%
10 Iowa	551	16%	20 Missouri	52	1%

[a]If totally utilized, the fraction of U.S. demand that wind could supply.
Source: Elliot et al. (1991)

much more difficult and environmentally damaging. Proximity to transmission lines and load centers affects the economic viability of projects, although in the future we could imagine wind generated electricity being converted, near the site, into hydrogen that could be pipelined to customers.

One attempt to incorporate land-use constraints into the estimate of U.S. wind energy potential was made by the Pacific Northwest Laboratory (Elliott et al., 1991). Assuming turbine efficiency of 25% and 25% array and system losses, the exploitable wind resource for the United States with no land-use restrictions is estimated to be 16,700 billion kWh/yr and 4600 billion kWh/yr under the most "severe" land use constraints. For comparison, the total amount of electricity generated in the United States in 2002 was about 3500 billion kWh, which means in theory that there is more than enough wind to supply all of U.S. electrical demand. Distances from windy sites to transmission lines and load centers, along with reliability issues, will constrain the total generation to considerably less than that, but nonetheless the statistic is impressive.

The top 20 states for wind energy potential are shown in Table 6.6. Notice that California, which in 2003 had the largest installed wind capacity, ranks only seventeenth among the states for wind potential. At the top of the list is North Dakota, with enough wind potential of its own to supply one-third of the total U.S. electrical demand.

6.9 SIMPLE ESTIMATES OF WIND TURBINE ENERGY

How much of the energy in the wind can be captured and converted into electricity? The answer depends on a number of factors, including the characteristics of the machine (rotor, gearbox, generator, tower, controls), the terrain (topography, surface roughness, obstructions), and, of course, the wind regime (velocity, timing, predictability). It also depends on the purpose behind the question. Are you an energy planner trying to estimate the contribution to overall electricity demand in a region that generic wind turbines might be able to make, or are you concerned about the performance of one wind turbine versus another? Is it for a homework question or are you investing millions of dollars in a wind farm somewhere? Some energy estimates can be made with "back of the envelope" calculations, and others require extensive wind turbine performance specifications and wind data for the site.

6.9.1 Annual Energy Using Average Wind Turbine Efficiency

Suppose that the wind power density has been evaluated for a site. If we make reasonable assumptions of the overall conversion efficiency into electricity by the wind turbine, we can estimate the annual energy delivered. We already know that the highest efficiency possible for the rotor itself is 59.3%. In optimum conditions, a modern rotor will deliver about three-fourths of that potential. To keep from overpowering the generator, however, the rotor must spill some of the most energetic high-speed winds, and low-speed winds are also neglected when they are too slow to overcome friction and generator losses. As Example 6.7 suggested, the gearbox and generator deliver about two-thirds of the shaft power created by the rotor. Combining all of these factors leaves us with an overall conversion efficiency from wind power to electrical power of perhaps 30%. Later in the chapter, more careful calculations of wind turbine performance will be made, but quick, simple estimates can be made based on wind classifications and overall efficiencies.

Example 6.11 Annual Energy Delivered by a Wind Turbine. Suppose that a NEG Micon 750/48 (750-kW generator, 48-m rotor) wind turbine is mounted on a 50-m tower in an area with 5-m/s average winds at 10-m height. Assuming standard air density, Rayleigh statistics, Class 1 surface roughness, and an overall efficiency of 30%, estimate the annual energy (kWh/yr) delivered.

Solution. We need to find the average power in the wind at 50 m. Since "surface roughness class" is given rather than the friction coefficient α, we need to use (6.16) to estimate wind speed at 50 m. From Table 6.4, we find the roughness length z for Class 1 to be 0.03 m. The average windspeed at 50 m is thus

$$v_{50} = v_{10}\frac{\ln(H_{50}/z)}{\ln(H_{10}/z)} = 5\ \text{m/s} \cdot \frac{\ln(50/0.03)}{\ln(10/0.03)} = 6.39\ \text{m/s}$$

Average power in the wind at 50 m is therefore (6.48)

$$\overline{P}_{50} = \frac{6}{\pi} \cdot \frac{1}{2}\rho\overline{v}^3 = 1.91 \times 0.5 \times 1.225 \times (6.39)^3 = 304.5\ \text{W/m}^2$$

Since this 48-m machine collects 30% of that, then, in a year with 8760 hours, the energy delivered would be

$$\begin{aligned}\text{Energy} &= 0.3 \times 304.5\ \text{W/m}^2 \times \frac{\pi}{4}(48\ \text{m})^2 \times 8760\ \text{h/yr} \times \frac{1\ \text{kW}}{1000\ \text{W}}\\ &= 1.45 \times 10^6\ \text{kWh/yr}\end{aligned}$$

6.9.2 Wind Farms

Unless it is a single wind turbine for a particular site, such as an off-grid home in the country, most often when a good wind site has been found it makes sense to install a large number of wind turbines in what is often called a *wind farm* or a *wind park*. Obvious advantages result from clustering wind turbines together at a windy site. Reduced site development costs, simplified connections to transmission lines, and more centralized access for operation and maintenance, all are important considerations.

So how many turbines can be installed at a given site? Certainly wind turbines located too close together will result in upwind turbines interfering with the wind received by those located downwind. As we know, the wind is slowed as some of its energy is extracted by a rotor, which reduces the power available to downwind machines. Eventually, however, some distance downwind, the wind speed recovers. Theoretical studies of square arrays with uniform, equal spacing illustrate the degradation of performance when wind turbines are too close together. For one such study, Figure 6.28 shows array efficiency (predicted output divided by the power that would result if there were no interference) as a function of tower spacing expressed in rotor diameters. The parameter is the number of turbines in an equally-spaced array. That is, for example, a 2 $\times$ 2 array consists of four wind turbines equally spaced within a square area, while an 8 $\times$ 8 array is 64 turbines in a square area. The larger the array, the greater the interference, so array efficiency drops.

Figure 6.28 shows that interference out to at least 9 rotor diameters for all of these square array sizes, but for small arrays performance degradation is modest, less than about 20% for 6-diameter spacing with 16 turbines. Intuitively, an array area should not be square, as was the case for the study shown in Fig. 6.28, but rectangular with only a few long rows perpendicular to the prevailing winds, with each row having many turbines. Experience has yielded some rough rules-of-thumb for tower spacing of such rectangular arrays. Recommended spacing is 3–5 rotor diameters separating towers within a row and 5–9 diameters between rows. The offsetting, or staggering, of one row of towers behind another, as illustrated in Fig. 6.29 is also common.

We can now make some preliminary estimates of the wind energy potential per unit of land area as the following example suggests.

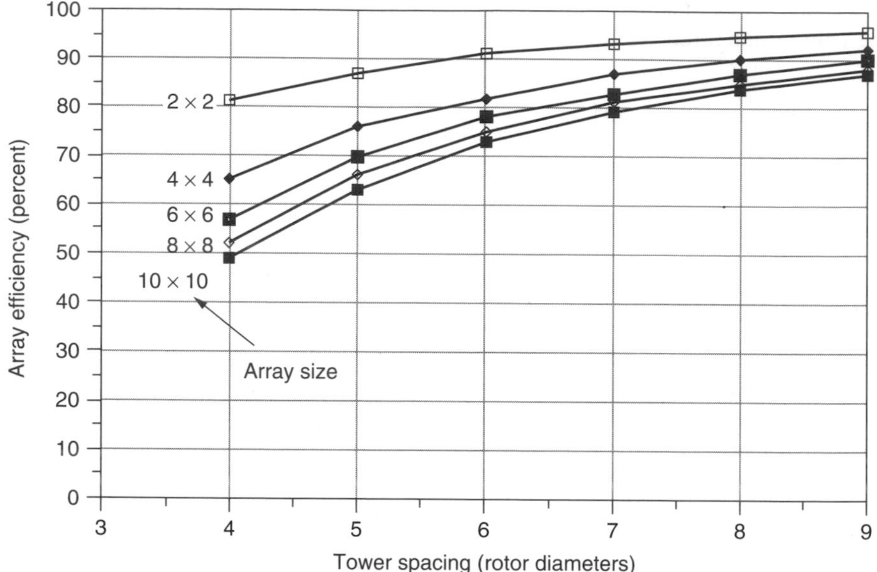

Figure 6.28 Impact of tower spacing and array size on performance of wind turbines. *Source*: Data in Milborrow and Surman (1987), presented in Grubb and Meyer (1993).

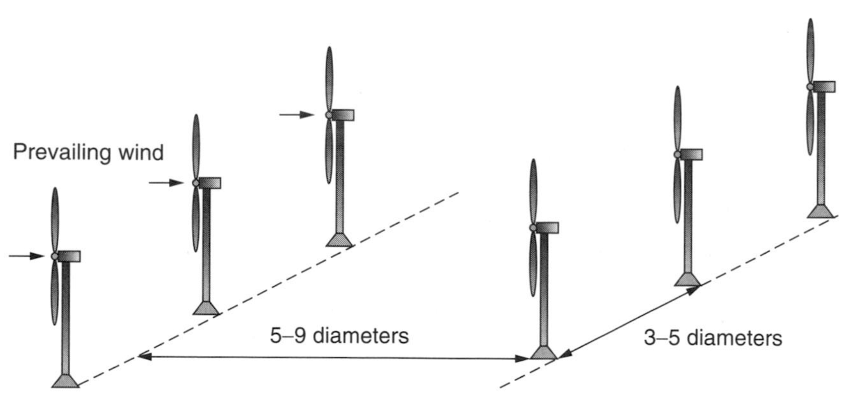

Figure 6.29 Optimum spacing of towers is estimated to be 3–5 rotor diameters between wind turbines within a row and 5–9 diameters between rows.

Example 6.12 Energy Potential for a Wind Farm. Suppose that a wind farm has 4-rotor-diameter tower spacing along its rows, with 7-diameter spacing between rows ($4D \times 7D$). Assume 30% wind turbine efficiency and an array efficiency of 80%.

a. Find the annual energy production per unit of land area in an area with 400-W/m^2 winds at hub height (the edge of 50 m, Class 4 winds).

b. Suppose that the owner of the wind turbines leases the land from a rancher for \$100 per acre per year (about 10 times what a Texas rancher makes on cattle). What does the lease cost per kWh generated?

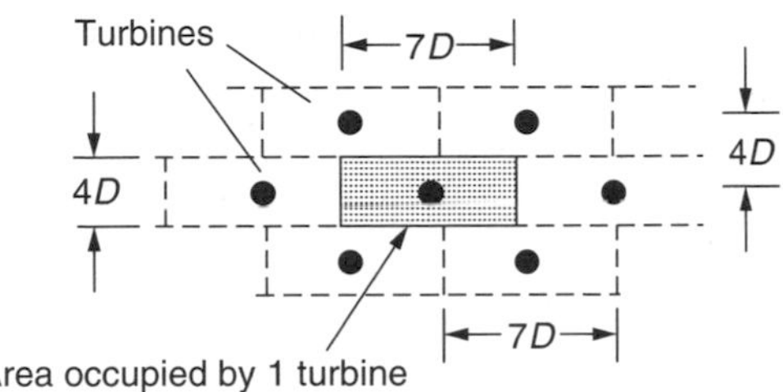

Solution

a. As the figure suggests, the land area occupied by one wind turbine is $4D \times 7D = 28D^2$, where D is the diameter of the rotor. The rotor area is $(\pi/4)D^2$. The energy produced per unit of land area is thus

$$\frac{\text{Energy}}{\text{Land area}} = \frac{1}{28D^2}\left(\frac{\text{Wind turbine}}{\text{m}^2\text{ land}}\right) \cdot \frac{\pi}{4}D^2\left(\frac{\text{m}^2\text{ rotor}}{\text{Wind turbine}}\right)$$

$$\times\, 400\left(\frac{\text{W}}{\text{m}^2\text{ rotor}}\right) \times 0.30 \times 0.80 \times 8760\frac{\text{h}}{\text{yr}}$$

$$\frac{\text{Energy}}{\text{Land area}} = 23,588\frac{\text{W}\cdot\text{h}}{\text{m}^2\cdot\text{yr}} = 23.588\frac{\text{kWh}}{\text{m}^2\cdot\text{yr}}$$

b. At 4047 m^2 per acre, the annual energy produced per acre is:

$$\frac{\text{Energy}}{\text{Land area}} = 23.588\frac{\text{kWh}}{\text{m}^2\cdot\text{yr}} \times \frac{4047\text{ m}^2}{\text{acre}} = 95,461\frac{\text{kWh}}{\text{acre}\cdot\text{yr}}$$

so, leasing the land costs the wind farmer:

$$\frac{\text{Land cost}}{\text{kWh}} = \frac{\$100}{\text{acre}\cdot\text{yr}} \times \frac{\text{acre}\cdot\text{yr}}{95,461\text{ kWh}} = \$0.00105/\text{kWh} = 0.1\ \text{¢}/\text{kWh}$$

The land leasing computation in the above example illustrates an important point. Wind farms are quite compatible with conventional farming, especially cattle ranching, and the added revenue a farmer can receive by leasing land to a wind park is often more than the value of the crops harvested on that same land.

As a result, ranchers and farmers are becoming some of the strongest proponents of wind power since it helps them to stay in their primary business while earning higher profits.

6.10 SPECIFIC WIND TURBINE PERFORMANCE CALCULATIONS

The techniques already described that help us go from power in the wind to electrical energy delivered have used only simple estimates of overall system efficiency linked to wind probability statistics. Now we will introduce techniques that can be applied to individual wind turbines based on their own specific performance characteristics.

6.10.1 Some Aerodynamics

In order to understand some aspects of wind turbine performance, we need a brief introduction to how rotor blades extract energy from the wind. Begin by considering the simple airfoil cross section shown in Fig. 6.30a. An airfoil, whether it is the wing of an airplane or the blade of a windmill, takes advantage of Bernoulli's principle to obtain lift. Air moving over the top of the airfoil has a greater distance to travel before it can rejoin the air that took the short cut under the foil. That means that the air pressure on top is lower than that under the airfoil, which creates the lifting force that holds an airplane up or that causes a wind turbine blade to rotate.

Describing the forces on a wind turbine blade is a bit more complicated than for a simple aircraft wing. A rotating turbine blade sees air moving toward it not only from the wind itself, but also from the relative motion of the blade as it rotates. As shown in Fig. 6.30b, the combination of wind and blade motion is

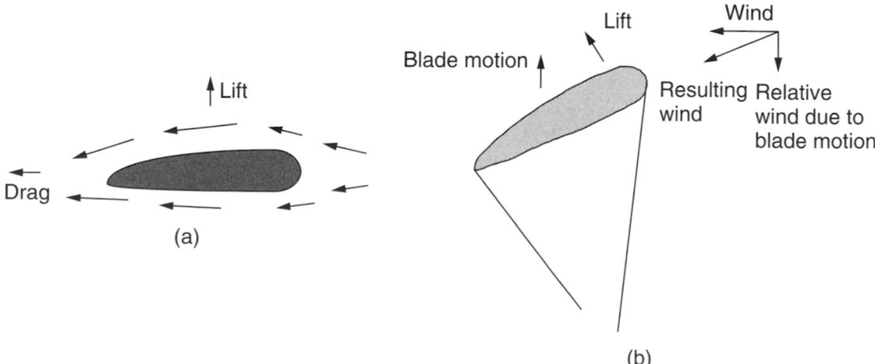

Figure 6.30 The lift in (a) is the result of faster air sliding over the top of the wind foil. In (b), the combination of actual wind and the relative wind due to blade motion creates a resultant that creates the blade lift.

like adding two vectors, with the resultant moving across the airfoil at the correct angle to obtain lift that moves the rotor along. Since the blade is moving much faster at the tip than near the hub, the blade must be twisted along its length to keep the angles right.

Up to a point, increasing the angle between the airfoil and the wind (called the angle of attack), improves lift at the expense of increased drag. As shown in Fig. 6.31, however, increasing the angle of attack too much can result in a phenomenon known as stall. When a wing stalls, the airflow over the top no longer sticks to the surface and the resulting turbulence destroys lift. When an aircraft climbs too steeply, stall can have tragic results.

6.10.2 Idealized Wind Turbine Power Curve

The most important technical information for a specific wind turbine is the power curve, which shows the relationship between windspeed and generator electrical output. A somewhat idealized power curve is shown in Fig. 6.32.

Cut-in Windspeed. Low-speed winds may not have enough power to overcome friction in the drive train of the turbine and, even if it does and the generator is

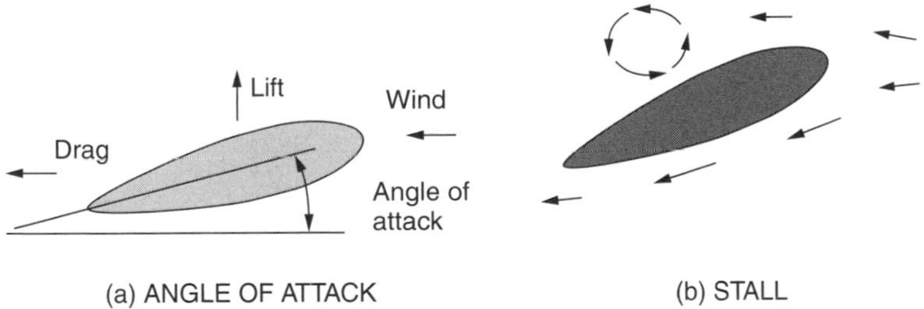

Figure 6.31 Increasing the angle of attack can cause a wing to stall.

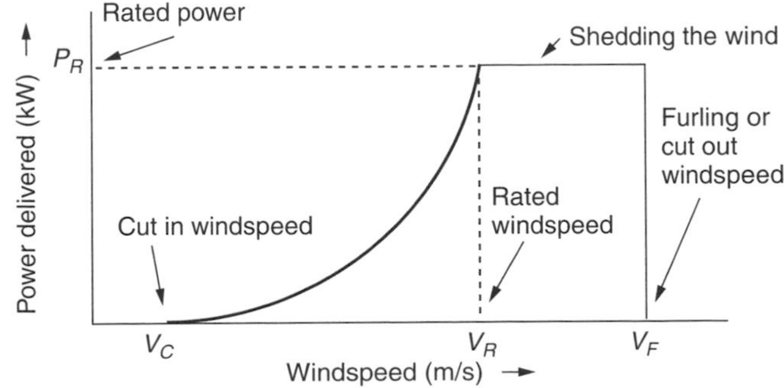

Figure 6.32 Idealized power curve. No power is generated at windspeeds below V_C; at windspeeds between V_R and V_F, the output is equal to the rated power of the generator; above V_F the turbine is shut down.

rotating, the electrical power generated may not be enough to offset the power required by the generator field windings. The cut-in windspeed V_C is the minimum needed to generate net power. Since no power is generated at windspeeds below V_C, that portion of the wind's energy is wasted. Fortunately, there isn't much energy in those low-speed winds anyway, so usually not much is lost.

Rated Windspeed. As velocity increases above the cut-in windspeed, the power delivered by the generator tends to rise as the cube of windspeed. When winds reach the rated windspeed V_R, the generator is delivering as much power as it is designed for. Above V_R, there must be some way to shed some of the wind's power or else the generator may be damaged. Three approaches are common on large machines: an active pitch-control system, a passive stall-control design, and a combination of the two.

For *pitch-controlled* turbines an electronic system monitors the generator output power; if it exceeds specifications, the pitch of the turbine blades is adjusted to shed some of the wind. Physically, a hydraulic system slowly rotates the blades about their axes, turning them a few degrees at a time to reduce or increase their efficiency as conditions dictate. The strategy is to reduce the blade's angle of attack when winds are high.

For *stall-controlled* machines, the blades are carefully designed to automatically reduce efficiency when winds are excessive. Nothing rotates—as it does in the pitch-controlled scheme—and there are no moving parts, so this is referred to as passive control. The aerodynamic design of the blades, especially their twist as a function of distance from the hub, must be very carefully done so that a gradual reduction in lift occurs as the blades rotate faster. The majority of modern, large wind turbines use this passive, stall-controlled approach.

For very large machines, above about 1 MW, an *active stall control* scheme may be justified. For these machines, the blades rotate just as they do in the active, pitch-control approach. The difference is, however, that when winds exceed the rated windspeed, instead of reducing the angle of attack of the blades, it is increased to induce stall.

Small, kilowatt-size wind turbines can have any of a variety of techniques to spill wind. Passive yaw controls that cause the axis of the turbine to move more and more off the wind as windspeeds increase are common. This can be accomplished by mounting the turbine slightly to the side of the tower so that high winds push the entire machine around the tower. Another simple approach relies on a wind vane mounted parallel to the plane of the blades. As winds get too strong, wind pressure on the vane rotate the machine away from the wind.

Cut-out or Furling Windspeed. At some point the wind is so strong that there is real danger to the wind turbine. At this windspeed V_F, called the cut-out windspeed or the furling windspeed ("furling" is the term used in sailing to describe the practice of folding up the sails when winds are too strong), the machine must be shut down. Above V_F, output power obviously is zero.

For pitch-controlled and active stall-controlled machines, the rotor can be stopped by rotating the blades about their longitudinal axis to create a stall. For stall-controlled machines, it is common on large turbines to have spring-loaded, rotating tips on the ends of the blades. When activated, a hydraulic system trips the spring and the blade tips rotate 90° out of the wind, stopping the turbine in a few rotor revolutions. If the hydraulic system fails, the springs automatically activate when rotor speed is excessive. Once a rotor has been stopped, by whatever control mechanism, a mechanical brake locks the rotor shaft in place, which is especially important for safety during maintenance.

6.10.3 Optimizing Rotor Diameter and Generator Rated Power

The idealized power curve of Fig. 6.32 provides a convenient framework within which to consider the trade-offs between rotor diameter and generator size as ways to increase the energy delivered by a wind turbine. As shown in Fig. 6.33a, increasing the rotor diameter, while keeping the same generator, shifts the power curve upward so that rated power is reached at a lower windspeed. This strategy increases output power for lower-speed winds. On the other hand, keeping the same rotor but increasing the generator size allows the power curve to continue upward to the new rated power. For lower-speed winds, there isn't much change, but in an area with higher wind speeds, increasing the generator rated power is a good strategy.

Manufacturers will sometimes offer a line of turbines with various rotor diameters and generator ratings so that customers can best match the distribution of windspeeds with an appropriate machine. In areas with relatively low windspeeds, a larger rotor diameter may be called for. In areas with relatively high windspeeds, it may be better to increase the generator rating.

6.10.4 Wind Speed Cumulative Distribution Function

Recall some of the important properties of a probability density function for wind speeds. The total area under a probability density function curve is equal to one,

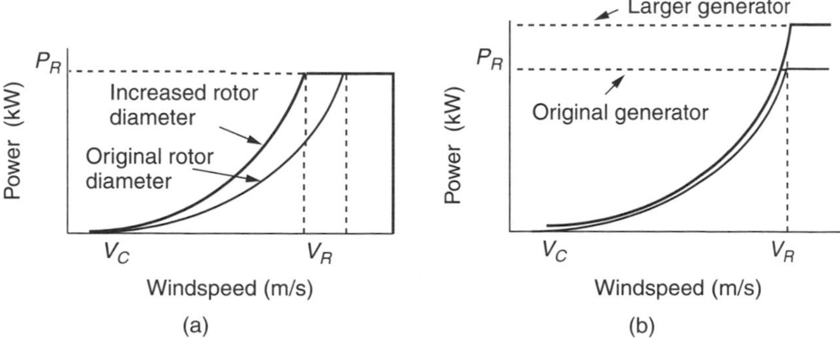

Figure 6.33 (a) Increasing rotor diameter reduces the rated windspeed, emphasizing lower speed winds. (b) Increasing the generator size increases rated power, emphasizing higher windspeeds.

and the area between any two windspeeds is the probability that the wind is between those speeds. Therefore, the probability that the wind is less than some specified windspeed V is given by

$$\text{prob}(v \le V) = F(V) = \int_0^V f(v)\, dv \tag{6.49}$$

The integral $F(V)$ in (6.49) is given a special name: the *cumulative distribution function*. The probability that the wind V is less than 0 is 0, and the probability that the wind is less than infinity is 1, so $F(V)$ has the following constraints:

$$F(V) = \text{probability } v \le V, \qquad F(0) = 0, \qquad \text{and} \qquad F(\infty) = 1 \tag{6.50}$$

In the field of wind energy, the most important p.d.f. is the Weibull function given before as (6.41):

$$f(v) = \frac{k}{c}\left(\frac{v}{c}\right)^{k-1} \exp\left[-\left(\frac{v}{c}\right)^k\right] \tag{6.41}$$

The cumulative distribution function for Weibull statistics is therefore

$$F(V) = \text{prob}(v \le V) = \int_0^V \frac{k}{c}\left(\frac{v}{c}\right)^{k-1} \exp\left[-\left(\frac{v}{c}\right)^k\right] dv \tag{6.51}$$

This integral looks pretty imposing. The trick to the integration is to make the change of variable:

$$x = \left(\frac{v}{c}\right)^k \qquad \text{so that } dx = \frac{k}{c}\left(\frac{v}{c}\right)^{k-1} dv \qquad \text{and} \qquad F(V) = \int_0^x e^{-x} dx \tag{6.52}$$

which results in

$$F(V) = \text{prob}(v \le V) = 1 - \exp\left[-\left(\frac{V}{c}\right)^k\right] \tag{6.53}$$

For the special case of Rayleigh statistics, $k = 2$, and from (6.44) $c = \frac{2\,\overline{v}}{\sqrt{\pi}}$, where $\overline{v}$ is the average windspeed, the probability that the wind is less than V is given by

$$F(V) = \text{prob}(v \le V) = 1 - \exp\left[-\frac{\pi}{4}\left(\frac{V}{\overline{v}}\right)^2\right] \qquad \text{(Rayleigh)} \tag{6.54}$$

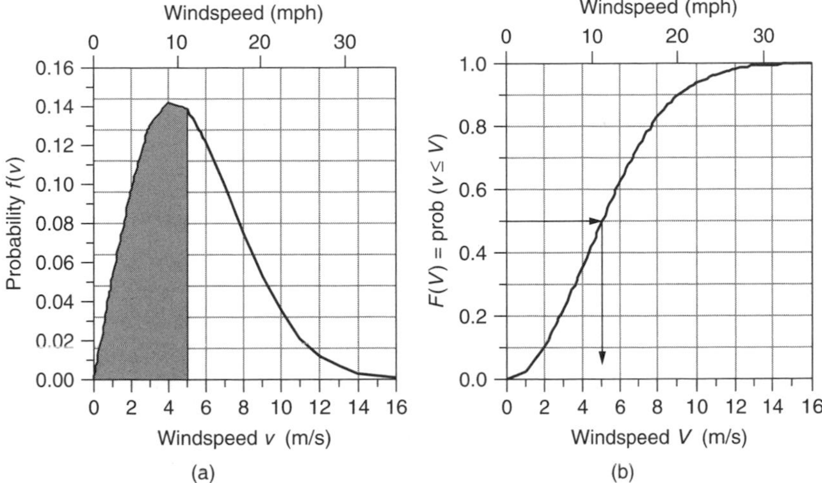

Figure 6.34 An example p.d.f. (a) and cumulative distribution function (b) for $k = 2$, $c = 6$ Weibull statistics. In this case, half the time the wind is less than or equal to 5 m/s; that is, half the area under the p.d.f. is to the left of $v = 5$ m/s.

A graph of a Weibull p.d.f. $f(v)$ and its cumulative distribution function, $F(V)$, is given in Fig. 6.34. The example shown there has $k = 2$ and $c = 6$, so it is actually a Rayleigh p.d.f. The figure shows that half of the time the wind is less than or equal to 5 m/s; that is, half the area under the $f(v)$ curve falls to the left of 5 m/s, and $F(5) = 0.5$. Note that this does not mean the average windspeed is 5 m/s. In fact, since this example is a Rayleigh p.d.f., the average windspeed is given by (6.44): $\overline{v} = c\sqrt{\pi}/2 = 6\sqrt{\pi}/2 = 5.32$ m/s.

Also of interest is the probability that the wind is greater than a certain value

$$\text{prob}(v \geq V) = 1 - \text{prob}(v \leq V) = 1 - F(V) \tag{6.55}$$

For Weibull statistics, (6.55) becomes

$$\text{prob}(v \geq V) = 1 - \left\{1 - \exp\left[-\left(\frac{V}{c}\right)^k\right]\right\} = \exp\left[-\left(\frac{V}{c}\right)^k\right] \tag{6.56}$$

and for Rayleigh statistics,

$$\text{prob}(v \geq V) = \exp\left[-\frac{\pi}{4}\left(\frac{V}{\overline{v}}\right)^2\right] \qquad \text{(Rayleigh)} \tag{6.57}$$

Example 6.13 Idealized Power Curve with Rayleigh Statistics. A NEG Micon 1000/54 wind turbine (1000-kW rated power, 54-m-diameter rotor) has a cut-in windspeed $V_C = 4$ m/s, rated windspeed $V_R = 14$ m/s, and a furling wind speed of $V_F = 25$ m/s. If this machine is located in Rayleigh winds with an average speed of 10 m/s, find the following:

a. How many h/yr is the wind below the cut-in wind speed?
b. How many h/yr will the turbine be shut down due to excessive winds?
c. How many kWh/yr will be generated when the machine is running at rated power?

Solution

a. Using (6.54), the probability that the windspeed is below cut-in 4 m/s is

$$F(V_C) = \text{prob}(v \le V_C) = 1 - \exp\left[-\frac{\pi}{4}\left(\frac{V_C}{\bar{v}}\right)^2\right]$$

$$= 1 - \exp\left[-\frac{\pi}{4}\left(\frac{4}{10}\right)^2\right] = 0.1181$$

In a year with 8760 hours (365 × 24), the number of hours the wind will be less than 4 m/s is

$$\text{Hours } (v \le 4 \text{ m/s}) = 8760 \text{ h/yr} \times 0.1181 = 1034 \text{ h/yr}$$

b. Using (6.57), the hours when the wind is higher than $V_F = 25$ m/s will be

$$\text{Hours}(v \ge V_F) = 8760 \cdot \exp\left[-\frac{\pi}{4}\left(\frac{V_F}{\bar{v}}\right)^2\right] = 8760 \cdot \exp\left[-\frac{\pi}{4}\left(\frac{25}{10}\right)^2\right]$$
$$= 65 \text{ h/yr}$$

that is, about 2.5 days per year the turbine will be shut down due to excessively high speed winds.

c. The wind turbine will deliver its rated power of 1000 kW any time the wind is between $V_R = 14$ m/s and $V_F = 25$ m/s. The number of hours that the wind is higher than 14 m/s is

$$\text{Hours}(v \ge 14) = 8760 \cdot \exp\left[-\frac{\pi}{4}\left(\frac{14}{10}\right)^2\right] = 1879 \text{ h/yr}$$

So, the number of hours per year that the winds blow between 14 m/s and 25 m/s is $1879 - 65 = 1814$ h/yr. The energy the turbine delivers from those winds will be

$$\text{Energy } (V_R \le v \le V_F) = 1000 \text{ kW} \times 1814 \text{ h/yr} = 1.814 \times 10^6 \text{ kWh/yr}$$

6.10.5 Using Real Power Curves with Weibull Statistics

Figure 6.35 shows power curves for three wind turbines: the NEG Micon 1500/64 (rated power is 1500 kW; rotor diameter is 64 m), the NEG Micon 1000/54, and the Vestas V42 600/42. Their resemblance to the idealized power curve is apparent, with most of the discrepancy resulting from the inability of wind-shedding techniques to precisely control output when winds exceed the rated windspeed. This is most pronounced in passive stall-controlled rotors. Notice how the rounding of the curve in the vicinity of the rated power makes it difficult to determine what an appropriate value of the rated windspeed V_R should be. As a result, rated windspeed is used less often these days as part of turbine product literature.

With the power curve in hand, we know the power delivered at any given wind speed. If we combine the power at any wind speed with the hours the wind blows at that speed, we can sum up total kWh of energy produced. If the site has data for hours at each wind speed, those would be used to calculate the energy delivered. Alternatively, when wind data are incomplete, it is customary to assume Weibull statistics with an appropriate shape parameter k, and scale parameter c. If only the average wind speed is known $\overline{v}$, we can use the simpler Rayleigh statistics with $k = 2$ and $c = \dfrac{2\,\overline{v}}{\sqrt{\pi}}$.

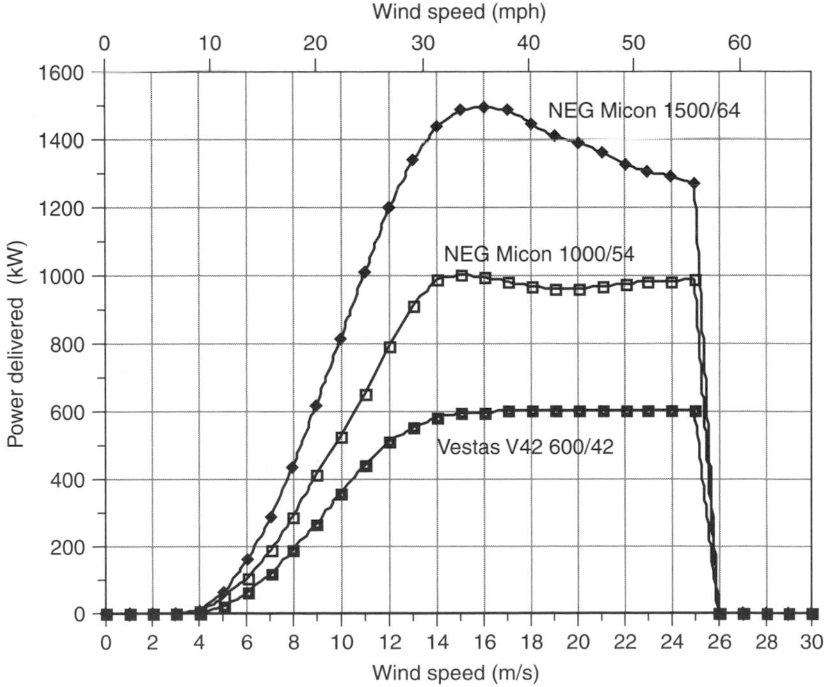

Figure 6.35 Power curves for three large wind turbines.

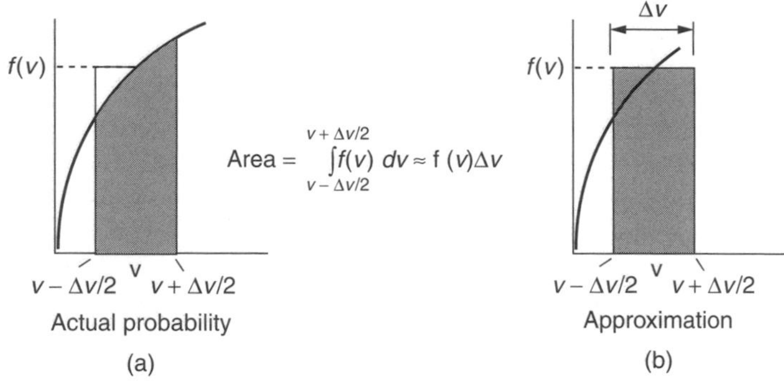

Figure 6.36 The probability that v is within $v \pm \Delta v/2$ is the shaded area in (a). A reasonable approximation is the shaded area in (b) $f(v)\Delta v$, as long as Δv is relatively small.

We started the description of wind statistics using discrete values of wind speed and hours per year at that wind speed, then moved on to continuous probability density functions. It is time to take a step backwards and modify the continuous p.d.f. to estimate hours at discrete wind speeds. With hours at any given speed and turbine power at that speed, we can easily do a summation to find energy produced.

Suppose we ask: What is the probability that the wind blows at some specified speed v? A statistician will tell you that the correct answer is zero. It never blows at *exactly* v m/s. The legitimate question is, What is the probability that the wind blows between $v - \Delta v/2$ and $v + \Delta v/2$? On a p.d.f., this probability is the area under the curve between $v - \Delta v/2$ and $v + \Delta v/2$ as shown in Fig. 6.36a. If Δv is small enough, then a reasonable approximation is the rectangular area shown in Fig. 6.36b. This suggests we can make the following approximation:

$$\text{prob}(v - \Delta v/2 \le V \le v + \Delta v/2) = \int_{v-\Delta v/2}^{v+\Delta v/2} f(v)\,dv \approx f(v)\Delta v \qquad (6.58)$$

While this may look complicated, it really makes life very simple. It says we can conveniently discretize a continuous p.d.f. by saying the probability that the wind blows at some windspeed V is just $f(V)$, and let the statisticians squirm. Let's us check the following example to see if this seems reasonable.

Example 6.14 Discretizing $f(v)$. For a wind site with Rayleigh winds having average speed $\overline{v} = 8$ m/s, what is the probability that the wind would blow between 6.5 and 7.5 m/s? How does this compare to the p.d.f. evaluated at 7 m/s?

Solution. Using (6.57), we obtain

$$\text{prob}(v \geq 6.5) = \exp\left[-\frac{\pi}{4}\left(\frac{6.5}{8}\right)^2\right] = 0.59542$$

$$\text{prob}(v \geq 7.5) = \exp\left[-\frac{\pi}{4}\left(\frac{7.5}{8}\right)^2\right] = 0.50143$$

So, the probability that the wind is between 6.5 and 7.5 m/s is

$$\text{prob}(6.5 \leq v \leq 7.5) = 0.59542 - 0.50143 - 0.09400$$

From (6.45), we will approximate the probability that the wind is at 7 m/s to be

$$f(v) = \frac{\pi}{2\bar{v}^2}\frac{v}{} \exp\left[-\frac{\pi}{4}\left(\frac{v}{\bar{v}}\right)^2\right]$$

so,

$$f(7\text{ m/s}) = \frac{\pi \cdot 7}{2 \cdot 8^2} \exp\left[-\frac{\pi}{4}\left(\frac{7}{8}\right)^2\right] = 0.09416$$

The approximation 0.09416 is only 0.2% higher than the correct value of 0.09400.

The above example is reassuring. It suggests that we can use the p.d.f. evaluated at integer values of windspeed to represent the probability that the wind blows at that speed. Combining power curve data supplied by the turbine manufacturer (examples are given in Table 6.7), with appropriate wind statistics, gives us a straightforward way to estimate annual energy production. This is most easily done using a spreadsheet. Example 6.15 demonstrates the process.

Example 6.15 Annual Energy Delivered Using a Spreadsheet. Suppose that a NEG Micon 60-m diameter wind turbine having a rated power of 1000 kW is installed at a site having Rayleigh wind statistics with an average windspeed of 7 m/s at the hub height.

a. Find the annual energy generated.
b. From the result, find the overall average efficiency of this turbine in these winds.
c. Find the productivity in terms of kWh/yr delivered per m^2 of swept area.

TABLE 6.7 Examples of Wind Turbine Power Specifications

Manufacturer:		NEG Micon	NEG Micon	NEG Micon	Vestas	Whisper	Wind World	Nordex	Bonus
Rated Power (kW):		1000	1000	1500	600	0.9	250	1300	300
Diameter (m):		60	54	64	42	2.13	29.2	60	33.4
Avg. Windspeed									
v (m/s)	v(mph)	kW	kW	kW	kW	kW	kW	kW	kW
0	0	0	0	0	0	0.00	0	0	0
1	2.2	0	0	0	0	0.00	0	0	0
2	4.5	0	0	0	0	0.00	0	0	0
3	6.7	0	0	0	0	0.03	0	0	4
4	8.9	33	10	9	0	0.08	0	25	15
5	11.2	86	51	63	22	0.17	12	78	32
6	13.4	150	104	159	65	0.25	33	150	52
7	15.7	248	186	285	120	0.35	60	234	87
8	17.9	385	291	438	188	0.45	92	381	129
9	20.1	535	412	615	268	0.62	124	557	172
10	22.4	670	529	812	356	0.78	153	752	212
11	24.6	780	655	1012	440	0.90	180	926	251
12	26.8	864	794	1197	510	1.02	205	1050	281
13	29.1	924	911	1340	556	1.05	224	1159	297
14	31.3	964	986	1437	582	1.08	238	1249	305
15	33.6	989	1006	1490	594	1.04	247	1301	300
16	35.8	1000	998	1497	598	1.01	253	1306	281
17	38.0	998	984	1491	600	1.00	258	1292	271
18	40.3	987	971	1449	600	0.99	260	1283	259
19	42.5	968	960	1413	600	0.97	259	1282	255
20	44.7	944	962	1389	600	0.95	256	1288	253
21	47.0	917	967	1359	600	0.00	250	1292	254
22	49.2	889	974	1329	600	0.00	243	1300	255
23	51.5	863	980	1307	600	0.00	236	1313	256
24	53.7	840	985	1288	600	0.00	230	1328	257
25	55.9	822	991	1271	600	0.00	224	1344	258
26	58.2	0	0	0	0	0.00	0	0	0

Source: Mostly based on data in www.windpower.dk.

Solution

a. To find the annual energy delivered, a spreadsheet solution is called for. Let's do a sample calculation for a 6-m/s windspeed to see how it goes, and then present the spreadsheet results.

From Table 6.7, at 6 m/s the NEG Micon 1000/60 generates 150 kW. From (6.45), the Rayleigh p.d.f. at 6 m/s in a regime with 7-m/s average windspeed is

$$f(v) = \frac{\pi v}{2\overline{v}^2} \exp\left[-\frac{\pi}{4}\left(\frac{v}{\overline{v}}\right)^2\right] = \frac{\pi \cdot 6}{2 \cdot 7^2} \exp\left[-\frac{\pi}{4}\left(\frac{6}{7}\right)^2\right] = 0.10801$$

In a year with 8760 h, our estimate of the hours the wind blows at 6 m/s is

$$\text{Hours @6 m/s} = 8760 \text{ h/yr} \times 0.10801 = 946 \text{ h/yr}$$

So the energy delivered by 6-m/s winds is

$$\text{Energy (@6 m/s)} = 150 \text{ kW} \times 946 \text{ h/yr} = 141,929 \text{ kWh/yr}$$

The rest of the spreadsheet is given below. The total energy produced is 2.85×10^6 kWh/yr.

Windspeed (m/s)	Power (kW)	Probability $f(v)$	Hrs/yr at v	Energy (kWh/yr)
0	0	0.000	0	0
1	0	0.032	276	0
2	0	0.060	527	0
3	0	0.083	729	0
4	33	0.099	869	28,683
5	86	0.107	941	80,885
6	150	0.108	946	141,929
7	248	0.102	896	222,271
8	385	0.092	805	310,076
9	535	0.079	690	369,126
10	670	0.065	565	378,785
11	780	0.051	444	346,435
12	864	0.038	335	289,551
13	924	0.028	243	224,707
14	964	0.019	170	163,779
15	989	0.013	114	113,101
16	1000	0.008	74	74,218
17	998	0.005	46	46,371
18	987	0.003	28	27,709
19	968	0.002	16	15,853
20	944	0.001	9	8,709
21	917	0.001	5	4,604
22	889	0.000	3	2,347
23	863	0.000	1	1,158
24	840	0.000	1	554
25	822	0.000	0	257
26	0	0.000	0	0
			Total:	2,851,109

b. The average efficiency is the fraction of the wind's energy that is actually converted into electrical energy. Since Rayleigh statistics are assumed, we can use (6.48) to find average power in the wind for a 60-m rotor diameter (assuming the standard value of air density equal to 1.225 kg/m^3):

$$\overline{P} = \frac{6}{\pi} \cdot \frac{1}{2}\rho A\overline{v}^3 = \frac{6}{\pi} \times 0.5 \times 1.225 \times \frac{\pi}{4}(60)^2 \times (7)^3$$
$$= 1.134 \times 10^6 \text{ W} = 1134 \text{ kW}$$

In a year with 8760 h, the energy in the wind is

$$\text{Energy in wind} = 8760 \text{ h/yr} \times 1134 \text{ kW} = 9.938 \times 10^6 \text{ kWh}$$

So the average efficiency of this machine in these winds is

$$\text{Average efficiency} = \frac{2.85 \times 10^6 \text{ kWh/yr}}{9.938 \times 10^6 \text{ kWh/yr}} = 0.29 = 29\%$$

c. The productivity (annual energy per swept area) of this machine is

$$\text{Productivity} = \frac{2.85 \times 10^6 \text{ kWh/yr}}{(\pi/4) \cdot 60^2 \text{ m}^2} = 1008 \text{ kWh/m}^2 \cdot \text{yr}$$

A histogram of hours per year and MWh per year at each windspeed for the above example is presented in Fig. 6.37. Notice how little energy is delivered at lower windspeeds in spite of the large number of hours of wind at those speeds. This is, of course, another example of the importance of the cubic relationship between power in the wind and wind speed.

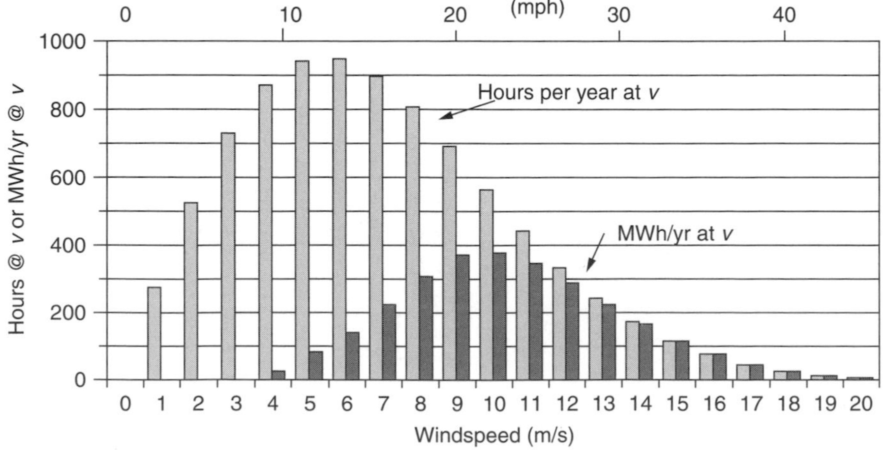

Figure 6.37 Hours per year and MWh per year at each windspeed for the NEG Micon (1000/60) turbine and Rayleigh winds with average speed 7 m/s.

6.10.6 Using Capacity Factor to Estimate Energy Produced

One of the most important characteristics of any electrical power system is its rated power; that is, how many kW it can produce on a continuous, full-power basis. If the system has a generator, the rated power is dictated by the rated output of the generator. If the generator were to deliver rated power for a full year, then the energy delivered would be the product of rated power times 8760 h/yr. Since power systems—especially wind turbines—don't run at full power all year, they put out something less than that maximum amount. The *capacity factor* CF is a convenient, dimensionless quantity between 0 and 1 that connects rated power to energy delivered:

$$\text{Annual energy (kWh/yr)} = P_R\ \text{(kW)} \times 8760\ \text{(h/yr)} \times \text{CF} \tag{6.59}$$

where P_R is the rated power (kW) and CF is the capacity factor. That is, the capacity factor is

$$\text{CF} = \frac{\text{Actual energy delivered}}{P_R \times 8760} \tag{6.60}$$

Or, another way to express it is

$$\text{CF} = \frac{\text{Actual energy delivered/8760 h/yr}}{P_R} = \frac{\text{Average power}}{\text{Rated power}} \tag{6.61}$$

Example 6.16 Capacity Factor for the NEG Micon 1000/60. What is the capacity factor for the NEG Micon 1000/60 in Example 6.14?

Solution

$$\text{CF} = \frac{\text{Actual energy delivered}}{P_R \times 8760} = \frac{2.851 \times 10^6\ \text{kWh/yr}}{1000\ \text{kW} \times 8760\ \text{h/yr}} = 0.325$$

Example 6.16 is quite artificial, in that a careful calculation of energy delivered was used to find capacity factor. The real purpose of introducing the capacity factor is to do just the opposite—that is, to use it to estimate energy delivered. The goal here is to find a simple way to estimate capacity factor when very little is known about a site and wind turbine.

Suppose we use the procedure just demonstrated in Examples 6.15 and 6.16 to work out the capacity factor for the above wind turbine while we vary the average wind speed. Figure 6.38 shows the result. For mid-range winds averaging from about 4 to 10 m/s (9 to 22 mph), the capacity factor for this machine is quite linear. These winds cover all the way from Class 2 to Class 7 winds, and so they are quite typical of sites for which wind power is attractive. For winds with higher averages, more and more of the wind is above the rated windspeed and

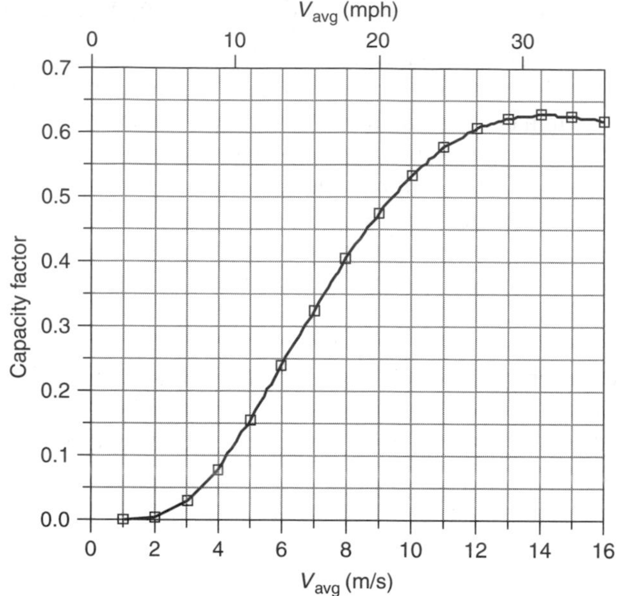

Figure 6.38 Capacity factor for the NEGMicon 1000/60 assuming Rayleigh wind statistics. For sites with average winds between about 4 and 10 m/s (9 to 22 mph), CF varies quite linearly with average windspeed. Rayleigh statistics are assumed.

capacity factor begins to level out and even drop some. A similar flattening of the curve occurs when the average windspeed is down near the cut-in windspeed and below, since much of the wind produces no electrical power.

The S-shaped curve of Fig. 6.38 was derived for a specific turbine operating in winds that follow Rayleigh statistics. As it turns out, all turbines show the same sort of curve with a sweet spot of linearity in the range of average wind speeds that are likely to be encountered in practice. This suggests the possibility of modeling capacity factor, in the linear region, with an equation of the form

$$\text{CF} = m\overline{V} + b \tag{6.62}$$

For the NEGMicon 1000/60, the linear fit shown in Fig. 6.39 leads to the following:

$$\text{CF} = 0.087\overline{V} - 0.278 \tag{6.63}$$

The rated power P_R of the NEG 1000/60 is 1000 kW and the rotor diameter D is 60 m. The ratio of rated power to the square of rotor diameter is

$$\frac{P_R}{D^2} = \frac{1000 \text{ kW}}{(60 \text{ m})^2} = 0.278 \qquad \text{for the NEG Micon 1000/60} \tag{6.64}$$

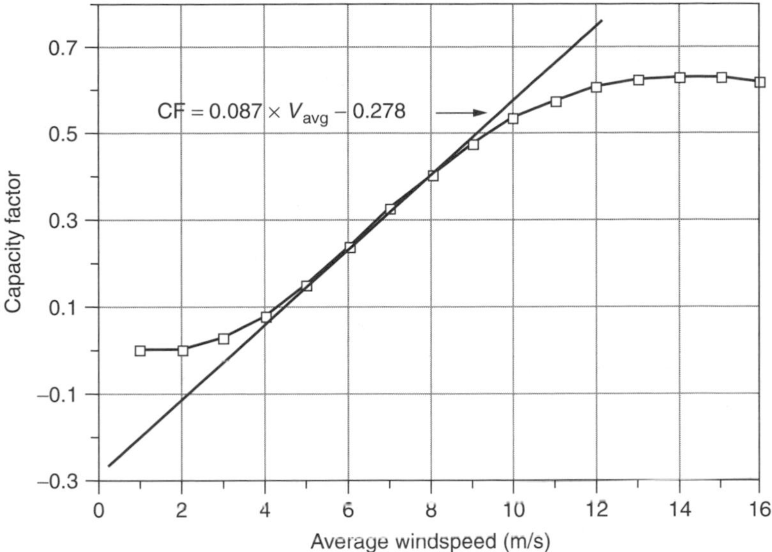

Figure 6.39 A line fitted to the linear portion of the NEGMicon 1000/60 wind turbine.

That's an interesting coincidence. For *this particular* wind turbine the y-axis intercept, b, equals P/D^2 so we can write the capacity factor as

$$\text{CF} = 0.087\overline{\text{V}} - \frac{P_R}{D^2} \qquad \text{(Rayleigh winds)} \tag{6.65}$$

where $\overline{V}$ is the average windspeed (m/s), P_R is the rated power (kW), and D is the rotor diameter (m).

Surprisingly, even though the estimate in (6.65) was derived for a single turbine, it seems to work quite well in general as a predictor of capacity factor. For example, when applied to all eight of the wind turbines in Table 6.7, Eq. (6.65) correlates very well with the correct capacity factors computed using the spreadsheet approach (Fig. 6.40). In fact, in the range of capacity factors of most interest, 0.2 to 0.5, Eq. (6.65) is accurate to within 10% for those eight turbines. This simple CF relationship is very handy since it only requires the rated power and rotor diameter for the wind turbine, along with average windspeed.

Using (6.65) for capacity factor gives the following simple estimate of energy delivered from a turbine with diameter D (m), rated power P_R (kW) in Rayleigh winds with average windspeed $\overline{\text{V}}$ (m/s)

$$\text{Annual energy (kWh/yr)} = 8760 \cdot P_R(\text{kW}) \left\{0.087\overline{V}(\text{m/s}) - \frac{P_R(\text{kW})}{[D(\text{m})]^2}\right\} \tag{6.66}$$

Of course, the spreadsheet approach has a solid theoretical basis and is the preferred method for determining annual energy, but (6.66) can be a handy one,

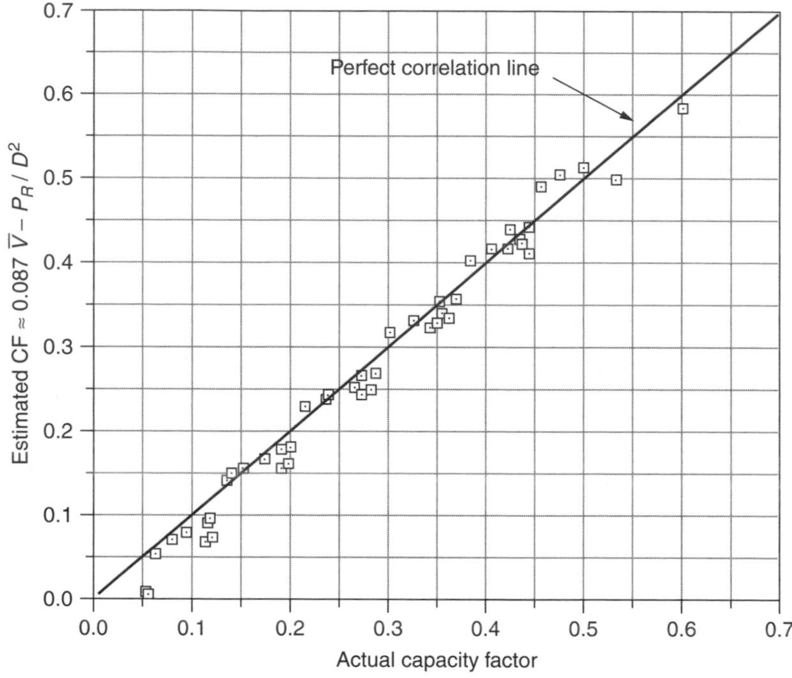

Figure 6.40 Correlation between actual capacity factor (using the spreadsheet approach) and an estimate given by (6.65) for the eight wind turbines in Table 6.7.

especially when little data for the wind and turbine are known (Jacobson and Masters, 2001).

Example 6.17 Energy Estimate Using the Capacity Factor Approach. The Whisper H900 wind turbine has a 900-W generator with 2.13-m blades. In an area with 6-m/s average windspeeds, estimate the energy delivered.

Solution. Using (6.65) for capacity factor gives

$$\text{CF} = 0.087\overline{V} - \frac{P_R}{D^2} = 0.087 \times 6 - \frac{0.90}{2.13^2} = 0.324$$

The energy delivered in a year's time would be

$$\text{Energy} = 8760\ \text{h/yr} \times 0.90\ \text{kW} \times 0.324 = 2551\ \text{kWh/yr}$$

Of course, we could have done this just by plugging into (6.66).

For comparison, the spreadsheet approach yields an answer of 2695 kWh/yr, just 6% higher.

In these winds, this little wind turbine puts out about 225 kWh/mo—probably enough for a small, efficient household.

It is reassuring to note that the capacity factor relationship in (6.65), which was derived for a very large 1000-kW wind turbine with 60-m blades, gives quite accurate answers for a very small 0.90-kW turbine with 2.13-m blades.

The question sometimes arises as to whether or not a high-capacity factor is a good thing. A high-capacity factor means that the turbine is deriving much of its energy in the flat, wind-shedding region of the power curve above the rated windspeed. This means that power production is relatively stable, which can have some advantages in terms of the interface with the local grid. On the other hand, a high-capacity factor means that a significant fraction of the wind's energy is not being captured since the blades are purposely shedding much of the wind to protect the generator. It might be better to have a larger generator to capture those higher-speed winds, in which case the capacity factor goes down while the energy delivered increases. A bigger generator, of course, costs more. In other words, the capacity factor itself is not a good indicator of the overall economics for the wind plant.

6.11 WIND TURBINE ECONOMICS

Wind turbine economics have been changing rapidly as machines have gotten larger and more efficient and are located in sites with better wind. In Fig. 6.41, the average rated power of new Danish wind turbines by year of sale shows a steady rise from roughly 50 kW in the early 1980s to 1200 kW in 2002 (Denmark accounts for more than half of world sales). The biggest machines currently being built are in the 2000-kW to 3000-kW size range. More efficient machines located in better sites with higher hub heights have doubled the average energy productivity from around 600 kWh/yr per square meter of blade area 20 years ago to around 1200 kWh/m^2-yr today.

6.11.1 Capital Costs and Annual Costs

While the rated power of new machines has increased year by year, the corresponding capital cost per kW dropped. As shown in Fig. 6.42, the capital cost of new installations has dropped from around \$1500/kW for 150-kW turbines in 1989 to about \$800/kW in 2000 for machines rated at 1650 kW. The impact of economies of scale is evident. The labor required to build a larger machine is not that much higher than for a smaller one; the cost of electronics are only moderately different; the cost of a rotor is roughly proportional to diameter while

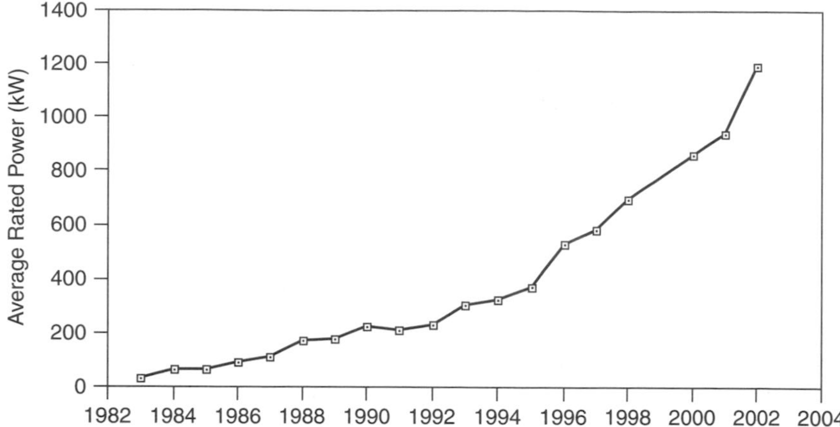

Figure 6.41 Average rated power of new wind turbines manufactured in Denmark (www.windpower.dk).

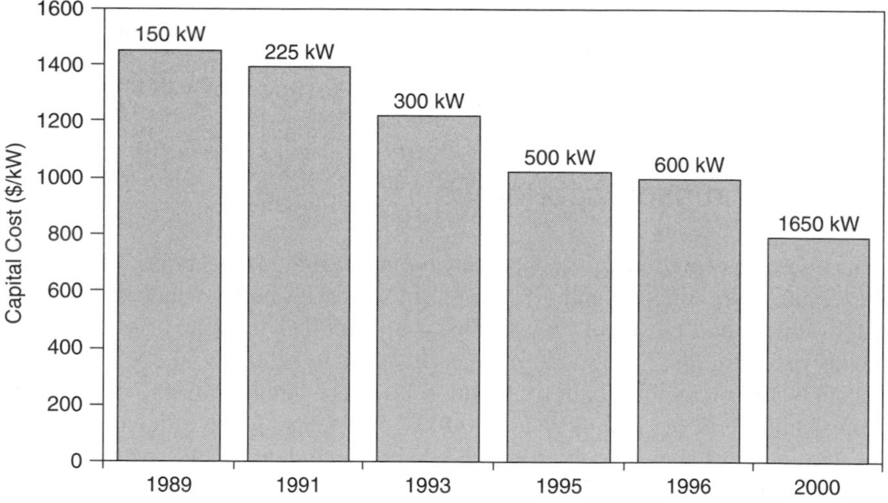

Figure 6.42 Capital costs of wind systems including turbine, tower, grid connection, site preparation, controls, and land. The 2000 cost is based on a wind farm rather than a single turbine. From Redlinger (1999) and AWEA.

power delivered is proportional to diameter squared; taller towers increase energy faster than costs increase; and so forth.

An example cost analysis for a 60-MW wind farm consisting of forty 1.5-MW turbines is given in Table 6.8. Included in the table is a cost breakdown for the initial capital costs and an estimate of the levelized annual cost of operations and maintenance (O&M). About three-fourths of the capital cost is associated with

TABLE 6.8 An Example Cost Analysis for a 60-MW Wind Park

Capital Costs	Amount ($)	Percentage
40 1.5-MW turbines @ $1.1 M, spare parts	46,640,000	76.6
Site prep, grid connections	9,148,000	15.0
Interest during construction, contingencies	3,514,000	5.8
Project development, feasibility study	965,000	1.6
Engineering	611,000	1.0
Total Capital Cost	60,878,000	100.0
Annual Costs	**Amount ($/yr)**	**Percentage**
Parts and labor	1,381,000	70.3
Insurance	135,000	6.9
Contingencies	100,000	5.1
Land lease	90,000	4.6
Property taxes	68,000	3.5
Transmission line maintenance	80,000	4.1
General and miscellaneous	111,000	5.6
Total Annual Costs	1,965,000	100.0

Source: Ministry of Natural Resources, Canada.

turbines, while the remaining portion covers costs related to turbine erection, grid connections, foundations, roads, and buildings. Operations and maintenance costs (O&M) include regular maintenance, repairs, stocking spare parts, insurance, land lease fees, insurance, and administration. Some of these are annual costs that don't particularly depend on the hours of operation of the wind turbines, such as insurance and administration, while others, those that involve wear and tear on parts, are directly related to annual energy produced. In this example, the annual O&M costs, which have already been levelized to include future cost escalations, are just over 3% of the initial capital cost of the wind farm.

In general, O&M costs depend not only on how much the machine is used in a given year, but also on the age of the turbine. That is, toward the end of the design life, more components will be subject to failure and maintenance will increase. Also, there are reasons to expect some economies of scale for O&M costs. A single turbine sitting somewhere will cost more to service than will a turbine located in a large wind park. Large turbines will also cost less to service, per kW of rated power, than a small one since labor costs will probably be comparable. Larger turbines are also newer-generation machines that have better components and designs to minimize the need for repairs.

6.11.2 Annualized Cost of Electricity from Wind Turbines

To find a levelized cost estimate for energy delivered by a wind turbine, we need to divide annual costs by annual energy delivered. To find annual costs, we must spread the capital cost out over the projected lifetime using an appropriate factor and then add in an estimate of annual O&M. Chapter 5 developed a number of techniques for doing such calculations, but let's illustrate one of the simpler approaches here.

To the extent that a wind project is financed by debt, we can annualize the capital costs using an appropriate capital recovery factor (CRF) that depends on the interest rate i and loan term n. The annual payments A on such a loan would be

$$A = P \cdot \left[\frac{i(1+i)^n}{(1+i)^n - 1}\right] = P \cdot \text{CRF}(i, n) \tag{6.67}$$

where A represents annual payments ($/yr), P is the principal borrowed ($), i is the interest rate (decimal fraction; e.g., 0.10 for a 10% interest rate), n is the loan term (yrs),
and

$$\text{CRF}(i, n) = \frac{i(1+i)^n}{(1+i)^n - 1} \tag{6.68}$$

A handy table of values for CRF(i, n) is given in Table 5.5.

Example 6.18 A Loan to Pay for a Small Wind Turbine. Suppose that a 900-W Whisper H900 wind turbine with 7-ft diameter (2.13 m) blade costs $1600. By the time the system is installed and operational, it costs a total of $2500, which is to be paid for with a 15-yr, 7 percent loan. Assuming O&M costs of $100/yr, estimate the cost per kWhr over the 15-year period if average windspeed at hub height is 15 mph (6.7 m/s).

Solution. The capital recovery factor for a 7%, 15-yr loan would be

$$\text{CRF}(0.07, 15\text{ yr}) = \frac{i(1+i)^n}{(1+i)^n - 1} = \frac{0.07(1+0.07)^{15}}{(1+0.07)^{15} - 1} = 0.1098/\text{yr}$$

which agrees with Table 5.5. So, the annual payments on the loan would be

$$A = P \times \text{CRF}(0.07, 15) = \$2500 \times 0.1098/\text{yr} = \$274.49/\text{yr}$$

The annual cost, including $100/yr of O&M, is therefore $274.49 + $100 = $374.49.

To estimate energy delivered by this machine in 6.7-m/s average wind, let us use the capacity factor approach (6.65):

$$\text{CF} = 0.087\overline{V}(\text{m/s}) - \frac{P_R(\text{kW})}{D^2(m^2)} = 0.087 \times 6.7 - \frac{0.90}{2.13^2} = 0.385$$

The annual energy delivered (6.59)

$$\text{kWh/yr} = 0.90\ \text{kW} \times 8760\ \text{h/yr} \times 0.385 = 3035\ \text{kWh/yr}$$

The average cost per kWh is therefore

$$\text{Average cost} = \frac{\text{Annual cost (\$/yr)}}{\text{Annual energy (kWh/yr)}} = \frac{\$374.49/\text{yr}}{3035\ \text{kWh/yr}} = \$0.123/\text{kWh}$$

That's a pretty good price of electricity for a small system—cheaper than grid electricity in many areas and certainly cheaper than any other off-grid, home-size generating system.

A sensitivity analysis of the cost of electricity from a 1500-kW, 64-m turbine, with a levelized O&M cost equal to 3% of capital costs, financed with a 7%, 20-year loan, is shown in Fig. 6.43. Again, taxes, depreciation, and the production tax credit are not included.

For large wind systems, capital costs are often divided into an equity portion, which comes out of the financial resources of the owner and must earn an appropriate annual rate of return, plus a debt portion that is borrowed over a loan term

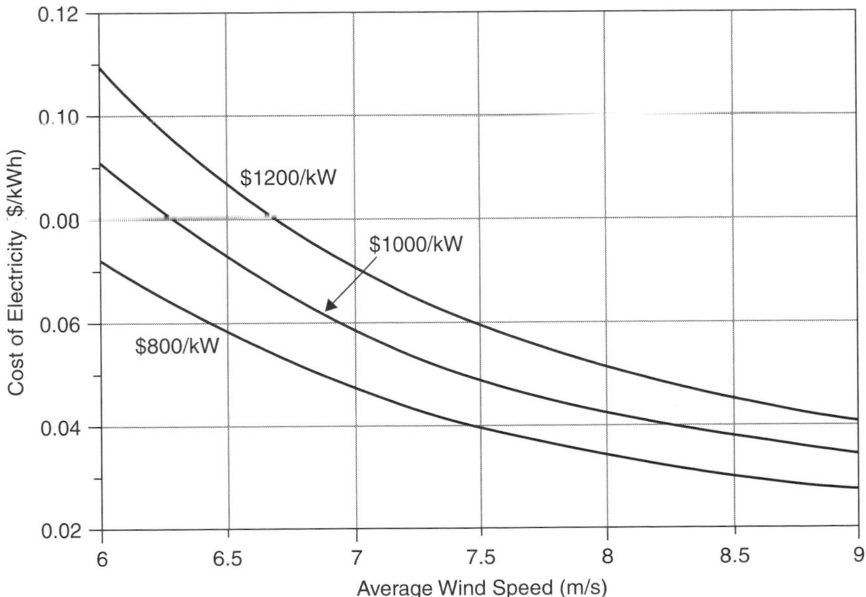

Figure 6.43 Sensitivity analysis of the levelized cost of a 1500-kW, 64-m wind turbine using (6.65) for capacity factor. Levelized O&M is 3% of capital cost, financing is 7%, 20 years. Depreciation, taxes, and government incentives are not included in this analysis.

at some interest rate. The price of electricity sold by the project must recover both the debt and equity portions of the financing.

Example 6.19 Price of Electricity from a Wind Farm. A wind farm project has 40 1500-kW turbines with 64-m blades. Capital costs are \$60 million and the levelized O&M cost is \$1.8 million/yr. The project will be financed with a \$45 million, 20-yr loan at 7% plus an equity investment of \$15 million that needs a 15% return. Turbines are exposed to Rayleigh winds averaging 8.5 m/s. What levelized price would the electricity have to sell for to make the project viable?

Solution. We can estimate the annual energy that will be delivered by starting with the capacity factor, (6.65):

$$\text{CF} = 0.087\overline{V}\ (\text{m/s}) - \frac{P_R(\text{kW})}{[D(\text{m})]^2} = 0.087 \times 8.5 - \frac{1500}{64^2} = 0.373$$

For 40 such turbines, the annual electrical production will be

$$\begin{aligned}\text{Annual energy} &= 40 \text{ turbines} \times 1500 \text{ kW} \times 8760 \text{ h/yr} \times 0.373 \\ &= 196 \times 10^6 \text{ kWh/yr}\end{aligned}$$

The debt payments will be

$$\begin{aligned}A = P \cdot \left[\frac{i(1+i)^n}{(1+i)^n - 1}\right] &= \$45{,}000{,}000 \cdot \left[\frac{0.07(1+0.07)^{20}}{(1+0.07)^{20} - 1}\right] \\ &= \$4.24 \times 10^6/\text{yr}\end{aligned}$$

The annual return on equity needs to be

$$\text{Equity} = 0.15/\text{yr} \times \$15{,}000{,}000 = \$2.25 \times 10^6/\text{yr}$$

The levelized O&M cost is \$1.8 million, so the total for O&M, debt, and equity is

$$\text{Annual cost} = \$(4.24 + 2.25 + 1.8) \times 10^6 = \$8.29 \times 10^6/\text{yr}$$

The levelized price at which electricity needs to be sold is therefore

$$\text{Selling price} = \frac{\$8.29 \times 10^6/\text{yr}}{196 \times 10^6 \text{ kWh/yr}} = \$0.0423 = 4.23¢/\text{kWh}$$

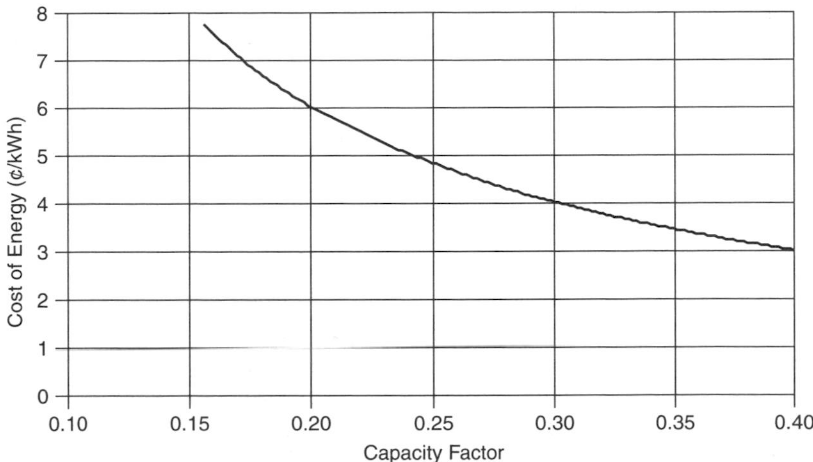

Figure 6.44 Cost of energy from a \$1000/ kW, 50-MW windfarm including PTC, depreciation, financing charges, and tax implications (O&M 1¢/kWh, inflation 2.3%, 60% equity with return 15%, debt interest 5%, income tax rate 40.7%, PTC 1.8¢/kWh for 10 yrs, CEC 5-yr incentive 0.75/kWh, discount rate 5%, MACRS 5-yr depreciation, property taxes 1.1%). Based on Bolinger et al. (2001).

Example 6.19 leaves out a number of other factors that affect the economic viability of doing the wind farm, including depreciation, income taxes, and a special tax incentive called the wind energy *production tax credit* (PTC). The production tax credit enacted in 1992 provides a 10-year, 1.5¢/kWh tax credit for electricity produced by wind energy systems installed by a certain date (inflation adjustable). The credit has been a mixed blessing to the wind industry. While it does provide a significant financial incentive, it has also caused a boom and bust cycle in the wind industry because the final deadline for projects to receive the credit is renewable at the pleasure of Congress. For example, when the credit expired in 1999, new installed capacity in the United States dropped from 661 MW in 1999 to only 53 MW in 2000. Then when it was renewed, installations jumped to 1696 MW in 2001, at which point it expired again, and in 2002 new wind power installations dropped back to 410 MW.

A careful analysis including PTC, equity and debt financing, depreciation, and inflation for a \$1000/ kW, 50-MW windfarm with a 0.30 capacity factor yields a cost of wind power of 4.03¢/kWh (Bolinger et al. (2001). Scaling that result for varying capacity factors yields the graph shown in Fig. 6.44.

6.12 ENVIRONMENTAL IMPACTS OF WIND TURBINES

Wind systems have negative as well as positive impacts on the environment. The negative ones relate to bird kills, noise, construction disturbances, aesthetic impacts, and pollution associated with manufacturing and installing the turbine. The positive impacts result from wind displacing other, more polluting energy systems.

Birds do collide with wind turbines, just as they collide with cars, cell-phone towers, glass windows, and high-voltage power lines. While the rate of deaths caused by wind turbines is miniscule compared to these other obstacles that humans put into their way, it is still an issue that can cause concern. Early wind farms had small turbines with fast-spinning blades and bird kills were more common but modern large turbines spin so slowly that birds now more easily avoid them. A number of European studies have concluded that birds almost always modify their flight paths well in advance of encountering a turbine, and very few deaths are reported. Studies of eider birds and offshore wind parks in Denmark concluded that the eiders avoided the turbines even when decoys to attract them were placed nearby. They also noted no change in the abundance of nearby eiders when turbines were purposely shut down to study their behavior.

People's perceptions of the aesthetics of wind farms are important in siting the machines. A few simple considerations have emerged, which can make them much more acceptable. Arranging same-size turbines in simple, uniform rows and columns seems to help, as does painting them a light gray color to blend with the sky. Larger turbines rotate more slowly, which makes them somewhat less distracting.

Noise from a wind turbine or a wind farm is another potentially objectionable phenomenon, and modern turbines have been designed specifically to control that noise. It is difficult to actually measure the sound level caused by turbines in the field because the ambient noise caused by the wind itself masks their noise. At a distance of only a few rotor diameters away from a turbine, the sound level is comparable to a person whispering.

The air quality advantages of wind are pretty obvious. Other than the very modest imbedded energy, wind systems emit none of the SO_x, NO_x, CO, VOCs, or particulate matter associated with fuel-fired energy systems. And, of course, since there are virtually no greenhouse gas emissions, wind economics will get a boost if and when carbon emitting sources begin to be taxed.

REFERENCES

Bolinger, M., R. Wiser, and W. Golove (2001). *Revisiting the "Buy versus Build" Decision for Publicly Owned Utilities in California Considering Wind and Geothermal Resources*, Lawrence Berkeley National Labs, LBNL 48831, October.

Cavallo, A. J., S. M. Hock, and D. R. Smith (1993). Wind Energy: Technology and Economics. Chapter 3 in *Renewable Energy, Sources for Fuels and Electricity*, L. Burnham, ed., Island Press, Washington, D.C.

Elliot, D. L., C. G. Holladay, W. R. Barchett, H. P. Foote, and W. F. Sandusky (1987). *Wind Energy Resource Atlas of the United States*, Solar Energy Research Institute, U.S. Department of Energy, DOE/CH 100934, Golden, CO.

Elliot, D. L., L. L. Windell, and G. I. Gower (1991). *An Assessment of the Available Windy Land Area and Wind Energy Potential in the Contiguous United States*, Pacific Northwest Laboratory, PNL-7789, August.

Grubb, M. J., and N. I. Meyer (1993). Wind Energy: Resources, Systems, and Regional Strategies. Chapter 4 in *Renewable Energy: Sources for Fuels and Electricity*, Island Press, Washington, D.C.

Jacobson, M. Z., and G. M. Masters (2001). Exploiting Wind Versus Coal, *Science*, 24 August, vol. 293, p. 1438.

Johnson, G. L. (1985). *Wind Energy Systems*, Prentice-Hall, Englewood Cliffs, N.J.

Krohn, S. (1997). The Energy Balance of Modern Wind Turbines, *Wind Power Note*, #10, December. Vindmolleindustrien, Denmark (available at www.windpower.dk).

Milborrow, D. J., and P. L. Surman (1987). Status of wake and array loss research. AWEA Windpower 91 Conference, Palm Springs, CA.

Pacific Northwest Laboratory (1991). *An Assessment of the Available Windy Land Area and Wind Energy Potential in the Contiguous United States*.

Patel, M. K. (1999). *Wind and Solar Power Systems*, CRC Press, New York.

Redlinger, R. Y., P. D. Andersen, and P. E. Morthorst (1999). *Wind Energy in the 21st Century*, Macmillan Press, London.

Troen, I., and E. L. Petersen (1991) *European Wind Atlas*, Risoe National Laboratory, Denmark.

PROBLEMS

6.1 A horizontal-axis wind turbine with rotor 20 meters in diameter is 30-% efficient in 10 m/s winds at 1 atmosphere of pressure and 15°C.

a. How much power would it produce in those winds?

b. Estimate the air density on a 2500-m mountaintop at 10°C.

b. Estimate the power the turbine would produce on that mountain with the same windspeed assuming its efficiency is not affected by air density.

6.2. An anemometer mounted at a height of 10 m above a surface with crops, hedges and shrubs, shows a windspeed of 5 m/s. Assuming 15°C and 1 atm pressure, determine the following for a wind turbine with hub height 60 m and rotor diameter of 60 m:

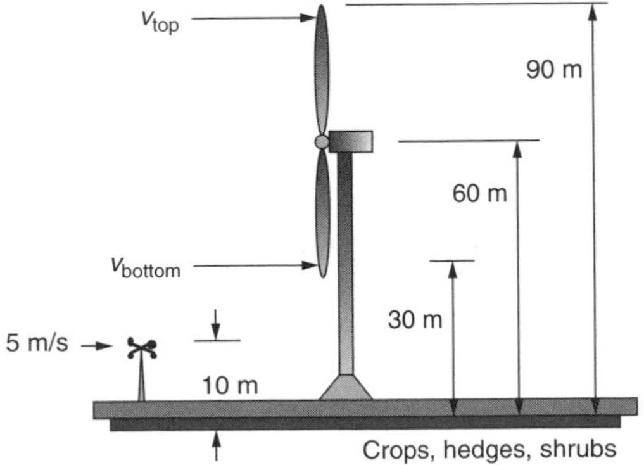

Figure P6.2

a. Estimate the windspeed and the specific power in the wind (W/m^2) at the highest point that a rotor blade reaches.
b. Repeat (a) at the lowest point at which the blade falls.
c. Compare the ratio of wind power at the two elevations using results of (a) and (b) and compare that with the ratio obtained using (6.17).

6.3 Consider the following probability density function for wind speed:

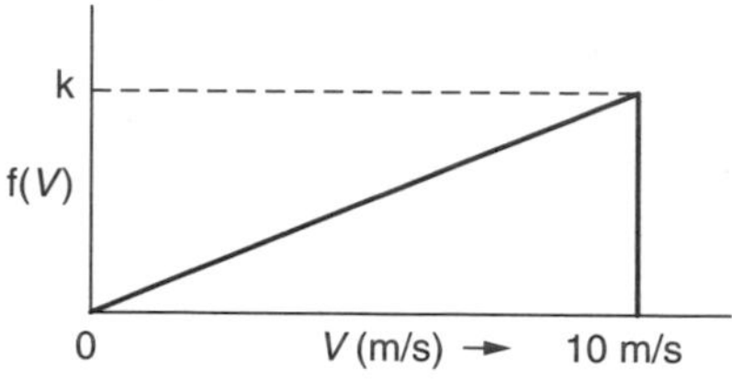

Figure P6.3

a. What is an appropriate value of k for this to be a legitimate probability density function?
b. What is the average power in these winds (W/m^2) under standard (15°C, 1 atm) conditions?

6.4 Suppose the wind probability density function is just a constant over the 5 to 20 m/s range of windspeeds, as shown below. The power curve for a small 1 kW windmill is also shown.

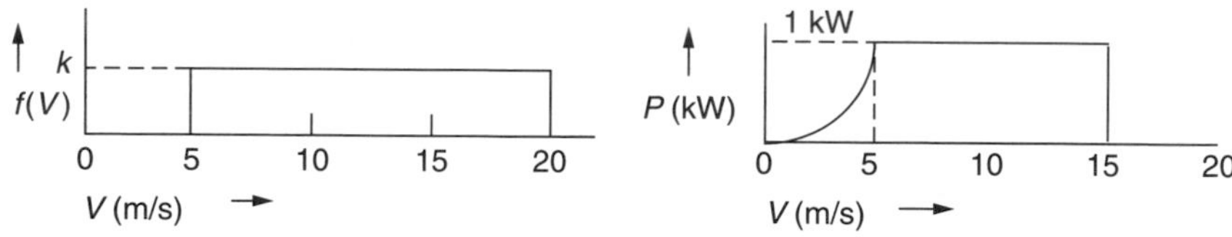

Figure P6.4

a. What is the probability that the wind is blowing between 5 and 15 m/s?
b. What is the annual energy that the wind turbine would generate?
c. What is the average power in the wind?

6.5 Suppose an anemometer mounted at a height of 10-m on a level field with tall grass shows an average windspeed of 6 m/s.

a. Assuming Rayleigh statistics and standard conditions (15°C, 1 atm), estimate the average wind power (W/m^2) at a height of 80 m.
b. Suppose a 1300-kW wind turbine with 60-m rotor diameter is located in those 80 m winds. Estimate the annual energy delivered (kWh/yr) if you assume the turbine has an overall efficiency of 30%.
c. What would the turbine's capacity factor be?

6.6 In the derivation of the cumulative distribution function, $F(V)$ for a Weibull function we had to solve the integral $F(V) = \int_0^V f(v)\,dv$. Show that $F(V)$ in (6.53) is the correct result by taking the derivative $f(v) = \dfrac{dF(V)}{dV}$ and seeing whether you get back to the Weibull probability density function given in (6.41).

6.7 The table below shows a portion of a spreadsheet that estimates the energy delivered by a NEG Micon 1000 kW/60 m wind turbine exposed to Rayleigh winds with an average speed of 8 m/s.

v (m/s)	kW	kWh/yr
0	0	0
1	0	0
2	0	0
3	0	0
4	33	23,321
5	86	?
6	...	...
etc.		

a. How many kWh/yr would be generated with 5 m/s winds?

b. Using Table 6.7, how many kWh/yr would be generated in 10 m/s winds for a Vestas 600/42 machine?

6.8 Consider the Nordex 1.3 MW, 60-m wind turbine with power specifications given in Table 6.7 located in an area with 8 m/s average wind speeds.

a. Find the average power in the wind (W/m^2) assuming Rayleigh statistics.

b. Create a spreadsheet similar to the one developed in Example 6.15 to determine the energy delivered (kWh/yr) from this machine.

c. What would be the average efficiency of the wind turbine?

d. If the turbine's rotor operates at 70% of the Betz limit, what is the efficiency of the gearing and generator?

6.9 For the following turbines and average Rayleigh wind speeds, set up a spreadsheet to find the total annual kWh delivered and compare that with an estimate obtained using the simple correlation given in (6.65):

a. Bonus 300 kW/33.4 m, 7 m/s average wind speed

b. NEG/Micon 1000 kW/60 m, 8 m/s average wind speed

c. Vestas 600 kW/42 m, 8 m/s average wind speed

d. Whisper 0.9 kW/2.13 m, 5 m/s average wind speed

6.10 Consider the design of a home-built wind turbine using a 350-W automobile dc generator. The goal is to deliver 70 kWh in a 30-day month.

a. What capacity factor would be needed for the machine?

b. If the average wind speed is 5 m/s, and Rayleigh statistics apply, what should the rotor diameter be if the correlation of (6.65) is used?

c. How fast would the wind have to blow to cause the turbine to put out its full 0.35 kW if the machine is 20% efficient at that point?

d. If the tip-speed-ratio is assumed to be 4, what gear ratio would be needed to match the rotor speed to the generator if the generator needs to turn at 1000 rpm to deliver its rated 350 W?

6.11 A 750-kW wind turbine with 45-m blade diameter operates in a wind regime that is well characterized by Rayleigh statistics with average windspeed equal to 7 m/s.

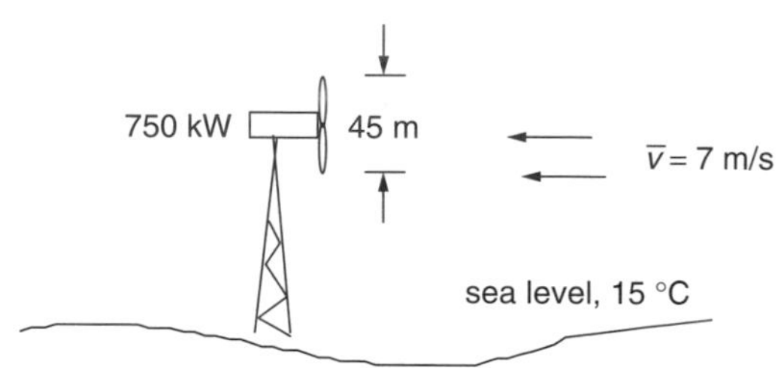

Figure P6.11

Assuming the capacity factor correlation (6.65), what is the average efficiency of this machine?

6.12 For Rayleigh winds with an average windspeed of 8 m/s:

a. How many hours per year do the winds blow at less than 13 m/s?

b. For how many hours per year are windspeeds above 25 m/s?

c. Suppose a 31-m, 340-kW turbine follows the idealized power curve shown in Figure 6.32. How many kWh/yr will it deliver when winds blow between its rated windspeed of 13 m/s and its furling windspeed of 25 m/s?

d. Using the capacity factor correlation given in (6.65), estimate the fraction of the annual energy delivered with winds that are above the rated windspeed?

6.13 Using the simple capacity factor correlation, derive an expression for the average (Rayleigh) windspeed that yields the highest efficiency for a turbine as a function of its rated power and blade diameter. What is the optimum windspeed for

a. The NEG/Micon 1000 kW/60 m turbine

b. The NEG/Micon 1000 kW/54 m turbine?

6.14 Consider a 64-m, 1.5 MW NEG Micon wind turbine (Table 6.7) located at a site with Rayleigh winds averaging 7.5 m/s.

a. Using the simple capacity factor correlation (6.65) estimate the annual energy delivered.

b. Suppose the total installed cost of the wind turbine is \$1.5 million (\$1/watt) and its annual cost is based on the equivalent of a 20-year, 6% loan to cover the capital costs. In addition, assume an annual operations and maintenance cost equal to 1-% of the capital cost. What would be the cost of electricity from this turbine (¢/kWh)?

c. If farmers are paid 0.1 ¢/kWh to put these towers on their land, what would their annual royalty payment be per turbine?

d. If turbines are installed with a density corresponding to $4D \times 7D$ separations (where D is rotor diameter), what would the annual payment be per acre?

6.15 This question has 4 different combinations of turbine, average wind speed, capital costs, return on equity, loan terms, and O&M costs. Using the capacity factor correlation, find their levelized costs of electricity.

	(a)	(b)	(c)	(d)
Turbine power (kW)	1500	600	250	1000
Rotor diameter (m)	64	42	29.2	60
Avg wind speed (m/s)	8.5	8.5	8.5	8.5
Capital cost (\$/kW)	800	1000	1200	900
Equity % of capital	25	25	25	25
Annual return on equity (%)	15	15	15	15
Loan interest (%)	7	7	7	7
Loan term (yrs)	20	20	20	20
Annual O&M percent of capital	3	3	3	3

CHAPTER 7

THE SOLAR RESOURCE

To design and analyze solar systems, we need to know how much sunlight is available. A fairly straightforward, though complicated-looking, set of equations can be used to predict where the sun is in the sky at any time of day for any location on earth, as well as the solar intensity (or *insolation: in*cident *sol*ar Radi*ation*) on a clear day. To determine average daily insolation under the combination of clear and cloudy conditions that exist at any site we need to start with long-term measurements of sunlight hitting a horizontal surface. Another set of equations can then be used to estimate the insolation on collector surfaces that are not flat on the ground.

7.1 THE SOLAR SPECTRUM

The source of insolation is, of course, the sun—that gigantic, 1.4 million kilometer diameter, thermonuclear furnace fusing hydrogen atoms into helium. The resulting loss of mass is converted into about 3.8×10^{20} MW of electromagnetic energy that radiates outward from the surface into space.

Every object emits radiant energy in an amount that is a function of its temperature. The usual way to describe how much radiation an object emits is to compare it to a theoretical abstraction called a *blackbody*. A blackbody is defined to be a perfect emitter as well as a perfect absorber. As a perfect emitter, it radiates more energy per unit of surface area than any real object at the same temperature. As a perfect absorber, it absorbs all radiation that impinges upon it; that is, none

Renewable and Efficient Electric Power Systems. By Gilbert M. Masters
ISBN 0-471-28060-7

is reflected and none is transmitted through it. The wavelengths emitted by a blackbody depend on its temperature as described by *Planck's law*:

$$E_\lambda = \frac{3.74 \times 10^8}{\lambda^5 \left[\exp\left(\dfrac{14{,}400}{\lambda T}\right) - 1\right]} \tag{7.1}$$

where E_λ is the emissive power per unit area of a blackbody (W/m^2 μm), T is the absolute temperature of the body (K), and λ is the wavelength (μm).

Modeling the earth itself as a 288 K (15°C) blackbody results in the emission spectrum plotted in Fig. 7.1.

The area under Planck's curve between any two wavelengths is the power emitted between those wavelengths, so the total area under the curve is the total radiant power emitted. That total is conveniently expressed by the *Stefan–Boltzmann law of radiation*:

$$E = A\sigma T^4 \tag{7.2}$$

where E is the total blackbody emission rate (W), σ is the Stefan–Boltzmann constant $= 5.67 \times 10^{-8}$ W/m^2-K^4, T is the absolute temperature of the blackbody (K), and A is the surface area of the blackbody (m^2).

Another convenient feature of the blackbody radiation curve is given by *Wien's displacement rule*, which tells us the wavelength at which the spectrum reaches its maximum point:

$$\lambda_{\max}(\mu\text{m}) = \frac{2898}{T(\text{K})} \tag{7.3}$$

where the wavelength is in microns (μm) and the temperature is in kelvins.

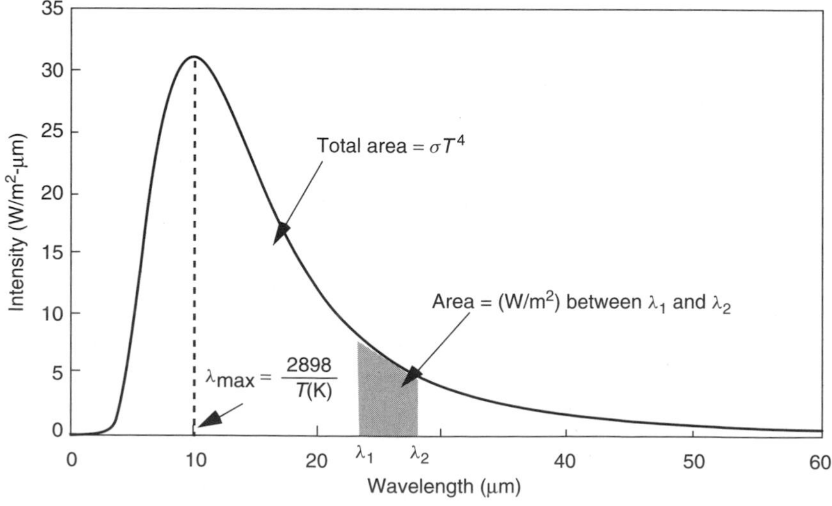

Figure 7.1 The spectral emissive power of a 288 K blackbody.

Example 7.1 The Earth's Spectrum. Consider the earth to be a blackbody with average surface temperature 15°C and area equal to 5.1×10^{14} m^2. Find the rate at which energy is radiated by the earth and the wavelength at which maximum power is radiated. Compare this peak wavelength with that for a 5800 K blackbody (the sun).

Solution. Using (7.2), the earth radiates:

$$E = \sigma A T^4 = (5.67 \times 10^{-8}\ \text{W/m}^2 \cdot \text{K}^4) \times (5.1 \times 10^{14}\ \text{m}^2) \times (15 + 273\ \text{K})^4$$
$$= 2.0 \times 10^{17}\ \text{W}$$

The wavelength at which the maximum power is emitted is given by (5.3):

$$\lambda_{\text{max}}(\text{earth}) = \frac{2898}{T(\text{K})} = \frac{2898}{288} = 10.1\ \mu\text{m}$$

For the 5800 K sun,

$$\lambda_{\text{max}}(\text{sun}) = \frac{2898}{5800} = 0.5\ \mu\text{m}$$

It is worth noting that earth's atmosphere reacts very differently to the much longer wavelengths emitted by the earth's surface (Fig. 7.1) than it does to the short wavelengths arriving from the sun (Fig. 7.2). This difference is the fundamental factor responsible for the greenhouse effect.

While the interior of the sun is estimated to have a temperature of around 15 million kelvins, the radiation that emanates from the sun's surface has a spectral distribution that closely matches that predicted by Planck's law for a 5800 K blackbody. Figure 7.2 shows the close match between the actual solar spectrum and that of a 5800 K blackbody. The total area under the blackbody curve has been scaled to equal 1.37 kW/m^2, which is the solar insolation just outside the earth's atmosphere. Also shown are the areas under the actual solar spectrum that corresponds to wavelengths within the ultraviolet UV (7%), visible (47%), and infrared IR (46%) portions of the spectrum. The visible spectrum, which lies between the UV and IR, ranges from 0.38 μm (violet) to 0.78 μm (red).

As solar radiation makes its way toward the earth's surface, some of it is absorbed by various constituents in the atmosphere, giving the terrestrial spectrum an irregular, bumpy shape. The terrestrial spectrum also depends on how much atmosphere the radiation has to pass through to reach the surface. The length of the path h_2 taken by the sun's rays as they pass through the atmosphere, divided by the minimum possible path length h_1, which occurs when

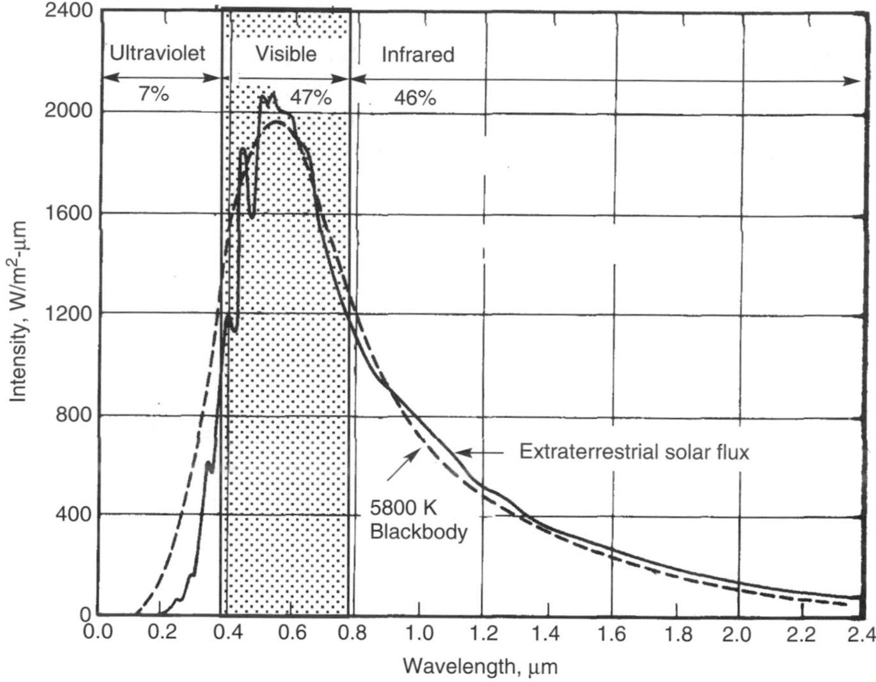

Figure 7.2 The extraterrestrial solar spectrum compared with a 5800 K blackbody.

the sun is directly overhead, is called the *air mass ratio*, m. As shown in Figure 7.3, under the simple assumption of a flat earth the air mass ratio can be expressed as

$$\text{Air mass ratio} \quad m = \frac{h_2}{h_1} = \frac{1}{\sin\beta} \tag{7.4}$$

where h_1 = path length through the atmosphere with the sun directly overhead, h_2 = path length through the atmosphere to reach a spot on the surface, and β = the altitude angle of the sun (see Fig. 7.3).

Thus, an air mass ratio of 1 (designated "AM1") means that the sun is directly overhead. By convention, AM0 means no atmosphere; that is, it is the extraterrestrial solar spectrum. Often, an air mass ratio of 1.5 is assumed for an average solar spectrum at the earth's surface. With AM1.5, 2% of the incoming solar energy is in the UV portion of the spectrum, 54% is in the visible, and 44% is in the infrared.

The impact of the atmosphere on incoming solar radiation for various air mass ratios is shown in Fig. 7.4. As sunlight passes through more atmosphere, less energy arrives at the earth's surface and the spectrum shifts some toward longer wavelengths.

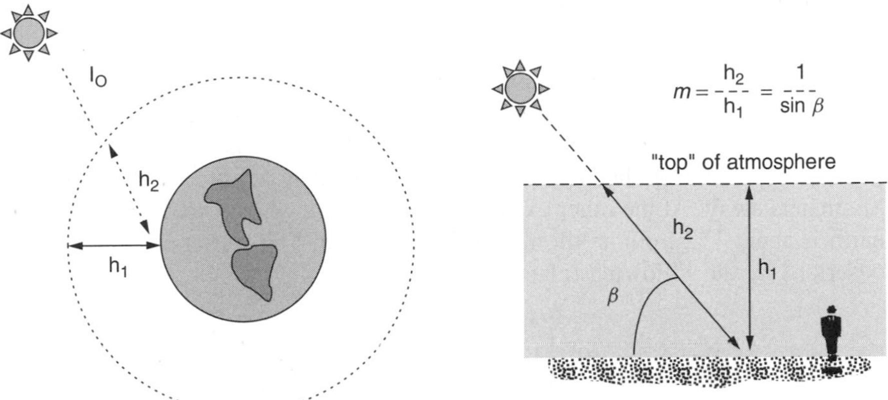

Figure 7.3 The air mass ratio m is a measure of the amount of atmosphere the sun's rays must pass through to reach the earth's surface. For the sun directly overhead, $m = 1$.

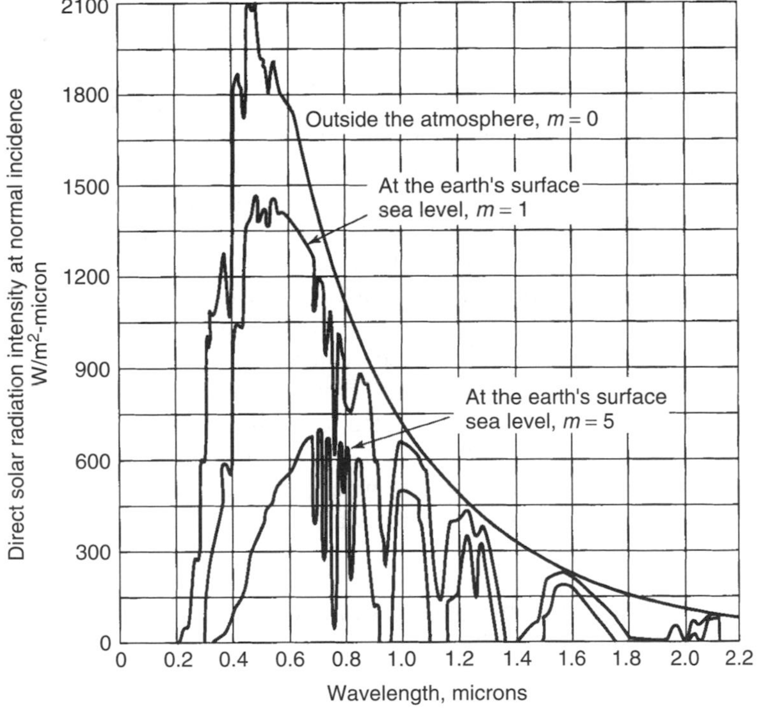

Figure 7.4 Solar spectrum for extraterrestrial ($m = 0$), for sun directly overhead ($m = 1$), and at the surface with the sun low in the sky, $m = 5$. From Kuen et al. (1998), based on *Trans. ASHRAE*, vol. 64 (1958), p. 50.

7.2 THE EARTH'S ORBIT

The earth revolves around the sun in an elliptical orbit, making one revolution every 365.25 days. The eccentricity of the ellipse is small and the orbit is, in fact, quite nearly circular. The point at which the earth is nearest the sun, the perihelion, occurs on January 2, at which point it is a little over 147 million kilometers away. At the other extreme, the aphelion, which occurs on July 3, the earth is about 152 million kilometers from the sun. This variation in distance is described by the following relationship:

$$d = 1.5 \times 10^8 \left\{ 1 + 0.017 \sin \left[\frac{360(n - 93)}{365} \right] \right\} \text{ km} \tag{7.5}$$

where n is the *day number*, with January 1 as day 1 and December 31 being day number 365. Table 7.1 provides a convenient list of day numbers for the first day of each month. *It should be noted that (7.5) and all other equations developed in this chapter involving trigonometric functions use angles measured in degrees, not radians.*

Each day, as the earth rotates about its own axis, it also moves along the ellipse. If the earth were to spin only 360° in a day, then after 6 months time our clocks would be off by 12 hours; that is, at noon on day 1 it would be the middle of the day, but 6 months later noon would occur in the middle of the night. To keep synchronized, the earth needs to rotate one extra turn each year, which means that in a 24-hour day the earth actually rotates 360.99°, which is a little surprising to most of us.

As shown in Fig. 7.5, the plane swept out by the earth in its orbit is called the ecliptic plane. The earth's spin axis is currently tilted 23.45° with respect to the ecliptic plane and that tilt is, of course, what causes our seasons. On March 21 and September 21, a line from the center of the sun to the center of the earth passes through the equator and everywhere on earth we have 12 hours of daytime and 12 hours of night, hence the term *equinox* (equal day and night). On December 21, the winter *solstice* in the Northern Hemisphere, the inclination of the North Pole reaches its highest angle away from the sun (23.45°), while on June 21 the opposite occurs. By the way, for convenience we are using the

TABLE 7.1 Day Numbers for the First Day of Each Month

January	$n = 1$	July	$n = 182$
February	$n = 32$	August	$n = 213$
March	$n = 60$	September	$n = 244$
April	$n = 91$	October	$n = 274$
May	$n = 121$	November	$n = 305$
June	$n = 152$	December	$n = 335$

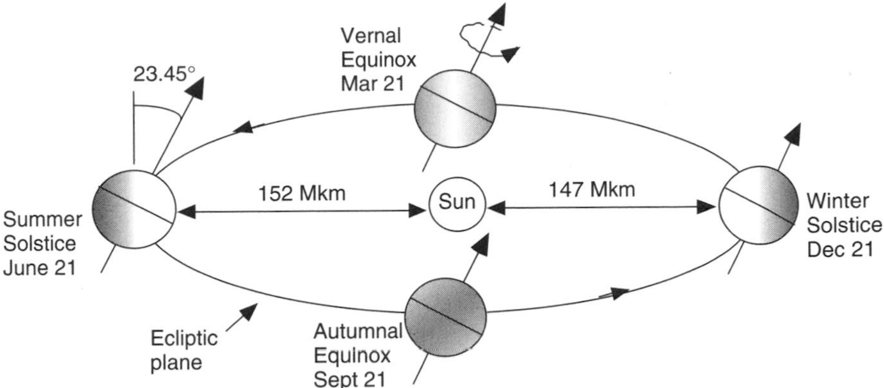

Figure 7.5 The tilt of the earth's spin axis with respect to the ecliptic plane is what causes our seasons. "Winter" and "summer" are designations for the solstices in the Northern Hemisphere.

twenty-first day of the month for the solstices and equinoxes even though the actual days vary slightly from year to year.

For solar energy applications, the characteristics of the earth's orbit are considered to be unchanging, but over longer periods of time, measured in thousands of years, orbital variations are extremely important as they significantly affect climate. The shape of the orbit oscillates from elliptical to more nearly circular with a period of 100,000 years (*eccentricity*). The earth's tilt angle with respect to the ecliptic plane fluctuates from 21.5° to 24.5° with a period of 41,000 years (*obliquity*). Finally, there is a 23,000-year period associated with the *precession* of the earth's spin axis. This precession determines, for example, where in the earth's orbit a given hemisphere's summer occurs. Changes in the orbit affect the amount of sunlight striking the earth as well as the distribution of sunlight both geographically and seasonally. Those variations are thought to be influential in the timing of the coming and going of ice ages and interglacial periods. In fact, careful analysis of the historical record of global temperatures does show a primary cycle between glacial episodes of about 100,000 years, mixed with secondary oscillations with periods of 23,000 years and 41,000 years that match these orbital changes. This connection between orbital variations and climate were first proposed in the 1930s by an astronomer named Milutin Milankovitch, and the orbital cycles are now referred to as *Milankovitch oscillations*. Sorting out the impact of human activities on climate from those caused by natural variations such as the Milankovitch oscillations is a critical part of the current climate change discussion.

7.3 ALTITUDE ANGLE OF THE SUN AT SOLAR NOON

We all know that the sun rises in the east and sets in the west and reaches its highest point sometime in the middle of the day. In many situations, it is quite useful to be able to predict exactly where in the sky the sun will be at any time, at any location on any day of the year. Knowing that information we can, for example, design an overhang to allow the sun to come through a window to help heat a house in the winter while blocking the sun in the summer. In the context of photovoltaics, we can, for example, use knowledge of solar angles to help pick the best tilt angle for our modules to expose them to the greatest insolation.

While Fig. 7.5 correctly shows the earth revolving around the sun, it is a difficult diagram to use when trying to determine various solar angles as seen from the surface of the earth. An alternative (and ancient!) perspective is shown in Fig. 7.6, in which the earth is fixed, spinning around its north–south axis; the sun sits somewhere out in space, slowly moving up and down as the seasons progress. On June 21 (the summer solstice) the sun reaches its highest point, and a ray drawn at that time from the center of the sun to the center of the earth makes an angle of 23.45° with the earth's equator. On that day, the sun is directly over the Tropic of Cancer at latitude 23.45°. At the two equinoxes, the sun is directly over the equator. On December 21 the sun is 23.45° below the equator, which defines the latitude known as the Tropic of Capricorn.

As shown in Fig. 7.6, the angle formed between the plane of the equator and a line drawn from the center of the sun to the center of the earth is called the solar declination, δ. It varies between the extremes of $\pm$ 23.45°, and a simple sinusoidal relationship that assumes a 365-day year and which puts the spring equinox on day $n = 81$ provides a very good approximation. Exact values of declination, which vary slightly from year to year, can be found in the annual publication *The American Ephemeris and Nautical Almanac*.

$$\delta = 23.45 \sin\left[\frac{360}{365}(n - 81)\right] \tag{7.6}$$

Computed values of solar declination on the twenty-first day of each month are given in Table 7.2.

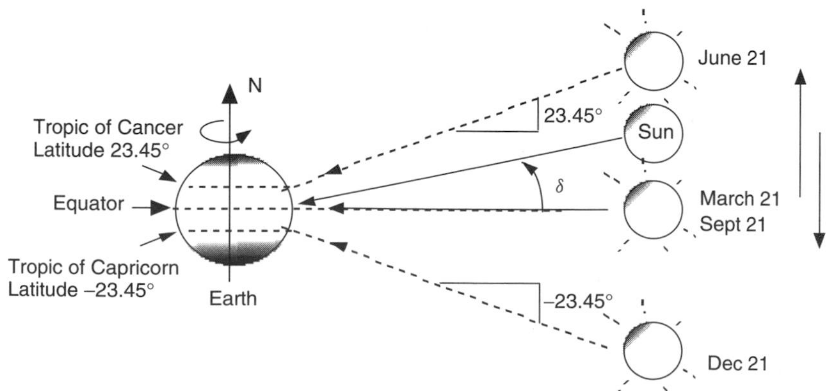

Figure 7.6 An alternative view with a fixed earth and a sun that moves up and down. The angle between the sun and the equator is called the solar declination δ.

TABLE 7.2 Solar Declination δ for the 21st Day of Each Month (degrees)

Month:	Jan	Feb	Mar	Apr	May	Jun	July	Aug	Sept	Oct	Nov	Dec
δ:	−20.1	−11.2	0.0	11.6	20.1	23.4	20.4	11.8	0.0	−11.8	−20.4	−23.4

While Fig. 7.6 doesn't capture the subtleties associated with the earth's orbit, it is entirely adequate for visualizing various latitudes and solar angles. For example, it is easy to understand the seasonal variation of daylight hours. As suggested in Fig. 7.7, during the summer solstice all of the earth's surface above latitude 66.55° (90° − 23.45°) basks in 24 hours of daylight, while in the Southern Hemisphere below latitude 66.55° it is continuously dark. Those latitudes, of course, correspond to the Arctic and Antarctic Circles.

It is also easy to use Fig. 7.6 to gain some intuition into what might be a good tilt angle for a solar collector. Figure 7.8 shows a south-facing collector on the earth's surface that is tipped up at an angle equal to the local latitude, L. As can be seen, with this tilt angle the collector is parallel to the axis of the earth. During an equinox, at *solar noon*, when the sun is directly over the local meridian (line of longitude), the sun's rays will strike the collector at the best possible angle; that is, they are perpendicular to the collector face. At other times of the year the sun is a little high or a little low for normal incidence, but on the average it would seem to be a good tilt angle.

Solar noon is an important reference point for almost all solar calculations. In the Northern Hemisphere, at latitudes above the Tropic of Cancer, solar noon occurs when the sun is due south of the observer. South of the Tropic of Capricorn, in New Zealand for example, it is when the sun is due north. And in the tropics, the sun may be either due north, due south, or directly overhead at solar noon.

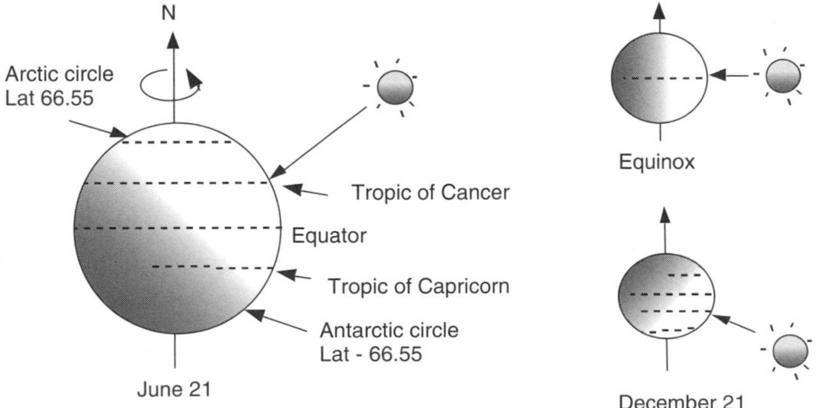

Figure 7.7 Defining the earth's key latitudes is easy with the simple version of the earth–sun system.

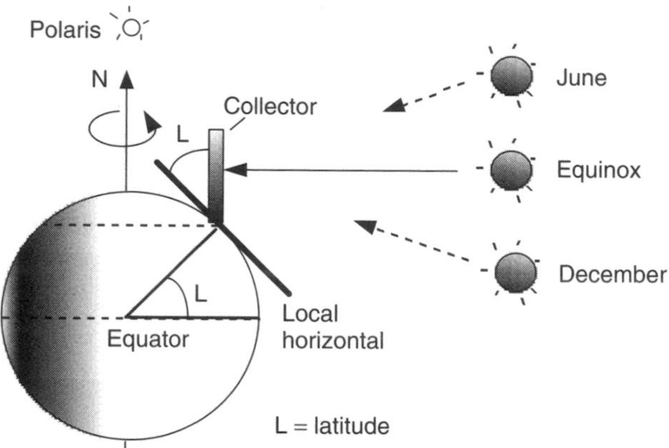

Figure 7.8 A south-facing collector tipped up to an angle equal to its latitude is perpendicular to the sun's rays at solar noon during the equinoxes.

On the average, facing a collector toward the equator (for most of us in the Northern Hemisphere, this means facing it south) and tilting it up at an angle equal to the local latitude is a good rule-of-thumb for annual performance. Of course, if you want to emphasize winter collection, you might want a slightly higher angle, and vice versa for increased summer efficiency.

Having drawn the earth-sun system as shown in Fig. 7.6 also makes it easy to determine a key solar angle, namely the *altitude angle* β_N of the sun at solar noon. The altitude angle is the angle between the sun and the local horizon directly beneath the sun. From Fig. 7.9 we can write down the following relationship by inspection:

$$\beta_N = 90^\circ - L + \delta \tag{7.7}$$

where L is the latitude of the site. Notice in the figure the term *zenith* is introduced, which refers to an axis drawn directly overhead at a site.

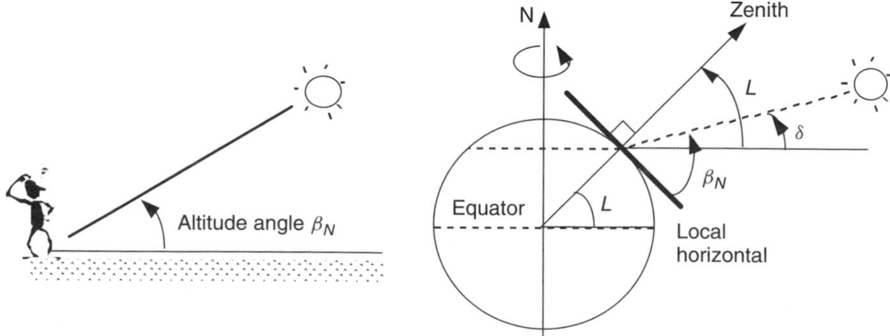

Figure 7.9 The altitude angle of the sun at solar noon.

Example 7.2 Tilt Angle of a PV Module. Find the optimum tilt angle for a south-facing photovoltaic module in Tucson (latitude 32.1°) at solar noon on March 1.

Solution. From Table 7.1, March 1 is the sixtieth day of the year so the solar declination (7.6) is

$$\delta = 23.45 \sin\left[\frac{360}{365}(n - 81)\right] = 23.45^\circ \sin\left[\frac{360}{365}(60 - 81)^\circ\right] = -8.3^\circ$$

which, from (7.7), makes the altitude angle of the sun equal to

$$\beta_N = 90^\circ - L + \delta = 90 - 32.1 - 8.3 = 49.6^\circ$$

The tilt angle that would make the sun's rays perpendicular to the module at noon would therefore be

$$\text{Tilt} = 90 - \beta_N = 90 - 49.6 = 40.4^\circ$$

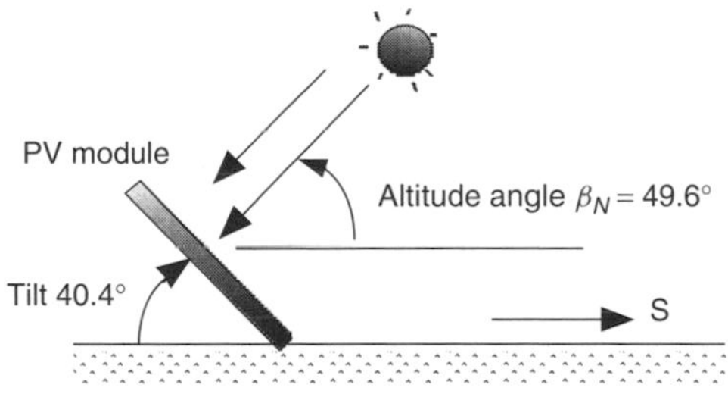

7.4 SOLAR POSITION AT ANY TIME OF DAY

The location of the sun at any time of day can be described in terms of its altitude angle β and its azimuth angle ϕ_s as shown in Fig. 7.10. The subscript s in the azimuth angle helps us remember that this is the azimuth angle of the sun. Later, we will introduce another azimuth angle for the solar collector and a different subscript c will be used. *By convention, the azimuth angle is positive in the morning with the sun in the east and negative in the afternoon with the sun in the west.* Notice that the azimuth angle shown in Fig. 7.10 uses true south as its reference, and this will be the assumption in this text unless otherwise stated. For solar work in the Southern Hemisphere, azimuth angles are measured relative to north.

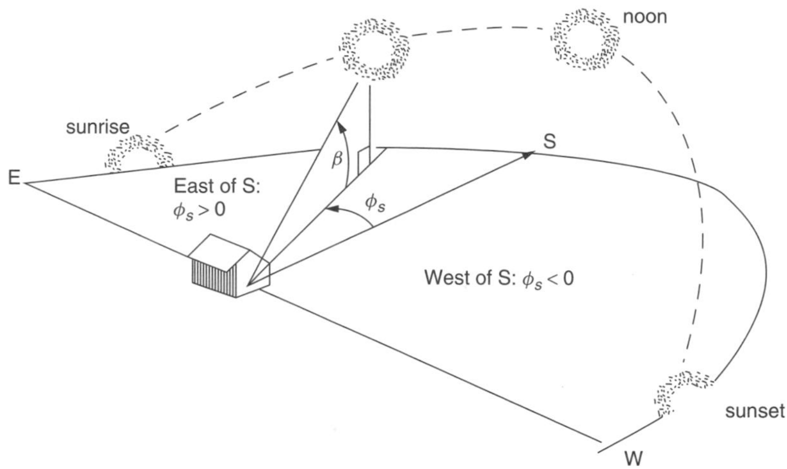

Figure 7.10 The sun's position can be described by its altitude angle β and its azimuth angle ϕ_S. By convention, the azimuth angle is considered to be positive before solar noon.

The azimuth and altitude angles of the sun depend on the latitude, day number, and, most importantly, the time of day. For now, we will express time as the number of hours before or after solar noon. Thus, for example, 11 A.M. *solar time* is one hour before the sun crosses your local meridian (due south for most of us). Later we will learn how to make the adjustment between solar time and local clock time. The following two equations allow us to compute the altitude and azimuth angles of the sun. For a derivation see, for example, T. H. Kuen et al. (1998):

$$\sin\beta = \cos L \cos\delta \cos H + \sin L \sin\delta \tag{7.8}$$

$$\sin\phi_S = \frac{\cos\delta \sin H}{\cos\beta} \tag{7.9}$$

Notice that time in these equations is expressed by a quantity called the *hour angle*, H. The hour angle is the number of degrees that the earth must rotate before the sun will be directly over your local meridian (line of longitude). As shown in Fig. 7.11, at any instant, the sun is directly over a particular line of longitude, called the sun's meridian. The difference between the local meridian and the sun's meridian is the hour angle, with positive values occurring in the morning before the sun crosses the local meridian.

Considering the earth to rotate 360° in 24 h, or 15°/h, the hour angle can be described as follows:

$$\text{Hour angle } H = \left(\frac{15^\circ}{\text{hour}}\right) \cdot \text{(hours before solar noon)} \tag{7.10}$$

Thus, the hour angle H at 11:00 A.M. solar time would be +15° (the earth needs to rotate another 15°, or 1 hour, before it is solar noon). In the afternoon, the hour angle is negative, so, for example, at 2:00 P.M. solar time H would be −30°.

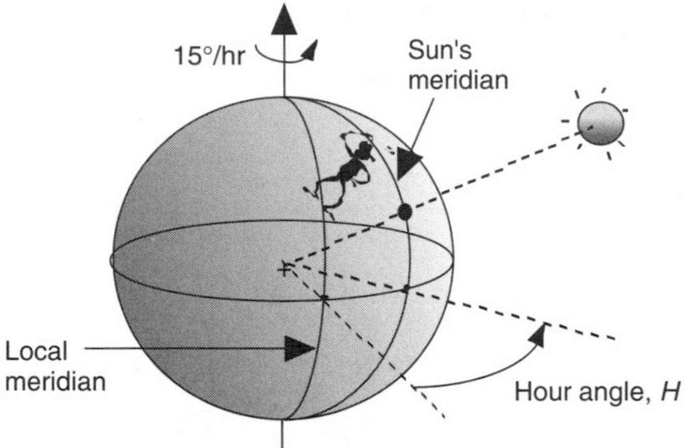

Figure 7.11 The hour angle is the number of degrees the earth must turn before the sun is directly over the local meridian. It is the difference between the sun's meridian and the local meridian.

There is a slight complication associated with finding the azimuth angle of the sun from (7.9). During spring and summer in the early morning and late afternoon, the magnitude of the sun's azimuth is liable to be more than 90° away from south (that never happens in the fall and winter). Since the inverse of a sine is ambiguous, sin x = sin $(180 - x)$, we need a test to determine whether to conclude the azimuth is greater than or less than 90° away from south. Such a test is

$$\text{if} \quad \cos H \geq \frac{\tan\delta}{\tan L}, \qquad \text{then } |\phi_S| \leq 90^\circ; \qquad \text{otherwise } |\phi_S| > 90^\circ \tag{7.11}$$

Example 7.3 Where Is the Sun? Find the altitude angle and azimuth angle for the sun at 3:00 P.M. solar time in Boulder, Colorado (latitude 40°) on the summer solstice.

Solution. Since it is the solstice we know, without computing, that the solar declination δ is 23.45°. Since 3:00 P.M. is three hours after solar noon, from (7.10) we obtain

$$H = \left(\frac{15^\circ}{\text{h}}\right) \cdot (\text{hours before solar noon}) = \frac{15^\circ}{\text{h}} \cdot (-3\ \text{h}) = -45^\circ$$

Using (7.8), the altitude angle is

$$\begin{aligned}\sin\beta &= \cos L \cos\delta \cos H + \sin L \sin\delta \\ &= \cos 40^\circ \cos 23.45^\circ \cos(-45^\circ) + \sin 40^\circ \sin 23.45^\circ = 0.7527 \\ \beta &= \sin^{-1}(0.7527) = 48.8^\circ\end{aligned}$$

From (7.9) the sine of the azimuth angle is

$$\sin\phi_S = \frac{\cos\delta\sin H}{\cos\beta}$$

$$= \frac{\cos 23.45^\circ \cdot \sin(-45^\circ)}{\cos 48.8^\circ} = -0.9848$$

But the arcsine is ambiguous and two possibilities exist:

$$\phi_S = \sin^{-1}(-0.9848) = -80^\circ \qquad (80^\circ \text{west of south})$$

$$\text{or} \quad \phi_S = 180 - (-80) = 260^\circ \qquad (100^\circ \text{west of south})$$

To decide which of these two options is correct, we apply (7.11):

$$\cos H = \cos(-45^\circ) = 0.707 \qquad \text{and} \qquad \frac{\tan\delta}{\tan L} = \frac{\tan 23.45^\circ}{\tan 40^\circ} = 0.517$$

Since $\cos H \geq \dfrac{\tan\delta}{\tan L}$ we conclude that the azimuth angle is

$$\phi_S = -80^\circ \qquad (80^\circ \text{west of south})$$

Solar altitude and azimuth angles for a given latitude can be conveniently portrayed in graphical form, an example of which is shown in Fig. 7.12. Similar sun path diagrams for other latitudes are given in Appendix B. As can be seen, in the spring and summer the sun rises and sets slightly to the north and our need for the azimuth test given in (7.11) is apparent; at the equinoxes, it rises and sets precisely due east and due west (everywhere on the planet); during the fall and winter the azimuth angle of the sun is never greater than 90°.

7.5 SUN PATH DIAGRAMS FOR SHADING ANALYSIS

Not only do sun path diagrams, such as that shown in Fig. 7.12, help to build one's intuition into where the sun is at any time, they also have a very practical application in the field when trying to predict shading patterns at a site—a very important consideration for photovoltaics, which are very shadow sensitive. The concept is simple. What is needed is a sketch of the azimuth and altitude angles for trees, buildings, and other obstructions along the southerly horizon that can be drawn on top of a sun path diagram. Sections of the sun path diagram that are covered by the obstructions indicate periods of time when the sun will be behind the obstruction and the site will be shaded.

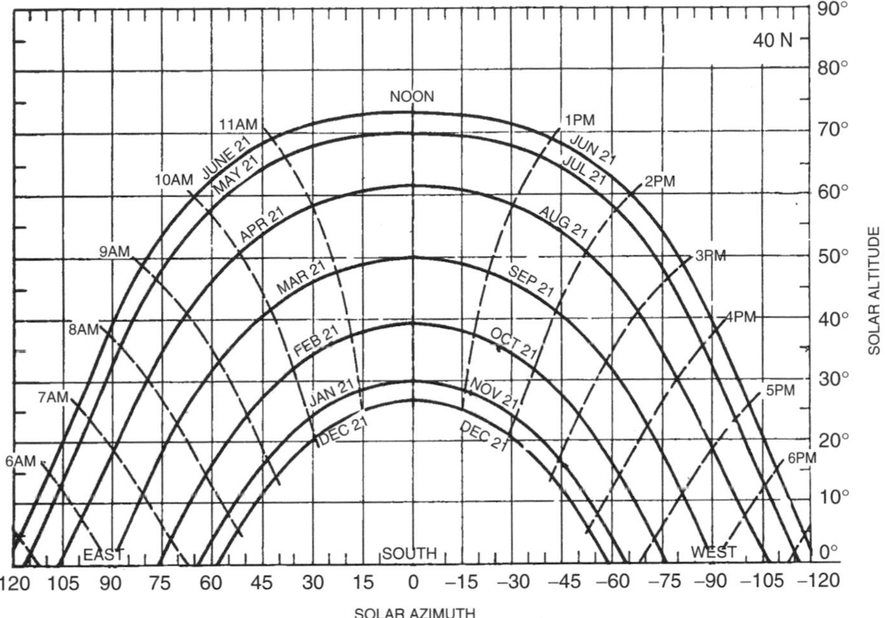

Figure 7.12 A sun path diagram showing solar altitude and azimuth angles for 40° latitude. Diagrams for other latitudes are in Appendix B.

There are several site assessment products available on the market that make the superposition of obstructions onto a sun path diagram pretty quick and easy to obtain. You can do just as good a job, however, with a simple compass, plastic protractor, and plumb bob, but the process requires a little more effort. The compass is used to measure azimuth angles of obstructions, while the protractor and plumb bob measure altitude angles.

Begin by tying the plumb bob onto the protractor so that when you sight along the top edge of the protractor the plumb bob hangs down and provides the altitude angle of the top of the obstruction. Figure 7.13 shows the idea. By standing at the site and scanning the southerly horizon, the altitude angles of major obstructions can be obtained reasonably quickly and quite accurately.

The azimuth angles of obstructions, which go along with their altitude angles, are measured using a compass. Remember, however, that a compass points to magnetic north rather than true north; this difference, called the magnetic declination or deviation, must be corrected for. In the continental United States, this deviation ranges anywhere from about 22°E in Seattle (the compass points 22° east of true north), to essentially zero along the east coast of Florida, to 22 °W at the northern tip of Maine. Figure 7.14 shows this variation in magnetic declination and shows an example, for San Francisco, of how to use it.

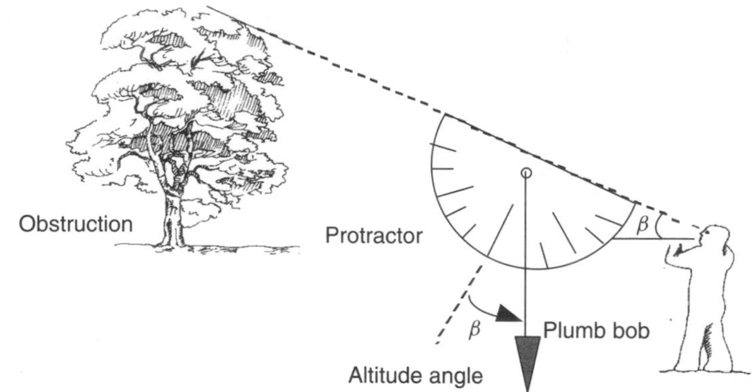

Figure 7.13 Measuring the altitude angle of a southerly obstruction using a plumb bob and protractor.

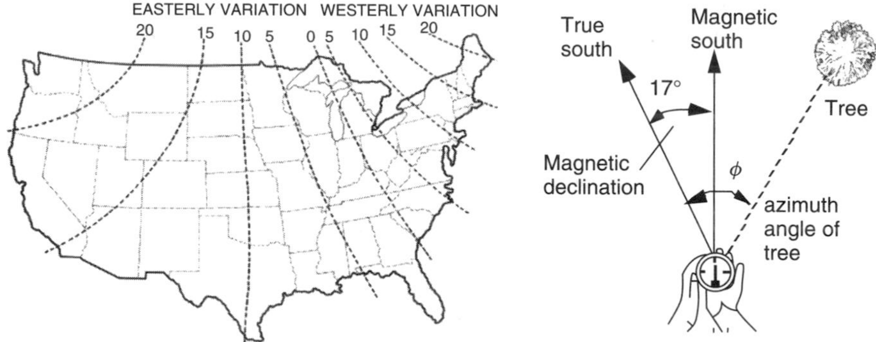

Figure 7.14 Lines of equal magnetic declination across the United States. The example shows the correction for San Francisco, which has a declination of 17°E.

Figure 7.15 shows an example of how the sun path diagram, with a superimposed sketch of potential obstructions, can be interpreted. The site is a proposed solar house with a couple of trees to the southeast and a small building to the southwest. In this example, the site receives full sun all day long from February through October. From November through January, the trees cause about one hour's worth of sun to be lost from around 8:30 A.M. to 9:30 A.M., and the small building shades the site after about 3 o'clock in the afternoon.

When obstructions plotted on a sun path diagram are combined with hour-by-hour insolation information, an estimate can be obtained of the energy lost due to shading. Table 7.3 shows an example of the hour-by-hour insolations available on a clear day in January at 40° latitude for south-facing collectors with fixed tilt angle, or for collectors mounted on 1-axis or 2-axis tracking systems. Later in this chapter, the equations that were used to compute this table will be presented, and in Appendix C there is a full set of such tables for a number of latitudes.

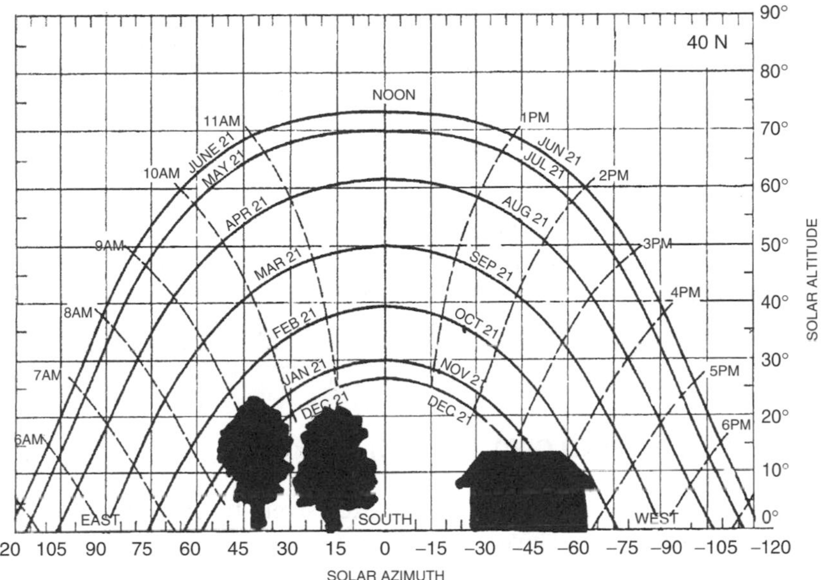

Figure 7.15 A sun path diagram with superimposed obstructions makes it easy to estimate periods of shading at a site.

TABLE 7.3 Clear Sky Beam Plus Diffuse Insolation at 40° Latitude in January for South-Facing Collectors with Fixed Tilt Angle and for Tracking Mounts (hourly W/m² and daily kWh/m²-day)[a]

Solar Time	Tracking		Tilt Angles						Latitude 40°
	One-Axis	Two-Axis	0	20	30	40	50	60	90
			January 21					(W/m²)	
7, 5	0	0	0	0	0	0	0	0	0
8, 4	439	462	87	169	204	232	254	269	266
9, 3	744	784	260	424	489	540	575	593	544
10, 2	857	903	397	609	689	749	788	803	708
11, 1	905	954	485	722	811	876	915	927	801
12	919	968	515	761	852	919	958	968	832
kWh/d:	6.81	7.17	2.97	4.61	5.24	5.71	6.02	6.15	5.47

[a] A complete set of tables is in Appendix C.

Example 7.4 Estimate the insolation available on a clear day in January on a south-facing collector with a fixed, 30° tilt angle at the site having the sun path and obstructions diagram shown in Fig. 7.15.

Solution. With no obstructions, Table 7.3 indicates that the panel would be exposed to 5.24 kWh/m^2-day. The sun path diagram shows loss of about 1 h of sun at around 9 A.M., which eliminates about 0.49 kWh. After about 3:30 P.M. there is no sun, which drops roughly another 0.20 kWh. The remaining insolation is

$$\text{Insolation} \approx 5.24 - 0.49 - 0.20 = 4.55\ \text{kWh/m}^2 \approx 4.6\ \text{kWh/m}^2\ \text{per day}$$

Notice it has been assumed that the insolations shown in Table 7.3 are appropriate averages covering the half-hour before and after the hour. Given the crudeness of the obstruction sketch (to say nothing of the fact that the trees are likely to grow anyway), a more precise calculation isn't warranted.

7.6 SOLAR TIME AND CIVIL (CLOCK) TIME

For most solar work it is common to deal exclusively in solar time (ST), where everything is measured relative to solar noon (when the sun is on our line of longitude). There are occasions, however, when local time, called civil time or clock time (CT), is needed. There are two adjustments that must be made in order to connect local clock time and solar time. The first is a longitude adjustment that has to do with the way in which regions of the world are divided into time zones. The second is a little fudge factor that needs to be thrown in to account for the uneven way in which the earth moves around the sun.

Obviously, it just wouldn't work for each of us to set our watches to show noon when the sun is on our own line of longitude. Since the earth rotates 15° per hour (4 minutes per degree), for every degree of longitude between one location and another, clocks showing solar time would have to differ by 4 minutes. The only time two clocks would show the same time would be if they both were due north/south of each other.

To deal with these longitude complications, the earth is nominally divided into 24 1-hour time zones, with each time zone ideally spanning 15° of longitude. Of course, geopolitical boundaries invariably complicate the boundaries from one zone to another. The intent is for all clocks within the time zone to be set to the same time. Each time zone is defined by a *Local Time Meridian* located, ideally, in the middle of the zone, with the origin of this time system passing through Greenwich, England, at 0° longitude. The local time meridians for the United States are given in Table 7.4.

The longitude correction between local clock time and solar time is based on the time it takes for the sun to travel between the local time meridian and the observer's line of longitude. If it is solar noon on the local time meridian, it will be solar noon 4 minutes later for every degree that the observer is west of that meridian. For example, San Francisco, at longitude 122°, will have solar noon 8 minutes after the sun crosses the 120° Local Time Meridian for the Pacific Time Zone.

The second adjustment between solar time and local clock time is the result of the earth's elliptical orbit, which causes the length of a *solar day* (solar noon

TABLE 7.4 Local Time Meridians for U.S. Standard Time Zones (Degrees West of Greenwich)

Time Zone	LT Meridian
Eastern	75°
Central	90°
Mountain	105°
Pacific	120°
Eastern Alaska	135°
Alaska and Hawaii	150°

to solar noon) to vary throughout the year. As the earth moves through its orbit, the difference between a 24-hour day and a solar day changes following an expression known as the *Equation of Time*, E:

$$E = 9.87 \sin 2B - 7.53 \cos B - 1.5 \sin B \qquad \text{(minutes)} \tag{7.12}$$

where

$$B = \frac{360}{364}(n - 81) \qquad \text{(degrees)} \tag{7.13}$$

As before, n is the day number. A graph of (7.12) is given in Fig. 7.16.

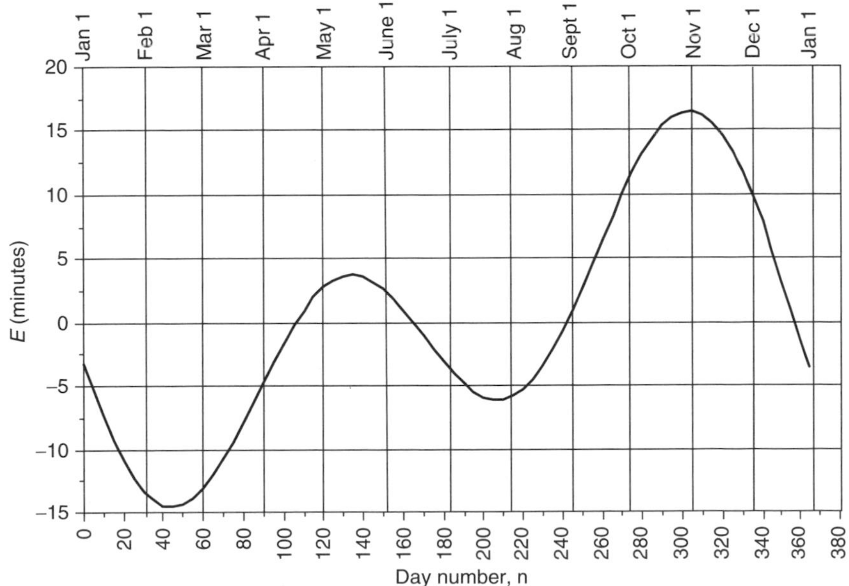

Figure 7.16 The Equation of Time adjusts for the earth's tilt angle and noncircular orbit.

Putting together the longitude correction and the Equation of Time gives us the final relationship between local standard clock time (CT) and solar time (ST).

$$\text{Solar Time (ST)} = \text{Clock Time (CT)} + \frac{4\ \text{min}}{\text{degree}}\ (\text{Local Time Meridian} - \text{Local longitude})^\circ + E(\text{min}) \tag{7.14}$$

When Daylight Savings Time is in effect, one hour must be added to the local clock time ("Spring ahead, Fall back").

Example 7.5 Solar Time to Local Time. Find Eastern Daylight Time for solar noon in Boston (longitude 71.1 °W) on July 1st.

Solution. From Table 7.1, July 1 is day number $n = 182$. Using (7.12) to (7.14) to adjust for local time, we obtain

$$B = \frac{360}{364}(n - 81) = \frac{360}{364}(182 - 81) = 99.89^\circ$$

$$E = 9.87 \sin 2B - 7.53 \cos B - 1.5 \sin B$$

$$= 9.87 \sin[2 \cdot (99.89)] - 7.53 \cos(99.89) - 1.5 \sin(99.89) = -3.5\ \text{min}$$

For Boston at longitude 71.7° in the Eastern Time Zone with local time meridian 75°

$$\text{CT} = \text{ST} - 4(\text{min}/^\circ)(\text{Local time meridian} - \text{Local longitude}) - E(\text{min})$$

$$\text{CT} = 12{:}00 - 4(75 - 71.1) - (-3.5) = 12{:}00 - 12.1\ \text{min}$$

$$= 11{:}47.9\ \text{A.M. EST}$$

To adjust for Daylight Savings Time add 1 h, so solar noon will be at about 12:48 P.M. EDT.

7.7 SUNRISE AND SUNSET

A sun path diagram, such as was shown in Fig. 7.12, can be used to locate the azimuth angles and approximate times of sunrise and sunset. A more careful estimate of sunrise/sunset can be found from a simple manipulation of (7.8). At sunrise and sunset, the altitude angle β is zero, so we can write

$$\sin\beta = \cos L\cos\delta\cos H + \sin L\sin\delta = 0 \tag{7.15}$$

$$\cos H = -\frac{\sin L\sin\delta}{\cos L\cos\delta} = -\tan L\tan\delta \tag{7.16}$$

Solving for the hour angle at sunrise, H_{SR}, gives

$$H_{SR} = \cos^{-1}(-\tan L\tan\delta) \qquad (+\text{ for sunrise}) \tag{7.17}$$

Notice in (7.17) that since the inverse cosine allows for both positive and negative values, we need to use our sign convention, which requires the positive value to be used for sunrise (and the negative for sunset).

Since the earth rotates 15°/h, the hour angle can be converted to time of sunrise or sunset using

$$\text{Sunrise(geometric)} = 12{:}00 - \frac{H_{SR}}{15^\circ/h} \tag{7.18}$$

Equations (7.15) to (7.18) are geometric relationships based on angles measured to the center of the sun, hence the designation *geometric sunrise* in (7.18). They are perfectly adequate for any kind of normal solar work, but they won't give you exactly what you will find in the newspaper for sunrise or sunset. The difference between weather service sunrise and our geometric sunrise (7.18) is the result of two factors. The first deviation is caused by atmospheric refraction; this bends the sun's rays, making the sun *appear* to rise about 2.4 min sooner than geometry would tell us and then set 2.4 min later. The second is that the weather service definition of sunrise and sunset is the time at which the upper limb (top) of the sun crosses the horizon, while ours is based on the center crossing the horizon. This effect is complicated by the fact that at sunrise or sunset the sun pops up, or sinks, much quicker around the equinoxes when it moves more vertically than at the solstices when its motion includes much more of a sideward component. An adjustment factor Q that accounts for these complications is given by the following (U.S. Department of Energy, 1978):

$$Q = \frac{3.467}{\cos L\cos\delta\sin H_{SR}} \qquad (min) \tag{7.19}$$

Since sunrise is earlier when it is based on the top of the sun rather than the middle, Q should be subtracted from geometric sunrise. Similarly, since the upper limb sinks below the horizon later than the middle of the sun, Q should be added to our geometric sunset. A plot of (7.19) is shown in Fig. 7.17. As

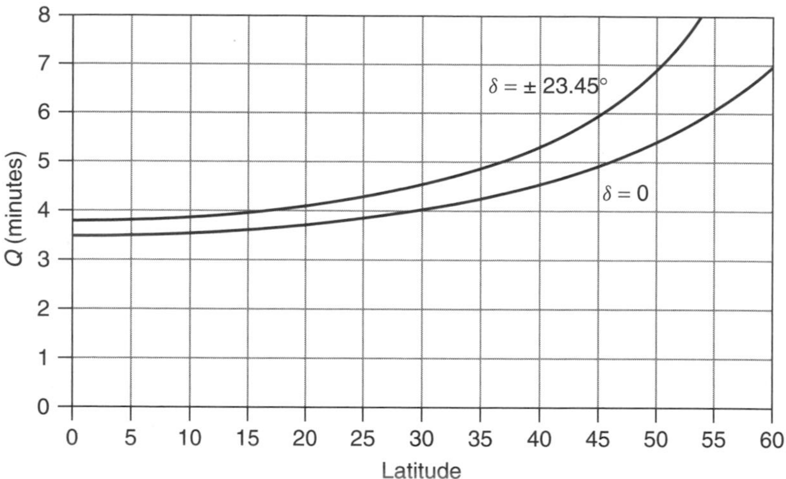

Figure 7.17 Sunrise/sunset adjustment factor to account for refraction and the upper-limb definition of sunrise. The range of solar declinations is shown.

can be seen, for mid-latitudes, the correction is typically in the range of about 4 to 6 min.

Example 7.6 Sunrise in Boston. Find the time at which sunrise (geometric and conventional) will occur in Boston (latitude 42.3°) on July 1 ($n = 182$). Also find conventional sunset.

Solution. From (7.6), the solar declination is

$$\delta = 23.45 \sin\left[\frac{360}{365}(n - 81)\right] = 23.45 \sin\left[\frac{360}{365}(182 - 81)\right] = 23.1^\circ$$

From (7.17), the hour angle at sunrise is

$$H_{SR} = \cos^{-1}(-\tan L \tan \delta) = \cos^{-1}(-\tan 42.3^\circ \tan 23.1^\circ) = 112.86^\circ$$

From (7.18) solar time of geometric sunrise is

$$\begin{aligned}\text{Sunrise (geometric)} &= 12{:}00 - \frac{H_{SR}}{15^\circ/h}\\ &= 12{:}00 - \frac{112.86^\circ}{15^\circ/h} = 12{:}00 - 7.524\text{ h}\\ &= 4{:}28.6 \text{ A.M. (solar time)}\end{aligned}$$

Using (7.19) to adjust for refraction and the upper-limb definition of sunrise gives

$$Q = \frac{3.467}{\cos L \cos \delta \sin H_{SR}} (\text{min})$$
$$= \frac{3.467}{\cos 42.3 \cos 23.1^\circ \sin 112.86^\circ} = 5.5 \text{ min}$$

The upper limb will appear 5.5 minutes sooner than our original geometric calculation indicated, so

$$\text{Sunrise} = 4{:}28.6 \text{ A.M.} - 5.5 \text{ min} = 4{:}23.1 \text{ A.M. (solar time)}$$

From Example 7.5, on this date in Boston, local clock time is 12.1 min earlier than solar time, so sunrise will be at

$$\text{Sunrise (upper limb)} = 4{:}23.1 - 12.1 = 4{:}11 \text{ A.M. Eastern Standard Time}$$

Similarly, geometric sunset is 7.524 h after solar noon, or 7:31.4 P.M. solar time. The upper limb will drop below the horizon 5.5 minutes later. Then adjusting for the 12.1 minutes difference between Boston time and solar time gives

$$\text{Sunset (upper limb)} = 7{:}31.4 + (5.5 - 12.1) \text{ min} = 7{:}24.8 \text{ P.M. EST}$$

There is a convenient website for finding sunrise and sunset times on the web at http://aa.usno.navy.mil/data/docs/RS_OneDay.html.

A fun, but fairly useless, application of these equations for sunrise and sunset is to work them in reverse order to navigate—that is, to find latitude and longitude, as the following example illustrates.

Example 7.7 Where in the World Are You? With your watch set for Pacific Standard Time (PST), you travel somewhere and when you arrive you note that the upper limb sunrise is at 1:11 A.M. (by your watch) and sunset is at 4:25 P.M. It is July 1st ($\delta = 23.1^\circ$, E $= -3.5$ min). Where are you?

Solution. Between 1:11 am and 4:25 P.M. there are 15 h and 14 min of daylight (15.233 h). With solar noon at the midpoint of that time—that is, 7 h and 37 min after sunrise—we have

$$\text{Solar noon} = 1{:}11 \text{ A.M.} + 7{:}37 = 8{:}48 \text{ A.M. PST}$$

Longitude can be determined from (7.14):

$$\text{Solar Time} = \text{Clock Time} + \frac{4\ \text{min}}{\text{degree}}(\text{Local Time Meridian} - \text{Local longitude})^\circ + E(\text{min})$$

Using the 120° Local Time Meridian for Pacific Time gives

$$12{:}00 - 8{:}48 = 192\ \text{min} = 4(120 - \text{Longitude}) + (-3.5)\ \text{min}$$

$$\text{Longitude} = \frac{480 - 3.5 - 192}{4} = 71.1^\circ$$

To find latitude, it helps to first ignore the correction factor Q. Doing so, the daylength is 15.233 h, which makes the hour angle at sunrise equal to

$$H_{SR} = \frac{15.233\ \text{hour}}{2} \cdot 15^\circ/\text{hour} = 114.25^\circ$$

A first estimate of latitude can now be found from (7.16)

$$L = \tan^{-1}\left(-\frac{\cos H_{SR}}{\tan\delta}\right) = \tan^{-1}\left(-\frac{\cos 114.25^\circ}{\tan 23.1^\circ}\right) = 43.9^\circ$$

Now we can find Q, from which we can correct our estimate of latitude

$$Q = \frac{3.467}{\cos L \cos\delta \sin H_{SR}} = \frac{3.467}{\cos 43.9^\circ \cos 23.1^\circ \sin 114.25^\circ} = 5.74\ \text{min}$$

Geometric daylength is therefore 2 × 5.74 min = 11.48 min = 0.191 h shorter than daylength based on the upper limb crossing the horizon. The geometric hour angle at sunrise is therefore

$$H_{SR} = \frac{(15.233 - 0.191)\text{h}}{2} \cdot 15^\circ/\text{h} = 112.81^\circ$$

Our final estimate of latitude is therefore

$$L = \tan^{-1}\left(-\frac{\cos H_{SR}}{\tan\delta}\right) = \tan^{-1}\left(-\frac{\cos 112.81^\circ}{\tan 23.1^\circ}\right) = 42.3^\circ$$

Notice we are back in Boston, latitude 42.3°, longitude 71.1°. Notice too, our watch worked fine even though it was set for a different time zone.

With so many angles to keep track of, it may help to summarize the terminology and equations for them all in one spot, which has been done in Box 7.1.

BOX 7.1 SUMMARY OF SOLAR ANGLES

δ = solar declination
n = day number
L = latitude
β = solar altitude angle, β_N = angle at solar noon
H = hour angle
H_{SR} = sunrise hour angle
ϕ_S = solar azimuth angle (+ before solar noon, − after)
ϕ_C = collector azimuth angle (+ east of south, − west of south)*
ST = solar time
CT = civil or clock time
E = equation of time
Q = correction for refraction and semidiameter at sunrise/sunset
Σ = collector tilt angle
θ = incidence angle between sun and collector face

$$\delta = 23.45 \sin\left[\frac{360}{365}(n - 81)\right]$$

$$\beta_N = 90^\circ - L + \delta$$

$$\sin\beta = \cos L \cos\delta \cos H + \sin L \sin\delta$$

$$\sin\phi_S = \frac{\cos\delta \sin H}{\cos\beta}$$

$$\text{If } \cos \text{H} \geq \frac{\tan\delta}{\tan L}, \quad \text{then} \quad |\phi_S| \leq 90^\circ; \qquad \text{otherwise } |\phi_S| > 90^\circ$$

$$\text{Hour angle } H = \left(\frac{15^\circ}{\text{hour}}\right) \cdot (\text{Hours before solar noon})$$

$$E = 9.87 \sin 2B - 7.53 \cos B - 1.5 \sin B \qquad (\text{min})$$

$$B = \frac{360}{364}(n - 81)$$

$$\text{Solar Time (ST)} = \text{Clock Time (CT)} + \frac{4\text{ min}}{\text{degree}}(\text{Local Time Meridian} - \text{Local Longitude})^\circ + E(\text{min})$$

$$H_{SR} = \cos^{-1}(-\tan L \tan\delta) \qquad (+\text{ for sunrise})$$

$$Q = \frac{3.467}{\cos L \cos\delta \sin H_{SR}} \qquad (\text{min})$$

$$\cos\theta = \cos\beta \cos(\phi_S - \phi_C) \sin\Sigma + \sin\beta \cos\Sigma$$

*Opposite signs in Southern Hemisphere.

7.8 CLEAR SKY DIRECT-BEAM RADIATION

Solar flux striking a collector will be a combination of *direct-beam* radiation that passes in a straight line through the atmosphere to the receiver, *diffuse* radiation that has been scattered by molecules and aerosols in the atmosphere, and *reflected* radiation that has bounced off the ground or other surface in front of the collector (Fig. 7.18). The preferred units, especially in solar–electric applications, are watts (or kilowatts) per square meter. Other units involving British Thermal Units, kilocalories, and langleys may also be encountered. Conversion factors between these units are given in Table 7.5.

Solar collectors that focus sunlight usually operate on just the beam portion of the incoming radiation since those rays are the only ones that arrive from a consistent direction. Most photovoltaic systems, however, don't use focusing devices, so all three components—beam, diffuse, and reflected—can contribute to energy collected. The goal of this section is to be able to estimate the rate at which just the beam portion of solar radiation passes through the atmosphere

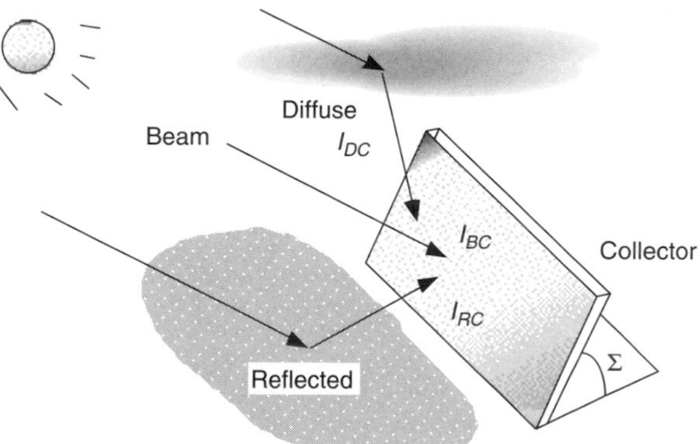

Figure 7.18 Solar radiation striking a collector, I_C, is a combination of direct beam, I_{BC}, diffuse, I_{DC}, and reflected, I_{RC}.

TABLE 7.5 Conversion Factors for Various Insolation Units

1 kW/m^2	=	316.95 Btu/h-ft^2
	=	1.433 langley/min
1 kWh/m^2	=	316.95 Btu/ft^2
	=	85.98 langleys
	=	3.60×10^6 joules/m^2
1 langley	=	1 cal/cm^2
	=	41.856 kjoules/m^2
	=	0.01163 kWh/m^2
	=	3.6878 Btu/ft^2

and arrives at the earth's surface on a clear day. Later, the diffuse and reflected radiation will be added to the clear day model. And finally, procedures will be presented that will enable more realistic average insolation calculations for specific locations based on empirically derived data for certain given sites.

The starting point for a clear sky radiation calculation is with an estimate of the extraterrestrial (ET) solar insolation, I_0, that passes perpendicularly through an imaginary surface just outside of the earth's atmosphere as shown in Fig. 7.19. This insolation depends on the distance between the earth and the sun, which varies with the time of year. It also depends on the intensity of the sun, which rises and falls with a fairly predictable cycle. During peak periods of magnetic activity on the sun, the surface has large numbers of cooler, darker regions called *sunspots*, which in essence block solar radiation, accompanied by other regions, called *faculae*, that are brighter than the surrounding surface. The net effect of sunspots that dim the sun, and faculae that brighten it, is an increase in solar intensity during periods of increased numbers of sunspots. Sunspot activity seems to follow an 11-year cycle. During sunspot peaks, the most recent of which was in 2001, the extraterrestrial insolation is estimated to be about 1.5% higher than in the valleys (U.S. Department of Energy, 1978).

Ignoring sunspots, one expression that is used to describe the day-to-day variation in extraterrestrial solar insolation is the following:

$$I_0 = \mathrm{SC} \cdot \left[1 + 0.034 \cos\left(\frac{360n}{365}\right)\right] \quad (\mathrm{W/m^2}) \tag{7.20}$$

where SC is called the *solar constant* and n is the day number. The solar constant is an estimate of the average annual extraterrestrial insolation. Based on early NASA measurements, the solar constant was often taken to be 1.353 kW/m^2, but 1.377 kW/m^2 is now the more commonly accepted value.

As the beam passes through the atmosphere, a good portion of it is absorbed by various gases in the atmosphere, or scattered by air molecules or particulate matter. In fact, over a year's time, less than half of the radiation that hits the top of the atmosphere reaches the earth's surface as direct beam. On a clear day, however, with the sun high in the sky, beam radiation at the surface can exceed 70% of the extraterrestrial flux.

Attenuation of incoming radiation is a function of the distance that the beam has to travel through the atmosphere, which is easily calculable, as well as factors

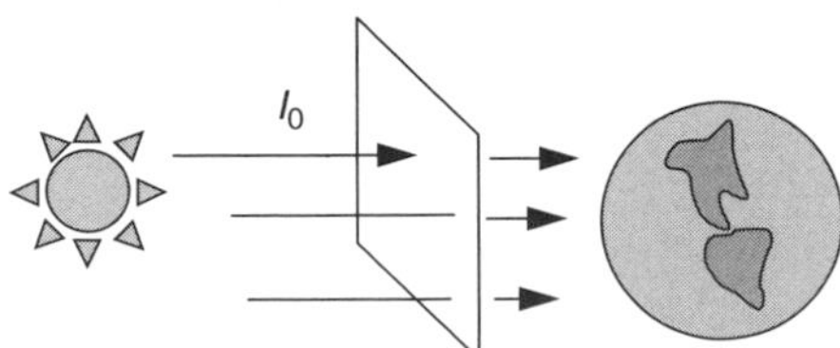

Figure 7.19 The extraterrestrial solar flux.

TABLE 7.6 Optical Depth k, Apparent Extraterrestrial Flux A, and the Sky Diffuse Factor C for the 21st Day of Each Month

Month:	Jan	Feb	Mar	Apr	May	June	July	Aug	Sept	Oct	Nov	Dec
A (W/m^2):	1230	1215	1186	1136	1104	1088	1085	1107	1151	1192	1221	1233
k:	0.142	0.144	0.156	0.180	0.196	0.205	0.207	0.201	0.177	0.160	0.149	0.142
C:	0.058	0.060	0.071	0.097	0.121	0.134	0.136	0.122	0.092	0.073	0.063	0.057

Source: ASHRAE (1993).

such as dust, air pollution, atmospheric water vapor, clouds, and turbidity, which are not so easy to account for. A commonly used model treats attenuation as an exponential decay function:

$$I_B = Ae^{-km} \tag{7.21}$$

where I_B is the beam portion of the radiation reaching the earth's surface (normal to the rays), A is an "apparent" extraterrestrial flux, and k is a dimensionless factor called the *optical depth*. The air mass ratio m was introduced earlier as (7.4)

$$\text{Air mass ratio} \quad m = \frac{1}{\sin\beta} \tag{7.4}$$

where β is the altitude angle of the sun.

Table 7.6 gives values of A and k that are used in the American Society of Heating, Refrigerating, and Air Conditioning Engineers (ASHRAE) Clear Day Solar Flux Model. This model is based on empirical data collected by Threlkeld and Jordan (1958) for a moderately dusty atmosphere with atmospheric water vapor content equal to the average monthly values in the United States. Also included is a diffuse factor, C, that will be introduced later.

For computational purposes, it is handy to have an equation to work with rather than a table of values. Close fits to the values of optical depth k and apparent extraterrestrial (ET) flux A given in Table 7.6 are as follows:

$$A = 1160 + 75\sin\left[\frac{360}{365}(n-275)\right] \qquad (\text{W/m}^2) \tag{7.22}$$

$$k = 0.174 + 0.035\sin\left[\frac{360}{365}(n-100)\right] \tag{7.23}$$

where again n is the day number.

Example 7.8 Direct Beam Radiation at the Surface of the Earth. Find the direct beam solar radiation normal to the sun's rays at solar noon on a clear day in Atlanta (latitude 33.7°) on May 21. Use (7.22) and (7.23) to see how closely they approximate Table 7.6.

Solution. Using Table 7.1 to help, May 21 is day number 141. From (7.22), the apparent extraterrestrial flux, A, is

$$A = 1160 + 75 \sin\left[\frac{360}{365}(n - 275)\right] = 1160 + 75 \sin\left[\frac{360}{365}(141 - 275)\right]$$
$$= 1104 \text{ W/m}^2$$

(which agrees with Table 7.6).

From (7.23), the optical depth is

$$k = 0.174 + 0.035 \sin\left[\frac{360}{365}(n - 100)\right]$$
$$= 0.174 + 0.035 \sin\left[\frac{360}{365}(141 - 100)\right] = 0.197$$

(which is very close to the value given in Table 7.6).

From Table 7.2, on May 21 solar declination is 20.1°, so from (7.7) the altitude angle of the sun at solar noon is

$$\beta_N = 90^\circ - L + \delta = 90 - 33.7 + 20.1 = 76.4^\circ$$

The air mass ratio (7.4) is

$$m = \frac{1}{\sin\beta} = \frac{1}{\sin(76.4^\circ)} = 1.029$$

Finally, using (7.21) the predicted value of clear sky beam radiation at the earth's surface is

$$I_B = Ae^{-km} = 1104\ e^{-0.197\times 1.029} = 902 \text{ W/m}^2$$

7.9 TOTAL CLEAR SKY INSOLATION ON A COLLECTING SURFACE

Reasonably accurate estimates of the clear sky, direct beam insolation are easy enough to work out and the geometry needed to determine how much of that will strike a collector surface is straightforward. It is not so easy to account for the diffuse and reflected insolation but since that energy bonus is a relatively small fraction of the total, even crude models are usually acceptable.

7.9.1 Direct-Beam Radiation

The translation of direct-beam radiation I_B (normal to the rays) into beam insolation striking a collector face I_{BC} is a simple function of the angle of incidence

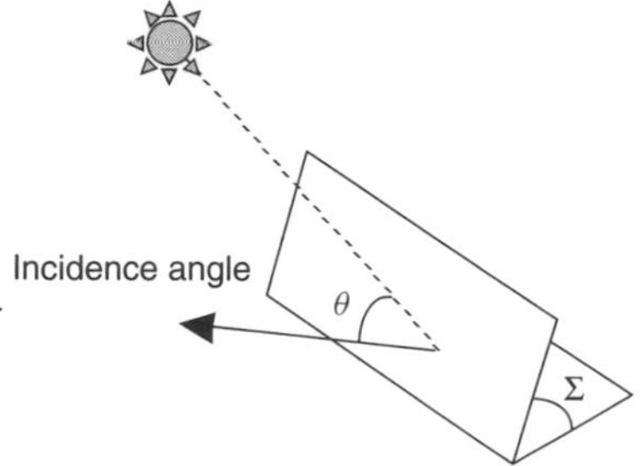

Figure 7.20 The incidence angle θ between a normal to the collector face and the incoming solar beam radiation.

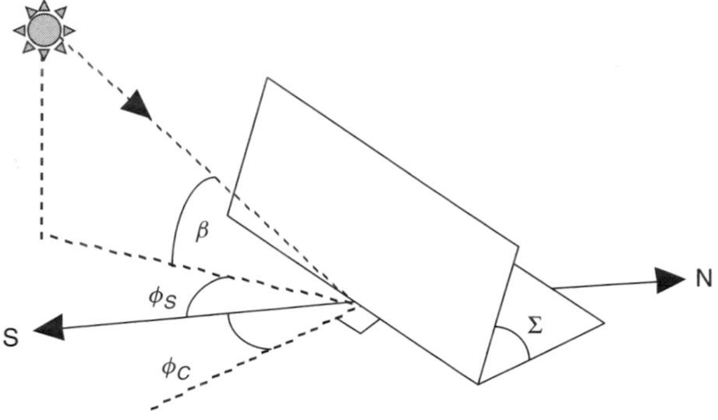

Figure 7.21 Illustrating the collector azimuth angle ϕ_C and tilt angle Σ along with the solar azimuth angle ϕ_S and altitude angle β. Azimuth angles are positive in the southeast direction and are negative in the southwest.

θ between a line drawn normal to the collector face and the incoming beam radiation (Fig. 7.20). It is given by

$$I_{BC} = I_B \cos\theta \tag{7.24}$$

For the special case of beam insolation on a horizontal surface I_{BH},

$$I_{BH} = I_B \cos(90^\circ - \beta) = I_B \sin\beta \tag{7.25}$$

The angle of incidence θ will be a function of the collector orientation and the altitude and azimuth angles of the sun at any particular time. Figure 7.21 introduces these important angles. The solar collector is tipped up at an angle Σ and faces in a direction described by its azimuth angle ϕ_C (measured relative to due south, *with positive values in the southeast direction and negative values in the southwest*). The incidence angle is given by

$$\cos\theta = \cos\beta\cos(\phi_S - \phi_C)\sin\Sigma + \sin\beta\cos\Sigma \tag{7.26}$$

Example 7.9 Beam Insolation on a Collector. In Example 7.8, at solar noon in Atlanta (latitude 33.7°) on May 21 the altitude angle of the sun was found to be 76.4° and the clear-sky beam insolation was found to be 902 W/m^2. Find the beam insolation at that time on a collector that faces 20° toward the southeast if it is tipped up at a 52° angle.

Solution. Using (7.26), the cosine of the incidence angle is

$$\begin{aligned}\cos\theta &= \cos\beta\cos(\phi_S - \phi_C)\sin\Sigma + \sin\beta\cos\Sigma \\ &= \cos 76.4^\circ \cos(0 - 20^\circ)\sin 52^\circ + \sin 76.4^\circ \cos 52^\circ = 0.7725\end{aligned}$$

From (7.24), the beam radiation on the collector is

$$I_{BC} = I_B \cos\theta = 902\ \text{W/m}^2 \cdot 0.7725 = 697\ \text{W/m}^2$$

7.9.2 Diffuse Radiation

The diffuse radiation on a collector is much more difficult to estimate accurately than it is for the beam. Consider the variety of components that make up diffuse radiation as shown in Fig. 7.22. Incoming radiation can be scattered from atmospheric particles and moisture, and it can be reflected by clouds. Some is reflected from the surface back into the sky and scattered again back to the ground. The simplest models of diffuse radiation assume it arrives at a site with equal intensity from all directions; that is, the sky is considered to be *isotropic*. Obviously, on hazy or overcast days the sky is considerably brighter in the vicinity of the sun, and measurements show a similar phenomenon on clear days as well, but these complications are often ignored.

The model developed by Threlkeld and Jordan (1958), which is used in the ASHRAE Clear-Day Solar Flux Model, suggests that diffuse insolation on a

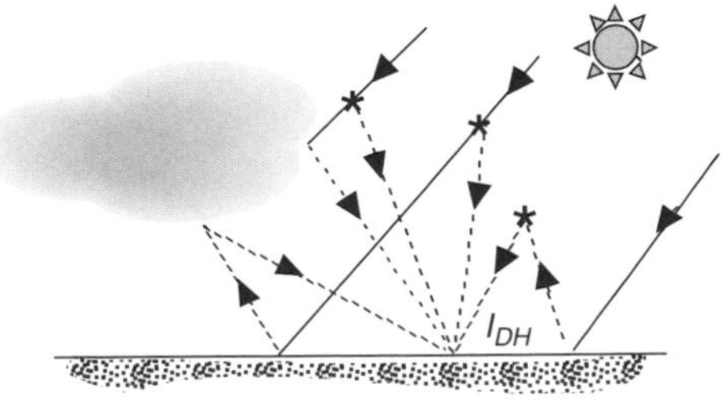

Figure 7.22 Diffuse radiation can be scattered by atmospheric particles and moisture or reflected from clouds. Multiple scatterings are possible.

horizontal surface I_{DH} is proportional to the direct beam radiation I_B no matter where in the sky the sun happens to be:

$$I_{DH} = C\, I_B \tag{7.27}$$

where C is a sky diffuse factor. Monthly values of C are given in Table 7.6, and a convenient approximation is as follows:

$$C = 0.095 + 0.04 \sin\left[\frac{360}{365}(n - 100)\right] \tag{7.28}$$

Applying (7.27) to a full day of clear skies typically predicts that about 15% of the total horizontal insolation on a clear day will be diffuse.

What we would like to know is how much of that horizontal diffuse radiation strikes a collector so that we can add it to the beam radiation. As a first approximation, it is assumed that diffuse radiation arrives at a site with equal intensity from all directions. This means that the collector will be exposed to whatever fraction of the sky the face of the collector points to, as shown in Fig. 7.23. When the tilt angle of the collector Σ is zero—that is, the panel is flat on the ground—the panel sees the full sky and so it receives the full horizontal diffuse radiation, I_{DH}. When it is a vertical surface, it sees half the sky and is exposed to half of the horizontal diffuse radiation, and so forth. The following expression for diffuse radiation on the collector, I_{DC}, is used when the diffuse radiation is idealized in this way:

$$I_{DC} = I_{DH}\left(\frac{1 + \cos\Sigma}{2}\right) = C\, I_B\left(\frac{1 + \cos\Sigma}{2}\right) \tag{7.29}$$

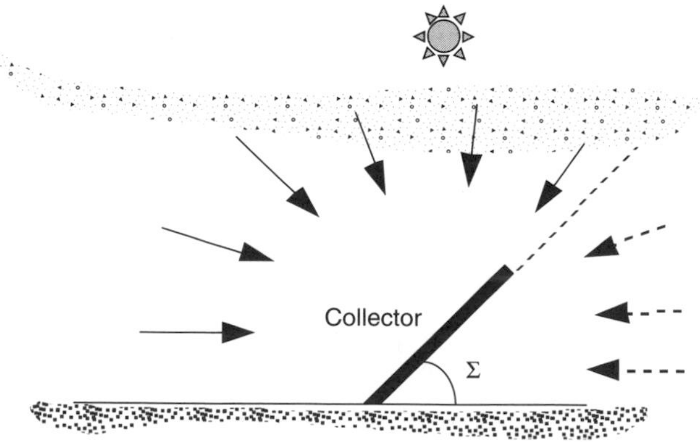

Figure 7.23 Diffuse radiation on a collector assumed to be proportional to the fraction of the sky that the collector "sees".

Example 7.10 Diffuse Radiation on a Collector. Continue Example 7.9 and find the diffuse radiation on the panel. Recall that it is solar noon in Atlanta on May 21 ($n = 141$), and the collector faces 20° toward the southeast and is tipped up at a 52° angle. The clear-sky beam insolation was found to be 902 W/m².

Solution. Start with (7.28) to find the diffuse sky factor, C:

$$C = 0.095 + 0.04 \sin\left[\frac{360}{365}(n - 100)\right]$$

$$C = 0.095 + 0.04 \sin\left[\frac{360}{365}(141 - 100)\right] = 0.121$$

And from (7.29), the diffuse energy striking the collector is

$$I_{DC} = CI_B\left(\frac{1 + \cos\Sigma}{2}\right)$$

$$= 0.121 \cdot 902 \text{ W/m}^2 \left(\frac{1 + \cos 52^\circ}{2}\right) = 88 \text{ W/m}^2$$

Added to the total beam insolation of 697 W/m² found in Example 7.9, this gives a total beam plus diffuse on the collector of 785 W/m².

7.9.3 Reflected Radiation

The final component of insolation striking a collector results from radiation that is reflected by surfaces in front of the panel. This reflection can provide a considerable boost in performance, as for example on a bright day with snow or water in front of the collector, or it can be so modest that it might as well be ignored. The assumptions needed to model reflected radiation are considerable, and the resulting estimates are very rough indeed. The simplest model assumes a large horizontal area in front of the collector, with a reflectance ρ that is diffuse, and it bounces the reflected radiation in equal intensity in all directions, as shown in Fig. 7.24. Clearly this is a very gross assumption, especially if the surface is smooth and bright.

Estimates of ground reflectance range from about 0.8 for fresh snow to about 0.1 for a bituminous-and-gravel roof, with a typical default value for ordinary ground or grass taken to be about 0.2. The amount reflected can be modeled as the product of the total horizontal radiation (beam I_{BH}, plus diffuse I_{DH}) times the ground reflectance ρ. The fraction of that ground-reflected energy that will

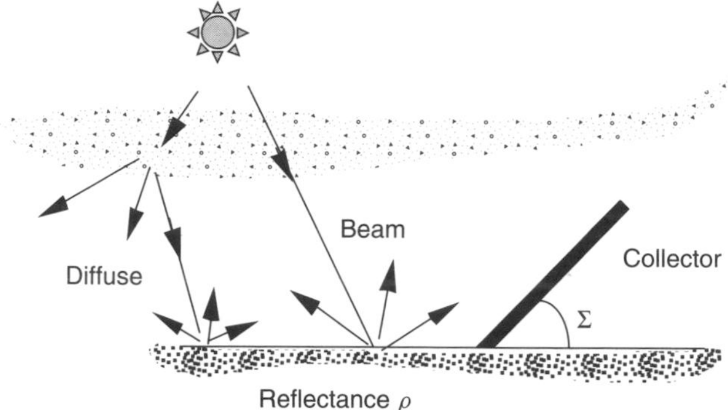

Figure 7.24 The ground is assumed to reflect radiation with equal intensity in all directions.

be intercepted by the collector depends on the slope of the panel Σ, resulting in the following expression for reflected radiation striking the collector I_{RC}:

$$I_{RC} = \rho(I_{BH} + I_{DH})\left(\frac{1 - \cos\Sigma}{2}\right) \tag{7.30}$$

For a horizontal collector (Σ = 0), Eq. (7.30) correctly predicts no reflected radiation on the collector; for a vertical panel, it predicts that the panel "sees" half of the reflected radiation, which also is appropriate for the model.

Substituting expressions (7.25) and (7.27) into (7.30) gives the following for reflected radiation on the collector:

$$I_{RC} = \rho I_B(\sin\beta + C)\left(\frac{1 - \cos\Sigma}{2}\right) \tag{7.31}$$

Example 7.11 Reflected Radiation Onto a Collector. Continue Examples 7.9 and 7.10 and find the reflected radiation on the panel if the reflectance of the surfaces in front of the panel is 0.2. Recall that it is solar noon in Atlanta on May 21, the altitude angle of the sun β is 76.4°, the collector faces 20° toward the southeast and is tipped up at a 52° angle, the diffuse sky factor C is 0.121, and the clear-sky beam insolation is 902 W/m^2.

Solution. From (7.31), the clear-sky reflected insolation on the collector is

$$I_{RC} = \rho I_B(\sin\beta + C)\left(\frac{1 - \cos\Sigma}{2}\right)$$

$$= 0.2 \cdot 902\ \text{W/m}^2(\sin 76.4^\circ + 0.121)\left(\frac{1 - \cos 52^\circ}{2}\right) = 38\ \text{W/m}^2$$

The total insolation on the collector is therefore

$$I_C = I_{BC} + I_{DC} + I_{RC} = 697 + 88 + 38 = 823 \text{ W/m}^2$$

Of that total, 84.7% is direct beam, 10.7% is diffuse, and 4.6% is reflected. The reflected portion is clearly modest and is often ignored.

Combining the equations for the three components of radiation, direct beam, diffuse and reflected gives the following for total rate at which radiation strikes a collector on a clear day:

$$I_C = I_{BC} + I_{DC} + I_{RC} \tag{7.32}$$

$$I_C = Ae^{-km}\left[\cos\beta\cos(\phi_S - \phi_C)\sin\Sigma + \sin\beta\cos\Sigma + C\left(\frac{1+\cos\Sigma}{2}\right) + \rho(\sin\beta + C)\left(\frac{1-\cos\Sigma}{2}\right)\right] \tag{7.33}$$

Equation (7.33) looks terribly messy, but it is a convenient summary, which can be handy when setting up a spreadsheet or other computerized calculation of clear sky insolation.

7.9.4 Tracking Systems

Thus far, the assumption has been that the collector is permanently attached to a surface that doesn't move. In many circumstances, however, racks that allow the collector to track the movement of the sun across the sky are quite cost effective. Trackers are described as being either *two-axis trackers*, which track the sun both in azimuth and altitude angles so the collectors are always pointing directly at the sun, or *single-axis trackers*, which track only one angle or the other.

Calculating the beam plus diffuse insolation on a two-axis tracker is quite straightforward (Fig. 7.25). The beam radiation on the collector is the full insolation I_B normal to the rays calculated using (7.21). The diffuse and reflected radiation are found using (7.29) and (7.31) with a collector tilt angle equal to the complement of the solar altitude angle, that is, $90° - \beta$.

Two-Axis Tracking:

$$I_{BC} = I_B \tag{7.34}$$

$$I_{DC} = CI_B\left[\frac{1+\cos(90° - \beta)}{2}\right] \tag{7.35}$$

$$I_{RC} = \rho(I_{BH} + I_{DH})\left[\frac{1-\cos(90° - \beta)}{2}\right] \tag{7.36}$$

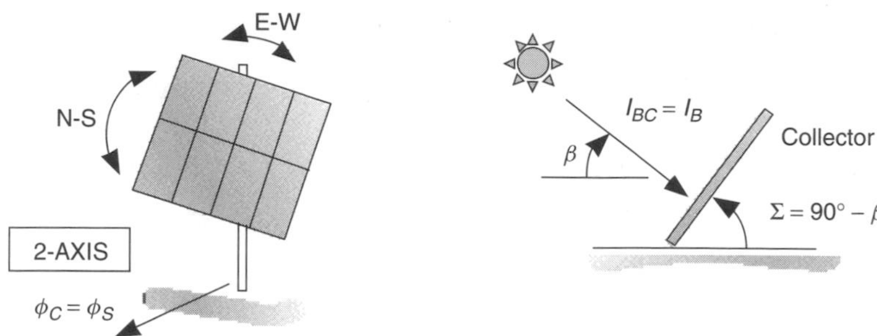

Figure 7.25 Two-axis tracking angular relationships.

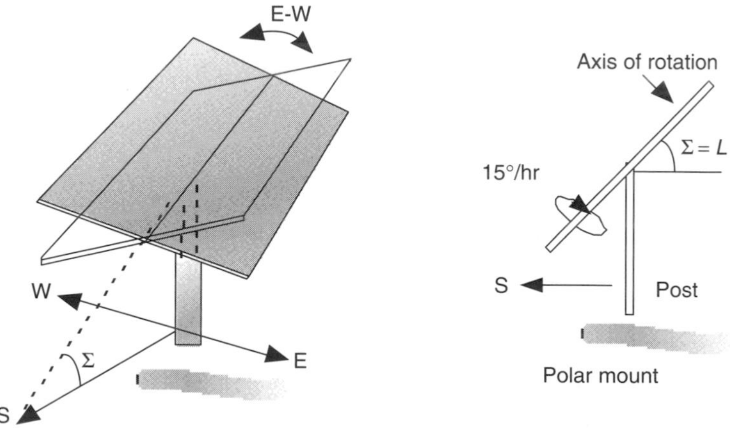

Figure 7.26 A single-axis tracking mount with east–west tracking. A polar mount has the axis of rotation facing south and tilted at an angle equal to the latitude.

Single-axis tracking for photovoltaics is almost always done with a mount having a manually adjustable tilt angle along a north-south axis, and a tracking mechanism that rotates the collector array from east-to-west, as shown in Fig. 7.26. When the tilt angle of the mount is set equal to the local latitude (called a *polar mount*), not only is that an optimum angle for annual collection, but the collector geometry and resulting insolation are fairly easy to evaluate as well.

As shown in Fig. 7.27, if a polar mount rotates about its axis at the same rate as the earth turns, 15°/h, then the centerline of the collector will always face directly into the sun. Under these conditions, the incidence angle θ between a normal to the collector and the sun's rays will be equal to the solar declination δ. That makes the direct-beam insolation on the collector just $I_B \cos \delta$. To evaluate diffuse and reflected radiation, we need to know the tilt angle of the collector. As can be seen in Fig. 7.26, while the axis of rotation has a fixed tilt of $\Sigma = L$, unless it is solar noon, the collector itself is cocked at an odd angle with respect

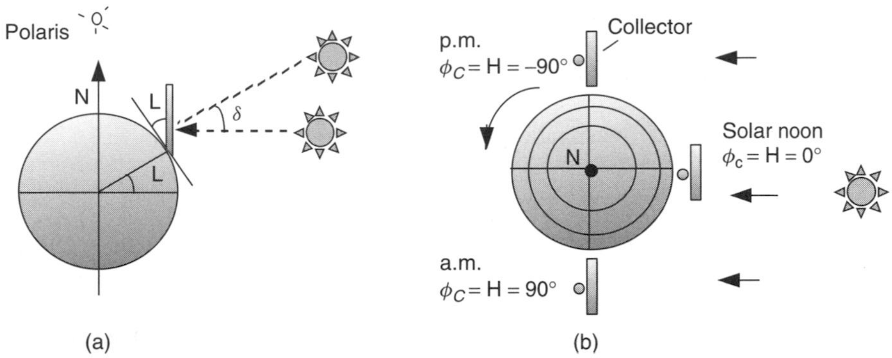

Figure 7.27 (a) Polar mount for a one-axis tracker showing the impact of a 15°/h angular rotation of the collector array. (b) Looking down on North Pole.

to the horizontal plane. The effective tilt, which is the angle between a normal to the collector and the horizontal plane, is given by

$$\Sigma_{\text{effective}} = 90 - \beta + \delta \tag{7.37}$$

The beam, diffuse, and reflected radiation on a polar mount, one-axis tracker are given by

One-Axis, Polar Mount:

$$I_{BC} = I_B \cos\delta \tag{7.38}$$

$$I_{DC} = CI_B\left[\frac{1+\cos(90^\circ - \beta + \delta)}{2}\right] \tag{7.39}$$

$$I_{RC} = \rho(I_{BH} + I_{DH})\left[\frac{1-\cos(90^\circ - \beta + \delta)}{2}\right] \tag{7.40}$$

Example 7.12 One-Axis and Two-Axis Tracker Insolation. Compare the 40° latitude, clear-sky insolation on a collector at solar noon on the summer solstice for a two-axis tracking mount versus a single-axis polar mount. Ignore ground reflectance.

Solution

1. *Two-Axis Tracker:* To find the beam insolation from (7.21) $I_B = Ae^{-km}$, we need the air mass ratio m, the apparent extraterrestrial flux A, and the optical depth k. To find m, we need the altitude angle of the sun. Using (7.7)

with a solstice declination of 23.45°,

$$\beta_N = 90^\circ - L + \delta = 90 - 40 + 23.45 = 73.45^\circ$$

$$\text{Air mass ratio } m = \frac{1}{\sin\beta} = \frac{1}{\sin 73.45^\circ} = 1.043$$

From Table 7.6, or Eqs. (7.22), (7.23) and (7.28), we find $A = 1088$ W/m^2, $k = 0.205$, and $C = 0.134$. The direct beam insolation on the collector is therefore

$$I_{BC} = I_B = Ae^{-km} = 1088\ (\text{W/m}^2) \cdot e^{-0.205x1.043} = 879\ \text{W/m}^2$$

Using (7.35) the diffuse radiation on the collector is

$$I_{DC} = CI_B\left[\frac{1+\cos(90^\circ - \beta)}{2}\right]$$

$$= 0.134 \cdot 879\left[\frac{1+\cos(90^\circ - 73.45^\circ)}{2}\right] = 115\ \text{W/m}^2$$

The total is $I_C = I_{BC} + I_{DC} = 879 + 115 = 994$ W/m^2

2. *One-Axis Polar Tracker:* The beam portion of insolation is given by (7.38)

$$I_{BC} = I_B\cos\delta = 879\ \text{W/m}^2\cos(23.45^\circ) = 806\ \text{W/m}^2$$

The diffuse portion, using (7.39), is

$$I_{DC} = CI_B\left[\frac{1+\cos(90^\circ - \beta + \delta)}{2}\right]$$

$$= 0.134 \cdot 879\ \text{W/m}^2\left[\frac{1+\cos(90 - 73.45 + 23.45)}{2}\right] = 104\ \text{W/m}^2$$

The total is $I_C = I_{BC} + I_{DC} = 806 + 104 = 910$ W/m^2

The two-axis tracker provides 994 W/m^2, which is only 9% higher than the single-axis mount.

To assist in keeping this whole set of clear-sky insolation relationships straight, Box 7.2 offers a helpful summary of nomenclature and equations. And, obviously, working with these equations is tedious until they have been put onto a spreadsheet. Or, for most purposes it is sufficient to look up values in a table and, if necessary, do some interpolation. In Appendix C there are tables of hour-by-hour clear-sky insolation for various tilt angles and latitudes, an example of which is given here in Table 7.7.

BOX 7.2 SUMMARY OF CLEAR-SKY SOLAR INSOLATION EQUATIONS

I_0 = extraterrestrial solar insolation
m = air mass ratio
I_B = beam insolation at earth's surface
A = apparent extraterrestrial solar insolation
k = atmospheric optical depth
C = sky diffuse factor
I_{BC} = beam insolation on collector
θ = incidence angle
Σ = collector tilt angle
I_H = insolation on a horizontal surface
I_{DH} = diffuse insolation on a horizontal surface
I_{DC} = diffuse insolation on collector
I_{RC} = reflected insolation on collector
ρ = ground reflectance
I_C = insolation on collector
n = day number
β = solar altitude angle
δ = solar declination
ϕ_S = solar azimuth angle (+ = AM)
ϕ_C = collector azimuth angle (+ = SE)

$$I_0 = 1370\left[1 + 0.034\cos\left(\frac{360n}{365}\right)\right]\ (\text{W/m}^2)$$

$$m = \frac{1}{\sin\beta}$$

$$I_B = Ae^{-km}$$

$$A = 1160 + 75\sin\left[\frac{360}{365}(n - 275)\right]\ (\text{W/m}^2)$$

$$k = 0.174 + 0.035\sin\left[\frac{360}{365}(n - 100)\right]$$

$$I_{BC} = I_B\cos\theta$$

$$\cos\theta = \cos\beta\cos(\phi_S - \phi_C)\sin\Sigma + \sin\beta\cos\Sigma$$

$$I_{BH} = I_B\cos(90^\circ - \beta) = I_B\sin\beta$$

$$I_{DH} = CI_B$$

$$C = 0.095 + 0.04 \sin\left[\frac{360}{365}(n - 100)\right]$$

$$I_{DC} = I_{DH}\left(\frac{1+\cos\Sigma}{2}\right) = I_B C\left(\frac{1+\cos\Sigma}{2}\right)$$

$$I_{RC} = \rho I_B(\sin\beta + C)\left(\frac{1-\cos\Sigma}{2}\right)$$

$$I_C = I_{BC} + I_{DC} + I_{RC}$$

$$I_C = Ae^{-km}\left[\cos\beta\cos(\phi_S - \phi_C)\sin\Sigma + \sin\beta\cos\Sigma + C\left(\frac{1+\cos\Sigma}{2}\right)\right.$$
$$\left. + \rho(\sin\beta + C)\left(\frac{1-\cos\Sigma}{2}\right)\right]$$

Two-Axis Tracking:

$$I_{BC} = I_B$$

$$I_{DC} = CI_B\left[\frac{1+\cos(90^\circ - \beta)}{2}\right]$$

$$I_{RC} = \rho(I_{BH} + I_{DH})\left[\frac{1-\cos(90^\circ - \beta)}{2}\right]$$

One-Axis, Polar Mount:

$$I_{BC} = I_B\cos\delta$$

$$I_{DC} = CI_B\left[\frac{1+\cos(90^\circ - \beta + \delta)}{2}\right]$$

$$I_{RC} = \rho(I_{BH} + I_{DH})\left[\frac{1-\cos(90^\circ - \beta + \delta)}{2}\right]$$

7.10 MONTHLY CLEAR-SKY INSOLATION

The instantaneous insolation equations just presented can be tabulated into daily, monthly and annual values that provide considerable insight into the impact of collector orientation. For example, Table 7.8 presents monthly and annual clear sky insolation on collectors with various azimuth and tilt angles, as well as for one- and two-axis tracking mounts, for latitude 40°N. They have been computed as the sum of just the beam plus diffuse radiation, which ignores the usually modest reflective contribution. Similar tables for other latitudes are given in

TABLE 7.7 Hour-by-Hour Clear-Sky Insolation in June for Latitude 40°

Solar Time	Tracking		Tilt Angles, Latitude 40°						
	One-Axis	Two-Axis	0	20	30	40	50	60	90
			June 21					(W/m^2)	
6, 6	471	524	188	128	93	57	53	48	32
7, 5	668	742	386	330	289	240	185	126	45
8, 4	772	855	572	538	498	445	380	305	51
9, 3	835	921	731	722	686	632	560	473	147
10, 2	875	961	853	865	834	780	703	607	233
11, 1	898	982	929	956	928	874	795	693	288
12	906	989	955	987	960	906	826	723	308
kWh/d:	9.94	10.96	8.27	8.06	7.62	6.96	6.18	5.23	1.90

Note: Similar tables for other months and latitudes are given in Appendix C

Appendix D. When plotted, as has been done in Fig. 7.28, it becomes apparent that *annual* performance is relatively insensitive to wide variations in collector orientation for nontracking systems. For this latitude, the annual insolation for south-facing collectors varies by less than 10% for collectors mounted with tilt angles ranging anywhere from 10° to 60°. And, only a modest degradation is noted for panels that don't face due south. For a 45° collector azimuth angle (southeast, southwest), the annual clear sky insolation available drops by less than 10% in comparison with south-facing panels at similar tilt angles.

While Fig. 7.28 seems to suggest that orientation isn't critical, remember that it has been plotted for *annual insolation* without regard to monthly distribution. For a grid-connected photovoltaic system, for example, this may be a valid way to consider orientation. Deficits in the winter are automatically offset by purchased utility power, and any extra electricity generated during the summer can simply go back onto the grid. For a stand-alone PV system, however, where batteries or a generator provide back-up power, it is quite important to try to smooth out the month-to-month energy delivered to minimize the size of the back-up system needed in those low-yield months.

A graph of monthly insolation, instead of the annual plots given in Fig. 7.28, shows dramatic variations in the pattern of monthly solar energy for different tilt angles. Such a plot for three different tilt angles at latitude 40°, each having nearly the same annual insolation, is shown in Fig. 7.29. As shown, a collector at the modest tilt angle of 20° would do well in the summer, but deliver very little in the winter, so it wouldn't be a very good angle for a stand-alone PV system. At 40° or 60°, the distribution of radiation is more uniform and would be more appropriate for such systems.

In Fig. 7.30, monthly insolation for a south-facing panel at a fixed tilt angle equal to its latitude is compared with a one-axis polar mount tracker and also

TABLE 7.8 Daily and Annual Clear-Sky Insolation (Beam plus Diffuse) for Various Fixed-Orientation Collectors, Along with one- and Two-Axis Trackers

Daily Clear-Sky Insolation (kWh/m^2) Latitude 40°N																					
Azim:		S						SE/SW						E, W						Tracking	
Tilt:	0	20	30	40	50	60	90	20	30	40	50	60	90	20	30	40	50	60	90	One-Axis	Two-Axis
Jan	3.0	4.6	5.2	5.7	6.0	6.2	5.5	4.1	4.5	4.7	4.9	4.9	4.0	2.9	2.8	2.7	2.6	2.4	1.7	6.8	7.2
Feb	4.2	5.8	6.3	6.6	6.8	6.7	5.4	5.3	5.6	5.7	5.7	5.5	4.2	4.1	3.9	3.7	3.5	3.3	2.2	8.2	8.3
Mar	5.8	6.9	7.2	7.3	7.1	6.8	4.7	6.5	6.6	6.6	6.4	6.0	4.1	5.5	5.3	5.0	4.6	4.3	2.8	9.5	9.5
Apr	7.2	7.7	7.7	7.4	6.9	6.2	3.3	7.5	7.4	7.1	6.6	6.1	3.7	6.9	6.6	6.2	5.7	5.2	3.3	10.3	10.6
May	8.1	8.0	7.7	7.1	6.4	5.5	2.3	8.0	7.6	7.2	6.5	5.8	3.2	7.7	7.3	6.8	6.2	5.5	3.5	10.2	11.0
Jun	8.3	8.1	7.6	7.0	6.2	5.2	1.9	8.0	7.6	7.1	6.4	5.6	3.0	7.8	7.4	6.9	6.3	5.6	3.4	9.9	11.0
July	8.0	7.9	7.6	7.0	6.3	5.5	2.2	7.9	7.5	7.1	6.4	5.7	3.2	7.6	7.2	6.7	6.1	5.5	3.4	10.0	10.7
Aug	7.1	7.5	7.5	7.2	6.7	6.0	3.2	7.3	7.2	6.9	6.5	5.9	3.6	6.7	6.4	6.0	5.5	5.0	3.2	9.8	10.1
Sept	5.6	6.7	6.9	7.0	6.9	6.5	4.5	6.3	6.4	6.3	6.1	5.8	4.0	5.4	5.2	4.9	4.5	4.1	2.7	9.0	9.0
Oct	4.1	5.5	6.0	6.3	6.4	6.4	5.1	5.0	5.3	5.4	5.4	5.2	4.0	3.9	3.7	3.6	3.3	3.1	2.1	7.7	7.8
Nov	2.9	4.5	5.1	5.5	5.8	5.9	5.3	3.9	4.3	4.6	4.7	4.7	3.9	2.8	2.7	2.6	2.5	2.3	1.6	6.5	6.9
Dec	2.5	4.1	4.7	5.2	5.5	5.7	5.2	3.6	3.9	4.2	4.4	4.4	3.8	2.4	2.3	2.2	2.1	2.0	1.4	6.0	6.5
Total	2029	2352	2415	2410	2342	2208	1471	2231	2249	2216	2130	1997	1357	1938	1848	1738	1612	1467	960	3167	3305

Tables for other latitudes are in Appendix D

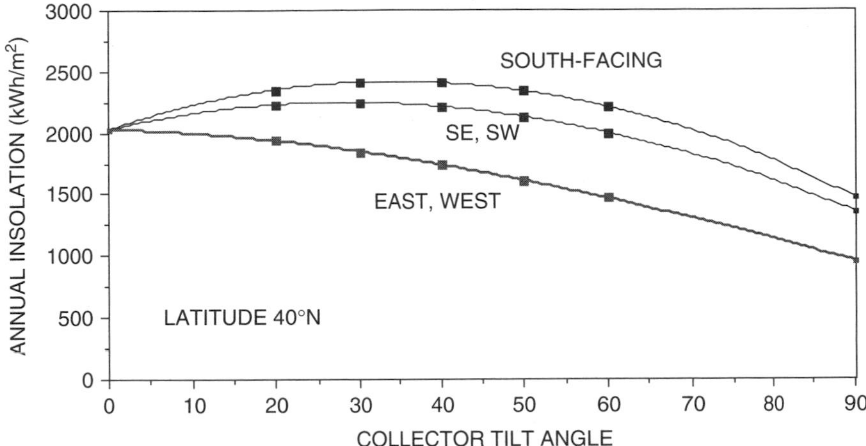

Figure 7.28 Annual insolation, assuming all clear days, for collectors with varying azimuth and tilt angles. Annual amounts vary only slightly over quite a range of collector tilt and azimuth angles.

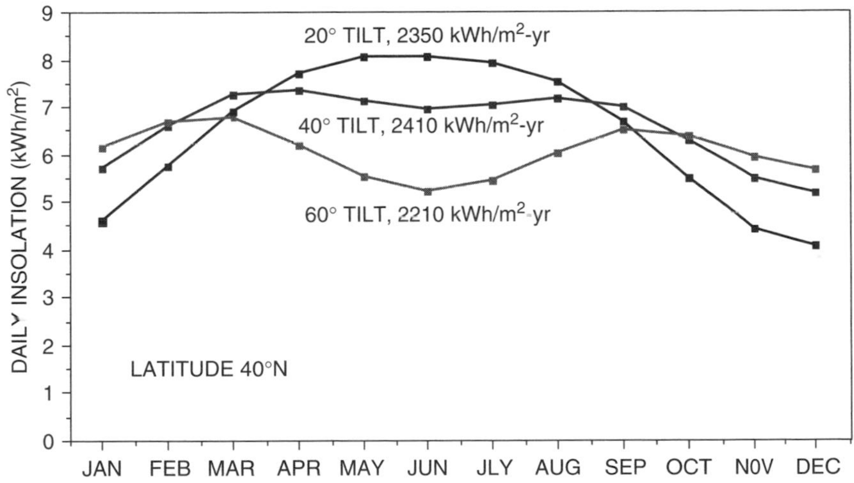

Figure 7.29 Daily clear-sky insolation on south-facing collectors with varying tilt angles. Even though they all yield roughly the same annual energy, the monthly distribution is very different.

a two-axis tracker. The performance boost caused by tracking is apparent: Both trackers are exposed to about one-third more radiation than the fixed collector. Notice, however, that the two-axis tracker is only a few percent better than the single-axis version, with almost all of this improvement occurring in the spring and summer months.

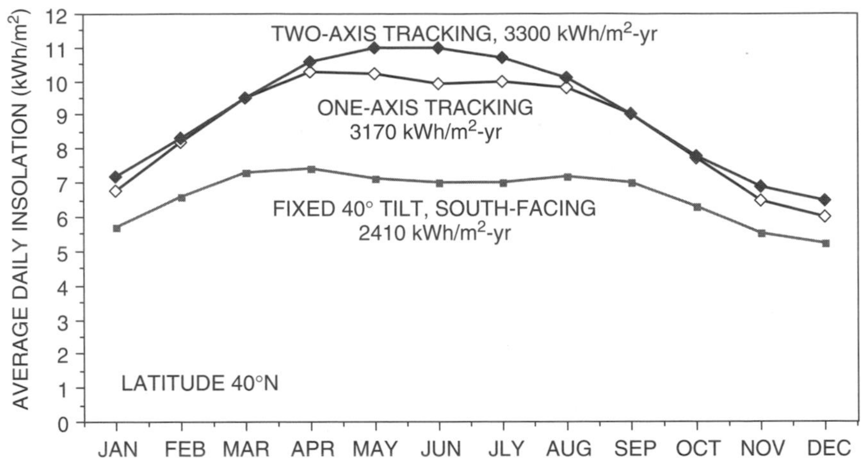

Figure 7.30 Clear sky insolation on a fixed panel compared with a one-axis, polar mount tracker and a two-axis tracker.

7.11 SOLAR RADIATION MEASUREMENTS

Creation of solar energy data bases began in earnest in the United States in the 1970s by the National Oceanic and Atmospheric Administration (NOAA) and later by the National Renewable Energy Laboratory (NREL). NREL has established the National Solar Radiation Data Base (NSRDB) for 239 sites in the United States. Of these, only 56 are primary stations for which long-term solar measurements have been made, while data for the remaining 183 sites are based on estimates derived from models incorporating meteorological data such as cloud cover. Figure 7.31 shows these 239 sites. The World Meteorological Organization (WMO), through its World Radiation Data Center in Russia, compiles data for hundreds of other sites around the world. Cloud mapping data taken by satellite are now a very important complement to the rather sparse global network of ground monitoring stations.

There are two principal types of devices used to measure solar radiation. The most widely used instrument, called a *pyranometer*, measures the total radiation arriving from all directions, including both direct and diffuse components. That is, it measures all of the radiation that is of potential use to a solar collecting system. The other device, called a *pyrheliometer*, looks at the sun through a narrow collimating tube, so it measures only the direct beam radiation. Data collected by pyrheliometers are especially important for focusing collectors since their solar resource is pretty much restricted to just the beam portion of incident radiation.

Pyranometers and pyrheliometers can be adapted to obtain other useful data. For example, as shall be seen in the next section, the ability to sort out the direct from the diffuse is a critical step in the conversion of measured insolation on a horizontal surface into estimates of radiation on tilted collectors. By temporarily affixing a shade ring to block the direct beam, a pyranometer can be used to

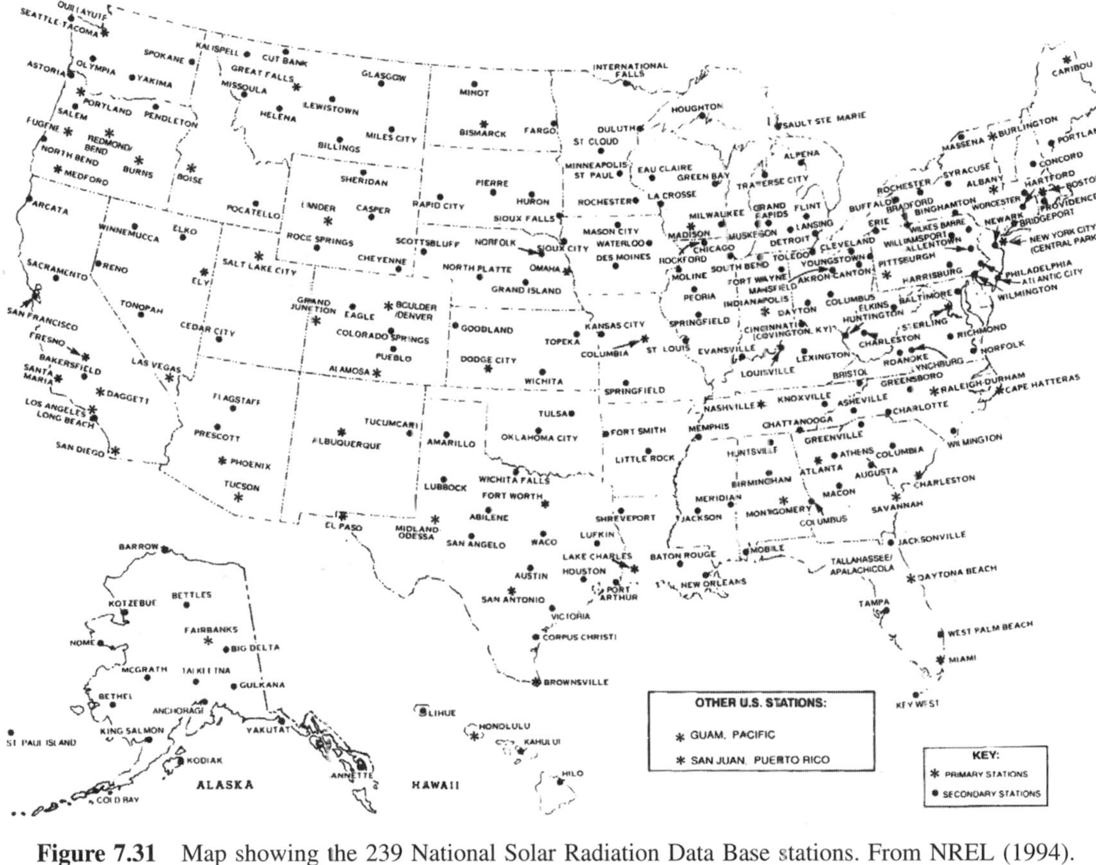

Figure 7.31 Map showing the 239 National Solar Radiation Data Base stations. From NREL (1994).

measure just diffuse radiation (Fig. 7.32). By subtracting the diffuse from the total, the beam portion can then be determined. In other circumstances, it is important to know not only how much radiation the sun provides, but also how much it provides within certain ranges of wavelengths. For example, newspapers now routinely report on the ultraviolet (UV) portion of the spectrum to warn us about skin cancer risks. This sort of data can be obtained by fitting pyranometers or pyrheliometers with filters to allow only certain wavelengths to be measured.

The most important part of a pyranometer or pyrheliometer is the detector that responds to incoming radiation. The most accurate detectors use a stack of thermocouples, called a thermopile, to measure how much hotter a black surface becomes when exposed to sunlight. The most accurate of these incorporate a sensor surface that consists of alternating black and white segments (Fig. 7.33). The thermopile measures the temperature difference between the black segments, which absorb sunlight, and the white ones, which reflect it, to produce a voltage that is proportional to insolation. Other thermopile pyranometers have sensors that are entirely black, and the temperature difference is measured between the case of the pyranometer, which is close to ambient, and the hotter, black sensor.

The alternative approach uses a photodiode sensor that sends a current through a calibrated resistance to produce a voltage proportional to insolation. These pyranometers are less expensive but are also less accurate than those based on thermopiles. Unlike thermopile sensors, which measure all wavelengths of incoming radiation, photoelectric sensors respond to only a limited portion of the solar spectrum. The most popular devices use silicon photosensors, which means that any photons with longer wavelengths than their band gap of 1100 μm don't contribute to the output. Photoelectric pyranometers are calibrated to produce very accurate results under clear skies, but if the solar spectrum is altered, as for

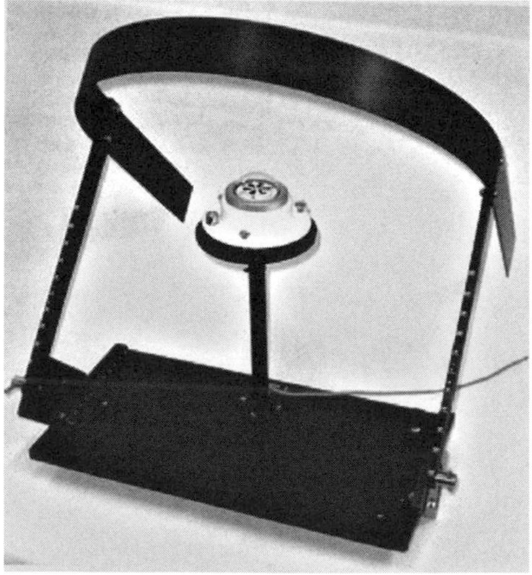

Figure 7.32 Pyranometer with a shade ring to measure diffuse radiation.

(a) (b)

Figure 7.33 (a) A thermopile-type, black-and-white pyranometer and (b) a Li-Cor silicon-cell pyranometer.

example when sunlight passes through glass or clouds, they won't be as accurate as a pyranometer that uses a thermopile sensor. Also, they don't respond accurately to artificial light.

7.12 AVERAGE MONTHLY INSOLATION

It is one thing to be able to compute the insolation on a tilted surface when the skies are clear, but what really is needed is a procedure for estimating the average insolation that can be expected to strike a collector under real conditions at a particular site. The starting point is site-specific, long-term radiation data, which is primarily insolation measured on a horizontal surface. Procedures used to convert these data into expected radiation on a tilted surface depend on being able to sort out what portion of the total measured horizontal insolation $\overline{I}_H$ is diffuse I_{DH} and what portion is direct beam, $\overline{I}_{BH}$.

$$\overline{I}_H = \overline{I}_{DH} + \overline{I}_{BH} \tag{7.41}$$

Once this decomposition has been estimated, adjusting the resulting horizontal diffuse radiation into diffuse and reflected radiation on a collecting surface is straightforward and uses equations already presented. Converting horizontal beam radiation is a little trickier.

Procedures for decomposing total horizontal insolation into its diffuse and beam components begin by defining a clearness index K_T, which is the ratio of the average horizontal insolation at the site $\overline{I}_H$ to the extraterrestrial insolation on a horizontal surface above the site and just outside the atmosphere, $\overline{I}_0$.

$$\text{Clearness index } K_T = \frac{\overline{I}_H}{\overline{I}_0} \tag{7.42}$$

A high clearness index corresponds to clear skies in which most of the radiation will be direct beam while a low one indicates overcast conditions having mostly diffuse insolation.

The average daily extraterrestrial insolation on a horizontal surface $\overline{I}_0$ (kWh/m^2-day) can be calculated by averaging the product of the normal radiation (7.20) and the sine of the solar altitude angle (7.8) from sunrise to sunset, resulting in

$$\overline{I}_0 = \left(\frac{24}{\pi}\right) \mathrm{SC}\left[1 + 0.034\cos\left(\frac{360n}{365}\right)\right](\cos L\cos\delta\sin H_{SR} + H_{SR}\sin L\sin\delta) \tag{7.43}$$

where SC is the solar constant and the sunrise hour angle H_{SR} is in radians.

Usually the clearness index is based on a monthly average, and (7.43) can be computed daily and those values averaged over the month or a day in the middle of the month can be used to represent the average monthly condition. The solar constant SC used here will be 1.37 kW/m^2.

A number of attempts to correlate clearness index and the fraction of horizontal insolation that is diffuse have been made, including Liu and Jordan (1961), and Collares-Pereira and Rabl (1979). The Liu and Jordan correlation is as follows:

$$\frac{\overline{I}_{DH}}{\overline{I}_H} = 1.390 - 4.027K_T + 5.531{K_T}^2 - 3.108{K_T}^3 \tag{7.44}$$

From (7.44), the diffuse portion of horizontal insolation can be estimated. Then, adjusting (7.29) and (7.30) to indicate average daylong values, the average diffuse and reflected radiation on a tilted collector surface can be found from

$$\overline{I}_{DC} = \overline{I}_{DH}\left(\frac{1+\cos\Sigma}{2}\right) \tag{7.45}$$

and

$$\overline{I}_{RC} = \rho\overline{I}_H\left(\frac{1-\cos\Sigma}{2}\right) \tag{7.46}$$

where Σ is the collector slope with respect to the horizontal. Equations (7.45) and (7.46) are sufficient for our purposes, but it should be noted that more complex models that don't require the assumption of an isotropic sky are available (Perez et al., 1990).

Average beam radiation on a horizontal surface can be found by subtracting the diffuse portion $\overline{I}_{DH}$ from the total $\overline{I}_H$. To convert the horizontal beam radiation into beam on the collector $\overline{I}_{BC}$, begin by combining (7.25)

$$I_{BH} = I_B\sin\beta \tag{7.25}$$

with (7.24)

$$I_{BC} = I_B\cos\theta \tag{7.24}$$

to get

$$I_{BC} = I_{BH}\left(\frac{\cos\theta}{\sin\beta}\right) = I_{BH}R_B \tag{7.47}$$

where θ is the incidence angle between the collector and beam, and β is the sun's altitude angle. The quantity in the parentheses is called the *beam tilt factor* R_B.

Equation (7.47) is correct on an instantaneous basis, but since we are working with monthly averages, what is needed is an average value for the beam tilt factor. In the Liu and Jordan procedure, the beam tilt factor is estimated by simply averaging the value of cos θ over those hours of the day in which the sun is in front of the collector and dividing that by the average value of sin β over those hours of the day when the sun is above the horizon. For south-facing collectors at tilt angle Σ, a closed-form solution for those averages can be found and the resulting average beam tilt factor becomes

$$\overline{R}_B = \frac{\cos(L-\Sigma)\cos\delta\sin H_{SRC} + H_{SRC}\sin(L-\Sigma)\sin\delta}{\cos L\cos\delta\sin H_{SR} + H_{SR}\sin L\sin\delta} \tag{7.48}$$

where H_{SR} is the sunrise hour angle (in radians) given in (7.17):

$$H_{SR} = \cos^{-1}(-\tan L\tan\delta) \tag{7.17}$$

H_{SRC} is the sunrise hour angle for the collector (when the sun first strikes the collector face, $\theta = 90°$):

$$H_{SRC} = \min\{\cos^{-1}(-\tan L\tan\delta), \cos^{-1}[-\tan(L-\Sigma)\tan\delta]\} \tag{7.49}$$

Recall that L is the latitude, Σ is the collector tilt angle, and δ is the solar declination (7.6).

To summarize the approach, once the horizontal insolation has been decomposed into beam and diffuse components, it can be recombined into the insolation striking a collector using the following:

$$\overline{I}_C = \overline{I}_H\left(1 - \frac{\overline{I}_{DH}}{\overline{I}_H}\right)\cdot\overline{R}_B + \overline{I}_{DH}\left(\frac{1+\cos\Sigma}{2}\right) + \rho\overline{I}_H\left(\frac{1-\cos\Sigma}{2}\right) \tag{7.50}$$

where $\overline{R}_B$ can be found for south-facing collectors using (7.48).

Example 7.13 Average Monthly Insolation on a Tilted Collector. Average horizontal insolation in Oakland, California (latitude 37.73°N) in July is 7.32 kWh/m^2-day. Estimate the insolation on a south-facing collector at a tilt angle of 30° with respect to the horizontal. Assume ground reflectivity of 0.2.

Solution. Begin by finding mid-month declination and sunrise hour angle for July 16 ($n = 197$):

$$\delta = 23.45 \sin\left[\frac{360}{365}(n-81)\right] = 23.45 \sin\left[\frac{360}{365}(197-81)\right]$$
$$= 21.35^\circ \qquad (7.6)$$
$$H_{SR} = \cos^{-1}(-\tan L \tan \delta) \qquad (7.17)$$
$$= \cos^{-1}(-\tan 37.73^\circ \tan 21.35^\circ) = 107.6^\circ = 1.878 \text{ radians}$$

Using a solar constant of 1.37 kW/m^2, the E.T. horizontal insolation from (7.43) is

$$\overline{I}_0 = \left(\frac{24}{\pi}\right) \text{SC}\left[1 + 0.034\cos\left(\frac{360n}{365}\right)\right](\cos L \cos \delta \sin H_{SR} + H_{SR} \sin L \sin \delta)$$
$$= \left(\frac{24}{\pi}\right) 1.37\left[1 + 0.034\cos\left(\frac{360 \cdot 197}{365}\right)^\circ\right](\cos 37.73 \cos 21.35^\circ \sin 107.6^\circ$$
$$+ 1.878 \sin 37.73^\circ \sin 21.35^\circ)$$
$$= 11.34 \text{ kWh/m}^2\text{-day}$$

From (7.42), the clearness index is

$$K_T = \frac{\overline{I}_H}{\overline{I}_0} = \frac{7.32 \text{ kWh/m}^2 \cdot \text{day}}{11.34 \text{ kWh/m}^2 \cdot \text{day}} = 0.645$$

From (7.44) the fraction diffuse is

$$\frac{\overline{I}_{DH}}{\overline{I}_H} = 1.390 - 4.027K_T + 5.531{K_T}^2 - 3.108{K_T}^3$$
$$= 1.390 - 4.027\ (0.645) + 5.531\ (0.645)^2 - 3.108\ (0.645)^3 = 0.258$$

So, the diffuse horizontal radiation is

$$\overline{I}_{DH} = 0.258 \cdot 7.32 = 1.89 \text{ kWh/m}^2\text{-day}$$

The diffuse radiation on the collector is given by (7.45)

$$\overline{I}_{DC} = \overline{I}_{DH}\left(\frac{1+\cos\Sigma}{2}\right) = 1.89\left(\frac{1+\cos 30^\circ}{2}\right) = 1.76 \text{ kWh/m}^2\text{-day}$$

The reflected radiation on the collector is given by (7.46)

$$\overline{I}_{RC} = \rho\,\overline{I}_H\left(\frac{1-\cos\Sigma}{2}\right) = 0.2 \cdot 7.32\left(\frac{1-\cos 30^\circ}{2}\right) = 0.10 \text{ kWh/m}^2\text{-day}$$

From (7.41), the beam radiation on the horizontal surface is

$$\overline{I}_{BH} = \overline{I}_H - \overline{I}_{DH} = 7.32 - 1.89 = 5.43 \text{ kWh/m}^2\text{-day}$$

To adjust this for the collector tilt, first find the sunrise hour angle on the collector from (7.49)

$$\begin{aligned} H_{SRC} &= \min\{\cos^{-1}(-\tan L \tan\delta),\ \cos^{-1}[-\tan(L-\Sigma)\tan\delta]\} \\ &= \min\{\cos^{-1}(-\tan 37.73^\circ \tan 21.35^\circ),\ \cos^{-1}[-\tan(37.73-30)^\circ \tan 21.35^\circ]\} \\ &= \min\{107.6^\circ, 93.0^\circ\} = 93.0^\circ = 1.624 \text{ radians} \end{aligned}$$

The beam tilt factor (7.48) is thus

$$\begin{aligned} \overline{R}_B &= \frac{\cos(L-\Sigma)\cos\delta \sin H_{SRC} + H_{SRC}\sin(L-\Sigma)\sin\delta}{\cos L \cos\delta \sin H_{SR} + H_{SR}\sin L \sin\delta} \\ &= \frac{\cos(37.73-30)^\circ \cos 21.35^\circ \sin 93^\circ + 1.624 \sin(37.73-30)^\circ \sin 21.35^\circ}{\cos 37.73^\circ \cos 21.35^\circ \sin 107.6^\circ + 1.878 \sin 37.73^\circ \sin 21.35^\circ} \\ &= 0.893 \end{aligned}$$

So the beam insolation on the collector is

$$\overline{I}_{BC} = \overline{I}_{BH}\overline{R}_B = 5.43 \cdot 0.893 = 4.85 \text{ kWh/m}^2\text{-day}$$

Total insolation on the collector is thus

$$\overline{I}_C = \overline{I}_{BC} + \overline{I}_{DC} + \overline{I}_{RC} = 4.85 + 1.76 + 0.10 = 6.7 \text{ kWh/m}^2\text{-day}$$

Clearly, with calculations that are this tedious it is worth spending the time to set up a spreadsheet or other computer analysis or, better still, use precomputed data available on the web or from publications such as the *Solar Radiation Data Manual for Flat-Plate and Concentrating Collectors* (NREL, 1994). An example of the sort of data available from NREL is shown in Table 7.9. Average total radiation data are given for south-facing collectors with various fixed-tilt angles as well as for one-axis and two-axis tracking mounts. In addition, the range of insolations each month is presented, which, along with the figure, gives a good sense of how variable insolation has been during the period in which the actual measurements were made. Also included are values for just the direct-beam portion of radiation for concentrating collectors that can't focus diffuse radiation. The direct-beam data are presented for horizontal collectors in which the tracking rotates about a north–south axis or an east–west axis as well as for tilted, tracking mounts. Horizontal mounts are common in solar–thermal systems that focus sunlight using parabolic troughs (Fig. 7.34).

TABLE 7.9 Average Solar Radiation for Boulder, CO (kWh/m²-day) for South-Facing, Fixed-Tilt Collectors, Tracking Collectors, and Tracking/Focusing Collectors that Operate on Just the Beam Portion of Insolation

Boulder, CO

WBAN NO. 94018

LATITUDE: 40.02° N
LONGITUDE: 105.25° W
ELEVATION: 1634 meters
MEAN PRESSURE: 836 millibars

STATION TYPE: Primary

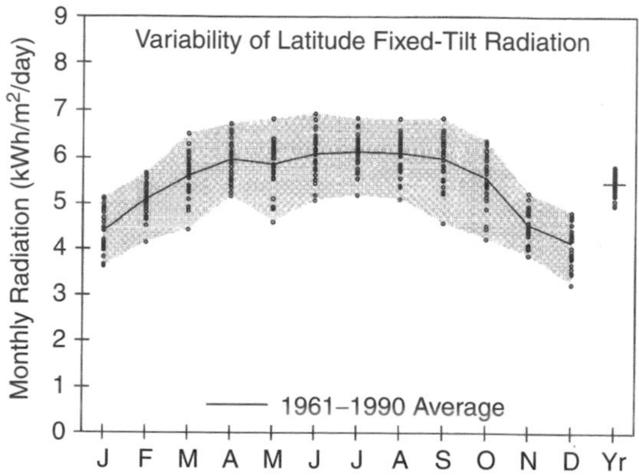

Solar Radiation for Flat-Plate Collectors Facing South at a Fixed Tilt (kWh/m² /day), Uncertainty ±9%

Tilt (°)		Jan	Feb	Mar	Apr	May	June	July	Aug	Sept	Oct	Nov	Dec	Year
0	Average	2.4	3.3	4.4	5.6	6.2	6.9	6.7	6.0	5.0	3.8	2.6	2.1	4.6
	Min/Max	2.1/2.7	2.8/3.5	3.7/5.0	4.8/6.1	5.1/7.2	5.7/7.8	5.6/7.4	5.2/6.6	4.0/5.5	3.1/4.2	2.3/2.8	1.9/2.3	4.3/4.8
Latitude −15	Average	3.8	4.6	5.4	6.1	6.2	6.6	6.6	6.3	5.9	5.1	4.0	3.5	5.4
	Min/Max	3.2/4.4	3.8/5.1	4.3/6.2	5.3/6.8	4.9/7.3	5.5/7.6	5.6/7.4	5.3/7.1	4.6/6.7	4.0/5.8	3.4/4.6	2.8/4.1	4.9/5.7
Latitude	Average	4.4	5.1	5.6	6.0	5.9	6.1	6.1	6.1	6.0	5.6	4.6	4.2	5.5
	Min/Max	3.6/5.1	4.2/5.7	4.4/6.5	5.2/6.7	4.6/6.8	5.1/6.9	5.2/6.8	5.1/6.8	4.6/6.8	4.2/6.4	3.9/5.2	3.2/4.8	5.0/5.8
Latitude +15	Average	4.8	5.3	5.6	5.6	5.2	5.2	5.3	5.5	5.8	5.7	4.8	4.5	5.3
	Min/Max	3.9/5.6	4.3/5.9	4.4/6.5	4.8/6.2	4.1/6.0	4.4/5.9	4.5/5.9	4.6/6.2	4.4/6.6	4.2/6.5	4.1/5.6	3.5/5.3	4.8/5.6
90	Average	4.5	4.6	4.3	3.6	2.8	2.6	2.7	3.2	4.0	4.6	4.4	4.3	3.8
	Min/Max	3.6/5.4	3.7/5.2	3.5/5.0	3.0/4.0	2.3/3.1	2.2/2.8	2.3/2.9	2.7/3.6	3.1/4.6	3.4/5.3	3.7/5.1	3.4/5.2	3.4/4.1

Solar Radiation for 1-Axis Tracking Flat-Plate Collectors with a North–South Axis (kWh/m² /day), Uncertainty ±9%

Axis Tilt (°)		Jan	Feb	Mar	Apr	May	June	July	Aug	Sept	Oct	Nov	Dec	Year
0	Average	3.7	4.9	6.2	7.6	8.2	9.1	9.0	8.2	7.1	5.7	4.0	3.3	6.4
	Min/Max	3.0/4.4	4.1/5.5	4.6/7.4	6.2/8.7	6.2/10.0	7.4/10.9	7.1/10.2	6.7/9.3	5.3/8.3	4.2/6.6	3.4/4.4	2.6/3.9	5.7/6.9
Latitude −15	Average	4.8	5.9	7.0	8.1	8.4	9.1	9.1	8.6	7.9	6.7	5.0	4.4	7.1
	Min/Max	3.8/5.6	4.8/6.7	5.1/8.4	6.6/9.2	6.3/10.2	7.4/10.9	7.1/10.3	7.0/9.8	5.8/9.2	4.8/7.8	4.2/5.7	3.3/5.2	6.2/7.6
Latitude	Average	5.2	6.2	7.2	8.0	8.1	8.8	8.7	8.4	7.9	7.1	5.5	4.9	7.2
	Min/Max	4.2/6.2	5.1/7.1	5.2/8.6	6.6/9.2	6.1/9.9	7.1/10.4	6.8/10.0	6.8/9.6	5.8/9.3	5.0/8.2	4.6/6.3	3.6/5.8	6.3/7.8
Latitude +15	Average	5.5	6.4	7.1	7.7	7.7	8.2	8.2	8.0	7.8	7.1	5.7	5.2	7.1
	Min/Max	4.4/6.6	5.2/7.3	5.2/8.6	6.3/8.9	5.8/9.4	6.6/9.8	6.4/9.3	6.5/9.2	5.6/9.1	5.0/8.3	4.8/6.6	3.8/6.2	6.1/7.6

Solar Radiation for 2-Axis Tracking Flat-Plate Collectors (kWh/m² /day), Uncertainty ±9%

Tracker		Jan	Feb	Mar	Apr	May	June	July	Aug	Sept	Oct	Nov	Dec	Year
2-Axis	Average	5.6	6.4	7.2	8.1	8.5	9.4	9.2	8.6	8.0	7.1	5.7	5.3	7.4
	Min/Max	4.5/6.7	5.2/7.3	5.2/8.6	6.7/9.3	6.4/10.4	7.6/11.1	7.2/10.5	7.0/9.8	5.8/9.3	5.1/8.3	4.8/6.6	3.9/6.3	6.5/8.0

Direct Beam Solar Radiation for Concentrating Collectors (kWh/m² /day), Uncertainty ±8%

Tracker		Jan	Feb	Mar	Apr	May	June	July	Aug	Sept	Oct	Nov	Dec	Year
1-Axis, E–W Horiz Axis	Average	3.5	3.7	3.7	4.0	4.2	5.0	4.9	4.5	4.4	4.3	3.6	3.4	4.1
	Min/Max	2.3/4.6	2.8/4.5	2.1/4.8	2.9/5.0	2.9/5.7	3.5/6.4	3.8/6.1	3.4/5.4	2.8/5.5	2.5/5.2	2.7/4.7	2.0/4.3	3.4/4.5
1-Axis, N–S Horiz Axis	Average	2.6	3.4	4.2	5.3	5.6	6.6	6.5	6.0	5.4	4.3	2.8	2.3	4.6
	Min/Max	1.6/3.4	2.5/4.2	2.2/5.7	3.6/6.4	3.8/7.6	4.8/8.5	4.8/8.1	4.5/7.1	3.4/6.7	2.4/5.3	2.2/3.6	1.3/3.0	3.7/5.1
1-Axis, N–S Tilt = Latitude	Average	3.9	4.5	5.0	5.6	5.5	6.2	6.2	6.1	6.0	5.5	4.1	3.6	5.2
	Min/Max	2.5/5.1	3.4/5.5	2.7/6.6	3.8/6.8	3.7/7.5	4.5/8.0	4.6/7.7	4.6/7.3	3.8/7.6	3.1/6.7	3.1/5.3	2.0/4.6	4.2/5.7
2-Axis	Average	4.1	4.6	5.0	5.7	5.8	6.8	6.7	6.3	6.1	5.6	4.3	4.0	5.4
	Min/Max	2.7/5.4	3.5/5.7	2.7/6.6	3.9/6.9	4.0/7.9	4.9/8.7	4.9/8.3	4.8/7.5	3.8/7.6	3.2/6.8	3.3/5.6	2.2/5.0	4.3/6.0

Note: Additional tables are in Appendix E.
Source: NREL (1994).

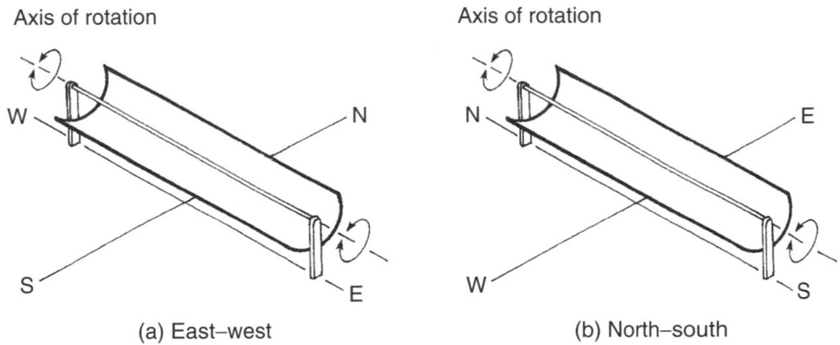

Figure 7.34 One-axis tracking parabolic troughs with horizontal axis oriented east–west or north–south. Most are oriented north–south.

TABLE 7.10 Sample of the Solar Data from Appendix E

	Los Angeles, CA: Latitude 33.93°N												
Tilt	Jan	Feb	Mar	Apr	May	June	July	Aug	Sept	Oct	Nov	Dec	Year
Lat − 15	3.8	4.5	5.5	6.4	6.4	6.4	7.1	6.8	5.9	5.0	4.2	3.6	5.5
Lat	4.4	5.0	5.7	6.3	6.1	6.0	6.6	6.6	6.0	5.4	4.7	4.2	5.6
Lat + 15	4.7	5.1	5.6	5.9	5.4	5.2	5.8	6.0	5.7	5.5	5.0	4.5	5.4
90	4.1	4.1	3.8	3.3	2.5	2.2	2.4	3.0	3.6	4.2	4.3	4.1	3.5
1-Axis (Lat)	5.1	6.0	7.1	8.2	7.8	7.7	8.7	8.4	7.4	6.6	5.6	4.9	7.0
Temp (°C)	18.7	18.8	18.6	19.7	20.6	22.2	24.1	24.8	24.8	23.6	21.3	18.8	21.3

Solar data from the NREL Solar Radiation Manual have been reproduced in Appendix E, a sample of which is shown in Table 7.10.

Radiation data for Boulder are plotted in Fig. 7.35. As was the case for clear-sky graphs presented earlier, there is little difference in annual insolation for fixed, south-facing collectors over a wide range of tilt angles, but the seasonal variation is significant. The boost associated with single-axis tracking is large, about 30%.

Maps of the seasonal variation in insolation, such as that shown in Fig. 7.36, provides a rough indication of the solar resource and are useful when more specific local data are not conveniently available. Analogous figures for the entire globe are included in Appendix F. The units in these figures are average kWh/m^2-day of insolation, but there is another way to interpret them. On a bright, sunny day with the sun high in the sky, the insolation at the earth's surface is roughly 1 kW/m^2. In fact, that convenient value, 1 kW/m^2, is defined to be *1-sun of insolation*. That means, for example, that an average daily insolation of say 5.5 kWh/m^2 is equivalent to 1 kW/m^2 (1-sun) for 5.5 h; that is, it is the same as

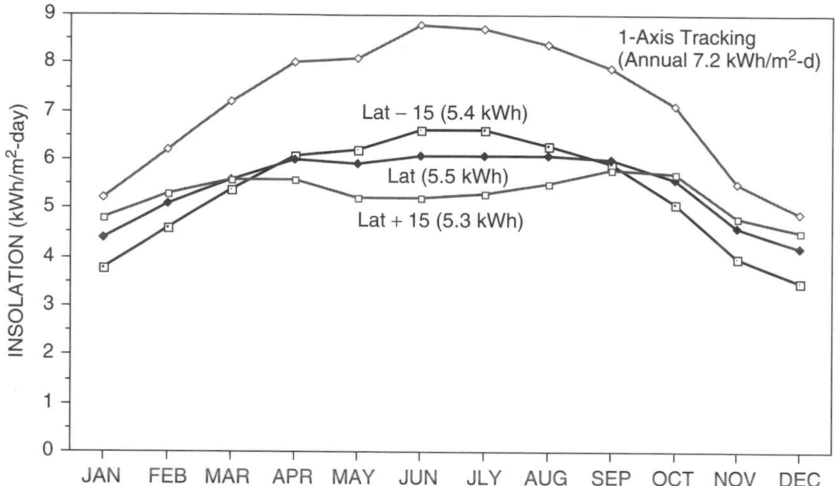

Figure 7.35 Insolation on south-facing collectors in Boulder, CO, at tilt angles equal to the latitude and latitude $\pm 15^\circ$. Values in parentheses are annual averages (kWh/m^2-day). The one-axis tracker with tilt equal to the latitude delivers about 30% more annual energy.

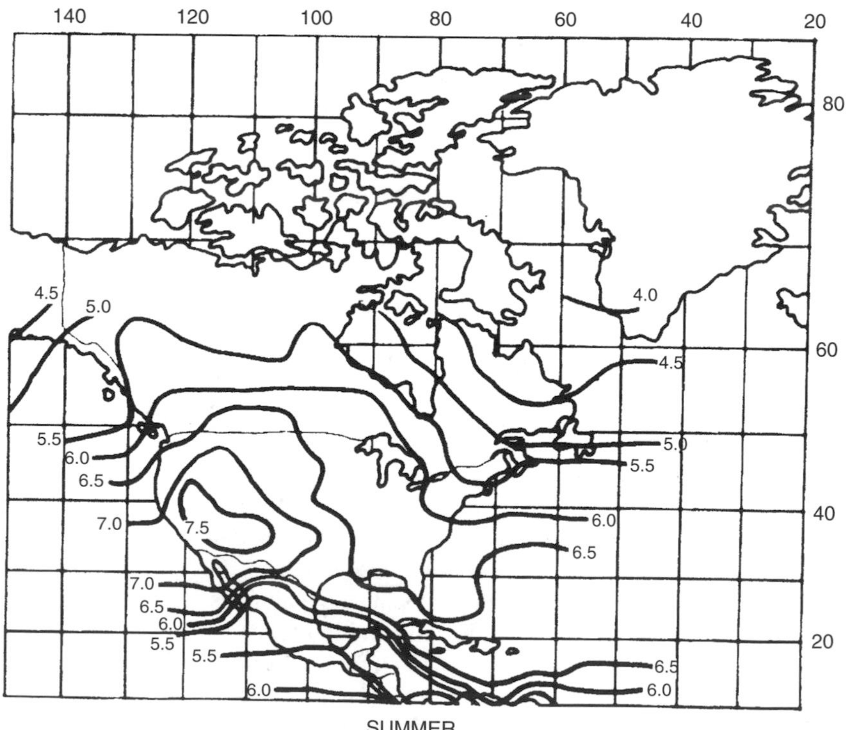

Figure 7.36 Solar radiation for south-facing collectors with tilt angle equal to L – 15° in summer (kWh/m^2-day). From Sandia National Laboratories (1987).

5.5 h of full sun. The units on these radiation maps can therefore be thought of as "hours of full sun." As will be seen in the next chapters on photovoltaics, the hours-of-full-sun approach is central to the analysis and design of PV systems.

REFERENCES

The American Ephemeris and Nautical Almanac, published annually by the Nautical Almanac Office, U.S. Naval Observatory, Washington, D.C.

ASHRAE, 1993, *Handbook of Fundamentals*, American Society of Heating, Refrigeration and Air Conditioning Engineers, Atlanta.

Collares Pereira, M., and A. Rabl (1979). The Average Distribution of Solar Radiation—Correlation Between Diffuse and Hemispherical, *Solar Energy*, vol. 22, pp. 155–166.

Kuen, T. H., J. W. Ramsey, and J. L. Threlkeld, 1998, *Thermal Environmental Engineering*, 3rd ed., Prentice-Hall, Englewood Cliffs, NJ.

Liu, B. Y. H., and R. C. Jordan (1961). Daily Insolation on Surfaces Tilted Toward the Equator, *Trans. ASHRAE*, vol. 67, pp. 526–541.

NREL, 1994. *Solar Radiation Data Manual for Flat-Plate and Concentrating Collectors*, NREL/TP-463-5607, National Renewable Energy Laboratory, Golden, CO.

Perez, R., P. Ineichen, R. Seals, J. Michalsky, and R. Stewart (1990). Modeling Daylight Availability and Irradiance Components from Direct and Global Irradiance, *Solar Energy*, vol. 44, no. 5, pp. 271–289.

Sandia National Laboratories (1987). *Water Pumping, The Solar Alternative*, SAND87-0804, M. Thomas, Albuquerque, NM.

Threlkeld, J. L., and R. C. Jordan (1958). *Trans. ASHRAE*, vol. 64, p. 45.

U.S. Department of Energy (1978). *On the Nature and Distribution of Solar Radiation*, Report No. HCP/T2552-01, UC-59, 62, 63A, March.

PROBLEMS

7.1 Using (7.5), determine the following:

a. The date on which the earth will be a maximum distance from the sun.

b. The date on which it will be a minimum distance from the sun.

7.2 What does (7.6) predict for the date of the following:

a. The two equinox dates

b. The two solstice dates

7.3 At what angle should a South-facing collector at 36° latitude be tipped up to in order to have it be normal to the sun's rays at solar noon on the following dates:

a. March 21

b. January 1

c. April 1

7.4 Consider June 21st (the solstice) in Seattle (latitude 47°).

a. Use (7.11) to help find the time of day (solar time) at which the sun will be due West.

b. At that time, what will the altitude angle of the sun be?

c. As a check on the validity of (7.11), use your answers from (a) and (b) in (7.8) and (7.9) to be sure they yield an azimuth angle of 90°.

7.5 Find the altitude angle and azimuth angle of the sun at the following (solar) times and places:

a. March 1st at 10:00 A.M. in New Orleans, latitude 30°N.

b. July 1st at 5:00 P.M. in San Francisco, latitude 38°

c. December 21st at 11 A.M. at latitude 68°

7.6 Suppose you are concerned about how much shading a tree will cause for a proposed photovoltaic system. Standing at the site with your compass and plumb bob, you estimate the altitude angle of the top of the tree to be about 30° and the width of the tree to have azimuth angles that range from about 30° to 45° West of South. Your site is at latitude 32°.
Using a sun path diagram (Appendix B), describe the shading problem the tree will pose (approximate shaded times each month).

7.7 Suppose you are concerned about a tall thin tree located 100 ft from a proposed PV site. You don't have a compass or protractor and plumb bob, but you do notice that an hour before solar noon on June 21, it casts a 30-ft shadow directly toward your site. Your latitude is 32°N.

a. How tall is the tree?

b. What is its azimuth angle with respect to your site?

c. What are the first and last days in the year when the shadow will land on the site?

7.8 Using Figure 7.16, what is the greatest difference between local standard time for the following locations and solar time? At approximately what date would that occur?

a. San Francisco, CA (longitude 122°, Pacific Time Zone)

b. Boston, MA (longitude 71.1°, Eastern Time Zone)

c. Boulder, CO (longitude 105.3°, Mountain Time Zone)

d. Greenwich, England (longitude 0°, Local time meridian 0°)

7.9 Using Figure 7.16, roughly what date(s) would local time be the same as solar time in the cities described in Problem 7.8?

7.10 Find the local Daylight Savings Time for geometric sunrise in Seattle (latitude 47°, longitude 123 °W) on the summer solstice ($n = 172$).

7.11 Find the local Daylight Savings time at which the upper limb of the sun will emerge at sunrise in Seattle (latitude 47°, longitude 123°) on the summer solstice.

7.12 Equations for solar angles, along with a few simple measurements and an accurate clock, can be used for rough navigation purposes. Suppose you know it is June 22 ($n = 172$, the solstice), your watch, which is set for Pacific Standard Time tells you that (geometric) sunrise occurred at 4:23 A.M. and sunset was 15 hs 42 min later. Ignoring refraction and assuming you measured geometric sunrise and sunset (mid-point in the sun), do the following calculations to find your latitude and longitude.

a. Knowing that solar noon occurs midway between sunrise and sunset, at what time would your watch tell you it is solar noon?

b. Use (7.14) to determine your longitude.

c. Use (7.17) to determine your latitude.

7.13 Following the procedure outlined in Example 7.7, you are to determine your location if on January 1, the newspaper says sunrise is 7:50 A.M. and sunset is at 3:50 P.M. (both Central Standard Time). Solar declination $\delta = -23.0^\circ$ and Figure 7.16 (or by calculation) $E = -3.6$ minutes.

a. What clock time is solar noon?

b. Use (7.14) to determine your longitude.

c. Using (7.16), estimate the latitude without using the Q correction

d. Estimate Q and from that find the clock time at which geometric sunrise occurs.

e. Use (7.17) to determine your latitude.

7.14 A south-facing collector at latitude 40° is tipped up at an angle equal to its latitude. Compute the following insolations for January 1st at solar noon:

a. The direct beam insolation normal to the sun's rays.

b. Beam insolation on the collector.

c. Diffuse radiation on the collector.

d. Reflected radiation on the collector with ground reflectivity 0.2.

7.15 Create a "Clear Sky Insolation Calculator" for direct and diffuse radiation using the following spreadsheet as a guide. In this example, the insolation has been computed to be 964 W/m^2 for a South-facing collector tipped up at 45° at noon on November 7 at latitude 37.5°. Note the third column simply adjusts angles measured in degrees to radians.
Use the calculator to compute clear sky insolation under the following conditions:

a. January 1, latitude 40°, horizontal insolation, solar noon

b. March 21, latitude 20°, South-facing collector with tilt 20°, 11:00 A.M. (solar time)

c. July 1, latitude 48°, South-East facing collector (azimuth 45°), tilt 20°, 2 P.M. (solar time)

Clear Sky Insolation Calculator:			
		radians	Radians = 0.017453292| × degrees
Day number n	311		ENTER (excel uses radians)
Latitude (L)	37.5	0.6545	ENTER
Collector azimuth (ϕ_C)	0	0	ENTER (+ is east of south)
Collector tilt (Σ)	45	0.7854	ENTER
Solar time (ST) (24 h)	12		ENTER
Hour angle H	0	0	$H = 15°/\text{h} \times (12 - \text{ST})$ (7.10)
Declination (δ)	−17.11	−0.2986	$\delta = 23.45 \sin(360/365\ (n - 81))$ (7.6)
Altitude angle (β)	35.39	0.6177	$\beta = \text{ASIN}((\cos L \cos\delta \cos H + \sin L \sin\delta)$ (7.8)
Solar azimuth (ϕ_S)	0.000	0	$\phi_S = \text{ASIN}(\cos\delta \sin H / \cos\beta)$ (7.9)
Air mass ratio (m)	1.727		$\text{m} = 1/\sin\beta$ (7.4)
A (W/m^2)	1204		$\text{A} = 1160 + 75 \sin(360/365\ (n - 275))$ (7.22)
k	0.158		$k = 0.174 + 0.035 \sin(360/365(n - 100))$ (7.23)
I_B (W/m^2)	917		$I_B = A \exp(-km)$ (7.21)
$\cos\theta$	0.986		$\cos\theta = \cos\beta \cos(\phi_s - \phi_c) \sin(\Sigma) + \sin\beta \cos(\Sigma)$ (7.26)
I_{BC} (W/m^2)	904		$I_{BC} = I_B \cos\theta$ (7.24)
C	0.076		$C = 0.095 + 0.04 \sin(360/365(n - 100))$ (7.28)
I_{DC} (W/m^2)	60		$I_{DC} = C I_B (1 + \cos\Sigma)/2$ (7.29)
$I_C = I_{BC} + I_{DC}$ (W/m^2)	964		$I_C = I_{BC} + I_{DC}$ (ignoring reflection) (7.32)

7.16 Air Mass AM1.5 is supposedly the basis for a standard 1-sun insolation of 1 kW/m^2. To see whether this is reasonable, compute the following for a clear day on March 21st:

a. What solar altitude angle gives AM1.5?

b. What would be the direct beam radiation normal to the sun's rays?

c. What would be the diffuse radiation on a collector normal to the rays?

d. What would be the reflected radiation on a collector normal to the rays with $\rho = 0.2$?

e. What would be the total insolation normal to the rays?

7.17 Consider a comparison between a south-facing photovoltaic (PV) array with a tilt equal to its latitude located in Los Angeles versus one with a polar-mount, single-axis tracker. Assuming the PVs are 10% efficient at converting sunlight into electricity:

Figure P7.17

a. For a house that needs 4000-kWh per year, how large would each array need to be?

b. If the PVs cost \$400/m^2 and everything else in the two systems has the same cost except for the extra cost of the tracker, how much can the tracker cost (\$) to make the systems cost the same amount? How much per unit area of tracker (\$/m^2)?

c. Derive a general expression for the justifiable extra cost of a tracker per unit area (\$/m^2) as a function of the PV cost (\$/m^2) and the ratio of tracker insolation I_T to fixed insolation I_F.

CHAPTER 8

PHOTOVOLTAIC MATERIALS AND ELECTRICAL CHARACTERISTICS

8.1 INTRODUCTION

A material or device that is capable of converting the energy contained in photons of light into an electrical voltage and current is said to be *photovoltaic*. A photon with short enough wavelength and high enough energy can cause an electron in a photovoltaic material to break free of the atom that holds it. If a nearby electric field is provided, those electrons can be swept toward a metallic contact where they can emerge as an electric current. The driving force to power photovoltaics comes from the sun, and it is interesting to note that the surface of the earth receives something like 6000 times as much solar energy as our total energy demand.

The history of photovoltaics (PVs) began in 1839 when a 19-year-old French physicist, Edmund Becquerel, was able to cause a voltage to appear when he illuminated a metal electrode in a weak electrolyte solution (Becquerel, 1839). Almost 40 years later, Adams and Day were the first to study the photovoltaic effect in solids (Adams and Day, 1876). They were able to build cells made of selenium that were 1% to 2% efficient. Selenium cells were quickly adopted by the emerging photography industry for photometric light meters; in fact, they are still used for that purpose today.

As part of his development of quantum theory, Albert Einstein published a theoretical explanation of the photovoltaic effect in 1904, which led to a Nobel

Renewable and Efficient Electric Power Systems. By Gilbert M. Masters
ISBN 0-471-28060-7

Prize in 1923. About the same time, in what would turn out to be a cornerstone of modern electronics in general, and photovoltaics in particular, a Polish scientist by the name of Czochralski began to develop a method to grow perfect crystals of silicon. By the 1940s and 1950s, the Czochralski process began to be used to make the first generation of single-crystal silicon photovoltaics, and that technique continues to dominate the photovoltaic (PV) industry today.

In the 1950s there were several attempts to commercialize PVs, but their cost was prohibitive. The real emergence of PVs as a practical energy source came in 1958 when they were first used in space for the Vanguard I satellite. For space vehicles, cost is much less important than weight and reliability, and solar cells have ever since played an important role in providing onboard power for satellites and other space craft. Spurred on by the emerging energy crises of the 1970s, the development work supported by the space program began to pay off back on the ground. By the late 1980s, higher efficiencies (Fig. 8.1) and lower costs (Fig. 8.2) brought PVs closer to reality, and they began to find application in many off-grid terrestrial applications such as pocket calculators, off-shore buoys, highway lights, signs and emergency call boxes, rural water pumping, and small home systems. While the amortized cost of photovoltaic power did drop dramatically in the 1990s, a decade later it is still about double what it needs to be to compete without subsidies in more general situations.

By 2002, worldwide production of photovoltaics had approached 600 MW per year and was increasing by over 40% per year (by comparison, global wind power sales were 10 times greater). However, as Fig. 8.3 shows, the U.S. share of this rapidly growing PV market has been declining and was, at the turn of the century,

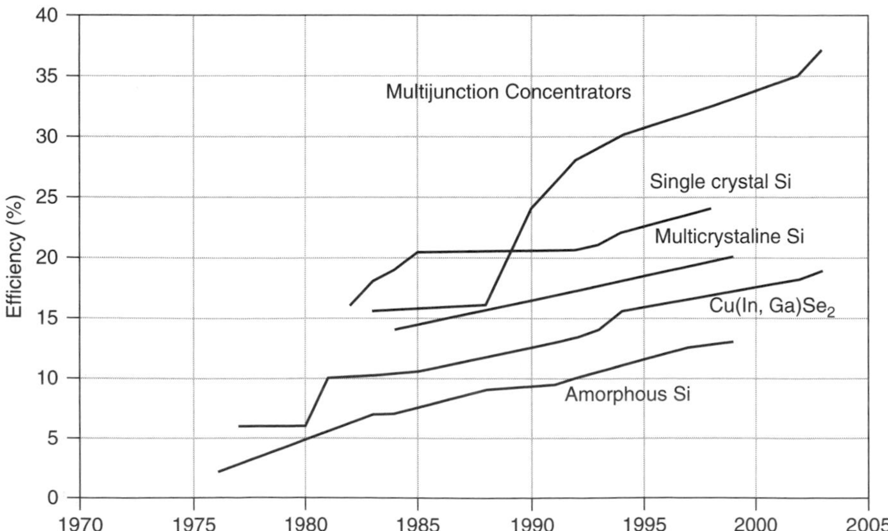

Figure 8.1 Best laboratory PV cell efficiencies for various technologies. (From National Center for Photovoltaics, www.nrel.gov/ncpv 2003).

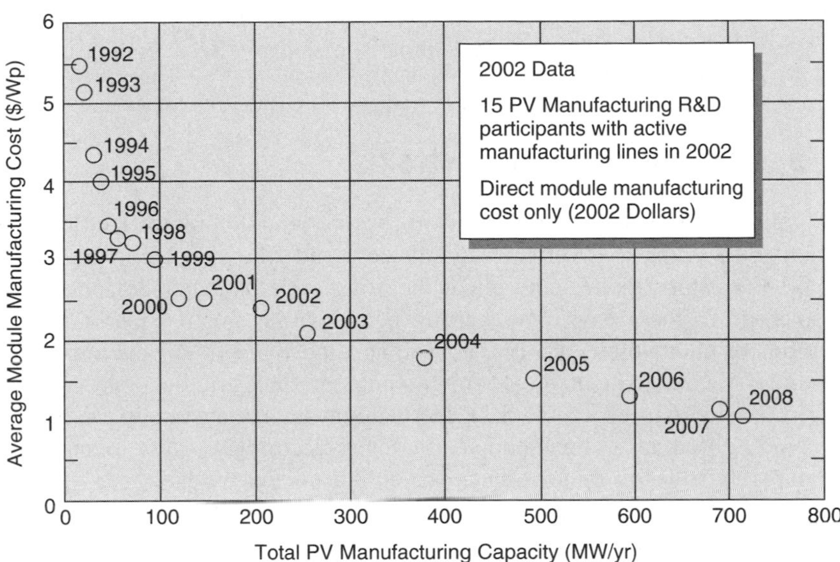

Figure 8.2 PV module manufacturing costs for DOE/US Industry Partners. Historical data through 2002, projections thereafter (www.nrel.gov/pvmat).

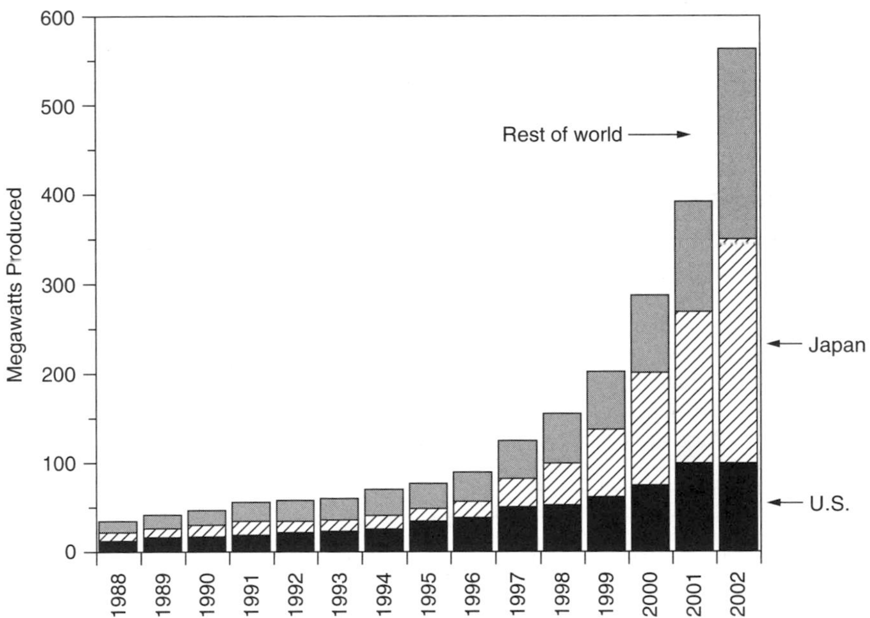

Figure 8.3 World production of photovoltaics is growing rapidly, but the U.S. share of the market is decreasing. Based on data from Maycock (2004).

less than 20% of the total. Critics of this decline point to the government's lack of enthusiasm to fund PV R&D. By comparison, Japan's R&D budget is almost an order of magnitude greater.

8.2 BASIC SEMICONDUCTOR PHYSICS

Photovoltaics use semiconductor materials to convert sunlight into electricity. The technology for doing so is very closely related to the solid-state technologies used to make transistors, diodes, and all of the other semiconductor devices that we use so many of these days. The starting point for most of the world's current generation of photovoltaic devices, as well as almost all semiconductors, is pure crystalline silicon. It is in the fourth column of the periodic table, which is referred to as Group IV (Table 8.1). Germanium is another Group IV element, and it too is used as a semiconductor in some electronics. Other elements that play important roles in photovoltaics are boldfaced. As we will see, boron and phosphorus, from Groups III and V, are added to silicon to make most PVs. Gallium and arsenic are used in GaAs solar cells, while cadmium and tellurium are used in CdTe cells.

Silicon has 14 protons in its nucleus, and so it has 14 orbital electrons as well. As shown in Fig. 8.4a, its outer orbit contains four valence electrons—that is, it is tetravalent. Those valence electrons are the only ones that matter in electronics, so it is common to draw silicon as if it has a +4 charge on its nucleus and four tightly held valence electrons, as shown in Fig. 8.4b.

In pure crystalline silicon, each atom forms covalent bonds with four adjacent atoms in the three-dimensional tetrahedral pattern shown in Fig. 8.5a. For convenience, that pattern is drawn as if it were all in a plane, as in Fig. 8.5b.

8.2.1 The Band Gap Energy

At absolute zero temperature, silicon is a perfect electrical insulator. There are no electrons free to roam around as there are in metals. As the temperature increases,

TABLE 8.1 The Portion of the Periodic Table of Greatest Importance for Photovoltaics Includes the Elements Silicon, Boron, Phosphorus, Gallium, Arsenic, Cadmium, and Tellurium

I	II	III	IV	V	VI
		5 B	6 C	7 N	8 O
		13 Al	**14 Si**	**15 P**	16 S
29 Cu	30 Zn	**31 Ga**	32 Ge	**33 As**	34 Se
47 Ag	**48 Cd**	49 In	50 Sn	51 Sb	**52 Te**

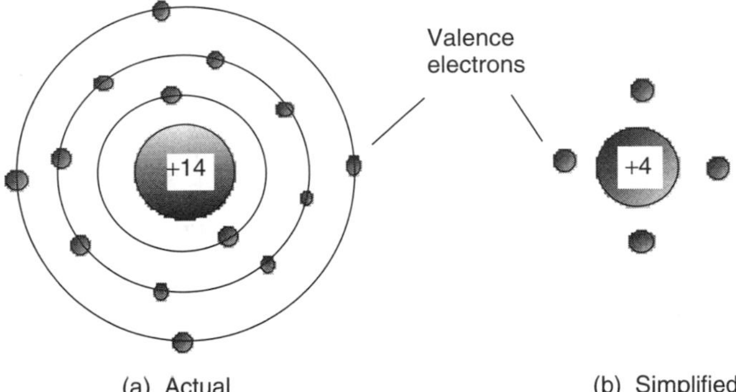

Figure 8.4 Silicon has 14 protons and electrons as in (a). A convenient shorthand is drawn in (b), in which only the four outer electrons are shown, spinning around a nucleus with a +4 charge.

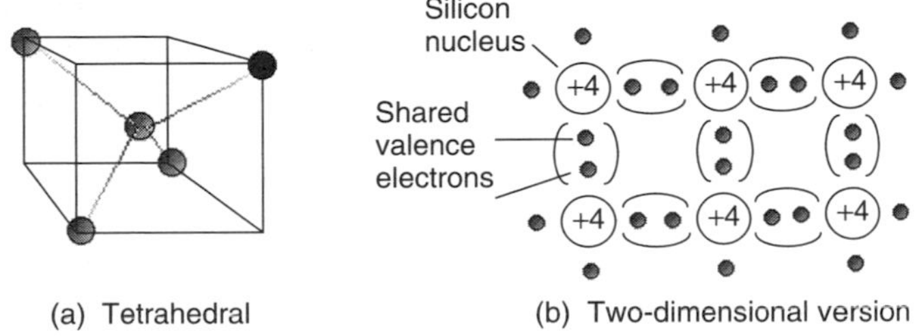

Figure 8.5 Crystalline silicon forms a three-dimensional tetrahedral structure (a); but it is easier to draw it as a two-dimensional flat array (b).

some electrons will be given enough energy to free themselves from their nuclei, making them available to flow as electric current. The warmer it gets, the more electrons there are to carry current, so its conductivity increases with temperature (in contrast to metals, where conductivity decreases). That change in conductivity, it turns out, can be used to advantage to make very accurate temperature sensors called *thermistors*. Silicon's conductivity at normal temperatures is still very low, and so it is referred to as a semiconductor. As we will see, by adding minute quantities of other materials, the conductivity of pure (intrinsic) semiconductors can be greatly increased.

Quantum theory describes the differences between conductors (metals) and semiconductors (e.g., silicon) using energy-band diagrams such as those shown in Fig. 8.6. Electrons have energies that must fit within certain allowable energy bands. The top energy band is called the conduction band, and it is electrons within this region that contribute to current flow. As shown in Fig. 8.6, the conduction band for metals is partially filled, but for semiconductors at absolute zero temperature, the conduction band is empty. At room temperature, only about one out of 10^{10} electrons in silicon exists in the conduction band.

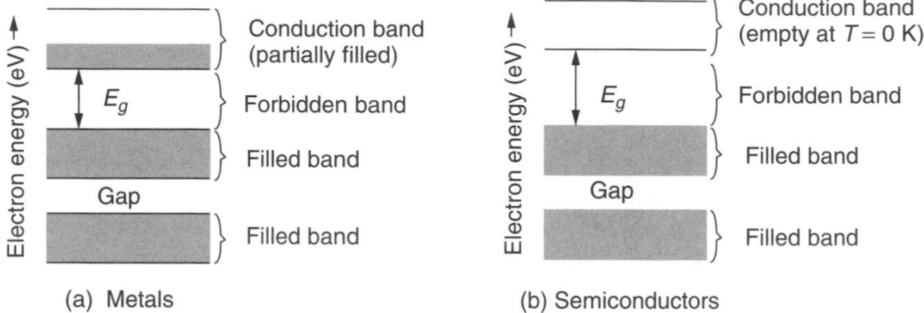

Figure 8.6 Energy bands for (a) metals and (b) semiconductors. Metals have partially filled conduction bands, which allows them to carry electric current easily. Semiconductors at absolute zero temperature have no electrons in the conduction band, which makes them insulators.

The gaps between allowable energy bands are called forbidden bands, the most important of which is the gap separating the conduction band from the highest filled band below it. The energy that an electron must acquire to jump across the forbidden band to the conduction band is called the band-gap energy, designated E_g. The units for band-gap energy are usually electron-volts (eV), where one electron-volt is the energy that an electron acquires when its voltage is increased by 1 V (1 eV $= 1.6 \times 10^{-19}$ J).

The band-gap E_g for silicon is 1.12 eV, which means an electron needs to acquire that much energy to free itself from the electrostatic force that ties it to its own nucleus—that is, to jump into the conduction band. Where might that energy come from? We already know that a small number of electrons get that energy thermally. For photovoltaics, the energy source is photons of electromagnetic energy from the sun. When a photon with more than 1.12 eV of energy is absorbed by a solar cell, a single electron may jump to the conduction band. When it does so, it leaves behind a nucleus with a +4 charge that now has only three electrons attached to it. That is, there is a net positive charge, called a *hole*, associated with that nucleus as shown in Fig. 8.7a. Unless there is some way to sweep the electrons away from the holes, they will eventually recombine, obliterating both the hole and electron as in Fig. 8.7b. When recombination occurs, the energy that had been associated with the electron in the conduction band is released as a photon, which is the basis for light-emitting diodes (LEDs).

It is important to note that not only is the negatively charged electron in the conduction band free to roam around in the crystal, but the positively charged hole left behind can also move as well. A valence electron in a filled energy band can easily move to fill a hole in a nearby atom, without having to change energy bands. Having done so, the hole, in essence, moves to the nucleus from which the electron originated, as shown in Fig. 8.8. This is analogous to a student leaving her seat to get a drink of water. A roaming student (electron) and a seat (hole) are created. Another student already seated might decide he wants that newly

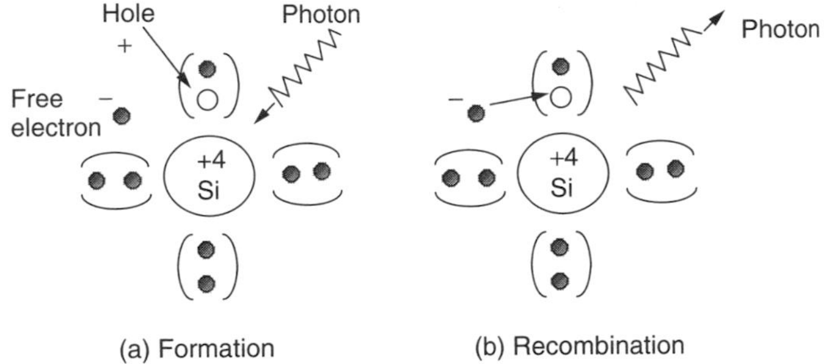

Figure 8.7 A photon with sufficient energy can create a hole–electron pair as in (a). The electron can recombine with the hole, releasing a photon of energy (b).

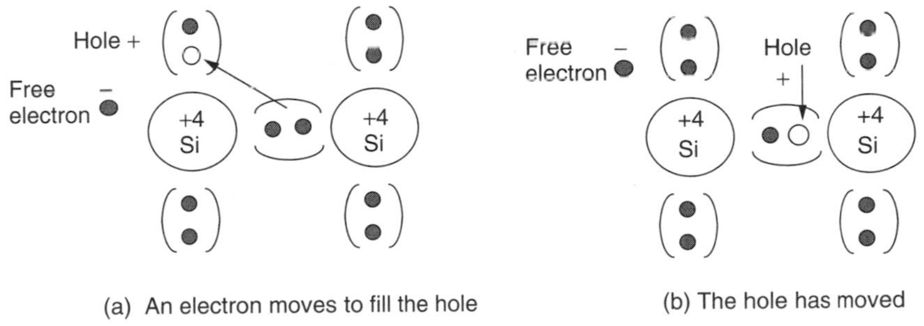

Figure 8.8 When a hole is filled by a nearby valence electron, the hole appears to move.

vacated seat, so he gets up and moves, leaving his seat behind. The empty seat appears to move around just the way a hole moves around in a semiconductor. The important point here is that electric current in a semiconductor can be carried not only by negatively charged electrons moving around, but also by positively charged holes that move around as well.

Thus, photons with enough energy create hole–electron pairs in a semiconductor. Photons can be characterized by their wavelengths or their frequency as well as by their energy; the three are related by the following:

$$c = \lambda v \tag{8.1}$$

where c is the speed of light (3×10^8 m/s), v is the frequency (hertz), λ is the wavelength (m), and

$$E = hv = \frac{hc}{\lambda} \tag{8.2}$$

where E is the energy of a photon (J) and h is Planck's constant (6.626×10^{-34} J-s).

Example 8.1 Photons to Create Hole–Electron Pairs in Silicon What maximum wavelength can a photon have to create hole–electron pairs in silicon? What minimum frequency is that? Silicon has a band gap of 1.12 eV and 1 eV = 1.6×10^{-19} J.

Solution. From (8.2) the wavelength must be less than

$$\lambda \leq \frac{hc}{E} = \frac{6.626 \times 10^{-34}\ \text{J} \cdot \text{s} \times 3 \times 10^{8}\ \text{m/s}}{1.12\ \text{eV} \times 1.6 \times 10^{-19}\text{J/eV}} = 1.11 \times 10^{-6}\ \text{m} = 1.11\ \mu\text{m}$$

and from (8.1) the frequency must be at least

$$\nu \geq \frac{c}{\lambda} = \frac{3 \times 10^{8}\ \text{m/s}}{1.11 \times 10^{-6}\ \text{m}} = 2.7 \times 10^{14}\ \text{Hz}$$

For a silicon photovoltaic cell, photons with wavelength greater than 1.11 μm have energy $h\nu$ less than the 1.12-eV band-gap energy needed to excite an electron. None of those photons create hole–electron pairs capable of carrying current, so all of their energy is wasted. It just heats the cell. On the other hand, photons with wavelengths shorter than 1.11 μm have more than enough energy to excite an electron. Since one photon can excite *only* one electron, any extra energy above the 1.12 eV needed is also dissipated as waste heat in the cell. Figure 8.9 uses a plot of (8.2) to illustrate this important concept. The band gaps for other photovoltaic materials—gallium arsenide (GaAs), cadmium telluride (CdTe), and indium phosphide (InP), in addition to silicon—are shown in Table 8.2.

These two phenomena relating to photons with energies above and below the actual band gap establish a maximum theoretical efficiency for a solar cell. To explore this constraint, we need to introduce the solar spectrum.

8.2.2 The Solar Spectrum

As was described in the last chapter, the surface of the sun emits radiant energy with spectral characteristics that well match those of a 5800 K blackbody. Just outside of the earth's atmosphere, the average radiant flux is about 1.377 kW/m^2, an amount known as the solar constant. As solar radiation passes through the atmosphere, some is absorbed by various constituents in the atmosphere, so that by the time it reaches the earth's surface the spectrum is significantly distorted.

The amount of solar energy reaching the ground, as well as its spectral distribution, depends very much on how much atmosphere it has had to pass through to get there. Recall that the length of the path taken by the sun's rays through the atmosphere to reach a spot on the ground, divided by the path length corresponding to the sun directly overhead, is called the *air mass ratio*, m. Thus, an

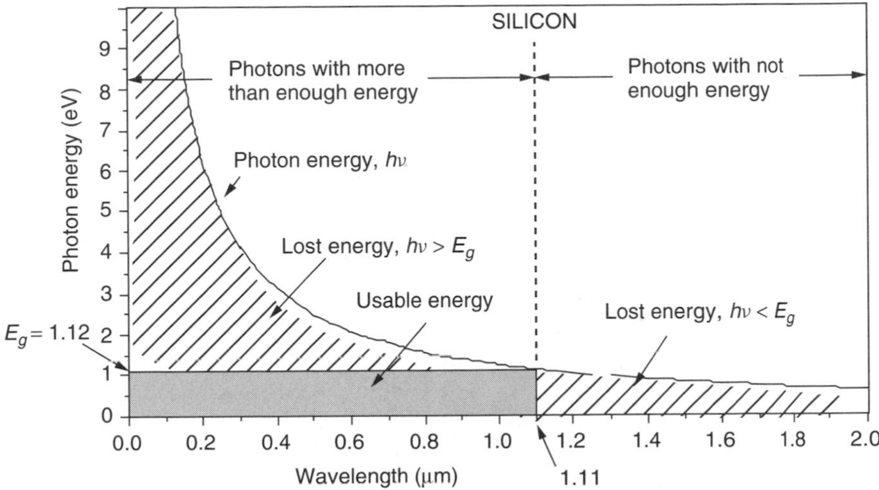

Figure 8.9 Photons with wavelengths above 1.11 μm don't have the 1.12 eV needed to excite an electron, and this energy is lost. Photons with shorter wavelengths have more than enough energy, but any energy above 1.12 eV is wasted as well.

TABLE 8.2 Band Gap and Cut-off Wavelength Above Which Electron Excitation Doesn't Occur

Quantity	Si	GaAs	CdTe	InP
Band gap (eV)	1.12	1.42	1.5	1.35
Cut-off wavelength (μm)	1.11	0.87	0.83	0.92

air mass ratio of 1 (designated "AM1") means that the sun is directly overhead. By convention, AM0 means no atmosphere; that is, it is the extraterrestrial solar spectrum. For most photovoltaic work, an air mass ratio of 1.5, corresponding to the sun being 42 degrees above the horizon, is assumed to be the standard. The solar spectrum at AM 1.5 is shown in Fig. 8.10. For an AM 1.5 spectrum, 2% of the incoming solar energy is in the UV portion of the spectrum, 54% is in the visible, and 44% is in the infrared.

8.2.3 Band-Gap Impact on Photovoltaic Efficiency

We can now make a simple estimate of the upper bound on the efficiency of a silicon solar cell. We know the band gap for silicon is 1.12 eV, corresponding to a wavelength of 1.11 μm, which means that any energy in the solar spectrum with wavelengths longer than 1.11 μm cannot send an electron into the conduction band. And, any photons with wavelength less than 1.11 μm waste their extra energy. If we know the solar spectrum, we can calculate the energy loss due to

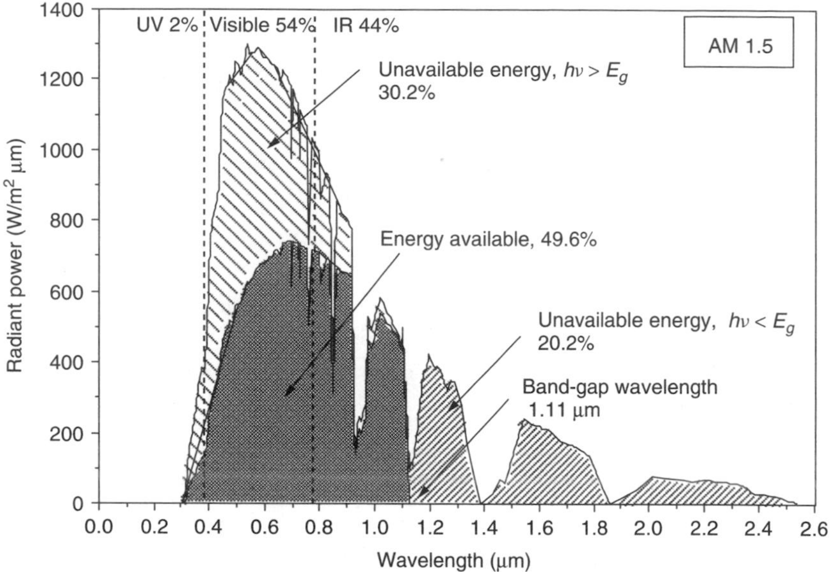

Figure 8.10 Solar spectrum at AM 1.5. Photons with wavelengths longer than 1.11 μm don't have enough energy to excite electrons (20.2% of the incoming solar energy); those with shorter wavelengths can't use all of their energy, which accounts for another 30.2% unavailable to a silicon photovoltaic cell. Spectrum is based on ERDA/NASA (1977).

these two fundamental constraints. Figure 8.10 shows the results of this analysis, assuming a standard air mass ratio AM 1.5. As is presented there, 20.2% of the energy in the spectrum is lost due to photons having less energy than the band gap of silicon ($h\nu < E_g$), and another 30.2% is lost due to photons with $h\nu > E_g$. The remaining 49.6% represents the maximum possible fraction of the sun's energy that could be collected with a silicon solar cell. That is, the constraints imposed by silicon's band gap limit the efficiency of silicon to just under 50%.

Even this simple discussion gives some insight into the trade-off between choosing a photovoltaic material that has a small band gap versus one with a large band gap. With a smaller band gap, more solar photons have the energy needed to excite electrons, which is good since it creates the charges that will enable current to flow. However, a small band gap means that more photons have surplus energy above the threshold needed to create hole–electron pairs, which wastes their potential. High band-gap materials have the opposite combination. A high band gap means that fewer photons have enough energy to create the current-carrying electrons and holes, which limits the current that can be generated. On the other hand, a high band gap gives those charges a higher voltage with less leftover surplus energy.

In other words, low band gap gives more current with less voltage while high band gap results in less current and higher voltage. Since power is the product of current and voltage, there must be some middle-ground band gap, usually estimated to be between 1.2 eV and 1.8 eV, which will result in the highest power and efficiency. Figure 8.11 shows one estimate of the impact of band gap on the theoretical maximum efficiency of photovoltaics at both AM0 and AM1. The figure includes band gaps and maximum efficiencies for many of the most promising photovoltaic materials being developed today.

Notice that the efficiencies in Fig. 8.11 are roughly in the 20–25% range—well below the 49.6% we found when we considered only the losses caused by (a) photons with insufficient energy to push electrons into the conduction band and (b) photons with energy in excess of what is needed to do so. Other factors that contribute to the drop in theoretical efficiency include:

1. Only about half to two-thirds of the full band-gap voltage across the terminals of the solar cell.
2. Recombination of holes and electrons before they can contribute to current flow.
3. Photons that are not absorbed in the cell either because they are reflected off the face of the cell, or because they pass right through the cell, or because they are blocked by the metal conductors that collect current from the top of the cell.
4. Internal resistance within the cell, which dissipates power.

8.2.4 The *p–n* Junction

As long as a solar cell is exposed to photons with energies above the band-gap energy, hole–electron pairs will be created. The problem is, of course, that

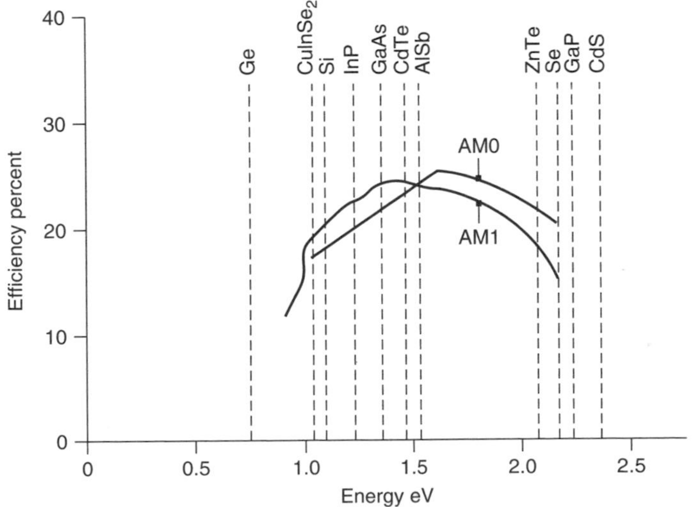

Figure 8.11 Maximum efficiency of photovoltaics as a function of their band gap. From Hersel and Zweibel (1982).

those electrons can fall right back into a hole, causing both charge carriers to disappear. To avoid that recombination, electrons in the conduction band must continuously be swept away from holes. In PVs this is accomplished by creating a built-in electric field within the semiconductor itself that pushes electrons in one direction and holes in the other. To create the electric field, two regions are established within the crystal. On one side of the dividing line separating the regions, pure (intrinsic) silicon is purposely contaminated with very small amounts of a trivalent element from column III of the periodic chart; on the other side, pentavalent atoms from column V are added.

Consider the side of the semiconductor that has been doped with a pentavalent element such as phosphorus. Only about 1 phosphorus atom per 1000 silicon atoms is typical. As shown in Fig. 8.12, an atom of the pentavalent impurity forms covalent bonds with four adjacent silicon atoms. Four of its five electrons are now tightly bound, but the fifth electron is left on its own to roam around the crystal. When that electron leaves the vicinity of its donor atom, there will remain a +5 donor ion fixed in the matrix, surrounded by only four negative valence electrons. That is, each donor atom can be represented as a single, fixed, immobile positive charge plus a freely roaming negative charge as shown in Fig. 8.12b. Pentavalent i.e., +5 elements donate electrons to their side of the semiconductor so they are called *donor* atoms. Since there are now negative charges that can move around the crystal, a semiconductor doped with donor atoms is referred to as an "*n-type material.*"

On the other side of the semiconductor, silicon is doped with a trivalent element such as boron. Again the concentration of dopants is small, something on the order of 1 boron atom per 10 million silicon atoms. These dopant atoms fall into place in the crystal, forming covalent bonds with the adjacent silicon atoms as shown in Fig. 8.13. Since each of these impurity atoms has only three electrons, only three of the covalent bonds are filled, which means that a positively charged hole appears next to its nucleus. An electron from a neighboring silicon atom can easily move into the hole, so these impurities are referred to as *acceptors* since they accept electrons. The filled hole now means there are four negative charges

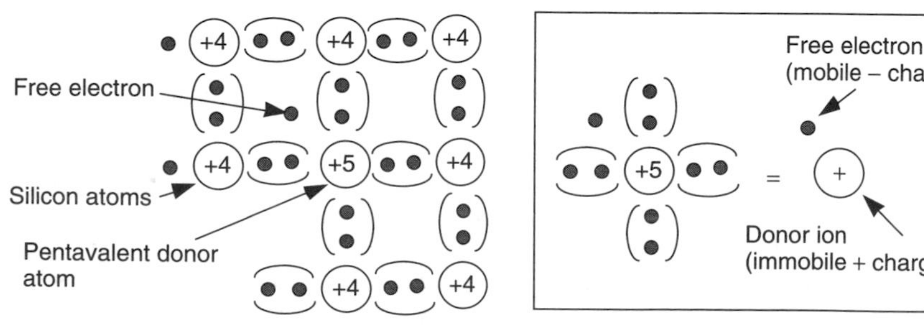

Figure 8.12 An n-type material. (a) The pentavalent donor. (b) The representation of the donor as a mobile negative charge with a fixed, immobile positive charge.

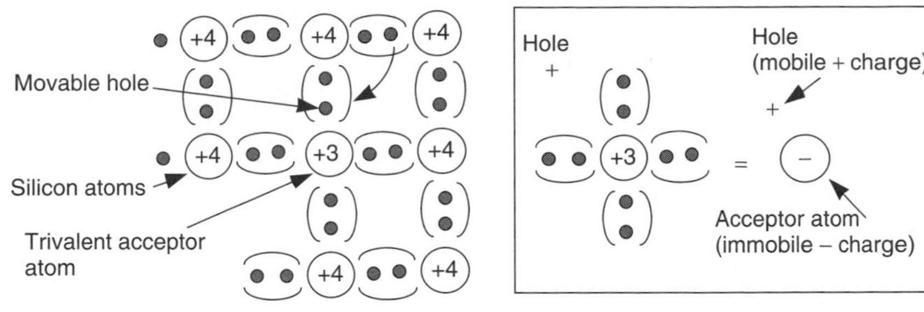

Figure 8.13 In a p-type material, trivalent acceptors contribute movable, positively charged holes leaving rigid, immobile negative charges in the crystal lattice.

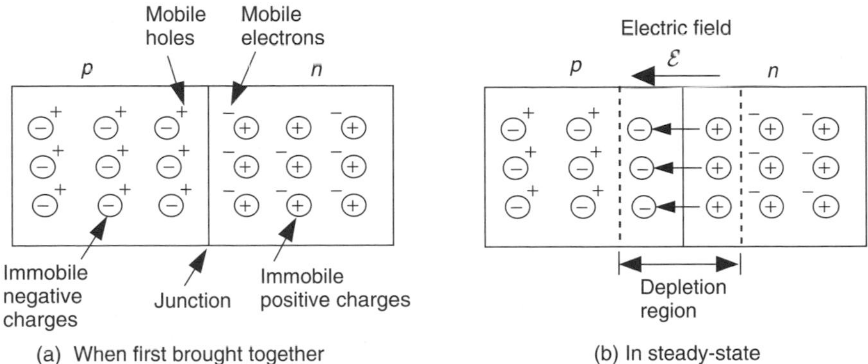

Figure 8.14 (a) When a p–n junction is first formed, there are mobile holes in the p-side and mobile electrons in the n-side. (b) As they migrate across the junction, an electric field builds up that opposes, and quickly stops, diffusion.

surrounding a +3 nucleus. All four covalent bonds are now filled creating a fixed, immobile net negative charge at each acceptor atom. Meanwhile, each acceptor has created a positively charged hole that is free to move around in the crystal, so this side of the semiconductor is called a *p-type* material.

Now, suppose we put an n-type material next to a p-type material forming a junction between them. In the n-type material, mobile electrons drift by diffusion across the junction. In the p-type material, mobile holes drift by diffusion across the junction in the opposite direction. As depicted in Fig. 8.14, when an electron crosses the junction it fills a hole, leaving an immobile, positive charge behind in the n-region, while it creates an immobile, negative charge in the p-region. These immobile charged atoms in the p and n regions create an electric field that works against the continued movement of electrons and holes across the junction. As the diffusion process continues, the electric field countering that movement increases until eventually (actually, almost instantaneously) all further movement of charged carriers across the junction stops.

The exposed immobile charges creating the electric field in the vicinity of the junction form what is called a *depletion region*, meaning that the mobile charges are depleted—gone—from this region. The width of the depletion region is only about 1 μm and the voltage across it is perhaps 1 V, which means the field strength is about 10,000 V/cm! Following convention, the arrows representing an electric field in Fig. 8.14b start on a positive charge and end on a negative charge. The arrow, therefore, points in the direction that the field would push a positive charge, which means that it holds the mobile positive holes in the *p*-region (while it repels the electrons back into the *n*-region).

8.2.5 The *p–n* Junction Diode

Anyone familiar with semiconductors will immediately recognize that what has been described thus far is just a common, conventional *p–n* junction diode, the characteristics of which are presented in Fig. 8.15. If we were to apply a voltage V_d across the diode terminals, forward current would flow easily through the diode from the *p*-side to the *n*-side; but if we try to send current in the reverse direction, only a very small ($\approx 10^{-12}$ A/cm^2) reverse saturation current I_0 will flow. This reverse saturation current is the result of thermally generated carriers with the holes being swept into the *p*-side and the electrons into the *n*-side. In the forward direction, the voltage drop across the diode is only a few tenths of a volt.

The symbol for a real diode is shown here as a blackened triangle with a bar; the triangle suggests an arrow, which is a convenient reminder of the direction in which current flows easily. The triangle is blackened to distinguish it from an "ideal" diode. Ideal diodes have no voltage drop across them in the forward direction, and no current at all flows in the reverse direction.

The voltage–current characteristic curve for the *p–n* junction diode is described by the following Shockley diode equation:

$$I_d = I_0(e^{qV_d/kT} - 1) \tag{8.3}$$

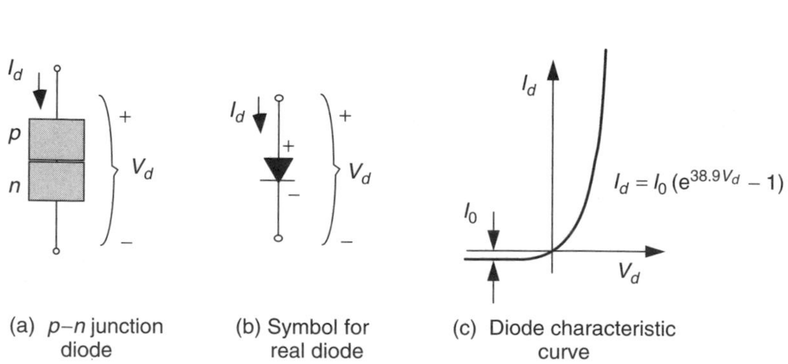

Figure 8.15 A *p–n* junction diode allows current to flow easily from the *p*-side to the *n*-side, but not in reverse. (a) *p–n* junction; (b) its symbol; (c) its characteristic curve.

where I_d is the diode current in the direction of the arrow (A), V_d is the voltage across the diode terminals from the p-side to the n-side (V), I_0 is the reverse saturation current (A), q is the electron charge (1.602×10^{-19}C), k is Boltzmann's constant (1.381×10^{-23} J/K), and T is the junction temperature (K).

Substituting the above constants into the exponent of (8.3) gives

$$\frac{qV_d}{kT} = \frac{1.602 \times 10^{-19}}{1.381 \times 10^{-23}} \cdot \frac{V_d}{T(\text{K})} = 11,600\frac{V_d}{T(\text{K})} \tag{8.4}$$

A junction temperature of 25°C is often used as a standard, which results in the following diode equation:

$$I_d = I_0(e^{38.9V_d} - 1) \qquad \text{(at 25°C)} \tag{8.5}$$

Example 8.2 A *p–n* Junction Diode. Consider a p–n junction diode at 25°C with a reverse saturation current of 10^{-9} A. Find the voltage drop across the diode when it is carrying the following:

a. no current (open-circuit voltage)
b. 1 A
c. 10 A

Solution

a. In the open-circuit condition, $I_d = 0$, so from (8.5) $V_d = 0$.
b. With $I_d = 1$ A, we can find V_d by rearranging (8.5):

$$V_d = \frac{1}{38.9}\ln\left(\frac{I_d}{I_0} + 1\right) = \frac{1}{38.9}\ln\left(\frac{1}{10^{-9}} + 1\right) = 0.532 \text{ V}$$

c. with $I_d = 10$ A,

$$V_d = \frac{1}{38.9}\ln\left(\frac{10}{10^{-9}} + 1\right) = 0.592 \text{ V}$$

Notice how little the voltage drop changes as the diode conducts more and more current, changing by only about 0.06 V as the current increased by a factor of 10. Often in normal electronic circuit analysis, the diode voltage drop when it is conducting current is assumed to be nominally about 0.6 V, which is quite in line with the above results.

While the Shockley diode equation (8.3) is appropriate for our purposes, it should be noted that in some circumstances it is modified with an "ideality factor" A, which accounts for different mechanisms responsible for moving carriers across the junction. The resulting equation is then

$$I_d = I_0(e^{qV_d/AkT} - 1) \tag{8.6}$$

where the ideality factor A is 1 if the transport process is purely diffusion, and $A \approx 2$ if it is primarily recombination in the depletion region.

8.3 A GENERIC PHOTOVOLTAIC CELL

Let us consider what happens in the vicinity of a p–n junction when it is exposed to sunlight. As photons are absorbed, hole-electron pairs may be formed. If these mobile charge carriers reach the vicinity of the junction, the electric field in the depletion region will push the holes into the p-side and push the electrons into the n-side, as shown in Fig. 8.16. The p-side accumulates holes and the n-side accumulates electrons, which creates a voltage that can be used to deliver current to a load.

If electrical contacts are attached to the top and bottom of the cell, electrons will flow out of the n-side into the connecting wire, through the load and back to the p-side as shown in Fig. 8.17. Since wire cannot conduct holes, it is only the electrons that actually move around the circuit. When they reach the p-side, they recombine with holes completing the circuit. By convention, positive current flows in the direction opposite to electron flow, so the current arrow in the figure shows current going from the p-side to the load and back into the n-side.

8.3.1 The Simplest Equivalent Circuit for a Photovoltaic Cell

A simple equivalent circuit model for a photovoltaic cell consists of a real diode in parallel with an ideal current source as shown in Fig. 8.18. The ideal current source delivers current in proportion to the solar flux to which it is exposed.

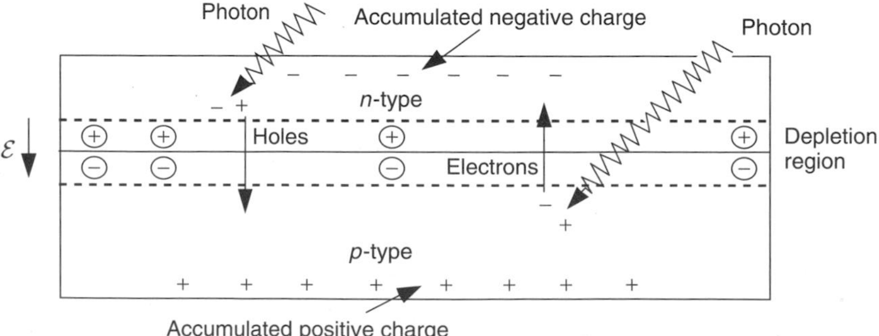

Figure 8.16 When photons create hole–electron pairs near the junction, the electric field in the depletion region sweeps holes into the p-side and sweeps electrons into the n-side of the cell.

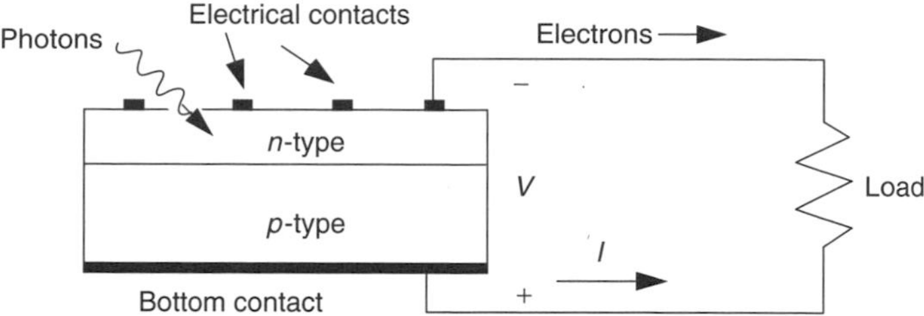

Figure 8.17 Electrons flow from the n-side contact, through the load, and back to the p-side where they recombine with holes. Conventional current I is in the opposite direction.

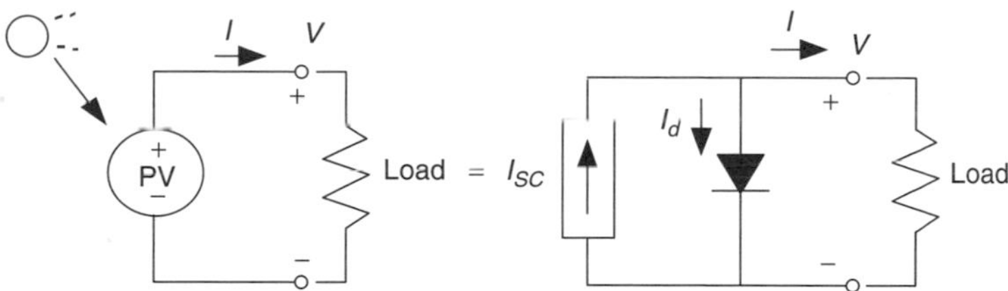

Figure 8.18 A simple equivalent circuit for a photovoltaic cell consists of a current source driven by sunlight in parallel with a real diode.

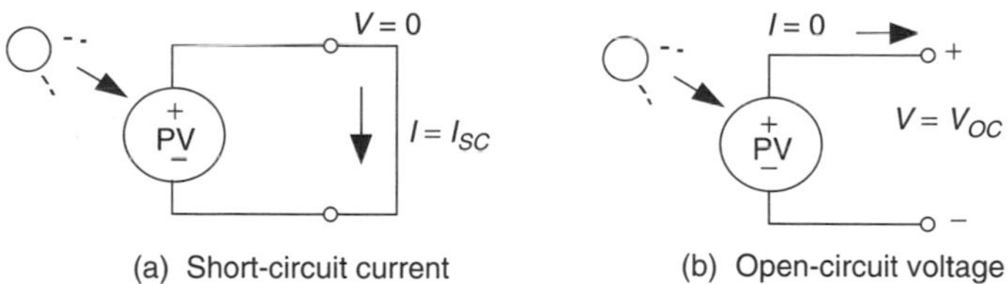

Figure 8.19 Two important parameters for photovoltaics are the short-circuit current I_{SC} and the open-circuit voltage V_{OC}.

There are two conditions of particular interest for the actual PV and for its equivalent circuit. As shown in Fig. 8.19, they are: (1) the current that flows when the terminals are shorted together (the short-circuit current, I_{SC}) and (2) the voltage across the terminals when the leads are left open (the open-circuit voltage, V_{OC}). When the leads of the equivalent circuit for the PV cell are shorted together, no current flows in the (real) diode since $V_d = 0$, so all of the current from the ideal source flows through the shorted leads. Since that short-circuit current must equal I_{SC}, the magnitude of the ideal current source itself must be equal to I_{SC}.

Now we can write a voltage and current equation for the equivalent circuit of the PV cell shown in Fig. 8.18b. Start with

$$I = I_{SC} - I_d \tag{8.7}$$

and then substitute (8.3) into (8.7) to get

$$I = I_{SC} - I_0\left(e^{qV/kT} - 1\right) \tag{8.8}$$

It is interesting to note that the second term in (8.8) is just the diode equation with a negative sign. That means that a plot of (8.8) is just I_{SC} added to the diode curve of Fig. 8.15c turned upside-down. Figure 8.20 shows the current–voltage relationship for a PV cell when it is dark (no illumination) and light (illuminated) based on (8.8).

When the leads from the PV cell are left open, $I = 0$ and we can solve (8.8) for the open-circuit voltage V_{OC} :

$$V_{OC} = \frac{kT}{q}\ln\left(\frac{I_{SC}}{I_0} + 1\right) \tag{8.9}$$

And at 25°C, (8.8) and (8.9) become

$$I = I_{SC} - I_0(e^{38.9\ V} - 1) \tag{8.10}$$

and

$$V_{OC} = 0.0257\ln\left(\frac{I_{SC}}{I_0} + 1\right) \tag{8.11}$$

In both of these equations, short-circuit current, I_{SC}, is directly proportional to solar insolation, which means that we can now quite easily plot sets of PV current–voltage curves for varying sunlight. Also, quite often laboratory specifications for the performance of photovoltaics are given per cm^2 of junction

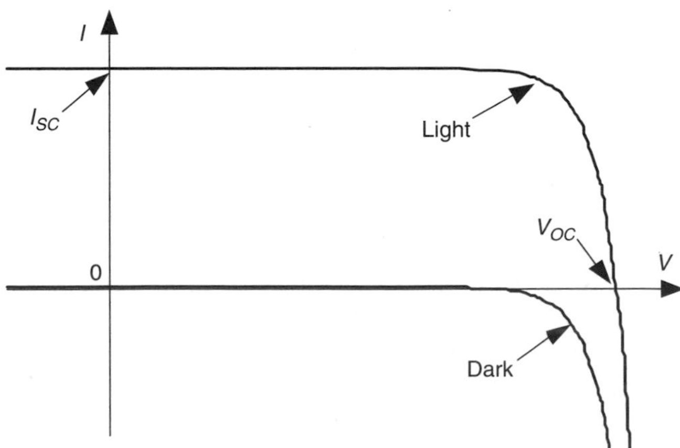

Figure 8.20 Photovoltaic current–voltage relationship for "dark" (no sunlight) and "light" (an illuminated cell). The dark curve is just the diode curve turned upside-down. The light curve is the dark curve plus I_{SC}.

area, in which case the currents in the above equations are written as current densities. Both of these points are illustrated in the following example.

Example 8.3 The *I*–*V* Curve for a Photovoltaic Cell. Consider a 100-cm^2 photovoltaic cell with reverse saturation current $I_0 = 10^{-12}$ A/cm^2. In full sun, it produces a short-circuit current of 40 mA/cm^2 at 25°C. Find the open-circuit voltage at full sun and again for 50% sunlight. Plot the results.

Solution. The reverse saturation current I_0 is 10^{-12} A/cm$^2 \times 100$ cm$^2 = 1 \times 10^{-10}$ A. At full sun I_{SC} is 0.040 A/cm$^2 \times 100$ cm$^2 = 4.0$ A. From (8.11) the open-circuit voltage is

$$V_{OC} = 0.0257 \ \ln\left(\frac{I_{SC}}{I_0} + 1\right) = 0.0257 \ln\left(\frac{4.0}{10^{-10}} + 1\right) = 0.627 \text{ V}$$

Since short-circuit current is proportional to solar intensity, at half sun $I_{SC} = 2$ A and the open-circuit voltage is

$$V_{OC} = 0.0257 \ln\left(\frac{2}{10^{-10}} + 1\right) = 0.610 \text{ V}$$

Plotting (8.10) gives us the following:

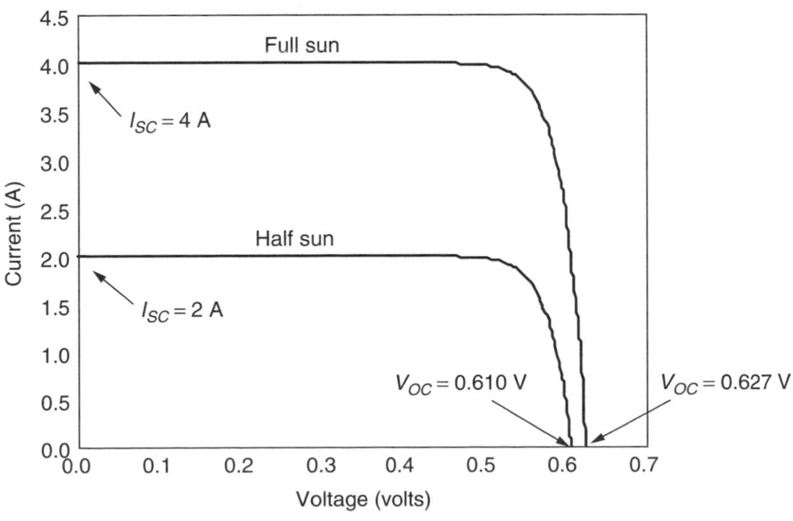

8.3.2 A More Accurate Equivalent Circuit for a PV Cell

There are times when a more complex PV equivalent circuit than the one shown in Fig. 8.18 is needed. For example, consider the impact of shading on a string of cells wired in series (Fig. 8.21 shows two such cells). If any cell in the string is in the dark (shaded), it produces no current. In our simplified equivalent circuit for the shaded cell, the current through that cell's current source is zero and its diode is back biased so it doesn't pass any current either (other than a tiny amount of reverse saturation current). This means that the simple equivalent circuit suggests that no power will be delivered to a load if any of its cells are shaded. While it is true that PV modules are very sensitive to shading, the situation is not quite as bad as that. So, we need a more complex model if we are going to be able to deal with realities such as the shading problem.

Figure 8.22 shows a PV equivalent circuit that includes some parallel leakage resistance R_p. The ideal current source I_{SC} in this case delivers current to the diode, the parallel resistance, and the load:

$$I = (I_{SC} - I_d) - \frac{V}{R_p} \tag{8.12}$$

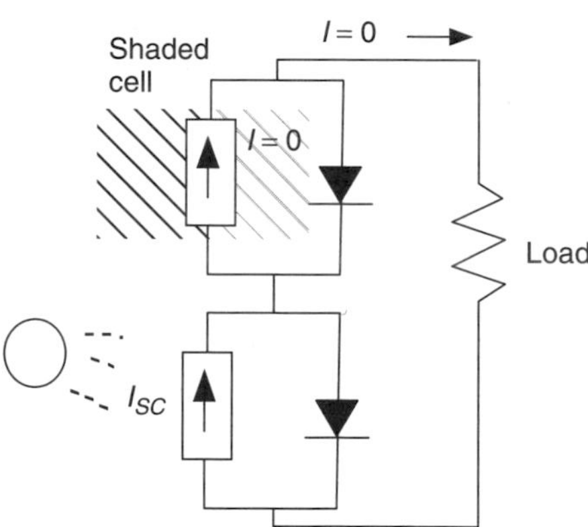

Figure 8.21 The simple equivalent circuit of a string of cells in series suggests no current can flow to the load if any cell is in the dark (shaded). A more complex model can deal with this problem.

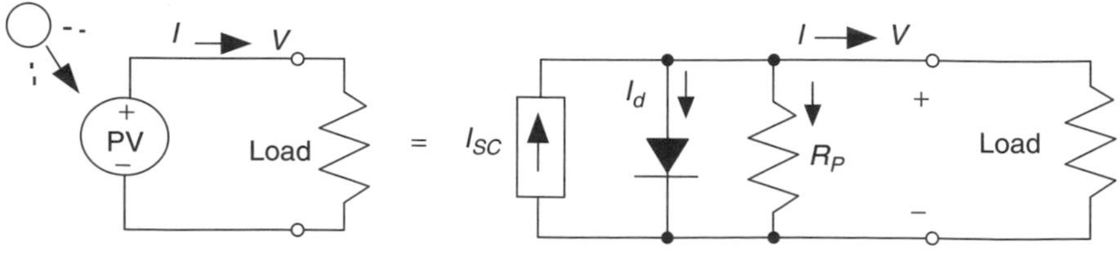

Figure 8.22 The simple PV equivalent circuit with an added parallel resistance.

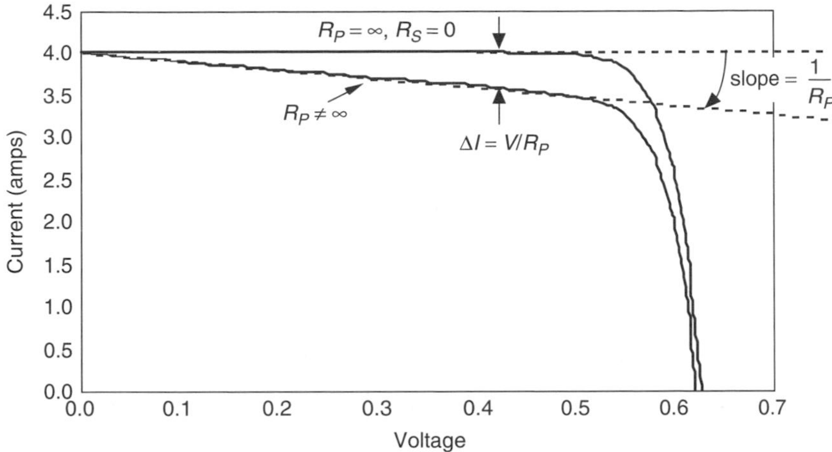

Figure 8.23 Modifying the idealized PV equivalent circuit by adding parallel resistance causes the current at any given voltage to drop by V/R_P.

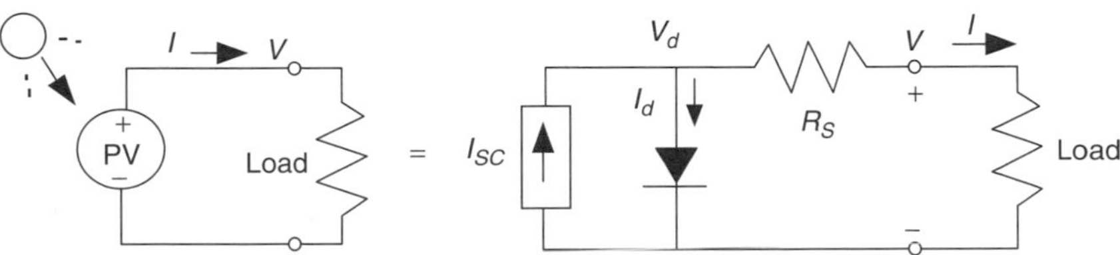

Figure 8.24 A PV equivalent circuit with series resistance.

The term in the parentheses of (8.12) is the same current that we had for the simple model. So, what (8.12) tells us is that at any given voltage, the parallel leakage resistance causes load current for the ideal model to be decreased by V/R_p as is shown in Fig. 8.23.

For a cell to have losses of less than 1% due to its parallel resistance, R_P should be greater than about

$$R_P > \frac{100V_{OC}}{I_{SC}} \tag{8.13}$$

For a large cell, I_{SC} might be around 7 A and V_{OC} might be about 0.6 V, which says its parallel resistance should be greater than about 9 Ω.

An even better equivalent circuit will include series resistance as well as parallel resistance. Before we can develop that model, consider Fig. 8.24 in which the original PV equivalent circuit has been modified to just include some series resistance, R_S. Some of this might be contact resistance associated with the bond between the cell and its wire leads, and some might be due to the resistance of the semiconductor itself.

To analyze Fig. 8.24, start with the simple equivalent circuit (8.8)

$$I = I_{SC} - I_d = I_{SC} - I_0\left(e^{qV_d/kT} - 1\right) \tag{8.8}$$

and then add the impact of R_S,

$$V_d = V + I \cdot R_S \tag{8.14}$$

to give

$$I = I_{SC} - I_0\left\{\exp\left[\frac{q(V + I \cdot R_S)}{kT}\right] - 1\right\} \tag{8.15}$$

Equation (8.15) can be interpreted as the original PV I–V curve with the voltage at any given current shifted to the left by $\Delta V = IR_S$ as shown in Fig. 8.25.

For a cell to have less than 1% losses due to the series resistance, R_S will need to be less than about

$$R_S < \frac{0.01V_{OC}}{I_{SC}} \tag{8.16}$$

which, for a large cell with $I_{SC} = 7$ A and $V_{OC} = 0.6$ V, would be less than 0.0009 Ω.

Finally, let us generalize the PV equivalent circuit by including both series and parallel resistances as shown in Fig. 8.26. We can write the following equation for current and voltage:

$$I = I_{SC} - I_0\left\{\exp\left[\frac{q(V + I \cdot R_S)}{kT}\right] - 1\right\} - \left(\frac{V + I \cdot R_S}{R_P}\right) \tag{8.17}$$

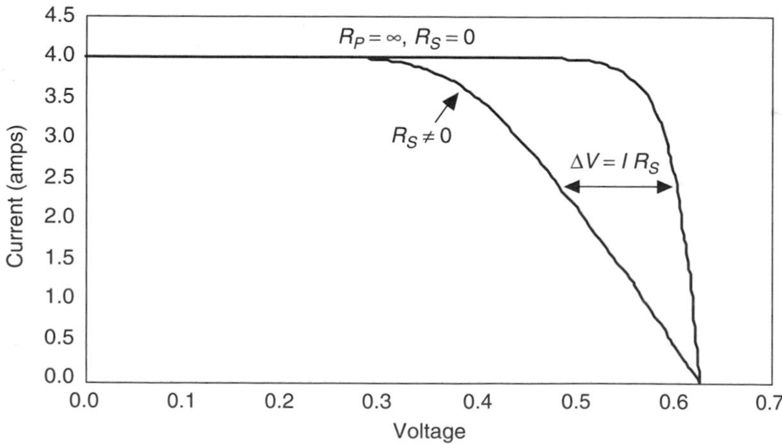

Figure 8.25 Adding series resistance to the PV equivalent circuit causes the voltage at any given current to shift to the left by $\Delta V = IR_S$.

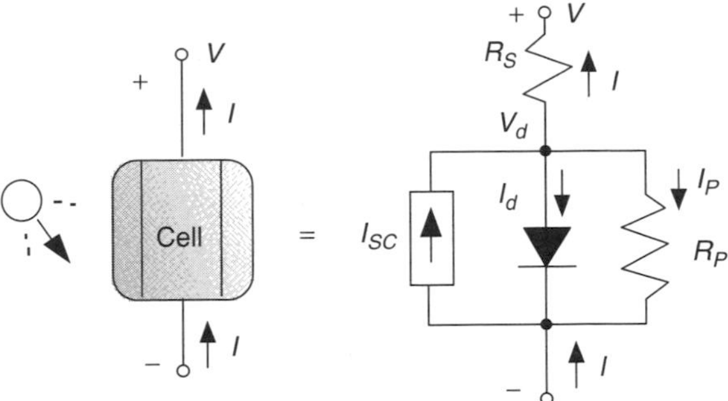

Figure 8.26 A more complex equivalent circuit for a PV cell includes both parallel and series resistances. The shaded diode reminds us that this is a "real" diode rather than an ideal one.

Under the standard assumption of a 25°C cell temperature, (8.17) becomes

$$I = I_{SC} - I_0\left[e^{38.9(V+IR_S)} - 1\right] - \frac{1}{R_P}(V + IR_S) \qquad \text{at } 25^\circ\text{C} \tag{8.18}$$

Unfortunately, (8.18) is a complex equation for which there is no explicit solution for either voltage V or current I. A spreadsheet solution, however, is fairly straightforward and has the extra advantage of enabling a graph of I versus V to be obtained easily. The approach is based on incrementing values of diode voltage, V_d, in the spreadsheet. For each value of V_d, corresponding values of current I and voltage V can easily be found.

Using the sign convention shown in Fig. 8.26 and applying Kirchhoff's Current Law to the node above the diode, we can write

$$I_{SC} = I + I_d + I_P \tag{8.19}$$

Rearranging, and substituting the Shockley diode equation (8.5) at 25°C gives

$$I = I_{SC} - \mathrm{I}_0(e^{38.9V_d} - 1) - \frac{V_d}{R_P} \tag{8.20}$$

With an assumed value of V_d in a spreadsheet, current I can be found from (8.20). Voltage across an individual cell then can be found from

$$V = V_d - IR_S \tag{8.21}$$

A plot of (8.18) obtained this way for an equivalent circuit with $R_S = 0.05\ \Omega$ and $R_P = 1\ \Omega$ is shown in Fig. 8.27. As might be expected, the graph combines features of Fig. 8.23 and 8.25.

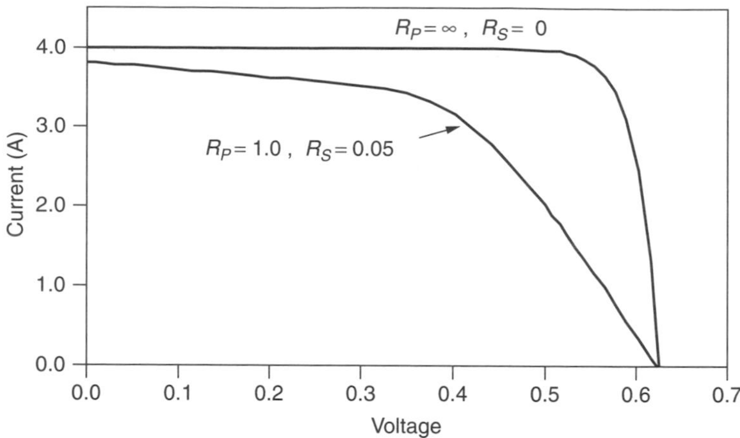

Figure 8.27 Series and parallel resistances in the PV equivalent circuit decrease both voltage and current delivered. To improve cell performance, high R_P and low R_S are needed.

8.4 FROM CELLS TO MODULES TO ARRAYS

Since an individual cell produces only about 0.5 V, it is a rare application for which just a single cell is of any use. Instead, the basic building block for PV applications is a *module* consisting of a number of pre-wired cells in series, all encased in tough, weather-resistant packages. A typical module has 36 cells in series and is often designated as a "12-V module" even though it is capable of delivering much higher voltages than that. Some 12-V modules have only 33 cells, which, as will be seen later may, be desirable in certain very simple battery charging systems. Large 72-cell modules are now quite common, some of which have all of the cells wired in series, in which case they are referred to as 24-V modules. Some 72-cell modules can be field-wired to act either as 24-V modules with all 72 cells in series or as 12-V modules with two parallel strings having 36 series cells in each.

Multiple modules, in turn, can be wired in series to increase voltage and in parallel to increase current, the product of which is power. An important element in PV system design is deciding how many modules should be connected in series and how many in parallel to deliver whatever energy is needed. Such combinations of modules are referred to as an *array*. Figure 8.28 shows this distinction between cells, modules, and arrays.

8.4.1 From Cells to a Module

When photovoltaics are wired in series, they all carry the same current, and at any given current their voltages add as shown in Fig. 8.29. That means we can continue the spreadsheet solution of (8.18) to find an overall module voltage

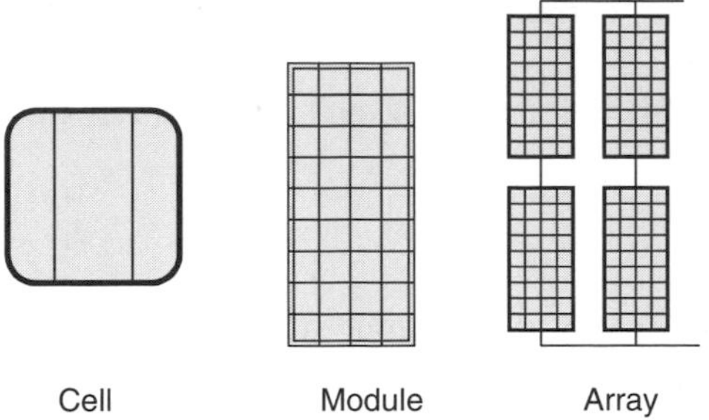

Figure 8.28 Photovoltaic cells, modules, and arrays.

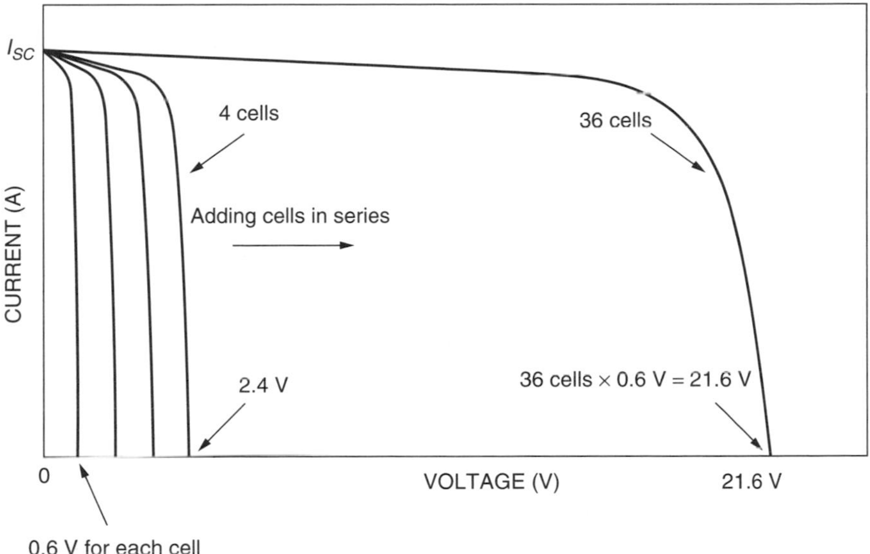

Figure 8.29 For cells wired in series, their voltages at any given current add. A typical module will have 36 cells.

V_{module} by multiplying (8.21) by the number of cells in the module n.

$$V_{\text{module}} = n(V_d - IR_S) \tag{8.22}$$

Example 8.4 Voltage and Current from a PV Module. A PV module is made up of 36 identical cells, all wired in series. With 1-sun insolation (1 kW/m^2), each cell has short-circuit current $I_{SC} = 3.4$ A and at 25°C its reverse saturation current is $I_0 = 6 \times 10^{-10}$ A. Parallel resistance $R_P = 6.6\ \Omega$ and series resistance $R_S = 0.005\ \Omega$.

a. Find the voltage, current, and power delivered when the junction voltage of each cell is 0.50 V.
b. Set up a spreadsheet for I and V and present a few lines of output to show how it works.

Solution.

a. Using $V_d = 0.50$ V in (8.20) along with the other data gives current:

$$I = I_{SC} - I_0(e^{38.9V_d} - 1) - \frac{V_d}{R_P}$$

$$= 3.4 - 6 \times 10^{-10}(e^{38.9 \times 0.50} - 1) - \frac{0.50}{6.6} = 3.16 \text{ A}$$

Under these conditions, (8.22) gives the voltage produced by the 36-cell module:

$$V_{\text{module}} = n(V_d - IR_S) = 36(0.50 - 3.16 \times 0.005) = 17.43 \text{ V}$$

Power delivered is therefore

$$P(\text{watts}) = V_{\text{module}}\, I = 17.43 \times 3.16 = 55.0 \text{ W}$$

b. A spreadsheet might look something like the following:

Number of cells, $n = 36$

Parallel resistance/cell R_P (ohms) = 6.6

Series resistance/cell R_S (ohms) = 0.005

Reverse saturation current I_0 (A) = 6.00E-10

Short-circuit current at 1-sun (A) = 3.4

V_d	$I = I_{SC} - I_0\left(e^{38.9V_d} - 1\right) - \frac{V_d}{R_p}$	$V_{\text{module}} = n(V_d - IR_S)$	P (watts) $= V_{\text{module}} I$
0.49	3.21	17.06	54.80
0.50	3.16	17.43	55.02
0.51	3.07	17.81	54.75
0.52	2.96	18.19	53.76
0.53	2.78	18.58	51.65
0.54	2.52	18.99	47.89
0.55	2.14	19.41	41.59

Notice that we have found the maximum power point for this module, which is at $I = 3.16$ A, $V = 17.43$ V, and $P = 55$ W. This would be described as a 55-W module.

8.4.2 From Modules to Arrays

Modules can be wired in series to increase voltage, and in parallel to increase current. Arrays are made up of some combination of series and parallel modules to increase power.

For modules in series, the I–V curves are simply added along the voltage axis. That is, at any given current (which flows through each of the modules), the total voltage is just the sum of the individual module voltages as is suggested in Fig. 8.30.

For modules in parallel, the same voltage is across each module and the total current is the sum of the currents. That is, at any given voltage, the I–V curve of the parallel combination is just the sum of the individual module currents at that voltage. Figure 8.31 shows the I–V curve for three modules in parallel.

When high power is needed, the array will usually consist of a combination of series and parallel modules for which the total I–V curve is the sum of the individual module I–V curves. There are two ways to imagine wiring a series/parallel combination of modules: The series modules may be wired as strings, and the strings wired in parallel as in Fig. 8.32a, or the parallel modules may be wired together first and those units combined in series as in 8.32b. The total I–V curve is just the sum of the individual module curves, which is the same in either case when everything is working right. There is a reason, however, to prefer the wiring of strings in parallel (Fig. 8.32a). If an entire string is removed from service for some reason, the array can still deliver whatever voltage is needed by the load, though the current is diminished, which is not the case when a parallel group of modules is removed.

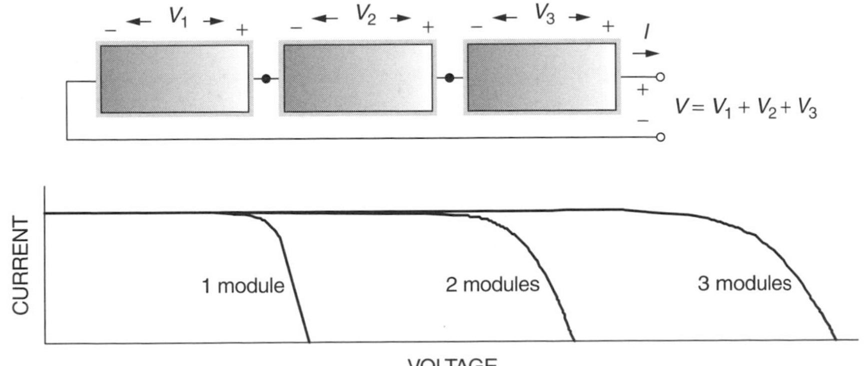

Figure 8.30 For modules in series, at any given current the voltages add.

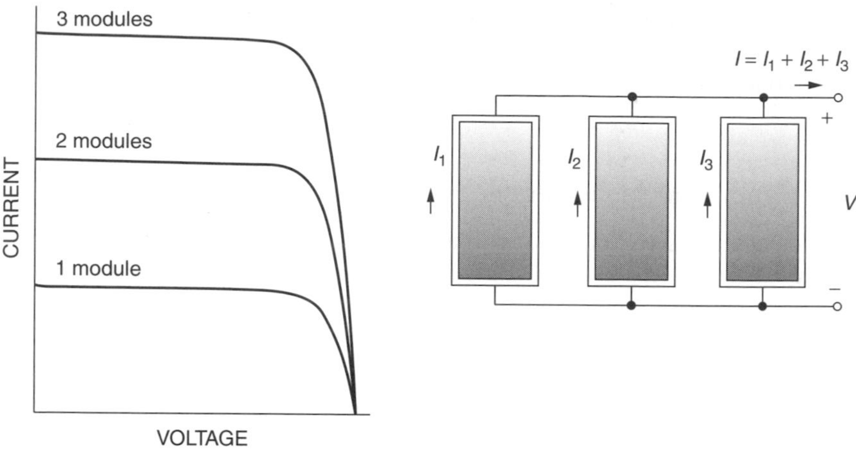

Figure 8.31 For modules in parallel, at any given voltage the currents add.

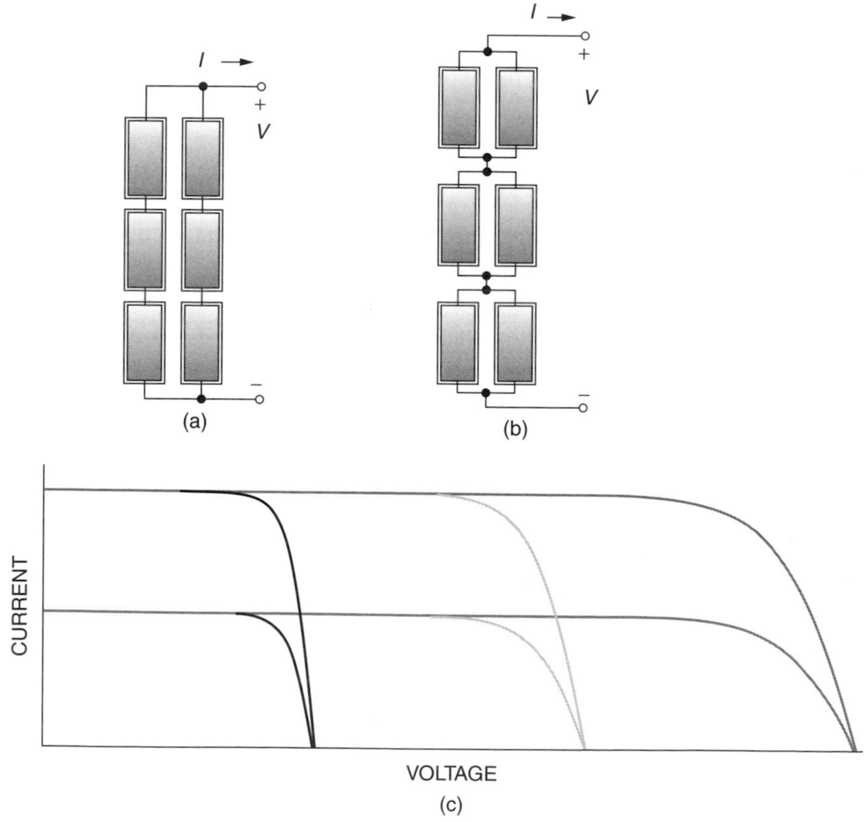

Figure 8.32 Two ways to wire an array with three modules in series and two modules in parallel. Although the $I-V$ curves for arrays are the same, two strings of three modules each (a) is preferred. The total $I-V$ curve of the array is shown in (c).

8.5 THE PV I–V CURVE UNDER STANDARD TEST CONDITIONS (STC)

Consider, for the moment, a single PV module that you want to connect to some sort of a load (Fig. 8.33). The load might be a dc motor driving a pump or it might be a battery, for example. Before the load is connected, the module sitting in the sun will produce an open-circuit voltage V_{OC}, but no current will flow. If the terminals of the module are shorted together (which doesn't hurt the module at all, by the way), the short-circuit current I_{SC} will flow, but the output voltage will be zero. In both cases, since power is the product of current and voltage, no power is delivered by the module and no power is received by the load. When the load is actually connected, some combination of current and voltage will result and power will be delivered. To figure out how much power, we have to consider the I–V characteristic curve of the module as well as the I–V characteristic curve of the load.

Figure 8.34 shows a generic I–V curve for a PV module, identifying several key parameters including the open-circuit voltage V_{OC} and the short-circuit current I_{SC}. Also shown is the product of voltage and current, that is, power delivered by the module. At the two ends of the I–V curve, the output power is zero since either current or voltage is zero at those points. The maximum power point (MPP) is that spot near the knee of the I–V curve at which the product of current and voltage reaches its maximum. The voltage and current at the MPP are sometimes designated as V_m and I_m for the general case and designated V_R and I_R (for *rated voltage* and *rated current*) under the special circumstances that correspond to idealized test conditions.

Another way to visualize the location of the maximum power point is by imagining trying to find the biggest possible rectangle that will fit beneath the I–V curve. As shown in Fig. 8.35, the sides of the rectangle correspond to current and voltage, so its area is power. Another quantity that is often used to characterize module performance is the *fill factor* (FF). The fill factor is the ratio of the power at the maximum power point to the product of V_{OC} and I_{SC}, so FF can be visualized as the ratio of two rectangular areas, as is suggested in

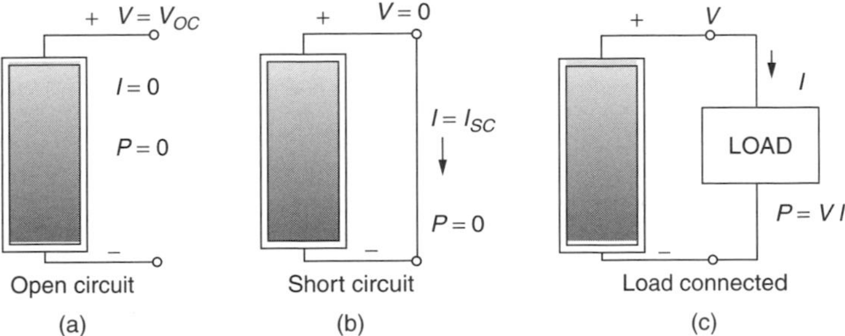

Figure 8.33 No power is delivered when the circuit is open (a) or shorted (b). When the load is connected (c), the same current flows through the load and module and the same voltage appears across them.

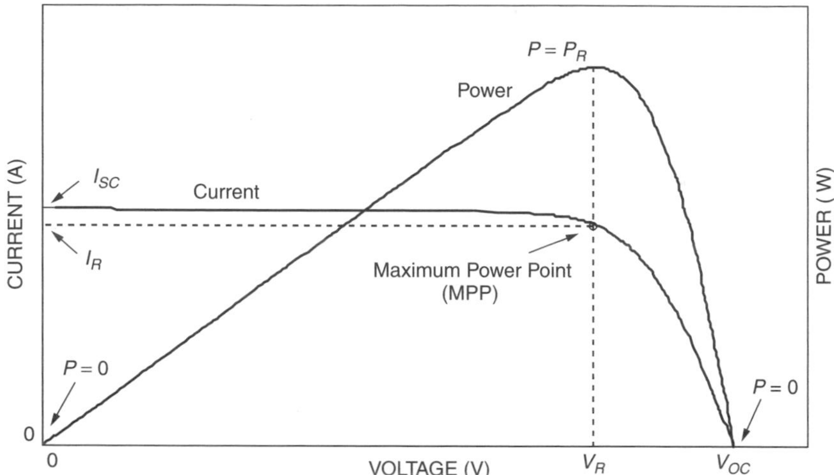

Figure 8.34 The $I-V$ curve and power output for a PV module. At the maximum power point (MPP) the module delivers the most power that it can under the conditions of sunlight and temperature for which the $I-V$ curve has been drawn.

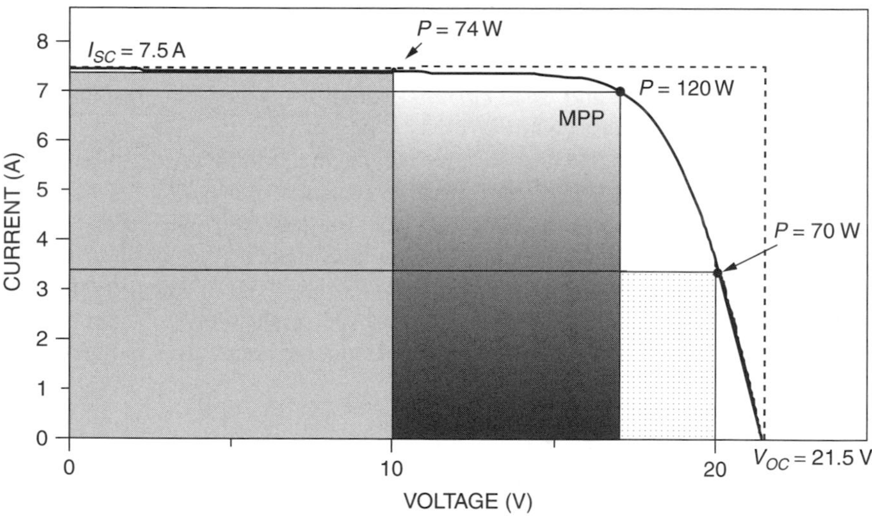

Figure 8.35 The maximum power point (MPP) corresponds to the biggest rectangle that can fit beneath the $I-V$ curve. The fill factor (FF) is the ratio of the area (power) at MPP to the area formed by a rectangle with sides V_{OC} and I_{SC}.

Fig. 8.35. Fill factors around 70–75% for crystalline silicon solar modules are typical, while for multijunction amorphous-Si modules, it is closer to 50–60%.

$$\text{Fill factor (FF)} = \frac{\text{Power at the maximum power point}}{V_{OC}\ I_{SC}} = \frac{V_R I_R}{V_{OC} I_{SC}} \tag{8.23}$$

TABLE 8.3 Examples of PV Module Performance Data Under Standard Test Conditions (1 kW/m², AM 1.5, 25°C Cell Temperature)

Manufacturer	Kyocera	Sharp	BP	Uni-Solar	Shell
Model	KC-120-1	NE-Q5E2U	2150S	US-64	ST40
Material	Multicrystal	Polycrystal	Monocrystal	Triple junction a-Si	CIS-thin film
Number of cells n	36	72	72		42
Rated Power $P_{DC,STC}$ (W)	120	165	150	64	40
Voltage at max power (V)	16.9	34.6	34	16.5	16.6
Current at rated power (A)	7.1	4.77	4.45	3.88	2.41
Open-circuit voltage V_{OC} (V)	21.5	43.1	42.8	23.8	23.3
Short-circuit current I_{SC} (A)	7.45	5.46	4.75	4.80	2.68
Length (mm/in.)	1425/56.1	1575/62.05	1587/62.5	1366/53.78	1293/50.9
Width (mm/in.)	652/25.7	826/32.44	790/31.1	741/29.18	329/12.9
Depth (mm/in.)	52/2.0	46/1.81	50/1.97	31.8/1.25	54/2.1
Weight (kg/lb)	11.9/26.3	17/37.5	15.4/34	9.2/20.2	14.8/32.6
Module efficiency	12.9%	12.7%	12.0%	6.3%	9.4%

Since PV $I-V$ curves shift all around as the amount of insolation changes and as the temperature of the cells varies, standard test conditions (STC) have been established to enable fair comparisons of one module to another. Those test conditions include a solar irradiance of 1 kW/m² (1 sun) with spectral distribution shown in Fig. 8.10, corresponding to an air mass ratio of 1.5 (AM 1.5). The standard cell temperature for testing purposes is 25°C (it is important to note that 25° is cell temperature, not ambient temperature). Manufacturers always provide performance data under these operating conditions, some examples of which are shown in Table 8.3. The key parameter for a module is its rated power; to help us remember that it is dc power measured under standard test conditions, it has been identified in Table 8.3 as $P_{DC,STC}$. Later we'll learn how to adjust rated power to account for temperature effects as well as see how to adjust it to give us an estimate of the actual ac power that the module and inverter combination will deliver.

8.6 IMPACTS OF TEMPERATURE AND INSOLATION ON *I–V* CURVES

Manufacturers will often provide $I-V$ curves that show how the curves shift as insolation and cell temperature changes. Figure 8.36 shows examples for the Kyocera 120-W multicrystal-silicon module described in Table 8.3. Notice as insolation drops, short-circuit current drops in direct proportion. Cutting insolation in half, for example, drops I_{SC} by half. Decreasing insolation also reduces

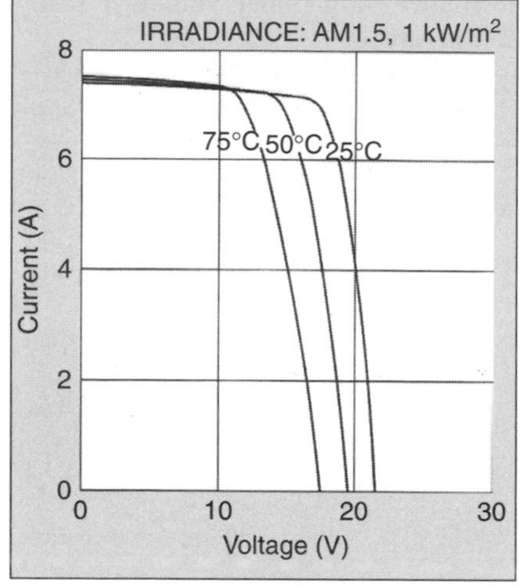

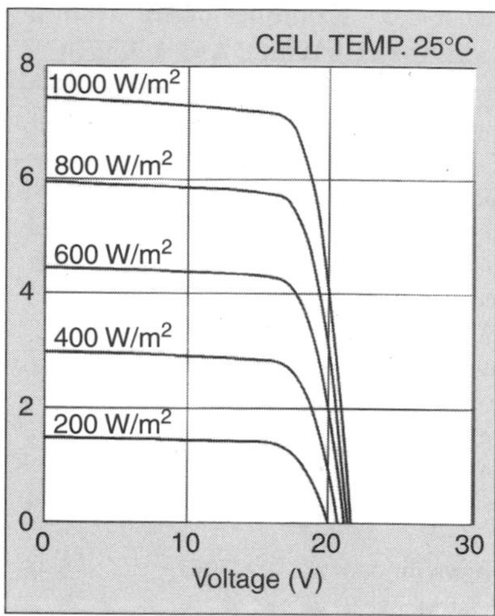

Figure 8.36 Current-voltage characteristic curves under various cell temperatures and irradiance levels for the Kyocera KC120-1 PV module.

V_{OC}, but it does so following a logarithmic relationship that results in relatively modest changes in V_{OC}.

As can be seen in Fig. 8.36, as cell temperature increases, the open-circuit voltage decreases substantially while the short-circuit current increases only slightly. Photovoltaics, perhaps surprisingly, therefore perform better on cold, clear days than hot ones. For crystalline silicon cells, V_{OC} drops by about 0.37% for each degree Celsius increase in temperature and I_{SC} increases by approximately 0.05%. The net result when cells heat up is the MPP slides slightly upward and toward the left with a decrease in maximum power available of about 0.5%/°C. Given this significant shift in performance as cell temperature changes, it should be quite apparent that temperature needs to be included in any estimate of module performance.

Cells vary in temperature not only because ambient temperatures change, but also because insolation on the cells changes. Since only a small fraction of the insolation hitting a module is converted to electricity and carried away, most of that incident energy is absorbed and converted to heat. To help system designers account for changes in cell performance with temperature, manufacturers often provide an indicator called the NOCT, which stands for nominal operating cell temperature. The NOCT is cell temperature in a module when ambient is 20°C, solar irradiation is 0.8 kW/m^2, and windspeed is 1 m/s. To account for other ambient conditions, the following expression may be used:

$$T_{\text{cell}} = T_{\text{amb}} + \left(\frac{\text{NOCT} - 20^\circ}{0.8}\right) \cdot S \tag{8.24}$$

where T_{cell} is cell temperature (°C), T_{amb} is ambient temperature, and S is solar insolation (kW/m^2).

Example 8.5 Impact of Cell Temperature on Power for a PV Module. Estimate cell temperature, open-circuit voltage, and maximum power output for the 150-W BP2150S module under conditions of 1-sun insolation and ambient temperature 30°C. The module has a NOCT of 47°C.

Solution. Using (8.24) with $S = 1$ kW/m^2, cell temperature is estimated to be

$$T_{cell} = T_{amb} + \left(\frac{\text{NOCT} - 20^\circ}{0.8}\right) \cdot S = 30 + \left(\frac{47 - 20}{0.8}\right) \cdot 1 = 64^\circ\text{C}$$

From Table 8.3, for this module at the standard temperature of 25°C, $V_{OC} =$ 42.8 V. Since V_{OC} drops by 0.37%/°C, the new V_{OC} will be about

$$V_{OC} = 42.8[1 - 0.0037(64 - 25)] = 36.7 \text{ V}$$

With maximum power expected to drop about 0.5%/°C, this 150-W module at its maximum power point will deliver

$$P_{max} = 150 \text{ W} \cdot [1 - 0.005(64 - 25)] = 121 \text{ W}$$

which is a rather significant drop of 19% from its rated power.

When the NOCT is not given, another approach to estimating cell temperature is based on the following:

$$T_{cell} = T_{amb} + \gamma\left(\frac{\text{Insolation}}{1 \text{ kW/m}^2}\right) \tag{8.25}$$

where γ is a proportionality factor that depends somewhat on windspeed and how well ventilated the modules are when installed. Typical values of γ range between 25°C and 35°C; that is, in 1 sun of insolation, cells tend to be 25–35°C hotter than their environment.

8.7 SHADING IMPACTS ON I–V CURVES

The output of a PV module can be reduced dramatically when even a small portion of it is shaded. Unless special efforts are made to compensate for shade problems, even a single shaded cell in a long string of cells can easily cut output power by more than half. External diodes, purposely added by the PV manufacturer or by the system designer, can help preserve the performance of PV modules. The main purpose for such diodes is to mitigate the impacts of shading on PV $I-V$ curves. Such diodes are usually added in parallel with modules or blocks of cells within a module.

8.7.1 Physics of Shading

To help understand this important shading phenomenon, consider Fig. 8.37 in which an n-cell module with current I and output voltage V shows one cell separated from the others (shown as the top cell, though it can be any cell in the string). The equivalent circuit of the top cell has been drawn using Fig. 8.26, while the other $(n-1)$ cells in the string are shown as just a module with current I and output voltage V_{n-1}.

In Fig. 8.37a, all of the cells are in the sun and since they are in series, the same current I flows through each of them. In Fig. 8.37b, however, the top cell is shaded and its current source I_{SC} has been reduced to zero. The voltage drop across R_P as current flows through it causes the diode to be reverse biased, so the diode current is also (essentially) zero. That means the entire current flowing through the module must travel through both R_P and R_S in the shaded cell on its way to the load. That means the top cell, instead of adding to the output voltage, actually reduces it.

Consider the case when the bottom $n-1$ cells still have full sun and still some how carry their original current I so they will still produce their original voltage V_{n-1}. This means that the output voltage of the entire module V_{SH} with

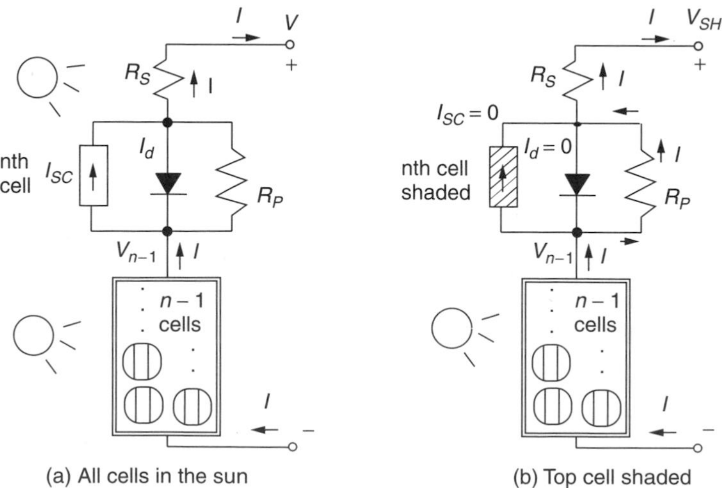

Figure 8.37 A module with n cells in which the top cell is in the sun (a) or in the shade (b).

one cell shaded will drop to

$$V_{SH} = V_{n-1} - I(R_P + R_S) \tag{8.26}$$

With all n cells in the sun and carrying I, the output voltage was V so the voltage of the bottom $n - 1$ cells will be

$$V_{n-1} = \left(\frac{n-1}{n}\right) V \tag{8.27}$$

Combining (8.26) and (8.27) gives

$$V_{SH} = \left(\frac{n-1}{n}\right) V - I(R_P + R_S) \tag{8.28}$$

The drop in voltage Δ V at any given current I, caused by the shaded cell, is given by

$$\Delta V = V - V_{SH} = V - \left(1 - \frac{1}{n}\right) V + I(R_P + R_S) \tag{8.29}$$

$$\Delta V = \frac{V}{n} + I(R_P + R_S) \tag{8.30}$$

Since the parallel resistance R_P is so much greater than the series resistance R_S, (8.30) simplifies to

$$\Delta V \cong \frac{V}{n} + IR_P \tag{8.31}$$

At any given current, the I–V curve for the module with one shaded cell drops by ΔV. The huge impact this can have is illustrated in Fig. 8.38.

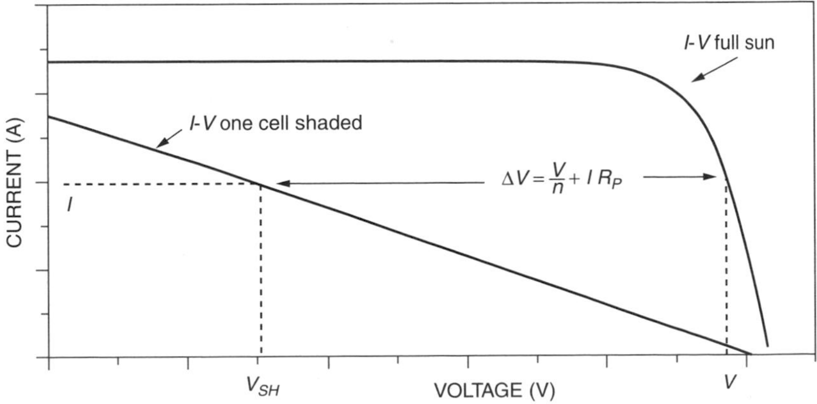

Figure 8.38 Effect of shading one cell in an n-cell module. At any given current, module voltage drops from V to $V - \Delta V$.

Example 8.6 Impacts of Shading on a PV Module. The 36-cell PV module described in Example 8.4 had a parallel resistance per cell of $R_P = 6.6\ \Omega$. In full sun and at current $I = 2.14$ A the output voltage was found there to be $V = 19.41$ V. If one cell is shaded and this current somehow stays the same, then:

a. What would be the new module output voltage and power?
b. What would be the voltage drop across the shaded cell?
c. How much power would be dissipated in the shaded cell?

Solution

a. From (8.31) the drop in module voltage will be

$$\Delta V = \frac{V}{n} + IR_P = \frac{19.41}{36} + 2.14 \times 6.6 = 14.66 \text{ V}$$

The new output voltage will be $19.41 - 14.66 = 4.75$ V.

Power delivered by the module with one cell shaded would be

$$P_{\text{module}} = VI = 4.75 \text{ V} \times 2.14 \text{ A} = 10.1 \text{ W}$$

For comparison, in full sun the module was producing 41.5 W.

b. All of that 2.14 A of current goes through the parallel plus series resistance (0.005 Ω) of the shaded cell, so the drop across the shaded cell will be

$$V_c = I(R_P + R_S) = 2.14(6.6 + 0.005) = 14.14\ V$$

(normally a cell in the sun will add about 0.5 V to the module; this shaded cell subtracts over 14 V from the module).

c. The power dissipated in the shaded cell is voltage drop times current, which is

$$P = V_c I = 14.14 \text{ V} \times 2.14 \text{ A} = 30.2 \text{ W}$$

All of that power dissipated in the shaded cell is converted to heat, which can cause a local hot spot that may permanently damage the plastic laminates enclosing the cell.

The procedures demonstrated in Examples 8.4 and 8.6 can be extended to develop $I-V$ curves under various conditions of shading. Figure 8.39 shows such curves for the example module under full-sun conditions and with one cell 50% shaded, one cell completely shaded, and two cells completely shaded. Also shown on the graph is a dashed vertical line at 13 V, which is a typical operating

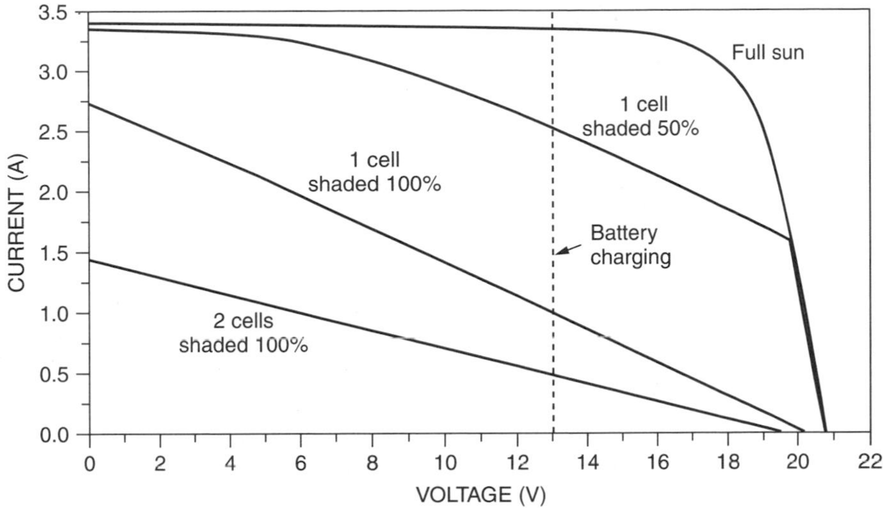

Figure 8.39 Effects of shading on the $I-V$ curves for a PV module. The dashed line shows a typical voltage that the module would operate at when charging a 12-V battery; the impact on charging current is obviously severe.

voltage for a module charging a 12-V battery. The reduction in charging current for even modest amounts of shading is severe. With just one cell shaded out of 36 in the module, the power delivered to the battery is decreased by about two-thirds!

8.7.2 Bypass Diodes for Shade Mitigation

Example 8.6 shows not only how drastically shading can shift the $I-V$ curve, but also how local, potentially damaging hot spots can be created in shaded cells. Figure 8.40 shows a typical situation. In Fig. 8.40a a solar cell in full sun operating in its normal range *contributes* about 0.5 V to the voltage output of the module, but in the equivalent circuit shown in 8.40b a shaded cell experiences a *drop* as current is diverted through the parallel and series resistances. This drop can be considerable (in Example 8.6 it was over 14 V).

The voltage drop problem in shaded cells could be to corrected by adding a *bypass diode* across each cell, as shown in Fig. 8.41. When a solar cell is in the sun, there is a voltage rise across the cell so the bypass diode is cut off and no current flows through it—it is as if the diode is not even there. When the solar cell is shaded, however, the drop that would occur if the cell conducted any current would turn on the bypass diode, diverting the current flow through that diode. The bypass diode, when it conducts, drops about 0.6 V. So, the bypass diode controls the voltage drop across the shaded cell, limiting it to a relatively modest 0.6 V instead of the rather large drop that may occur without it.

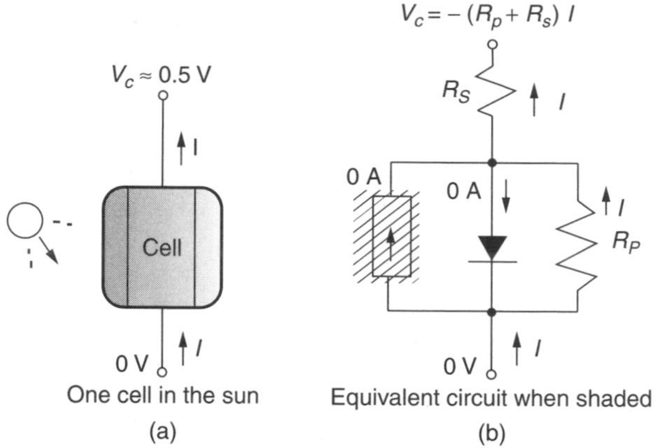

Figure 8.40 In full sun a cell may contribute around 0.5 V to the module output; but when a cell is shaded, it can have a large voltage drop across it.

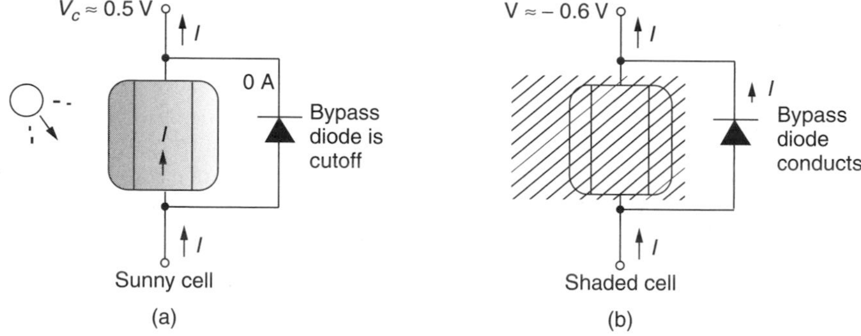

Figure 8.41 Mitigating the shade problem with a bypass diode. In the sun (a), the bypass diode is cut off and all the normal current goes through the solar cell. In shade (b), the bypass diode conducts current around the shaded cell, allowing just the diode drop of about 0.6 V to occur.

In real modules, it would be impractical to add bypass diodes across every solar cell, but manufacturers often do provide at least one bypass diode around a module to help protect arrays, and sometimes several such diodes around groups of cells within a module. These diodes don't have much impact on shading problems of a single module, but they can be very important when a number of modules are connected in series. Just as cells are wired in series to increase module voltage, modules can be wired in series to increase array voltage. Also, just as a single cell can drag down the current within a module, a few shaded cells in a single module can drag down the current delivered by the entire string in an array. The benefit already demonstrated for a bypass diode on a single cell also applies to a diode applied across a complete module.

To see how bypass diodes wired in parallel with modules can help mitigate shading problems, consider Fig. 8.42, which shows $I-V$ curves for a string of five modules (the same modules that were used to derive Fig. 8.39). The graph shows the modules in full sun as well as the $I-V$ curve that results when one module has two cells completely shaded. Imagine the PVs delivering charging current at about 65 V to a 60-V battery bank. As can be seen, in full sun about 3.3 A are delivered to the batteries. However, when just two cells in one module are shaded, the current drops by one-third to about 2.2 A. With a bypass diode across the shaded module, however, the $I-V$ curve is improved considerably as shown in the figure.

Figure 8.43 helps explain how the bypass diodes do their job. Imagine five modules, wired in series, connected to a battery that forces the modules to operate at 65 V. In full sun the modules deliver 3.3 A at 65 V. When any of the cells are shaded, they cease to produce voltage and instead begin to act like resistors (6.6 Ω per cell in this example) that cause voltage to drop as the other modules continue to try to push current through the string. Without a bypass diode to divert the current, the shaded module loses voltage and the other modules try to compensate by increasing voltage, but the net effect is that current in the whole string drops. If, however, bypass diodes are provided, as shown in Fig. 8.43c, then current will go around the shaded module and the charging current bounces back to nearly the same level that it was before shading occurred.

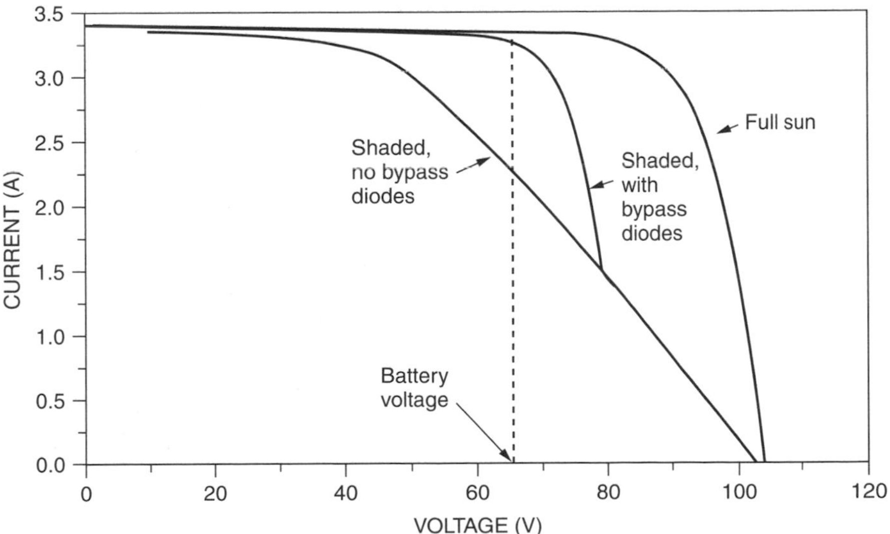

Figure 8.42 Impact of bypass diodes. Drawn for five modules in series delivering 65 V to a battery bank. With one module having two shaded cells, charging current drops by almost one-third when there are no bypass diodes. With the module bypass diodes there is very little drop.

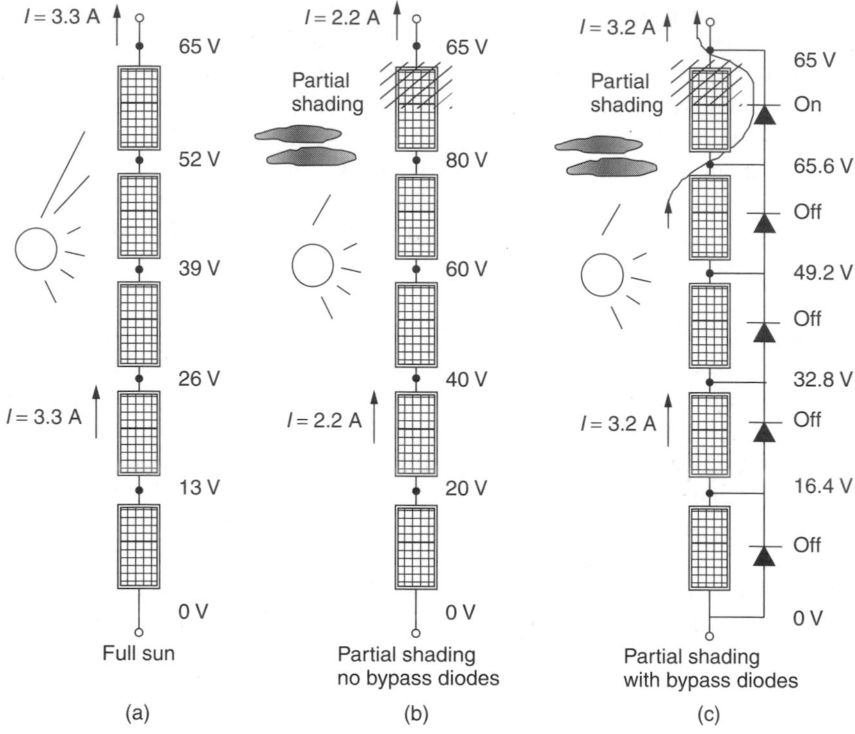

Figure 8.43 Showing the ability of bypass diodes to mitigate shading when modules are charging a 65 V battery. Without bypass diodes, a partially shaded module constricts the current delivered to the load (b). With bypass diodes, current is diverted around the shaded module.

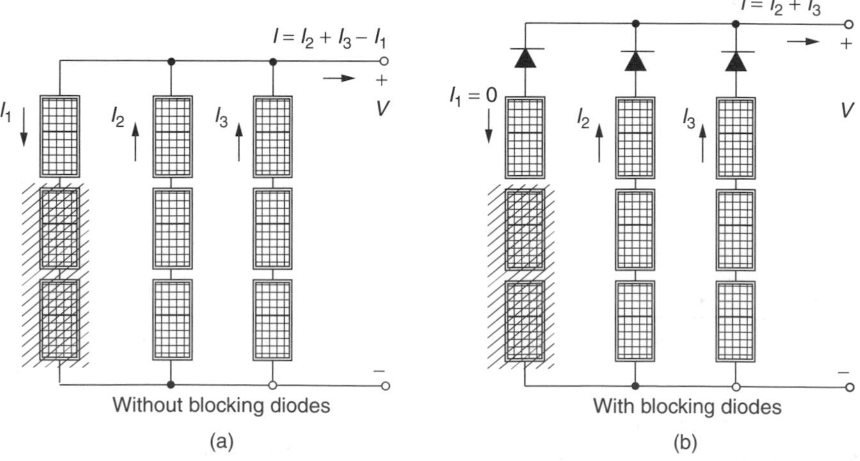

Figure 8.44 Blocking diodes prevent reverse current from flowing down malfunctioning or shaded strings.

8.7.3 Blocking Diodes

Bypass diodes help current go around a shaded or malfunctioning module within a string. This not only improves the string performance, but also prevents hot spots from developing in individual shaded cells. When strings of modules are wired in parallel, a similar problem may arise when one of the strings is not performing well. Instead of supplying current to the array, a malfunctioning or shaded string can withdraw current from the rest of the array. By placing *blocking diodes* (also called *isolation diodes*) at the top of each string as shown in Fig. 8.44, the reverse current drawn by a shaded string can be prevented.

8.8 CRYSTALLINE SILICON TECHNOLOGIES

Thus far, the discussion of photovoltaics has been quite generic; that is, it hasn't particularly depended upon what technology was used to fabricate the cells. The circuit concepts just developed will be used in the next chapter when we explore PV system design, but before we get there it will be helpful to explore the different types of technologies currently used to manufacture photovoltaics.

There are a number of ways to categorize photovoltaics. One dichotomy is based on the thickness of the semiconductor. Conventional crystalline silicon solar cells are, relatively speaking, very thick—on the order of 200–500 μm (0.008–0.020 in.). An alternative approach to PV fabrication is based on thin films of semiconductor, where "thin" means something like 1–10 μm. Thin-film cells require much less semiconductor material and are easier to manufacture, so they have the potential to be cheaper than thick cells. The first generation of thin-film PVs were only about half as efficient as conventional thick silicon cells; they were less reliable over time, yet they were no cheaper per watt, so they really weren't competitive. All three of these negative attributes have been addressed, more or less successfully, and thin-film PVs are beginning to become more competitive. In the near future they may even dominate PV sales. Currently, however, about 80% of all photovoltaics are thick cells and the remaining 20% are thin-film cells used mostly in calculators, watches, and other consumer electronics.

Photovoltaic technologies can also be categorized by the extent to which atoms bond with each other in individual crystals. As described by Bube (1998), there is a "family tree" of PVs based on the size of these crystals. The historically generic name "polycrystalline" can be broken down into the following more specific terms: (1) *single crystal*, the dominant silicon technology; (2) *multicrystalline*, in which the cell is made up of a number of relatively large areas of single crystal grains, each on the order of 1 mm to 10 cm in size, including multicrystalline silicon (mc-Si); (3) *polycrystalline*, with many grains having dimensions on the order of 1 μm to 1 mm, as is the case for cadmium telluride (CdTe) cells, copper indium diselenide ($CuInSe_2$,) and polycrystalline, thin-film silicon; (4) *microcrystalline* cells with grain sizes less than 1 μm; and (5) *amorphous*, in which there are no single-crystal regions, as in amorphous silicon (a-Si).

Another way to categorize photovoltaic materials is based on whether the p and n regions of the semiconductor are made of the same material (with different dopings, of course)—for example, silicon. These are called *homojunction* photovoltaics. When the p–n junction is formed between two different semiconductors, they are called *heterojunction* PVs. For example, one of the most promising heterojunction combinations uses cadmium sulfide (CdS) for the n-type layer and copper indium diselenide ($CuInSe_2$, also known as "CIS") for the p-type layer.

Other distinctions include multiple junction solar cells (also known as cascade or tandem cells) made up of a stack of p–n junctions with each junction designed to capture a different portion of the solar spectrum. The shortest-wavelength, highest-energy photons are captured in the top layer while most of the rest pass through to the next layer. Subsequent layers have lower and lower band gaps, so they each pick off the most energetic photons that they see, while passing the rest down to the next layer. Very high efficiencies are possible using this approach.

Finally, some cells are specifically designed to work best with concentrated sunlight while others are used in nonconcentrating flat-plate systems. Figure 8.45 provides a rough road map for our subsequent discussion of each technology.

8.8.1 Single-Crystal Czochralski (CZ) Silicon

Silicon is the second most abundant element on earth, comprising approximately 20% of the earth's crust. Pure silicon almost immediately forms a layer of SiO_2 on its surface when exposed to air, so it exists in nature mostly in SiO_2-based minerals such as quartzite or in silicates such as mica, feldspars, and zeolites.

The raw material for photovoltaics and other semiconductors could be common sand, but it is usually naturally purified, high-quality silica or quartz (SiO_2) from mines. The first processing step to convert silica into crystalline silicon is an energy-intensive process in which SiO_2 is reduced to an impure Si using carbon in an arc furnace. It is then transformed to a liquid trichlorosilane ($SiHCl_3$),

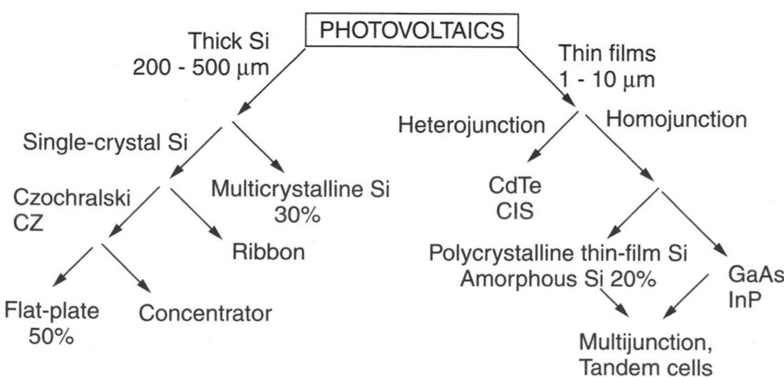

Figure 8.45 One way to organize the discussion of photovoltaic technologies. Percentages represent fraction of PV sales in the late 1990s.

which can be purified by fractional distillation. The purified trichlorosilane can then be reduced with hydrogen to form extremely pure silicon:

$$SiHCl_3 + H_2 + \text{heat} \rightarrow Si + 3HCl \tag{8.32}$$

The result is rock-like chunks of 99.9999% pure silicon. When heated to over 1400°C, the rocks can be melted in a quartz crucible to form a molten vat of silicon.

The most commonly used technique for forming single-crystal silicon from the crucible of molten silicon is the *Czochralski*, or CZ, method (Fig. 8.46a), in which a small seed of solid, crystalline silicon about the size of a pencil is dipped into the vat and then slowly withdrawn using a combination of pulling (10^{-4}–10^{-2} cm/s) and rotating (10–40 rpm). As it is withdrawn, the molten silicon atoms bond with atoms in the crystal and then solidify (freeze) in place. The result is a large cylindrical ingot or "boule" of single-crystal silicon perhaps a meter long and as large as about 20 cm in diameter. By adding proper amounts of a dopant to the melt, the resulting ingot can be fabricated as an n- or p-type material. Usually the dopant is boron and the ingot is therefore a p-type semiconductor. An alternative to the Czochralski method is called the *float-zone* (FZ) process, in which a solid ingot of silicon is locally melted and then solidified by an RF field that passes slowly along the ingot.

After the cylindrical ingot is formed, four sides may be sliced off, resulting in a block of silicon that has an approximately square cross section. This "square-ness" allows greater packing density when the cells are assembled into a photovoltaic module. The silicon block must then be sawed into thin slices called *wafers*. This can be done with a saw blade (Fig. 8.46b) or with a diamond impregnated wire that cuts through the wafer. In either case, a large fraction of the ingot is lost as kerf (like sawdust)—as much as 50% when a saw blade is used, less if the wire saw is used. The wafers are then etched to remove some of the surface damage and to expose the microscopic crystalline structure at the top of the cell. The surface is made up of a jumble of four-sided pyramids, which

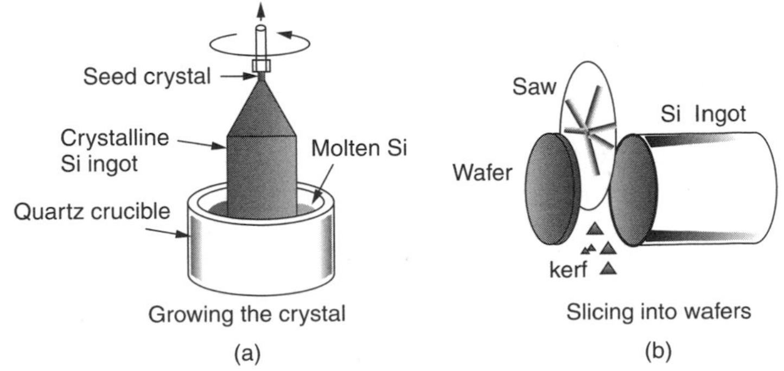

Figure 8.46 The Czochralski method for growing single-crystal silicon.

helps reflect light down into the crystal. After polishing, the wafers are ready to be doped to make the p–n junction.

During the above wafer fabrication, the crystalline silicon is usually doped with acceptor atoms, making it p-type throughout its 200- to- 500-μm thickness. To form the junction, a thin 0.1- to 0.5-μm n-type layer is created by diffusing enough donor atoms into the top of the cell to overwhelm the already existing acceptors. The wafers are placed in long tubes of silica glass for the diffusion process. The impurities, in gaseous form, flow through the tubes, thereby exposing the wafers under carefully controlled exposure time and temperature conditions. For most crystalline silicon, the donor atoms are phosphorus from phosphine gas (PH_3) and the acceptors are boron (from diborane, B_2H_6).

Since silicon is naturally quite reflective to solar wavelengths, some sort of surface treatment is required to reduce those losses. An antireflection (AR) coating of some transparent material such as tin oxide (SnO_2) is applied. These coatings tend to readily transmit the green, yellow, and red light into the cell, but some of the shorter-wavelength blue light is reflected, which gives the cells their characteristic dark blue color.

The next step is the attachment of electrical contacts to the cell. For many years, the bottom contacts were formed by vacuum deposition of a layer of aluminum that covered the back side of the cell. Aluminum is a Group III element, so it not only serves as a conductor but also can contribute to the concentration of holes in the bottom of the p-layer, forming what is called a p^+ layer. Those extra holes help reduce the contact resistance between the silicon and aluminum, and the gradient of holes that they create helps reduce recombination at the contact by driving holes away from the back surface. A cross section of a typical 1970s vintage, single-crystal silicon cell is shown in Fig. 8.47a.

The front-surface contacts in most cells have been formed by depositing a grid of metal conductors that covers on the order of 5–10% of the total area. That coverage, of course, reduces the amount of sunlight reaching the junction and hence reduces the overall cell efficiency. Some newer cells, called *back-point contact cells*, put both contacts on the bottom to avoid that shading effect. Another approach involves use of lasers to dig deep, narrow grooves into the cell. The deep grooves in these *laser-grooved, buried-contact cells* are filled with metal, forming a large contact area while minimizing the top-surface area shaded by the contact. The bottom contacts can also be formed using this laser technique, resulting in what is called a "double-sided" laser-grooved photovoltaic cell (Fig. 8.47b).

In newer cells, a number of other techniques may be incorporated into the top surface to improve performance. One of the most advanced crystalline silicon photovoltaics is called *the passive-emitter, rear locally diffused* (PERL) cell (Fig. 8.47c). In a PERL cell, inverted pyramids on the front surface, covered with an antireflection coating, help capture and bounce light into the cell. These cells not only direct more sunlight into the cell, they also reflect back into the cell photons that were reflected off of the bottom oxide layer covered by aluminum. Efficiencies approaching 25% have been achieved.

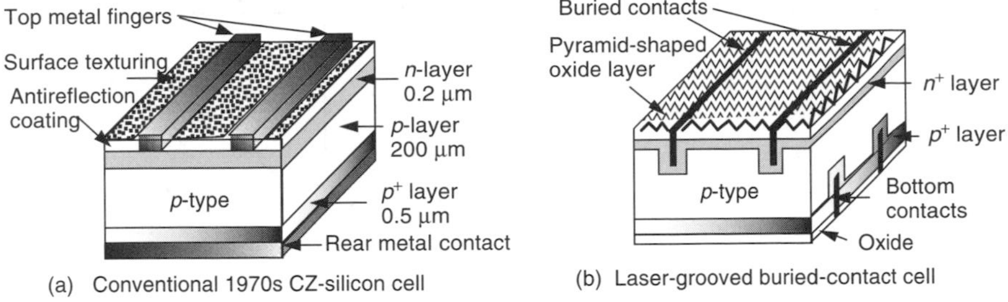

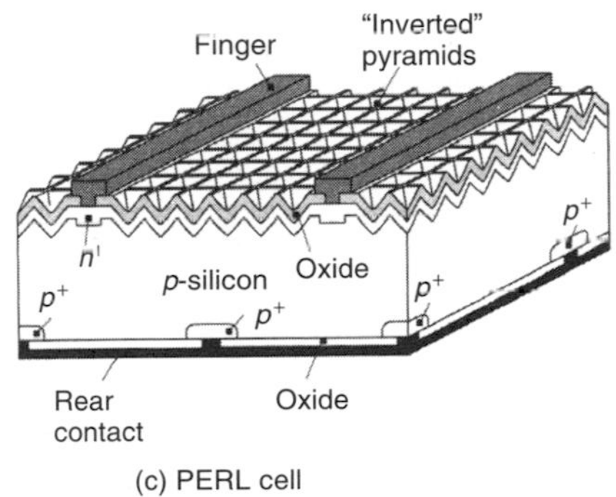

Figure 8.47 Evolution of CZ-silicon solar cells. (a) A conventional 1970s cell, showing typical thicknesses of semiconductor. (b) Double-sided, laser-grooved, buried-contact cell. (c) Passive-emitter, rear locally diffused (PERL) cell. Based on Green (1993).

Great progress has been made in improving the maximum efficiency achieved in laboratory cells. Figure 8.48 illustrates the rise from less than 1% in the 1940s, to just over 15% in the early 1970s, to over 25% by 2000. Actual production module efficiencies in 2003 were in the range of 14 to 17%.

8.8.2 Ribbon Silicon Technologies

The fact that the CZ and FZ processes for wafer production were developed for the semiconductor industry has been both a blessing and a curse. Without question, the photovoltaics industry has benefited greatly by being able to piggyback onto technology developed for the immense microelectronics industry. Also, since PVs don't need the very highest quality wafers, they can use some of the rejected crystalline silicon originally meant for the semiconductor industry. On the other hand, a wafer that is fabricated into thousands of high-priced integrated circuits can generate a lot more revenue than one that ends up as a photovoltaic cell, which means an electronics company can pay a lot more for wafers, keeping their price high.

With the cost of sliced-and-polished wafers being a significant fraction of the cost of PVs, attempts have been made to find other ways to fabricate crystalline

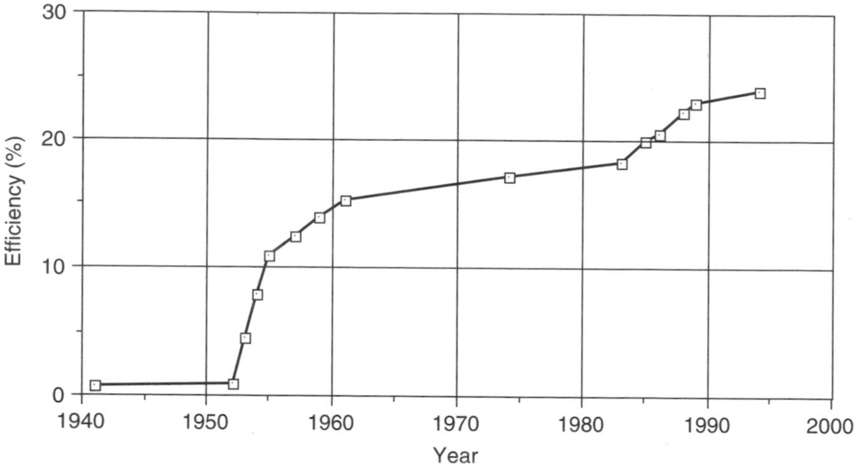

Figure 8.48 Increasing efficiency of single-crystal silicon, laboratory-scale cells. From Bube (1998).

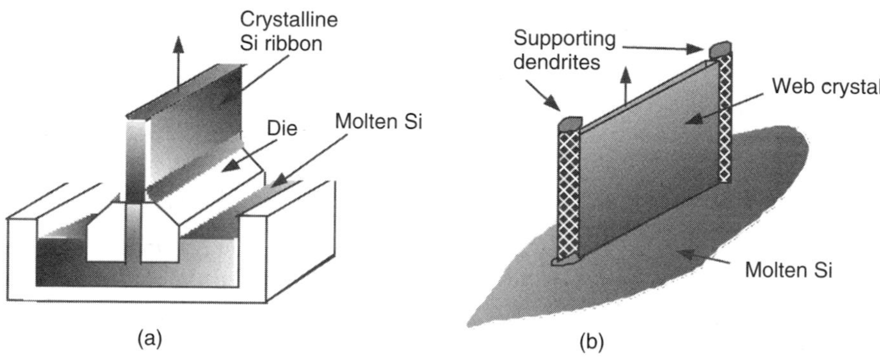

Figure 8.49 Ribbons of crystalline silicon can be grown using the edge-defined film-fed growth process (EFG) in (a) or using the dendritic web process (b).

silicon. Several such technologies are based on growing crystalline silicon that emerges as a long, thin, continuous ribbon from the silicon melt. The ribbons can then be scribed and broken into rectangular cells without the wastefulness of sawing an ingot and without the need for separate polishing steps.

One such approach, called "edge-defined, film-fed growth (EFG)," is illustrated in Fig. 8.49a. It is based on a carbon die that is partially immersed in molten silicon. Within the slot-shaped aperture in the die, molten silicon solidifies and emerges as a frozen ribbon of crystalline silicon. An earlier ribbon process was based on growth of a thin sheet of crystal between two parallel dendrite supports (Fig. 8.49b). Very precise temperature control is required for the dendrites, and the solidifying silicon between them, to form good crystalline material as

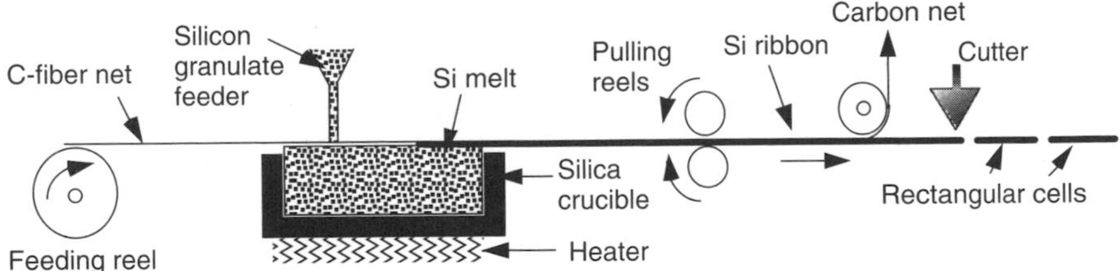

Figure 8.50 The S-Web process produces a continuous ribbon of silicon, which can be doped and cut into individual rectangular cells. Based on Schmela (2000).

the ribbon is pulled from the melt. A similar process is called the String Ribbon technique, in which two high-temperature vertical strings are pulled vertically through a shallow silicon melt. The molten silicon spans and freezes between the strings, forming a long, thin single-crystal ribbon. The silicon ribbon is then cut to length without interrupting the continuous growth of new ribbon.

A third ribbon technology was introduced by Siemens Solar in 1990, but was never fully developed until the German company, Brandl AG, picked it up in 2000. The S-Web process utilizes a carbon-fiber net that is pulled horizontally along the surface of a silicon melt. The silicon solidifies along the lower side of the net, forming a crystal that can be drawn out into a continuous ribbon, which can then be doped and cut into rectangular, cell-sized pieces. A sketch of the process is shown in Fig. 8.50.

8.8.3 Cast Multicrystalline Silicon

Another way to avoid the costly Czochralski and float-zone processes is based on carefully cooling and solidifying a crucible of molten silicon, yielding a large, solid rectangular ingot. Since these ingots may be quite large, on the order of $40 \times 40 \times 40$ cm and weighing over 100 kg, the ingot may need to be cut into smaller, more manageable blocks, which are then sliced into silicon wafers using either the saw or wire-cutting techniques. Sawing can waste a significant fraction of the ingot, but since this casting method is itself cheaper and it utilizes less expensive, less pure silicon than the CZ process, the waste is less important.

Casting silicon in a mold and then carefully controlling its rate of solidification results in an ingot that is not a single, large crystal. Instead, it consists of many regions, or grains, that are individually crystalline and which tend to have grain boundaries that run perpendicularly to the plane of the cell. Defective atomic bonds at these boundaries increase recombination and diminish current flow, resulting in cell efficiencies that tend to be a few percentage points below CZ cells. Figure 8.51 illustrates the casting, cutting, slicing, and grain boundary structure of these multicrystalline silicon (mc-Si) cells.

8.8.4 Crystalline Silicon Modules

The photovoltaic technologies described thus far result in individual thick cells that must be wired together to create modules with the desired voltage and current

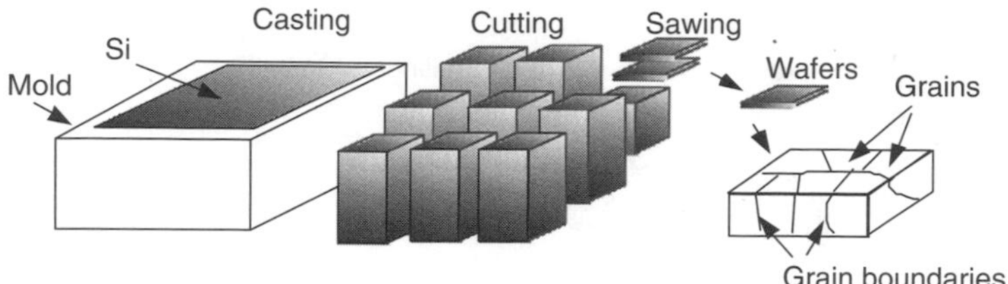

Figure 8.51 Casting, cutting and sawing of silicon results in wafers with individual grains of crystalline silicon separated by grain boundaries.

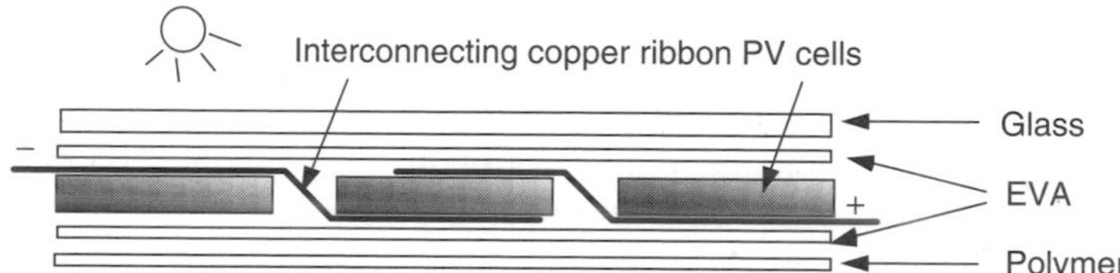

Figure 8.52 Thick crystalline cells must be wired together in series and then sandwiched in layers of glass, EVA, and polymers for structural support and weather protection.

characteristics. This wiring is done with automated soldering machines, which connect the cells in series—that is, with the front of one cell connected to the back of the next, as shown in Fig. 8.52. After soldering, the cells are laminated into a sandwich of materials that offer structural support as well as weather protection. The upper surface is tempered glass, and the cells are encapsulated in two layers of ethylene vinyl acetate (EVA). Finally, the back is covered with sheets of polymer that prevent moisture penetration.

8.9 THIN-FILM PHOTOVOLTAICS

Conventional crystalline silicon technologies (x-Si) require considerable amounts of expensive material with additional complexity and costs needed to wire individual cells together. Competing technologies, however, are based on depositing extremely thin films of photovoltaic materials onto glass or metal substrates. Thin-film devices use relatively little material (their thickness is in the micron range rather than the hundreds-of-microns range needed by crystalline silicon),* they do not require the complexity of cell interconnections, and they are particularly well suited to mass-production techniques. Their thinness allows photons that aren't absorbed to pass completely through the photovoltaic material, which offers two special opportunities. Their semitransparency means that they can be deposited onto windows, making building glass a provider of both light and

*For comparison, a human hair is roughly 100 μm in diameter.

electricity. They also lend themselves to multiple-junction, tandem cells in which photons of different wavelengths are absorbed in different layers of the device.

In exchange for these highly desirable properties, thin-film cells are not as efficient as x-Si—especially when they are not used in tandem devices. While the likelihood of significant reductions in module costs are modest for conventional crystalline silicon, many opportunities remain to increase efficiency and dramatically reduce costs using thin-film technologies. Many believe that thin films will be the dominant photovoltaic technology in the future.

8.9.1 Amorphous Silicon

Almost all of today's thin-film technology is based on amorphous (glassy) silicon (a-Si)—that is, silicon in which there is very little order to the arrangement of atoms. Since it is not crystalline, the organized tetrahedral structure in which one silicon atom bonds to its four adjacent neighbors in a precisely defined manner does not apply. While almost all of the atoms do form bonds with four other silicon atoms, there remain numerous "dangling bonds" where nothing attaches to one of the valence electrons. These dangling-bond defects act as recombination centers so that photogenerated electrons recombine with holes before they can travel very far. The key to making a-Si into a decent photovoltaic material was first discovered, somewhat by accident, in 1969, by a British team that noted a glow when silane gas SiH_4 was bombarded with a stream of electrons (Chittick et al., 1969). This led to their critical discovery that by alloying amorphous silicon with hydrogen the concentration of defects could be reduced by about three orders of magnitude. The concentration of hydrogen atoms in these alloys is roughly 1 atom in 10, so their chemical composition is approximately $Si_{0.9}H_{0.1}$. Moreover, the silicon–hydrogen alloy that results, designated as a-Si:H, is easily doped to make n-type and p-type materials for solar cells.*

The first a-Si:H solar cells were described in the literature in 1976 (Carlson and Wronski, 1976). While only 1% efficient, their potential was quickly recognized. The first commercial use of a-Si:H came in 1980 when Sanyo introduced a line of solar-powered pocket calculators. And by the mid-1980s, amorphous silicon PV modules for use outdoors were on the market. Their efficiency was only about 5% or 6% when new; within the first few months of use, this dropped to about 3% or 4%. Understanding this efficiency instability, and dealing with it, has remained a significant challenge. By the early 1990s, single-junction cells had stabilized efficiencies of about 10%, while considerably higher efficiencies have been reached for multijunction cells.

So, how can a p–n junction be formed in an amorphous material with very little organization among its atoms? Figure 8.53 shows a cross section of a simple a-Si:H cell that uses glass as the supporting superstrate. On the underside of the

*It is interesting to note that amorphous silicon is finding application in other products besides photovoltaics. The fact that it is an insulator in the dark but readily conducts current when exposed to light has led to its incorporation in photoelectronic products such as photocopiers, laser printers, and fax machines.

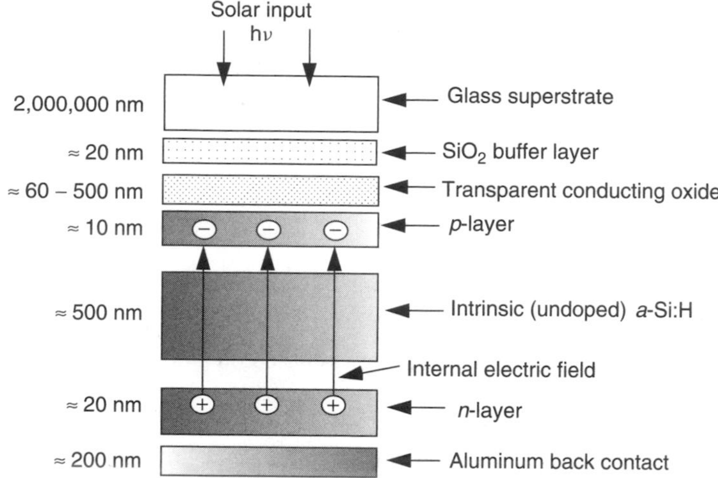

Figure 8.53 Cross section of an amorphous silicon p–i–n cell. The example thicknesses are in nanometers (10^{-9} m) and are not drawn to scale.

glass a buffer layer of SiO_2 may be deposited in order to prevent subsequent layers of atoms from migrating into the glass. Next comes the electrical contact for the top of the cell, which is usually a transparent conducting oxide (TC) such as tin oxide, indium–tin oxide, or zinc oxide. The solar cell itself is often formed by subjecting gaseous silicon in the form of silane SiH_4 (or disilane Si_2H_6) to bombardment by a stream of electrons. The bombardment creates reactive silane radicals, which can attach themselves onto what becomes a growing layer of amorphous silicon. During this process, the excited silane can emit photons of light, in what is called a glow discharge. The n-layer is formed by adding phosphene gas PH_3 to the mix, while diborane gas B_2H_6 is used to form the p-layer.

The actual p–n junction, whose purpose is of course to create the internal electric field in the cell to separate holes and electrons, consists of three layers consisting of the p-layer and n-layers separated by an undoped (intrinsic) region of a-Si:H As shown, the p-layer is only 10 nm thick; the intrinsic layer, or i-layer, is 500 nm; and the n-layer is 20 nm. Notice that the electric field created between the rigid positive charges in the n-layer and the rigid negative charges in the p-layer spans almost the full depth of the cell. This means that light-induced hole–electron pairs created almost anywhere within the cell will be swept across the intrinsic layer by the internal field. These amorphous silicon PVs are referred to as p–i–n cells.

A variation on the a-Si p–i–n cell uses a thin, flexible stainless steel substrate located at the bottom of the cell instead of the glass superstrate shown in Fig. 8.53. The stainless steel provides the mechanical strength needed while its flexibility allows modules of these cells to be literally rolled up and stored (Fig. 8.54). These flexible modules are often fitted with grommets so they can be lashed to irregular surfaces such as those found on boats.

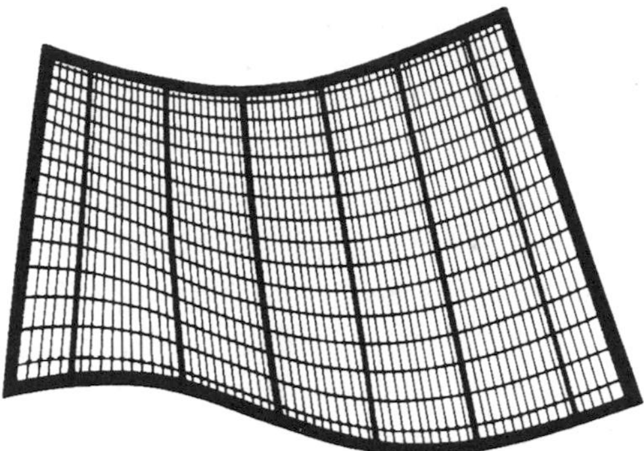

Figure 8.54 Flexible a-Si modules can be rolled up and stored when not in use. From SERI (1985).

The theoretical maximum efficiency for single-junction a-Si is 28%. The best laboratory cells today have a stabilized efficiency of just over 13 percent but commercial modules have stabilized efficiencies of only 5 to 8%.

Amorphous Silicon Processing. An important advantage of thin-film photovoltaics over conventional crystalline silicon is the ease with which they can be manufactured. Starting with rather ordinary glass, layer upon layer of materials are deposited onto the glass in a decomposition chamber. The raw materials are gaseous rather than solid, and whole modules (many cells connected in series) are created at one time rather than separately fabricating individual cells that must be soldered together to form a module.

A simplified step-by-step a-Si manufacturing process is described in Fig. 8.55. It begins with the deposition of a transparent conductor covering the entire underside of the glass superstrate (not shown is an SiO_2 buffer layer that might also be included). Then long narrow metal "stitch" bars, which run the length of the module, are printed onto the transparent conductor. After scribing the transparent conductor, the bars will connect the top of one cell to the bottom of the next one. Now the three layers of silicon are deposited, forming the p–i–n structure of the cell itself. The final layer to be deposited is the bottom conductor, which is then scribed with long, narrow cuts that run the length of the module to complete the separation of one cell from the next. The bottom conductor may be a transparent oxide or it may be metal. If it is a transparent oxide, then photons that aren't captured in the cells can pass through the entire system, delivering some light to whatever is beneath the module.

Figure 8.56 shows how the individual cells run the entire length of the module with scribe lines that separate each cell from adjacent cells.

Multijunction or Tandem a-Si The band gap of a-Si:H is 1.75 eV, which is quite a bit higher than the 1.12 eV for crystalline silicon. Recall that higher band gaps increase voltage at the expense of lower currents (a smaller fraction of

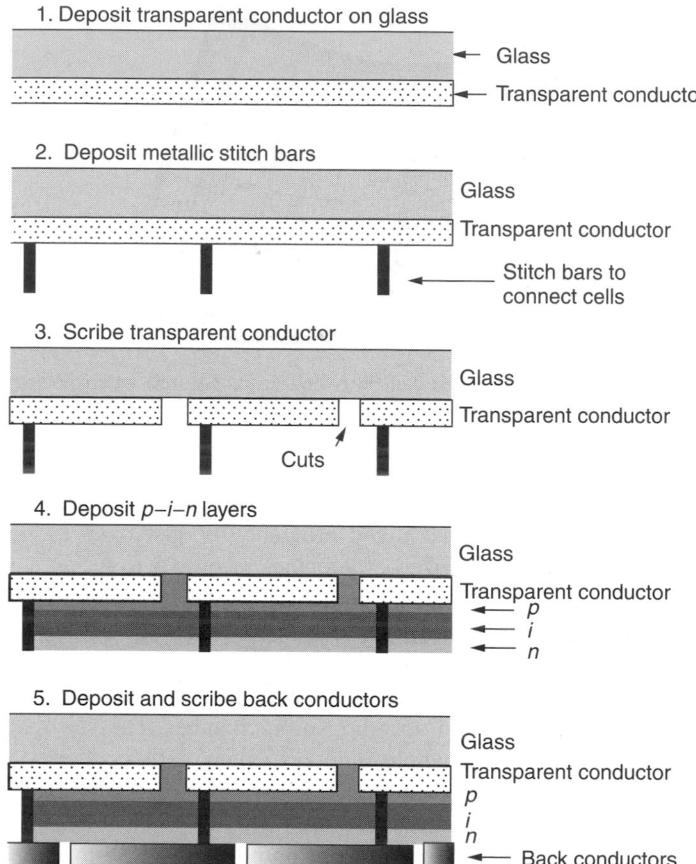

Figure 8.55 The sequence of steps taken to create a module of amorphous silicon cells.

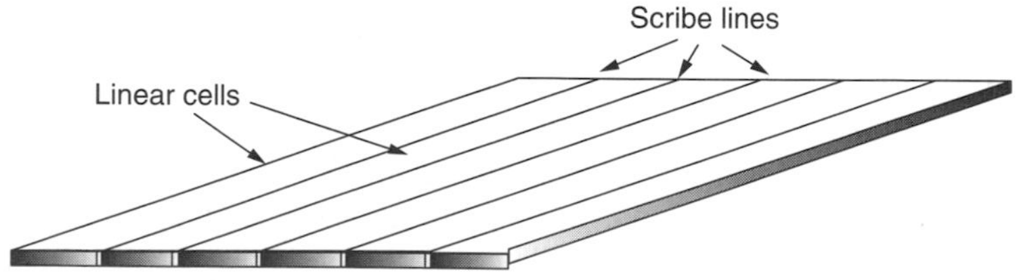

Figure 8.56 Individual cells can run the entire length of an amorphous silicon module.

solar photons have sufficient energy to create hole–electron pairs). Since power is the product of voltage and current, there will be some optimum band gap for a single-junction cell, which will theoretically result in the most efficient device. As was shown in Fig. 8.11, that optimum band gap (at AM1 corresponding to the sun directly overhead) is about 1.4 eV. Thus crystalline silicon has a band gap somewhat too low, while a-Si has one too high to be optimum. As it turns out, however, amorphous silicon has the handy property that alloys made with other Group IV elements will cause the band gap to change. As a general rule, moving up a row in the Periodic Table increases band gap, while moving down a row decreases band gap. Referring to the portion of the Periodic Table given in Table 8.1, this rule would suggest that carbon (directly above silicon) would have a higher band gap than silicon, while germanium (directly below silicon) would have a lower band gap. To lower the 1.75-eV band gap of amorphous silicon toward the 1.4 eV optimum suggests that an alloy of silicon with the right amount of germanium (forming *a*-Si:H:Ge) can help improve cell efficiency. And that is the case.

The above discussion on *a*-Si alloys leads to an even more important opportunity, however. When *a*-Si is alloyed with carbon, for example, the band gap can be increased (to about 2 eV), and when alloyed with germanium the gap will be reduced (to about 1.3 eV). That suggests that multijunction photovoltaic devices can be fabricated by layering p–n junctions of different alloys. The idea behind a multijunction cell is to create junctions with decreasing band gaps as photons penetrate deeper and deeper into the cell. As shown in Fig. 8.57a, the top junction should capture the most energetic photons while allowing photons with less energy to pass through to the next junction below, and so forth. In Fig. 8.57b, an example amorphous silicon, three-junction photovoltaic device is shown in which advantage is taken of the ability of germanium and carbon to increase or decrease the *a*-Si:H band gap.

The theoretical maximum efficiency of an ideal multijunction *a*-Si:H cell is 42%, and some estimate a practically achievable efficiency of about 24% (Carlson and Wagner, 1993). By the turn of the century, multijunction, amorphous silicon modules had a stabilized efficiency of almost 11%.

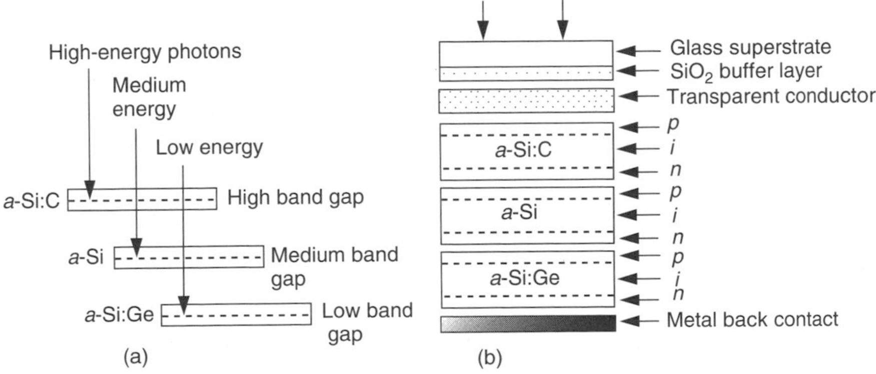

Figure 8.57 Multijunction amorphous silicon solar cells can be made by alloying *a*-Si:H (band gap ≈1.75 eV) with carbon *a*-Si:C in the top layer (≈2.0 eV) to capture the highest-energy photons and germanium *a*-Si:Ge (≈1.3 eV) in the bottom layer to capture the lowest-energy photons.

8.9.2 Gallium Arsenide and Indium Phosphide

While silicon dominates the photovoltaic industry, there is emerging competition from thin films made of compounds of two or more elements. Referring back to the portion of the Periodic Table of the elements shown in Table 8.1, recall that silicon is in the fourth column, and it is referred to as a Group IV element. These other compounds are often made up of pairs of elements from the third and fifth columns (called III–V materials), or pairs from the second and sixth columns (II–VI materials). For example, gallium, which is a Group III element, paired with arsenic, which is Group V, can be used to make gallium arsenide (GaAs) photovoltaics. Similarly, indium (Group III) and phosphorus (Group V) can be made into indium phosphide (InP) cells. Later we will consider II–VI materials such as cadmium (Group II) and tellurium (Group VI) in CdTe ("cad-telluride") cells.

Compounds such as GaAs can be grown as crystals and doped with acceptor (p-type) and donor (n-type) impurities. Common donors include Group VI elements such as Se and Te, while Group II elements such as Zn and Cd can be used as acceptors. It is even possible for elements from Group IV such as C, Si, Ge, and Sn to act as donors or acceptors, depending on which element they displace. For example, when Ge substitutes for Ga on a particular site in the lattice, it acts as a donor, but when it substitutes for As it acts as an acceptor.

As shown in Fig. 8.11, the GaAs band gap of 1.43 eV is very near the optimum value of 1.4 eV. It should not be surprising, therefore, to discover that GaAs cells are among the most efficient single-junction solar cells around. In fact, the theoretical maximum efficiency of single-junction GaAs solar cells, without solar concentration, is a very high 29%, and with concentration it is all the way up to 39% (Bube, 1998). GaAs cells with efficiencies exceeding 20% have been reported since the mid-1970s; and when used in concentrator systems in which solar energy is focused onto the cells, efficiencies approaching 30% have been realized.

In contrast to silicon cells, the efficiency of GaAs is relatively insensitive to increasing temperature, which helps them perform better than x-Si under concentrated sunlight. They are also less affected by cosmic radiation, and as thin films they are lightweight, which gives them an advantage in space applications. On the other hand, gallium is much less abundant in the earth's crust and it is a very expensive material. When coupled with the much more difficult processing required to fabricate GaAs cells, they have been too expensive for all but space applications and, potentially, for concentrator systems in which expensive cells are offset by cheap optical concentrators. Ongoing work with alloys and multijunction cells may, however, change that prognosis. Of particular interest are cells in which GaAs is coupled with other photovoltaic materials. A multijunction cell consisting of layers of GaAs and GaInP has achieved efficiencies of 29.5% for nonconcentrating AM1.5 conditions, while a hybrid, multijunction, solar concentrating cell of GaAs and Si has reached 31% efficiency.

8.9.3 Cadmium Telluride

Cadmium telluride (CdTe) is the most successful example of a II–VI photovoltaic compound. Although it can be doped in both p-type and n-type forms, it is most often used as the p-layer in heterojunction solar cells. Recall the distinction between homojunctions (the same material on each side of the junction) and heterojunctions (different materials). One difficulty associated with heterojunctions is the mismatch between the size of the crystalline lattice of the two materials, which leads to dangling bonds as shown in Fig. 8.58.

One way to sort out the best materials to use for the n-layer of CdTe cells is based on the mismatch of their lattice dimensions as expressed by their *lattice constants* (the a_1, a_2 dimensions shown in Fig. 8.58). One compound that is often used for the n-layer is cadmium sulfide CdS, which has a lattice mismatch of 9.7% with CdTe.

The band gap for CdTe is 1.44 eV, which puts it very close to the optimum for terrestrial cells. Thin-film laboratory cells using the n-CdS/p-CdTe heterojunction have efficiencies approaching 16% and prototype modules are reaching efficiencies beyond 9%. The equipment needed to manufacture these cells is orders of magnitude cheaper than that required for x-Si, and their relatively high efficiency makes them attractive candidates for mass production. While CdTe cells were used for years on pocket calculators made by Texas Instruments, full-scale modules have not yet successfully entered the marketplace.

One aspect of CdS/CdTe cells that needs to be considered carefully is the potential hazard to human health and the environment associated with cadmium. Cadmium is a very toxic substance, and it is categorized as a probable human carcinogen. Use of cadmium during the manufacture of CdTe cells needs to be carefully monitored and controlled to protect worker health, but apparently necessary safety precautions are relatively straightforward. Waste cadmium produced during the manufacturing process needs to be kept out of the environment and

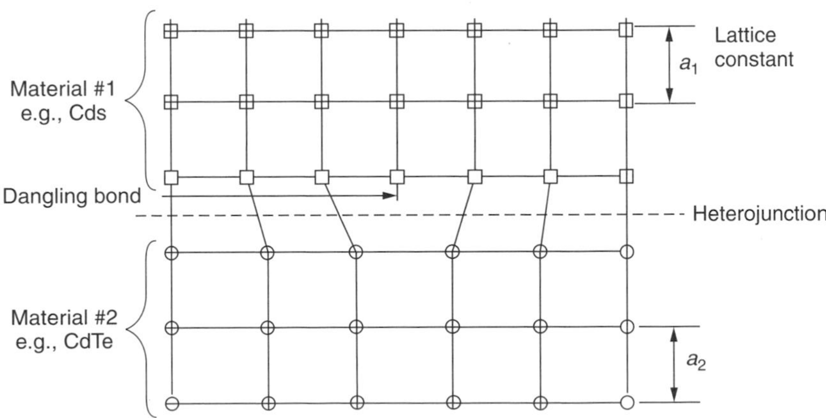

Figure 8.58 The mismatch between heterojunction materials leads to dangling bonds as shown.

should be recycled. The question then arises as to what precautions are necessary once modules have been manufactured and installed. CdS/CdTe modules contain about 6 g of cadmium per square meter of surface area, but it is completely sealed inside of the module so it should pose no risk under normal circumstances. If all of the cadmium in a rooftop PV system were to vaporize in a fire, however, and be inhaled by an individual, it would be pose a very serious health risk. But the likelihood of someone inhaling enough cadmium to cause harm without also having inhaled a lethal dose of smoke is considered to be insignificantly small.

8.9.4 Copper Indium Diselenide (CIS)

The goal in exploring compounds made up of a number of elements is to find combinations with band gaps that approach the optimum value while minimizing inefficiencies associated with lattice mismatch. Copper indium diselenide, $CuInSe_2$, better known as "CIS," is a *ternary* compound consisting of one element, copper, from the first column of the Periodic Table, another from the third column, indium, along with selenium from the sixth column. It is therefore referred to as a I–III–VI material. A simplistic way to think about this complexity is to imagine that the average properties of Cu (Group I) and In (Group III) are somewhat like those of an element from the second column (Group II), so the whole molecule might be similar to a II–VI compound such as CdTe (Bube, 1998).

While the crystal structure of silicon is a simple tetrahedral that is easy to understand and visualize, crystalline $CuInSe_2$ is much more complicated. As shown in Fig. 8.59, each selenium atom serves as the center of a tetrahedron of

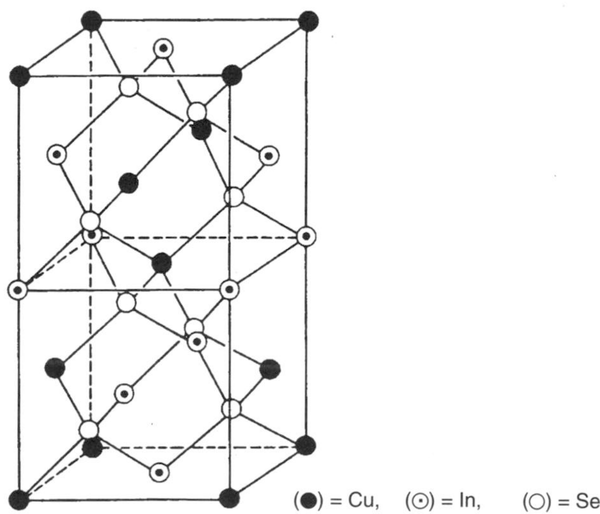

Figure 8.59 The crystalline structure of $CuInSe_2$ or "CIS." From Bube (1998) based on Kazmerski and Wagner (1985).

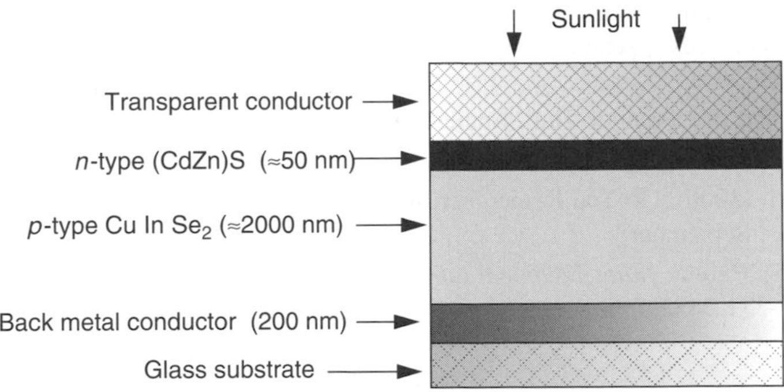

Figure 8.60 Structure of a simple, thin-film copper indium diselenide (CIS) cell.

two Cu and two In atoms, and each Cu atom is the center of a tetrahedron of selenium atoms.

CIS cells often use a thin layer of cadmium and zinc sulfide (CdZn)S for the n-layer as shown in Fig. 8.60.

With the substitution of gallium for some of the indium in the CIS material, the relatively low 1.04-eV band gap of CIS is increased and efficiency is improved. This is consistent with our interpretation of the Periodic Table in which band gap increases for elements in higher rows of the table (Ga is above In). The resulting $CuIn_{1-x}Ga_xSe_2$ alloy is called "CIGS" for short. By 2003, laboratory CIGS cells had achieved efficiencies of almost 20% and production modules had efficiencies in the range of 8 to 10%.

REFERENCES

Adams, W. G., and R. E. Day (15 June 1876). The Action of Light on Selenium, *Proceedings of the Royal Society*, vol. A25, pp. 113–117.

Becquerel, E. (4 November 1839). Memoire sur les effets electriques produits sous l'influence des rayons solaires, *Comptes Rendus*, vol. 9, pp. 561–564.

Bube, R. H. (1998). *Photovoltaic Materials*, Imperial College Press, London.

Carlson, D. E., and C. R. Wronski (1976). Amorphous Silicon Solar Cell, *Applied Physics Letters*, vol. 28:11: 671–773.

Chittick, R. C., J. H. Alexander, and H. F. Sterling (1969). The Preparation and Properties of Amorphous Silicon, *Journal of the Electrochemical Society*, vol. 116; pp. 77–81.

ERDA/NASA (1977). *Terrestrial Photovoltaic Measurement Procedures*. ERDA/NASA NASA/1022-77/16, NASA TM 73702, Cleveland, Ohio.

Green, M. (1993). Crystalline-and-Polycrystalline Solar Cells, *Renewable Energy: Sources for Fuels and Electricity*, T. B. Johansson, H. Kelly, A. K. N. Reddy, and R. H. Williams (eds.), Island Press, Washington, D.C.

Hersel, P., and K. Zweibel (1982). *Basic Photovoltaic Principles and Methods*, Solar Energy Research Institute, Golden, CO.

Kazmerski, L., and S. Wagner (1985). *Current Topics in Photovoltaics*, T. J. Coutts and J. D. Meakin (eds.), Academic Press, London, p. 41.

Kuehn, T. H., J. W. Ramsey, and J. L. Threlkeld (1998). *Thermal Environmental Engineering*, 3rd ed., Prentice-Hall, Englewood Cliffs, NJ.

Maycock, P. (2004). The State of the PV Market, *Solar Today*, Jan/Feb pp 32–35.

Schmela, M. (2000). Do You Remember S-Web?, *Photon International, The Photovoltaic Magazine*, September.

SERI (1985). *Photovoltaics Technical Information Guide*, Solar Energy Research Institute, SERI/SP-271-2452, U.S. Department of Energy, Washington, D.C.

PROBLEMS

8.1 For the following materials, determine the maximum wavelength of solar energy capable of creating hole-electron pairs:

a. Gallium arsenide, GaAs, band gap 1.42 eV.

b. Copper indium diselenide, $CuInSe_2$, band gap 1.01 eV

c. Cadmium sulfide, CdS, band gap 2.42 eV.

8.2 A *p-n* junction diode at 25°C carries a current of 100 mA when the diode voltage is 0.5 V. What is the reverse saturation current, I_0?

8.3 For the simple equivalent circuit for a 0.005 m^2 photovoltaic cell shown below, the reverse saturation current is $I_0 = 10^{-9}$ A and at an insolation of 1-sun the short-circuit current is $I_{SC} = 1$ A,. At 25°C, find the following:

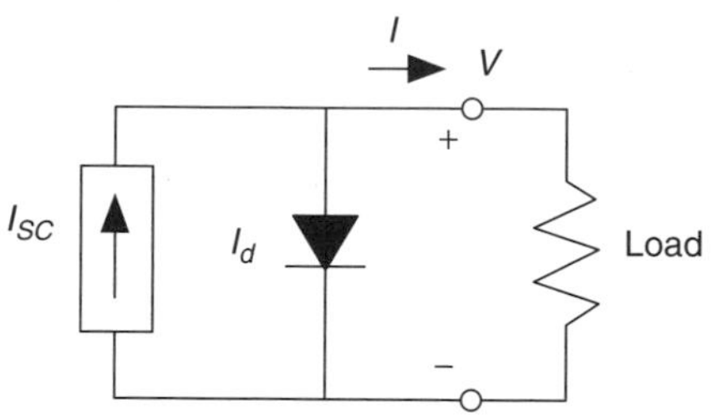

Figure P8.3

a. The open-circuit voltage.

b. The load current when the output voltage is $V = 0.5$ V.

c. The power delivered to the load when the output voltage is 0.5 V.

d. The efficiency of the cell at $V = 0.5$ V.

8.4 The equivalent circuit for a PV cell includes a parallel resistance of $R_P = 10\ \Omega$. The cell has area 0.005 m^2, reverse saturation current of $I_0 = 10^{-9}$ A and at an insolation of 1-sun the short-circuit current is $I_{SC} = 1$ A, At 25°C, with an output voltage of 0.5 V, find the following:

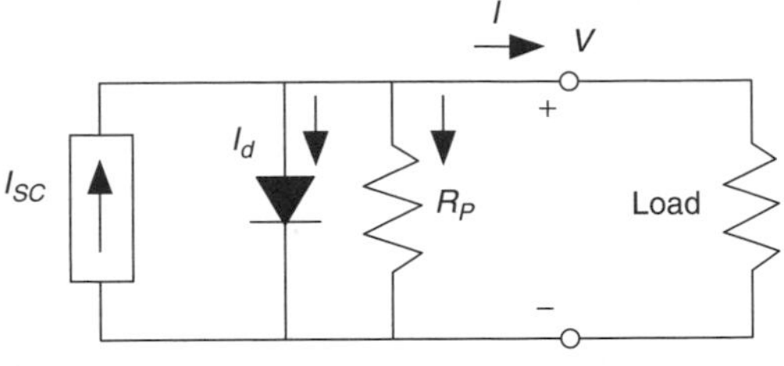

Figure P8.4

a. The load current.

b. The power delivered to the load.

c. The efficiency of the cell.

8.5 The following figure shows two I-V curves. One is for a PV cell with an equivalent circuit having an infinite parallel resistance (and no series resistance). What is the parallel resistance in the equivalent circuit of the other cell?

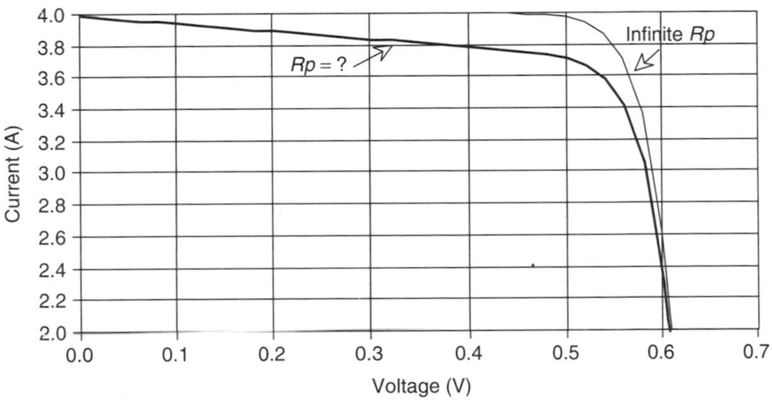

Figure P8.5

8.6 The following figure shows two I-V curves. One is for a PV cell with an equivalent circuit having no series resistance (and infinite parallel resistance). What is the series resistance in the equivalent circuit of the other cell?

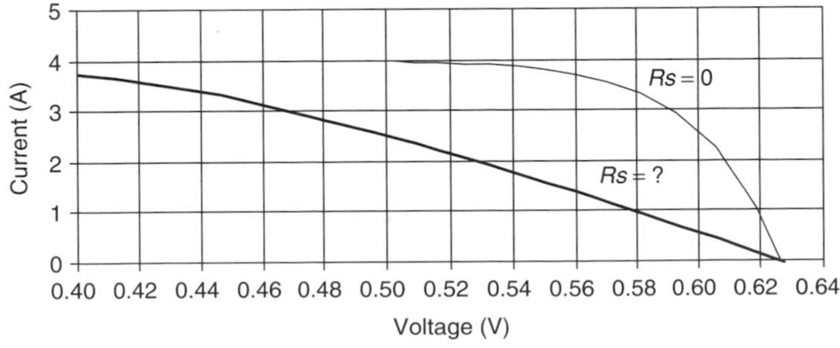

Figure P8.6

8.7 Estimate the cell temperature and power delivered by a 100-W PV module with the following conditions. Assume 0.5%/°C power loss.

a. NOCT = 50°C, ambient temperature of 25°C, insolation of 1-sun.

b. NOCT = 45°C, ambient temperature of 0°C, insolation of 500 W/m^2.

c. NOCT = 45°C, ambient temperature of 30°C, insolation of 800 W/m^2.

8.8 A module with 40 cells has an idealized, rectangular I-V curve with $I_{SC} =$ 4 A and $V_{OC} = 20$ V. If a single cell has a parallel resistance of 5 Ω and negligible series resistance, draw the I-V curve if one cell is completely shaded. What current would it deliver to a 12-V battery (vertical I-V load at 12 V)?

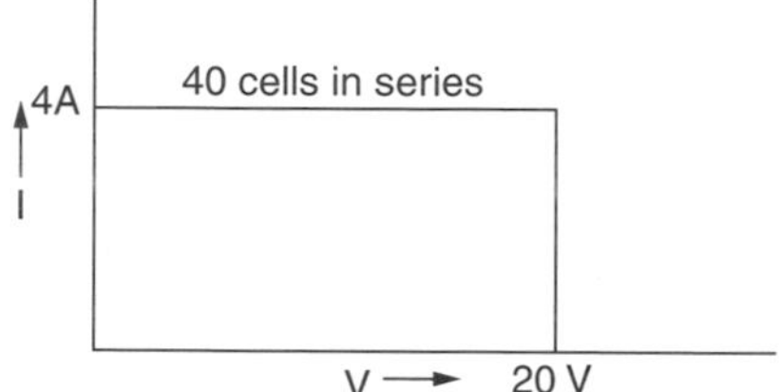

Figure P8.8

8.9 Suppose a PV module has the 1-sun I-V curve shown below. Within the module itself, the manufacturer has provided a pair of bypass diodes to help the panel deliver some power even when many of the cells are shaded. Each diode bypasses half of the cells, as shown. You may consider the diodes to be "ideal;" that is, they have no voltage drop across them when conducting.

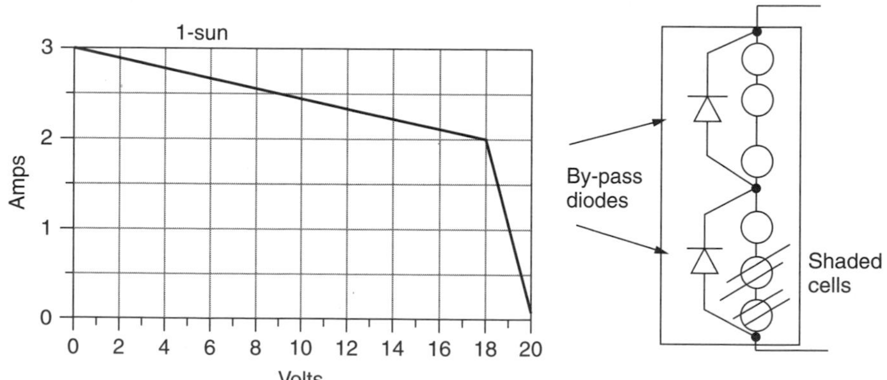

Figure P8.9

Suppose there is enough shading on the bottom cells to cause the lower diode to start conducting. Draw the new "shaded" I-V curve for the module.

CHAPTER 9

PHOTOVOLTAIC SYSTEMS

9.1 INTRODUCTION TO THE MAJOR PHOTOVOLTAIC SYSTEM TYPES

The focus of this chapter is on the analysis and design of photovoltaic (PV) systems in their three most commonly encountered configurations: systems that feed power directly into the utility grid, stand-alone systems that charge batteries, perhaps with generator back-up, and applications in which the load is directly connected to the PVs as is the case for most water-pumping systems.

Figure 9.1 shows a simplified diagram of the first of these systems—a grid-connected or *utility interactive* (UI) system in which PVs are supplying power to a building. The PV array may be pole-mounted, or attached externally to the roof, or it may become an integral part of the skin of the building itself. PV roofing shingles and thin-film PVs applied to glazing serve dual purposes, power, and building structure, and when that is the case the system is referred to as *building-integrated photovoltaics* (BIPV).

The photovoltaics in a grid-connected system deliver dc power to a power conditioning unit (PCU) that converts dc to ac and sends power to the building. If the PVs supply less than the immediate demand of the building, the PCU draws supplementary power from the utility grid, so demand is always satisfied. If, at any moment, the PVs supply more power than is needed, the excess is sent back onto the grid, potentially spinning the electric meter backwards. The system is relatively simple since failure-prone batteries are not needed for back-up power, although sometimes they may be included if utility outages are problematic. The

Renewable and Efficient Electric Power Systems. By Gilbert M. Masters
ISBN 0-471-28060-7

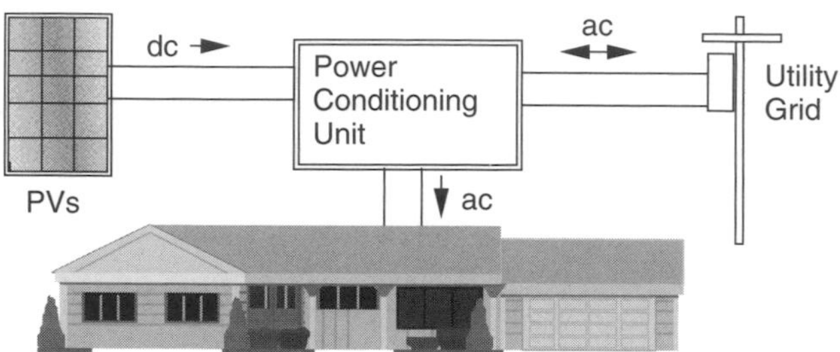

Figure 9.1 Simplified grid-connected PV system.

power-conditioning unit also helps keep the PVs operating at the most efficient point on their $I-V$ curves as conditions change.

Grid-connected PV systems have a number of desirable attributes. Their relative simplicity can result in high reliability; their maximum-power-tracking unit assures high PV efficiency; their potential to be integrated into the structure of the building means that there are no additional costs for land and, in some cases, materials displaced by PVs in such systems may offset some of their costs; and finally, their ability to deliver power during the middle of the day, when utility rates are highest, increases the economic value of their kilowatt-hours. All of these attributes contribute to the cost effectiveness of these systems. On the other hand, they have to compete with the relatively low price of utility power.

Figure 9.2 shows the second system, which is an off-grid, stand-alone system with battery storage and a generator for back-up power. In this particular system, an inverter converts battery dc voltages into ac for conventional household electricity, but in very simple systems everything may be run on dc and no inverter may be necessary. The charging function of the inverter allows the generator to top up the batteries when solar is insufficient.

Stand-alone PV systems can be very cost effective in remote locations where the only alternatives may be noisy, high-maintenance generators burning relatively expensive fuel, or extending the existing utility grid to the site, which can cost thousands of dollars per mile. These systems suffer from several inefficiencies, however, including battery losses and the fact that the PVs usually operate well off of the their most efficient operating point. Moreover, inefficiencies are often increased by mounting the array at an overly steep tilt angle to supply relatively uniform amounts of energy through the seasons, rather than picking an angle that results in the maximum possible annual energy delivery. These systems also require much more attention and care than stand-alone systems; and if generator usage is to be minimized (or eliminated), those using the energy may need to modify their lifestyles to accommodate the uneven availability of power as the seasons change or the weather deteriorates.

The third system type that we will pay close attention to has photovoltaics directly coupled to their loads, without any batteries or major power conditioning equipment. The most common example is PV water pumping in which the wires from the array are connected directly to the motor running a pump (Fig. 9.3).

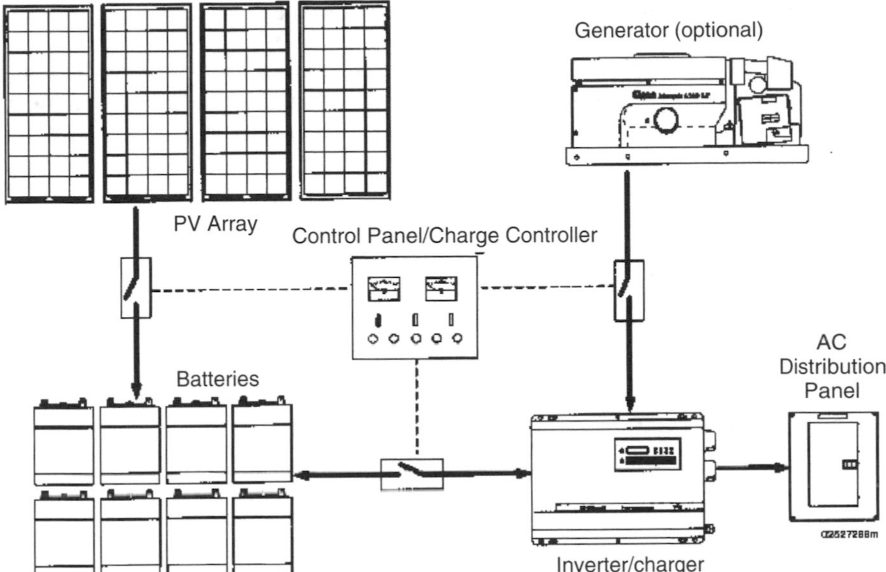

Figure 9.2 Example of a stand-alone PV system with optional generator for back-up.

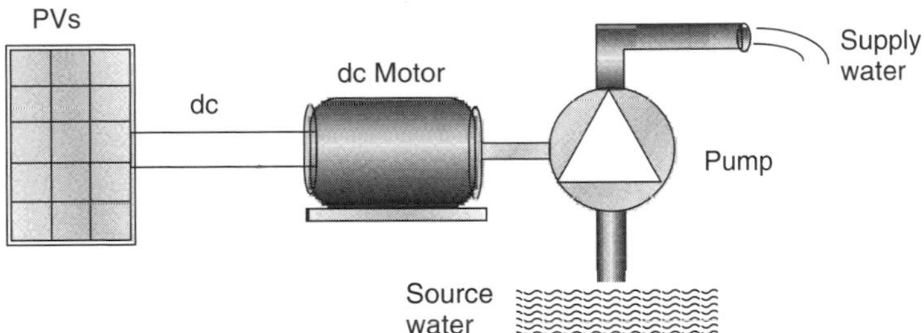

Figure 9.3 Conceptual diagram of a photovoltaic-powered water pumping system.

When the sun shines, water is pumped. There is no electric energy storage, but potential energy may be stored in a tank of water up the hill for use whenever it is needed. These systems are the ultimate in simplicity and reliability and are the least costly as well. But they need to be carefully designed to be efficient.

Our goal in this chapter is to try to learn how to properly size photovoltaic systems to provide for these various types of loads. Power delivered by a photovoltaic system will be a function of not only ambient conditions—especially solar intensity, spectral variations associated with overcast conditions, ambient temperature, and windspeed—but also what type of load the photovoltaics are supplying. As we shall see, very different analysis procedures apply to grid-connected systems, battery-charging stand-alone systems, and directly coupled water pumping systems.

9.2 CURRENT–VOLTAGE CURVES FOR LOADS

While the $I-V$ curve for a photovoltaic cell, module, or array defines the combinations of voltage and current that are permissible under the existing ambient conditions, it does not by itself tell us anything about just where on that curve the system will actually be operating. This determination is a function of the load into which the PVs deliver their power. Just as PVs have an $I-V$ curve, so do loads. As shown in Fig. 9.4, the same voltage is across both the PVs and load, and the same current runs through the PVs and load. Therefore, when the $I-V$ curve for the load is plotted onto the same graph that has the $I-V$ curve for the PVs, the intersection point is the one spot at which both the PVs and load are satisfied. This is called the operating point.

9.2.1 Simple Resistive-Load *I–V* Curve

To illustrate the importance and need for load curves, consider a simple resistive load as shown in Fig. 9.5. For the load,

$$V = IR \qquad \text{or} \qquad I = \left(\frac{1}{R}\right)V \tag{9.1}$$

which, when plotted on current versus voltage axes, is a straight line with slope $1/R$. As R increases, the operating point where the PV and resistance $I-V$ curves intersect moves along the PV $I-V$ curve from left to right. In fact, that suggests a simple way to actually measure the $I-V$ curve for PV module. By using a variable resistance, called a potentiometer, or *pot,* as the load, and then varying its resistance, pairs of current and voltage can be obtained, which can be plotted to give the module $I-V$ curve.

Since power delivered to any load is the product of current and voltage, there will be one particular value of resistance that will result in maximum power:

$$R_m = \frac{V_m}{I_m} \tag{9.2}$$

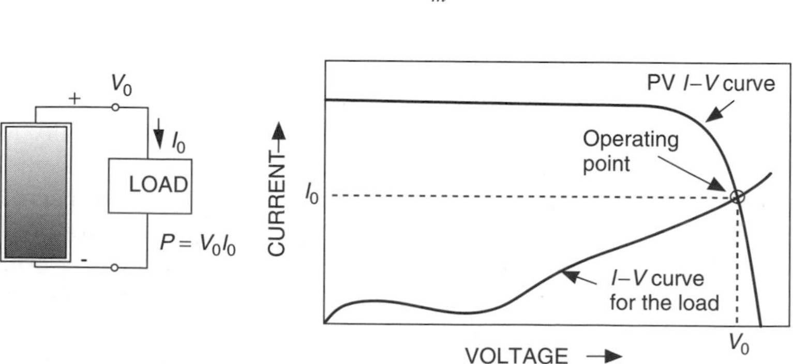

Figure 9.4 The operating point is the intersection of the current–voltage curves for the load and the PVs.

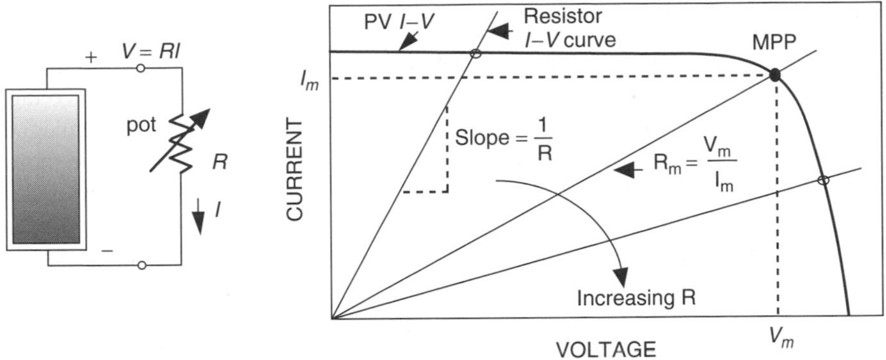

Figure 9.5 A module supplying power to a resistive load. As resistance changes, the operating point moves around on the PV $I-V$ curve.

where V_m and I_m are the voltage and current at the maximum power point (MPP). Under the special conditions at which modules are tested, the MPP corresponds to the rated voltage V_R and current I_R of the module. That means the best value of resistance, for maximum power transfer, should be V_R/I_R under 1-sun, 25°C, AM 1.5 conditions. As Fig. 9.6 shows, however, with a fixed resistance the operating point slips off the MPP as conditions change and the module becomes less and less efficient. Later, a device called a *maximum power point tracker* (MPPT) will be introduced, the purpose of which is to keep the PVs operating at their highest efficiency point at all times.

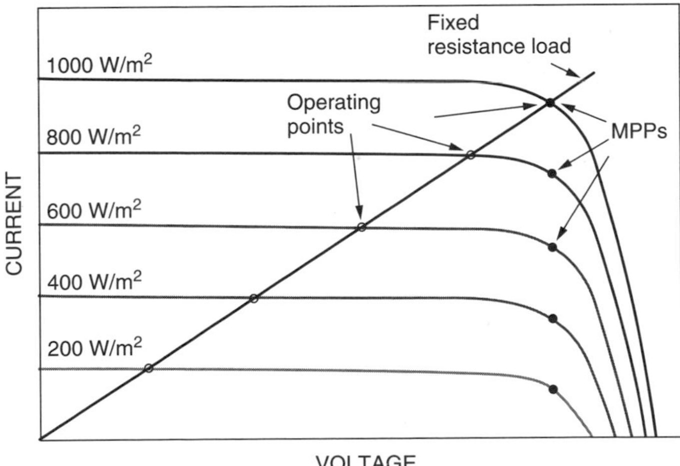

Figure 9.6 The efficiency of a PV module with a fixed resistance load designed for 1-sun conditions will decline with changing insolation. The solid maximum power point (MPP) dots show the operating points that would result in maximum PV efficiency.

9.2.2 DC Motor *I–V* Curve

While it is not often that a load would be an actual resistor, dc motors, such as those often used in PV-water-pumping systems, do exhibit a current–voltage relationship that is quite similar to that of a resistor. Most are permanent-magnet dc motors, which can be modeled as shown in Fig. 9.7. Notice that as the motor spins, it develops a *back electromotive force e*, which is a voltage proportional to the speed of the motor (ω) that opposes the voltage supplied by the photovoltaics. From the equivalent circuit, the voltage–current relationship for the dc motor is simply

$$V = IR_a + k\omega \tag{9.3}$$

where back emf $e = k\omega$ and R_a is the armature resistance.

A dc motor runs at nearly constant speed for any given applied voltage even though the torque requirement of its load may change. For example, as the torque requirement increases, the motor slows slightly, which drops the back emf and allows more armature current to flow. Since motor torque is proportional to armature current, the slowing motor draws more current, delivers more torque to the load, and regains almost all of its lost speed.

Based on (9.3), the electrical characteristic curve of a dc motor will appear to be something like the one shown in Fig. 9.8. Notice that at start-up, while $\omega = 0$, the current rises rapidly with increasing voltage until current is sufficient to create enough starting torque to break the motor loose from static friction. Once the motor starts to spin, back emf drops the current and thereafter I rises more slowly with increasing voltage. Notice that if you stall a dc motor while the voltage is way above the starting voltage, the current may be so high that the armature windings will burn out. That is why you should never leave the power on a dc motor if the armature is mechanically stuck for some reason.

A dc motor $I-V$ curve is superimposed on a set of photovoltaic $I-V$ curves in Fig. 9.9. The mismatch of operating points with the ideal MPP is apparent. Notice in this somewhat exaggerated example that the motor doesn't have enough current to overcome static friction until insolation reaches at least 400 W/m^2. Once it starts spinning, however, it only needs about 200 W/m^2 to keep running.

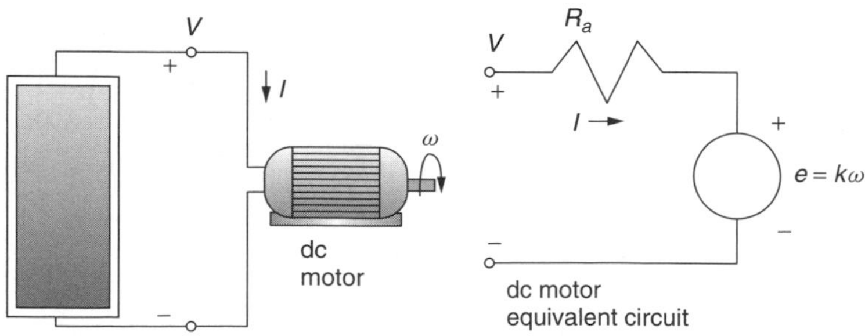

Figure 9.7 Electrical model of a permanent magnet dc motor.

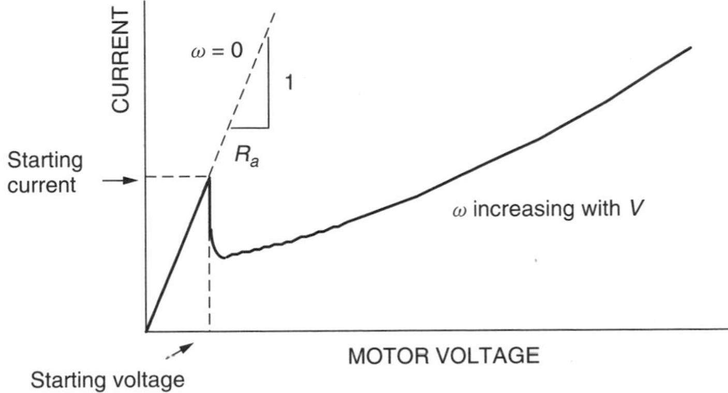

Figure 9.8 Electrical characteristics of a permanent-magnet dc motor.

This could mean that a fair amount of insolation is unusable in the morning while the motor struggles to break loose, which adds to the inefficiency of this simple PV–motor setup.

There is a device, called a *linear current booster* (LCB), that is designed to help overcome this loss of potentially usable insolation when current delivered to the motor is insufficient to overcome friction (Fig. 9.10). Notice from the I–V curves of Fig. 9.9 that the operating point in the morning is nowhere near the knee of the insolation curve where maximum power is available. Out by the knee of the curve, the PVs may be able to supply enough power to overcome friction, but without some clever electronics, this power would be delivered with

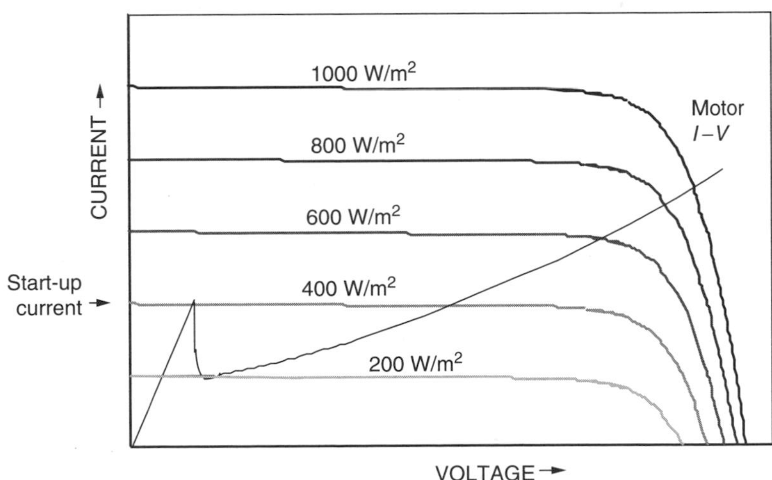

Figure 9.9 DC motor I–V curve on photovoltaic I–V curves for varying insolation. In this example (somewhat exaggerated), the motor won't start spinning until insolation reaches 400 W/m^2, but after that it only needs 200 W/m^2 to keep running.

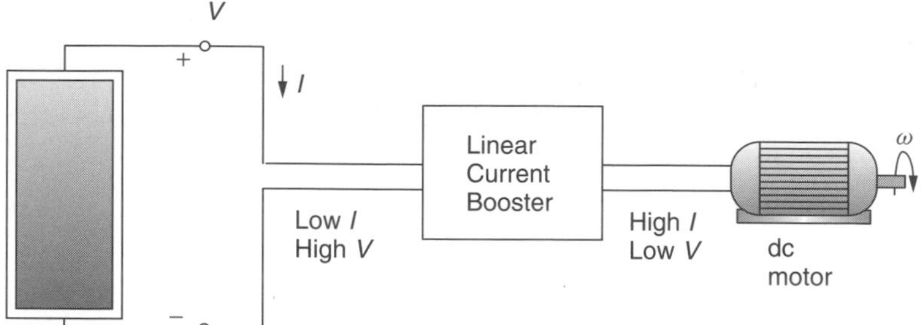

Figure 9.10 A linear current booster (LCB) increases current to help start or keep the motor running in low sunlight.

relatively low current and relatively high voltage and still wouldn't start the motor. What an LCB does is to shift this relationship around. By converting low-current, high-voltage power into high-current, low-voltage power, they can get the motor started earlier in the morning. The lower voltage, however, means that the motor will spin at a slower rate, but at least it is working. In addition, the motor with an LCB will not stall as early in the afternoon, though it will slow down. So there are additional gains.

9.2.3 Battery *I–V* Curves

Since PVs only provide power during the daylight hours and many applications require energy when the sun isn't shining, some method of energy storage often is needed. For a water pumping system, this might be the potential energy of water stored in a tank. For grid-connected systems, the utility lines themselves can be thought of as the storage mechanism: PV energy is put onto the grid during the day and taken back at night. For most off-grid applications, however, energy is stored in batteries for use whenever it is needed.

An ideal battery is one in which the voltage remains constant no matter how much current is drawn. This means that it will have an $I-V$ curve that is simply a straight up-and-down line as shown in Fig. 9.11. A real battery, on the other hand, has some internal resistance and is often modeled with an equivalent circuit consisting of an ideal battery of voltage V_B in series with some internal resistance R_i as shown in Fig. 9.12. During the charge cycle, with positive current flow into the battery, we can write

$$V = V_B + R_i I \tag{9.4}$$

which plots as a slightly-tilted, straight line with slope equal to $1/R_i$. During charging, the applied voltage needs to be greater than V_B; as the process continues, V_B itself increases so the $I-V$ line slides to the right as shown in Fig. 9.12a.

During discharge, the output voltage of the battery is less than V_B, the slope of the $I-V$ line flips, and the $I-V$ curve moves back to the left as shown in Fig. 9.12b.

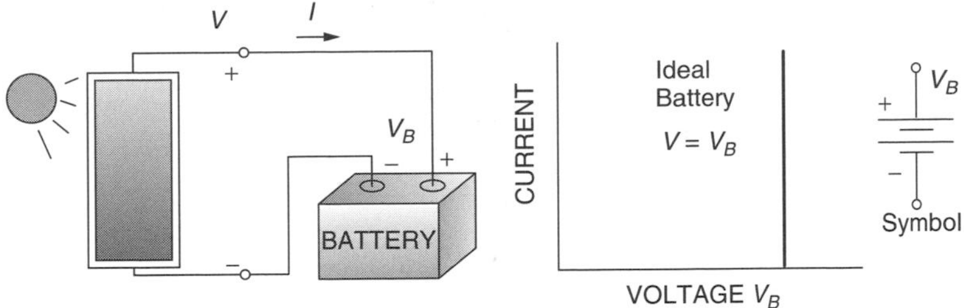

Figure 9.11 An ideal battery has a vertical current–voltage characteristic curve.

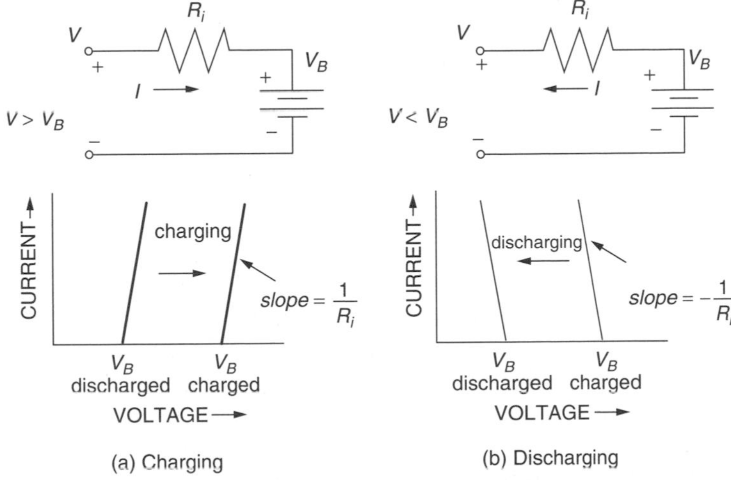

Figure 9.12 A real battery can be modeled as an ideal battery in series with its internal resistance, with current flowing in opposite directions during charging (a) and discharging (b). During charging/discharging, the slightly tilted I–V curve slides right or left.

The simple equivalent circuit representation of Fig. 9.12 is complicated by a number of factors, including the fact that the open-circuit voltage (V_B) depends not only on the state of charge but also on battery temperature and how long it has been resting without any current flowing. For a conventional 12-V lead-acid battery at 78°F, which has been allowed to rest for a few hours, V_B ranges from 12.7 V for a fully charged battery to about 11.7 V for one that has only a few percent of its charge remaining. Internal resistance is also a function of temperature and state of charge, as well the age and condition of the battery.

Example 9.1 Charging a 12-volt Battery. Suppose that a nearly depleted 12-V lead-acid battery has an open-circuit voltage of 11.7 V and an internal resistance of 0.03 Ω.

a. What voltage would a PV module operate at if it is delivering 6 A to the battery?
b. If 20 A is drawn from a fully charged battery with open-circuit voltage 12.7 V, what voltage would the PV module operate at?

Solution

a. Using (9.4), the PV voltage would be

$$V = V_B + R_i I = 11.7 + 0.03 \times 6 = 11.88 \text{ V}$$

b. While drawing 20 A after V_B has reached 12.7 V, the output voltage of the battery would be

$$V_{\text{load}} = V_B - IR_i = 12.7 - 20 \times 0.03 = 12.1 \text{ V}$$

and since the voltage that the PVs operate at is determined by the battery voltage, they would also be at 12.1 V.

Since the $I-V$ curve for a battery moves toward the right as the battery gains charge during the day, there is a chance that the PV operating point will begin to slide off the edge of the knee—especially late in the day when the knee itself is moving toward the left. This may not be a bad thing, however, since current has to be slowed or stopped anyway when a battery reaches full charge. If the PV–battery system has a charge controller, it will automatically prevent overcharging of the batteries. For very small battery charging systems, however, the charge controller can sometimes be omitted if modules with fewer cells in series are used. Such self-regulating modules sometimes have 33, or even 30, cells instead of the usual 36 to purposely cause the current to drop off as the battery approaches full charge as shown in Fig. 9.13.

9.2.4 Maximum Power Point Trackers

Clearly, significant efficiency gains could be realized if the operating points for resistive, dc motor, and battery loads could somehow be kept near the knee of the PV $I-V$ curves throughout the ever-changing daily conditions. Devices to do just that, called *maximum power trackers* (MPPTs), are available and are a standard part of many PV systems—especially those that are grid-connected.

There are some very clever, quite simple circuits that are at the heart of not only MPPTs but also linear current boosters (LCBs) as well as a number of other important power devices. The key is to be able to convert dc voltages from one level to another—something that was very difficult to do efficiently before high-power, field-effect transistors (FETs) became available in the 1980s and insulated-gate bipolar transistors (IGBTs) became available in the 1990s. At

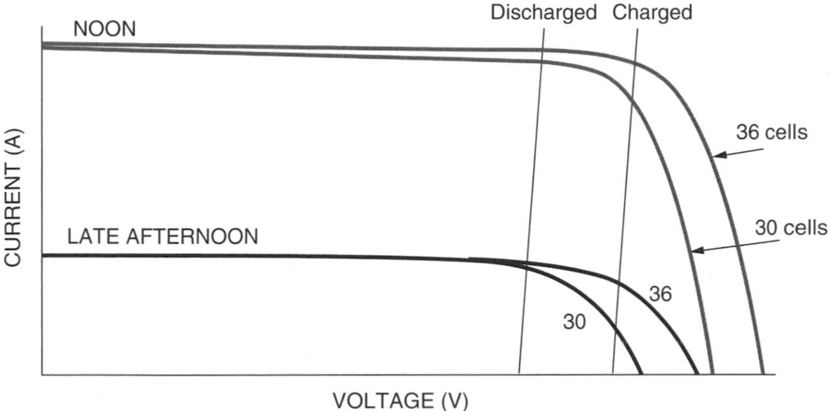

Figure 9.13 A self-regulating PV module with fewer cells can automatically reduce charging current as the batteries approach full charge.

the heart of modern *switched-mode* dc-to-dc converters is one of these transistors used as a simple on–off switch that either allows current to pass or blocks it.

A *boost converter* is a commonly used circuit to step up the voltage from a dc source, while a *buck converter* is often used to step down voltage. The circuit of Fig. 9.14 is a combination of these two circuits and is called a *buck-boost converter*. A buck-boost converter is capable of raising or lowering a dc voltage from its source to whatever dc voltage is needed by the load. The source in this case is shown as being a PV module and the load is shown as a dc motor, but the basic concept is used for a wide variety of electric power applications. The transistor switch flips on and off at a rapid rate (on the order of 20 kHz) under control of some sensing and logic circuitry that isn't shown. Also not shown is a capacitor across the PVs that helps smooth the voltage supplied by the PVs.

To analyze the buck-boost converter, we have to go back to first principles. Conventional dc or ac circuit analysis doesn't help much and instead the analysis

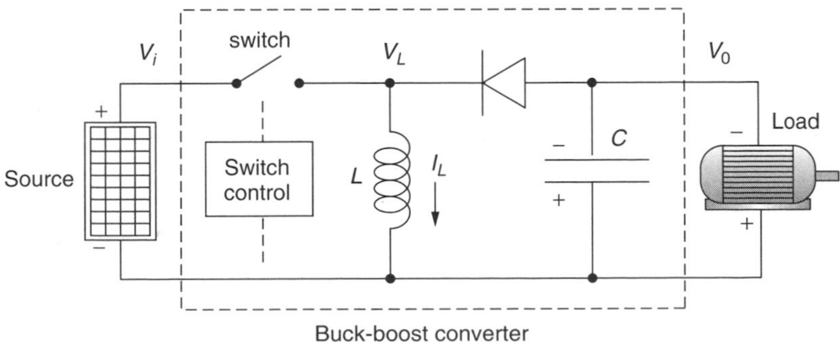

Figure 9.14 A buck-boost converter used as a the heart of a maximum power tracker.

is based on an energy balance for the magnetic field of the inductor. Basically there are two situations to consider: the circuit with the switch closed and the circuit with the switch open.

When the switch is closed, the input voltage V_i is applied across the inductor, driving current I_L through the inductor. All of the source current goes through the inductor since the diode blocks any flow to the rest of the circuit. During this portion of the cycle, energy is being added to the magnetic field in the inductor as current builds up. If the switch stayed closed, the inductor would eventually act like a short-circuit and the PVs would deliver short-circuit current at zero volts.

When the switch is opened, current in the inductor continues to flow as the magnetic field begins to collapse (remember that current through an inductor cannot be changed instantaneously—to do so would require infinite power). Inductor current now flows through the capacitor, the load, and the diode. Inductor current charging the capacitor provides a voltage (with a polarity reversal) across the load that will help keep the load powered after the switch closes again.

If the switch is cycled quickly enough, the current through the inductor doesn't have a chance to drop much while the switch is open before the next jolt of current from the source. With a fast enough switch and a large enough inductor, the circuit can be designed to have nearly constant inductor current. That's our first important insight into how this circuit works: Inductor current is essentially constant.

If the switch is cycled quickly enough, the voltage across the capacitor doesn't have a chance to drop much while the switch is closed before the next jolt of current from the inductor charges it back up again. Capacitors, recall, can't have their voltage change instantaneously so if the switch is cycling fast enough and the capacitor is sized large enough, the output voltage across the capacitor and load is nearly constant. We now have our second insight into this circuit: Output voltage V_o is essentially constant (and opposite in sign to V_i).

Finally, we need to introduce the duty cycle of the switch itself. This is what controls the relationship between the input and output voltages of the converter. The duty cycle D ($0 < D < 1$) is the fraction of the time that the switch is closed, as illustrated in Fig. 9.15. This variation in the fraction of time the switch is in one state or the other is referred to as *pulse-width modulation* (PWM).

For this simple description, all of the components in the converter will be considered to be ideal. As such, the inductor, diode and capacitor do not consume any net energy over a complete cycle of the switch. Therefore the average power into the converter is equal to the average power delivered by the converter; that is, it has 100% efficiency. Real MPPTs have efficiencies in the mid-90% range, so this isn't a bad assumption.

Now focus on the inductor. While the switch is closed, from time $t = 0$ to $t = \text{DT}$, the voltage across the inductor is a constant V_i. The average power put into the magnetic field of the inductor during one complete cycle is given by

$$\overline{P}_{L,\text{in}} = \frac{1}{T}\int_0^{DT} V_i I_L \, dt = \frac{1}{T} V_i \int_0^{DT} I_L \, dt \tag{9.5}$$

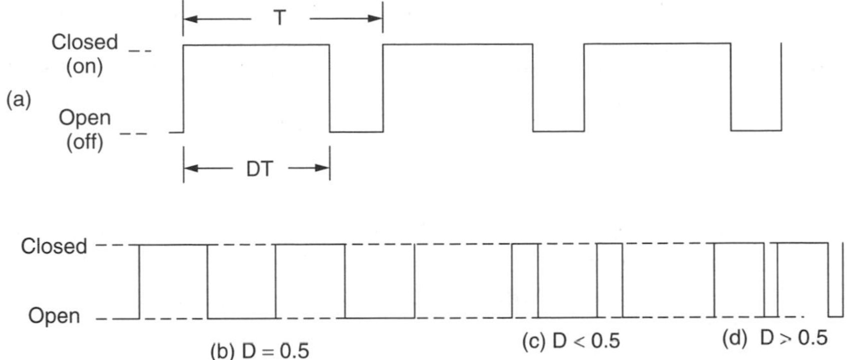

Figure 9.15 The duty cycle D is the fraction of the time the switch is closed (a). Examples: (b) 50% duty cycle; (c) D < 0.5; (d) D > 0.5.

Under the assumption that inductor current is constant, the average power into the inductor is

$$\overline{P}_{L,\text{in}} = \frac{1}{T} V_i I_L \int_0^{DT} dt = V_i I_L D \tag{9.6}$$

When the switch opens, the inductor's magnetic field begins to collapse, returning the energy it just acquired. The diode conducts, which means that the voltage across the inductor V_L is the same as the voltage across the load V_0. The average power delivered by the inductor is therefore

$$\overline{P}_{L,\text{out}} = \frac{1}{T} \int_{DT}^{T} V_L I_L \, dt = \frac{1}{T} \int_{DT}^{T} V_0 I_L \, dt \tag{9.7}$$

With good design, both V_0 and I_L are essentially constant, so average power from the inductor is

$$\overline{P}_{L,\text{out}} = \frac{1}{T} V_0 I_L (T - DT) = V_0 I_L (1 - D) \tag{9.8}$$

Over a complete cycle, average power into the inductor equals average power out of the inductor. So, from (9.6) and (9.8), we get

$$\frac{V_0}{V_i} = -\left(\frac{D}{1 - D}\right) \tag{9.9}$$

Equation (9.9) is pretty interesting. It tells us we can bump dc voltages up or down (there is a sign change) just by varying the duty cycle of the buck-boost converter. Longer duty cycles allow more time for the capacitor to charge up and less time for it to discharge, so the output voltage increases as D increases. For a duty cycle of 1/2, the output voltage is the same as the input voltage. A duty cycle of 2/3 results in a doubling of voltage, while $D = 1/3$ cuts voltage in half.

An actual MPP tracker needs some way for the dc-to-dc converter to know the proper duty cycle to provide at any given instant. This can be done with a microprocessor that periodically varies the duty cycle up and down a bit while monitoring the output power to see whether any improvement can be achieved.

Example 9.2 Duty Cycle for a MPPT. Under certain ambient conditions, a PV module has its maximum power point at $V_m = 17$ volts and $I_m = 6$ A. What duty cycle should an MPPT have if the module is delivering power to a 10 ohm resistance?

Solution. The maximum power delivered by the PVs is $P = 17\text{ V} \times 6\text{ A} = 102$ W. To deliver all of that 102 W to the 10 Ω resistor means that the resistor needs a voltage of

$$P = \frac{{V_R}^2}{R} = 102 = \frac{{V_R}^2}{10}$$

$$V_R = \sqrt{102 \cdot 10} = 31.9\text{ V}$$

The MPPT must bump the 17-V PV voltage to the desired 31.9-V resistor voltage. Using (9.9) and ignoring the sign change (it doesn't matter), we obtain

$$\frac{31.9}{17} = \left(\frac{D}{1 - D}\right) = 1.88$$

solving

$$D = 1.88 - 1.88D$$

$$D = \frac{1.88}{2.88} = 0.65$$

For a 100% efficient MPPT, the product of current and voltage (power) from the PVs is the same as the current-voltage product delivered by the MPPT to the load (Fig. 9.16). One way to visualize the impact of the MPPT is to redraw the PV $I-V$ curves using D as a parameter. For the MPPT's output voltage and current, one goes up and the other goes down compared with the original PV $I-V$ curve as shown in Fig. 9.17.

9.2.5 Hourly *I–V* Curves

As a typical solar day progresses, ambient temperature and available insolation are constantly changing. That means, of course, that the $I-V$ curve for a PV array is constantly shifting and the operating point for any given load is constantly

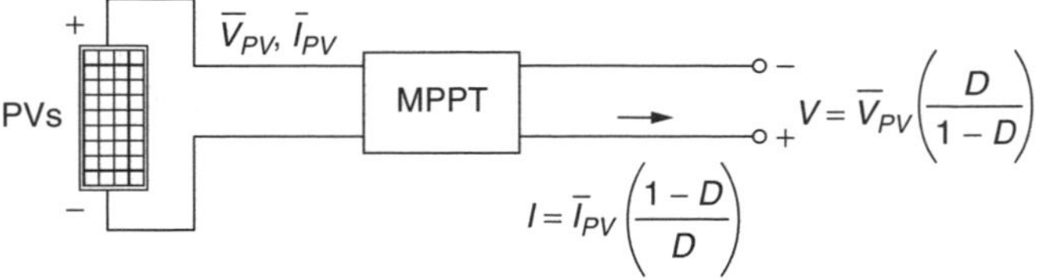

Figure 9.16 The MPPT bumps the PV voltages and currents to appropriate values for the load (one goes up, the other down).

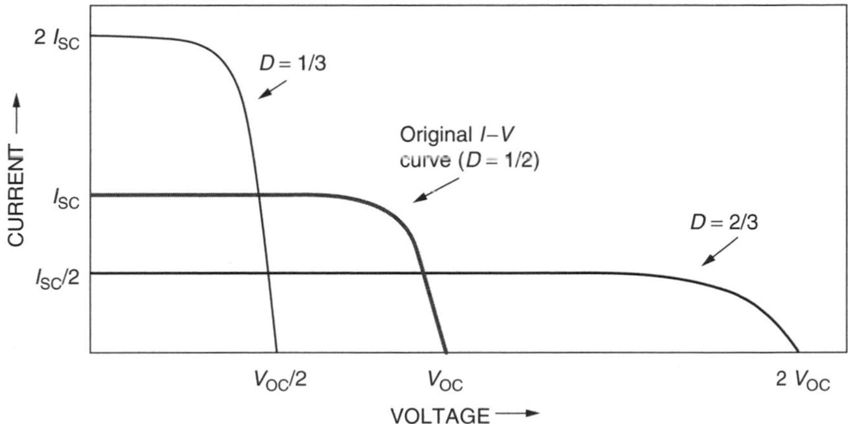

Figure 9.17 Redrawing the PV $I-V$ curves with an MPPT.

moving around as well. Manufacturers provide $I-V$ curves for various temperatures and solar intensity (e.g., Fig. 8.36), but there are times when hour-by-hour curves are helpful.

Over most of a PV $I-V$ curve, current at any voltage is directly proportional to insolation. That suggests we can simply scale the 1-sun (1000 W/m^2) $I-V$ curve by moving it up or down in proportion to the anticipated insolation. This generalization is completely true for short-circuit current I_{SC} (i.e., $V = 0$). Recall, however, that open-circuit voltage V_{OC} decreases somewhat as insolation decreases, so the simple assumption of current being proportional to insolation breaks down near V_{OC}. Under most circumstances, however, the operating voltage of a system is around the knee, or even lower, where current is very close to being proportional to insolation. Figure 9.18 illustrates this point. In it, a 1-sun $I-V$ curve having $I_{SC} = 6$ A has been drawn along with two $I-V$ curves that would be expected if insolation happened to be 677 W/m^2. One of the curves uses the assumption that current is proportional to insolation, the other properly accounts for the drop in V_{OC} as insolation decreases. As can be seen, there is very little difference between the 677-W/m^2 curves as long as the module doesn't operate below the knee.

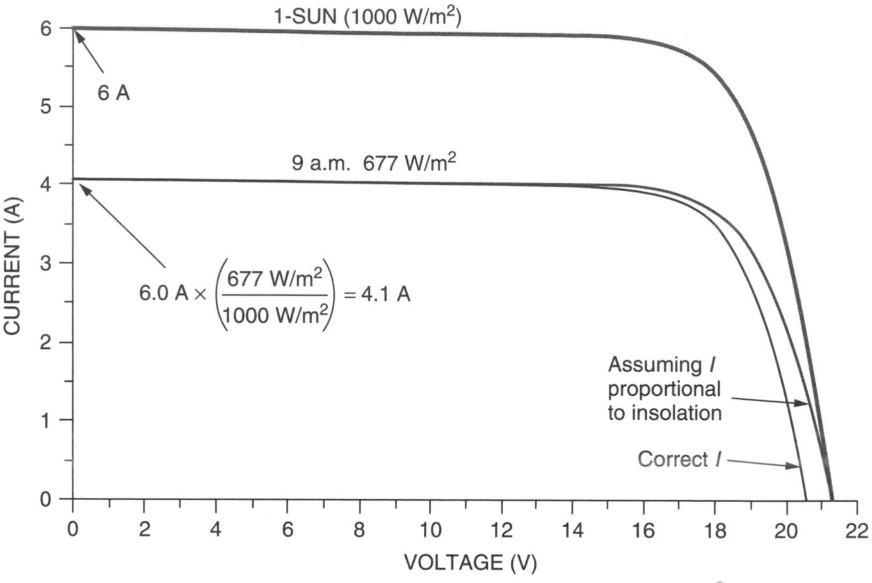

Figure 9.18 The 1-sun $I-V$ curve with two $I-V$ curves when insolation is 677 W/m^2: One is drawn under the simplifying assumption that I is proportional to insolation; the other accounts for the drop in V_{OC} as insolation decreases. For voltages below the knee, there is very little difference.

The simple assumption that current is proportional to insolation makes it easy to draw hour-by-hour $I-V$ curves for clear days. Techniques for estimating hourly insolation on clear days were presented in Chapter 7, and Appendix C has tables of values as well, so all we need to do is scale the 1-sun (1 kW/m^2) $I-V$ curve in direct proportion to those estimated hourly solar intensities. Since the 1-sun $I-V$ curve itself depends on cell temperature, and cell temperature depends on insolation and ambient temperature, we could imagine adjusting the 1-sun reference curve on an hour-by-hour basis as well. But since our purpose is to illustrate certain principles, that degree of refinement will be ignored here.

In Fig. 9.19, hourly PV $I-V$ curves have been drawn using insolations corresponding to a south-facing collector in April at 40° latitude with tilt angle of 40° using clear-sky values given in Appendix C. The module is the same one used in the example in Fig. 9.18. Superimposed onto these $I-V$ curves are example $I-V$ curves for three different kinds of loads: a dc motor, a 12-V battery with a constant charging voltage of 13.5 V, and a maximum power point tracker (MPPT). As can be seen, the dc motor has been well matched to the 1-sun $I-V$ curve, but does poorly in the early morning and late afternoon. The 12-V battery is consistently somewhat below the maximum power point. Table 9.1 provides a compilation of the hourly performance of each of these loads. The dc motor loses about 15% of the available daily energy because it doesn't operate at the maximum power point while the 12-V battery loses 17%.

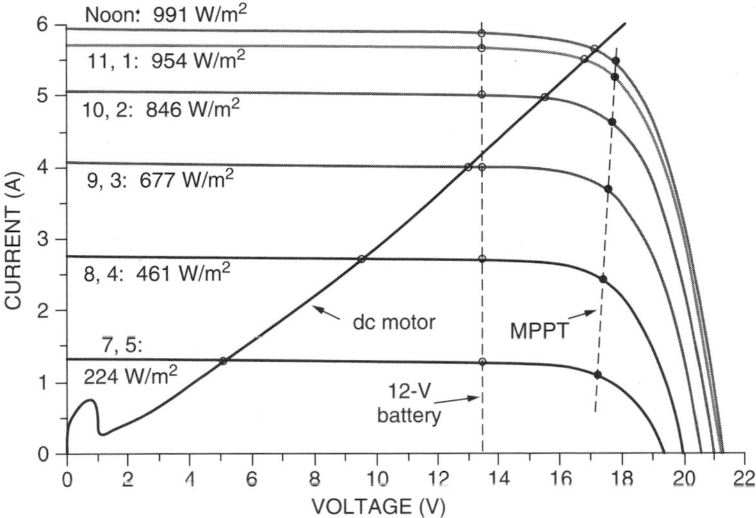

Figure 9.19 Hour-by-hour PV $I-V$ curves with examples of three different load types: dc motor, 12-V battery, MPPT.

TABLE 9.1 Daily Energy Delivered to Three Loads[a]

		dc Motor			12-V Battery			MPPT		
Time	Insolation (W/m^2)	Amps	Volts	Watts	Amps	Volts	Watts	Amps	Volts	Watts
7	224	1.3	5.0	6.5	1.3	13.5	17.6	1.1	17.3	19.0
8	461	2.7	9.6	25.9	2.7	13.5	36.5	2.5	17.4	43.5
9	677	4.0	13.0	52.0	4.0	13.5	54.0	3.7	17.5	64.8
10	846	5.0	15.6	78.0	5.0	13.5	67.5	4.7	17.6	82.7
11	954	5.3	16.9	89.6	5.6	13.5	75.6	5.2	17.7	92.0
12	991	5.5	17.1	94.1	5.9	13.5	79.7	5.4	17.8	96.1
1	954	5.3	16.9	89.6	5.6	13.5	75.6	5.2	17.7	92.0
2	846	5.0	15.6	78.0	5.0	13.5	67.5	4.7	17.6	82.7
3	677	4.0	13.0	52.0	4.0	13.5	54.0	3.7	17.5	64.8
4	461	2.7	9.6	25.9	2.7	13.5	36.5	2.5	17.4	43.5
5	224	1.3	5.0	6.5	1.3	13.5	17.6	1.1	17.3	19.0
W-h:	7315			598			582			700
Eff. vs. MPPT:				85%			83%			100%

[a]Without an MPPT, the dc motor is unable to collect 15% of the available energy and the 12-V battery loses 17%

9.3 GRID-CONNECTED SYSTEMS

Photovoltaic systems mounted on buildings are becoming increasingly popular as prices decrease and the installation infrastructure becomes increasingly mature.

As shown in Fig. 9.20, the principal components in a grid-connected, home-size PV system consists of the array itself with the two leads from each string sent to a combiner box that includes blocking diodes, individual fuses for each string, and usually a lightning surge arrestor. Two heavy-gauge wires from the combiner box deliver dc power to a fused array disconnect switch, which allows the PVs to be completely isolated from the system. The inverter sends ac power, usually at 240 V, through a breaker to the utility service panel. By tying each end of the inverter output to opposite sides of the service panel, 120-V power is delivered to each household circuit. Additional components not shown include the maximum power point tracker (MPPT), a ground-fault circuit interrupter (GFCI) that shuts the system down if any currents flow to ground, and circuitry to disconnect the PV system from the grid if the utility loses power. The system may also include a small battery bank to provide back-up power in case the grid is down. The inverter, some of the fuses and switches, the MPPT, GFCI, and other power management devices are usually integrated into a single power conditioning unit (PCU).

An alternative approach to the single inverter system shown in Fig. 9.20 is based on each PV module having its own small inverter mounted directly onto the backside of the panel. These ac modules allow simple expansion of the system, one module at a time, as the needs or budget dictate. Another advantage is that the connections from modules to the house distribution panel can all be done with relatively inexpensive, conventional 120- or 240-V ac switches, breakers, and wiring. Currently available module-mounted inverters are designed to work with individual 24-V modules, or with pairs of 12-V modules wired in series. Figure 9.21 suggests how simple this approach can be.

For large grid-connected systems, strings of PV modules may be tied into inverters in a manner analogous to the individual inverter/module concept (Fig. 9.22a). By doing so, the system is modularized, making it easier to service portions of the system without taking the full array off line. Expensive dc cabling

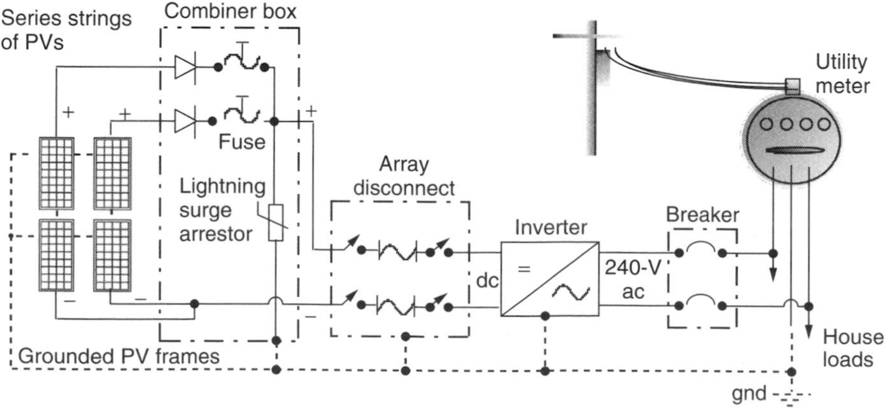

Figure 9.20 Principal components in a grid-connected PV system using a single inverter.

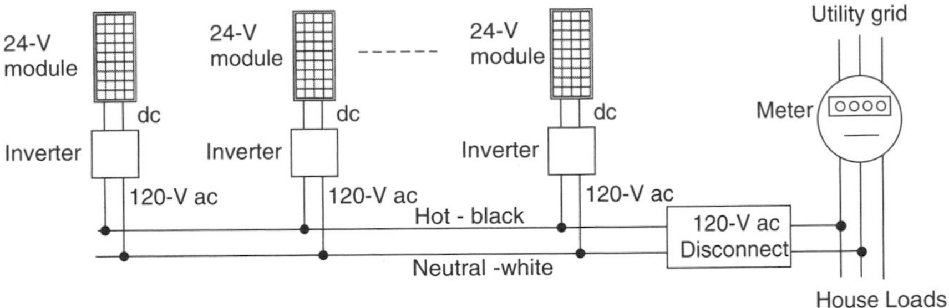

Figure 9.21 AC modules each have their own inverters mounted on the backside of the collector, allowing simple system expansion at any time.

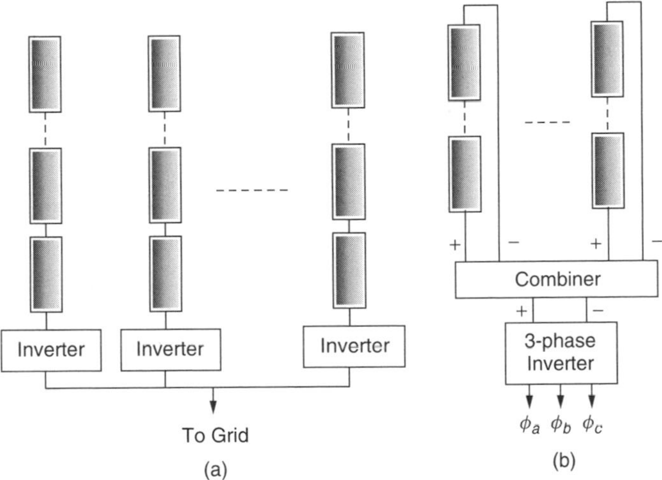

Figure 9.22 Large grid connected systems may use an individual inverter for each string (a) or may incorporate a large, central inverter system to provide three-phase power (b).

is also minimized making the installation potentially cheaper than a large, central inverter. Large, central inverter systems providing three-phase power to the grid are also an option (Fig. 9.22b).

9.3.1 Interfacing with the Utility

The ac output of a grid-connected PV system is fed into the main electrical distribution panel of the house, from which it can provide power to the house or put power back onto the grid as shown in Fig. 9.23. In most cases, whenever the PV system delivers more power than the home needs at that moment, the electric meter runs backwards, building up a credit with the utility. At other times, when demand exceeds that supplied by the PVs, the grid provides supplementary power. This arrangement, in which a single electric meter runs in both directions,

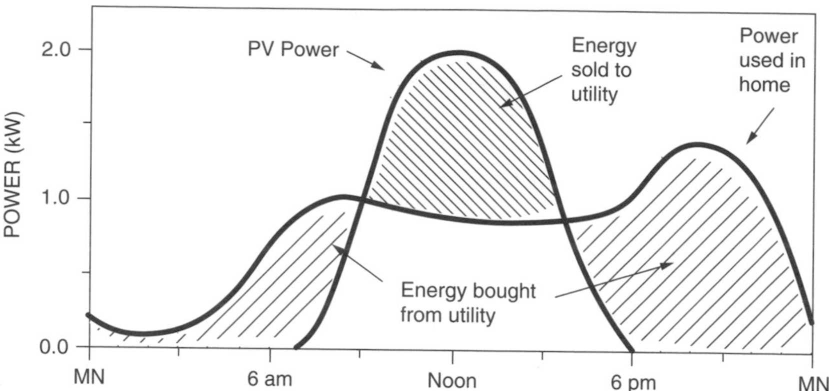

Figure 9.23 During the day, excess power from the array is sold to the utility; at night, the deficit is purchased from the utility.

is called *net metering* since the customer's monthly electric bill is only for that net amount of energy that the PV system is unable to supply. For an example of one version of net metering, refer back to Example 5.2 in Chapter 5. It is also possible to use two ratcheted meters, one to measure power you sell to the grid and the other to measure power you buy back from the grid. The two-meter arrangement is not only more cumbersome, it also can mean that the customer may have to sell electricity at a wholesale price and buy it back at the utility's more expensive retail price. From a PV owner's perspective, the net metering approach is usually preferred.

The power conditioning unit absolutely must be designed to quickly and automatically drop the PV system from the grid in the event of a utility power outage. When there is an outage, breakers automatically isolate a section of the utility lines in which the fault has occurred, creating what is referred to as an "island." A number of very serious problems may occur if, during such an outage, a self-generator, such as a grid-connected PV system, supplies power to that island.

Most faults are transient in nature, such as a tree branch brushing against the lines, and so utilities have automatic procedures that are designed to limit the amount of time the outage lasts. When there is a fault, breakers trip to isolate the affected lines, and then they are automatically reclosed a few tenths of a second later. It is hoped that in the interim the fault clears and customers are without power for just a brief moment. If that doesn't work, the procedure is repeated with somewhat longer intervals until finally, if the fault doesn't clear, workers are dispatched to the site to take care of the problem. If a self-generator is still on the line during such an incident, even for less than one second, it may interfere with the automatic reclosing procedure, leading to a longer-than-necessary outage. And if a worker attempts to fix a line that has supposedly been disconnected from all energy sources, but it is not, then a serious hazard has been created.

When a grid-connected system must provide power to its owners during a power outage, a small battery back-up system may be included. If the users really need uninterruptible power for longer periods of time, the battery system can be augmented with a generator such as has been suggested in Fig. 9.2.

9.3.2 DC and AC Rated Power

Grid-connected systems consist of an array of modules and a power conditioning unit that includes an inverter to convert dc from the PVs into ac required by the grid. A good starting point to estimate system performance is the rated dc power output of an individual module under standard test conditions (STC)—that is, 1-sun, AM 1.5 and 25°C cell temperature. Then we can try to estimate the actual ac power output under varying conditions.

When a PV system is put into the field, the actual ac power delivered at 1-sun, call it P_{ac}, can be represented as the following product:

$$P_{ac} = P_{dc,STC} \times \text{(Conversion Efficiency)} \tag{9.10}$$

where $P_{dc(STC)}$ is the dc power of the array obtained by simply adding the individual module ratings under standard test conditions. The conversion efficiency accounts for inverter efficiency, dirty collectors, mismatched modules, and differences in ambient conditions. Even in full sun, the impact of these losses can easily derate the power output by 20–40%.

Consider first the impact of slight variations in I–V curves for modules in an array. Figure 9.24 shows a simple example consisting of two mismatched

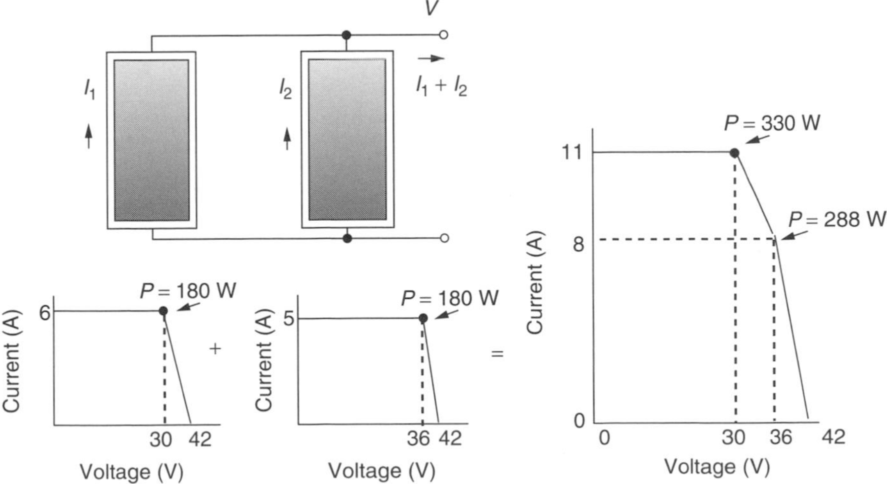

Figure 9.24 Illustrating the loss due to mismatched modules. Each module is rated at 180 W, but the parallel combination yields only 330 W at the maximum power point.

180-W modules wired in parallel. Their somewhat idealized $I-V$ curves have been drawn so that one produces 180 W at 30 V and the other does so at 36 V. As shown, the sum of their $I-V$ curves shows that the maximum power of the combined modules is only 330 W instead of the 360 W that would be expected if their $I-V$ curves were identical. In addition, not all modules coming off the very same production line will have exactly the same rated output. Some 100-W modules may really be 103 W and others 97 W, for example. In other words, production tolerances can reduce array output as well. These two module mismatch factors can easily drop the array output by several percent.

An even more important factor that reduces module power below the rated value is cell temperature. In the field, the cells are likely to be much hotter than the 25°C at which they are rated and we know that as temperature increases, power decreases. To help account for the change in module power caused by elevated cell temperatures, another rating system has been evolving that is based on field tests performed as part of an extensive monitoring program called PVUSA. The PVUSA test conditions (PTC) are defined as 1-sun irradiance in the plane of the array, 20°C ambient temperature, and a wind-speed of 1 m/s. The ac output of an array under PTC conditions $P_{ac(PTC)}$ is a much better indicator of the actual power delivered to the building in full sun than is the more commonly used $P_{dc(STC)}$. For its PV rebate programs, California, for example, has chosen to use the PTC rating of collectors.

Finally there is the efficiency of the inverter itself, which varies depending on the load, as is suggested in Fig. 9.25. Good grid-connect inverters have efficiencies above 90% when operating at all but very low loads.

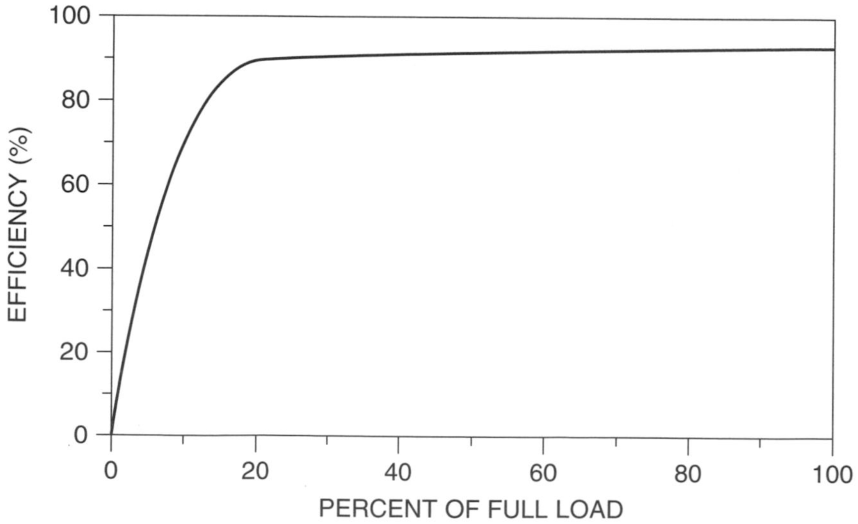

Figure 9.25 The efficiency of an inverter depends on the fraction of its rated power at which it operates.

Example 9.3 Derating a PV Array to a PTC, AC Rating. Consider a PV array rated at 1 kW under standard test conditions. Module nominal operating cell temperature (NOCT) is 47°C (see Section 8.6). DC power output at the MPP drops by 0.5%/°C above the STC temperature of 25°C.

Estimate its ac output under PTC conditions if there is a 3% array loss due to mismatched modules, dirt loss is 4%, and the inverter has an efficiency of 90%.

Solution. Inserting PTC conditions (ambient 20°C, insolation $S = 1\ \text{kW}(\text{m}^2)$ into (8.24) gives us an estimated cell temperature of

$$T_{\text{cell}} = T_{\text{amb}} + \left(\frac{\text{NOCT} - 20}{0.8}\right) \cdot S$$

$$= 20 + \left(\frac{47 - 20}{0.8}\right) \cdot 1 = 53.8^\circ\text{C}$$

With power loss at 0.5% per degree above 25°C, the dc rated power of the array would be

$$P_{dc(PTC)} = 1\ \text{kW}[1 - 0.005(53.8 - 25)] = 0.856\ \text{kW}$$

Including mismatch, dirt, and inverter efficiencies will result in an estimated ac rated power at PTC of

$$P_{ac(PTC)} = 0.856\ \text{kW} \times 0.97 \times 0.96 \times 0.90 = 0.72\ \text{kW}$$

That's a pretty sizably reduced indicator of rated power, but it is much more realistic in terms of actual field experience than the 1-kW dc STC rating that would often be quoted.

PV systems have been traditionally described in terms of their dc output under standard test conditions. In the above example, the system would likely be sold as a "1-kW system." But in that example the array will deliver only 72% of that as ac power to the load under the much more realistic PTC conditions. It should probably be called a 0.72-kW(ac) system. And in fact studies of real systems in the field under PTC conditions suggest that the ratio of $P_{ac(PTC)}$ to $P_{dc(STC)}$ may be lower still. Scheuermann et al. (2002) measured 19 PV systems in California and found the ratio to be between 53% and 70%.

Whether the ac PTC rating system will be adopted as the standard remains to be seen. It is much more realistic than the usual STC rating, but it has the disadvantage of being less easy to define since it depends on the specific inverter that has been chosen. In the meantime, it should be clear that describing a system based on its dc performance under STC without also including corrections for temperature and inverter is misleading. Even the PTC system is crude since it too is based on arbitrary ambient conditions. In cold climates, for example, where

PVs perform better, the PTC would underestimate energy delivered, and in hot climates the opposite would occur.

9.3.3 The "Peak-Hours" Approach to Estimating PV Performance

Predicting performance is a matter of combining the characteristics of the major components—the PV array and the inverter—with local insolation and temperature data. After having adjusted dc power under STC to expected ac from the inverter, the second key factor is the amount of sun available at the site. Chapter 7 was devoted to developing equations for clear sky insolation, and tables of values are given in Appendices C and D. In Appendix E there are tables of estimated average insolations for a number of locations in the United States and Appendix F presents seasonal averages around the globe.

When the units for daily, monthly, or annual average insolation are specifically kWh/m^2-day, then there is a very convenient way to interpret that number. Since 1-sun of insolation is defined as 1 kW/m^2, we can think of an insolation of say 5.6 kWh/m^2-day as being the same as 5.6 h/day of 1-sun, or 5.6 h of "peak sun." So, if we know the ac power delivered by an array under 1-sun insolation (P_{ac}), we can just multiply that rated power by the number of hours of peak sun to get daily kWh delivered.

To see whether this simple approach is reasonable, consider the following analysis. We can write the energy delivered in a day's time as

$$\text{Energy (kWh/day)} = \text{Insolation}\left(\frac{\text{kWh/m}^2}{\text{day}}\right) \cdot A\ (\text{m}^2) \cdot \overline{\eta} \tag{9.11}$$

where A is the area of the PV array and $\overline{\eta}$ is the average system efficiency over the day.

When exposed to 1-sun of insolation, we can write for ac power from the system

$$P_{ac}(\text{kW}) = \left(\frac{1\ \text{kW}}{\text{m}^2}\right) \cdot A\ (\text{m}^2) \cdot \eta_{1-\text{sun}} \tag{9.12}$$

where η_{1-sun} is the system efficiency at 1-sun. Combining (9.11) and (9.12) gives

$$\text{Energy (kWh/day)} = P_{ac}(\text{kW}) \cdot \left[\frac{\text{Insolation (kWh/m}^2\text{/day)}}{1\ \text{kW/m}^2}\right] \cdot \left(\frac{\overline{\eta}}{\eta_{1-\text{sun}}}\right) \tag{9.13}$$

If we assume that the average efficiency of the system over a day's time is the same as the efficiency when it is exposed to 1-sun, then the energy collected is what we hoped it would be

$$\text{Energy (kWh/day)} = P_{ac}(\text{kW}) \cdot (\text{h/day of "peak sun"}) \tag{9.14}$$

The key assumption in (9.14) is that system efficiency remains pretty much constant throughout the day. The main justification is that these grid-connected systems have maximum power point trackers that keep the operating point near the knee of the $I-V$ curve all day long. Since power at the maximum point is nearly directly proportional to insolation, system efficiency should be reasonably constant. Cell temperature also plays a role, but it is less important. Efficiency might be a bit higher than average in the morning, when it is cooler and there is less insolation, but all that will do is make (9.14) slightly conservative.

The value of something like the ac PTC rating system is now quite apparent. Using it to indicate system ac power at 1-sun, coupled with the interpretation of kWh/m^2-day as hours of peak sun, gives us an easy way to estimate energy production.

Example 9.4 Annual Energy Using the Peak-Sun Approach. Estimate the annual energy delivered by the 1-kW (dc, STC) array described in Example 9.3 if it located in Madison, WI, is south-facing, and has a tilt angle equal to its latitude minus 15°. Use the PTC ac rating.

Solution. Appendix E shows the annual insolation in Madison at L-15 is 4.5 kWh/m^2-day. Using the de-rated ac output of 0.72 kW (ac, PTC) that was found in Example 9.3, along with 4.5 h/day of peak sun, gives

$$\text{Energy} = 0.72\text{ kW} \times 4.5\text{ h/day} \times 365\text{ day/yr} = 1183\text{ kWh/yr}$$

The PTC rating assumes a nominal ambient of 20°C, which is a pretty good average estimate for many locations in the United States. We could expect it to overpredict performance when it is hotter than that, and underpredict when it is cooler. To test this, let us rework Example 9.4 on a monthly basis using Madison ambient temperature instead of the assumed 20°C.

Example 9.5 Correcting Predicted Performance for Temperature Effects. Estimate the energy that the 1-kW (dc, STC) array described in Example 9.3 would deliver in Madison in January. Assume south-facing with tilt = L-15 and use the average daily maximum temperature instead of the 20°C assumed by PTC. The nominal operating cell temperature (NOCT) was given as 47°C for this array.

Solution. In Appendix E, the average daily maximum temperature for Madison in January is given as −4.0°C. When it is that cold, (8.24) estimates cell temperature at 1-sun to be

$$T_{\text{cell}} = T_{\text{amb}} + \left(\frac{\text{NOCT} - 20}{0.8}\right) \cdot S$$

$$= -4.0 + \left(\frac{47 - 20}{0.8}\right) \cdot 1 = 29.8^\circ\text{C}$$

With power loss at 0.5% per degree above 25°C, the dc rated power of the array without dirt and mismatched modules would be

$$P_{dc} = 1\ \text{kW}[1 - 0.005(29.8 - 25)] = 0.98\ \text{kW}$$

For comparison, in Example 9.3 the cell temperature at PTC was a much warmer 53.8°C and the dc power was

$$P_{dc(PTC)} = 1\ \text{kW}[1 - 0.005(53.8 - 25)] = 0.856\ \text{kW}$$

In other words, the array will perform considerably better than PTC would have predicted because it is so cold in Madison.

Including mismatch, dirt and inverter efficiencies given in Example 9.3, yields an estimated ac rated power at of

$$P_{ac} = 0.98\ \text{kW} \times 0.97 \times 0.96 \times 0.90 = 0.82\ \text{kW}$$

Appendix E gives January insolation at L-15 in Madison as 3.0 kWh/m^2 or 3.0 h/day of 1-sun. So we estimate this 1 kW array will deliver

$$\text{Energy} = 0.82\ \text{kW} \times 3.0\ \text{h/day} \times 31\ \text{day/mo} = 76\ \text{kWh/mo}$$

It is easy to systematize calculations of the sort demonstrated in Example 9.5 so that actual site temperature data can be used instead of the 20°C assumed in PTC. A spreadsheet for the 1-kW (dc) Madison system using the same dirt, mismatch, and inverter assumptions is demonstrated in Table 9.2. Notice that the PTC estimate for annual energy found in Example 9.4 was 1183 kWh/yr, and this more carefully done spreadsheet gives 1202 kWh. The difference is negligible. A similar comparison for Phoenix, which is a much hotter place than Madison, results in a 7% overestimate using PTC. These differences are so small that it

TABLE 9.2 Estimated Energy Delivered by a 1-kW (dc, STC) PV Array in Madison, WI, Using Average Maximum Monthly Temperatures to Compute Performance Degradation[a]

Madison, WI, South L-15

dc Power	1 kW at STC
Temp. coef.	0.5%/°C
Mismatch	0.03
Dirt	0.04
Inverter	0.90
NOCT	47°C

Month	Insolation (kWh/m^2-day)	Avg Max Temp. (°C)	Cell Temp. (°C)	Array dc Power (kW)	Array ac Power (kW)	Energy (kWh/mo)
Jan	3.0	−4.0	29.8	0.98	0.82	76
Feb	3.9	−1.1	32.7	0.96	0.81	88
Mar	4.5	5.3	39.1	0.93	0.78	109
Apr	5.1	13.7	47.5	0.89	0.74	114
May	5.8	20.5	54.3	0.85	0.72	129
Jun	6.2	25.7	59.5	0.83	0.69	129
July	6.2	28.0	61.8	0.82	0.68	131
Aug	5.7	26.4	60.2	0.82	0.69	122
Sept	4.8	21.9	55.7	0.85	0.71	102
Oct	3.8	15.5	49.3	0.88	0.74	87
Nov	2.5	6.7	40.5	0.92	0.77	58
Dec	2.3	−1.2	32.6	0.96	0.81	57
Avg:	4.5	13.2			kWh/yr = 1202	

[a] Inverter, mismatch, and dirt losses from Example 9.3 are included.

would appear that using the simpler PTC approach is quite reasonable under a wide range of temperature conditions.

One of the most useful ways to describe energy production from a PV array is in terms of kWh/yr delivered per kW of STC rated power. Table 9.3 presents such data for a number of U.S. cities. The table has been derived following the methods presented in Example 9.5. Temperature and insolation data are from Appendix E. A 90% efficient inverter, along with 3% module mismatch and 4% dirt depreciation were assumed. Annual energy delivered for south-facing modules tilted at the latitude −15° ranges from about 1000 kWh per kW (dc, STC) in Seattle, to over 1600 kWh/kW in Albuquerque. When a single-axis E–W tracker with polar tilt angle is used, annual production increases by 24–36%.

Table 9.3 suggests a very crude rule-of-thumb for annual performance of a grid-connected system in a pretty good location:

> Each 1 kW of dc, STC-rated PV will deliver an average of roughly
>
> 1400 kWh per year.

TABLE 9.3 Annual Energy Production in Various Cities per kW (dc, STC) of Installed PV Capacity[a]

	South Facing, L-15 Fixed			1-Axis, Polar Mount		
Location	Average High Temp. (°C)	Insolation (kWh/ m^2-d)	Annual kWh/kW	Insolation (kWh/ m^2-d)	Annual kWh/kW	Ratio 1-axis/ Fixed
Seattle, WA	15.3	3.8	1006	4.7	1247	1.24
New York, NY	16.8	4.5	1195	5.6	1479	1.24
Madison, WI	13.2	4.5	1202	5.7	1519	1.26
Boston, MA	15.0	4.5	1209	5.7	1529	1.26
Atlanta, GA	21.8	5.0	1294	6.4	1639	1.27
Honolulu, HI	29.1	5.5	1373	7.4	1834	1.34
Boulder, CO	17.9	5.3	1404	7.2	1885	1.34
Los Angeles, CA	21.3	5.5	1420	7.0	1808	1.27
El Paso, TX	25.3	6.3	1583	8.6	2159	1.36
Albuquerque, NM	21.2	6.3	1618	8.5	2199	1.36

[a] Assumed inverter efficiency 90%, mismatching loss 3%, dirt loss 4%

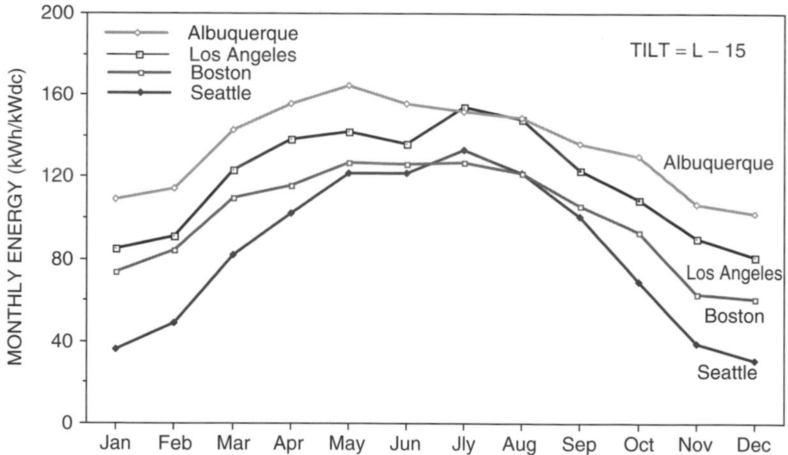

Figure 9.26 Monthly energy production for four cities in kWh per kW (dc, STC) for fixed south-facing, L-15 tilt. Assumed inverter efficiency 90%, mismatch loss 3%, dirt loss 4%. Includes local temperature impacts.

Figure 9.26 shows monthly values of kWh/kW for four U.S. cities that pretty much represent the range of likely performance across the coterminous states. A couple of features are worth pointing out. In Albuquerque, the decline in output after it peaks in May is due in large part to the high summer temperatures. At the other extreme, Seattle, with latitude about as high as it gets in the coterminous

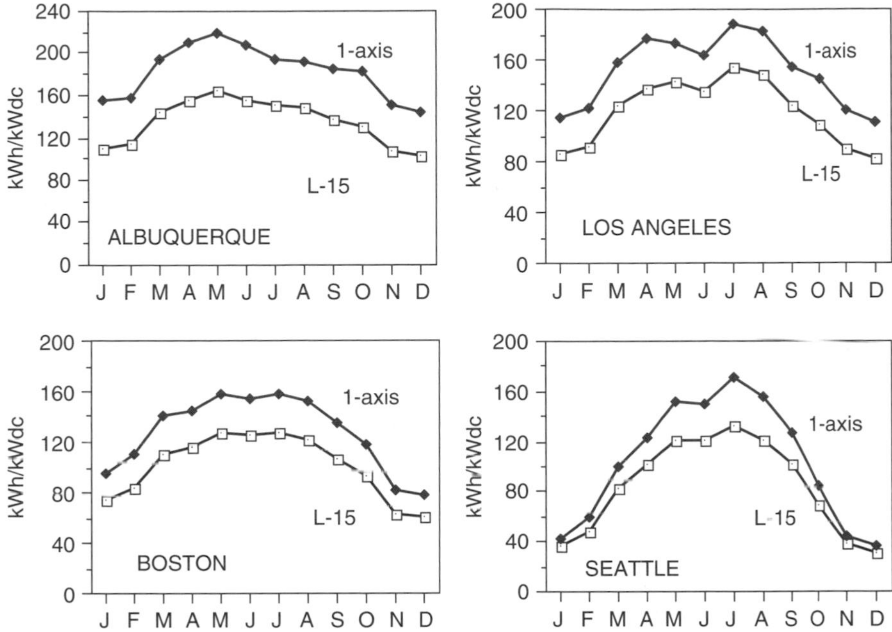

Figure 9.27 Comparing energy delivered from fixed L-15 tilt with single-axis polar tracking.

states, does quite well in the summer due to long, clear, relatively cool days. Figure 9.27 shows the improvement in monthly performance for those four cities when a single-axis polar mount tracker is incorporated into the system.

9.3.4 Capacity Factors for PV Grid-Connected Systems

A simple way to present the energy delivered by any electric power generation system is in terms of its rated ac power and its capacity factor (CF). If the system delivered full, rated power continuously, the CF would be unity. A CF of 0.4, for example, could mean that the system delivers full-rated power 40% of the time and no power at all the rest of the time, but that is not the only interpretation. It could also deliver 40% of rated power all of the time and still have CF = 0.4, or any of a number of other combinations.

The governing equation for annual performance in terms of CF is simply

$$\text{Energy (kWh/yr)} = P_{ac}\text{(kW)} \cdot \text{CF} \cdot 8760\text{(h/yr)} \tag{9.15}$$

where 8760 is the product of 24 hours per day times 365 days per year. Monthly or daily capacity factors are similarly defined.

Combining (9.14) and (9.15) leads to the simple interpretation of capacity factor for grid-connected PV systems:

$$\text{Capacity factor (CF)} = \frac{\text{(h/day of "peak sun")}}{\text{24 h/day}} \tag{9.16}$$

Notice that the complication associated with temperature is included in the definition of P_{ac} so it doesn't affect CF. Capacity factors for a number of U.S. cities are shown in Fig. 9.28. They range from 0.16 to 0.26 for fixed, south-facing panels at tilt L-15 and range from 0.20 to 0.36 for single-axis polar-mount trackers.

9.3.5 Grid-Connected System Sizing

With the utility there to provide energy storage and back-up power, sizing grid-connected systems is not nearly as critical as it is for stand-alone systems. Moreover, there are few economies of scale with these systems—it costs roughly twice as much to install a system that will deliver twice as much energy. Sizing grid-connected systems therefore is more a matter of how much area is conveniently available on the building, and the budget of the buyer, than it is trying to match supply to demand. It is, nonetheless, very important to be able to predict as accurately as possible the annual energy delivered by the system in order to decide whether it makes economic sense. Certain components will dictate some of the details, but what has already been developed on rated ac power and "peak hours" of insolation provides a good start to system design.

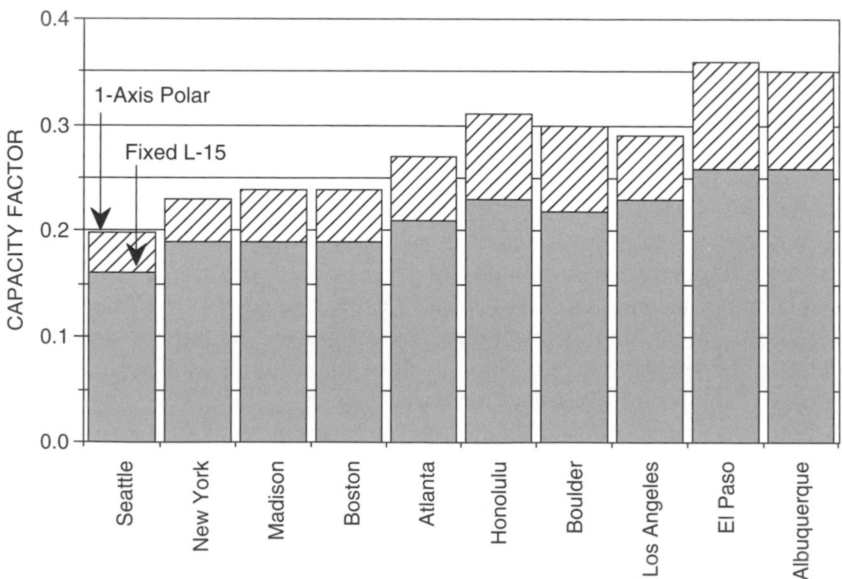

Figure 9.28 AC photovoltaic capacity factors for a number of U.S. cities.

As an academic exercise, system sizing is straightforward. How many kWh/yr are required? How many peak watts of dc PV power are needed to provide that amount? How much area will that system require? The realities of design, however, revolve around real components, which are available only in certain sizes and which have their own design constraints. It is hard to find a collector that is rated at 103.45 W. Available rooftop areas and orientations, whether a pole-mount is physically and aesthetically acceptable, is a collector rack in the yard viable, all affect system sizing. Some decisions require not only technical data but cost data as well, such as whether a tracking system is more cost effective than a fixed array. And certainly budget constraints dominate every decision.

Example 9.6 System Sizing in Fresno, CA: A First Cut. An energy efficient house in Fresno is to be fitted with a rooftop PV array that will annually displace all of the 3600 kWh/yr of electricity that the home uses. How many kW (dc, STC) of panels will be required and what area will be needed? Make assumptions as needed.

Solution. We'll assume the roof is south-facing with a moderate tilt angle. Aesthetically, the array will look best if it is mounted at the same angle as the roof, which means a shallow pitch. Data in Appendix E indicate 5.7 kWh/m^2-day of annual insolation for L-15, which at Fresno's latitude of 37° means a tilt of 22°. That's probably a good estimate for the roof insolation.

Using the peak hour approach, we can write

$$\text{Energy (kWh/yr)} = P_{ac}(\text{kW}) \cdot (\text{h/day @ 1-sun}) \cdot 365 \text{ days/yr}$$

Solving

$$P_{ac} = \frac{3600 \text{ kWh/yr}}{5.7 \text{ h/day} \times 365 \text{ days/yr}} = 1.73 \text{ kW}$$

Previous examples have shown how to individually estimate the impacts of temperature, inverter efficiency, module mismatch, and dirt to come up with conversion efficiency from dc to ac. Those results suggests that a de-rating of about 25%, or an efficiency of 75%, is in the ballpark, so we'll use that to estimate the STC rated dc power of the array:

$$P_{dc,STC} = \frac{P_{ac}}{\text{Conversion efficiency}} = \frac{1.73 \text{ kW}}{0.75} = 2.3 \text{ kW}$$

If we can estimate collector efficiency, we can find collector area from the following:

$$P_{dc,STC} = 1 \text{ kW/m}^2 \text{ insolation} \cdot A \text{ (m}^2) \cdot \eta$$

Assuming crystalline silicon modules, Table 8.3 suggests that an efficiency of about 12.5% is reasonable, resulting in an area estimate of

$$A = \frac{2.3 \text{ kW}}{1 \text{ kW/m}^2 \cdot 0.125} = 18.4 \text{ m}^2 \text{ (198 ft}^2\text{)}$$

The procedure outlined in Example 9.6 illustrates the first step in grid-connected system design, which is to estimate the rated power and area required for the PV array. Local insolation in kWh/m²-day was a key site parameter, and the resulting calculations included rough estimates for a de-rating from dc to ac power along with the module efficiency. Figure 9.29 shows how the estimate of annual energy production depends on the de-rating and local insolation, while Fig. 9.30 shows how area required depends on module efficiency.

Example 9.7 Fresno House with a 1-Axis Polar Tracker. Use Figs. 9.29 and 9.30 to estimate the module rated power and area needed to deliver 3600 kWh/yr in Fresno if a single-axis, polar mount tracker is used.

Solution. From Appendix E, annual insolation on a 1-axis tracker in Fresno is 7.6 kWh/m²-day. From Fig. 9.29 with the 75% dc-to-ac efficiency, the energy delivered per kW (dc, STC) is about 2100 kWh/yr. This suggests that we need

$$P_{dc,STC} = \frac{3600 \text{ kWh/yr}}{2100 \text{ kWh/yr/kW}_{dc,STC}} = 1.7 \text{ kW}$$

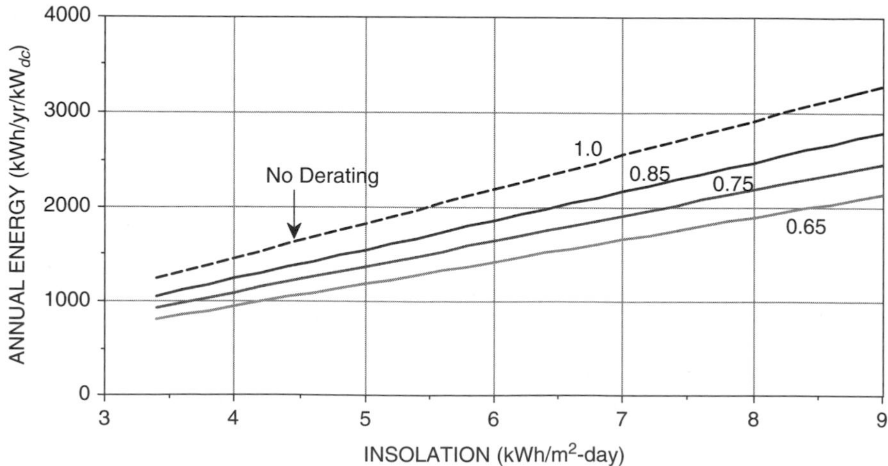

Figure 9.29 Annual energy delivered by a 1 kW(dc, STC) PV array, with dc to ac conversion efficiency as a parameter.

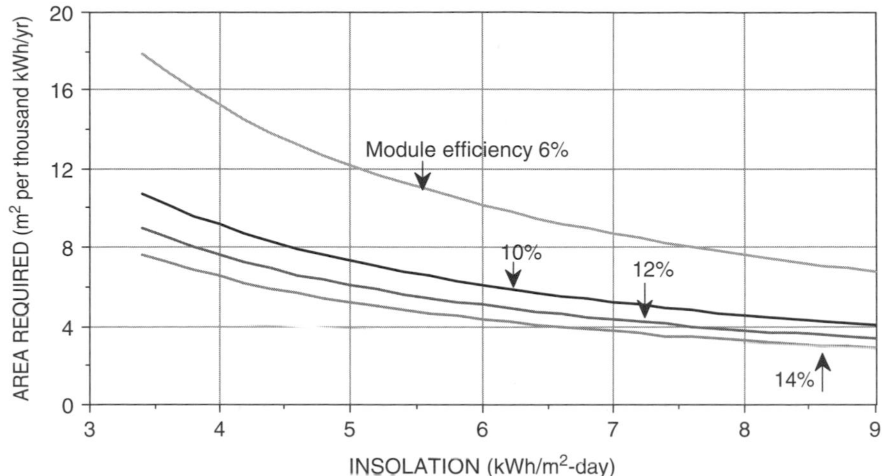

Figure 9.30 Area required to deliver 1000 kWh/yr with module efficiency as a parameter. Assumes a conversion efficiency from dc to ac of 75%.

This is considerably less than the 2.3 kW needed for the nontracking array in Example 9.6. The extra cost of the tracker will need to be compared to the reduction in cost of both the modules and the inverter.

From Fig. 9.30, the area of collectors at the assumed efficiency of 12.5% looks like about 3.9 m^2 per 1000 kWh/yr. That is, the area would be about

$$\text{Area} = 3600\ \text{kWh/yr} \times 3.9\ \text{m}^2/(1000\ \text{kWh/yr}) = 14\ \text{m}^2$$

Examples 9.6 and 9.7 illustrate the first step in grid-connected system design, which is to estimate the rated power and area required for the PV array. The next step is to explore the interactions between the choice of PV modules and inverters and how those impact the layout of the PV array. Finally, we need to consider details about voltage and current ratings for fuses, switches, and conductors.

Most traditional collectors on the market have 36 or 72 series cells in order to satisfy 12- or 24-V battery charging applications. Higher-voltage, higher-power modules are now becoming popular in grid-connected systems, for which battery-voltage constraints no longer apply. The key characteristics for a number of high-power modules intended for grid connections are given in Table 9.4.

Similarly, inverters for grid-connected systems are also different from those designed for battery-charging applications. Grid-connected inverters, for example, accept much higher input voltages and, as we shall see, those voltage constraints very much affect how the PV array is configured. The most important parameters for a number of inverters intended for grid-connected applications are given in Table 9.5.

TABLE 9.4 Important Characteristics of Several High-Power PV Modules

Module:	Sharp NE-K125U2	Kyocera KC158G	Shell SP150	Uni-Solar SSR256
Material:	Poly Crystal	Multicrystal	Monocrystal	Triple junction a-Si
Rated power $P_{dc,STC}$:	125 W	158 W	150 W	256 W
Voltage at max power:	26.0 V	23.2 V	34 V	66.0 V
Current at max power:	4.80 A	6.82 A	4.40 A	3.9
Open-circuit voltage V_{OC}:	32.3 V	28.9 V	43.4 V	95.2
Short-circuit current I_{SC}:	5.46 A	7.58 A	4.8 A	4.8
Length:	1.190 m	1.290 m	1.619 m	11.124 m
Width:	0.792 m	0.990 m	0.814 m	0.420 m
Efficiency:	13.3%	12.4%	11.4%	5.5%

TABLE 9.5 Example Inverter Characteristics for Grid-Connected Systems

Manufacturer:	Xantrex	Xantrex	Xantrex	Sunny Boy	Sunny Boy
Model:	STXR1500	STXR2500	PV 10	SB2000	SB2500
AC power:	1500 W	2500 W	10,000 W	2000 W	2500 W
AC voltage:	211–264 V	211–264 V	208 V, 3Φ	198–251 V	198–251 V
PV voltage range MPPT:	44–85 V	44–85 V	330–600 V	125–500 V	250–550 V
Max input voltage:	120 V	120 V	600 V	500 V	600 V
Max input current:	—	—	31.9 A	10 A	11 A
Maximum efficiency:	92%	94%	95%	96%	94%

To explore the interactions between modules, inverters, and the PV array, let us continue the design started in Example 9.6. For this example, let us try the Kyocera KC158G 158-W module with the Xantrex STXR2500 inverter. Begin by determining the number of modules required. Since they are rated at 158 W each and we need 2300 dc, STC watts, we have

$$\text{Number of modules} = \frac{2300\text{ W}}{158\text{ W/module}} = 14.6$$

To help us decide between 14 modules or 15 modules, consider how they might be arranged into an array. With two modules per string, the STC rated voltage would be $2 \times 23.2 = 46.4$ V, which just barely falls into the MPPT range of 44–85 V for the inverter picked. At higher temperatures, the module voltage could drop below 44 V, which isn't so good. At three modules per string, the rated voltage becomes $3 \times 23.2 = 69.6$ V, which fits nicely with the MPPT range. This suggests using an array with five strings of three modules each, for a total of 15 modules.

It is important to estimate the maximum open-circuit voltage of the array to be sure that it doesn't violate the highest dc voltage that the inverter can accept, which in this case is 120 V. With three modules in series, each having a V_{OC} at STC of 28.9 V, the string voltage could reach $3 \times 28.9 = 86.7$ V. This is well below the 120-V limit. But, remember that V_{OC} increases when cell temperature is below the STC assumption of 25°C. We could imagine that on a cold morning, with a strong, cold wind and low sunlight, cell temperature might be close to ambient, and that might be well below 25°C. With V_{OC} increasing by 0.38%/°C below 25°C (for crystalline silicon), the open-circuit voltage could then be well above its STC value.

Suppose that it is −5°C on the coldest morning in Fresno, and assume that cell temperature and ambient temperature are the same. The three-module string V_{OC} would now be

$$V_{OC,max} = 86.7\ V \cdot [1 + 0.0038(25 + 5)] = 86.7 \times 1.114 = 97\ \text{V}$$

which is still below the 120-V limit of this inverter. That 1.114 multiplier is, of course, temperature-dependent, and Fig. 9.31 shows the relationship.

Another reason for checking the open-circuit voltage of a residential grid-connected array under the coldest possible ambient conditions at the site is that the National Electrical Code restricts all voltages in one- and two-family dwellings to no more than 600 V (Wiles, 2001). Our array easily satisfies this constraint, so we will choose a design with 15 modules arranged in five strings of three modules each.

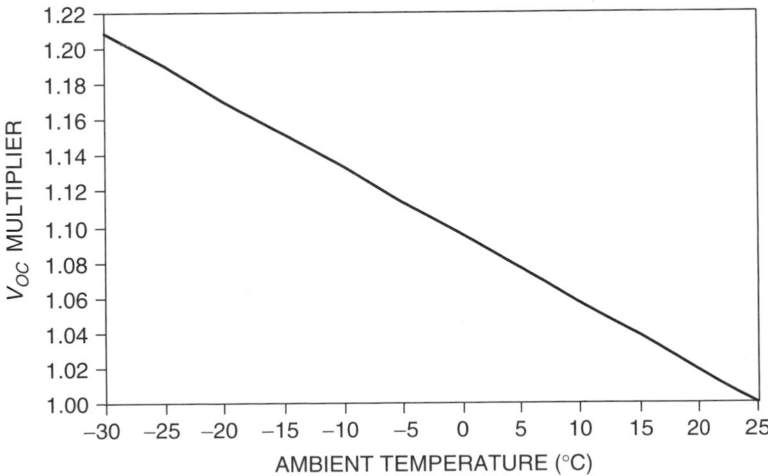

Figure 9.31 The V_{OC} multiplier for crystalline silicon assuming cells are at ambient temperature.

Example 9.8 Fresno House Example: Second Pass. Find the roof area required for the 15-module Fresno system with the fixed orientation collector, and estimate the annual energy delivered.

Solution. Collector dimensions are given in Table 9.4. With 15 modules, the collector area will be

$$\text{Area} = 15 \text{ modules} \times 1.29 \text{ m} \times 0.99 \text{ m} = 19.1 \text{ m}^2 \ (206 \text{ ft}^2)$$

which is just a bit larger than our original estimate of 18.4 m^2.

The dc STC rated power of the array will be

$$P_{dc,STC} = 158 \text{ W/module} \times 15 \text{ modules} = 2370 \text{ W}$$

Assuming a 25% de-rating for ac and using the 5.7-kWh/m^2-day average insolation for Fresno means that the system is predicted to deliver

$$\text{Energy} = 2.37 \text{ kW} \times 0.75 \times 5.7 \text{ h/day} \times 365 \text{ day/yr} = 3698 \text{ kWh/yr}$$

So, our goal of 3600 kWh/yr has been met.

In addition to the 600-V maximum voltage limit for one- and two-family dwellings, the proposed National Electrical Code for PV systems specifies other constraints on the choice of wires, fuses, and disconnect switches. They must be capable of withstanding 1.25 times the expected dc voltage. The NEC also recommends multiplying all PV currents by 1.25 to account for two possibilities: (1) the potential for insolation to exceed the 1-sun level of 1000 W/m^2 and (2) the increase in short-circuit current (approximately 0.1%/°C) caused by cell temperatures above the 25°C standard. In addition, the NEC requires that the continuous current of any circuit be multiplied by 1.25 to ensure that fuses, switches, and other disconnects as well as conductors are not operated above 80% of their rating. The two 125% current factors are independent and must be multiplied ($1.25 \times 1.25 = 1.56$) to properly size conductors and switchgear.

Finally, the allowable current that a conductor can handle (called its *ampacity*) is related to the type of insulation it has and the ambient temperature to which it is exposed. Standard ampacity ratings for cables assume an ambient temperature of 30°C. Given the possibility of higher ambients, exposure to sunlight, and nearness to hot collector surfaces, cable ampacity should be de-rated by factors of 0.33 to 0.58 depending on cable type, installation method (free air or conduit), and the temperature rating of the insulation (Wiles, 2001).

Example 9.9 Current and Voltage Ratings for the Fresno House Breakers. For our PV array consisting of five parallel strings of three Kyocera KC158G modules each, what minimum current rating should fuses have in the combiner, the array disconnect switch, and the utility breaker from the inverter? What maximum voltage should switchgear be rated for if the minimum daytime ambient temperature is expected to be −5°C?

Solution. From Table 9.4, the STC short-circuit current for each string of panels is 7.58 A. Applying the 125% factor for the effects of higher insolation and lower temperature, plus the 125 percent factor to afford a safe oversizing margin, results in combiner fuses that must allow at least

$$\text{Combiner fuses} > 7.58\ \text{A} \times 1.25 \times 1.25 = 11.8\ \text{A}$$

The array disconnect fuse must accommodate five such strings, so it must handle

$$\text{Array disconnect fuse} > 11.8\ \text{A} \times 5 = 59.2\ \text{A}$$

Applying the NEC 1.25 current multiplier to the 2500-W, 240-V inverter says its fuse must be rated to handle

$$\text{Inverter fuse} > 1.25 \times \frac{2500\ \text{W}}{240\ \text{V}} = 13\ \text{A}$$

Notice only a single 1.25 multiplier is used for the inverter since the PV temperature and insolation adjustments don't apply to the ac portions of the system.

The final system design is for the Fresno house is shown in Fig. 9.32.

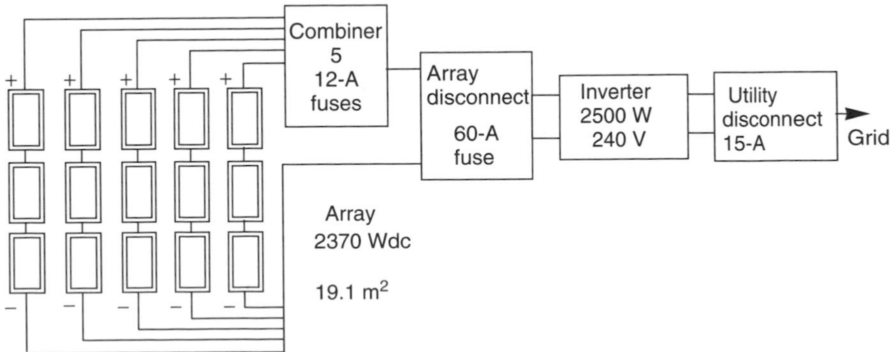

Figure 9.32 System design for the Fresno house example.

TABLE 9.6 Performance Adjustment Factors for Various Roof Pitches and Directions for 40° Latitude

Tilt Angle:	Flat	20	30	40	50	90
South:	0.84	0.97	1.00	1.00	0.97	0.61
SE, SW:	0.84	0.92	0.93	0.92	0.88	0.56
E, W:	0.84	0.80	0.77	0.72	0.67	0.40

Reference condition is 30° tilt, south facing.

Since most residential PV systems are mounted at the same pitch as the roof, and since those roofs seldom face due south, it is handy to have an adjustment factor to account for various collector orientations. The clear sky insolation tables in Appendix D can be used to estimate these adjustments. An example presentation in which a due south array at a 30° pitch is the reference condition is given in Table 9.6.

9.4 GRID-CONNECTED PV SYSTEM ECONOMICS

We now have the tools to allow us to estimate the energy delivered by a grid-connected PV system, so the next step is to explore its economic viability. Two types of economic analyses need to be made. One helps to make decisions between different system options—for example, whether to use a tracking system or a fixed mount, which collectors and which inverter are most cost effective, and so forth—and the other helps a buyer decide whether the investment is worthwhile.

9.4.1 System Trade-offs

To illustrate the decision between system options, consider the trade-off between the benefits of higher irradiance with a tracking mount compared to a simple fixed, roof mount. The following example demonstrates a process that can be used, but realistic decisions depend on current and accurate cost estimates for the equipment and installation.

Example 9.10 Should a House in Boulder Use a 1-Axis Tracker? A PV system for a house in Boulder, CO, is to be designed to deliver about 4000 kWh/yr. Given the following costs, decide whether to recommend a fixed array at tilt L-15 or a single-axis tracker. Assume 12%-efficient PVs and a 0.75 dc-to-ac efficiency factor.

Component	Cost
PVs	\$4.20/Wdc
Inverter	\$1.20/ W
Tracker	\$400 + \$100/m^2
Installation, BOS	\$3800

Solution
1-Axis Tracker: From Appendix E, annual insolation with a 1-axis tracker is 7.2 kWh/m^2-day. Using the peak-hours approach combined with the derating from dc, STC to ac:

$$\text{kWh/yr ac} = P_{dc,STC}(\text{kW}) \cdot (\text{Conversion efficiency}) \cdot \left(\frac{\text{h}}{\text{day @1-sun}}\right) \cdot 365 \text{ days/yr}$$

$$P_{dc,STC} = \frac{4000 \text{ kWh/yr}}{0.75 \times 7.2 \text{ h/day} \times 365 \text{ days/yr}} = 2.03 \text{ kW}$$

which would cost about \$4.20/W $\times$ 2030 W = \$8524 for the PVs.

At \$1.20/W, a 2.03-kW inverter would cost \$2435.

The area of collector needed at 12% efficiency would be

$$A(\text{m}^2) = \frac{P_{dc,STC}}{1 \ (\text{kW/m}^2) \cdot \eta} = \frac{2.03 \text{ kW}}{1 \times 0.12} = 16.9 \text{ m}^2$$

The extra cost of the tracker would be \$400 + \$100/m^2 $\times$ 16.9 m^2 = \$2091.

Fixed Array: The insolation at L-15 is 5.4 kWh/m^2-day (Appendix E).

$$P_{dc,STC} = \frac{4000 \text{ kWh/yr}}{0.75 \times 5.4 \text{ h/day} \times 365 \text{ day/yr}} = 2.71 \text{ kW}$$

costing \$4.20/W $\times$ 2710 W = \$11,365 for the PVs. Notice that a bigger inverter is needed to accommodate the higher power rating of the collectors. A 2.71-kW inverter would cost \$1.20/W $\times$ 2710 W = \$3247.

The cost summary is thus:

Component	Tracker	Fixed Tilt
PVs	\$8,524	\$11,365
Inverter	\$2,435	\$3,247
Tracker	\$2,091	\$0
Installation, BOS	\$3,800	\$3,800
Total	\$16,850	\$18,412

It looks like there might some advantage to using the tracker, but only a more careful analysis based on real components could affirm the validity of that conclusion.

9.4.2 Dollar-per-Watt Ambiguities

The most important inputs to any economic analysis of a PV system are the initial cost of the system and the amount of energy it will deliver each year. Whether the system is economically viable depends on other factors—most especially, the price of the energy displaced by the system, whether there are any tax credits or other economic incentives, and how the system is to be paid for. A detailed economic analysis will include: estimates of operation and maintenance costs; future costs of utility electricity; loan terms and income tax implications if the money is to be borrowed, or personal discount rates if the owner purchases it outright; system lifetime; costs or residual value when the system is ultimately removed; and so forth.

Begin with the installed cost of the system. For individual buyers, it is total dollars for their system that matters, but when an overall snapshot of the industry is being presented, it is common practice to describe installed costs in dollars per watt of peak power. There are two ambiguities with the $/W indicator, however, which must be made clear for the parameter to have any meaning. One is whether the watts are based on dc power from the PVs or ac power from the inverter. The other is whether or not a tracker has been used. To illustrate these differences, Table 9.7 summarizes the costs for the tracker versus fixed-array analysis just derived in Example 9.10. As can be seen, even though the tracker delivers the same kWh/yr and it is cheaper than the fixed-tilt array, it appears to have a higher cost when expressed as $/Wdc or $/Wac. Clearly, some factor is needed to account for the greater energy delivered by a module in a tracking system.

When a PV system uses tracking, an *Energy Production Factor* (EPF) must be included in order to make the simple $/Wdc or $/Wac descriptors directly comparable to those systems that have a fixed orientation. The EPF is essentially the ratio of insolation on the tracking surface to that on a fixed surface (for the example in Table 9.7, EPF = 7.2/5.4 = 1.333), which leads to the following:

$$\text{Tracker}(\$/\text{W}) = \frac{\$/\text{W}}{\text{EPF}} = \frac{\$/\text{W}}{(\text{Tracking insolation/Fixed insolation})} \tag{9.17}$$

When the EPF is included, the tracking array in Table 9.7 does correctly appear to be the least-cost system.

So, having resolved the $/W ambiguities, how much do systems cost? Data gathered by the Utility Photovoltaic Group (UPVG) on actual costs of over 600 residential PV systems installed between 1994 and 2000 has been summarized in Fig. 9.33. Of the total $6.80/Wdc, a little over 60% is modules with the remainder roughly equally divided between other components and installation.

TABLE 9.7 Illustrating the Ambiguities Associated with \$/W Cost Indicators[a]

Parameter	Tracker	Fixed Tilt
Energy (kWh/yr)	4000	4000
Insolation (kWh/m^2-day)	7.2	5.4
System cost (\$)	\$16,850	\$18,412
$P_{dc,STC}$ (W)	2029	2706
P_{ac} (W)	1522	2029
Cost \$/Wdc	\$8.30	\$6.80
Cost \$/Wac	\$11.07	\$9.07
Cost \$/Wdc (EPF)	\$6.23	
Cost \$/Wac (EPF)	\$8.30	

[a] Without an energy production factor (EPF) correction in \$/W, the tracking system incorrectly appears to be more expensive. (Data are based on Example 9.10.)

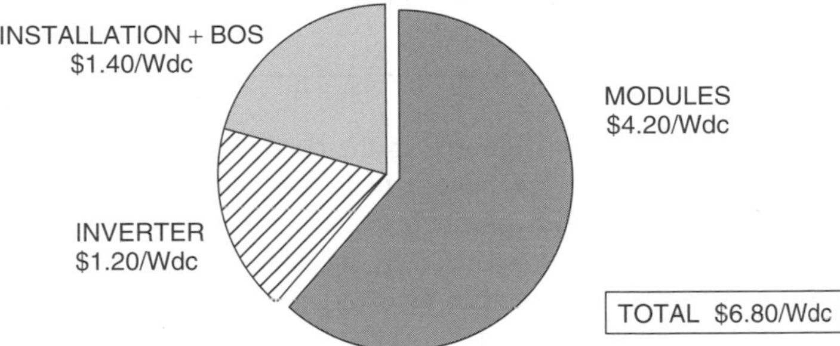

Figure 9.33 Average installed cost for 625 residential PV systems installed between 1994 and 2000. From Chang (2000).

9.4.3 Amortizing Costs

A simple way to estimate the cost of electricity generated by a PV system is to imagine taking out a loan to pay for the system and then using annual payments divided by annual kWh delivered to give ¢/kWh. If an amount of money, or principal, P (\$), is borrowed over a period of n (years) at an interest rate of i (decimal fraction/yr), then the annual loan payments, A (\$/yr), will be

$$A = P \cdot \text{CRF}(i, n) \tag{9.18}$$

where CRF(i, n) is the capital recovery factor given by

$$\text{CRF}(i, n) = \frac{i(1+i)^n}{(1+i)^n - 1} \tag{9.19}$$

Example 9.11 Cost of PV Electricity for the Boulder House. The tracking PV system of Example 9.10 costs \$16,850 to deliver 4000 kWh/yr. If the system is paid for with a 6%, 30-year loan, what would be the cost of electricity?

Solution. The capital recovery factor for this loan would be

$$\text{CRF}(i, n) = \frac{i(1+i)^n}{(1+i)^n - 1} = \frac{0.06(1.06)^{30}}{(1.06)^{30} - 1} = 0.07265/\text{yr}$$

So the annual payments would be

$$\text{A} = \text{P CRF}(i, n) = \$16{,}850 \times 0.07265/\text{yr} = \$1224/\text{yr}$$

The cost per kWh is therefore

$$\text{Cost of electricity} = \frac{\$1224/\text{yr}}{4000 \text{ kWh/yr}} = \$0.306/\text{kWh}$$

A significant factor that was ignored in the cost calculation of Example 9.11 is the impact of the income tax benefit that goes with a home loan. As was described in Chapter 5, interest on such loans is tax deductible, which means that a person's gross income is reduced by the loan interest, and it is only the resulting net income that is subject to income taxes. The tax benefit that results depends on the marginal tax bracket (MTB) of the homeowner. For married couples filing jointly, the 2002 MTB for federal taxes is shown in Table 9.8. For example, a married couple earning \$120,000 per year is in the 30.5% marginal tax bracket, which means that a one-dollar tax deduction reduces their income tax by 30.5 cents. The value of the tax deduction will be even greater when it reduces the homeowner's state income tax as well. For example, the same taxpayer in California would be in the 37% marginal tax bracket when both state and federal taxes are considered.

TABLE 9.8 Federal Income Tax for Married Couples Filing Jointly (2002)

Income Over	But Not Over	Tax Is	of the Amount Over
\$0	\$45,200	15%	\$0
45,200	109,250	\$6,780 + 27.5%	45,200
109,250	166,500	24,394 + 30.5%	109,250
166,500	297,350	41,855 + 35.5%	166,500
297,350	—	88,307 + 39.1%	297,350

During the first years of a long-term loan, almost all of the annual payments will be interest, with very little left to reduce the principal, while the opposite occurs toward the end of the loan. This means that the tax benefit of interest payments varies from year to year. For example, in the first year, interest is owed on the entire amount borrowed and the tax benefit is

$$\text{First-year tax benefit} = i \times P \times \text{MTB} \tag{9.20}$$

where MTB is the marginal tax bracket.

Example 9.12 Cost of PV Energy Including Income Tax Benefit. The PV system for the house in Boulder costs \$16,850 and annual payments on its 6%, 30-year loan are \$1224. If the homeowner is in the 30.5% marginal tax bracket, what is the cost of PV energy in the first year?

Solution. From (9.20) the reduction in federal income tax will be

$$\text{First-year tax benefit} = 0.06 \times \$16{,}850 \times 0.305 = \$308$$

The net cost of the PV system in that first year will therefore be

$$\text{First-year cost of PV} = \$1224 - \$308 = \$916$$

which means that its electricity cost will be

$$\text{Cost of electricity} = \frac{\$916/\text{yr}}{4000\ \text{kWh/yr}} = \$0.23/\text{kWh}$$

Figure 9.34 shows how the first-year cost of electricity varies as a function of annual insolation and system cost. A 30-year, 6% loan for a family in the 30.5% tax bracket has been used to derive the figure. For a reasonably good site, with annual insolation above 5.5 kWh/m^2-day, the installed cost of a PV system needs to drop to below about \$3/Wdc to make PV electricity competitive with 10¢/kWh electricity from the grid. At the \$6.80/Wdc average cost at the turn of the century, significant cost reductions are needed to make PVs competitive with the local utility. To help encourage the PV industry, some states have initiated sizable rebates and tax credits. California, for example, has a buy-down program funded by utility ratepayers that offers a \$4.50/Wac rebate (about \$3.50/Wdc) which pretty much closes the gap making PVs cost effective right now.

A very effective way to present the economics of a residential grid-connected PV system is with a spreadsheet that incorporates the price of avoided electricity

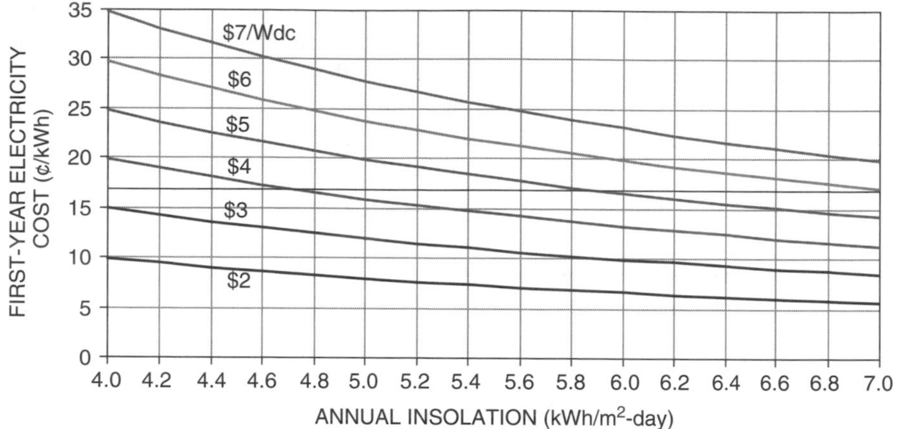

Figure 9.34 First-year cost of electricity with \$/Wdc (STC) as the parameter. Assumptions: 6%, 30-yr loan, MTB 0.305, 75% conversion from dc (STC) to ac (PTC).

purchases, loan terms, including the homeowner MTB and income tax benefits, and any rebates or other monetary features that affect the decision.

Example 9.13 Financing a PV System. A 3-kWac PV system is predicted to deliver 6000 kWh/yr to a house that currently pays \$0.12/kWh for electricity. The system, which costs \$27,000, is eligible for a rebate of \$4.50/Wac. If the balance is paid for with a 6%, 30-year loan and the owner is in the 37% tax bracket (combined state and federal), what is the cost of PV electricity in the first year and what would be the net economic benefit in the first year?

Solution. The net cost of the system after the rebate is

$$P = \$27{,}000 - \$4.50/\text{W} \times 3000\ \text{W} = \$13{,}500$$

From Example 9.11, CRF(0.06, 30 yr) is 0.07265/yr, so the annual loan payment is

$$\text{Loan} = 0.07265/\text{yr} \times \$13{,}500 = \$980.78$$

The reduction in income taxes on the first-year's interest portion of the loan payment is

$$\text{Tax savings} = 0.06 \times \$13{,}500 \times 0.37 = \$299.70$$

The first-year cost of electricity from the PV system is therefore

$$\text{Cost of electricity} = \frac{\$980.78 - \$299.70/\text{yr}}{6000\ \text{kWh/yr}} = \$0.1135/\text{kWh}$$

The net economic benefit in the first year is

$$\text{Benefit} = 6000\ \text{kWh/yr} \times (0.12 - 0.1135)\$/\text{kWh} = \$39/\text{yr}$$

It is easy to set up a spreadsheet to show the year-by-year economic value of a PV system as has been done in Table 9.9 for Example 9.13. In that spreadsheet, the price of utility electricity is assumed to escalate at 2%/yr, with the result that

TABLE 9.9 Annual Cash Flow for the PV System of Example 9.13[a]

Year	Loan Balance	Loan Payment	Loan Interest	Delta P	Delta tax	Annual Cost	PV cost ¢/kWh	Utility ¢/kWh	Savings $/yr
0	13,500	981	810	171	300	681	11.4	12.0	39
1	13,329	981	800	181	296	685	11.4	12.2	50
2	13,148	981	789	192	292	689	11.5	12.5	60
3	12,956	981	777	203	288	693	11.6	12.7	71
4	12,753	981	765	216	283	698	11.6	13.0	82
5	12,537	981	752	229	278	702	11.7	13.2	93
6	12,309	981	739	242	273	708	11.8	13.5	103
7	12,067	981	724	257	268	713	11.9	13.8	114
8	11,810	981	709	272	262	719	12.0	14.1	125
9	11,538	981	692	288	256	725	12.1	14.3	136
10	11,249	981	675	306	250	731	12.2	14.6	147
11	10,943	981	657	324	243	738	12.3	14.9	157
12	10,619	981	637	344	236	745	12.4	15.2	168
13	10,276	981	617	364	228	753	12.5	15.5	179
14	9,911	981	595	386	220	761	12.7	15.8	189
15	9,525	981	572	409	211	769	12.8	16.2	200
16	9,116	981	547	434	202	778	13.0	16.5	210
17	8,682	981	521	460	193	788	13.1	16.8	220
18	8,223	981	493	487	183	798	13.3	17.1	230
19	7,735	981	464	517	172	809	13.5	17.5	240
20	7,218	981	433	548	160	821	13.7	17.8	249
21	6,671	981	400	581	148	833	13.9	18.2	259
22	6,090	981	365	615	135	846	14.1	18.6	268
23	5,475	981	328	652	122	859	14.3	18.9	276
24	4,823	981	289	691	107	874	14.6	19.3	284
25	4,131	981	248	733	92	889	14.8	19.7	292
26	3,398	981	204	777	75	905	15.1	20.1	300
27	2,622	981	157	823	58	923	15.4	20.5	306
28	1,798	981	108	873	40	941	15.7	20.9	313
29	925	981	56	925	21	960	16.0	21.3	318
30	0	0	0	0	0	0	0.0	21.7	1304

[a] Buyer is in the 37% tax bracket and utility electricity is projected to grow at 2%/yr.

annual savings with the PV system grows year-by-year, reaching \$318/yr in the last year of the 30-year loan.

9.5 STAND-ALONE PV SYSTEMS

Since grid-connected systems use the utility for back-up, there is no need for battery storage unless power outages are a problem. This makes them relatively simple and relatively inexpensive. Having the grid right there, however, means that they have to compete with relatively inexpensive utility power, which makes it hard to justify the PV system unless significant subsidies are provided. When the grid isn't nearby, electricity suddenly becomes much more valuable and the extra cost and complexity of a totally self-sufficient, stand-alone power system can provide enormous benefit. Instead of competing with 10-cent utility power, a PV–battery system competes with 50-cent gasoline or diesel-powered generators. Or it competes with the cost of bringing the grid to the site, which may run many thousands of dollars per mile. In developing countries, a panel and battery to run a few lights and a radio can change people's lives.

Figure 9.35 suggests a quite general system that includes a generator backup as well as the possibility for some loads to be served directly with more-efficient dc and others with ac. A combination charger–inverter is shown, which has the capability to convert ac to dc or vice versa. As a charger, it converts ac from the generator into dc to charge the batteries; as an inverter, it converts dc from the batteries into ac needed by the load. The charger–inverter unit may include an automatic transfer switch that allows the generator to supply ac loads directly whenever it is running.

Off-grid systems must be designed with great care to assure satisfactory performance. Users must be willing to check and maintain batteries, they must be willing to adjust their energy demands as weather and battery charge vary, they may have to fuel and fix a noisy generator, and they must take responsibility for the safe operation of the system. The reward is electricity that is truly valued.

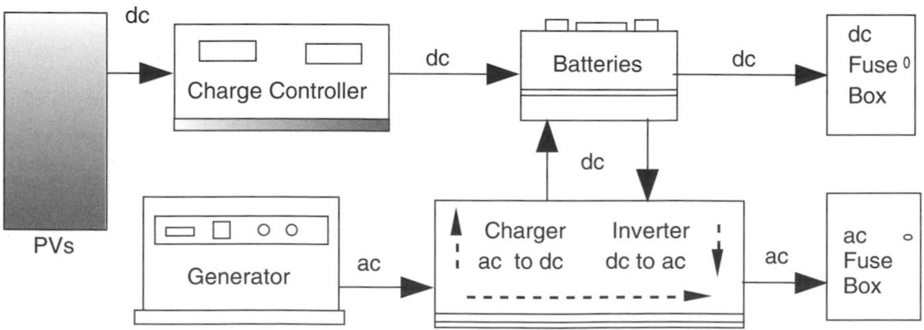

Figure 9.35 A stand-alone system with back-up generator and separate outputs for dc and ac loads.

9.5.1 Estimating the Load

The design process for stand-alone systems begins with an estimate of the loads that are to be provided for. As with all design processes, a number of iterations may be required. On the first pass, the user may try to provide the capability to power anything and everything that normal, grid-connected living allows. Various iterations will follow in which trade-offs are made between more expensive, but more efficient, appliances and devices in exchange for fewer PVs and batteries. Lifestyle adjustments need to be considered in which some loads are treated as essentials that must be provided for, and others are luxuries to be used only when conditions allow. A key decision involves whether to use all dc loads to avoid the inefficiencies associated with inverters, or whether the convenience of an all ac system is worth the extra cost, or perhaps a combination of the two is best. Another important decision is whether to include a generator back-up system and, if so, what fraction of the load it will have to supply.

The simplest of systems will incorporate only devices that run directly on dc. A rather large market already exists for such dc equipment to meet the needs of the boating and recreational vehicle communities. In addition, catalogs of such equipment are readily available. For example, Real Goods Trading Corporation in Ukiah, California, publishes its continuously updated *Solar Living Source Book*, which not only provides detailed descriptions of specific equipment, but also presents theory and perspectives on system design and equipment selection. At the other extreme, off-grid homes can be as conventional as any other, with ac appliances and devices purchased from mainstream suppliers.

Power needed by a load, as well as energy required over time by that load, is important for system sizing. In the simplest case, energy (watt-hours or kilowatt-hours) is just the product of some nominal power rating of the device multiplied by the hours that it is in use. The situation is often more complicated, however. For example, an amplifier needs more power when the volume is increased, and many appliances, such as refrigerators and washing machines, use different amounts of power during different portions of their operating cycle. An especially important consideration for household electronic devices—TVs, VCRs, computers, portable phones, and so on—is the power consumed when the device is in its standby or charging modes. Many devices, such as TVs, use power even while they are turned off since some circuits remain energized awaiting the turn-on signal from the remote. Consumer electronics now account for about 10% of all U.S. residential electricity, and researchers at Lawrence Berkeley National Labs conclude that almost two-thirds of this energy occurs when these devices are not actually being used (Rosen and Meier, 2000). Major appliances and shop tools have another complication caused by the surge of power required to start their electric motors. While that large initial spike doesn't add much to the energy used by a motor, it has important implications for sizing inverters, wires, fuses, and other ancillary electrical components in the system.

Table 9.10 lists examples of power used by a number of household electrical loads. Some of these are simply watts of power, which can be multiplied by hours of use to get watt-hours of energy. Many of the devices listed in the

TABLE 9.10 Power Requirements of Typical Loads

Kitchen Appliances	*Power*
Refrigerator: ac EnergyStar, 14 cu. ft	300 W, 1080 Wh/day
Refrigerator: ac EnergyStar, 19 cu. ft	300 W, 1140 Wh/day
Refrigerator: ac EnergyStar, 22 cu. ft	300 W, 1250 Wh/day
Refrigerator: dc Sun Frost, 12 cu. ft	58 W, 560 Wh/day
Freezer: ac 7.5 cu. ft	300 W, 540 Wh/day
Freezer: dc Sun Frost, 10 cu. ft	88 W, 880 Wh/day
Electric range (small burner)	1250 W
Electric range (large burner)	2100 W
Dishwasher: cool dry	700 W
Dishwasher: hot dry	1450 W
Microwave oven	750–1100 W
Coffeemaker (brewing)	1200 W
Coffeemaker (warming)	600 W
Toaster	800–1400 W
General Household	
Clothes washer: vertical axis	500 W
Clothes washer: horizontal axis	250 W
Dryer (gas)	500 W
Vacuum cleaner	1000–1400 W
Furnace fan: 1/4 hp	600 W
Furnace fan: 1/3 hp	700 W
Furnace fan: 1/2 hp	875 W
Ceiling fan	65–175 W
Whole house fan	240–750 W
Air conditioner: window, 10,000 Btu	1200 W
Heater (portable)	1200–1875 W
Compact fluorescent lamp (100-W equivalent)	30 W
Compact fluorescent lamp (60-W equivalent)	16 W
Electric blanket, single/double	60/100 W
Clothes iron	1000–1800 W
Electric clock	4 W
Consumer Electronics	
TV: >39-in. (active/standby)	142/3.5 W
TV: 25 to 27-in. color (active/standby)	90/4.9 W
TV: 19 to 20-in. color (active/standby)	68/5.1 W
Analog cable box (active/standby)	12/11 W
Satellite receiver (active/standby)	17/16 W
VCR (active/standby)	17/5.9 W
Component stereo (active/standby)	44/3 W
Compact stereo (active/standby)	22/9.8 W
Cordless phone	4 W
Clock radio (active/standby)	2.0/1.7 W
Computer, desktop (active/idle/standby)	125/80/2.2 W
Laptop computer	20 W

TABLE 9.10 (*continued*)

Ink-jet printer	35 W
Dot-matrix printer	200 W
Laser printer	900 W
Shop	
Circular saw, 7 1/4″	900 W
Table saw, 10-in.	1800 W
Hand drill, 1/4″	250 W
Water Pumping	
Centrifugal pump: 36 Vdc, 50-ft @ 10 gpm	450 W
Submersible pump: 24 Vdc, 100-ft @ 1.6 gpm	100 W
Submersible pump: 48 Vdc, 300-ft @ 1.5 gpm	180 W
DC pump (house pressure system), typical use 1–2 h/day	60 W

Source: Rosen and Meier (2000) and others.

consumer electronics category show power while they are being used (active) and power consumed the rest of the time (standby), both of which must be considered when determining energy consumption. Refrigerators are also unusual since they are always turned on, but their power demand varies throughout the day. The data given on refrigerator labels are average watt-hours per day based on measurements made with the refrigerator located in a 90°F room. This means that they are likely to overstate actual demand in someone's home—perhaps by as much as 20%. Tables like this one are useful for average values, but the best source of power and energy data are actual measurements that can easily be performed with readily available meters. Another source is the device nameplate itself, but those tend to overstate power since they are meant to describe maximum demand rather than the likely average. Some nameplates provide only amperage and voltage; and while it is tempting to multiply the two to get power, this can also be an overestimate since it ignores the phase angle, or power factor, between current and voltage.

Example 9.14 A Modest Household Demand. Estimate the monthly energy demand for a cabin with all ac appliances, consisting of a 19-cu. ft refrigerator, six 30-W compact fluorescents (CFLs) used 5 h/day, a 19-in. TV turned on 3 h/day and connected to a satellite, a cordless phone, a 1000-W microwave used 6 min/day, and a 100-ft deep well that supplies 120 gallons/day.

Solution. Using data from Table 9.10, we can put together the following table of power and energy demands. The total is just over 3.11 kWh/day.

Notice how little of the energy for TV/satellite is used during the 3 h that it is actually turned on (255 Wh out of a total of 698 Wh, or 36.5%). This household really needs to provide a power strip to allow manual complete shut off of these electricity vampires.

Appliance	Power (W)	Hours	Watt-hours/day	Percentage
Refrigerator, 19 cu. ft	300		1140	37%
Lights (6 @ 30 W)	180	5	900	29%
TV, 19-in., active mode	68	3	204	7%
TV, 19-in., standby mode	5.1	21	107	3%
Satellite, active mode	17	3	51	2%
Satellite, standby mode	16	21	336	11%
Cordless phone	4	24	96	3%
Microwave	1000	0.1	100	3%
Washing machine	250	0.2	50	2%
Well pump, 100 ft, 1.6 gpm	100	1.25	125	4%
Total			3109	100%

9.5.2 The Inverter and the System Voltage

In Example 9.14, an example calculation was made for the average daily energy consumption for appliances and loads that were all assumed to run on ac power. To figure out how much power the batteries must supply, the calculation needs to be modified to account for losses in the dc-to-ac inverter. This can be tricky to do accurately since the inverter's efficiency is a function of the magnitude of the load it happens to be supplying at that particular instant. Most inverters now operate at around 90% efficiency over most of their range, as is suggested in Fig. 9.36. For calculations, an overall inverter efficiency of about 85% is considered to be a conservative default assumption.

When no load is present, a good inverter will power down to less than 1 watt of standby power while it waits for something to be turned on that needs ac. When it senses a load, the inverter powers up and while it runs uses on the order of 5–20 W of its own. That means those standby losses associated with so many of our electronic devices may keep the inverter running continuously, even though no real energy service is being delivered. In that case, that 5-to-20 watts of inverter loss adds to the other standby losses, which emphasizes the need for a manual shutdown of turned-off electronic equipment.

Example 9.15 Accounting for Inverter Losses. Suppose that a dc refrigerator that uses 800 Wh/day is being considered instead of the 1140 Wh/d ac one given in Example 9.14. Estimate the dc load that the batteries must provide if an 85% efficient inverter is used (a) with all loads running on ac and (b) with everything but the refrigerator running on ac.

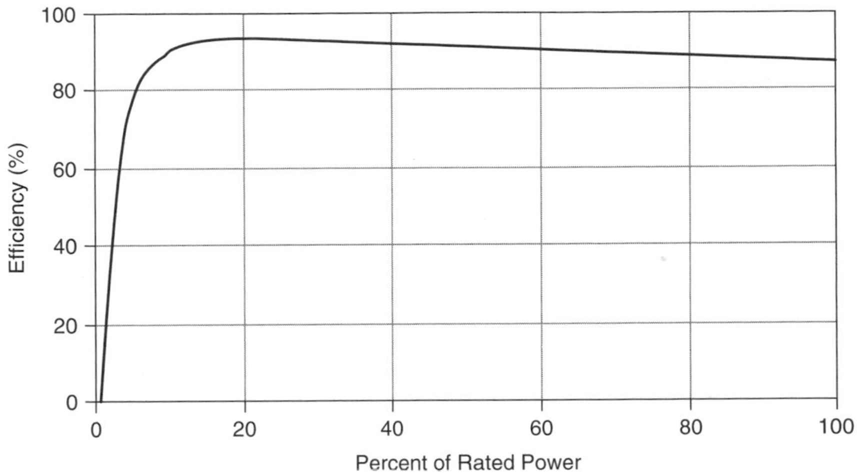

Figure 9.36 Typical efficiency of a stand-alone system inverter.

Solution

a. With all 3109 Wh/day running on an 85% efficient inverter, the dc load that the batteries must supply would be

$$\text{dc battery load} = \frac{3109 \text{ Wh/day}}{0.85} = 3658 \text{ Wh/day}$$

b. With the 1140-Wh/day refrigerator removed, the remaining ac load is

$$\text{ac load} = (3109 - 1140) \text{ Wh/day} = 1969 \text{ Wh/day}$$

accounting for the inverter efficiency, that ac load would be supplied by

$$\text{dc for ac loads} = \frac{1969 \text{ Wh/day}}{0.85} = 2316 \text{ Wh/day}$$

Adding in the 800-Wh/day dc refrigerator, the total dc load becomes

$$\text{dc battery load} = (2316 + 800) \text{ Wh/day} = 3116 \text{ Wh/day}$$

That's a 15% decrease in energy needed. Figure 9.37 shows these data.

Swapping out the ac refrigerator in the above example for a dc one reduces the total dc load that the batteries have to supply by about 15%. This can translate into a 15% reduction in the size and cost of the photovoltaic array as well as the batteries. Moreover, the inverter itself can be smaller and cheaper since it doesn't have to supply as much ac power. Offsetting those gains is the additional

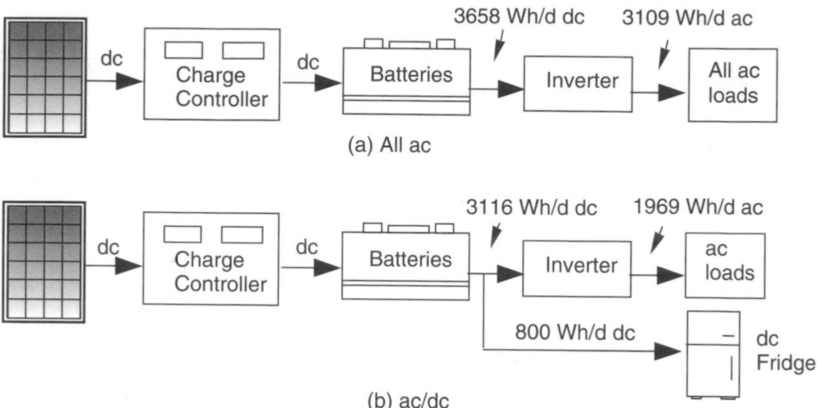

Figure 9.37 Switching out the ac refrigerator with a more efficient dc one. Numbers are based on Example 9.15.

cost of a dc refrigerator plus the added complexity and cost of wiring a house to provide for some ac and some dc loads. An economic analysis would be needed to determine the right decision.

System Voltage Inverters are specified by their dc input voltage as well as by their ac output voltage, continuous power handling capability, and the amount of surge power they can supply for brief periods of time. The inverter's dc input voltage, which is the same as the voltage of the battery bank and the PV array, is called the *system voltage*. The system voltage is usually 12 V, 24 V, or 48 V. Higher voltages need less current, making it easier to minimize wire losses. On the other hand, higher voltage means more batteries wired in series, which impacts the number of batteries that may be needed to supply the load.

One guideline that can be used to pick the system voltage is based on keeping the maximum steady-state current drawn below around 100 A so that readily available electrical hardware and wire sizes can be used. Using this guideline results in the system-voltage suggestions given in Table 9.11.

The maximum ac power that the inverter needs to deliver can be estimated by adding the power demand of all of the loads that will ever be anticipated to

TABLE 9.11 Suggested System Voltages Based on Limiting Current to 100 A

Maximum ac Power	System dc Voltage
<1200 W	12 V
1200–2400 W	24 V
2400–4800 W	48 V

TABLE 9.12 The Maximum Continuous Power Demand for The House in Example 9.14

Load	Watts
Refrigerator	300
Lights	180
TV/satellite, active mode	85
Cordless phone	4
Microwave	1000
Washing machine	250
Well pump	100
Maximum ac power demand	1919

be operating simultaneously. Table 9.10 provides representative values of steady-state power for a number of common household devices, which can be used as a starting point in the analysis. For the house described in Example 9.14, the total ac power demand with everything turned on at once would be 1919 W (Table 9.12), which would draw 160 A if the system voltage is only 12 V. A 24-V inverter would allow plenty of growth in demand without exceeding the 100-A guideline and so should be chosen for this system.

The most important specification for an inverter is the amount of ac power that it can supply on a continuous basis. But it is also critically important that the inverter be able to supply surges of current that occur when electric motors are started. Until the rotor starts spinning, there is no back emf to limit motor current and surges many times higher than steady state can occur. Table 9.13 provides estimates of steady-state and surge power requirements for typical household loads.

9.5.3 Batteries

Stand-alone systems obviously need some method to store energy gathered during good times to be able to use it during the bad. While various exotic technologies are possible—including flywheels, compressed air, or even hydrogen production—it is the lowly battery that makes the most sense today. And, among the many possible battery technologies, it is the familiar lead-acid battery that continues to be the workhorse of PV systems. In addition to energy storage, batteries provide several other important energy services for PV systems, including the ability to provide surges of current that are much higher than the instantaneous current available from the array, as well as the inherent and automatic property of controlling the output voltage of the array so that loads receive voltages that are within their own range of acceptability.

Competitors to conventional lead-acid batteries include nickel–cadmium, nickel–metal hydride, lithium–ion, lithium–polymer, and nickel–zinc technologies. Of these, only nickel–cadmium "Nicads" are even barely competitive with

TABLE 9.13 Steady-State and Surge Power Requirements for Example Loads

Load	Steady State (watts)	Surge (watts)
Refrigerator (ac)	300	1500
Refrigerator (dc)	58	700
Dishwasher	700	1400
Jet pump (1/3 hp) (ac)	750	1400
Submersible pump (ac)	1000	6000
Clothes washer (vertical axis)	650	1150
Clothes washer (horizontal axis)	250	750
Dryer (gas)	500	1800
Furnace fan 1/4 hp	600	1000
Furnace fan 1/3 hp	700	1400
Furnace fan 1/2 hp	875	2350
Air conditioner, window 10 kBtu	1200	1500
Worm drive 7 1/4″ saw	1800	3000
Table saw, 10″	1800	4500

Source: Real Goods (2002).

TABLE 9.14 Rough Comparison of Battery Characteristics[a]

Battery	Max Depth Discharge	Energy Density (Wh/kg)	Cycle Life (cycles)	Calendar Life (years)	Efficiencies Ah %	Efficiencies Wh %	Cost ($/kWh)
Lead-acid, SLI	20%	50	500	1–2	90	75	50
Lead-acid, golf cart	80%	45	1000	3–5	90	75	60
Lead-acid, deep-cycle	80%	35	2000	7–10	90	75	100
Nickel–cadmium	100%	20	1000–2000	10–15	70	60	1000
Nickel–metal hydride	100%	50	1000–2000	8–10	70	65	1200

[a] Actual performance depends greatly on how they are used.

Source: Linden (1995) and Patel (1999).

lead-acid batteries, but this may change in the near future due to the surge of interest and development in new battery technologies for electric and hybrid vehicles. Table 9.14 summarizes typical values of some of the important characteristics of these battery technologies. Lead-acid batteries are listed in three categories: conventional automobile batteries for engine starting, vehicle lighting, and engine ignition (SLI); low-cost, deep-cycle batteries typically used in golf carts; and longer-lifetime, true deep-cycle batteries. Two other battery types are shown, nickel–cadmium (or Nicads) and nickel–metal hydride batteries, which are beginning to be used in some hybrid-electric vehicles. As can be seen, lead-acid batteries are by far the least expensive option, they have the highest efficiencies, and the more expensive ones, when used properly, can last nearly as long as their competitors. Nicads are much more expensive, but they last longer. Nicads also perform better in harsh climates; and since they can be discharged nearly 100% without damage, they are far more forgiving when abused.

9.5.4 Basics of Lead-Acid Batteries

Lead-acid batteries date back to the 1860s when inventor Raymond Gaston Planté fabricated the first practical cells made with corroded lead-foil electrodes and a dilute solution of sulfuric acid and water. Many advances since then have lead to a global market that now exceeds $30 billion in annual retail sales, with about three fourths of that being starting, lighting, and ignition (SLI) automobile batteries. Lead-acid batteries are used in everything from small electronic devices such as cell phones, to car batteries, to enormous utility battery banks, the largest of which, in Chino, California, is capable of delivering 4 h of 10 MW power (5000 A at 2000 V) into the grid.

Automobile SLI batteries have been highly refined to perform their most important task, which is to start your engine. To do so, they have to provide short bursts of very high current (400–600 A!). Once the engine has started, its alternator quickly recharges the battery, which means that under normal circumstances the battery is almost always at or near full charge. SLI batteries are not designed to withstand deep discharges, and in fact they will fail after only a few complete-discharge cycles. This makes them inappropriate for most PV systems, in which slow, but deep, discharges are the norm. If they must be used, as is sometimes the case in developing countries where they may be the only batteries available, daily discharges of less than about 20% can yield approximately 500 cycles, or a year or two of operation.

In comparison with SLI batteries, deep discharge batteries have thicker plates, which are housed in bigger cases that provide greater space both above and beneath the plates. Greater space below allows more debris to accumulate without shorting out the plates, and greater space above lets there be more electrolyte in the cell to help keep water losses from exposing the plates. Thicker plates and larger cases mean that these batteries are big and heavy. A single 12-V deep-discharge battery can weigh several hundred pounds. They are designed to be discharged repeatedly by 80% of their capacity without harm, although such deep discharges result in a lower lifetime number of cycles. Figure 9.38 suggests that a typical deep-cycle, lead-acid battery can be cycled about 4000 times when discharged by 25% of its rated capacity, which would give it a lifetime of over 10 years. At a daily discharge of 80%, about 1800 cycles could be expected, which suggests a lifetime of around 5 years. While Fig. 9.38 provides a rough indication of battery life, other factors, including quality of battery, frequency of maintenance, charge rates, and final charging cut-off voltages, are also important.

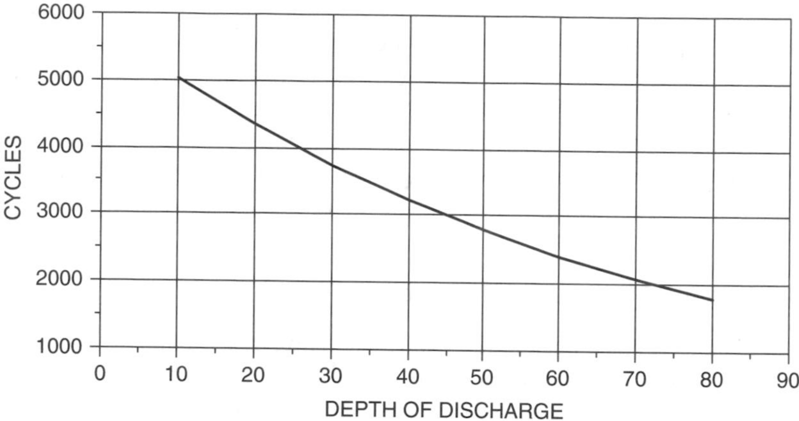

Figure 9.38 Impact of depth of discharge on the number of cycles a typical deep-cycle lead-acid battery might be able to provide. An automobile SLI battery delivers only around 500 cycles at 20% discharge.

To understand some of the subtleties in sizing battery systems, we need a basic understanding of their chemistry. Very simply, an individual 2-V cell in a lead-acid battery consists of a positive electrode made of lead dioxide (PbO_2) and an negative electrode made of a highly porous, metallic lead (Pb) structure, both of which are completely immersed in an electrolyte consisting of a dilute solution of sulfuric acid and water. Thin lead plates are structurally very weak and would not hold up well to physical abuse unless alloyed with a strengthening material. Automobile SLI batteries use calcium for strengthening, but calcium does not tolerate discharges of more than about 25 percent very well. Deep discharge batteries use antimony instead, and so are often referred to as lead-antimony batteries.

The chemical reactions taking place while the battery discharges are as follows:

$$\text{Positive plate}: \quad PbO_2 + 4H^+ + SO_4^{2-} + 2e^- \rightarrow PbSO_4 + 2H_2O \tag{9.21}$$

$$\text{Negative plate}: \quad Pb + SO_4^{2-} \rightarrow PbSO_4 + 2e^- \tag{9.22}$$

It is, by the way, simpler to refer to the terminals by their charge (positive or negative) rather than as the anode and cathode. Strictly speaking, the anode is the electrode at which oxidation occurs, which means that during discharge the anode is the negative terminal, but during charging the anode is the positive terminal.

As can be seen from (9.22), during discharge the electrons are released at the negative electrode, which then flow through the load to the positive plate where they enter into the reaction given by (9.21). The key feature of both reactions is that sulfate ions (SO_4^{2-}) that start out in the electrolyte when the battery is fully charged end up being deposited onto each of the two electrodes as lead sulfate ($PbSO_4$) during discharge. This lead sulfate, which is an electrical insulator, blankets the electrodes, leaving less and less active area for the reactions to take place.

As the battery approaches its fully discharged state, the cell voltage drops sharply while its internal resistance rises abruptly. Meanwhile, during discharge the specific gravity of the electrolyte drops as sulfate ions leave solution, providing an accurate indicator of the battery's state of charge. The battery is more vulnerable to freezing in its discharged state since the anti-freeze action of the sulfuric acid is diminished when there is less of it present. A fully discharged lead-acid battery will freeze at around −8°C (17°F), while a fully charged one won't freeze until the electrolyte drops below −57°C (−71°F). In very cold conditions, concern for freezing may limit the maximum allowable depth of discharge, as shown in Fig. 9.39.

The opposite reactions occur during charging. Battery voltage and specific gravity rise, while freeze temperature and internal resistance drop. Sulfate is removed from the plates and reenters the electrolyte as sulfate ions (Fig. 9.40). Unfortunately, not all of the lead sulfate returns to solution, and each battery charge/discharge cycle leaves a little more sulfate permanently attached to the

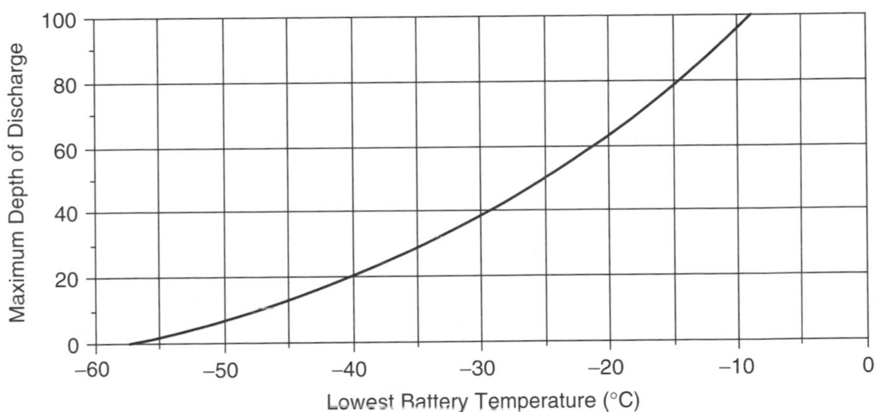

Figure 9.39 Concern for battery freezing may limit the allowable depth of discharge of a lead-acid battery.

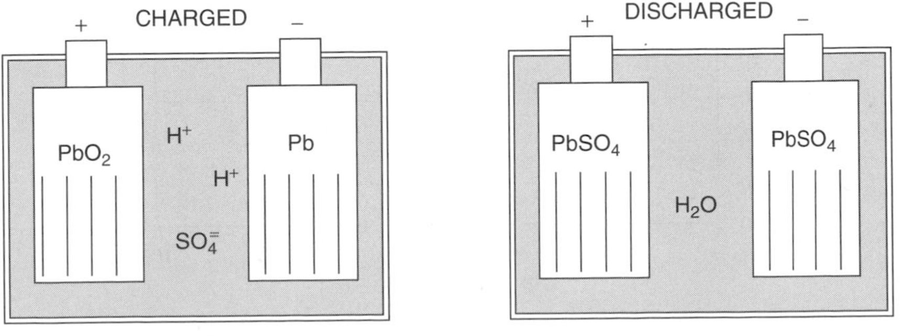

Figure 9.40 A lead-acid battery in its charged and discharged states.

plates. This sulfation is a primary cause of a battery's finite lifetime. The amount of lead sulfate that permanently bonds to the electrodes depends on the length of time that it is allowed to exist, which means that for good battery longevity it is important to keep batteries as fully charged as possible and to completely charge them on a regular basis. This suggests that a generator back-up system to top up batteries is an important consideration.

As batteries cycle between their charged and partially discharged states, the voltage as measured at the terminals and the specific gravity of the electrolyte changes. While either may be used as an indication of the state of charge (SOC) of the battery, both are tricky to measure correctly. To make an accurate voltage reading, the battery must be at rest, which means that at least several hours must elapse after any charging or discharging. Specific gravity is also difficult to measure since stratification of the electrolyte means that a sample taken from the liquid above the plates may not be an accurate average value. Bearing those complications in mind, Fig. 9.41 shows voltage for a nominal lead-acid 12-V battery at rest along with the specific gravity for a well-mixed electrolyte, as a function of the state of charge. It is interesting to note that a 12-V battery is only about 20% charged when its terminal voltage is 12 V.

9.5.5 Battery Storage Capacity

Energy storage in a battery is typically given in units of amp-hours (Ah) at some nominal voltage and at some specified discharge rate. A lead-acid battery, for example, has a nominal voltage of 2 V per cell (e.g., 6 cells for a 12-V battery), and manufacturers typically specify the amp-hour capacity at a discharge rate that would drain the battery down to 1.75 V over a specified period of time at a temperature of 25°C. For example, a fully charged 12-V battery that is specified to have a 10-h, 200-Ah capacity could deliver 20 A for 10 h, at which point the

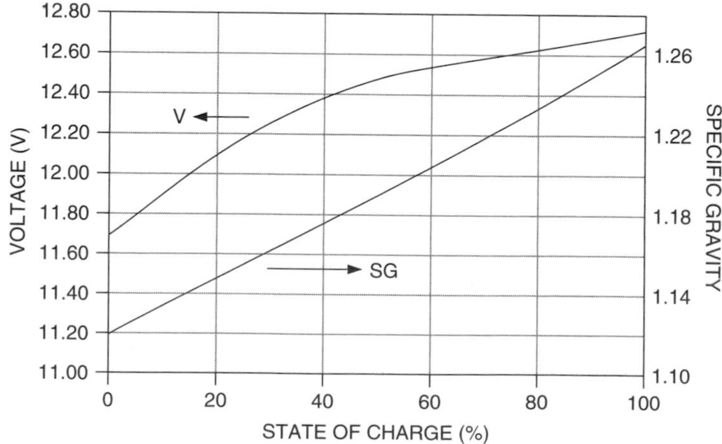

Figure 9.41 Voltage and specific gravity for a typical deep-cycle lead-acid 12-V battery. Data from Sandia National Laboratories (1991)..

battery would have a voltage of 10.5 V ($6 \times 1.75 = 10.5$) and be considered to be fully discharged. Notice how tricky it would be to specify how much energy the battery delivered during its discharge. Energy is volts × amps × hours, but since voltage varies throughout the discharge period, we can't just say 12 V × 20 A × 10 h = 2400 Wh. To avoid that ambiguity, almost everything having to do with battery storage capacity is specified in amp-hours rather than watt-hours.

A 200-Ah battery that is delivering 20 A is said to be discharging at a $C/10$ rate, where the C refers to Ah of capacity and the 10 is hours it would take to deplete ($C/10 = 200$ Ah/10 h = 20 A). That same 200-Ah battery won't be able to deliver 50 A for a full 4 h ($C/4$), however, and it will actually deliver 10 A for more than 20 h ($C/20$). In other words, the amp-hour capacity depends on the rate at which current is withdrawn. Rapid draw-down of a battery results in lower Ah capacity, while long discharge times result in higher Ah capacity. Deep-cycle batteries intended for photovoltaic systems are often specified in terms of their 20-h discharge rate ($C/20$), which is more or less of a standard, as well as in terms of the much longer $C/100$ rate that is more representative of how they are actually used. Table 9.15 provides some examples of such batteries, including their $C/20$ and $C/100$ rates as well as their voltage and weight.

The amp-hour capacity of a battery is not only rate-dependent, but also depends on temperature. Figure 9.42 captures both of these phenomena by comparing capacity under varying temperature and discharge rates to a reference condition of $C/20$ and 25°C. These curves are approximate for typical deep-cycle lead-acid batteries, so specific data available from the battery manufacturer should be used whenever possible. As shown in Fig. 9.42, battery capacity decreases dramatically in colder conditions. At −30°C (−22°F), for example, a battery that is discharged at the $C/20$ rate will have only half of its rated capacity. The combination of cold temperature effects on battery performance—decreased capacity, decreased output voltage, and increased vulnerability to freezing when discharged—mean that lead-acid batteries need to be well protected in cold climates. Nicad batteries, by the way, don't suffer from these cold weather effects, which is the main reason they are sometimes used instead of lead-acid batteries in cold climates. Also, the apparent improvement in battery capacity at high temperatures does not mean that heat is good for a battery. In fact, a rule-of-thumb estimate is that battery life is shortened by 50% for every 10°C above the optimum 25°C operating temperature.

TABLE 9.15 Example Deep-Cycle Lead-Acid Battery Characteristics

BATTERY	Voltage	Weight (lbs)	Ah @ $C/20$	Ah @ $C/100$
Concorde PVX 5040T	2	57	495	580
Trojan T-105	6	62	225	250
Trojan L16	6	121	360	400
Concorde PVX 1080	12	70	105	124
Surette 12CS11PS	12	272	357	503

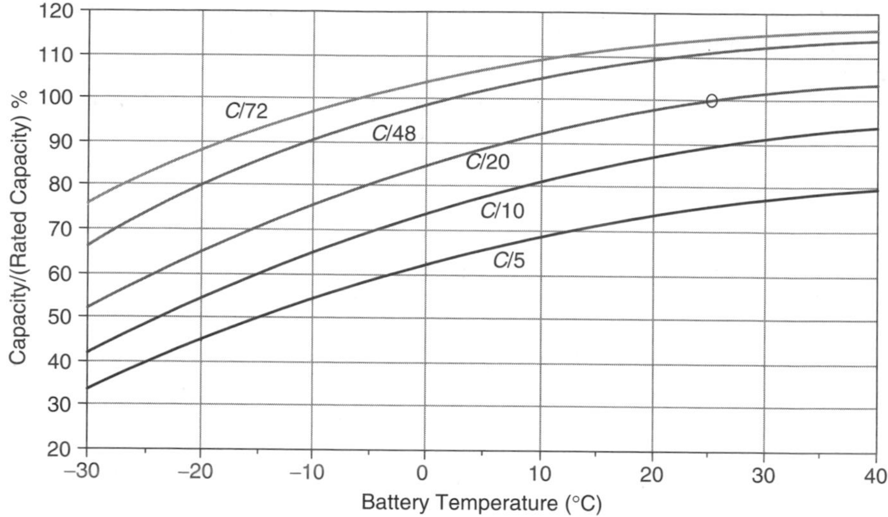

Figure 9.42 Lead-acid battery capacity depends on discharge rate and temperature. Ratio is based on a rated capacity at *C*/20 and 25°C.

Example 9.16 Battery Storage Calculation in a Cold Climate. Suppose that batteries located at a remote telecommunications site may drop to −20°C. If they must provide 2 days of storage for a load that needs 500 Ah/day at 12 V, how many amp-hours of storage should be specified for the battery bank?

Solution. From Fig. 9.39, to avoid freezing, the maximum depth of discharge at −20°C is about 60%. For 2 days of storage, with a discharge of no more than 60%, the batteries need to store

$$\text{Battery storage} = \frac{500\ \text{Ah/day} \times 2\ \text{days}}{0.60} = 1667\ \text{Ah}$$

Since the rated capacity of batteries is likely to be specified at an assumed temperature of 25°C at a *C*/20 rate, we need to adjust the battery capacity to account for our different temperature and discharge period. From Fig. 9.42, the actual capacity of batteries at −20°C discharged over a 48-h period is about 80% of their rated capacity. This means that we need to specify batteries with rated capacity

$$\text{Battery storage (25°C, 10-hour rate)} = \frac{1667\ \text{Ah}}{0.8} = 2083\ \text{Ah}$$

Most PV–battery systems are based on 6-V or 12-V batteries, which may be wired in series and parallel combinations to achieve the needed Ah capacity and voltage rating. For batteries wired in series, the voltages add, but since the same current flows through each battery, the amp-hour rating of the string is the same as it is for each battery. For batteries wired in parallel the voltage across each battery is the same, but since currents add, the amp-hour capacity is additive. Figure 9.43 illustrates these notions.

Since there is no difference in energy stored in the two-battery series and parallel example shown in Fig. 9.43, the question arises as to which is better. The key difference between the two is the amount of current that flows to deliver a given amount of power. Batteries in series have higher voltage and lower current, which means more manageable wire sizes without excessive voltage and power losses, along with smaller fuses and switches, and slightly easier connections between batteries. On the other hand, a storage system consisting of batteries in parallel is easy to expand, one battery at a time. In series, a whole new string of batteries must be added to increase storage.

9.5.6 Coulomb Efficiency Instead of Energy Efficiency

As mentioned earlier, almost everything having to do with batteries is described in terms of currents rather than voltage or energy. Battery capacity C is given in amp-hours rather than watt-hours; charging and discharging are expressed in C/T rates, which are amps. And, as we shall see, even battery efficiency is more easily expressed in terms of current efficiency than in terms of energy efficiency. The reason, of course, is that battery voltage is so ambiguous without specifying whether it is a "rest" voltage measured some time after charging or discharging, a voltage during charging, or a voltage during discharge. And even those charging and discharging voltages depend on the rate at which current is entering or leaving the battery as well as the state of charge of the battery, its temperature, age, and general condition.

Imagine charging a battery with a constant current I_C over a period of time ΔT_C during which time applied voltage is V_C. The energy input to the battery

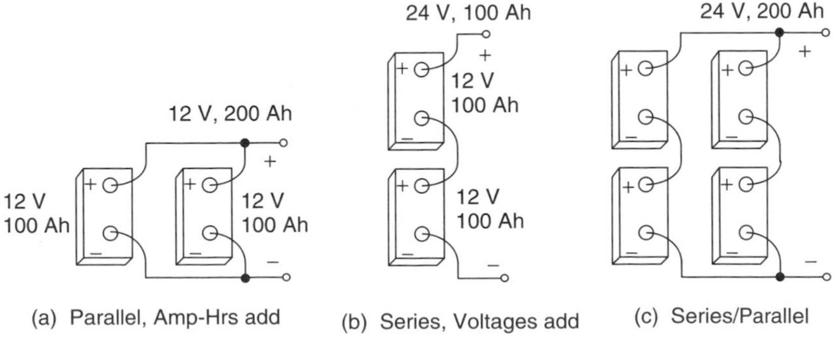

Figure 9.43 For batteries wired in parallel, amp-hours add (a). For batteries in series, voltages add (b). For series/parallel combinations, both add.

is thus

$$E_{\text{in}} = V_C I_C \; \Delta T_C \tag{9.23}$$

Suppose that the battery is discharged at current I_D and voltage V_D over a period of time ΔT_D, delivering energy

$$E_{\text{out}} = V_D I_D \; \Delta T_D \tag{9.24}$$

The energy efficiency of the battery would be

$$\text{Energy efficiency} = \frac{E_{\text{out}}}{E_{\text{in}}} = \frac{V_D I_D \; \Delta T_D}{V_C I_C \; \Delta T_C} \tag{9.25}$$

If we recognize that current (A) times time (h) is Coulombs of charge expressed as Ah, then

$$\text{Energy efficiency} = \left(\frac{V_D}{V_C}\right)\left(\frac{I_D \; \Delta T_D}{I_C \; \Delta T_C}\right) = \left(\frac{V_D}{V_C}\right)\left(\frac{\text{coulombs out, Ah}_{\text{out}}}{\text{coulombs in, Ah}_{\text{in}}}\right) \tag{9.26}$$

The ratio of discharge voltage to charge voltage is called the *voltage efficiency* of the battery, and the ratio of Ah_{out} to Ah_{in} is called the *Coulomb efficiency*.

$$\text{Energy efficiency} = (\text{Voltage efficiency}) \times (\text{Coulomb efficiency}) \tag{9.27}$$

A typical 12-V lead-acid battery might be charged at a voltage of around 14 V, and its discharge voltage might be around 12 V. Its voltage efficiency would therefore be

$$\text{Voltage efficicncy} = \frac{12 \text{ V}}{14 \text{ V}} = 0.86 = 86\% \tag{9.28}$$

The Coulomb efficiency is the ratio of coulombs of charge out of the battery to coulombs that went in. If they don't all come back out, where did they go? When a battery approaches full charge, its cell voltage gets high enough to electrolyze water, creating hydrogen and oxygen gases that may be released. Among the negative effects of this *gassing* is loss of some of those charging electrons along with the escaping gases. While the battery state of charge (SOC) is low, little gassing occurs and the Coulomb efficiency is nearly 100%, but it can drop below 90% during the final stages of charging. Over a full charge cycle, it is typically between 90% and 95%. As we shall see later, when it comes to sizing batteries, the Coulomb efficiency will be the measure that is most appropriate.

Assuming a 90% Coulomb efficiency, the overall energy efficiency of a lead-acid battery with 86% voltage efficiency would be about

$$\text{Energy efficiency} = 0.86 \times 0.90 = 0.77 = 77\% \tag{9.29}$$

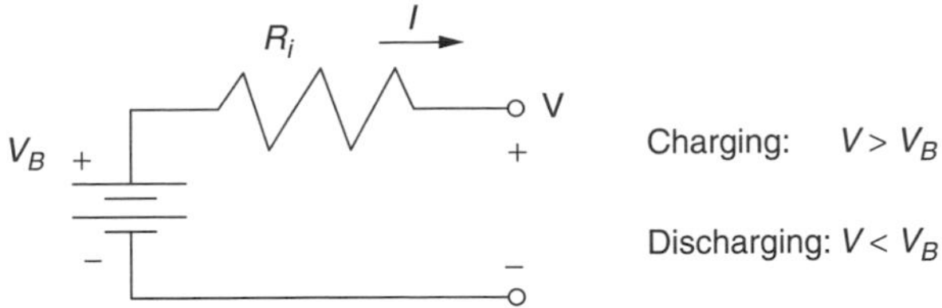

Figure 9.44 Thevenin equivalent circuit for a battery.

which is close to the commonly quoted estimate of 75% for lead-acid battery energy efficiency.

To help understand where that energy loss occurs, consider the simple Thevenin equivalent for a battery consisting of an ideal battery of voltage V_B in series with an internal resistance, R_i (Fig. 9.44). Voltage V_B can be considered to be the open-circuit "rest" voltage of the battery as measured some hours after either charging or discharging has occurred. To charge the battery, the voltage applied to the terminals must be greater than V_B; when the battery is discharging, the output voltage will be less than V_B. During those charge and discharge times, there are I^2R power losses in the internal resistance. Since those losses go as the square of current, faster charge or discharge times result in much higher losses.

Example 9.17 Losses at High and Low Charging Rates. A 100-Ah, 12-V battery with a rest voltage of 12.5 V (at its current SOC) is charged at a *C*/5 rate, during which time the applied voltage is 13.2 V. Using a simple Thevenin equivalent:

a. Estimate the internal resistance of the battery.
b. What fraction of the input power is lost in the internal resistance of the battery?
c. If the charging is done at a *C*/20 rate, what fraction of the input power would be lost due to the internal resistance?

Solution

a. At *C*/5, the current is 100 Ah/5 h = 20 A. The internal resistance must have been

$$R_i = \frac{V_{in} - V_B}{I} = \frac{13.2 - 12.5}{20} = 0.035\ \Omega$$

b. So the I^2R losses as a fraction of the input power would be

$$\frac{\text{Power lost in } R_i}{\text{Input power}} = \frac{I^2R}{V_{in}I} = \frac{(20)^2 \times 0.035}{13.2 \times 20} = 0.053 = 5.3\%$$

c. At $C/20$, the current is 100 Ah/20 h = 5 A. The input voltage needed to drive 5 A through the 0.035-Ω resistance is

$$V_{\text{in}} = V_B + IR = 12.5 + 5 \times 0.035 = 12.68 \text{ V}$$

So at the $C/20 = 5$-A rate, the losses are now only

$$\frac{\text{Power lost in } R_i}{\text{Input power}} = \frac{I^2 R}{V_{\text{in}} I} = \frac{(5)^2 \times 0.035}{12.68 \times 5} = 0.014 = 1.4\%$$

The simple Thevenin model of a battery is complicated by the fact that both V_B and R_i depend on the battery's state of charge (SOC), its temperature, and its history. But since the model is so simple, it does help provide an intuitive understanding of the actual charging and discharging curves for a battery. Figure 9.45 shows representative values of battery voltage for differing charging and discharging currents as a function of SOC.

During charging, notice the sudden rise in cell voltage around the 14-V level as the battery nears full charge. This is when charging is most inefficient and when gassing occurs. The release of generated hydrogen and oxygen gases removes water from the battery, which has to be replaced or the plates may be damaged. It also creates a potentially dangerous situation since hydrogen gas is so explosive. In addition, very small quantities of the poisonous gases arsine (AsH_3) and stibine (SbH_3) can be released when hydrogen comes in contact with the lead alloys arsenic and antimony. Proper ventilation is clearly an important consideration in the design of a safe battery storage system.

To reduce the need for water replacement, some lead-acid batteries now use pressure-relief valves that allow most of the oxygen formed during overcharging to recombine with lead rather than being released. Hydrogen gas releases are also suppressed. While they do open as necessary to relieve the pressure built up by hydrogen and oxygen gases, these valve-regulated lead-acid batteries (VRLA) can reduce gas emissions by over 95% (Linden, 1995).

Gassing and charging losses can be minimized, by using a charge controller that has been designed to slow the charging rate as the battery approaches its fully charged condition. Charge controllers also protect batteries from overcharging by completely disconnecting the PV array at some predetermined battery voltage, usually around 14 V (for a 12-V battery). They also keep the batteries from being overly discharged by disconnecting the load when battery voltage drops below another set point, usually around 11.5 V.

9.5.7 Battery Sizing

If good weather could be counted on, battery sizing might mean simply providing enough storage to carry the load through the night and into the next day until

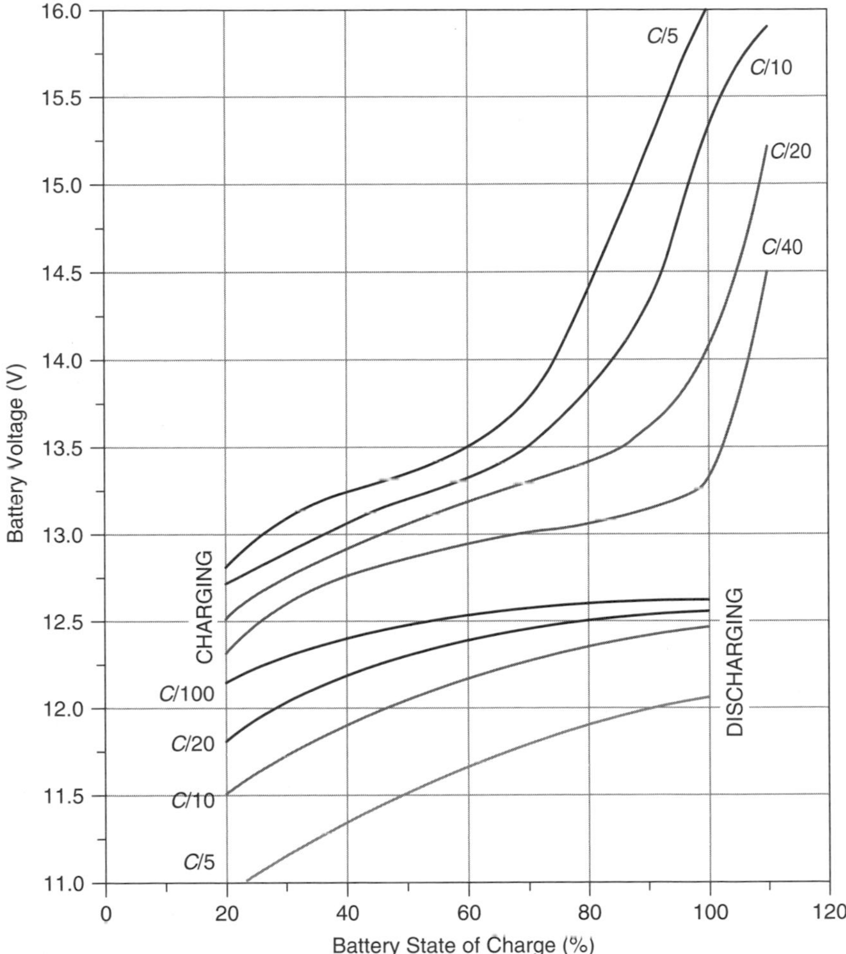

Figure 9.45 Terminal voltage and state of charge for 12-V lead-acid batteries for various rates of charging and discharging. Based on Sandia National Laboratories (1991).

the sun picks up the load once again. The usual case, of course, is one in which there are periods of time when little or no sunlight is available and the batteries might have to be relied on to carry the load for some number of days. During those periods, there may be some flexibility in the strategy to be taken. Some noncritical loads, for example, might be reduced or eliminated; and if a generator is part of the system, a trade-off between battery storage and generator run times will be part of the design.

Given the statistical nature of weather and the variability of responses to inclement conditions, there are no set rules about how best to size battery storage. The key trade-off will be cost. Sizing a storage system to meet the demand 99% of the time can easily cost triple that of one that meets demand only 95%

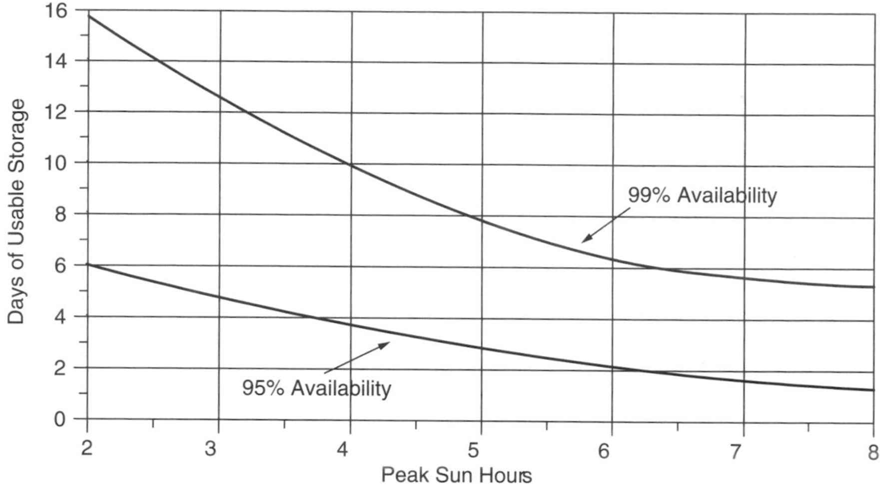

Figure 9.46 Days of battery storage needed for a stand-alone system with 95% and 99% system availability. Peak sun hours are for the worst month of the year and availability is on an annual basis. Based on Sandia National Laboratories (1995).

of the time. As a starting point for estimating the number of days of storage to be provided, consider Fig. 9.46, which is based on the excellent guidebook *Stand-Alone Photovoltaic Systems Handbook of Recommended Design Practices* (Sandia National Laboratories, 1995). The graph gives an estimate for days of battery storage needed to supply a load as a function of the peak sun hours per day in the *design month*, which is the month with the worst combination of insolation and load. To account for a range of load criticality, two curves are given: one for loads that must be satisfied during 99% of the 8760 h in a year and one for less critical loads, for which a 95% system availability is satisfactory.

When doing designs with a computer, it is handy to have equations for the curves given in Fig. 9.46. The following match pretty well.

$$\text{Storage days (99\%)} \approx 24.0 - 4.73\,(\text{Peak sun hours}) + 0.3\,(\text{Peak sun hours})^2 \tag{9.30}$$

$$\text{Storage days (95\%)} \approx 9.43 - 1.9\,(\text{Peak sun hours}) + 0.11\,(\text{Peak sun hours})^2 \tag{9.31}$$

Figure 9.46 refers to days of "usable storage," which means after accounting for impacts associated with maximum allowable battery discharge, Coulomb efficiency, battery temperature, and discharge rate. The relationship between usable storage and nominal, rated storage (at $C/20$, 25°C) is given by

$$\text{Nominal } (C/20, 25^\circ\text{C}) \text{ battery capacity} = \frac{\text{Usable battery capacity}}{(\text{MDOD})(\text{T, DR})} \tag{9.32}$$

where MDOD stands for maximum depth of discharge (default: 0.8 for lead-acid, deep-discharge batteries, 0.25 for auto SLI; subject to freeze constraints given in Fig. 9.39) and T, DR stands for temperature and discharge-rate factor (Fig. 9.42).

The following example illustrates the process.

Example 9.18 Battery Sizing for an Off-Grid Cabin. A cabin near Salt Lake City, Utah, has an ac demand of 3000 Wh/day in the winter months. A decision has been made to size the batteries such that a 95% system availability will be provided, and a back-up generator will be kept in reserve to cover the other 5%. The batteries will be kept in a ventilated shed whose temperature may reach as low as -10°C. The system voltage is to be 24 V, and an inverter with overall efficiency of 85% will be used.

Solution. With an 85% efficient inverter, the dc load is

$$\text{DC load} = \frac{\text{AC load}}{\text{Inverter efficiency}} = \frac{3000\ \text{Wh/day}}{0.85} = 3529\ \text{Wh/day}$$

With a 24-V system voltage the batteries need to supply

$$\text{Load} = \frac{3529\ \text{Wh/day}}{24\ V} = 147\ \text{Ah/day @ 24 V}$$

From Appendix E the following monthly insolation data are found for Salt Lake City:

Tilt	Jan	Feb	Mar	Apr	May	Jun	July	Aug	Sept	Oct	Nov	Dec	Year
Lat − 15	2.9	4.0	5.0	5.9	6.6	7.2	7.3	7.0	6.3	5.0	3.3	2.5	5.2
Lat	3.2	4.3	5.2	5.8	6.2	6.6	6.7	6.7	6.4	5.4	3.7	2.9	5.3
Lat + 15	3.4	4.4	5.1	5.4	5.5	5.6	5.8	6.1	6.1	5.5	3.9	3.1	5.0

Even though annual insolation is highest for a tilt angle equal to the local latitude, we'll use an L + 15 tilt to help meet winter needs. December has the lowest insolation so that will be our design month. From Fig. 9.46 at 95% availability and 3.1 peak sun hours in December, it looks like we need about 4.6 days of storage.

$$\text{Usable storage} = 147\ \text{Ah/day} \times 4.6\ \text{day} = 676\ \text{Ah}$$

We'll pick deep discharge lead-acid batteries that can be routinely discharged by 80%. But, we need to check to see whether that depth of discharge will expose the batteries to a potential freeze problem. From Fig. 9.39, at -10°C the batteries could be discharged to over 95% without freezing the electrolyte, so an 80% discharge is acceptable.

Batteries that are nominally rated at $C/20$ and 25°C will be operated in much colder conditions which degrades storage capacity, but they will be discharged at a much slower rate, which increases capacity. Figure 9.42 suggests that at −10°C and a $C/72$ rate (the 5-day rate for the cabin is even slower, so this is conservative), storage capacity would be about 0.97 times the nominal capacity.

Applying the 0.80 factor for maximum discharge and the 0.97 factor for discharge rate and temperature in (9.32) gives

$$\text{Nominal } (C/20, 25^{\circ}\text{C}) \text{ battery capacity} = \frac{676 \text{ Ah}}{0.80 \times 0.97} = 871 \text{ Ah (at 24 V)}$$

Table 9.15 can help us pick a battery. None of those batteries comes close to providing that many Ah, so we're going to have to have parallel strings.

Partly to keep the weight of each battery low enough to be able to handle, and partly to keep the number of batteries under control, suppose we choose the Trojan 6 V, 225 Ah, T-105. Three parallel strings would give us 675 Ah, four would give 900 Ah, and our goal is 871 Ah. Suppose we slightly oversize with four strings.

To get 24 V, we need each string to have four batteries, so the total battery bank will have four parallel strings of four batteries each as shown in Fig. 9.47.

9.5.8 Blocking Diodes

The simplest PV–battery system consists of just a single module connected to a battery and a load, with no charge controller, inverter, or anything else to

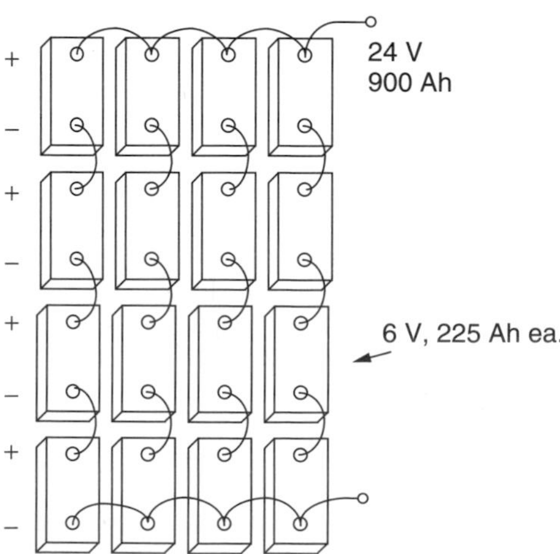

Figure 9.47 The 24-V, 900-Ah battery bank for the cabin in Example 9.18.

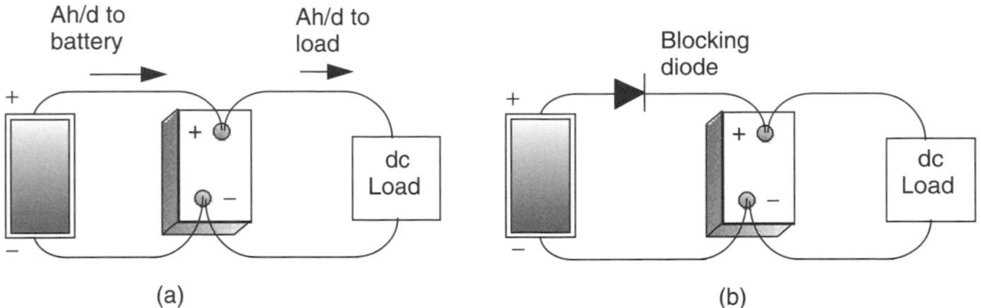

Figure 9.48 Simplest PV–battery system (a). Adding a blocking diode to prevent losses from the battery through the PV at night (b).

complicate things (Fig. 9.48a). Such a system might provide someone with a bit of light at night and maybe a few other simple amenities. As long as the user is careful about not letting the batteries discharge too deeply, or be overcharged, the system will perform well. There is another concern, however. The system shown in Fig. 9.48a allows the battery to leak current back through the PV module at night, which raises the question of whether it might be worthwhile to add a blocking diode as shown in Fig. 9.48b to prevent that nightly discharge.

The equivalent circuit of a single PV cell shown in Fig. 9.49a will help us analyze the potential nighttime battery loss problem. Remember that the blackened symbol for a diode means that it is a real diode, rather than an ideal one. Ignoring the insignificant impact of the very small series resistance and eliminating the ideal current source I_{SC} because the cell is in the dark at night leaves the simple circuit of Fig. 9.49b.

Current through the diode in the equivalent circuit for a cell (at 25°C) is given by the Shockley equation introduced in Chapter 8:

$$I_d = I_0(e^{38.9V_d} - 1) \tag{9.33}$$

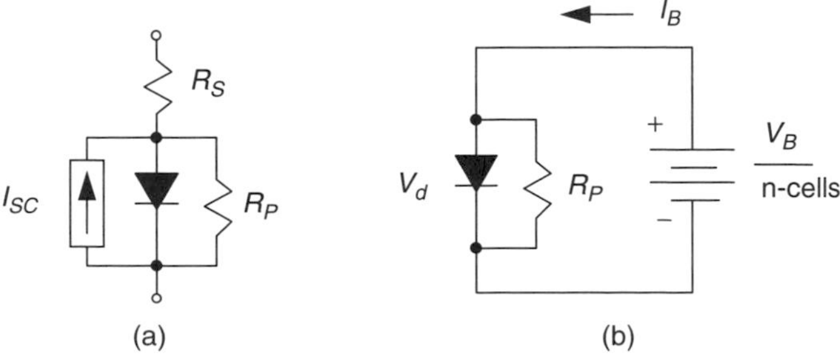

Figure 9.49 Nighttime leakage from a battery back through a PV module with n cells. (a) Equivalent circuit of one PV cell. (b) A simplified equivalent circuit at night for one cell having V_B/n volts from the battery across it.

The nighttime current from the battery through each cell will be

$$I_B = I_d + I_{R_P} = I_0(e^{38.9V_d} - 1) + \frac{V_d}{R_p} \tag{9.34}$$

where the voltage V_d across the diode will be equal to the battery voltage V_B divided by the number of cells n in the PV module.

With this simple nighttime equivalent circuit, we can decide how much leakage will occur from the battery through the PVs. The following example shows how to evaluate the potential advantage of using a blocking diode to prevent that current from flowing.

Example 9.19 Impact of a Blocking Diode to Control Nighttime Battery Leakage. A PV module is made up of 36 cells, each having a reverse saturation current I_0 of 1×10^{-10} A and a parallel resistance of 8 Ω. The PVs provide the equivalent of 5 A for 6 h each day. The module is connected without a blocking diode to a battery with voltage 12.5 V.

a. How many Ah will be discharged from the battery over a 15-h night?
b. How much energy will be lost due to this discharge?
c. If a blocking diode is added, how much energy will be dissipated through the diode during the daytime. Assume the diode while conducting has a voltage drop of 0.6 V.

Solution. The voltage across each PV cell will be about 12.5 V/36 cells = 0.347 V. From (9.34) the current discharged from the battery while the PV is in the dark will be

$$I_B = 10^{-10}(e^{38.9\times0.347} - 1) + \frac{0.347}{8} = 0.000073 + 0.043 = 0.043 \text{ A} = 43 \text{ mA}$$

a. Over a 15-h nighttime period, the loss in Ah from the battery will be

$$\text{Nighttime loss} = 0.043 \text{ A} \times 15 \text{ h} = 0.65 \text{ Ah}$$

b. At a nominal 12.5 V, the energy loss at night will be

$$\text{Nighttime loss} = 0.65 \text{ Ah} \times 12.5 \text{ V} = 8.1 \text{ Wh}$$

c. During the day the PVs will deliver

$$\text{PV output} = 6 \text{ h} \times 5 \text{ A} = 30 \text{ Ah}$$

The nighttime loss without a blocking diode is 0.65 Ah/30 Ah = 0.02; that is, 2% of the daytime gains.

With the blocking diode dropping 0.6 V, the daytime loss caused by that diode is

$$\text{Blocking diode loss} = 30\text{ Ah} \times 0.6\text{ V} = 18\text{ Wh}$$

In the example just presented, the blocking diode loses more energy during the day while it is conducting (18 Wh) than it saves overnight (8.1 Wh). Another way to look at it is that without a blocking diode, only about 2% of the Ah of daytime solar gains are lost overnight. *In spite of these arguments, blocking diodes still make some sense.* Since the operating point of the battery–PV system is generally some distance to the left of the knee of the I–V curve, shifting the battery I–V curve approximately 0.6 V to the right to cover the diode drop barely changes current delivered by the PVs. That is, in terms of amp-hours, the diode doesn't lose any during battery charging, but does stop nighttime battery discharge, so in terms of amp-hours it does offer a modest net benefit.

9.5.9 Sizing the PV Array

Designing stand-alone PV–battery systems is clearly much more demanding than sizing grid-connected systems. Month-by-month load estimates and solar resource evaluations, making trade-offs between ac and dc loads, choosing a system voltage, and determining battery storage with or without a back-up generator are things that simply don't apply to grid-connected systems. Having addressed those topics, we can now deal with the most important part of the system: the PV array itself.

In Fig. 9.50 a 1-sun, PV I–V curve has been drawn along with a vertical I–V line for a battery. As shown, during battery charging, the operating point of the PVs is almost always above the knee of the PV I–V curve, which means that charging current will exceed the rated current of the PVs. It is a fairly conservative estimate therefore to simply use the rated current of the PVs as an indication of the battery charging current at 1-sun insolation. There are circumstances in which this assumption should be checked—as, for example, when a 12-V battery is charged in a high-temperature environment with a "self-regulating" PV module having fewer than the usual 36 cells in series. Fewer cells and higher temperatures move the maximum power point (MPP) toward the battery I–V curve, and the conservatism of the assumption decreases.

Our simple sizing procedure will be based on the same "peak hours" approach used for grid-connected systems, except that it will be applied to current rather than power. So, for example, an area with 6 kWh/m^2-day of insolation is treated as if it has 6 h/day of 1-sun, 1-kW/m^2 radiation. Then using the rated current I_R at 1-sun, times peak hours of sun, gives us amp-hours of current provided to the batteries.

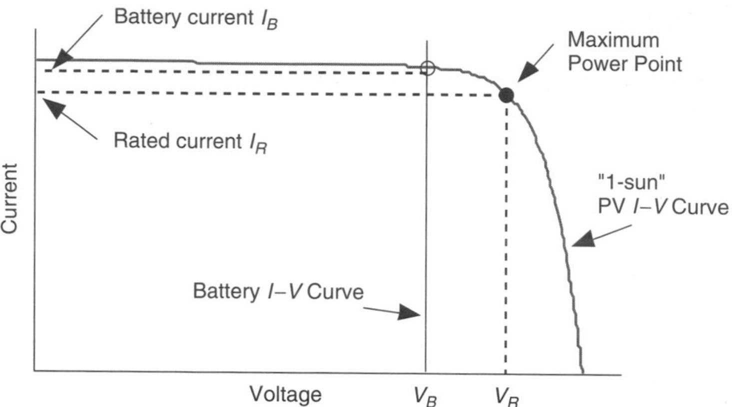

Figure 9.50 Estimating battery charging at 1-sun to be the rated current of the PVs is a fairly conservative assumption.

Notice that the operating point for battery charging is usually some distance away from the MPP. This means that a considerable fraction of power that the PVs could provide based on the rated power P_R of the module is not being delivered to the batteries. *It would not be appropriate, therefore, to multiply the peak hours of sunlight times the rated power of the module to estimate energy to the batteries.*

The product of rated current I_R times peak hours of insolation provides a good starting-point estimate for Ah delivered to the batteries. It is common practice to apply a de-rating of about 10% to account for dirt and gradual aging of the modules. The temperature and module-mismatch factors that were quite important for grid-connected systems with MPP trackers are usually ignored for PV–battery systems. This is because the operating point for battery systems is far enough away from the knee of the $I-V$ curve that those variations are minimum and are to some extent offset by the conservatism associated with assuming that charging current is only I_R.

Here is another important feature of stand-alone PV sizing. It will be based on (a) amp-hours from the PVs into the batteries and (b) amp-hours from the batteries to the load. That is, the appropriate battery efficiency measure will be the Coulomb efficiency. The current delivered to the batteries needs to multiplied by the Coulomb efficiency ($\text{Ah}_{\text{out}}/\text{Ah}_{\text{in}}$) to give Ah delivered from the batteries to the load:

$$\text{Ah to load} = I_R \times \text{Peak sun hours} \times \text{Coulomb efficiency} \times \text{De-rating factor} \tag{9.35}$$

The only other thing to keep track of is the system voltage. For a 12-V system voltage and 12-V modules, modules are added in parallel until sufficient Ah are provided for the load. For a 24-V system voltage and 12-V modules, two modules in series are needed to provide the 24 V, and then parallel strings of two series modules each are added to deliver the Ah needed by the load. Or, if 24 V is the system voltage, it may make more sense to simply choose 24-V PV modules rather than using two 12-V versions.

Example 9.20 PVs for the Cabin in Salt Lake City. The cabin from Example 9.18 needs 3000 Wh/day of ac delivered from an 85%-efficient inverter. For a 24-V system voltage, a 90% Coulomb efficiency, and 10% de-rating (de-rating factor = 0.90), size a PV array using Kyocera KC120 modules.

Solution. From Table 9.3, the Kyocera KC120 is a 120-W module with its maximum power point at a current of 7.1 A and a voltage of 16.9 V. From Example 9.18, the worst solar month is December, with 3.1 peak hours of sunlight at a tilt of L + 15. Using (9.35), one string of modules will therefore deliver in December about

$$\text{Ah to inverter} = 7.1\ \text{A} \times 3.1\ \text{h/day} \times 0.90 \times 0.90 = 17.83\ \text{Ah/day per string}$$

For an 85% efficient inverter to deliver 3000 Wh/day of 120 V ac, it needs a 24-V dc input of

$$\text{Inverter dc input} = \frac{3000\ \text{Wh/day}}{0.85 \times 24\ \text{V}} = 147\ \text{Ah/day @ 24 V}$$

Since these modules have a rated voltage of 16.9 V, they are nominally "12-V modules." Therefore two modules are needed in series to provide a single 24-V string.

The number of parallel strings of modules needed is

$$\text{Number of parallel strings} = \frac{147\ \text{Ah/day}}{17.83\ \text{Ah/day-string}} = 8.25\ \text{strings}$$

Suppose that we undersize it slightly and use eight parallel strings with two modules per string, for a total of 16 modules.

Including the 0.90 de-rating factor, the PVs will deliver

$$\begin{aligned}\text{PV output} &= 8\ \text{strings} \times 7.1\ \text{A/string} \times 3.1\ \text{h/day} \times 0.90 \\ &= 158\ \text{Ah/day @ 24 Vdc}\end{aligned}$$

The batteries with 0.90 Coulomb efficiency will deliver

$$\text{Battery output} = 158\ \text{Ah/day} \times 0.90 = 142\ \text{Ah/day @ 24 Vdc}$$

The 85% efficient inverter will deliver

$$\text{Inverter output} = 142 \text{ Ah/day} \times 24 \text{ V} \times 0.85 = 2900 \text{ Wh/day @ 120 Vac}$$

So, the design is just a bit shy of its goal of 3000 Wh/day ac. The system diagram for this cabin, including some of the energy flows for December, is shown in Fig. 9.51.

The month-by-month energy that the system just designed will deliver is shown in Fig. 9.52. The system delivers far more energy in the summer than was needed in the winter.

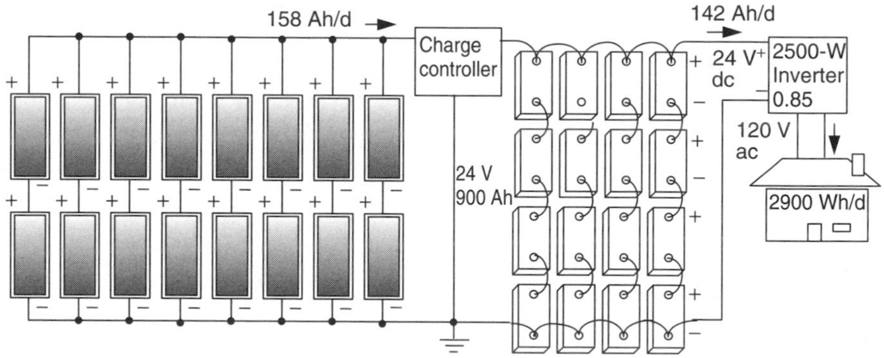

Figure 9.51 Design for the cabin in Salt Lake City (additional fuses, breakers, and diodes not shown). Energy values are for the design month, December.

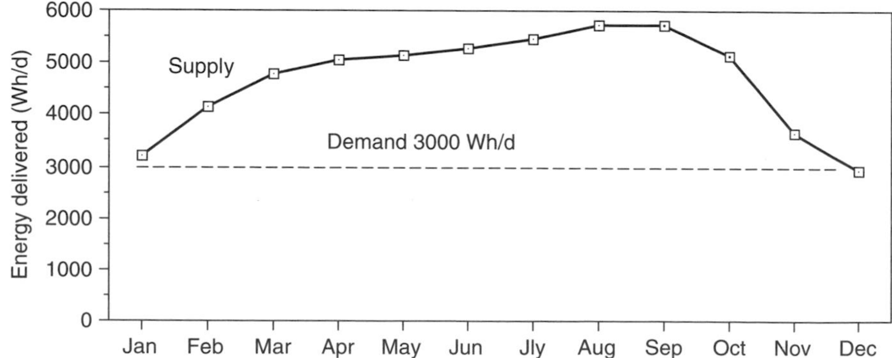

Figure 9.52 A PV–battery system sized to cover the worst month delivers much more energy than is needed during the rest of the year. Data are for the system designed in Example 9.20.

9.5.10 Hybrid PV Systems

A PV system designed to supply the entire load in the worst month (the "design month") usually delivers much more energy than is needed during the rest of the year. Outside of the tropics, it is not at all unusual for the energy supplied during the best month to be nearly double that of the design month. Figure 9.52 shows that to be the case for the cabin in Salt Lake City designed in Example 9.20.

After figuring out the cost of a system that has been designed to be completely solar, a buyer may very well decide that a hybrid system with most of the load covered by PVs and the remainder supplied by a generator is worth considering. The key to that decision is estimating the relationship between shrinking the PV system and increasing the fraction of the load carried by the generator. An analysis based on month-by-month calculations of energy supplied by the solar portion of the system while varying the area of PVs is quite straightforward. Figure 9.53 shows a pretty typical result (drawn for Salt Lake City with south-facing array at a tilt of $L + 15°$). As can be seen, significant reductions in the size of the solar system may still cover a high fraction of the annual load. For example, Fig. 9.53 suggests that a PV system designed to deliver only 50% of the load in the design month will still cover about 80% of the annual load.

An approximation to Fig. 9.53 that is helpful for computer analysis is the following:

$$\text{Design mo. solar fraction} = 0.625 \times \text{Annual solar fraction} \qquad (\text{Annual} \le 0.80) \qquad (9.36)$$

$$\text{Design mo. solar fraction} = 0.50 + 28(\text{Annual solar fraction} - 0.80)^{2.5} \qquad (\text{Annual} > 0.80) \qquad (9.37)$$

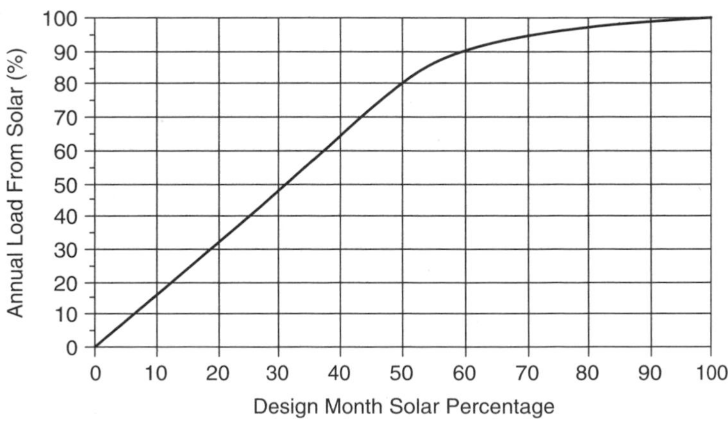

Figure 9.53 The fraction of the annual load supplied by solar as a function of the percent of the load covered in the worst month of the year. Derived for Salt Lake City.

TABLE 9.16 Characteristics of Generators for Hybrid PV Systems

Type	Size Range (kW)	Applications	Cost ($/W)	Maintenance Intervals (hours)		
				Oil Change	Tune-up	Engine Rebuild
Gasoline (3600 rpm)	1–20	Cabin Light use	$0.50	25	300	2000–5000
Gasoline (1800 rpm)	5–20	Residence Heavy use	$0.75	50	300	2000–5000
Diesel	3–100	Industrial	$1.00	125–750	500–1500	6000

Source: Sandia National Laboratories (1995).

When a generator is included in the system, the most versatile inverter is an inverter-charger capable of converting dc from the batteries into ac for the load, as well as converting ac from the generator into dc to charge the batteries. Switching from one mode to the other can be done manually or with an automatic transfer switch in the unit itself. The generator can be sized just to charge the batteries, which is the usual case, or it can be sized large enough to charge batteries and simultaneously run the entire household.

With a hybrid system, the battery storage bank can be smaller since the generator can charge the batteries during prolonged periods of poor weather. One constraint on how small storage can be is to check to be sure that the load can't discharge the batteries at too fast a rate—certainly no faster than $C/5$. A nominal 3-day storage system is often recommended since it will avoid discharging too rapidly, while at the same time keeping the number of times the generator has to be fired up to a reasonable level. Finally, the generator should be sized so that it doesn't charge the batteries too rapidly—again, certainly no faster than $C/5$.

Generators are somewhat costly, depending on the quality of the machine. They require periodic oil changes, tune-ups, and major overhauls. Home-size generators burn fuel at varying rates, but a rough estimate is that they will generate about 5 kWh per gallon of fuel. Table 9.16 summarizes some of the cost and maintenance characteristics of generators.

9.5.11 Stand-Alone System Design Summary

While the above series of examples and sets of equations for stand-alone sizing of PV systems may at first appear daunting, it is actually quite straightforward and can fairly easily be set up in a spreadsheet, which allows multiple design options to be quickly explored. The following is a summary of the design approach for the principal components of the system.

Load Analysis

1. Analyze the load to determine daily watt-hours of demand. See Table 9.10 for representative load power. Adjust ac loads using an inverter efficiency

(default 85%) to find their dc equivalent and add it to the load that runs directly on dc:

$$\text{Total dc load (Wh/day)} = \text{dc load (Wh/day)} + \frac{\text{ac load (Wh/day)}}{\text{Inverter efficiency}} \tag{9.38}$$

2. Use Tables 9.10 and 9.13 to estimate the maximum steady-state power demand (W) and, with the help of Table 9.11, pick an appropriate system voltage (12 V, 24 V, or 48 V).
3. Convert total dc load to a load expressed as Ah @ the system voltage:

$$\text{Total load (Ah/day @ System voltage)} = \frac{\text{Total dc load (Wh/day)}}{\text{System voltage (V)}} \tag{9.39}$$

4. Use Fig. 9.53 [or Eqs. (9.36) and (9.37)] to help decide what annual solar fraction to design for and to determine from that the solar fraction in the design month. The design-month load is then

$$\text{Design-month load (Ah/day)} = \text{Design-month fraction} \times \text{Total load (Ah/day)} \tag{9.40}$$

PV Sizing

5. Using insolation data for the site (Appendix E), find the insolation (kWh/m^2-day = hours @ 1-sun) in the worst month of the year (the design month).
6. Pick a PV module and use its rated current I_R along with an estimated Coulomb efficiency (default 0.90), de-rating factor (default 0.90), and the design month insolation to determine design-month Ah/day delivered per string:

$$\text{Ah/day-string} = \text{Insolation (h/day @ 1-sun)} \times I_R \text{ (A)} \times \text{Coulomb.} \times \text{De-rating} \tag{9.41}$$

7. Find an integer value for the number of parallel strings of modules based on

$$\text{Strings in parallel} = \frac{\text{Design-month load (Ah/day)}}{\text{Ah/day per module in design month}} \tag{9.42}$$

8. Find the number of modules in series from

$$\text{Modules in series} = \frac{\text{System voltage (V)}}{\text{Nominal module voltage (V)}} \tag{9.43}$$

Battery Sizing

9. Decide the number of days of storage needed. For a totally solar system, use design-month insolation in Fig. 9.46 as a guide. For a hybrid system, the choice depends on how inconvenient running the generator is perceived to be (3 days of storage is a reasonable default).
10. Find the usable storage needed from the Total load (Ah/day) found in Step 3 and the days of storage found in step 9:

$$\text{Usable storage (Ah)} = \text{Total load (Ah/day)} \times \text{Days of storage (days)} \tag{9.44}$$

11. Using the days of storage as an indicator of the discharge rate C/T, along with anticipated minimum battery temperature, find the temperature and discharge rate factor (T, DR) from Fig. 9.42.
12. Pick the maximum depth of discharge (MDOD defaults: deep-cycle lead-acid 0.80, auto SLI 0.25, NiCad 0.90) subject to the freeze constraint given in Fig. 9.39.
13. Find the total storage capacity from

$$\begin{matrix}\text{Total storage capacity} \\ (\text{Ah @ } C/20, 25^\circ\text{C})\end{matrix} = \frac{\text{Usable storage capacity (Ah)}}{(\text{MDOD}) \times (T, \text{DR})} \tag{9.45}$$

14. Check to be sure the battery capacity is sufficient to avoid a too-rapid discharge rate (default $C/5$ rate). Use the maximum power demand found in step 2 along with a maximum discharge rate of $C/5$ to find the minimum storage capacity:

$$\begin{matrix}\text{Minimum storage} \\ \text{capacity (Ah)}\end{matrix} = \frac{\text{Maximum load power (W)} \times 5\text{ h}}{\text{System voltage (V)} \times \text{Max depth of discharge}} \tag{9.46}$$

Choose the tentative storage capacity to be the larger of the total and minimum.

15. Choose a battery (voltage and Ah) and determine the number of batteries in series in each string and the number of parallel strings:

$$\text{Number of batteries in series} = \frac{\text{System voltage}}{\text{Nominal battery voltage}} \tag{9.47}$$

$$\begin{matrix}\text{Number of strings of batteries} \\ \text{in parallel}\end{matrix} = \frac{\text{Total storage capacity (Ah)}}{\text{Capacity of a single battery (Ah)}} \tag{9.48}$$

Actual battery capacity is then

$$\text{Actual battery capacity (Ah)} = \text{Ah/battery} \times \text{No. of parallel strings} \tag{9.49}$$

Generator Sizing

16. A generator should be able to replenish the batteries in a reasonable amount of time, but no faster than at a *C*/5 rate. Including a charger efficiency factor (default 0.80), the generator output to the battery charger should be

$$\text{Generator (W)} = \frac{\text{Total storage capacity (Ah)} \times \text{System voltage (V)}}{\text{Charge time (h)} \times \text{Charger efficiency}} \tag{9.50}$$

The actual generator size will be determined by what is available on the market.

17. The generator output and run time can be estimated from the annual solar fraction (step 4) along with the annual dc load and the charger efficiency (default 0.80):

$$\text{Generator (kWh/yr)} = \frac{\text{Total dc load (Wh/day)} \times 365\ \text{day/yr} \times (1 - \text{Annual solar fraction})}{\text{Charger efficiency} \times 1000\ \text{(W/kW)}} \tag{9.51}$$

$$\text{Generator run time (h/yr)} = \frac{\text{Generator (kWh/yr)}}{\text{Generator rated power (kW)}} \tag{9.52}$$

System Costs

18. In lieu of actual data on component costs, the following can be used for preliminary estimates of total system cost:

Photovoltaics:	\$4.00/W
Batteries:	Golf cart \$0.07/Wh; True Deep cycle \$0.10/Wh
Inverter-charger:	\$0.60/W
Generator:	Light duty \$0.50/W; heavy duty \$0.75/W
Installation + BOS:	20% × (PV + batteries + inverter + generator \$)

19. Annual fuel costs can be estimated based on unit cost and use rate (default: \$1.50/gal × 0.2 gal/kWh = \$0.30/kWh)

$$\text{Fuel (\$/yr)} = \text{Generator (kWh/yr)} \times \text{Fuel cost (\$/gal)} \times \text{Fuel rate (gal/kWh)} \tag{9.53}$$

Maintenance costs depend on hours of use, maintenance intervals and unit costs. (Defaults: \$100 tune-ups @ 300 h, \$400 overhauls @ 4000 h):

$$\text{Maintenance (\$/yr)} = \frac{\text{Generator run time (h/yr)} \times \text{Rate (\$/service)}}{\text{Maintenance interval (h/service)}} \tag{9.54}$$

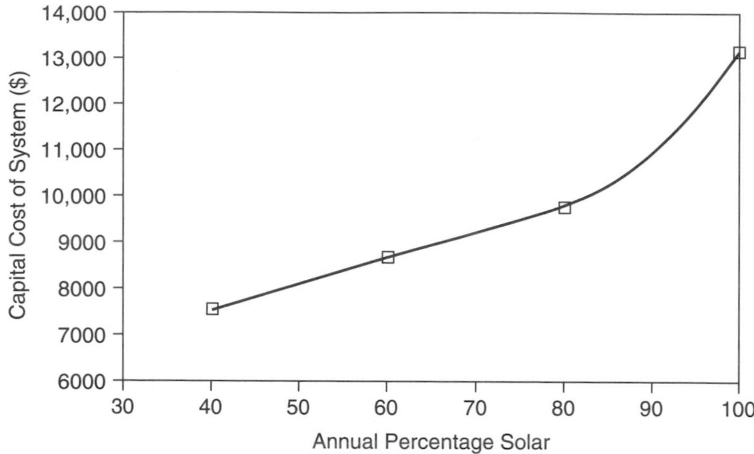

Figure 9.54 The capital cost of the Salt Lake City off-grid cabin drops quickly when it includes a backup generator to supply even a small portion of the annual load.

While these 19 design steps may imply a linear flow from top to bottom, a real design will of course involve many iterations as different modules and batteries are analyzed and as the economics of various solar fractions are determined.

Figure 9.54 is based on applying the above sequence of steps to the cabin already analyzed in previous examples, and it does so for a system that is 100% solar for comparison with hybrid systems having annual solar fractions of 80% 60% and 40%. Even though the cost estimates used are crude, the conclusions drawn are quite robust. That is, the capital cost of a system drops rather dramatically when a generator provides even a small fraction of the annual energy.

9.6 PV-POWERED WATER PUMPING

One of the most economically viable photovoltaic applications today is for water pumping in remote areas. For an off-grid home, a simple PV system can raise water from a well or spring and store it in either a pressurized or an unpressurized tank, or it can circulate water through a solar water heating system. Water for irrigation, cattle watering, or village water supply—especially in developing countries—can be critically important, and the value of a PV water-pumping system in these circumstances can far exceed its costs.

The simplest PV water-pumping systems consist of just a PV array attached to a dc pump. Water that is pumped when the sun is out may be used at that time or stored in a tank for use later, so the disadvantages of battery storage can be avoided. The result can be a system that combines simplicity, low cost, and reliability. Matching photovoltaics and pumps in such directly coupled systems (without battery storage), along with predicting their daily performance, is

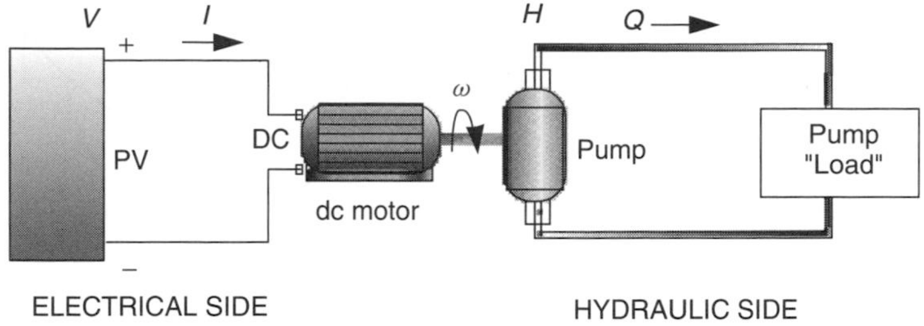

Figure 9.55 The electrical characteristics of the PV–motor combination need to be matched to the hydraulic characteristics of the pump and its load.

actually a quite challenging task. More complex systems may include a battery and inverter to run a conventional ac pump, along with a linear current booster (LCB) to improve performance in low-light conditions.

As suggested in Fig. 9.55, a simple, directly coupled PV–pump system has (a) an electrical side in which PVs create a voltage V that drives current I through wires to a motor load and (b) a hydraulic side in which a pump creates a pressure H (for *head*) that drives water at some flow rate Q through pipes to some destination. The figure suggests that the hydraulic side is a *closed* loop with water circulating back to the pump, but it may also be an *open* system in which water is raised from one level to the next and then released. On the electrical side of the system, the voltage and current delivered at any instant are determined by the intersection of the PV I–V curve and the motor I–V curve. In the hydraulic system, H is analogous to voltage while Q is analogous to current; as we shall see, the role of H–Q curves in determining a hydraulic operating point is exactly analogous to the role of I–V curves that determine the electrical operating point.

9.6.1 Hydraulic System Curves

Figure 9.56a shows an open system in which water is to be raised from one level to the next. The vertical distance between the lower water surface and the elevation of the discharge point is referred to as the *static head* (or gravity head), and in the United States it is usually given in "feet of water." Head can also be measured in units of pressure, such as pounds per square inch (psi) or pascals (1 psi = 6895 Pa). To convert between these two equivalent approaches to units, just picture the pressure that a cube of water exerts on its base. For example, the pressure that a 1-ft cube weighing 62.4 lb would exert on its 144 square inches of base would be

$$1 \text{ ft of head} = 62.4 \text{ lb}/144 \text{ in.}^2 = 0.433 \text{ psi} \tag{9.55}$$

Conversely, 1 psi = 2.31 ft of water. Typical city water pressure is about 60 psi which corresponds to a column of water roughly 140 ft high.

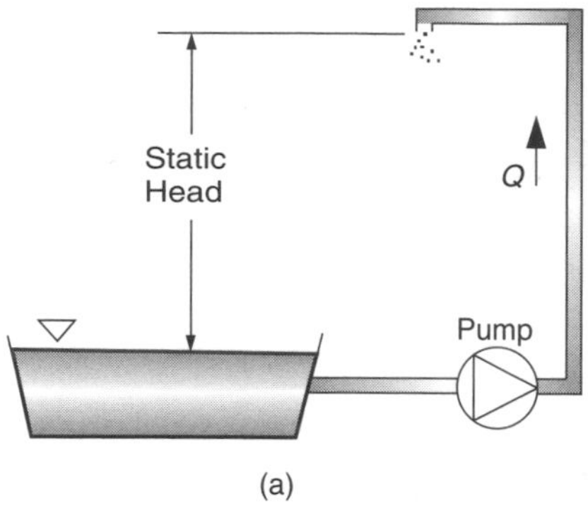

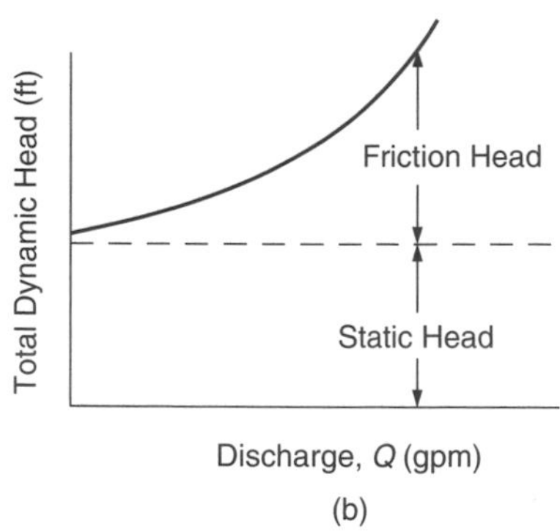

Figure 9.56 An "open" system and the resulting "system curve" showing the static and friction head components.

If the pump is capable of supplying only enough pressure to the column of water to overcome the static head, the water would rise in the pipe and just make it to the discharge point and then stop. In order to create flow, the pump must provide an extra amount of head to overcome friction losses in the piping system. These friction losses rise roughly as the square of the flow velocity (as is suggested in Figure 9.56); they depend on roughness of the inside of the pipe and on the numbers of bends and valves in the system. For example, the pressure drop per 100 ft of plastic water pipe for various flow rates and diameters is presented in Table 9.17. In keeping with U.S. tradition, flow rates are given in gallons per minute (gpm), pipe diameters in inches, and head in feet of water.

Table 9.18 gives pressure drop of various plumbing fittings expressed as equivalent lengths of pipe. For example, each 3/4″ 90° elbow in a plumbing run adds to the pressure drop the same amount as would 2.0 ft of straight pipe. So we can

TABLE 9.17 Pressure Loss Due to Friction in Plastic Pipe, Feet of Water per 100 ft of Tube for Various Nominal Tube Diameters

gpm	0.5 in.	0.75 in.	1 in.	1.5 in.	2 in.	3 in.
1	1.4	0.4	0.1	0.0	0.0	0.0
2	4.8	1.2	0.4	0.0	0.0	0.0
3	10.0	2.5	0.8	0.1	0.0	0.0
4	17.1	4.2	1.3	0.2	0.0	0.0
5	25.8	6.3	1.9	0.2	0.0	0.0
6	36.3	8.8	2.7	0.3	0.1	0.0
8	63.7	15.2	4.6	0.6	0.2	0.0
10	97.5	26.0	6.9	0.8	0.3	0.0
15		49.7	14.6	1.7	0.5	0.0
20		86.9	25.1	2.9	0.9	0.1

TABLE 9.18 Friction Loss in Valves and Elbows Expressed as Equivalent Lengths of Tube[a]

Fitting	0.5 in.	0.75 in.	1 in.	1.5 in.	2 in.	3 in.
90-degree ell	1.5	2.0	2.7	4.3	5.5	8.0
45-degree ell	0.8	1.0	1.3	2.0	2.5	3.8
Long sweep ell	1.0	1.4	1.7	2.7	3.5	5.2
Close return bend	3.6	5.0	6.0	10.0	13.0	18.0
Tee—straight run	1.0	2.0	2.0	3.0	4.0	
Tee—side inlet or outlet	3.3	4.5	5.7	9.0	12.0	17.0
Globe valve, open	17.0	22.0	27.0	43.0	55.0	82.0
Gate valve, open	0.4	0.5	0.6	1.0	1.2	1.7
Check valve, swing	4.0	5.0	7.0	11.0	13.0	20.0

[a]Units are feet of pipe for various nominal pipe diameters.

add up all the bends and valves in a pipe run and find what equivalent length of straight pipe would have the same pressure drop.

The sum of the friction head and the static head is known as the *total dynamic head (H)*.

Example 9.21 Total Dynamic Head for a Well. What pumping head would be required to deliver 4 gpm from a depth of 150 ft. The well is 80 ft from the storage tank, and the delivery pipe rises another 10 ft. The piping is 3/4-in. diameter plastic, and there are three 90° elbows, one swing-type check valve, and one gate valve in the line.

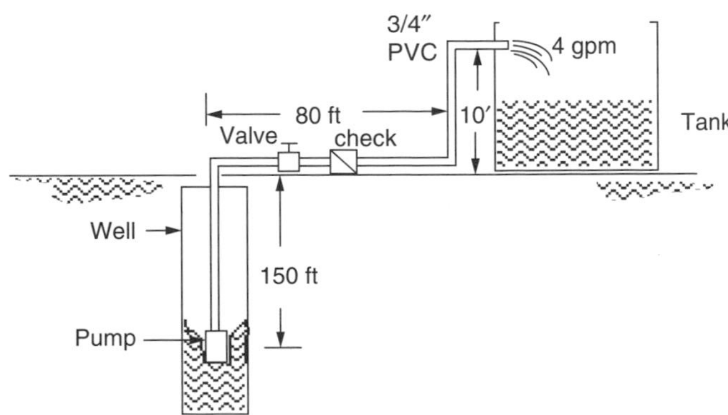

Solution. The total length of pipe is 150 + 80 + 10 = 240 ft. From Table 9.18, the three ells add the equivalent of 3 × 2.0 = 6 ft of pipe; the check valve adds the equivalent of 5.0 ft of pipe; the gate valve (assuming it is totally open) adds the equivalent of 0.5 ft of pipe. The total equivalent length of pipe is therefore $240 + 6 + 5 + 0.5 = 251.5$ ft.

From Table 9.17, 100 ft of 3/4-in. pipe at 4 gpm has a pressure drop of 4.2 ft/100′ of tube. Our friction-head requirement is therefore $4.2 \times 251.5/100 =$ 10.5 ft of water.

The water must be lifted $150 + 10 = 160$ ft (static head). Total head requirement is the sum of static and friction heads, or $160 + 10.5 = 170.5$ ft of water pressure.

If the process followed in Example 9.21 is repeated for varying flow rates, a plot of total dynamic head H (static plus friction) versus flow rate, called the *hydraulic system curve*, can be derived. The hydraulic system curve for Example 9.21 is given in Fig. 9.57.

9.6.2 Hydraulic Pump Curves

The hydraulic system curve tells us the amount of head that the pump must provide to supply a given flow rate Q. To determine the actual flow that a given pump will provide, we need to know something about the characteristics of the pump that will be used. Pumps suitable for PV-powered systems generally fall into one of two categories: *centrifugal* and *positive displacement* pumps.

Centrifugal pumps have fast-spinning impellers that literally throw the water out of the pump, creating suction on the input side of the pump and creating pressure on the delivery side. When these are installed above the water, they are limited by the ability of atmospheric pressure to push water up into the suction side of the pump—that is, to a theoretical maximum of about 32 ft. In

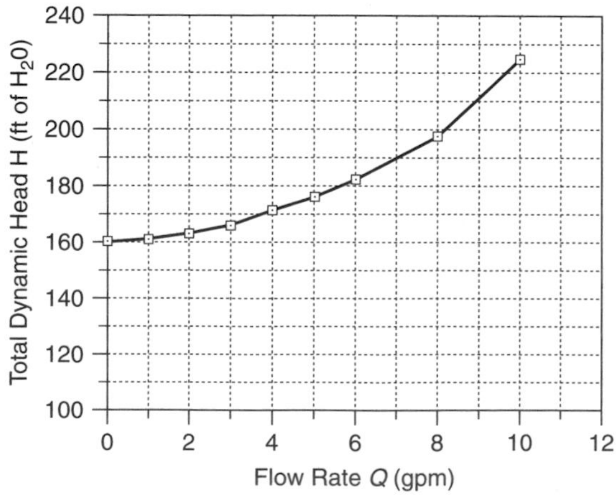

Figure 9.57 The hydraulic system curve for Example 9.21.

practice, this is more like 20 ft. When the pump is installed below the water line, however, the pump can push water up hundreds of feet. Submersible pumps with waterproof housings for the motor are suspended in a well using the same pipe that the water is pumped through. In this configuration, centrifugal pumps can push water over 1000 vertical feet. One of the disadvantages of centrifugal pumps, however, is that their speedy impellers are susceptible to abrasion and clogging by grit in the water. When powered by PVs, they are also particularly sensitive to changes in solar intensity during the day.

Positive displacement pumps come in several types, including *helical pumps*, which use a rotating shaft to push water up a cavity, *jack pumps*, which have an above-ground oscillating arm that pulls a long drive shaft up and down (like the classic oil-rig pumper), and *diaphragm pumps*, which use a rotating cam to open and close valves. The traditional hand pump as well as the wind-powered water pumps are versions of jack pumps. Jack pumps use simple flap valves that work very much like hydraulic diodes. During each upward stroke of the shaft, a flap closes and a gulp of water is carried upward; during the downward stroke, the valve opens and new water enters a chamber to be carried upward on the next stroke. In general, positive displacement pumps pump at slower rates so they are most useful in low volume applications. They easily handle high heads, however, and they are much less susceptible to gritty water problems than centrifugal pumps. They also are less sensitive to changes in solar intensity. A brief comparison of the two types of pumps is presented in Table 9.19.

The graphical relationship between head and flow is called the *hydraulic pump curve*, two examples of which are shown in Fig. 9.58. Notice that the fundamental difference in processes that produce the pumping for centrifugal and positive displacement pumps yield quite different shapes to their pump curves. For a centrifugal pump, as the height of the water column above the pump increases, more and more of the pump's energy is devoted to simply holding up the water so flow rates rapidly diminish. For example, imagine a small centrifugal pump connected to a hose. Raising the open end of the hose higher and higher (increasing the head) will result in less and less flow until a point is reached at which there is no flow at all. On the other hand, the flapper valve, diaphragm, or rotating screw in

TABLE 9.19 A Comparison Between Centrifugal and Positive-Displacement Pumps

Centrifugal	Positive Displacement
High-speed impellers	Volumetric movement
Large flow rates	Lower flow rates
Loss of flow with higher heads	Flow rate less affected by head
Low irradiance reduces ability to achieve head	Low irradiance has little effect on head
Potential grit abrasion	Unaffected by grit

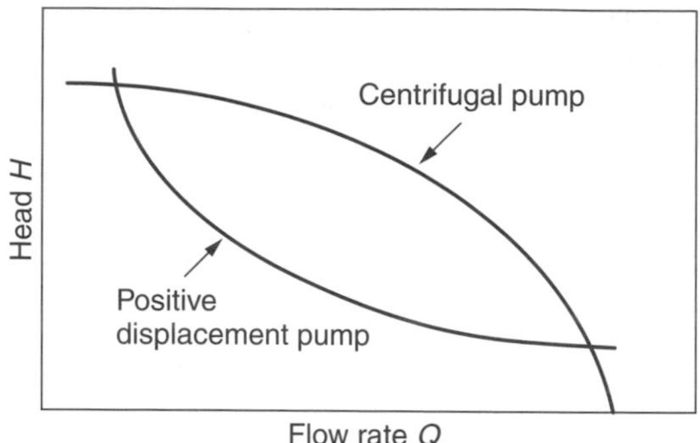

Figure 9.58 The pump curves for positive displacement pumps and centrifugal pumps have quite different shapes.

a positive displacement pump holds up the water column mechanically, so their flow rates are much less affected by increasing head.

Electrical $I-V$ curves and hydraulic $Q-H$ curves share many similar features. For example, recall that the electrical power delivered by a PV is the product of I times V and the maximum power point is at the knee of the $I-V$ curve. For the hydraulic side, the power delivered by the pump to the fluid is given by

$$P = \rho H Q \tag{9.56}$$

where ρ is fluid density. In American units,

$$P(\text{watts}) = 8.34 \text{ lb/gal} \times H(\text{ft}) \times Q(\text{gal/min}) \times (1 \text{ min}/60 \text{ s}) \times 1.356 \text{ W/(ft-lb/s)}$$

$$P(\text{watts}) = 0.1885 \times H(\text{ft}) \times Q(\text{gpm}) \tag{9.57}$$

In SI units,

$$P(\text{watts}) = 9.81 \times H(\text{m}) \times Q(\text{L/s}) \tag{9.58}$$

When Q is zero, there is no power delivered to the fluid; when the head H is zero, there is no power delivered either.

To complicate matters, for directly coupled PV-to-pump systems the voltage delivered to the pump will vary as insolation changes. In turn, the pump curve will shift as the pump voltage changes, which means that the *pump curves vary with insolation*. Many manufacturers of pumps intended for solar applications will supply pump curves for voltages corresponding to nominal 12-V module voltages. Figure 9.59 shows an example of a set of pump curves for the Jacuzzi SJ1C11 dc centrifugal pump, which is intended for use with photovoltaics. Individual curves have been given for 15-V, 30-V, 45-V, and 60-V inputs. A typical "12-V" PV module operating near the knee of its $I-V$ curve delivers about 15 V,

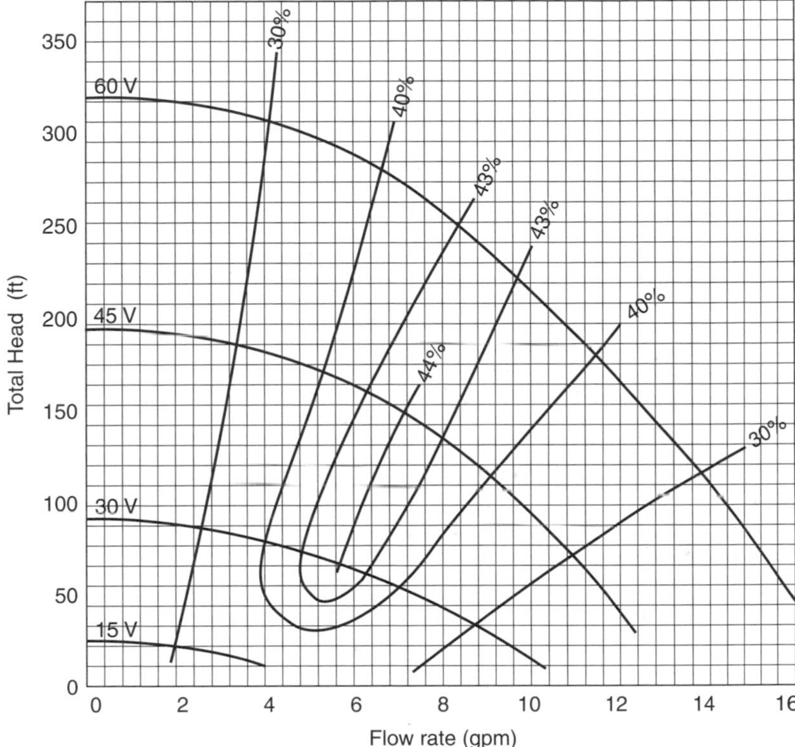

Figure 9.59 Pump curves for the Jacuzzi SJ1C11 pump showing pump efficiency for various input voltages. Pump efficiencies are also shown, with the peak along the knee of the curves.

so these pump voltages are meant to correspond to 1, 2, 3, and 4, typical "12-V" PV modules wired in series. Also shown are indications of the efficiency of the pump as a function of flow rate and head. Notice that the peak in efficiency (about 44%) occurs along the knee of the pump curves, which is exactly analogous to the case for a PV $I-V$ curve.

9.6.3 Hydraulic System Curve and Pump Curve Combined

Just as an $I-V$ curve for a PV load is superimposed onto the $I-V$ curves for the photovoltaics, so too is the $Q-H$ system curve superimposed onto the $Q-H$ pump curve to determine the hydraulic operating point. For example, superimposing the system curve of Fig. 9.57 onto the pump curves in Fig. 9.59 gives us the diagram in Fig. 9.60. A glance at the figure tells us a lot. For example, this pump will not deliver any water unless the voltage applied to the pump is at least about 36 V. At 45 V, about 5 gpm would be pumped, while at 60 V the flow would be about 9.5 gpm.

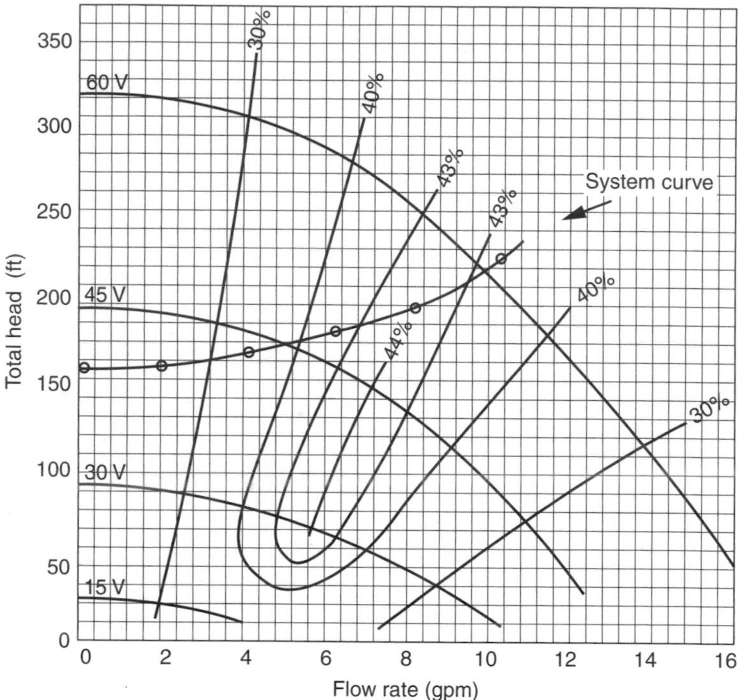

Figure 9.60 The system curve for the example, superimposed onto the pump curves for the Jacuzzi SJ1C11. No flow occurs until pump voltage exceeds about 36 V.

9.6.4 A Simple Directly Coupled PV–Pump Design Approach

The easiest approach to estimating average performance of directly coupled PV–pump systems is based on the familiar concept of "peak sun hours." That is, insolation expressed as kWh/m^2-day is treated as if it is numerically equivalent to "peak hours" at 1-sun. This lets us base the analysis of PV performance on its 1-sun rated voltage, current, and power. And it lets us assume that the flow rates on a pump curve are deliverable for the number of peak sun hours per day.

This procedure assumes that a linear current booster (LCB) is included in the system to help start the pump in the morning and keep it running under conditions of low insolation. Starting a pump requires high current at low voltage, but under low-light conditions the maximum power point on the PV I–V curve has just the opposite characteristic. An LCB is the clever dc-to-dc converter described in Section 9.2.2 that enables the PVs to operate at their highest efficiency in low light—that is, at low current and relatively high voltage—while providing the pump with what it needs to start or keep running—that is, high current and low voltage.

Usually manufacturers provide a nominal voltage and power for their pump curves. From pump power (W) and 1-sun PV power (W/module) we can determine the needed number of photovoltaic modules. If pump voltage and efficiency are given, as is the case with the pump curves in Fig. 9.60, pump power can be determined using (9.57).

$$P_{\text{in}}\ (\text{W})\ \text{to pump} = \frac{\text{Power to fluid}}{\text{Pump efficiency}} = \frac{0.1885 \times H(\text{ft}) \times Q(\text{gpm})}{\eta_P} \tag{9.59}$$

The sizing procedure is based on the following simple steps (Thomas, 1987):

1. Determine the water production goal (gallons/day) in the design month (highest water need and lowest insolation).
2. Use design-month insolation (hours @ 1-sun) as the hours of pumping to find the pumping rate:

$$Q(\text{gpm}) = \frac{\text{Daily demand (gal/day)}}{\text{Insolation(h/day@1-sun)} \times 60\ \text{min/h}} \tag{9.60}$$

3. Find the total dynamic head H @ Q (gpm). As a default, the friction head may be assumed to be 5% of the static head.
4. Find a pump capable of delivering the desired head H and flow Q. Note its input power P_{in} and nominal voltage. Pump input power can also be estimated from (9.59) along with estimated pump efficiency η_P (defaults: suction pumps 25%; submersible pumps 35%)
5. The number of PV modules in series (assuming that modules will operate at about 15 V) is an integer number based on

$$\text{Modules in series} = \frac{\text{Pump voltage(V)}}{15\ \text{V/module}} \tag{9.61}$$

6. The number of PV strings in parallel will be an integer number based on

$$\#\ \text{strings} = \frac{\text{Pump input power } P_{\text{in}}(\text{W})}{\#\ \text{mods in series} \times 15\ \text{V/mod} \times I_R(\text{A}) \times \text{de-rating}} \tag{9.62}$$

   where I_R is the PV rated current at STC. The de-rating factor takes into account dirt and temperature effects. A reasonable default value is 0.80.
7. After having sized the system, the water pumped can be estimated by rearranging (9.59) and adding in the de-rating factor:

$$\begin{aligned} Q(\text{gal/day}) &= 15\ \text{V/mod} \times I_R\ (\text{A}) \times (\#\ \text{mods}) \times (\text{Peak h/day}) \times 60\ \text{min/h} \\ &\quad \times \text{de-rating} \times \eta_P/[0.1885 \times H(\text{ft})] \end{aligned} \tag{9.63}$$

Example 9.22 Sizing an Array for a 150-ft Well in Santa Maria, California. Suppose that the goal is to pump at least 1200 gallons per day from the

150-ft well described in Example 9.21 using the Jacuzzi SJ1C11 pump. Size the PV array based on Siemens SR100 modules with rated current 5.9 A, mounted at an L + 15° tilt.

Solution. From the solar radiation tables given in Appendix E, the worst month is December, with an insolation of 4.9 kwh/m^2-day. From (9.60)

$$Q = \frac{1200 \text{ gal/day}}{4.9 \text{ (h/day @1-sun)} \times 60 \text{ min/h}} = 4.1 \text{ gpm}$$

From Fig. 9.60, at 4.1 gpm the total dynamic head is about 170 ft and the pump efficiency is about 34%. From (9.59), the estimated pump input power is

$$P_{\text{in}}(\text{W}) = \frac{0.1885 \times H(\text{ft}) \times Q(\text{gpm})}{\text{Pump efficiency}} = \frac{0.1885 \times 170 \times 4.1}{0.34} = 386 \text{ W}$$

From Fig. 9.60, at 4.1 gpm and 170 ft of head, the pump voltage is a little under 45 V, which means that at 15 V per module, three modules in series should be sufficient.

Using (9.62), we can decide upon the number of parallel strings of modules

$$\# \text{ strings} = \frac{386 \text{ W}}{3 \text{ modules string} \times 15 \text{ V/module} \times 5.9\ A \times 0.80} = 1.8$$

so choose two parallel strings.

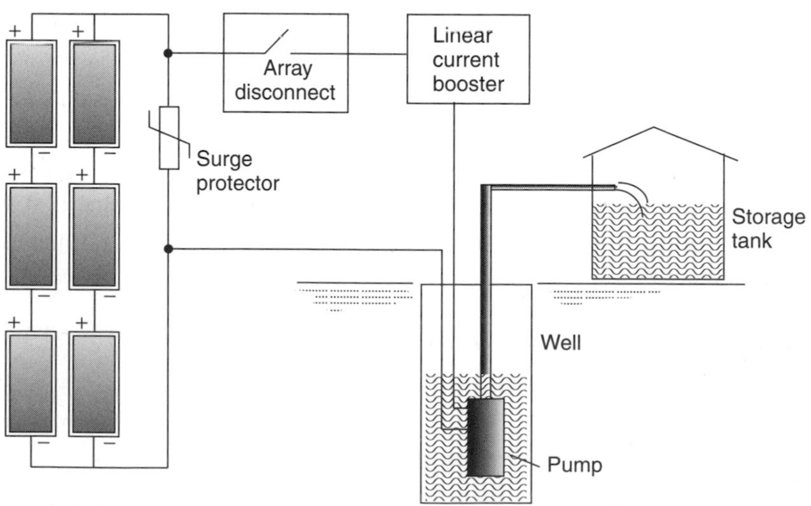

Figure 9.61 PV water pumping system for Example 9.22.

From (9.63), estimated delivery in January with two strings of three modules would be

$$Q \approx \frac{15\text{ V} \times 5.9\text{ A} \times 6\text{ modules} \times 4.9\text{ h/day} \times 60\ \text{min/h} \times 0.80 \times 0.34}{0.1885 \times 170\text{ ft}}$$

$$= 1325\text{ gal/day}$$

A system diagram is shown in Fig. 9.61.

REFERENCES

Applebaum, J. (1981). Performance Analysis of D.C. Motor Photovoltaic Converter System II, *Solar Energy,* vol 27, no. 5, pp. 421–431.

Chang, A. (2000). *Residential Systems Summary Report: Analysis of TEAM-UP Residential Installations*, Utility Photovoltaic Group, Washington, D.C.

Linden, D. (1995). *Handbook of Batteries*, McGraw-Hill, New York.

Patel, M. R. (1999). *Wind and Solar Power Systems*, CRC Press, Boca Raton, FL.

Real Goods Trading Corporation (2002). *Solar Living Source Book*, Ukiah, CA.

Roger, J. A. (1979). Theory of the Direct Coupling Between D. C. Motors and Photovoltaic Solar Arrays, *Solar Energy*, vol 23, pp 193–198.

Rosen, K., and A. Meier (1999). *Energy Use of Televisions and Videocasette Recorders in the U.S.*, Lawrence Berkeley National Labs, LBNL-42393.

Rosen, K., and A. Meier (2000). *Energy Use of U.S. Consumer Electronics At the End of the 20th Century*, Lawrence Berkeley National Labs, LBNL-46212.

Sandia National Laboratories (1991). *Maintenance and Operation of Stand-Alone Photovoltaic Systems*, U.S. Department of Energy, Albuquerque, NM.

Sandia National Laboratories (1995). Stand-Alone. *Photovoltaic Systems Handbook of Recommended Design Practices*, U.S. Department of Energy, Albuquerque, NM.

Scheuermann, K., D. R. Boleyn, P. N. Lilly, and S. Miller (2002). Measured Output for nineteen residential PV systems: Updated analysis of actual system Performance and net metering impacts, *Proceedings of the 31st American Solar Energy Society Annual Conference*, Reno, NV.

Thomas, M. G. (1987), *Water Pumping, the Solar Alternative*, Sandia National Laboratories, SAND87-0804, Albuquerque, NM.

Wiles, J. (2001). *Photovoltaic Power Systems and the National Electrical Code: Suggested Practices*, Sandia National Laboratories, SAND2001-0674, Albuquerque, NM.

PROBLEMS

9.1 Suppose the $I-V$ curve for a PV module exposed to 1-sun (1 kW/m^2) of insolation is as shown below:

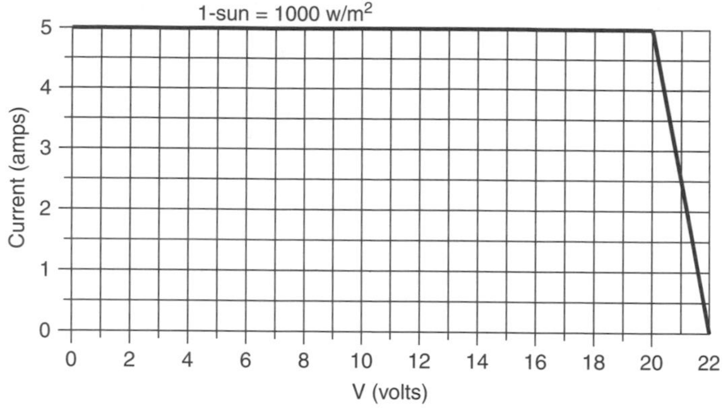

Figure P9.1

a. What load resistance R would result in the maximum power delivered to the panel at 1-sun? How much power would that be (watts)?

b. Suppose the hour-by-hour insolation striking this single panel is given in the following table (e.g., from 7:30 A.M. to 8:30 A.M. the insolation averages 200 W/m^2):

Time	Insolation (W/m^2)
8A.M., 4P.M.	200
9, 3	400
10, 2	600
11, 1	800
noon	1000
Daily total	5000 W-hr/m^2

If current (at any given voltage) is directly proportional to insolation, carefully draw the hour-by-hour $I-V$ curves for the module. For the resistive load determined in (a), what would be the energy (watt-hr) delivered by this panel in a day's time?

c. If the module is equipped with a maximum-power-point tracker, what energy would be delivered in a day's time? What percentage improvement does it provide?

9.2 Consider a single PV module supplying power to a resistance load. The hour-by-hour $I-V$ curves for the PV module are shown below along with the $I-V$ curve for the resistance. For simplicity, consider each $I-V$ curve to apply for one hour (e.g., the "8-A.M." curve is from 7:30 A.M. to 8:30 A.M.).

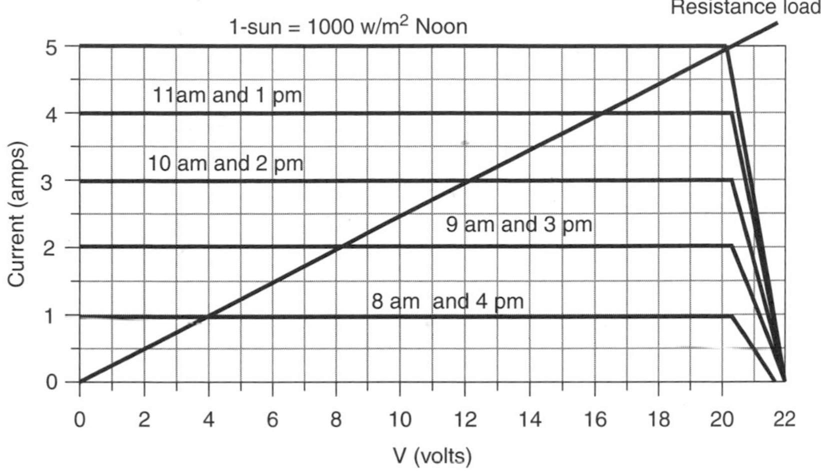

Figure P9.2a

a. What is the resistance of the load?

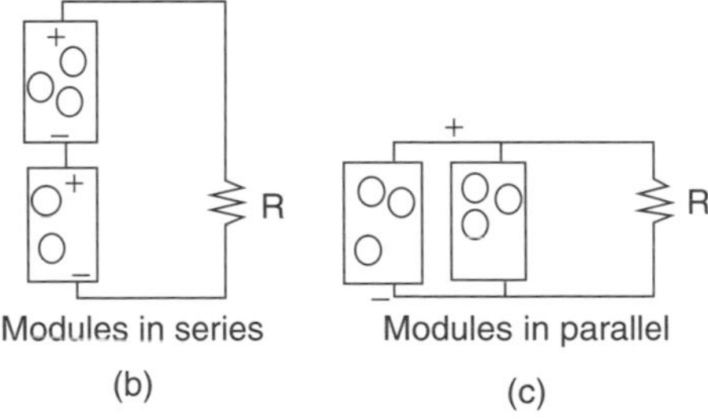

Figure P9.2b

b. Suppose the same resistive load is connected to two such modules wired together in series (so their voltages add). Draw the $I-V$ curves for the pair of modules and from that, determine the daily energy delivered (W-hr). Compare that with the energy that would have been delivered with just a single module.

c. If two modules are connected in parallel (so their currents add), what would be the daily energy delivered to the same resistive load?

9.3 A dc pump connected to a PV module runs a garden water fountain. The pump $I-V$ curve and the hourly $I-V$ curves for the module are shown. Once the pump starts running, it needs 8 V to keep spinning. At what time in the morning will the pump start running? At what time will it stop running?

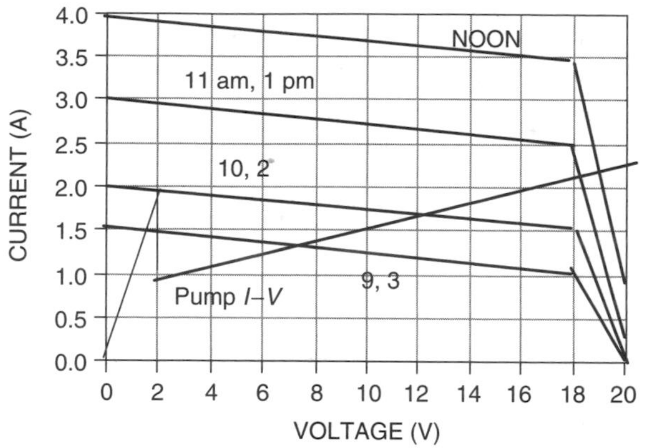

Figure P9.3

9.4 Imagine a very simple, reliable solar water heater using PVs directly connected to the electric resistance heating element of a small electric water heater. The water heater is one that would deliver 2.88 kW if it were wired to normal 120 V ac.

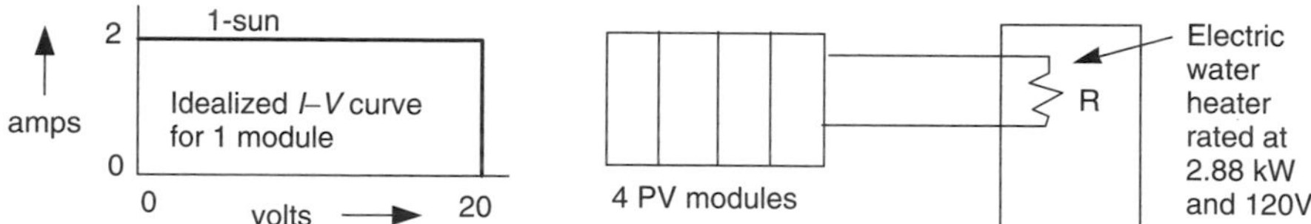

Figure P9.4a

Suppose you have 4 identical PV modules, each with an idealized, 1-sun, $I-V$ curve as shown above. Three ways that you can wire up the PVs are A, B, and C as shown.

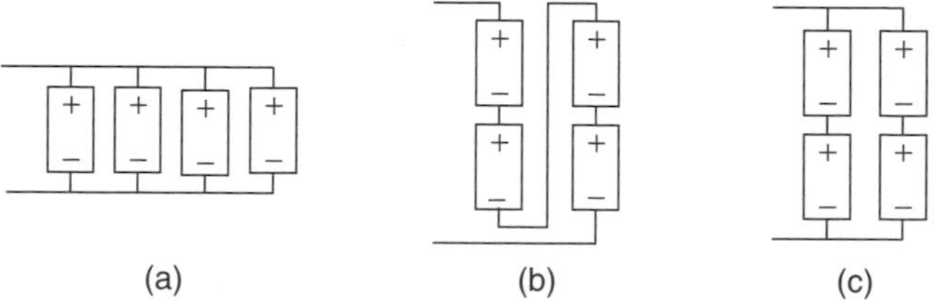

Figure P9.4b

Plot the $I-V$ curves for the three combinations (A, B, and C) and decide which one delivers the most energy in a day's time. Explain your answer.

9.5 A *clean*, 15-%-efficient module (STC), 1 m^2 in size, has its own 90% efficient inverter. Its NOCT is 44°C and its rated power degrades by 0.5%/°C above the 25°C STC.

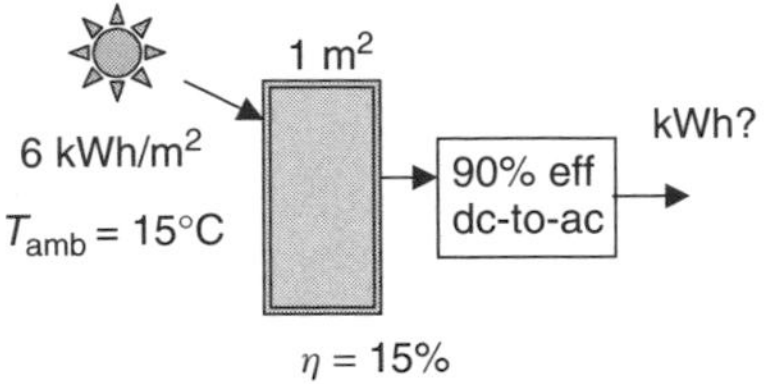

Figure P9.5

a. What would its STC rated power be?

b. Find the kWh that it would be expected to deliver on a 15°C day with daylong insolation equal to 6 kWh/m^2?

9.6 You have decided to enter the business of selling single-axis, polar mount, trackers and want to know how much you can sell them for on a dollar-per-square meter basis to make the resulting PV system no more expensive than when using a fixed mount system. Using simple dc, STC ratings along with the following assumptions, how much can the trackers cost \$/m^2?

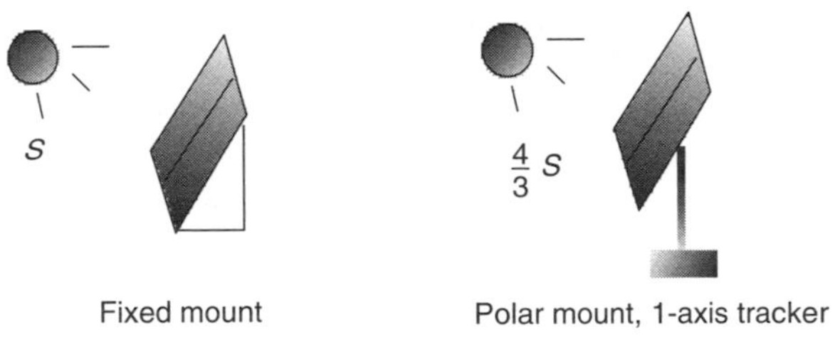

Figure P9.6

1. Trackers see one-third more insolation.
2. Modules cost \$4/W and have conversion efficiencies of 10%.
3. Inverters cost \$1 per W of module.
4. Installation cost is the same for fixed mount and trackers.

9.7 You are to size a grid-connected south-facing PV system for Las Vegas to deliver 4000 kWh/yr. Use a tilt equal to latitude minus 15°.

a. What should be the ac-rated power of the PV/inverter system?

b. What temperature derating should you use if the modules have an NOCT of 45°C, the maximum power drops by 0.36%/°C, and for ambient the average annual daily high given in Appendix E is appropriate.

c. For 3% dirt, 3% mismatched modules, and a 92% efficient inverter, what should the DC, STC rated power of the modules?

d. If the modules are 13% efficient, what area of PVs would be required?

e. If the installed system cost is \$6 per dc, STC W, what is the total cost of the system?

f. Suppose the system is paid for with a 6%, 30-year loan with interest being tax deductible, The customer is married and has a taxable income around $200,000/yr, which puts them in the 35.5% marginal tax bracket. Nevada has no state income tax. There is a renewable energy credit (Green Tag program) that will pay the owner 5¢/kWh generated. What would be the first year cost of PV electricity for this customer (¢/kWh)?

9.8 A grid-connected PV array consisting of sixteen Shell SP150 modules (Table 9.4) can be arranged in a number of series and parallel combinations: (16S, 1P), (8S, 2P), (4S, 4P), (2S, 8P), (1S, 16P). The array delivers power to a Sunny Boy SB2500 inverter (Table 9.5). Using the input voltage range of the maximum power point tracker (MPPT) and the maximum input voltage of the inverter as design constraints, what series/parallel combination of modules would best match the PVs to the inverter?

9.9 A 1.5 V AA battery that costs $1 is rated at 1.8 Ah. What is its cost per kWh?

9.10 Suppose a 12-V battery bank rated at 200 Ah under standard conditions needs to deliver 600 Wh over a 12-h period each day. If they operate at −10°C, how many days of use would they be able to supply?

9.11 A small cabin needs 2.4-kWh per day, all of which is used in the evening from 7 pm to 11 pm. If you decide to supply only enough 12-V battery storage to cover each evening, what should the minimum rated A-hr of batteries be if they shouldn't be discharged more than 80 percent? Assume battery temperature is 25°C.

9.12 Analyze this PV/battery system (ignore battery discharge rate and temperature constraints):

1. PV Module characteristics: $P_R = 100$ W, $V_R = 20.0$ V, $V_{OC} = 22$ V, $I_R = 5.0$ A, $I_{SC} = 6$ A
2. Battery characteristics: 6 V, 100 Ah (each)
3. Coulomb efficiency 90%; Maximum discharge 80%
4. Insolation: 6 kWh/m²day
5. Inverter efficiency: 80%;
6. Dirt, etc, efficiency: 90% (10% loss)

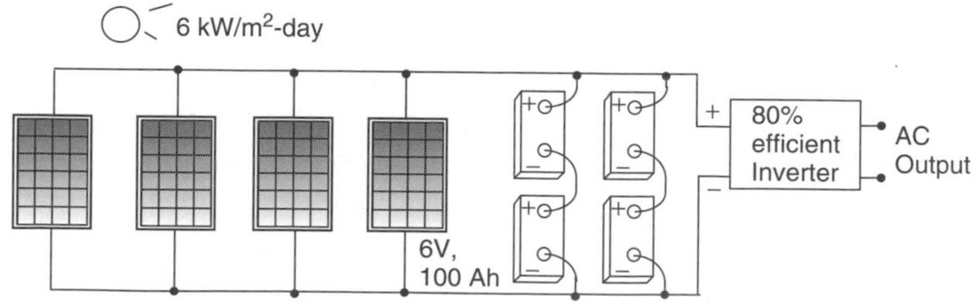

Figure P9.12

a. Find the watt-hours/day that would reach the loads (assume all PV current passes through the batteries):

b. If the load requires 600 Wh/day, for how many cloudy days in a row can fully-charged batteries supply the load?

9.13 Consider the design of a small PV-powered light-emitting-diode (LED) light. The PV array consists of 8 series cells, each with rated current 0.3 A @ 0.6 V. Storage is provided by three series AA batteries that each store 2 Ah at 1.2 V when fully charged. The LED provides full brightness when it draws 0.4 A @ 3.6 V.

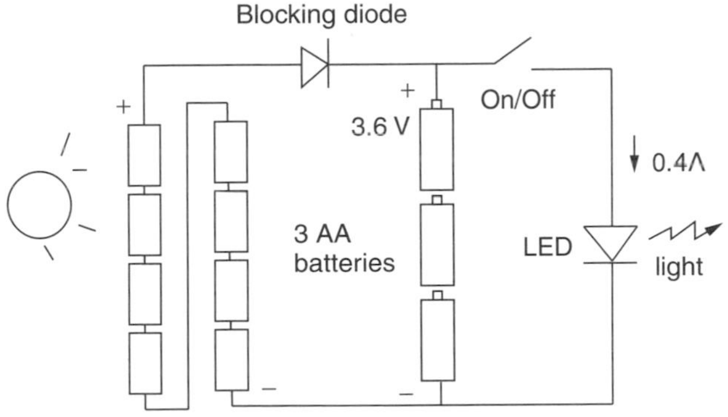

Figure P9.13

The batteries have a Coulomb efficiency of 90% and for maximum cycle life can be discharged by up to 80%. Ignore any complications associated with temperature, dirt, discharge rate, etc.

a. How many hours of light could this design provide each evening if the batteries are fully charged during the day?

b. How many kWh/m^2-day of insolation would be needed to provide the amount of light found in (a)?

c. With 12-%-efficiency cells, what PV area would be required?

9.14 Consider a PV array charging batteries for an off-the-grid, stand-alone system near Boulder, Colorado You plan to run everything in the house on 120 V ac.

a. Suppose the load is estimated to consist of the following 120 V ac appliances. Supplement load data given below with values from Table 9.10.

A 19-cu. ft. refrigerator/freezer.

A 1000-W microwave used 15 min./day.

Five 20-W compact fluorescent lamps, each used 8 hrs/day.

A horizontal-axis washing machine used 3 hrs/week

A 24-V well pump that delivers 288 gal/day at 1.6 gpm of water from 100-ft depth

A 19-inch color TV used 2 hrs/day and drawing standby power the other 22 hrs/day

A satellite receiver system for the TV, used 2 hours/day and in standby mode 22 hrs.

A laptop computer used 4 hrs/day,

Six 3-W transformer units for chargers that run all day long

Find the total watt-hrs per day needed by these appliances.

b. Pick an appropriate system voltage.

c. How many amp-hours/day would the battery bank have to deliver if the loads are all ac provided by an 85% efficient inverter?

d. How many Ah/d would have to be delivered to the batteries if they have a Coulomb efficiency of 90%?

e. Suppose 5% of the PV output is lost due to dirt, etc. How many Ah/d should the PVs provide before that derating?

f. Using the worst month in Boulder with south-facing panels tilted at L + 15, how many AstroPower APex 90-W modules with DC, STC rated current 5.3 A and rated voltage 17.1 volts would be needed in series and parallel? If they cost $400 each, what is the cost of PVs?

g. What is the maximum current that you might expect in the wires connecting the array to battery system (just use the rated current)? What gage wire would you suggest using (Table 1.3)?

h. If your design goal is to provide needed electricity 95% of the time, about how many days of usable battery storage would you need?

i. If the coldest temperature the batteries might experience is −20°C, what is the maximum depth of discharge that lead-acid batteries could tolerate without freezing (Figure 9.39)?

j. If a C/72 discharge rate is assumed along with −20°C and the results from (c), what should be the rated (nominal) Ah capacity of the battery bank? (use Figure 9.42 and Eq. 9.32)?

k. Suppose you use Concorde PVX 1080 batteries (Table 9.15). How many batteries in series and how many in parallel would you recommend (round up if necessary). At $160 each, how much would the batteries cost?

l. Assume a power control unit with inverter costs $1/watt and they come in 500 W increments (1 kW, 1.5 kW, 2 kW, 2.5 kW, etc). You want one big enough to cover all your appliances on at once. Pick an inverter and how much would it cost?

m. Draw the system "wiring" diagram showing series/parallel combinations of the PV modules and batteries, similar to Figure 9.51.

n. What would the total system cost be?

o. Using the annual insolation in Boulder, estimate the average Amp-hours/day (at the system voltage) that the PVs could deliver all year long. If all of that electricity is run through the batteries and inverter and ends up being used in the house, how many kWh/year would be delivered (i.e., in the better months you use all of the kWh not just the design load specified in (a))?

p. Paying for the system with a 15-year, 5% loan, find the first-year cost of electricity ($/kWh) for someone in the 36% marginal tax bracket if the loan interest is tax deductible (see Section 5.3.8).

9.15 A PV module is directly connected to a water pump that needs to raise water 10-ft high through 70 ft of 1/2″ plastic tubing with twenty 90° ells (Figure P9.15a).

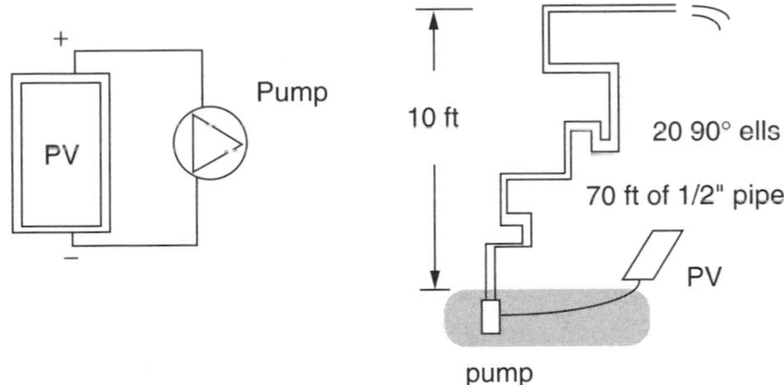

Figure P9.15a

The pump curves of head H versus flow Q as a function of input voltage are shown in Figure P9.15b.

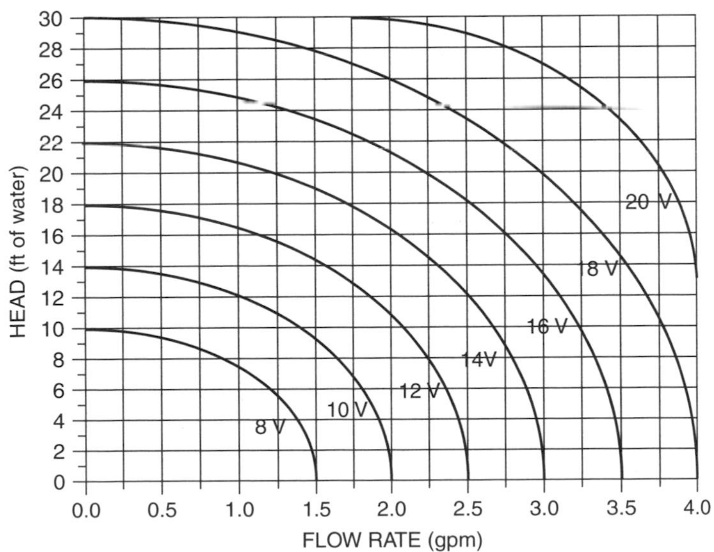

Figure P9.15b Pump curves.

The hour-by-hour $I-V$ curves for the PV module are shown in Figure P9.19c along with the pump $I-V$ curve:

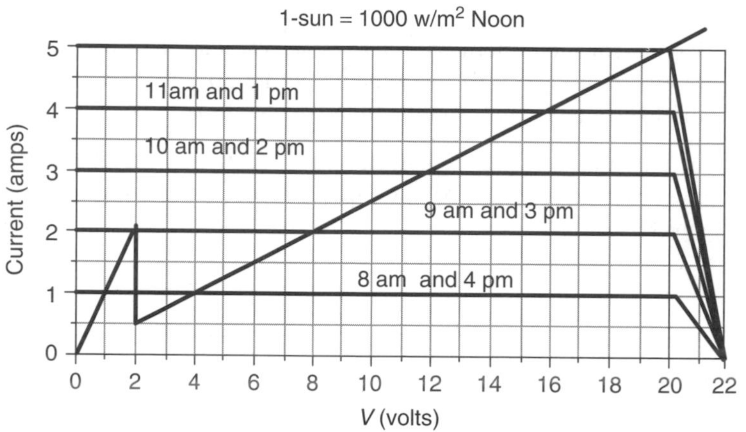

Figure P9.15c

Using Table 9.19, find the equivalent length of the pipe and ells to create a hydraulic system curve similar to that in Figure 9.56. Remember to include the 10 feet of static head. Plot that system curve on top of the pump curves in Figure P9.15b. Now find the hour-by-hour pumping rate Q (gpm) that will result.

9.16 Your client wants a quick design for a system to pump on the order of 1800 gallons per day from a depth of 200 ft. You decide to use a 45-V submersible pump with an estimated efficiency of 35% along with PV modules having rated current of 5 A at a rated voltage of 16 V. The insolation is estimated to be 6 kWh/m^{2}day in the design month. Since this is only a quick estimate, piping friction losses will be estimated at 5% of the static head. Design the PV system.

APPENDICES

Renewable and Efficient Electric Power Systems. By Gilbert M. Masters
ISBN 0-471-28060-7

APPENDIX A

USEFUL CONVERSION FACTORS

LENGTH

1 inch	= 2.540 cm
1 foot	= 0.3048 m
1 yard	= 0.9144 m
1 mile	= 1.6093 km
1 meter	= 3.2808 ft
	= 39.37 in.
1 kilometer	= 0.6214 mile

AREA

1 square inch	= 6.452 cm^2
	= 0.0006452 m^2
1 square foot	= 0.0929 m^2
1 acre	= 43,560 ft^2
	= 0.0015625 sq mile
	= 4046.85 m^2
	= 0.404685 ha
1 square mile	= 640 acre
	= 2.604 km^2
	= 259 ha

Renewable and Efficient Electric Power Systems. By Gilbert M. Masters
ISBN 0-471-28060-7

1 square meter	= 10.764 ft^2
1 hectare	= 2.471 acre
	= 0.00386 sq mile
	= 10,000 m^2

VOLUME

1 cubic foot	= 0.03704 cu yd
	= 7.4805 gal (U.S.)
	= 0.02832 m^3
	= 28.32 L
1 acre foot	= 43,560 ft^3
	= 1233.49 m^3
	= 325,851 gal (U.S.)
1 gallon (U.S.)	= 0.134 ft^3
	= 0.003785 m^3
	= 3.785 L
1 cubic meter	= 8.11×10^{-4} Ac ft
	= 35.3147 ft^3
	= 264.172 gal (U.S.)
	= 1000 L
	= 10^6 cm^3

LINEAR VELOCITY

1 foot per second	= 0.6818 mph
	= 0.3048 m/s
1 mile per hour	= 1.467 ft/s
	= 0.4470 m/s
	= 1.609 km/h
1 meter per second	= 3.280 ft/s
	= 2.237 mph

MASS

1 pound (avoirdupois)	= 0.453592 kg
1 kilogram	= 2.205 lb (avoirdupois)
	= 35.27396 oz (avoirdupois)
1 ton (short)	= 2000 lb (avoirdupois)
	= 907.2 kg
	= 0.9072 ton (metric)

1 ton (metric)	= 1000 kg
	= 2204.622 lb (avoirdupois)
	= 1.1023 ton (short)

FLOWRATE

1 cubic foot per second	= 0.028316 m^3/s
	= 448.8 gal (U.S.)/min (gpm)
1 cubic foot per minute	= 4.72×10^{-4} m^3/s
	= 7.4805 gpm
1 gallon (U.S.) per minute	= 6.31×10^{-5} m^3/s
1 million gallons per day	= 0.0438 m^3/s
1 million acre feet per year	= 39.107 m^3/s
1 cubic meter per second	= 35.315 ft^3/s (cfs)
	= 2118.9 ft^3/min (cfm)
	= 22.83×10^6 gal/d
	= 70.07 Ac-ft/d

DENSITY

1 pound per cubic foot	= 16.018 kg/m^3
1 pound per gallon	= 1.2×10^5 mg/L
1 kilogram per cubic meter	= 0.062428 lb/ft^3
1 gram per cubic centimeter	= 62.427961 lb/ft^3

CONCENTRATION

1 milligram per liter in water (specific gravity = 1.0)	= 1.0 ppm
	= 1000 ppb
	= 1.0 g/m^3
	= 8.34 lb per million gal

PRESSURE

1 atmosphere	= 76.0 cm Hg
	= 14.696 $lb/in.^2$ (psia)
	= 29.921 in. of Hg (32°F)
	= 33.8995 ft of H_2O (32°F)
	= 101.325 kPa

1 pound per square inch	= 2.307 ft of H_2O
	= 2.036 in. Hg
	= 0.06805 atm
1 Pascal (Pa)	= 1 N/m^2
	= 1.45×10^{-4} psia
1 inch of mercury (32°F)	= 3386.4 Pa
(60°F)	= 3376.9 Pa

ENERGY

1 British Thermal Unit	= 778 ft-lb
	= 252 cal
	= 1055 J
	= 0.2930 Wh
1 quadrillion Btu	= 10^{15} Btu
	= 1055×10^{15} J
	= 2.93×10^{11} kWh
	= 172×10^6 barrels (42-gal) of oil equivalent
	= 36.0×10^6 metric tons of coal equivalent
	= 0.93×10^{12} cubic feet of natural gas equivalent
1 joule	= 1 N-m
	= 9.48×10^{-4} Btu
	= 0.73756 ft-lb
1 kilowatt-hour	= 3600 kJ
	= 3412 Btu
	= 860 kcal
1 kilocalorie	= 4.185 kJ

POWER

1 kilowatt	= 1000 J/s
	= 3412 Btu/h
	= 1.340 hp
1 horsepower	= 746 W
	= 550 ft-lb/s
1 quadrillion Btu per year	= 0.471 million barrels of oil per day
	= 0.03345 TW

APPENDIX B

SUN-PATH DIAGRAMS

Renewable and Efficient Electric Power Systems. By Gilbert M. Masters
ISBN 0-471-28060-7

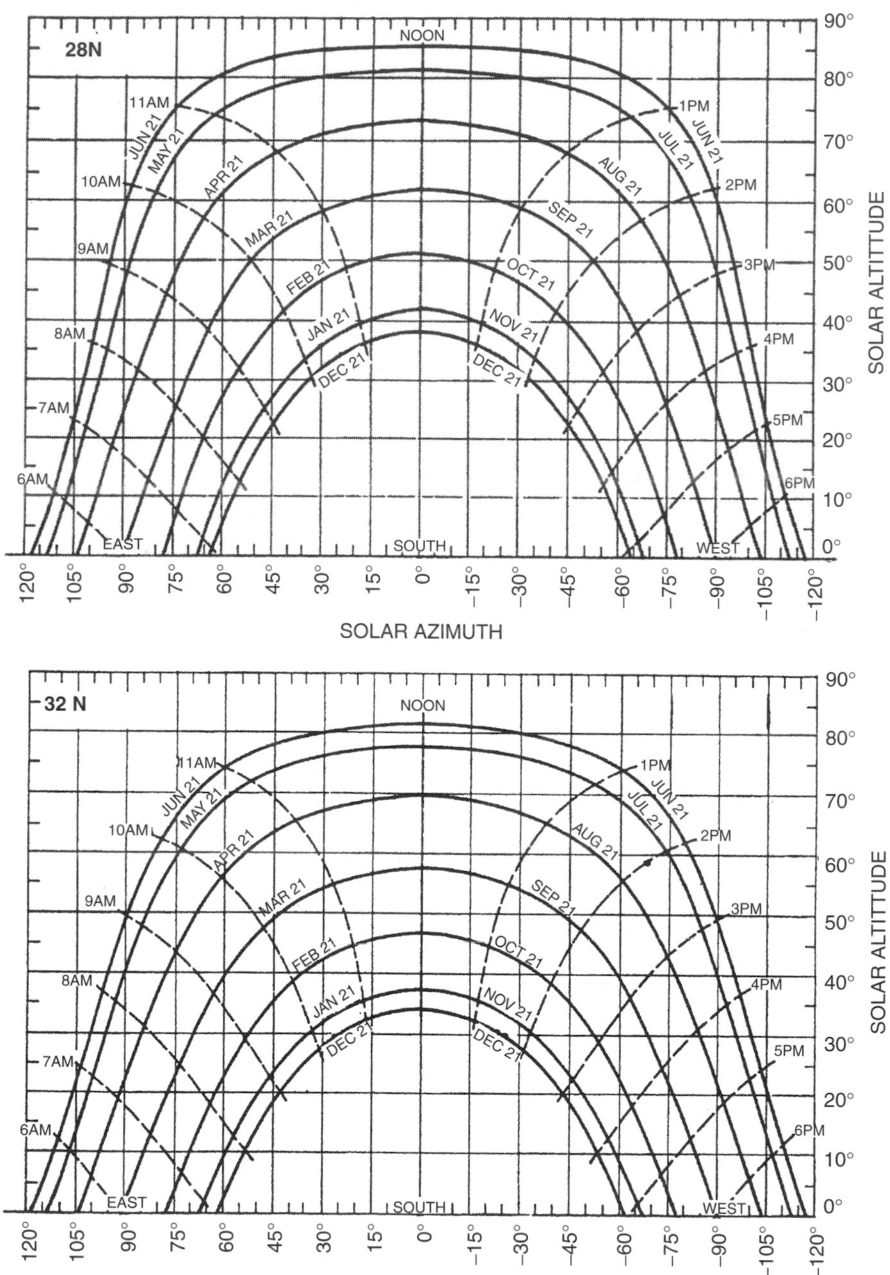
28N
NOON
11AM
10AM
9AM
8AM
7AM
6AM
1PM
2PM
3PM
4PM
5PM
6PM
JUN 21
MAY 21
APR 21
MAR 21
FEB 21
JAN 21
DEC 21
JUL 21
AUG 21
SEP 21
OCT 21
NOV 21
EAST
SOUTH
WEST
90°
80°
70°
60°
50°
40°
30°
20°
10°
0°
SOLAR ALTITUDE
120°
105°
90°
75°
60°
45°
30°
15°
0°
−15°
−30°
−45°
−60°
−75°
−90°
−105°
−120°
SOLAR AZIMUTH
32 N
NOON
11AM
10AM
9AM
8AM
7AM
6AM
1PM
2PM
3PM
4PM
5PM
6PM
JUN 21
MAY 21
APR 21
MAR 21
FEB 21
JAN 21
DEC 21
JUL 21
AUG 21
SEP 21
OCT 21
NOV 21
EAST
SOUTH
WEST
90°
80°
70°
60°
50°
40°
30°
20°
10°
0°
SOLAR ALTITUDE
120°
105°
90°
75°
60°
45°
30°
15°
0°
−15°
−30°
−45°
−60°
−75°
−90°
−105°
−120°
SOLAR AZIMUTH

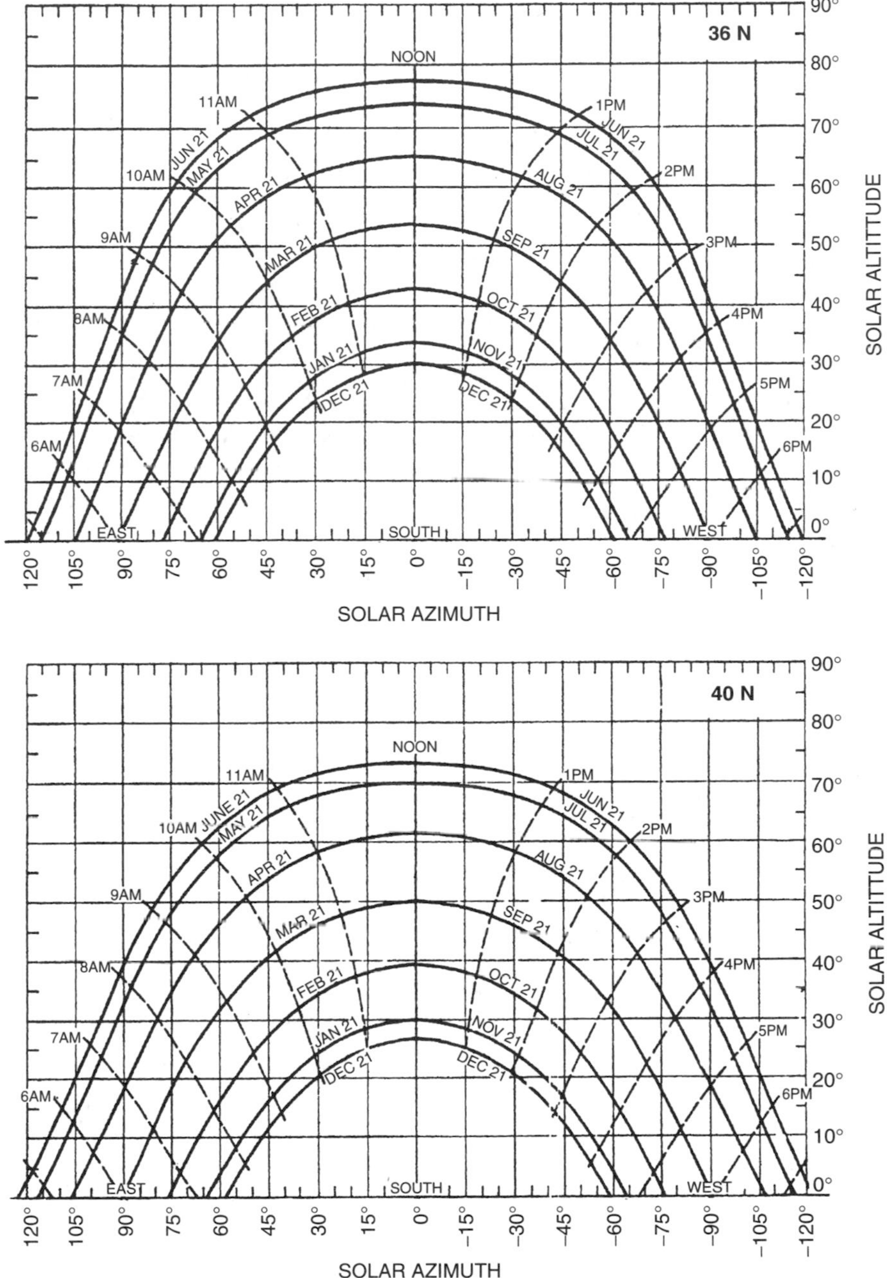

36 N
NOON
11AM
1PM
10AM
2PM
9AM
3PM
8AM
4PM
7AM
5PM
6AM
6PM
JUN 21
MAY 21
APR 21
MAR 21
FEB 21
JAN 21
DEC 21
JUL 21
AUG 21
SEP 21
OCT 21
NOV 21
EAST
SOUTH
WEST
SOLAR ALTITUDE
SOLAR AZIMUTH
40 N
JUNE 21

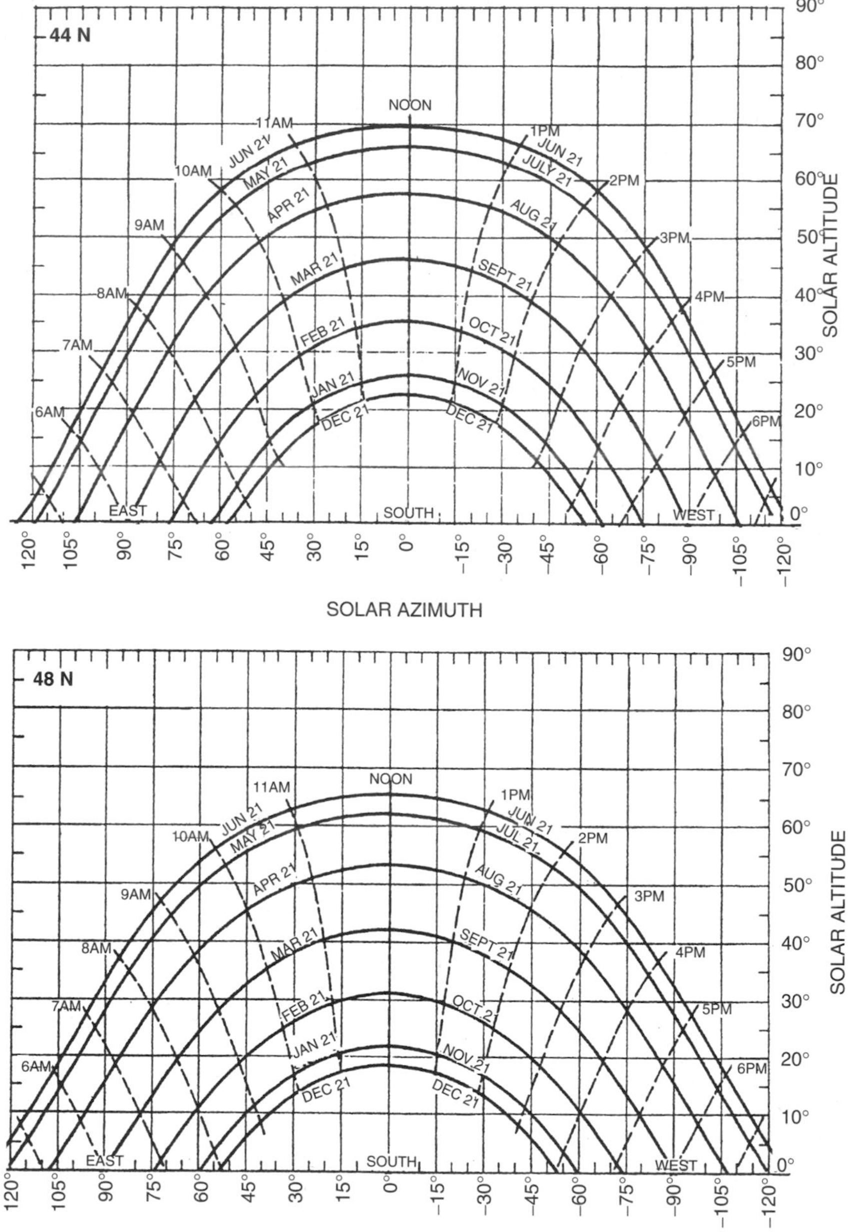
44 N
NOON
11AM
10AM
9AM
8AM
7AM
6AM
1PM
2PM
3PM
4PM
5PM
6PM
JUN 21
MAY 21
APR 21
MAR 21
FEB 21
JAN 21
DEC 21
JUN 21
JULY 21
AUG 21
SEPT 21
OCT 21
NOV 21
DEC 21
EAST
SOUTH
WEST
90°
80°
70°
60°
50°
40°
30°
20°
10°
0°
SOLAR ALTITUDE
120°
105°
90°
75°
60°
45°
30°
15°
0°
−15°
−30°
−45°
−60°
−75°
−90°
−105°
−120°
SOLAR AZIMUTH
48 N
NOON
11AM
10AM
9AM
8AM
7AM
6AM
1PM
2PM
3PM
4PM
5PM
6PM
JUN 21
MAY 21
APR 21
MAR 21
FEB 21
JAN 21
DEC 21
JUN 21
JUL 21
AUG 21
SEPT 21
OCT 21
NOV 21
DEC 21
EAST
SOUTH
WEST
90°
80°
70°
60°
50°
40°
30°
20°
10°
0°
SOLAR ALTITUDE
120°
105°
90°
75°
60°
45°
30°
15°
0°
−15°
−30°
−45°
−60°
−75°
−90°
−105°
−120°
SOLAR AZIMUTH

APPENDIX C

HOURLY CLEAR-SKY INSOLATION TABLES

Renewable and Efficient Electric Power Systems. By Gilbert M. Masters
ISBN 0-471-28060-7

Clear-Sky Insolation: Beam + Diffuse — Latitude 20°N

Solar	Tracking		Tilt Angles Latitude 20°						
Time	1-Axis	2-Axis	0	20	30	40	50	60	90
			JANUARY 21 (W/m^2)						
7, 5	341	359	58	103	122	138	149	156	151
8, 4	786	828	303	418	459	486	499	498	409
9, 3	923	973	525	670	715	739	740	721	540
10, 2	983	1037	696	860	905	923	914	878	624
11, 1	1010	1067	804	978	1023	1037	1021	975	673
12	1018	1076	841	1018	1063	1076	1057	1007	689
kWh/day	9.10	9.60	5.61	7.08	7.51	7.72	7.70	7.46	5.48
			FEBRUARY 21 (W/m^2)						
7, 5	523	530	121	162	176	185	189	188	154
8, 4	856	867	381	460	479	486	478	456	318
9, 3	973	986	610	710	730	728	705	662	425
10, 2	1028	1043	786	901	919	910	875	814	499
11, 1	1053	1069	897	1020	1037	1024	981	908	545
12	1061	1078	935	1061	1078	1063	1017	940	560
kWh/day	9.93	10.07	6.53	7.57	7.76	7.73	7.47	7.00	4.44
			MARCH 21 (W/m^2)						
7, 5	630	630	193	203	199	190	177	158	81
8, 4	881	881	456	481	474	452	418	373	182
9, 3	983	983	684	724	712	680	629	559	266
10, 2	1034	1034	860	910	896	855	790	702	331
11, 1	1060	1059	970	1027	1011	965	892	792	371
12	1067	1067	1008	1067	1051	1003	926	823	385
kWh/day	10.24	10.24	7.34	7.76	7.63	7.29	6.74	5.99	2.85
			APRIL 21 (W/m^2)						
6, 6	84	86	14	8	8	7	7	6	4
7, 5	654	673	258	223	196	164	128	90	32
8, 4	847	871	510	476	439	389	330	261	41
9, 3	937	962	727	700	657	595	516	423	88
10, 2	986	1011	893	874	827	756	663	552	141
11, 1	1013	1036	998	984	934	858	756	633	175
12	1022	1043	1034	1022	971	892	788	661	187
kWh/day	10.06	10.32	7.84	7.55	7.09	6.43	5.59	4.59	1.15
			MAY 21 (W/m^2)						
6, 6	205	222	50	24	23	22	21	19	13
7, 5	625	677	292	225	183	136	86	57	38
8, 4	787	850	528	456	400	334	259	178	47
9, 3	870	936	731	662	599	519	425	320	51
10, 2	919	984	886	822	754	665	558	435	53
11, 1	946	1009	984	923	853	759	643	509	55
12	958	1017	1017	958	887	790	672	534	55
kWh/day	9.66	10.37	7.96	7.18	6.51	5.66	4.66	3.57	0.57
			JUNE 21 (W/m^2)						
6, 6	229	255	64	31	29	28	26	24	16
7, 5	599	666	300	224	176	125	71	61	41
8, 4	752	833	528	443	382	311	232	147	50
9, 3	833	918	723	640	571	486	388	280	55
10, 2	881	967	873	793	719	625	514	389	57
11, 1	909	992	967	891	814	714	594	459	58
12	921	1000	999	924	846	745	622	483	59
kWh/day	9.33	10.26	7.91	6.97	6.23	5.32	4.27	3.20	0.61

Solar	Tracking		Tilt Angles Latitude 20°						
Time	1-Axis	2-Axis	0	20	30	40	50	60	90
			DECEMBER 21 (W/m^2)						
7, 5	235	252	35	72	87	100	110	118	118
8, 4	738	794	269	393	439	473	492	497	428
9, 3	887	955	486	645	697	729	740	729	572
10, 2	951	1025	654	832	886	914	915	889	660
11, 1	980	1057	760	949	1003	1027	1021	985	711
12	988	1066	796	988	1042	1065	1057	1017	728
kWh/day	8.57	9.23	5.20	6.77	7.27	7.55	7.61	7.45	5.71
			NOVEMBER 21 (W/m^2)						
7, 5	299	315	54	94	110	124	134	140	135
8, 4	755	795	296	407	446	472	484	483	398
9, 3	900	948	515	657	701	724	726	707	531
10, 2	964	1017	685	845	889	907	899	864	616
11, 1	993	1049	792	962	1006	1020	1005	960	665
12	1002	1058	828	1002	1045	1059	1041	992	682
kWh/day	8.82	9.31	5.51	6.93	7.35	7.55	7.53	7.30	5.37
			OCTOBER 21 (W/m^2)						
7, 5	456	461	112	148	161	169	172	171	141
8, 4	808	818	368	443	462	468	461	441	311
9, 3	937	949	593	690	709	708	686	645	419
10, 2	997	1012	766	877	895	888	854	796	494
11, 1	1026	1042	875	995	1012	1000	958	889	539
12	1035	1051	912	1035	1051	1038	994	921	554
kWh/day	9.48	9.62	6.34	7.34	7.53	7.50	7.26	6.81	4.36
			SEPTEMBER 21 (W/m^2)						
6, 6	0	0	0	0	0	0	0	0	0
7, 5	569	569	187	194	190	181	168	150	78
8, 4	833	833	445	467	458	437	404	360	176
9, 3	945	945	668	703	691	659	609	542	259
10, 2	1002	1002	840	886	870	830	767	681	322
11, 1	1030	1030	947	1000	983	938	866	769	362
12	1039	1039	984	1039	1022	975	900	799	376
kWh/day	9.80	9.80	7.16	7.54	7.41	7.07	6.53	5.80	2.77
			AUGUST 21 (W/m^2)						
6, 6	65	67	12	8	7	7	6	6	4
7, 5	607	627	251	217	192	162	128	91	36
8, 4	809	833	498	465	429	381	323	257	48
9, 3	905	931	711	684	642	581	505	415	91
10, 2	959	984	874	854	807	738	648	540	142
11, 1	988	1011	976	961	912	837	739	619	175
12	998	1020	1011	998	948	871	770	646	187
kWh/day	9.66	9.93	7.65	7.38	6.93	6.28	5.47	4.50	1.18
			JULY 21 (W/m^2)						
6, 6	188	204	48	25	24	23	21	19	13
7, 5	600	652	288	223	181	136	87	60	40
8, 4	765	829	521	449	395	330	256	176	50
9, 3	851	918	720	652	590	511	419	316	55
10, 2	901	968	874	810	743	655	549	428	58
11, 1	930	994	970	909	840	746	632	500	59
12	943	1003	1003	943	873	778	660	525	59
kWh/day	9.41	10.13	7.84	7.08	6.42	5.58	4.59	3.52	0.61

Clear-Sky Insolation: Beam + Diffuse — Latitude 25°N

Solar	Tracking		Tilt Angles Latitude 25°						
Time	1-Axis	2-Axis	0	20	30	40	50	60	90
			JANUARY 21 (W/m^2)						
7, 5	185	195	25	50	61	71	78	83	84
8, 4	735	774	251	367	411	442	461	466	402
9, 3	894	943	464	621	674	706	718	710	562
10, 2	962	1015	629	811	867	898	902	880	664
11, 1	993	1048	733	928	987	1015	1014	983	724
12	1002	1058	769	969	1027	1055	1052	1018	744
kWh/day	8.54	9.01	4.97	6.52	7.03	7.32	7.40	7.26	5.62
			FEBRUARY 21 (W/m^2)						
7, 5	454	460	95	133	147	157	162	163	140
8, 4	827	838	343	429	455	467	466	451	334
9, 3	955	968	564	679	708	717	704	671	463
10, 2	1013	1028	733	869	900	904	882	834	553
11, 1	1041	1056	840	989	1020	1022	993	935	609
12	1049	1065	877	1029	1061	1061	1031	969	627
kWh/day	9.63	9.76	6.03	7.23	7.52	7.59	7.45	7.08	4.82
			MARCH 21 (W/m^2)						
7, 5	613	613	183	196	195	189	178	162	91
8, 4	869	869	436	473	471	457	429	389	212
9, 3	973	973	656	714	712	690	648	588	316
10, 2	1025	1025	825	899	898	870	817	740	395
11, 1	1051	1051	932	1016	1014	983	923	836	445
12	1059	1059	968	1056	1054	1022	959	869	462
kWh/day	10.12	10.12	7.03	7.65	7.64	7.40	6.95	6.30	3.38
			APRIL 21 (W/m^2)						
6, 6	139	143	25	16	13	12	11	10	7
7, 5	664	683	267	237	212	181	147	108	33
8, 4	847	871	510	489	457	413	358	293	62
9, 3	935	960	719	712	678	624	553	467	143
10, 2	983	1007	880	885	849	789	707	604	210
11, 1	1009	1032	981	994	958	894	804	691	253
12	1017	1039	1015	1032	995	930	838	721	267
kWh/day	10.17	10.43	7.78	7.70	7.33	6.76	6.00	5.07	1.68
			MAY 21 (W/m^2)						
6, 6	282	306	76	42	32	31	28	26	17
7, 5	645	698	314	251	210	163	112	60	39
8, 4	794	857	541	481	431	369	297	217	47
9, 3	872	938	737	686	632	559	472	371	51
10, 2	919	984	887	846	789	710	610	494	81
11, 1	945	1008	981	947	889	805	699	573	119
12	956	1016	1013	981	923	838	730	600	132
kWh/day	9.87	10.60	8.08	7.49	6.89	6.11	5.17	4.08	0.84
			JUNE 21 (W/m^2)						
6, 6	307	342	95	52	39	37	35	32	21
7, 5	622	691	327	253	206	154	99	63	42
8, 4	761	842	546	473	416	349	272	188	51
9, 3	837	923	734	669	608	530	437	333	55
10, 2	883	968	879	822	758	673	569	449	57
11, 1	910	993	969	919	853	764	653	524	70
12	922	1001	1000	952	886	795	682	550	83
kWh/day	9.56	10.52	8.10	7.33	6.65	5.81	4.81	3.73	0.67

Solar	Tracking		Tilt Angles Latitude 25°						
Time	1-Axis	2-Axis	0	20	30	40	50	60	90
			DECEMBER 21 (W/m^2)						
7, 5	59	64	6	16	20	24	27	29	30
8, 4	673	723	213	334	381	417	441	452	407
9, 3	853	918	421	588	648	688	708	708	585
10, 2	927	998	582	776	841	880	894	881	693
11, 1	960	1034	684	893	959	997	1006	985	755
12	970	1045	719	932	999	1037	1044	1020	775
kWh/day	7.91	8.52	4.53	6.15	6.70	7.05	7.19	7.13	5.72
			NOVEMBER 21 (W/m^2)						
7, 5	149	157	21	42	51	58	64	68	68
8, 4	700	737	244	355	396	426	444	449	388
9, 3	869	916	455	607	658	690	702	693	550
10, 2	941	993	618	795	850	880	885	863	653
11, 1	974	1028	721	912	969	997	996	966	713
12	984	1039	756	952	1009	1037	1034	1001	733
kWh/day	8.25	8.70	4.87	6.37	6.86	7.14	7.21	7.08	5.48
			OCTOBER 21 (W/m^2)						
7, 5	384	388	85	118	130	138	143	144	124
8, 4	776	786	330	411	435	447	446	433	323
9, 3	916	928	546	658	686	694	683	652	453
10, 2	981	995	713	845	875	879	858	813	544
11, 1	1012	1027	818	962	993	994	967	912	599
12	1021	1037	854	1002	1033	1034	1004	946	617
kWh/day	9.16	9.28	5.84	6.99	7.27	7.34	7.20	6.85	4.70
			SEPTEMBER 21 (W/m^2)						
6, 6	0	0	0	0	0	0	0	0	0
7, 5	553	553	177	187	186	179	168	153	86
8, 4	820	820	425	458	456	440	413	375	204
9, 3	934	934	641	694	691	668	627	568	305
10, 2	992	992	806	875	872	844	792	717	383
11, 1	1021	1021	910	989	986	954	895	810	431
12	1030	1030	946	1028	1024	992	930	842	448
kWh/day	9.67	9.67	6.86	7.43	7.40	7.16	6.72	6.09	3.27
			AUGUST 21 (W/m^2)						
6, 6	112	116	23	15	13	12	11	10	7
7, 5	617	637	260	230	207	178	145	109	37
8, 4	809	834	498	477	446	404	350	288	65
9, 3	903	928	703	695	661	609	540	456	143
10, 2	955	980	860	864	829	770	690	590	208
11, 1	983	1007	959	971	935	872	785	675	249
12	993	1015	993	1007	971	907	817	704	264
kWh/day	9.75	10.02	7.60	7.51	7.15	6.60	5.86	4.96	1.68
			JULY 21 (W/m^2)						
6, 6	263	286	75	43	33	31	29	27	18
7, 5	621	674	310	248	208	162	113	62	41
8, 4	772	836	534	475	425	364	293	215	51
9, 3	853	921	727	676	622	551	464	365	55
10, 2	902	969	875	833	777	698	600	486	80
11, 1	930	994	968	933	875	792	687	563	117
12	941	1002	999	966	908	824	717	590	130
kWh/day	9.62	10.36	7.98	7.38	6.79	6.02	5.09	4.03	0.86

Clear-Sky Insolation: Beam + Diffuse — Latitude 30°N

Solar Time	Tracking 1-Axis	Tracking 2-Axis	Tilt 0	Tilt 20	Tilt 30	Tilt 40	Tilt 50	Tilt 60	Tilt 90
			JANUARY 21 (W/m^2)						
7, 5	31	33	3	7	9	11	13	14	14
8, 4	666	701	197	310	354	387	410	421	380
9, 3	858	904	399	564	623	664	685	686	572
10, 2	936	987	556	752	819	861	878	869	693
11, 1	971	1024	656	870	940	982	995	979	764
12	981	1035	690	909	981	1023	1035	1016	787
kWh/day	7.90	8.33	4.31	5.92	6.47	6.83	7.00	6.95	5.64
			FEBRUARY 21 (W/m^2)						
7, 5	369	373	68	101	113	123	129	132	118
8, 4	791	802	302	394	424	441	447	439	343
9, 3	932	944	512	641	679	697	695	672	493
10, 2	995	1009	674	830	872	889	880	845	600
11, 1	1025	1040	776	948	993	1009	995	952	664
12	1034	1049	811	988	1034	1049	1034	988	686
kWh/day	9.26	9.39	5.48	6.81	7.20	7.37	7.32	7.07	5.12
			MARCH 21 (W/m^2)						
7, 5	592	592	170	188	189	185	176	163	99
8, 4	853	853	412	460	464	456	435	401	239
9, 3	960	960	622	698	706	693	661	610	361
10, 2	1014	1014	784	881	892	876	836	771	454
11, 1	1040	1040	885	996	1009	992	946	872	513
12	1048	1048	920	1035	1048	1031	983	907	533
kWh/day	9.97	9.97	6.67	7.48	7.57	7.44	7.09	6.54	3.87
			APRIL 21 (W/m^2)						
6, 6	193	199	38	25	18	17	16	14	10
7, 5	670	690	273	248	226	197	164	127	33
8, 4	845	868	505	497	472	433	383	323	99
9, 3	930	955	705	717	692	648	585	506	197
10, 2	977	1002	858	888	864	816	744	651	276
11, 1	1002	1026	955	996	973	922	845	743	327
12	1010	1033	988	1033	1010	959	880	775	345
kWh/day	10.24	10.51	7.66	7.77	7.50	7.03	6.35	5.50	2.23
			MAY 21 (W/m^2)						
6, 6	348	377	103	62	40	38	35	32	21
7, 5	662	716	333	275	235	189	139	85	40
8, 4	798	861	550	503	458	400	332	256	48
9, 3	872	938	737	705	659	595	514	419	80
10, 2	917	982	880	863	817	748	658	549	149
11, 1	942	1006	970	962	917	845	750	633	194
12	951	1013	1001	997	951	879	781	662	210
kWh/day	10.03	10.77	8.15	7.74	7.20	6.51	5.64	4.61	1.27
			JUNE 21 (W/m^2)						
6, 6	372	414	127	76	48	45	42	38	25
7, 5	641	712	350	281	235	184	128	69	43
8, 4	767	849	560	499	447	384	310	229	51
9, 3	838	924	740	693	639	569	482	383	55
10, 2	882	968	877	843	790	715	619	506	102
11, 1	908	991	964	939	886	808	707	586	145
12	919	999	993	972	919	839	737	613	159
kWh/day	9.74	10.72	8.23	7.63	7.01	6.25	5.31	4.24	1.00

Solar Time	Tracking 1-Axis	Tracking 2-Axis	Tilt 0	Tilt 20	Tilt 30	Tilt 40	Tilt 50	Tilt 60	Tilt 90
			DECEMBER 21 (W/m^2)						
7, 5	0	0	0	0	0	0	0	0	0
8, 4	584	627	155	266	310	346	372	388	365
9, 3	808	869	351	523	587	635	663	672	584
10, 2	896	964	505	711	784	834	860	860	713
11, 1	934	1006	602	827	905	955	978	972	786
12	945	1018	636	866	945	996	1018	1009	810
kWh/day	7.39	7.95	3.86	5.52	6.12	6.54	6.76	6.79	5.71
			NOVEMBER 21 (W/m^2)						
7, 5	17	18	2	4	5	6	7	8	8
8, 4	628	661	190	296	338	369	391	401	362
9, 3	830	874	390	549	606	645	666	667	557
10, 2	913	963	546	736	801	842	858	850	679
11, 1	950	1003	644	852	921	962	975	960	750
12	961	1014	677	892	961	1003	1014	996	773
kWh/day	7.64	8.05	4.22	5.77	6.30	6.65	6.81	6.77	5.49
			OCTOBER 21 (W/m^2)						
7, 5	296	300	59	85	96	103	108	111	99
8, 4	736	745	288	374	402	419	424	417	328
9, 3	890	902	495	619	655	672	670	649	479
10, 2	960	974	654	804	845	861	853	820	585
11, 1	994	1008	754	920	963	979	966	925	650
12	1004	1018	789	959	1004	1019	1004	961	672
kWh/day	8.76	8.88	5.29	6.56	6.92	7.09	7.05	6.80	4.95
			SEPTEMBER 21 (W/m^2)						
6, 6	0	0	0	0	0	0	0	0	0
7, 5	532	532	165	179	179	175	166	153	93
8, 4	804	804	402	445	448	439	417	385	229
9, 3	920	920	608	677	683	670	638	588	347
10, 2	979	979	766	856	865	849	808	745	438
11, 1	1009	1009	865	969	979	961	916	844	496
12	1018	1018	899	1007	1018	999	952	877	515
kWh/day	9.50	9.50	6.51	7.26	7.33	7.19	6.84	6.31	3.72
			AUGUST 21 (W/m^2)						
6, 6	161	166	35	24	18	17	16	15	10
7, 5	624	644	266	242	220	193	161	126	37
8, 4	806	831	494	485	460	423	374	315	99
9, 3	897	923	690	700	675	632	570	494	194
10, 2	949	974	840	867	843	795	725	635	271
11, 1	976	1001	934	972	949	899	823	724	321
12	985	1009	966	1008	985	934	857	755	338
kWh/day	9.81	10.09	7.48	7.59	7.32	6.85	6.20	5.37	2.20
			JULY 21 (W/m^2)						
6, 6	327	356	102	63	41	39	36	33	22
7, 5	638	692	329	272	233	188	139	86	43
8, 4	777	841	543	496	451	395	328	252	51
9, 3	854	921	727	695	649	585	506	412	79
10, 2	900	967	869	850	804	736	647	540	146
11, 1	927	991	957	948	903	832	737	622	190
12	936	999	987	981	936	864	768	650	206
kWh/day	9.78	10.53	8.04	7.63	7.10	6.41	5.55	4.54	1.27

Clear-Sky Insolation: Beam + Diffuse — Latitude 35°N

Solar	Tracking		Tilt Angles Latitude 35°						
Time	1-Axis	2-Axis	0	20	30	40	50	60	90
			JANUARY 21 (W/m²)						
7, 5	0	0	0	0	0	0	0	0	0
8, 4	572	602	142	244	286	319	343	357	338
9, 3	810	853	331	498	562	609	638	649	568
10, 2	902	951	479	685	760	812	841	844	709
11, 1	942	994	573	801	882	936	963	961	791
12	954	1006	605	840	923	978	1004	1001	818
kWh/day	7.41	7.81	3.65	5.30	5.90	6.33	6.57	6.62	5.63
			FEBRUARY 21 (W/m²)						
7, 5	265	268	42	67	77	85	91	94	88
8, 4	746	755	258	353	386	408	419	417	343
9, 3	903	915	456	596	642	669	676	664	515
10, 2	973	986	609	781	835	863	867	845	637
11, 1	1005	1019	706	897	955	985	985	957	711
12	1015	1029	739	937	996	1026	1026	995	736
kWh/day	8.80	8.92	4.88	6.33	6.79	7.05	7.10	6.95	5.32
			MARCH 21 (W/m²)						
7, 5	566	566	156	176	180	178	172	161	105
8, 4	834	834	384	442	452	450	435	408	263
9, 3	944	944	583	674	692	689	667	625	401
10, 2	1000	999	735	854	877	874	846	793	508
11, 1	1027	1027	831	967	993	990	959	899	575
12	1035	1035	864	1005	1033	1030	997	935	598
kWh/day	9.78	9.77	6.24	7.23	7.42	7.39	7.15	6.71	4.30
			APRIL 21 (W/m²)						
6, 6	243	250	52	36	27	21	20	18	12
7, 5	674	694	276	257	238	212	180	144	33
8, 4	840	863	496	500	482	449	405	349	134
9, 3	923	948	685	716	701	666	612	541	248
10, 2	969	994	830	883	872	835	775	692	340
11, 1	994	1018	921	989	980	942	878	788	399
12	1002	1025	952	1025	1017	979	913	821	419
kWh/day	10.29	10.56	7.47	7.79	7.61	7.23	6.65	5.89	2.75
			MAY 21 (W/m²)						
6, 6	404	437	130	84	58	44	41	37	25
7, 5	674	729	348	297	259	214	165	111	41
8, 4	800	863	554	520	481	429	365	292	48
9, 3	870	936	730	718	681	625	552	463	135
10, 2	913	979	866	872	838	780	700	600	215
11, 1	937	1001	951	970	937	878	794	687	268
12	945	1008	980	1003	971	912	826	717	286
kWh/day	10.14	10.90	8.14	7.92	7.48	6.85	6.06	5.10	1.75
			JUNE 21 (W/m²)						
6, 6	426	474	158	102	70	51	48	44	29
7, 5	656	729	370	307	263	212	157	98	44
8, 4	771	853	568	520	475	416	346	268	51
9, 3	838	924	738	710	666	603	523	430	93
10, 2	880	966	869	858	816	750	664	559	168
11, 1	904	988	951	951	911	844	754	642	218
12	913	995	979	983	943	876	785	671	235
kWh/day	9.86	10.86	8.29	7.88	7.34	6.63	5.77	4.75	1.44

Solar	Tracking		Tilt Angles Latitude 35°						
Time	1-Axis	2-Axis	0	20	30	40	50	60	90
			DECEMBER 21 (W/m²)						
7, 5	0	0	0	0	0	0	0	0	0
8, 4	456	490	97	188	226	257	282	298	293
9, 3	748	804	279	449	515	567	602	619	563
10, 2	855	920	424	636	716	775	811	823	716
11, 1	901	969	516	751	838	900	935	943	802
12	914	984	547	790	879	942	976	983	830
kWh/day	6.83	7.35	3.18	4.84	5.47	5.94	6.24	6.35	5.58
			NOVEMBER 21 (W/m²)						
7, 5	0	0	0	0	0	0	0	0	0
8, 4	529	557	135	229	268	298	321	334	315
9, 3	779	820	322	483	543	588	616	627	549
10, 2	877	924	469	668	741	791	819	822	691
11, 1	920	970	561	783	861	914	940	939	773
12	932	983	593	822	902	956	981	978	800
kWh/day	7.14	7.52	3.57	5.15	5.73	6.14	6.37	6.42	5.46
			OCTOBER 21 (W/m²)						
7, 5	194	197	33	52	59	65	69	72	67
8, 4	686	694	244	332	362	383	393	391	323
9, 3	858	869	439	572	615	641	648	636	496
10, 2	935	948	589	754	805	833	836	816	618
11, 1	971	985	684	868	924	952	953	926	691
12	982	996	717	907	964	993	993	964	716
kWh/day	8.27	8.38	4.70	6.06	6.50	6.74	6.79	6.65	5.11
			SEPTEMBER 21 (W/m²)						
6, 6	0	0	0	0	0	0	0	0	0
7, 5	506	506	151	168	170	168	161	150	97
8, 4	783	783	375	427	435	432	416	390	250
9, 3	902	902	569	654	669	665	642	601	384
10, 2	963	963	719	829	849	845	817	765	488
11, 1	994	994	813	939	963	959	927	868	554
12	1003	1003	845	977	1002	997	964	903	576
kWh/day	9.30	9.30	6.10	7.01	7.17	7.13	6.89	6.45	4.12
			AUGUST 21 (W/m²)						
6, 6	207	214	48	35	27	22	21	19	13
7, 5	628	648	269	251	232	207	177	142	37
8, 4	801	826	485	488	469	438	394	340	133
9, 3	890	916	670	698	683	648	595	526	243
10, 2	940	966	812	861	849	813	754	673	331
11, 1	967	992	902	965	955	918	855	767	389
12	976	1000	932	1000	991	954	889	799	408
kWh/day	9.84	10.12	7.31	7.59	7.42	7.04	6.48	5.73	2.70
			JULY 21 (W/m²)						
6, 6	382	416	129	85	60	46	42	39	26
7, 5	651	706	344	294	257	213	164	112	43
8, 4	779	843	547	513	474	423	360	288	51
9, 3	851	919	722	708	671	615	543	455	132
10, 2	896	963	855	859	825	767	688	589	211
11, 1	921	986	939	955	923	864	780	675	262
12	930	994	967	988	956	897	812	704	280
kWh/day	9.89	10.66	8.04	7.82	7.37	6.75	5.97	5.02	1.73

Clear-Sky Insolation: Beam + Diffuse — Latitude 40°N

Solar Time	Tracking 1-Axis	Tracking 2-Axis	Tilt 0	20	30	40	50	60	90
			JANUARY 21 (W/m²)						
7, 5	0	0	0	0	0	0	0	0	0
8, 4	439	462	87	169	204	232	254	269	266
9, 3	744	784	260	424	489	540	575	593	544
10, 2	857	903	397	609	689	749	788	803	708
11, 1	905	954	485	722	811	876	915	927	801
12	919	968	515	761	852	919	958	968	832
kWh/day	6.81	7.17	2.97	4.61	5.24	5.71	6.02	6.15	5.47
			FEBRUARY 21 (W/m²)						
7, 5	146	147	19	34	39	44	48	50	49
8, 4	687	696	211	306	341	367	382	386	332
9, 3	866	877	396	544	596	630	646	644	526
10, 2	944	957	539	725	787	827	842	833	663
11, 1	980	993	629	838	907	949	963	950	747
12	990	1004	660	876	947	990	1005	989	775
kWh/day	8.24	8.35	4.25	5.77	6.29	6.62	6.77	6.71	5.41
			MARCH 21 (W/m²)						
7, 5	534	534	140	163	168	168	164	156	108
8, 4	809	809	352	419	435	438	429	409	282
9, 3	924	924	538	644	671	677	665	633	435
10, 2	982	982	681	819	853	862	847	807	554
11, 1	1010	1010	771	928	968	979	962	916	629
12	1019	1019	801	966	1007	1019	1001	953	655
kWh/day	9.54	9.54	5.76	6.91	7.19	7.27	7.13	6.79	4.67
			APRIL 21 (W/m²)						
6, 6	288	297	65	47	37	25	24	22	14
7, 5	674	694	277	264	247	224	194	160	39
8, 4	832	856	482	499	487	461	422	372	169
9, 3	913	938	659	708	702	677	632	570	296
10, 2	959	984	794	870	871	846	798	726	399
11, 1	983	1007	880	973	977	954	903	826	465
12	991	1015	909	1008	1014	991	939	860	488
kWh/day	10.29	10.56	7.22	7.73	7.66	7.36	6.88	6.21	3.25
			MAY 21 (W/m²)						
6, 6	450	488	155	107	78	48	45	41	27
7, 5	684	740	360	316	281	238	190	137	41
8, 4	799	862	552	532	500	454	395	326	78
9, 3	866	932	718	724	697	650	585	503	188
10, 2	906	972	844	874	852	805	735	645	279
11, 1	929	994	924	969	950	903	831	735	338
12	937	1001	951	1001	983	937	864	766	359
kWh/day	10.21	10.98	8.06	8.05	7.70	7.13	6.43	5.54	2.26
			JUNE 21 (W/m²)						
6, 6	471	524	188	128	93	57	53	48	32
7, 5	668	742	386	330	289	240	185	126	45
8, 4	772	855	572	538	498	445	380	305	51
9, 3	835	921	731	722	686	632	560	473	147
10, 2	875	961	853	865	834	780	703	607	233
11, 1	898	982	929	956	928	874	795	693	288
12	906	989	955	987	960	906	826	723	308
kWh/day	9.94	10.96	8.27	8.06	7.62	6.96	6.18	5.23	1.90

Solar Time	Tracking 1-Axis	Tracking 2-Axis	Tilt 0	20	30	40	50	60	90
			DECEMBER 21 (W/m²)						
7, 5	0	0	0	0	0	0	0	0	0
8, 4	271	292	43	99	123	144	161	172	178
9, 3	663	712	205	364	429	481	519	542	516
10, 2	799	860	339	551	634	699	744	766	699
11, 1	855	920	425	665	758	828	874	895	799
12	871	937	454	703	799	871	918	937	831
kWh/day	6.05	6.51	2.48	4.06	4.69	5.18	5.51	5.69	5.21
			NOVEMBER 21 (W/m²)						
7, 5	0	0	0	0	0	0	0	0	0
8, 4	392	412	80	154	184	210	229	242	239
9, 3	709	746	251	407	469	517	550	567	520
10, 2	828	873	388	591	668	726	763	777	686
11, 1	880	927	474	704	789	852	890	901	780
12	894	942	504	742	830	894	932	943	811
kWh/day	6.51	6.86	2.89	4.45	5.05	5.50	5.80	5.92	5.26
			OCTOBER 21 (W/m²)						
7, 5	87	88	13	21	25	28	30	31	30
8, 4	622	629	198	283	315	338	352	355	307
9, 3	816	827	379	518	567	599	615	612	502
10, 2	903	915	520	696	756	793	808	799	639
11, 1	943	956	608	807	873	914	928	915	721
12	955	968	638	845	913	955	968	954	749
kWh/day	7.70	7.80	4.07	5.50	5.98	6.30	6.43	6.38	5.15
			SEPTEMBER 21 (W/m²)						
6, 6	0	0	0	0	0	0	0	0	0
7, 5	475	475	135	154	158	158	153	145	99
8, 4	757	757	344	404	417	419	410	389	266
9, 3	881	881	526	624	647	652	639	607	416
10, 2	944	944	666	794	825	832	816	776	531
11, 1	975	975	754	901	937	946	928	882	604
12	985	985	784	938	975	985	966	919	629
kWh/day	9.05	9.05	5.63	6.69	6.95	7.00	6.86	6.52	4.46
			AUGUST 21 (W/m²)						
6, 6	249	257	62	46	37	27	25	23	15
7, 5	629	649	270	257	241	218	190	157	42
8, 4	793	818	472	486	474	448	411	362	165
9, 3	880	905	645	690	684	658	614	553	288
10, 2	929	955	778	848	848	823	775	705	388
11, 1	955	980	861	948	952	928	877	802	452
12	964	989	890	983	987	964	912	835	474
kWh/day	9.83	10.12	7.06	7.54	7.46	7.17	6.70	6.04	3.17
			JULY 21 (W/m²)						
6, 6	428	466	154	107	80	51	47	43	29
7, 5	661	717	357	313	278	237	189	137	44
8, 4	778	843	547	526	493	447	389	321	77
9, 3	847	914	710	714	687	640	575	495	185
10, 2	889	957	834	861	838	792	722	633	273
11, 1	913	979	913	954	935	888	816	722	331
12	921	986	940	986	968	921	849	752	351
kWh/day	9.95	10.74	7.97	7.94	7.59	7.03	6.33	5.45	2.23

Clear-Sky Insolation: Beam + Diffuse									
Solar	Tracking		Tilt Angles Latitude 45°						
Time	1-Axis	2-Axis	0	20	30	40	50	60	90
			JANUARY 21 (W/m²)						
7, 5	0	0	0	0	0	0	0	0	0
8, 4	248	261	37	85	106	124	138	148	153
9, 3	651	686	187	338	401	451	489	512	491
10, 2	795	837	313	521	604	670	715	740	683
11, 1	854	900	393	633	727	799	848	872	790
12	871	918	420	670	768	843	893	916	825
kWh/day	5.97	6.29	2.28	3.82	4.44	4.93	5.27	5.46	5.06
			FEBRUARY 21 (W/m²)						
7, 5	36	36	4	7	9	10	11	12	12
8, 4	610	618	163	253	288	315	333	341	308
9, 3	818	829	332	484	540	581	605	611	524
10, 2	907	919	464	659	729	778	804	807	676
11, 1	948	960	547	768	847	901	928	928	768
12	959	973	575	805	887	942	970	969	800
kWh/day	7.60	7.70	3.59	5.15	5.71	6.11	6.33	6.37	5.38
			MARCH 21 (W/m²)						
7, 5	496	496	123	147	153	156	154	148	108
8, 4	779	779	318	391	411	421	418	403	295
9, 3	899	899	489	608	642	657	654	632	462
10, 2	960	959	620	775	820	841	838	810	593
11, 1	989	989	703	881	932	957	953	922	675
12	998	998	731	917	971	996	993	960	703
kWh/day	9.24	9.24	5.24	6.52	6.89	7.06	7.03	6.79	4.97
			APRIL 21 (W/m²)						
6, 6	327	337	78	59	47	34	27	24	16
7, 5	672	692	274	268	254	234	207	175	57
8, 4	822	845	464	493	487	468	436	391	201
9, 3	901	925	627	693	697	682	646	593	341
10, 2	946	971	752	849	862	849	812	752	454
11, 1	969	994	831	948	966	956	918	854	526
12	977	1001	858	981	1001	992	954	889	551
kWh/day	10.25	10.53	6.91	7.60	7.63	7.44	7.05	6.47	3.74
			MAY 21 (W/m²)						
6, 6	489	529	179	129	99	67	49	45	30
7, 5	690	747	369	332	300	261	214	163	42
8, 4	796	859	546	540	514	474	421	357	117
9, 3	859	925	699	725	707	669	612	539	240
10, 2	898	964	816	868	857	822	764	683	340
11, 1	919	985	889	959	953	920	860	776	405
12	927	992	914	990	986	953	893	808	428
kWh/day	10.23	11.01	7.91	8.10	7.85	7.38	6.73	5.93	2.78
			JUNE 21 (W/m²)						
6, 6	508	565	216	155	118	79	57	52	35
7, 5	677	751	399	351	313	266	213	155	46
8, 4	772	854	570	550	517	470	410	339	84
9, 3	830	916	717	727	701	656	592	512	200
10, 2	868	954	830	864	845	802	736	649	295
11, 1	889	975	900	952	937	895	828	737	356
12	896	981	924	981	968	927	859	768	377
kWh/day	9.98	11.01	8.19	8.18	7.83	7.26	6.53	5.66	2.41

Latitude 45°N									
Solar	Tracking		Tilt Angles Latitude 45°						
Time	1-Axis	2-Axis	0	20	30	40	50	60	90
			DECEMBER 21 (W/m²)						
7, 5	0	0	0	0	0	0	0	0	0
8, 4	45	48	5	14	18	22	25	27	30
9, 3	536	576	131	265	322	369	406	431	429
10, 2	720	774	252	453	536	603	652	682	651
11, 1	792	852	330	566	661	737	791	822	768
12	812	874	357	604	703	782	837	868	806
kWh/day	5.00	5.37	1.79	3.20	3.78	4.24	4.58	4.79	4.56
			NOVEMBER 21 (W/m²)						
7, 5	0	0	0	0	0	0	0	0	0
8, 4	201	212	31	70	87	101	113	121	125
9, 3	610	642	179	320	378	425	460	482	462
10, 2	762	802	303	502	581	643	687	710	656
11, 1	825	869	383	613	703	773	820	843	764
12	843	888	410	650	744	816	864	887	799
kWh/day	5.64	5.94	2.20	3.66	4.24	4.70	5.02	5.20	4.81
			OCTOBER 21 (W/m²)						
7, 5	11	11	1	2	3	3	3	4	4
8, 4	539	545	150	229	260	283	299	306	276
9, 3	763	773	316	457	509	547	569	574	494
10, 2	861	872	445	629	695	742	766	769	646
11, 1	906	918	527	737	811	862	888	888	737
12	920	932	555	773	850	903	929	929	768
kWh/day	7.08	7.17	3.43	4.88	5.41	5.78	5.98	6.01	5.08
			SEPTEMBER 21 (W/m²)						
6, 6	0	0	0	0	0	0	0	0	0
7, 5	437	437	118	139	144	145	143	136	98
8, 4	726	726	310	376	394	401	397	382	277
9, 3	854	854	478	588	618	631	627	604	439
10, 2	919	919	607	751	792	810	805	777	566
11, 1	952	952	688	854	901	923	918	886	646
12	962	962	716	889	939	961	956	923	673
kWh/day	8.74	8.74	5.12	6.30	6.64	6.78	6.73	6.49	4.73
			AUGUST 21 (W/m²)						
6, 6	287	296	74	57	47	35	28	26	17
7, 5	627	647	268	261	247	227	202	171	58
8, 4	782	807	454	480	474	454	423	379	195
9, 3	866	892	614	676	678	662	627	574	330
10, 2	915	941	737	828	838	825	788	729	439
11, 1	941	966	814	924	940	929	891	828	510
12	949	974	840	957	974	964	927	862	534
kWh/day	9.78	10.07	6.76	7.41	7.42	7.23	6.85	6.28	3.63
			JULY 21 (W/m²)						
6, 6	467	507	179	130	101	70	52	47	31
7, 5	668	725	366	329	298	258	213	162	44
8, 4	776	840	541	533	507	468	415	351	115
9, 3	840	907	691	715	696	658	602	529	235
10, 2	880	948	807	856	844	809	750	671	333
11, 1	903	970	879	945	938	904	845	761	396
12	910	977	904	976	970	937	877	793	418
kWh/day	9.98	10.77	7.83	7.99	7.74	7.27	6.63	5.84	2.73

Clear-Sky Insolation: Beam + Diffuse — Latitude 50°N

Solar Time	Tracking 1-Axis	Tracking 2-Axis	Tilt 0	Tilt 20	Tilt 30	Tilt 40	Tilt 50	Tilt 60	Tilt 90
			JANUARY 21 (W/m^2)						
7, 5	0	0	0	0	0	0	0	0	0
8, 4	30	32	3	9	12	14	16	18	19
9, 3	512	539	114	239	291	336	371	394	396
10, 2	705	743	226	420	501	567	616	647	626
11, 1	782	824	298	530	625	701	757	790	750
12	803	846	323	567	666	746	803	837	790
kWh/day	4.86	5.12	1.60	2.96	3.52	3.98	4.32	4.54	4.37
			FEBRUARY 21 (W/m^2)						
7, 5	0	0	0	0	0	0	0	0	0
8, 4	507	514	115	193	225	251	269	280	264
9, 3	754	764	266	416	474	518	548	561	505
10, 2	858	870	385	584	659	716	751	765	672
11, 1	905	917	460	689	774	838	876	889	773
12	919	931	486	724	814	879	919	931	807
kWh/day	6.97	7.06	2.93	4.49	5.08	5.52	5.81	5.92	5.24
			MARCH 21 (W/m^2)						
7, 5	450	450	104	129	136	140	140	136	105
8, 4	742	742	280	358	383	396	399	390	301
9, 3	868	868	435	564	605	629	634	622	481
10, 2	932	932	554	723	778	809	818	802	621
11, 1	963	963	630	824	887	923	933	916	709
12	973	973	655	858	924	962	973	955	740
kWh/day	8.88	8.88	4.66	6.05	6.50	6.76	6.82	6.69	5.17
			APRIL 21 (W/m^2)						
6, 6	361	372	90	70	58	44	30	27	18
7, 5	666	686	269	270	259	241	217	187	75
8, 4	808	831	442	482	483	470	444	405	230
9, 3	885	910	590	672	685	679	653	609	381
10, 2	929	954	704	820	844	843	819	770	502
11, 1	953	978	775	914	945	948	924	873	580
12	960	985	800	946	979	984	960	909	607
kWh/day	10.17	10.45	6.54	7.40	7.53	7.43	7.13	6.65	4.18
			MAY 21 (W/m^2)						
6, 6	521	564	202	152	121	87	52	48	32
7, 5	694	751	374	346	317	281	237	187	42
8, 4	791	854	535	543	524	490	444	385	155
9, 3	850	916	674	718	710	681	634	569	289
10, 2	887	953	780	854	855	832	785	715	397
11, 1	907	973	847	941	948	928	881	809	467
12	914	980	870	971	980	961	914	841	492
kWh/day	10.21	11.00	7.69	8.08	7.93	7.56	6.98	6.27	3.26
			JUNE 21 (W/m^2)						
6, 6	538	599	241	181	143	103	60	55	37
7, 5	683	758	408	369	334	290	239	183	46
8, 4	769	851	563	558	532	491	437	371	125
9, 3	823	909	697	725	709	673	618	546	251
10, 2	858	944	799	856	848	817	762	685	353
11, 1	878	964	864	939	937	908	853	774	420
12	885	970	886	967	967	939	885	805	443
kWh/day	9.98	11.02	8.03	8.22	7.97	7.50	6.82	6.03	2.90

Solar Time	Tracking 1-Axis	Tracking 2-Axis	Tilt 0	Tilt 20	Tilt 30	Tilt 40	Tilt 50	Tilt 60	Tilt 90
			DECEMBER 21 (W/m^2)						
7, 5	0	0	0	0	0	0	0	0	0
8, 4	0	0	0	0	0	0	0	0	0
9, 3	337	363	59	147	185	218	244	264	275
10, 2	599	644	164	339	413	475	524	557	556
11, 1	699	751	233	452	543	619	676	714	698
12	726	780	257	490	586	666	726	764	743
kWh/day	4.00	4.29	1.17	2.37	2.87	3.29	3.61	3.83	3.80
			NOVEMBER 21 (W/m^2)						
7, 5	0	0	0	0	0	0	0	0	0
8, 4	14	14	1	4	5	6	7	8	9
9, 3	464	489	107	219	266	307	338	359	360
10, 2	666	701	217	400	475	537	584	613	592
11, 1	748	787	288	509	599	671	724	756	717
12	771	811	313	546	640	716	771	803	758
kWh/day	4.56	4.79	1.54	2.81	3.33	3.76	4.08	4.27	4.11
			OCTOBER 21 (W/m^2)						
7, 5	0	0	0	0	0	0	0	0	0
8, 4	429	434	101	168	195	216	232	241	227
9, 3	693	701	250	387	440	480	507	520	468
10, 2	807	817	367	552	623	675	708	721	635
11, 1	859	870	441	656	736	796	832	844	735
12	874	885	466	691	775	837	874	886	769
kWh/day	6.45	6.53	2.78	4.22	4.76	5.17	5.43	5.54	4.90
			SEPTEMBER 21 (W/m^2)						
6, 6	0	0	0	0	0	0	0	0	0
7, 5	393	393	100	121	127	129	128	124	94
8, 4	687	687	273	343	365	376	377	368	282
9, 3	820	820	426	544	581	602	605	592	455
10, 2	889	889	543	700	749	777	783	767	590
11, 1	924	923	617	798	855	888	896	877	676
12	934	934	642	831	892	926	934	915	706
kWh/day	8.36	8.36	4.56	5.84	6.25	6.47	6.51	6.37	4.90
			AUGUST 21 (W/m^2)						
6, 6	319	330	86	69	57	45	32	29	19
7, 5	622	641	263	262	252	234	211	182	74
8, 4	768	792	432	469	469	456	430	392	222
9, 3	850	875	578	655	666	658	633	589	368
10, 2	897	923	689	799	820	818	793	745	485
11, 1	923	948	759	890	918	920	896	845	560
12	931	956	783	922	952	955	931	880	586
kWh/day	9.69	9.98	6.40	7.21	7.32	7.22	6.92	6.45	4.04
			JULY 21 (W/m^2)						
6, 6	499	542	201	152	122	90	55	50	34
7, 5	672	729	371	343	315	278	235	186	45
8, 4	770	834	531	536	517	483	437	378	152
9, 3	831	898	667	708	699	670	623	558	282
10, 2	869	936	772	842	842	818	771	702	388
11, 1	890	958	838	927	933	912	865	793	457
12	897	964	860	956	964	944	897	825	480
kWh/day	9.96	10.76	7.62	7.98	7.82	7.45	6.87	6.16	3.19

Clear-Sky Insolation: Beam + Diffuse — Latitude 55°N

Solar Time	Tracking 1-Axis	Tracking 2-Axis	Tilt 0	20	30	40	50	60	90
			JANUARY 21 (W/m²)						
7, 5	0	0	0	0	0	0	0	0	0
8, 4	0	0	0	0	0	0	0	0	0
9, 3	293	308	46	120	152	180	202	219	230
10, 2	566	596	139	301	371	430	476	508	514
11, 1	672	708	201	410	499	572	629	667	663
12	701	739	223	447	541	619	679	719	710
kWh/day	3.76	3.96	0.99	2.11	2.58	2.98	3.29	3.51	3.52
			FEBRUARY 21 (W/m²)						
7, 5	0	0	0	0	0	0	0	0	0
8, 4	366	371	67	127	151	172	188	198	195
9, 3	666	675	197	337	394	439	472	491	462
10, 2	792	802	302	498	575	636	679	701	646
11, 1	847	858	369	598	688	758	805	829	756
12	864	875	392	632	726	799	848	873	793
kWh/day	6.21	6.29	2.26	3.75	4.34	4.81	5.14	5.31	4.91
			MARCH 21 (W/m²)						
7, 5	395	395	85	108	116	121	122	121	97
8, 4	695	695	239	320	347	365	373	370	300
9, 3	829	829	377	513	560	590	604	601	489
10, 2	897	897	484	663	725	766	785	781	637
11, 1	930	930	551	758	830	878	900	896	731
12	940	940	574	790	866	916	939	935	764
kWh/day	8.43	8.43	4.05	5.51	6.02	6.36	6.51	6.47	5.27
			APRIL 21 (W/m²)						
6, 6	390	402	101	81	68	54	39	29	19
7, 5	658	677	262	268	260	246	225	198	91
8, 4	791	814	415	467	473	466	447	415	256
9, 3	866	890	547	644	666	669	652	618	415
10, 2	909	934	649	783	818	828	815	779	544
11, 1	932	957	713	871	914	930	919	882	626
12	940	965	735	901	947	965	955	918	655
kWh/day	10.03	10.31	6.11	7.13	7.35	7.35	7.15	6.76	4.56
			MAY 21 (W/m²)						
6, 6	547	592	222	173	142	108	71	50	33
7, 5	696	752	376	356	331	298	257	210	43
8, 4	783	845	520	541	529	502	461	408	191
9, 3	838	903	643	705	706	687	649	593	334
10, 2	873	939	738	833	845	833	798	739	449
11, 1	892	958	798	914	934	926	893	833	524
12	899	965	818	942	964	958	925	866	549
kWh/day	10.16	10.95	7.41	7.98	7.94	7.67	7.18	6.53	3.70
			JUNE 21 (W/m²)						
6, 6	563	626	264	206	168	127	83	57	38
7, 5	686	762	413	383	352	312	264	209	46
8, 4	763	845	552	560	541	507	459	399	164
9, 3	814	899	671	718	711	684	638	574	298
10, 2	846	932	762	840	844	823	780	714	408
11, 1	865	951	820	918	928	912	870	803	478
12	871	957	839	945	957	943	901	834	502
kWh/day	9.95	10.99	7.80	8.19	8.05	7.67	7.09	6.35	3.37

Solar Time	Tracking 1-Axis	Tracking 2-Axis	Tilt 0	20	30	40	50	60	90
			DECEMBER 21 (W/m²)						
7, 5	0	0	0	0	0	0	0	0	0
8, 4	0	0	0	0	0	0	0	0	0
9, 3	56	60	6	21	28	33	38	42	46
10, 2	400	430	78	199	252	298	335	362	378
11, 1	548	589	136	315	392	458	510	547	560
12	588	632	157	353	438	509	565	605	616
kWh/day	2.60	2.79	0.60	1.42	1.78	2.09	2.33	2.51	2.58
			NOVEMBER 21 (W/m²)						
7, 5	0	0	0	0	0	0	0	0	0
8, 4	0	0	0	0	0	0	0	0	0
9, 3	241	254	40	100	127	149	168	182	190
10, 2	520	547	130	279	342	396	438	468	473
11, 1	631	664	192	387	470	538	591	627	622
12	662	697	214	423	512	585	641	678	670
kWh/day	3.45	3.63	0.94	1.96	2.39	2.75	3.04	3.23	3.24
			OCTOBER 21 (W/m²)						
7, 5	0	0	0	0	0	0	0	0	0
8, 4	283	286	55	101	120	136	148	156	153
9, 3	596	604	182	306	357	397	426	443	417
10, 2	733	742	285	465	536	592	630	651	600
11, 1	794	804	351	563	647	711	756	778	710
12	812	822	374	597	684	752	798	821	746
kWh/day	5.63	5.70	2.12	3.47	4.00	4.42	4.72	4.88	4.51
			SEPTEMBER 21 (W/m²)						
6, 6	0	0	0	0	0	0	0	0	0
7, 5	341	341	81	101	107	111	111	109	86
8, 4	639	639	234	306	330	345	350	346	278
9, 3	779	778	369	493	535	562	574	569	460
10, 2	851	851	474	640	696	733	749	744	603
11, 1	888	888	540	732	798	841	860	855	694
12	899	899	562	764	833	878	898	893	725
kWh/day	7.89	7.89	3.96	5.31	5.77	6.06	6.19	6.14	4.97
			AUGUST 21 (W/m²)						
6, 6	347	359	98	80	68	55	40	31	21
7, 5	613	632	256	261	253	238	218	192	90
8, 4	751	774	407	454	459	451	432	400	246
9, 3	829	855	536	627	646	648	631	596	399
10, 2	876	902	636	762	794	803	789	752	523
11, 1	901	927	699	848	887	901	890	852	603
12	909	935	720	877	919	935	924	887	631
kWh/day	9.54	9.83	5.98	6.94	7.13	7.13	6.92	6.53	4.39
			JULY 21 (W/m²)						
6, 6	525	571	221	174	143	110	74	53	35
7, 5	673	731	373	353	329	296	255	208	45
8, 4	762	826	515	534	522	495	454	402	187
9, 3	819	885	637	695	695	676	638	582	326
10, 2	855	922	731	821	832	819	783	725	439
11, 1	875	942	790	901	919	910	876	817	511
12	881	949	810	928	949	942	908	848	536
kWh/day	9.90	10.70	7.34	7.88	7.83	7.55	7.07	6.42	3.62

Clear-Sky Insolation: Beam + Diffuse — Latitude 60°N

Solar Time	Tracking 1-Axis	Tracking 2-Axis	Tilt 0	Tilt 20	Tilt 30	Tilt 40	Tilt 50	Tilt 60	Tilt 90
			JANUARY 21 (W/m²)						
7, 5	0	0	0	0	0	0	0	0	0
8, 4	0	0	0	0	0	0	0	0	0
9, 3	19	20	2	7	9	11	13	14	15
10, 2	332	349	56	154	198	236	267	290	306
11, 1	490	516	106	264	333	392	440	475	493
12	534	562	124	300	377	443	496	534	551
kWh/day	2.22	2.33	0.45	1.15	1.46	1.72	1.93	2.09	2.18
			FEBRUARY 21 (W/m²)						
7, 5	0	0	0	0	0	0	0	0	0
8, 4	178	181	25	55	68	79	88	94	97
9, 3	539	546	128	247	297	338	370	390	384
10, 2	696	705	218	399	474	535	580	609	586
11, 1	765	775	275	494	583	655	708	741	707
12	785	795	295	526	620	696	752	785	747
kWh/day	5.14	5.21	1.59	2.91	3.46	3.91	4.24	4.45	4.29
			MARCH 21 (W/m²)						
7, 5	330	330	65	86	94	99	101	101	85
8, 4	636	635	197	277	306	326	338	339	289
9, 3	778	778	316	454	505	541	562	566	484
10, 2	851	851	409	593	662	710	739	746	639
11, 1	888	887	467	682	761	818	851	860	737
12	899	898	487	712	795	855	890	899	771
kWh/day	7.86	7.86	3.40	4.90	5.45	5.84	6.07	6.12	5.24
			APRIL 21 (W/m²)						
6, 6	414	426	111	92	79	64	48	31	21
7, 5	646	665	251	263	259	248	230	206	106
8, 4	770	792	385	446	458	457	444	418	277
9, 3	842	866	500	610	639	651	644	618	442
10, 2	885	909	589	738	782	804	802	778	576
11, 1	907	932	644	820	873	901	904	880	662
12	915	940	663	848	904	935	938	915	692
kWh/day	9.84	10.12	5.62	6.78	7.08	7.18	7.08	6.78	4.86
			MAY 21 (W/m²)						
6, 6	569	616	239	193	163	128	91	52	35
7, 5	694	750	374	363	342	313	275	231	68
8, 4	773	834	499	533	528	508	474	428	225
9, 3	824	888	607	685	695	686	657	610	374
10, 2	856	922	690	803	826	826	802	755	495
11, 1	874	940	742	878	910	915	894	848	573
12	880	946	759	904	939	946	926	880	600
kWh/day	10.06	10.85	7.06	7.82	7.87	7.70	7.31	6.73	4.14
			JUNE 21 (W/m²)						
6, 6	584	649	284	229	192	151	106	59	40
7, 5	687	762	414	394	367	331	286	234	51
8, 4	755	837	535	557	545	518	477	423	200
9, 3	801	886	639	703	705	688	651	597	342
10, 2	831	917	719	816	831	822	790	735	457
11, 1	848	935	769	888	911	907	878	823	530
12	854	940	786	913	938	937	908	854	556
kWh/day	9.87	10.91	7.51	8.09	8.04	7.77	7.29	6.60	3.80

Solar Time	Tracking 1-Axis	Tracking 2-Axis	Tilt 0	Tilt 20	Tilt 30	Tilt 40	Tilt 50	Tilt 60	Tilt 90
			DECEMBER 21 (W/m²)						
7, 5	0	0	0	0	0	0	0	0	0
8, 4	0	0	0	0	0	0	0	0	0
9, 3	0	0	0	0	0	0	0	0	0
10, 2	81	87	9	35	46	56	65	71	78
11, 1	283	304	46	141	183	220	251	274	294
12	345	370	61	180	233	279	316	345	367
kWh/day	1.07	1.15	0.17	0.53	0.69	0.83	0.95	1.04	1.11
			NOVEMBER 21 (W/m²)						
7, 5	0	0	0	0	0	0	0	0	0
8, 4	0	0	0	0	0	0	0	0	0
9, 3	6	7	1	2	3	4	4	5	5
10, 2	279	294	49	131	168	199	225	244	258
11, 1	441	464	97	239	300	353	396	427	443
12	485	511	115	275	344	404	451	485	501
kWh/day	1.94	2.04	0.41	1.02	1.29	1.52	1.70	1.84	1.91
			OCTOBER 21 (W/m²)						
7, 5	0	0	0	0	0	0	0	0	0
8, 4	106	107	16	34	41	48	53	57	58
9, 3	459	465	114	214	256	291	318	335	330
10, 2	628	636	202	364	430	485	525	550	530
11, 1	703	712	258	457	538	604	652	681	650
12	725	734	278	489	574	644	695	725	690
kWh/day	4.52	4.57	1.46	2.62	3.11	3.50	3.79	3.97	3.83
			SEPTEMBER 21 (W/m²)						
6, 6	0	0	0	0	0	0	0	0	0
7, 5	281	281	62	80	86	89	91	90	74
8, 4	578	578	192	263	288	305	314	315	265
9, 3	725	725	310	435	481	513	530	533	452
10, 2	803	803	401	570	632	676	701	706	600
11, 1	842	841	458	656	729	781	810	816	695
12	853	853	477	686	762	816	847	853	728
kWh/day	7.31	7.31	3.32	4.69	5.19	5.54	5.74	5.77	4.90
			AUGUST 21 (W/m²)						
6, 6	371	383	108	90	78	64	49	33	22
7, 5	601	620	246	256	251	240	222	199	103
8, 4	729	751	377	433	443	442	428	403	265
9, 3	804	829	490	592	619	629	620	595	424
10, 2	850	875	577	717	758	777	774	749	553
11, 1	874	900	632	797	846	872	872	848	636
12	882	908	650	824	877	904	906	882	665
kWh/day	9.34	9.62	5.51	6.59	6.87	6.95	6.84	6.54	4.67
			JULY 21 (W/m²)						
6, 6	547	594	239	194	164	130	93	55	37
7, 5	672	729	371	360	340	310	273	229	68
8, 4	752	815	495	527	521	501	467	420	220
9, 3	804	870	602	676	685	674	645	598	365
10, 2	837	904	683	792	813	812	787	740	483
11, 1	856	923	735	866	895	899	878	831	559
12	862	930	752	891	923	929	909	862	585
kWh/day	9.80	10.60	7.00	7.72	7.76	7.58	7.19	6.61	4.05

APPENDIX D

MONTHLY CLEAR-SKY INSOLATION TABLES

Daily Clear-Sky Insolation (kWh/m^2)														Latitude 20°N							
Azimuth:		South						SE, SW						East, West						Tracking	
Tilt:	0	20	30	40	50	60	90	20	30	40	50	60	90	20	30	40	50	60	90	1-Axis	2-Axis
Jan	5.61	7.08	7.51	7.72	7.70	7.46	5.48	7.85	7.52	7.06	6.45	5.73	3.18	7.57	7.17	6.67	6.10	5.47	3.41	9.10	9.60
Feb	6.53	7.57	7.76	7.73	7.47	7.00	4.44	7.32	7.19	6.90	6.46	5.89	3.61	6.73	6.40	6.01	5.54	5.01	3.22	9.93	10.07
Mar	7.34	7.76	7.63	7.29	6.74	5.99	2.85	6.30	6.41	6.34	6.13	5.77	3.95	5.39	5.15	4.85	4.51	4.13	2.73	10.24	10.24
Apr	7.84	7.55	7.09	6.43	5.59	4.59	1.15	5.02	5.29	5.41	5.40	5.25	3.98	3.90	3.73	3.56	3.33	3.08	2.09	10.06	10.32
May	7.96	7.18	6.51	5.66	4.66	3.57	0.57	3.95	4.31	4.56	4.68	4.67	3.85	2.79	2.71	2.58	2.47	2.29	1.62	9.66	10.37
Jun	7.91	6.97	6.23	5.32	4.27	3.20	0.61	3.56	3.95	4.23	4.40	4.43	3.78	2.39	2.34	2.24	2.15	2.00	1.44	9.33	10.26
July	7.84	7.08	6.42	5.58	4.59	3.52	0.61	7.85	7.52	7.06	6.45	5.73	3.18	7.57	7.17	6.67	6.10	5.47	3.41	9.41	10.13
Aug	7.65	7.38	6.93	6.28	5.47	4.50	1.18	7.32	7.19	6.90	6.46	5.89	3.61	6.73	6.40	6.01	5.54	5.01	3.22	9.66	9.93
Sept	7.16	7.54	7.41	7.07	6.53	5.80	2.77	6.30	6.41	6.34	6.13	5.77	3.95	5.39	5.15	4.85	4.51	4.13	2.73	9.80	9.80
Oct	6.34	7.34	7.53	7.50	7.26	6.81	4.36	5.02	5.29	5.41	5.40	5.25	3.98	3.90	3.73	3.56	3.33	3.08	2.09	9.48	9.62
Nov	5.51	6.93	7.35	7.55	7.53	7.30	5.37	3.95	4.31	4.56	4.68	4.67	3.85	2.79	2.71	2.58	2.47	2.29	1.62	8.82	9.31
Dec	5.20	6.77	7.27	7.55	7.61	7.45	5.71	3.56	3.95	4.23	4.40	4.43	3.78	2.39	2.34	2.24	2.15	2.00	1.44	8.57	9.23
kWh/m^2-yr	2522	2650	2603	2483	2292	2041	1065	2067	2108	2098	2038	1930	1359	1750	1672	1576	1465	1336	882	3469	3615

Renewable and Efficient Electric Power Systems. By Gilbert M. Masters
ISBN 0-471-28060-7

Daily Clear-Sky Insolation (kWh/m²)														Latitude 25°N							
Azimuth:		South						SE, SW						East, West						Tracking	
Tilt:	0	20	30	40	50	60	90	20	30	40	50	60	90	20	30	40	50	60	90	1-Axis	2-Axis
Jan	4.97	6.52	7.03	7.32	7.40	7.26	5.62	6.00	6.26	6.34	6.23	5.98	4.31	4.74	4.53	4.26	3.93	3.59	2.32	8.54	9.01
Feb	6.03	7.23	7.52	7.59	7.45	7.08	4.82	6.79	6.91	6.83	6.56	6.17	4.12	5.76	5.50	5.15	4.76	4.32	2.78	9.63	9.76
Mar	7.03	7.65	7.64	7.40	6.95	6.30	3.38	7.35	7.26	6.96	6.51	5.92	3.57	6.70	6.34	5.91	5.43	4.87	3.04	10.12	10.12
Apr	7.78	7.70	7.33	6.76	6.00	5.07	1.68	7.62	7.27	6.75	6.13	5.36	2.86	7.39	6.99	6.50	5.92	5.26	3.21	10.17	10.43
May	8.08	7.49	6.89	6.11	5.17	4.08	0.84	7.58	7.07	6.42	5.69	4.86	2.36	7.68	7.27	6.73	6.10	5.43	3.27	9.87	10.60
Jun	8.10	7.33	6.65	5.81	4.81	3.73	0.67	7.47	6.91	6.21	5.45	4.61	2.14	7.70	7.27	6.72	6.08	5.41	3.22	9.56	10.52
July	7.98	7.38	6.79	6.02	5.09	4.03	0.86	7.47	6.97	6.32	5.60	4.78	2.32	7.58	7.16	6.63	6.01	5.35	3.21	9.62	10.36
Aug	7.60	7.51	7.15	6.60	5.86	4.96	1.68	7.43	7.09	6.58	5.97	5.22	2.79	7.22	6.82	6.34	5.76	5.12	3.11	9.75	10.02
Sept	6.86	7.43	7.40	7.16	6.72	6.09	3.27	7.15	7.05	6.75	6.30	5.72	3.44	6.54	6.19	5.76	5.28	4.73	2.94	9.67	9.67
Oct	5.84	6.99	7.27	7.34	7.20	6.85	4.70	6.56	6.68	6.60	6.34	5.95	3.98	5.58	5.32	4.97	4.59	4.16	2.65	9.16	9.28
Nov	4.87	6.37	6.86	7.14	7.21	7.08	5.48	5.86	6.11	6.19	6.08	5.83	4.20	4.65	4.43	4.17	3.83	3.50	2.25	8.25	8.70
Dec	4.53	6.15	6.70	7.05	7.19	7.13	5.72	5.60	5.90	6.03	5.98	5.77	4.26	4.31	4.11	3.87	3.55	3.25	2.09	7.91	8.52
kWh/m²-yr	2424	2608	2591	2502	2341	2116	1175	2521	2477	2370	2215	2011	1226	2308	2189	2039	1864	1673	1037	3413	3558

Daily Clear-Sky Insolation (kWh/m²)														Latitude 30°N							
Azimuth:		South						SE, SW						East, West						Tracking	
Tilt:	0	20	30	40	50	60	90	20	30	40	50	60	90	20	30	40	50	60	90	1-Axis	2-Axis
Jan	4.31	5.92	6.47	6.83	7.00	6.95	5.64	5.38	5.68	5.82	5.79	5.63	4.22	4.11	3.91	3.71	3.43	3.15	2.06	7.90	8.33
Feb	5.48	6.81	7.20	7.37	7.32	7.07	5.12	6.35	6.53	6.53	6.36	6.04	4.22	5.24	5.02	4.73	4.38	4.01	2.63	9.26	9.39
Mar	6.67	7.48	7.57	7.44	7.09	6.54	3.87	7.13	7.11	6.90	6.53	6.01	3.79	6.36	6.04	5.63	5.21	4.70	2.98	9.97	9.97
Apr	7.66	7.77	7.50	7.03	6.35	5.50	2.23	7.65	7.37	6.93	6.36	5.64	3.16	7.29	6.90	6.44	5.89	5.26	3.27	10.24	10.51
May	8.15	7.74	7.20	6.51	5.64	4.61	1.27	7.77	7.32	6.72	6.03	5.21	2.65	7.75	7.33	6.80	6.18	5.51	3.36	10.03	10.77
Jun	8.23	7.63	7.01	6.25	5.31	4.24	1.00	7.72	7.20	6.56	5.82	4.98	2.43	7.82	7.38	6.83	6.19	5.52	3.32	9.74	10.72
July	8.04	7.63	7.10	6.41	5.55	4.54	1.27	7.66	7.21	6.62	5.93	5.12	2.60	7.65	7.23	6.70	6.09	5.43	3.29	9.78	10.53
Aug	7.48	7.59	7.32	6.85	6.20	5.37	2.20	7.46	7.19	6.75	6.18	5.49	3.06	7.12	6.74	6.28	5.73	5.11	3.16	9.81	10.09
Sept	6.51	7.26	7.33	7.19	6.84	6.31	3.72	6.93	6.90	6.68	6.31	5.80	3.64	6.21	5.89	5.49	5.06	4.56	2.88	9.50	9.50
Oct	5.29	6.56	6.92	7.09	7.05	6.80	4.95	6.12	6.29	6.28	6.11	5.80	4.05	5.05	4.84	4.55	4.20	3.84	2.50	8.76	8.88
Nov	4.22	5.77	6.30	6.65	6.81	6.77	5.49	5.25	5.54	5.67	5.64	5.48	4.10	4.02	3.83	3.62	3.34	3.07	2.00	7.64	8.05
Dec	3.86	5.52	6.12	6.54	6.76	6.79	5.71	4.97	5.31	5.50	5.53	5.42	4.19	3.69	3.52	3.34	3.11	2.86	1.90	7.39	7.95
kWh/m²-yr	2310	2545	2555	2497	2368	2172	1289	2445	2423	2341	2207	2025	1280	2200	2088	1951	1790	1613	1015	3346	3489

Daily Clear-Sky Insolation (kWh/m²)														Latitude 35°N							
Azimuth:		South						SE, SW						East, West						Tracking	
Tilt:	0	20	30	40	50	60	90	20	30	40	50	60	90	20	30	40	50	60	90	1-Axis	2-Axis
Jan	3.65	5.30	5.90	6.33	6.57	6.62	5.63	4.76	5.11	5.31	5.37	5.29	4.15	3.50	3.36	3.20	3.00	2.77	1.89	7.41	7.81
Feb	4.88	6.33	6.79	7.05	7.10	6.95	5.32	5.84	6.08	6.15	6.07	5.83	4.26	4.67	4.49	4.26	3.95	3.66	2.45	8.80	8.92
Mar	6.24	7.23	7.42	7.39	7.15	6.71	4.30	6.84	6.90	6.77	6.48	6.04	3.96	5.96	5.68	5.32	4.94	4.50	2.92	9.78	9.77
Apr	7.47	7.79	7.61	7.23	6.65	5.89	2.75	7.61	7.41	7.04	6.53	5.89	3.46	7.12	6.76	6.32	5.81	5.22	3.31	10.29	10.56
May	8.14	7.92	7.48	6.85	6.06	5.10	1.75	7.90	7.50	6.98	6.31	5.53	2.93	7.74	7.32	6.81	6.20	5.55	3.42	10.14	10.90
Jun	8.29	7.88	7.34	6.63	5.77	4.75	1.44	7.90	7.43	6.85	6.14	5.32	2.71	7.87	7.43	6.88	6.25	5.58	3.39	9.86	10.86
July	8.04	7.82	7.37	6.75	5.97	5.02	1.73	7.79	7.40	6.87	6.21	5.44	2.87	7.64	7.23	6.72	6.12	5.47	3.36	9.89	10.66
Aug	7.31	7.59	7.42	7.04	6.48	5.73	2.70	7.42	7.22	6.86	6.35	5.71	3.35	6.96	6.60	6.17	5.66	5.08	3.20	9.84	10.12
Sept	6.10	7.01	7.17	7.13	6.89	6.45	4.12	6.64	6.69	6.54	6.25	5.82	3.80	5.82	5.54	5.19	4.81	4.36	2.81	9.30	9.30
Oct	4.70	6.06	6.50	6.74	6.79	6.65	5.11	5.60	5.83	5.88	5.80	5.57	4.06	4.49	4.30	4.08	3.77	3.48	2.31	8.27	8.38
Nov	3.57	5.15	5.73	6.14	6.37	6.42	5.46	4.63	4.96	5.16	5.21	5.13	4.03	3.42	3.28	3.12	2.93	2.70	1.83	7.14	7.52
Dec	3.18	4.84	5.47	5.94	6.24	6.35	5.58	4.30	4.68	4.93	5.04	5.01	4.07	3.06	2.96	2.81	2.66	2.46	1.71	6.83	7.35
kWh/m²-yr	2178	2462	2500	2470	2372	2207	1393	2350	2349	2292	2182	2024	1326	2078	1977	1853	1708	1547	992	3271	3412

Daily Clear-Sky Insolation (kWh/m²)													**Latitude 40°N**								
Azimuth:		South						SE, SW						East, West						Tracking	
Tilt:	0	20	30	40	50	60	90	20	30	40	50	60	90	20	30	40	50	60	90	1-Axis	2-Axis
Jan	2.97	4.61	5.24	5.71	6.02	6.15	5.47	4.08	4.47	4.72	4.86	4.85	4.00	2.87	2.79	2.67	2.55	2.37	1.69	6.81	7.17
Feb	4.25	5.77	6.29	6.62	6.77	6.71	5.41	5.26	5.56	5.69	5.69	5.53	4.21	4.08	3.91	3.74	3.51	3.26	2.23	8.24	8.35
Mar	5.76	6.91	7.19	7.27	7.13	6.79	4.67	6.49	6.62	6.57	6.36	6.01	4.14	5.51	5.28	4.98	4.64	4.27	2.84	9.54	9.54
Apr	7.22	7.73	7.66	7.36	6.88	6.21	3.25	7.51	7.38	7.10	6.65	6.07	3.74	6.89	6.55	6.16	5.70	5.16	3.34	10.29	10.56
May	8.06	8.05	7.70	7.13	6.43	5.54	2.26	7.96	7.62	7.17	6.55	5.82	3.24	7.66	7.25	6.76	6.18	5.55	3.47	10.21	10.98
Jun	8.27	8.06	7.62	6.96	6.18	5.23	1.90	8.02	7.61	7.09	6.41	5.64	2.99	7.85	7.41	6.87	6.26	5.61	3.45	9.94	10.96
July	7.97	7.94	7.59	7.03	6.33	5.45	2.23	7.85	7.52	7.06	6.45	5.73	3.18	7.57	7.17	6.67	6.10	5.47	3.41	9.95	10.74
Aug	7.06	7.54	7.46	7.17	6.70	6.04	3.17	7.32	7.19	6.90	6.46	5.89	3.61	6.73	6.40	6.01	5.54	5.01	3.22	9.83	10.12
Sept	5.63	6.69	6.95	7.00	6.86	6.52	4.46	6.30	6.41	6.34	6.13	5.77	3.95	5.39	5.15	4.85	4.51	4.13	2.73	9.05	9.05
Oct	4.07	5.50	5.98	6.30	6.43	6.38	5.15	5.02	5.29	5.41	5.40	5.25	3.98	3.90	3.73	3.56	3.33	3.08	2.09	7.70	7.80
Nov	2.89	4.45	5.05	5.50	5.80	5.92	5.26	3.95	4.31	4.56	4.68	4.67	3.85	2.79	2.71	2.58	2.47	2.29	1.62	6.51	6.86
Dec	2.48	4.06	4.69	5.18	5.51	5.69	5.21	3.56	3.95	4.23	4.40	4.43	3.78	2.39	2.34	2.24	2.15	2.00	1.44	6.05	6.51
kWh/m²-yr	2029	2352	2415	2410	2342	2208	1471	2231	2249	2216	2130	1997	1357	1938	1848	1738	1612	1467	960	3167	3305

Daily Clear-Sky Insolation (kWh/m²)													**Latitude 45°N**								
Azimuth:		South						SE, SW						East, West						Tracking	
Tilt:	0	20	30	40	50	60	90	20	30	40	50	60	90	20	30	40	50	60	90	1-Axis	2-Axis
Jan	2.28	3.82	4.44	4.93	5.27	5.46	5.06	3.33	3.73	4.02	4.19	4.25	3.67	2.21	2.17	2.11	2.03	1.90	1.41	5.97	6.29
Feb	3.59	5.15	5.71	6.11	6.33	6.37	5.38	4.64	4.97	5.16	5.23	5.15	4.08	3.46	3.34	3.20	3.04	2.83	1.99	7.60	7.70
Mar	5.24	6.52	6.89	7.06	7.03	6.79	4.97	6.07	6.27	6.29	6.17	5.90	4.26	5.02	4.83	4.60	4.30	4.00	2.74	9.24	9.24
Apr	6.91	7.60	7.63	7.44	7.05	6.47	3.74	7.33	7.28	7.08	6.71	6.21	3.98	6.60	6.30	5.95	5.54	5.05	3.35	10.25	10.53
May	7.91	8.10	7.85	7.38	6.73	5.93	2.78	7.94	7.68	7.29	6.73	6.07	3.53	7.52	7.12	6.65	6.11	5.51	3.51	10.23	11.01
Jun	8.19	8.18	7.83	7.26	6.53	5.66	2.41	8.06	7.73	7.26	6.63	5.92	3.29	7.76	7.33	6.81	6.22	5.59	3.48	9.98	11.01
July	7.83	7.99	7.74	7.27	6.63	5.84	2.73	7.84	7.58	7.19	6.63	5.97	3.46	7.44	7.05	6.57	6.04	5.43	3.45	9.98	10.77
Aug	6.76	7.41	7.42	7.23	6.85	6.28	3.63	7.15	7.09	6.88	6.50	6.01	3.84	6.46	6.15	5.80	5.39	4.91	3.23	9.78	10.07
Sept	5.12	6.30	6.64	6.78	6.73	6.49	4.73	5.88	6.06	6.06	5.93	5.66	4.05	4.90	4.71	4.47	4.17	3.87	2.63	8.74	8.74
Oct	3.43	4.88	5.41	5.78	5.98	6.01	5.08	4.40	4.71	4.88	4.94	4.86	3.85	3.30	3.18	3.04	2.88	2.67	1.87	7.08	7.17
Nov	2.20	3.66	4.24	4.70	5.02	5.20	4.81	3.20	3.57	3.84	4.00	4.05	3.49	2.13	2.09	2.02	1.94	1.82	1.33	5.64	5.94
Dec	1.79	3.20	3.78	4.24	4.58	4.79	4.56	2.76	3.13	3.42	3.60	3.68	3.27	1.73	1.68	1.65	1.57	1.48	1.08	5.00	5.37
kWh/m²-yr	1866	2216	2299	2317	2273	2167	1514	2088	2124	2111	2046	1938	1362	1782	1704	1610	1499	1372	915	3027	3160

Daily Clear-Sky Insolation (kWh/m²)													**Latitude 50°N**								
Azimuth:		South						SE, SW						East, West						Tracking	
Tilt:	0	20	30	40	50	60	90	20	30	40	50	60	90	20	30	40	50	60	90	1-Axis	2-Axis
Jan	1.60	2.96	3.52	3.98	4.32	4.54	4.37	2.54	2.91	3.19	3.38	3.47	3.14	1.56	1.53	1.51	1.45	1.39	1.04	4.86	5.12
Feb	2.93	4.49	5.08	5.52	5.81	5.92	5.24	3.98	4.34	4.60	4.72	4.71	3.91	2.84	2.78	2.68	2.58	2.42	1.78	6.97	7.06
Mar	4.66	6.05	6.50	6.76	6.82	6.69	5.17	5.58	5.85	5.94	5.91	5.72	4.32	4.48	4.34	4.18	3.95	3.70	2.62	8.88	8.88
Apr	6.54	7.40	7.53	7.43	7.13	6.65	4.18	7.09	7.11	7.00	6.70	6.28	4.21	6.26	6.00	5.68	5.34	4.92	3.34	10.17	10.45
May	7.69	8.08	7.93	7.56	6.98	6.27	3.26	7.86	7.68	7.36	6.85	6.26	3.81	7.31	6.93	6.50	6.00	5.44	3.53	10.21	11.00
Jun	8.03	8.22	7.97	7.50	6.82	6.03	2.90	8.04	7.78	7.38	6.80	6.14	3.58	7.60	7.18	6.69	6.14	5.53	3.51	9.98	11.02
July	7.62	7.98	7.82	7.45	6.87	6.16	3.19	7.76	7.58	7.25	6.75	6.15	3.73	7.23	6.86	6.42	5.93	5.36	3.47	9.96	10.76
Aug	6.40	7.21	7.32	7.22	6.92	6.45	4.04	6.91	6.92	6.79	6.49	6.07	4.05	6.12	5.86	5.54	5.19	4.77	3.22	9.69	9.98
Sept	4.56	5.84	6.25	6.47	6.51	6.37	4.90	5.41	5.64	5.71	5.66	5.46	4.09	4.37	4.23	4.06	3.82	3.57	2.50	8.36	8.36
Oct	2.78	4.22	4.76	5.17	5.43	5.54	4.90	3.75	4.08	4.31	4.43	4.42	3.66	2.69	2.63	2.53	2.44	2.28	1.66	6.45	6.53
Nov	1.54	2.81	3.33	3.76	4.08	4.27	4.11	2.41	2.75	3.02	3.19	3.27	2.96	1.50	1.46	1.44	1.38	1.32	0.98	4.56	4.79
Dec	1.17	2.37	2.87	3.29	3.61	3.83	3.80	2.00	2.33	2.60	2.79	2.90	2.72	1.15	1.14	1.13	1.09	1.06	0.80	4.00	4.29
kWh/m²-yr	1692	2059	2157	2194	2169	2089	1520	1928	1978	1982	1937	1851	1343	1618	1552	1473	1380	1272	866	2863	2990

Daily Clear-Sky Insolation (kWh/m^2)															Latitude 55°N						
Azimuth:		South						SE, SW						East, West						Tracking	
Tilt:	0	20	30	40	50	60	90	20	30	40	50	60	90	20	30	40	50	60	90	1-Axis	2-Axis
Jan	0.99	2.11	2.58	2.98	3.29	3.51	3.52	1.77	2.08	2.34	2.53	2.64	2.52	0.98	0.99	0.98	0.97	0.95	0.74	3.76	3.96
Feb	2.26	3.75	4.34	4.81	5.14	5.31	4.91	3.28	3.66	3.94	4.12	4.17	3.62	2.21	2.19	2.16	2.10	2.01	1.54	6.21	6.29
Mar	4.05	5.51	6.02	6.36	6.51	6.47	5.27	5.04	5.35	5.51	5.56	5.45	4.30	3.92	3.81	3.71	3.56	3.36	2.46	8.43	8.43
Apr	6.11	7.13	7.35	7.35	7.15	6.76	4.56	6.77	6.88	6.84	6.62	6.28	4.42	5.86	5.65	5.39	5.10	4.75	3.32	10.03	10.31
May	7.41	7.98	7.94	7.67	7.18	6.53	3.70	7.70	7.60	7.35	6.93	6.40	4.05	7.03	6.68	6.29	5.84	5.33	3.53	10.16	10.95
Jun	7.80	8.19	8.05	7.67	7.09	6.35	3.37	7.95	7.77	7.42	6.93	6.32	3.84	7.38	6.98	6.52	6.01	5.44	3.52	9.95	10.99
July	7.34	7.88	7.83	7.55	7.07	6.42	3.62	7.61	7.50	7.25	6.82	6.29	3.97	6.97	6.62	6.22	5.77	5.26	3.47	9.90	10.70
Aug	5.98	6.94	7.13	7.13	6.92	6.53	4.39	6.60	6.68	6.63	6.41	6.07	4.24	5.73	5.51	5.25	4.95	4.60	3.19	9.54	9.83
Sept	3.96	5.31	5.77	6.06	6.19	6.14	4.97	4.86	5.14	5.28	5.30	5.18	4.05	3.82	3.71	3.60	3.44	3.23	2.34	7.89	7.89
Oct	2.12	3.47	4.00	4.42	4.72	4.88	4.51	3.04	3.38	3.64	3.79	3.83	3.32	2.07	2.04	2.01	1.94	1.85	1.40	5.63	5.70
Nov	0.94	1.96	2.39	2.75	3.04	3.23	3.24	1.64	1.93	2.16	2.34	2.44	2.33	0.92	0.93	0.92	0.91	0.88	0.68	3.45	3.63
Dec	0.60	1.42	1.78	2.09	2.33	2.51	2.58	1.17	1.41	1.62	1.77	1.87	1.85	0.59	0.59	0.59	0.59	0.57	0.44	2.60	2.79
kWh/m^2-yr	1511	1878	1984	2034	2026	1965	1477	1749	1808	1826	1799	1732	1293	1447	1393	1330	1255	1165	811	2664	2784

Daily Clear-Sky Insolation (kWh/m^2)															Latitude 60°N						
Azimuth:		South						SE, SW						East, West						Tracking	
Tilt:	0	20	30	40	50	60	90	20	30	40	50	60	90	20	30	40	50	60	90	1-Axis	2-Axis
Jan	0.45	1.15	1.46	1.72	1.93	2.09	2.18	0.94	1.15	1.32	1.46	1.55	1.56	0.45	0.45	0.46	0.46	0.45	0.36	2.22	2.33
Feb	1.59	2.91	3.46	3.91	4.24	4.45	4.29	2.50	2.87	3.15	3.34	3.43	3.12	1.58	1.58	1.58	1.55	1.52	1.20	5.14	5.21
Mar	3.40	4.90	5.45	5.84	6.07	6.12	5.24	4.42	4.77	5.00	5.11	5.08	4.19	3.32	3.27	3.20	3.13	2.97	2.27	7.86	7.86
Apr	5.62	6.78	7.08	7.18	7.08	6.78	4.86	6.39	6.57	6.61	6.48	6.22	4.58	5.41	5.25	5.06	4.81	4.54	3.28	9.84	10.12
May	7.06	7.82	7.87	7.70	7.31	6.73	4.14	7.48	7.45	7.28	6.94	6.47	4.30	6.69	6.38	6.03	5.64	5.20	3.52	10.06	10.85
Jun	7.51	8.09	8.04	7.77	7.29	6.60	3.80	7.80	7.67	7.40	6.99	6.44	4.09	7.10	6.71	6.30	5.85	5.33	3.51	9.87	10.91
July	7.00	7.72	7.76	7.58	7.19	6.61	4.05	7.40	7.36	7.17	6.83	6.36	4.20	6.64	6.32	5.97	5.58	5.13	3.46	9.80	10.60
Aug	5.51	6.59	6.87	6.95	6.84	6.54	4.67	6.22	6.37	6.40	6.25	5.99	4.38	5.29	5.12	4.93	4.68	4.40	3.15	9.34	9.62
Sept	3.32	4.69	5.19	5.54	5.74	5.77	4.90	4.26	4.57	4.76	4.85	4.80	3.92	3.23	3.18	3.09	3.01	2.85	2.14	7.31	7.31
Oct	1.46	2.62	3.11	3.50	3.79	3.97	3.83	2.26	2.58	2.82	2.99	3.07	2.78	1.44	1.43	1.43	1.39	1.35	1.05	4.52	4.57
Nov	0.41	1.02	1.29	1.52	1.70	1.84	1.91	0.83	1.02	1.17	1.29	1.37	1.37	0.41	0.41	0.41	0.41	0.40	0.31	1.94	2.04
Dec	0.17	0.53	0.69	0.83	0.95	1.04	1.11	0.42	0.53	0.63	0.71	0.76	0.79	0.17	0.18	0.18	0.18	0.18	0.14	1.07	1.15
kWh/m^2-yr	1327	1670	1774	1828	1830	1780	1366	1552	1612	1635	1620	1569	1195	1273	1228	1179	1119	1046	743	2404	2514

APPENDIX E

SOLAR INSOLATION TABLES BY CITY

Tilt	Fairbanks, AK						Latitude 64.82°N						
	Jan	Feb	Mar	Apr	May	Jun	July	Aug	Sept	Oct	Nov	Dec	Year
Lat − 15	0.7	2.2	4.5	5.6	5.7	5.7	5.4	4.5	3.4	1.9	1.0	0.2	3.4
Lat	0.7	2.4	4.7	5.6	5.3	5.2	4.9	4.2	3.4	2.0	1.1	0.3	3.3
Lat + 15	0.8	2.5	4.7	5.3	4.6	4.5	4.3	3.8	3.2	2.0	1.1	0.3	3.1
90	0.8	2.5	4.5	4.9	4.1	3.9	3.7	3.3	3.0	2.0	1.1	0.3	2.8
1-Axis (Lat)	0.8	2.7	5.8	7.7	8.0	8.0	7.4	5.8	4.4	2.3	1.2	0.3	4.5
Temp. (°C)	−18.7	−13.8	−4.6	5.0	15.2	21.2	22.4	19.1	12.7	0.0	−11.7	−16.8	2.5

Tilt	Montgomery, AL						Latitude 32.30°N						
	Jan	Feb	Mar	Apr	May	Jun	July	Aug	Sept	Oct	Nov	Dec	Year
Lat − 15	3.4	4.2	5.0	5.9	6.2	6.3	6.0	5.9	5.3	4.9	3.8	3.3	5.0
Lat	3.8	4.6	5.2	5.8	5.8	5.8	5.6	5.7	5.4	5.3	4.3	3.7	5.1
Lat + 15	4.0	4.7	5.1	5.4	5.2	5.1	4.9	5.2	5.2	5.4	4.5	4.0	4.9
90	3.5	3.7	3.4	2.9	2.3	2.1	2.2	2.6	3.2	4.0	3.8	3.5	3.1
1-Axis (Lat)	4.5	5.5	6.5	7.5	7.6	7.5	7.0	7.1	6.7	6.5	5.1	4.4	6.3
Temp. (°C)	13.5	16.0	20.3	24.7	28.3	31.9	32.8	32.4	30.6	25.7	20.4	15.7	24.3

Radiation data measured in kWh/m^2-day, average daily maximum temperature (°C).

Renewable and Efficient Electric Power Systems. By Gilbert M. Masters
ISBN 0-471-28060-7

Tilt	Phoenix, AZ Jan	Feb	Mar	Apr	May	Jun	Latitude 33.43°N July	Aug	Sept	Oct	Nov	Dec	Year
Lat − 15	4.4	5.4	6.4	7.5	8.0	8.1	7.5	7.3	6.8	6.0	4.9	4.2	6.4
Lat	5.1	6.0	6.7	7.4	7.5	7.3	6.9	7.1	7.0	6.5	5.6	4.9	6.5
Lat + 15	5.5	6.2	6.6	6.9	6.6	6.3	6.0	6.4	6.7	6.7	5.9	5.3	6.3
90	4.9	5.0	4.5	3.7	2.7	2.3	2.4	3.1	4.2	5.1	5.1	4.8	4.0
1-Axis (Lat)	6.2	7.5	8.7	10.3	10.7	10.8	9.6	9.6	9.3	8.4	6.8	5.8	8.6
Temp. (°C)	18.8	21.5	24.2	29.2	34.2	39.7	41.1	39.8	36.8	31.2	23.8	19.0	29.9

Tilt	Tucson, AZ Jan	Feb	Mar	Apr	May	Jun	Latitude 32.12°N July	Aug	Sept	Oct	Nov	Dec	Year
Lat − 15	4.6	5.5	6.4	7.5	7.8	7.8	6.9	6.6	6.6	6.1	5.0	4.3	6.3
Lat	5.4	6.2	6.7	7.3	7.3	7.1	6.4	6.6	6.8	6.6	5.8	5.1	6.5
Lat + 15	5.9	6.4	6.6	6.8	6.4	6.1	5.6	6.0	6.6	6.8	6.2	5.6	6.3
90	5.2	5.1	4.4	3.5	2.5	2.1	2.2	2.8	4.0	5.1	5.3	5.1	3.9
1-Axis (Lat)	6.7	7.8	9.0	10.4	10.7	10.6	8.8	9.1	9.1	8.6	7.3	6.2	8.7
Temp. (°C)	17.7	19.9	22.7	27.3	32.2	37.6	37.4	36.0	34.1	29.1	22.6	17.9	27.9

Tilt	Daggett, CA Jan	Feb	Mar	Apr	May	Jun	Latitude 34.87°N July	Aug	Sept	Oct	Nov	Dec	Year
Lat − 15	4.6	5.4	6.5	7.5	7.9	8.1	7.8	7.6	7.1	6.2	5.0	4.4	6.5
Lat	5.3	6.0	6.8	7.4	7.4	7.4	7.2	7.3	7.3	6.8	5.8	5.2	6.6
Lat + 15	5.7	6.2	6.7	6.8	6.5	6.3	6.2	6.6	7.0	6.9	6.2	5.6	6.4
90	5.2	5.1	4.7	3.8	2.8	2.4	2.5	3.3	4.4	5.4	5.4	5.2	4.2
1-Axis (Lat)	6.5	7.5	9.0	10.3	10.9	11.2	10.7	10.6	10.1	8.8	7.2	6.3	9.1
Temp. (°C)	15.9	18.9	21.4	25.6	30.8	36.6	39.9	38.6	34.3	28.2	20.8	15.8	27.2

Tilt	Fresno, CA Jan	Feb	Mar	Apr	May	Jun	Latitude 36.77°N July	Aug	Sept	Oct	Nov	Dec	Year
Lat − 15	2.8	4.1	5.5	6.8	7.6	7.8	7.9	7.5	6.8	5.5	3.6	2.5	5.7
Lat	3.1	4.4	5.7	6.7	7.1	7.2	7.3	7.3	6.9	6.0	4.1	2.8	5.7
Lat + 15	3.2	4.5	5.6	6.2	6.3	6.1	6.3	6.6	6.7	6.1	4.2	3.0	5.4
90	2.8	3.7	4.0	3.6	3.0	2.6	2.7	3.5	4.4	4.8	3.7	2.6	3.4
1-Axis (Lat)	3.4	5.2	7.2	8.9	10.1	10.5	10.8	10.3	9.2	7.4	4.7	3.1	7.6
Temp. (°C)	12.3	16.5	19.2	23.9	29.0	33.7	37.0	35.9	32.3	26.5	18.2	12.1	24.7

Tilt	Los Angeles, CA Jan	Feb	Mar	Apr	May	Jun	Latitude 33.93°N July	Aug	Sept	Oct	Nov	Dec	Year
Lat − 15	3.8	4.5	5.5	6.4	6.4	6.4	7.1	6.8	5.9	5.0	4.2	3.6	5.5
Lat	4.4	5.0	5.7	6.3	6.1	6.0	6.6	6.6	6.0	5.4	4.7	4.2	5.6
Lat + 15	4.7	5.1	5.6	5.9	5.4	5.2	5.8	6.0	5.7	5.5	5.0	4.5	5.4
90	4.1	4.1	3.8	3.3	2.5	2.2	2.4	3.0	3.6	4.2	4.3	4.1	3.5
1-Axis (Lat)	5.1	6.0	7.1	8.2	7.8	7.7	8.7	8.4	7.4	6.6	5.6	4.9	7.0
Temp. (°C)	18.7	18.8	18.6	19.7	20.6	22.2	24.1	24.8	24.8	23.6	21.3	18.8	21.3

Radiation data measured in kWh/m^2-day, average daily maximum temperature (°C).

Tilt	San Diego, CA						Latitude 32.73°N						
	Jan	Feb	Mar	Apr	May	Jun	July	Aug	Sept	Oct	Nov	Dec	Year
Lat − 15	4.1	4.8	5.6	6.4	6.3	6.3	6.8	6.7	6.0	5.3	4.5	3.9	5.6
Lat	4.7	5.3	5.8	6.3	5.9	5.8	6.4	6.5	6.1	5.7	5.1	4.6	5.7
Lat + 15	5.1	5.5	5.7	5.9	5.2	5.1	5.6	5.9	5.8	5.8	5.4	5.0	5.5
90	4.5	4.3	3.9	3.2	2.4	2.1	2.2	2.9	3.6	4.4	4.6	4.5	3.5
1-Axis (Lat)	5.7	6.5	7.4	8.2	7.5	7.4	8.4	8.4	7.7	7.1	6.2	5.5	7.2
Temp. (°C)	18.8	19.2	19.1	20.2	20.6	22.0	24.6	25.4	25.1	23.7	21.1	18.9	21.6

Tilt	Santa Maria, CA						Latitude 34.90°N						
	Jan	Feb	Mar	Apr	May	Jun	July	Aug	Sept	Oct	Nov	Dec	Year
Lat − 15	4.0	4.7	5.7	6.6	7.0	7.2	7.4	7.1	6.3	5.4	4.4	3.9	5.8
Lat	4.6	5.2	5.9	6.5	6.6	6.6	6.9	6.9	6.4	5.8	5.0	4.5	5.9
Lat + 15	4.9	5.4	5.8	6.0	5.8	5.7	6.0	6.2	6.2	6.0	5.3	4.9	5.7
90	4.4	4.4	4.0	3.4	2.7	2.3	2.5	3.2	3.9	4.6	4.6	4.5	3.7
1-Axis (Lat)	5.5	6.4	7.5	8.6	8.8	9.0	9.2	8.9	8.2	7.3	6.0	5.4	7.6
Temp. (°C)	17.7	18.2	17.9	19.4	19.9	21.7	22.9	23.4	23.8	23.3	20.4	17.9	20.6

Tilt	Alamosa, CO						Latitude 37.45°N						
	Jan	Feb	Mar	Apr	May	Jun	July	Aug	Sept	Oct	Nov	Dec	Year
Lat − 15	4.7	5.5	6.2	6.9	7.1	7.4	7.0	6.8	6.5	5.9	4.9	4.4	6.1
Lat	5.5	6.2	6.5	6.8	6.6	6.8	6.5	6.5	6.7	6.4	5.6	5.2	6.3
Lat + 15	6.0	6.5	6.4	6.3	5.8	5.7	5.6	5.9	6.4	6.5	6.0	5.7	6.1
90	5.7	5.6	4.7	3.7	2.9	2.5	2.6	3.2	4.3	5.2	5.5	5.5	4.3
1-Axis (Lat)	6.8	7.9	8.7	9.6	9.8	10.3	9.5	9.2	9.1	8.4	7.0	6.3	8.5
Temp. (°C)	0.7	4.4	9.3	14.8	20.0	25.4	27.8	26.2	22.6	16.9	8.6	1.9	14.9

Tilt	Boulder, CO						Latitude 40.02°N						
	Jan	Feb	Mar	Apr	May	Jun	July	Aug	Sept	Oct	Nov	Dec	Year
Lat − 15	3.8	4.6	5.4	6.1	6.2	6.6	6.6	6.3	5.9	5.1	4.0	3.5	5.4
Lat	4.4	5.1	5.6	6.0	5.9	6.1	6.1	6.1	6.0	5.6	4.6	4.2	5.5
Lat + 15	4.8	5.3	5.6	5.6	5.2	5.2	5.3	5.5	5.8	5.7	4.8	4.5	5.3
90	4.5	4.6	4.3	3.6	2.8	2.6	2.7	3.2	4.0	4.6	4.4	4.3	3.8
1-Axis (Lat)	5.2	6.2	7.2	8.0	8.1	8.8	8.7	8.4	7.9	7.1	5.5	4.9	7.2
Temp. (°C)	6.2	8.1	11.2	16.6	21.6	27.4	31.2	29.9	24.9	19.1	11.4	6.9	17.9

Tilt	Grand Junction, CO						Latitude 39.12°N						
	Jan	Feb	Mar	Apr	May	Jun	July	Aug	Sept	Oct	Nov	Dec	Year
Lat − 15	3.8	4.7	5.5	6.5	7.0	7.5	7.3	7.0	6.5	5.4	4.1	3.6	5.7
Lat	4.4	5.2	5.7	6.4	6.6	6.8	6.7	6.7	6.6	5.9	4.6	4.1	5.8
Lat + 15	4.7	5.4	5.6	6.0	5.8	5.8	5.8	6.1	6.4	6.0	4.9	4.5	5.6
90	4.4	4.7	4.2	3.7	3.0	2.7	2.8	3.5	4.4	4.8	4.4	4.3	3.9
1-Axis (Lat)	5.2	6.4	7.3	8.7	9.3	10.1	9.8	9.4	9.0	7.5	5.6	4.9	7.8
Temp. (°C)	1.9	7.4	13.1	18.8	24.4	30.9	34.2	32.5	27.3	19.8	10.8	3.7	18.8

Radiation data measured in kWh/m^2-day, average daily maximum temperature (°C).

	Daytona Beach, FL					Latitude 29.18°N							
Tilt	Jan	Feb	Mar	Apr	May	Jun	July	Aug	Sept	Oct	Nov	Dec	Year
Lat − 15	3.8	4.5	5.5	6.4	6.4	6.0	5.9	5.8	5.2	4.7	4.1	3.6	5.2
Lat	4.3	4.9	5.7	6.3	6.0	5.5	5.5	5.6	5.3	5.0	4.6	4.1	5.2
Lat + 15	4.6	5.1	5.6	5.9	5.4	4.8	4.9	5.1	5.1	5.1	4.8	4.4	5.1
90	3.9	3.8	3.6	2.9	2.1	1.8	1.9	2.4	3.0	3.6	4.0	3.9	3.1
1-Axis (Lat)	5.2	6.0	7.2	8.3	7.9	7.1	7.1	7.0	6.5	6.1	5.5	4.9	6.6
Temp. (°C)	20.0	20.8	23.8	26.7	29.2	31.1	32.1	31.7	30.4	27.5	24.2	21.3	26.6

	Miami, FL					Latitude 25.80°N							
Tilt	Jan	Feb	Mar	Apr	May	Jun	July	Aug	Sept	Oct	Nov	Dec	Year
Lat − 15	4.1	4.7	5.5	6.2	5.9	5.5	5.7	5.6	5.1	4.7	4.2	3.9	5.1
Lat	4.7	5.2	5.7	6.1	5.6	5.1	5.4	5.5	5.1	5.1	4.7	4.5	5.2
Lat + 15	5.0	5.4	5.6	5.7	5.0	4.5	4.8	5.0	4.9	5.1	4.9	4.9	5.1
90	4.1	3.9	3.4	2.6	1.9	1.6	1.7	2.1	2.7	3.5	3.9	4.1	3.0
1-Axis (Lat)	5.7	6.4	7.2	7.8	7.2	6.3	6.7	6.7	6.2	6.1	5.6	5.4	6.5
Temp. (°C)	24.0	24.7	26.2	28.0	29.6	30.9	31.7	31.7	31.0	29.2	26.9	24.8	28.2

	Tallahassee, FL					Latitude 30.38°N							
Tilt	Jan	Feb	Mar	Apr	May	Jun	July	Aug	Sept	Oct	Nov	Dec	Year
Lat − 15	3.6	4.3	5.2	6.1	6.2	6.0	5.7	5.6	5.3	5.0	4.1	3.4	5.0
Lat	4.0	4.7	5.4	6.0	5.9	5.6	5.4	5.4	5.3	5.4	4.6	4.0	5.1
Lat + 15	4.3	4.9	5.3	5.6	5.2	4.9	4.7	4.9	5.1	5.5	4.8	4.2	5.0
90	3.6	3.7	3.4	2.9	2.2	1.9	2.0	2.4	3.1	4.0	4.0	3.7	3.1
1-Axis (Lat)	4.8	5.8	6.7	7.8	7.7	7.1	6.8	6.8	6.6	6.7	5.5	4.6	6.4
Temp. (°C)	17.2	19.1	23.1	26.9	30.2	32.7	32.9	32.8	31.4	27.5	22.8	18.8	26.3

	Atlanta, GA					Latitude 33.65°N							
Tilt	Jan	Feb	Mar	Apr	May	Jun	July	Aug	Sept	Oct	Nov	Dec	Year
Lat − 15	3.4	4.2	5.1	6.0	6.2	6.3	6.1	5.9	5.3	4.9	3.8	3.2	5.0
Lat	3.8	4.6	5.3	5.8	5.8	5.8	5.7	5.7	5.4	5.2	4.2	3.7	5.1
Lat + 15	4.1	4.7	5.1	5.4	5.2	5.1	5.0	5.2	5.1	5.3	4.9	3.9	4.9
90	3.5	3.7	3.5	3.0	2.4	2.2	2.2	2.7	3.2	4.0	3.8	3.5	3.1
1-Axis (Lat)	4.5	5.5	6.6	7.6	7.7	7.6	7.3	7.2	6.7	6.4	5.0	4.3	6.4
Temp. (°C)	10.2	12.8	17.9	22.6	26.4	29.9	31.1	30.6	27.7	22.6	17.4	12.2	21.8

	Savannah, GA					Latitude 32.13°N							
Tilt	Jan	Feb	Mar	Apr	May	Jun	July	Aug	Sept	Oct	Nov	Dec	Year
Lat − 15	3.5	4.3	5.2	6.1	6.2	6.1	6.0	5.6	5.1	4.8	3.9	3.4	5.0
Lat	4.0	4.7	5.4	6.0	5.8	5.7	5.6	5.4	5.1	5.1	4.4	3.9	5.1
Lat + 15	4.3	4.8	5.3	5.6	5.2	4.9	4.9	5.0	4.9	5.2	4.6	4.2	4.9
90	3.7	3.8	3.6	3.0	2.3	2.0	2.1	2.5	3.0	3.8	3.9	3.7	3.1
1-Axis (Lat)	4.8	5.7	6.8	7.9	7.7	7.4	7.1	6.8	6.3	6.3	5.3	4.5	6.4
Temp. (°C)	15.4	16.9	21.2	25.3	28.9	31.6	32.8	32.1	29.6	25.3	21.1	16.8	24.7

Radiation data measured in kWh/m^2-day, average daily maximum temperature (°C).

Tilt	Honolulu, HI Jan	Feb	Mar	Apr	May	Jun	Latitude 21.33°N July	Aug	Sept	Oct	Nov	Dec	Year
Lat − 15	4.3	5.0	5.6	5.9	6.3	6.4	6.5	6.5	6.1	5.3	4.5	4.1	5.5
Lat	4.9	5.5	5.8	5.9	5.9	5.9	6.0	6.2	6.2	5.7	5.1	4.8	5.7
Lat + 15	5.3	5.8	5.8	5.5	5.3	5.1	5.3	5.7	6.0	5.8	5.4	5.2	5.5
90	4.2	4.0	3.2	2.2	1.5	1.4	1.4	1.8	2.8	3.7	4.1	4.3	2.9
1-Axis (Lat)	6.1	7.0	7.5	7.7	8.0	8.1	8.3	8.5	8.2	7.3	6.2	5.9	7.4
Temp. (°C)	26.7	26.9	27.6	28.2	29.3	30.3	30.8	31.5	31.4	30.5	28.9	27.3	29.1

Tilt	Boise, ID Jan	Feb	Mar	Apr	May	Jun	Latitude 43.57°N July	Aug	Sept	Oct	Nov	Dec	Year
Lat − 15	2.5	3.5	4.7	5.9	6.7	7.1	7.6	7.1	6.3	4.9	2.9	2.3	5.1
Lat	2.8	3.8	4.9	5.8	6.2	6.5	7.0	6.8	6.5	5.2	3.2	2.6	5.1
Lat + 15	2.9	3.9	4.8	5.4	5.5	5.5	6.0	6.2	6.2	5.3	3.3	2.8	4.8
90	2.7	3.4	3.7	3.6	3.2	3.0	3.3	3.8	4.6	4.5	3.0	2.6	3.5
1-Axis (Lat)	3.1	4.5	6.2	7.8	9.0	9.6	10.7	9.9	8.8	6.6	3.7	2.9	6.9
Temp. (°C)													

Tilt	Indianapolis, IN Jan	Feb	Mar	Apr	May	Jun	Latitude 39.73°N July	Aug	Sept	Oct	Nov	Dec	Year
Lat − 15	2.8	3.6	4.3	5.2	5.9	6.3	6.2	5.9	5.2	4.2	2.8	2.3	4.6
Lat	3.1	3.9	4.4	5.1	5.6	5.8	5.8	5.7	5.3	4.5	3.1	2.6	4.6
Lat + 15	3.3	4.0	4.3	4.7	4.9	5.0	5.1	5.2	5.1	4.5	3.2	2.7	4.3
90	3.0	3.5	3.2	3.0	2.7	2.6	2.7	3.1	3.5	3.6	2.8	2.5	3.0
1-Axis (Lat)	3.6	4.6	5.3	6.5	7.4	7.9	7.8	7.5	6.7	5.5	3.5	2.9	5.8
Temp. (°C)	0.9	3.5	10.5	17.4	23.2	28.2	29.7	28.7	25.3	18.8	11.1	3.6	16.7

Tilt	Dodge City, KS Jan	Feb	Mar	Apr	May	Jun	Latitude 37.77°N July	Aug	Sept	Oct	Nov	Dec	Year
Lat − 15	4.0	4.8	5.5	6.3	6.5	7.0	7.1	6.6	5.9	5.2	4.0	3.6	5.5
Lat	4.6	5.3	5.8	6.2	6.1	6.4	6.5	6.4	6.0	5.6	4.6	4.2	5.6
Lat + 15	5.0	5.5	5.7	5.8	5.4	5.5	5.7	5.8	5.7	5.7	4.8	4.6	5.4
90	4.6	4.6	4.2	3.5	2.7	2.5	2.7	3.2	3.9	4.5	4.3	4.3	3.7
1-Axis (Lat)	5.6	6.5	7.4	8.3	8.4	9.0	9.2	8.7	7.8	7.0	5.5	5.0	7.4
Temp. (°C)	5.3	8.5	13.7	20.0	24.6	30.6	33.9	32.6	27.4	21.5	12.9	6.6	19.8

Tilt	Lake Charles, LA Jan	Feb	Mar	Apr	May	Jun	Latitude 30.12°N July	Aug	Sept	Oct	Nov	Dec	Year
Lat − 15	3.3	4.1	4.9	5.5	6.0	6.2	5.9	5.7	5.4	5.0	3.9	3.2	4.9
Lat	3.7	4.5	5.1	5.4	5.6	5.7	5.5	5.6	5.4	5.4	4.3	3.7	5.0
Lat + 15	3.9	4.6	4.9	5.1	5.0	5.0	4.9	5.1	5.2	5.4	4.6	3.9	4.8
90	3.2	3.5	3.2	2.6	2.1	1.9	2.0	2.4	3.1	3.9	3.7	3.4	2.9
1-Axis (Lat)	4.3	5.4	6.2	6.9	7.2	7.5	7.0	7.0	6.8	6.6	5.2	4.3	6.2
Temp. (°C)	15.4	17.4	21.5	25.5	28.9	31.7	32.7	32.7	30.4	26.8	21.6	17.4	25.2

Radiation data measured in kWh/m^2-day, average daily maximum temperature (°C).

Tilt	Boston, MA						Latitude 42.37°N						
	Jan	Feb	Mar	Apr	May	Jun	July	Aug	Sept	Oct	Nov	Dec	Year
Lat − 15	3.0	3.8	4.6	5.2	5.7	6.0	6.0	5.7	5.0	4.1	2.8	2.5	4.5
Lat	3.4	4.2	4.7	5.0	5.3	5.5	5.6	5.5	5.1	4.3	3.1	2.9	4.6
Lat + 15	3.6	4.3	4.6	4.7	4.7	4.8	4.9	5.0	4.9	4.4	3.3	3.1	4.4
90	3.4	3.9	3.7	3.1	2.8	2.6	2.8	3.1	3.5	3.6	3.0	2.9	3.2
1-Axis (Lat)	3.9	5.0	5.9	6.5	7.1	7.4	7.5	7.1	6.4	5.2	3.6	3.2	5.7
Temp. (°C)	2.1	3.1	7.7	13.3	19.2	24.6	27.7	26.6	22.7	17.1	11.2	4.7	15.0

Tilt	Caribou, ME						Latitude 46.87°N						
	Jan	Feb	Mar	Apr	May	Jun	July	Aug	Sept	Oct	Nov	Dec	Year
Lat − 15	2.9	3.9	5.0	5.1	5.3	5.6	5.6	5.2	4.3	3.1	2.2	2.2	4.2
Lat	3.3	4.3	5.2	5.0	4.9	5.1	5.2	4.9	4.3	3.3	2.4	2.5	4.2
Lat + 15	3.5	4.5	5.2	4.7	4.4	4.5	4.6	4.5	4.1	3.3	2.5	2.7	4.0
90	3.4	4.2	4.6	3.7	2.9	2.8	2.9	3.0	3.2	2.8	2.3	2.7	3.2
1-Axis (Lat)	3.7	5.1	6.5	6.5	6.7	7.1	7.2	6.6	5.6	3.9	2.7	2.8	5.4
Temp. (°C)	−7.0	−5.0	1.3	8.2	16.5	22.2	24.7	23.1	17.8	11.1	3.1	−4.4	9.3

Tilt	Columbia, MO						Latitude 38.82°N						
	Jan	Feb	Mar	Apr	May	Jun	July	Aug	Sept	Oct	Nov	Dec	Year
Lat − 15	3.3	4.0	4.7	5.6	6.0	6.4	6.6	6.2	5.3	4.5	3.2	2.8	4.9
Lat	3.8	4.4	4.9	5.5	5.6	5.9	6.1	5.9	5.4	4.9	3.6	3.2	4.9
Lat + 15	4.0	4.5	4.8	5.1	5.0	5.1	5.3	5.4	5.1	4.9	3.8	3.4	4.7
90	3.7	3.9	3.5	3.2	2.7	2.5	2.7	3.1	3.5	3.9	3.3	3.2	3.3
1-Axis (Lat)	4.4	5.2	6.0	7.1	7.6	8.1	8.4	7.9	6.9	6.0	4.2	3.7	6.3
Temp. (°C)	2.6	5.2	11.8	18.7	23.4	28.2	31.4	30.4	26.0	19.8	12.0	4.6	17.8

Tilt	Glascow, MT						Latitude 48.22°N						
	Jan	Feb	Mar	Apr	May	Jun	July	Aug	Sept	Oct	Nov	Dec	Year
Lat − 15	2.7	3.7	4.8	5.4	5.8	6.4	6.8	6.4	5.2	4.1	2.8	2.3	4.7
Lat	3.1	4.1	5.0	5.3	5.5	5.9	6.3	6.1	5.3	4.3	3.1	2.6	4.7
Lat + 15	3.3	4.2	4.9	4.9	4.8	5.0	5.5	5.5	5.0	4.4	3.3	2.8	4.5
90	3.2	3.9	4.2	3.6	3.2	3.1	3.4	3.8	3.9	3.8	3.1	2.8	3.5
1-Axis (Lat)	3.5	4.8	6.3	7.2	7.7	8.6	9.4	8.7	6.9	5.3	3.6	3.0	6.3
Temp. (°C)	−6.7	−2.7	4.3	13.6	19.6	25.3	29.3	28.6	21.3	14.8	4.2	−3.9	12.3

Tilt	Cape Hatteras, NC						Latitude 35.27°N						
	Jan	Feb	Mar	Apr	May	Jun	July	Aug	Sept	Oct	Nov	Dec	Year
Lat − 15	3.3	4.1	5.1	6.0	6.1	6.2	6.1	5.8	5.3	4.6	3.7	3.2	5.0
Lat	3.8	4.5	5.2	5.9	5.8	5.7	5.7	5.6	5.4	4.9	4.2	3.6	5.0
Lat + 15	4.0	4.6	5.2	5.5	5.1	5.0	5.0	5.1	5.1	4.9	4.5	3.9	4.8
90	3.6	3.7	3.6	3.1	2.5	2.2	2.3	2.7	3.3	3.8	3.9	3.6	3.2
1-Axis (Lat)	4.5	5.4	6.6	7.7	7.6	7.6	7.4	7.2	6.7	6.0	5.0	4.3	6.3
Temp. (°C)	11.3	11.9	15.3	19.4	23.6	27.1	29.2	29.3	27.1	22.4	18.2	13.8	20.7

Radiation data measured in kWh/m^2-day, average daily maximum temperature (°C).

	Raleigh, NC					Latitude 35.87°N							
Tilt	Jan	Feb	Mar	Apr	May	Jun	July	Aug	Sept	Oct	Nov	Dec	Year
Lat − 15	3.4	4.1	5.0	5.8	6.0	6.2	6.0	5.7	5.1	4.6	3.7	3.1	4.9
Lat	3.8	4.5	5.2	5.7	5.7	5.7	5.6	5.5	5.2	4.9	4.1	3.6	5.0
Lat + 15	4.1	4.6	5.1	5.3	5.0	4.9	4.9	5.0	5.0	5.0	4.3	3.8	4.8
90	3.6	3.8	3.6	3.1	2.5	2.3	2.4	2.8	3.3	3.8	3.7	3.5	3.2
1-Axis (Lat)	4.5	5.5	6.5	7.5	7.4	7.5	7.2	6.9	6.4	6.0	4.9	4.2	6.2
Temp. (°C)	9.4	11.4	16.7	22.1	25.9	29.4	31.1	30.4	27.3	22.0	17.0	11.5	21.2

	Bismarck, ND					Latitude 46.77°N							
Tilt	Jan	Feb	Mar	Apr	May	Jun	July	Aug	Sept	Oct	Nov	Dec	Year
Lat − 15	3.1	4.0	5.0	5.6	6.1	6.5	6.8	6.4	5.3	4.2	2.9	2.6	4.9
Lat	3.5	4.4	5.2	5.5	5.7	5.9	6.3	6.1	5.4	4.5	3.2	3.0	4.9
Lat + 15	3.7	4.5	5.1	5.1	5.1	5.1	5.5	5.5	5.1	4.5	3.4	3.2	4.7
90	3.7	4.2	4.3	3.7	3.3	3.1	3.4	3.7	3.9	3.9	3.2	3.1	3.6
1-Axis (Lat)	4.0	5.2	6.4	7.3	8.0	8.6	9.2	8.5	7.0	5.5	3.7	3.3	6.4
Temp. (°C)	−18.7	−14.9	−7.9	−0.6	5.7	10.9	13.6	12.2	6.2	0.3	−7.9	−15.9	−1.4

	Omaha, NE					Latitude 41.37°N							
Tilt	Jan	Feb	Mar	Apr	May	Jun	July	Aug	Sept	Oct	Nov	Dec	Year
Lat − 15	3.3	4.0	4.7	5.5	6.0	6.5	6.5	6.1	5.2	4.4	3.2	2.7	4.9
Lat	3.8	4.4	4.9	5.3	5.6	6.0	6.0	5.8	5.3	4.7	3.5	3.2	4.9
Lat + 15	4.1	4.6	4.8	5.0	5.0	5.2	5.3	5.3	5.1	4.7	3.7	3.4	4.7
90	3.9	4.1	3.8	3.2	2.8	2.7	2.8	3.2	3.6	3.9	3.3	3.2	3.4
1-Axis (Lat)	4.5	5.3	6.1	7.0	7.5	8.3	8.3	7.8	6.7	5.7	4.1	3.6	6.3
Temp. (°C)	−1.3	1.7	8.7	16.9	22.7	28.0	30.3	28.9	23.8	17.8	8.7	0.5	15.6

	Albuquerque, NM					Latitude 35.05°N							
Tilt	Jan	Feb	Mar	Apr	May	Jun	July	Aug	Sept	Oct	Nov	Dec	Year
Lat − 15	4.6	5.4	6.3	7.3	7.7	7.8	7.4	7.2	6.6	5.9	4.8	4.3	6.3
Lat	5.3	6.0	6.5	7.2	7.2	7.1	6.9	6.9	6.8	6.5	5.5	5.0	6.4
Lat + 15	5.8	6.2	6.5	6.6	6.3	6.1	6.0	6.3	6.5	6.6	5.9	5.5	6.2
90	5.2	5.1	4.5	3.7	2.8	2.4	2.5	3.2	4.2	5.1	5.2	5.1	4.1
1-Axis (Lat)	6.5	7.5	8.6	9.9	10.3	10.4	9.5	9.3	9.0	8.3	6.8	6.1	8.5
Temp. (°C)	8.2	11.9	16.3	21.6	26.5	32.2	33.6	31.7	27.7	21.7	14.1	8.6	21.2

	Ely, NV					Latitude 39.28°N							
Tilt	Jan	Feb	Mar	Apr	May	Jun	July	Aug	Sept	Oct	Nov	Dec	Year
Lat − 15	4.0	4.7	5.5	6.3	6.6	7.2	7.2	6.9	6.6	5.5	4.1	3.6	5.7
Lat	4.6	5.2	5.7	6.2	6.2	6.6	6.6	6.6	6.7	6.0	4.7	4.3	5.8
Lat + 15	5.0	5.5	5.6	5.7	5.5	5.6	5.7	6.0	6.5	6.1	5.0	4.7	5.6
90	4.8	4.8	4.3	3.6	2.9	2.6	2.8	3.4	4.4	4.9	4.5	4.5	4.0
1-Axis (Lat)	5.5	6.5	7.4	8.4	8.9	9.9	9.9	9.5	9.2	7.7	5.6	5.1	7.8
Temp. (°C)	4.3	6.4	9.1	13.9	19.6	25.7	30.6	29.1	24.0	17.5	9.6	4.8	16.2

Radiation data measured in kWh/m^2-day, average daily maximum temperature (°C).

	LAS Vegas, NV						Latitude 36.08°N						
Tilt	Jan	Feb	Mar	Apr	May	Jun	July	Aug	Sept	Oct	Nov	Dec	Year
Lat − 15	4.4	5.3	6.4	7.5	7.8	8.1	7.7	7.5	7.1	6.1	4.8	4.2	6.4
Lat	5.1	5.9	6.7	7.4	7.3	7.4	7.1	7.2	7.2	6.6	5.5	4.9	6.5
Lat + 15	5.6	6.1	6.6	6.8	6.5	6.3	6.2	6.5	7.0	6.8	5.9	5.4	6.3
90	5.0	5.1	4.7	3.9	3.0	2.6	2.6	3.4	4.5	5.3	5.2	5.0	4.2
1-Axis (Lat)	6.2	7.3	8.8	10.2	10.6	11.1	10.4	10.3	9.8	8.6	6.7	5.9	8.8
Temp. (°C)	14.1	17.4	20.4	25.3	31.0	37.9	41.1	39.6	34.8	27.8	19.7	14.2	26.9

	Albany, NY						Latitude 42.75°N						
Tilt	Jan	Feb	Mar	Apr	May	Jun	July	Aug	Sept	Oct	Nov	Dec	Year
Lat − 15	2.7	3.6	4.4	5.0	5.5	5.8	6.0	5.5	4.8	3.7	2.4	2.1	4.3
Lat	3.0	3.9	4.5	4.9	5.1	5.4	5.5	5.2	4.8	3.9	2.6	2.4	4.3
Lat + 15	3.2	4.1	4.4	4.5	4.6	4.6	4.8	4.8	4.6	3.9	2.7	2.5	4.1
90	3.1	3.7	3.5	3.1	2.7	2.6	2.8	3.0	3.3	3.2	2.4	2.4	3.0
1-Axis (Lat)	3.5	4.6	5.5	6.3	6.8	7.2	7.5	6.9	6.0	4.6	2.9	2.6	5.4
Temp. (°C)	−11.7	−10.1	−4.2	1.7	7.4	12.6	15.3	14.3	9.7	3.7	−0.7	−7.7	2.6

	New York City, NY						Latitude 40.78°N						
Tilt	Jan	Feb	Mar	Apr	May	Jun	July	Aug	Sept	Oct	Nov	Dec	Year
Lat − 15	2.9	3.7	4.6	5.3	5.8	6.0	6.0	5.7	5.0	4.1	2.9	2.4	4.5
Lat	3.2	4.0	4.8	5.2	5.4	5.5	5.6	5.5	5.0	4.4	3.2	2.8	4.6
Lat + 15	3.4	4.1	4.6	4.8	4.8	4.8	4.9	5.0	4.8	4.4	3.3	3.0	4.3
90	3.2	3.6	3.5	3.1	2.7	2.6	2.7	3.0	3.4	3.6	3.0	2.7	3.1
1-Axis (Lat)	3.7	4.7	5.8	6.6	6.9	7.2	7.2	6.9	6.2	5.3	3.7	3.1	5.6
Temp. (°C)	3.1	4.6	10.0	16.2	22.1	26.7	29.6	28.7	24.6	18.5	12.2	5.8	16.8

	Burns, OR						Latitude 43.58°N						
Tilt	Jan	Feb	Mar	Apr	May	Jun	July	Aug	Sept	Oct	Nov	Dec	Year
Lat − 15	2.8	3.7	4.7	5.8	6.5	6.9	7.5	7.0	6.3	4.8	2.8	2.4	5.1
Lat	3.1	4.0	4.9	5.7	6.1	6.3	6.9	6.8	6.4	5.2	3.1	2.7	5.1
Lat + 15	3.3	4.1	4.8	5.3	5.4	5.4	6.0	6.1	6.1	5.2	3.3	2.9	4.8
90	3.2	3.7	3.8	3.6	3.2	3.0	3.3	3.8	4.5	4.4	2.9	2.8	3.5
1-Axis (Lat)	3.5	4.8	6.1	7.6	8.7	9.4	10.5	9.8	8.7	6.5	3.6	3.1	6.9
Temp. (°C)	0.9	4.2	8.7	13.6	18.7	23.6	29.5	28.5	23.1	16.6	7.3	1.8	14.7

	Eugene, OR						Latitude 44.12°N						
Tilt	Jan	Feb	Mar	Apr	May	Jun	July	Aug	Sept	Oct	Nov	Dec	Year
Lat − 15	1.8	2.6	3.8	4.8	5.6	6.0	6.7	6.3	5.4	3.6	1.9	1.4	4.2
Lat	2.0	2.7	3.8	4.7	5.3	5.5	6.2	6.0	5.5	3.8	2.1	1.6	4.1
Lat + 15	2.0	2.8	3.7	4.3	4.6	4.8	5.4	5.5	5.2	3.8	2.1	1.6	3.8
90	1.8	2.3	2.9	3.0	2.8	2.7	3.1	3.5	3.9	3.2	1.8	1.5	2.7
1-Axis (Lat)	2.1	3.1	4.6	5.9	7.1	7.8	9.1	8.4	7.1	4.6	2.3	1.7	5.3
Temp. (°C)	8.0	10.8	13.3	15.8	19.5	23.4	27.6	27.7	24.6	18.1	11.3	7.9	17.3

Radiation data measured in kWh/m^2-day, average daily maximum temperature (°C).

Tilt	Medford, OR Jan	Feb	Mar	Apr	May	Jun	Latitude 42.37°N July	Aug	Sept	Oct	Nov	Dec	Year
Lat − 15	2.1	3.2	4.5	5.7	6.6	7.1	7.7	7.2	6.3	4.5	2.3	1.7	4.9
Lat	2.3	3.5	4.6	5.6	6.2	6.5	7.1	6.9	6.4	4.8	2.4	1.9	4.9
Lat + 15	2.4	3.5	4.5	5.2	5.5	5.6	6.2	6.3	6.2	4.8	2.5	2.0	4.5
90	2.2	3.0	3.4	3.5	3.1	2.9	3.2	3.8	4.4	4.0	2.2	1.8	3.1
1-Axis (Lat)	2.6	4.0	5.6	7.3	8.7	9.5	10.7	9.9	8.6	5.9	2.7	2.0	6.5
Temp. (°C)	7.6	11.8	14.7	18.1	22.7	27.8	32.5	32.2	28.2	20.8	11.4	6.8	19.6

Tilt	Portland, OR Jan	Feb	Mar	Apr	May	Jun	Latitude 45.60°N July	Aug	Sept	Oct	Nov	Dec	Year
Lat − 15	1.7	2.5	3.6	4.6	5.4	5.8	6.3	5.9	5.1	3.4	1.9	1.4	4.0
Lat	1.9	2.6	3.7	4.5	5.0	5.3	5.8	5.6	5.1	3.6	2.1	1.6	3.9
Lat + 15	1.9	2.7	3.5	4.1	4.4	4.6	5.1	5.1	4.9	3.6	2.1	1.6	3.6
90	1.8	2.3	2.8	2.9	2.8	2.7	3.1	3.4	3.7	3.0	1.9	1.5	2.6
1-Axis (Lat)	2.0	3.0	4.3	5.5	6.5	7.2	8.1	7.5	6.5	4.3	2.2	1.7	4.9
Temp. (°C)	7.4	10.6	13.3	15.9	19.5	23.3	26.6	26.8	23.7	17.8	11.4	7.6	17.0

Tilt	Redmond, OR Jan	Feb	Mar	Apr	May	Jun	Latitude 44.27°N July	Aug	Sept	Oct	Nov	Dec	Year
Lat − 15	2.6	3.5	4.8	5.9	6.6	7.0	7.6	7.1	6.3	4.7	2.9	2.4	5.1
Lat	3.0	3.8	4.9	5.7	6.2	6.4	7.0	6.9	6.4	5.1	3.2	2.7	5.1
Lat + 15	3.1	3.9	4.8	5.3	5.5	5.5	6.1	6.2	6.1	5.1	3.3	2.9	4.8
90	2.9	3.4	3.8	3.7	3.3	3.1	3.4	3.9	4.5	4.3	3.0	2.7	3.5
1-Axis (Lat)	3.4	4.4	6.2	7.7	8.8	9.5	10.6	9.9	8.6	6.3	3.6	3.0	6.9
Temp. (°C)	5.1	8.4	11.5	15.1	19.6	24.8	29.6	28.8	24.1	17.9	9.4	5.4	16.7

Tilt	Pittsburgh, PA Jan	Feb	Mar	Apr	May	Jun	Latitude 40.50°N July	Aug	Sept	Oct	Nov	Dec	Year
Lat − 15	2.4	3.2	4.1	4.9	5.5	5.9	5.9	5.5	4.8	3.8	2.4	1.9	4.2
Lat	2.6	3.4	4.2	4.8	5.2	5.4	5.5	5.3	4.8	4.1	2.6	2.1	4.2
Lat + 15	2.7	3.5	4.1	4.4	4.6	4.7	4.8	4.8	4.6	4.1	2.7	2.2	3.9
90	2.5	3.0	3.1	2.9	2.6	2.5	2.6	2.9	3.2	3.3	2.3	2.0	2.7
1-Axis (Lat)	2.9	3.9	5.0	6.0	6.6	7.2	7.2	6.8	6.0	4.8	2.9	2.3	5.1
Temp. (°C)	0.9	2.7	9.4	15.7	21.4	26.1	28.1	27.1	23.5	16.9	10.2	3.7	15.5

Tilt	Guam, PI Jan	Feb	Mar	Apr	May	Jun	Latitude 13.55°N July	Aug	Sept	Oct	Nov	Dec	Year
Lat − 15	4.3	4.8	5.4	5.8	5.7	5.5	5.1	4.9	4.8	4.6	4.3	4.1	4.9
Lat	5.0	5.2	5.7	5.7	5.4	5.1	4.8	4.7	4.9	4.9	4.9	4.8	5.1
Lat + 15	5.3	5.4	5.6	5.3	4.9	4.5	4.3	4.4	4.7	4.9	5.1	5.1	5.0
90	3.8	3.3	2.6	1.7	1.5	1.5	1.5	1.5	2.0	2.8	3.5	3.8	2.5
1-Axis (Lat)	6.1	6.4	7.1	7.3	7.0	6.6	6.0	5.7	6.0	5.9	5.9	5.8	6.3
Temp. (°C)	28.7	28.6	29.2	29.9	30.4	30.5	30.2	29.9	30.1	29.9	29.6	29.0	29.7

Radiation data measured in kWh/m^2-day, average daily maximum temperature (°C).

Tilt	San Juan, PR Jan	Feb	Mar	Apr	May	Jun	Latitude 18.43°N July	Aug	Sept	Oct	Nov	Dec	Year
Lat − 15	4.5	5.1	5.8	6.1	5.7	6.0	6.0	6.0	5.6	5.0	4.5	4.1	5.4
Lat	5.1	5.6	6.1	6.1	5.4	5.5	5.6	5.8	5.7	5.4	5.1	4.8	5.5
Lat + 15	5.5	5.8	6.0	5.7	4.9	4.8	5.0	5.3	5.5	5.5	5.4	5.2	5.4
90	4.2	3.9	3.1	2.0	1.5	1.5	1.5	1.7	2.5	3.3	3.9	4.1	2.8
1-Axis (Lat)	6.3	6.9	7.7	7.8	6.9	7.2	7.3	7.4	7.1	6.7	6.1	5.7	6.9
Temp. (°C)	28.4	28.7	29.1	29.9	30.7	31.4	31.4	31.5	31.6	31.3	29.9	28.8	30.2

Tilt	Charleston, SC Jan	Feb	Mar	Apr	May	Jun	Latitude 32.90°N July	Aug	Sept	Oct	Nov	Dec	Year
Lat − 15	3.5	4.3	5.3	6.2	6.2	6.1	6.0	5.6	5.1	4.8	3.9	3.4	5.0
Lat	4.0	4.7	5.5	6.1	5.8	5.6	5.6	5.4	5.2	5.2	4.5	3.9	5.1
Lat + 15	4.3	4.9	5.4	5.7	5.2	4.9	4.9	4.9	5.0	5.2	4.7	4.2	4.9
90	3.7	3.9	3.6	3.1	2.4	2.1	2.2	2.5	3.1	3.9	4.0	3.8	3.2
1-Axis (Lat)	4.8	5.8	6.9	8.0	7.7	7.3	7.2	6.8	6.4	6.3	5.3	4.6	6.4
Temp. (°C)	14.3	16.1	20.3	24.3	28.2	30.9	32.3	31.7	29.4	25.1	20.8	16.4	24.2

Tilt	Nashville, TN Jan	Feb	Mar	Apr	May	Jun	Latitude 36.12°N July	Aug	Sept	Oct	Nov	Dec	Year
Lat − 15	3.1	3.9	4.7	5.7	6.0	6.4	6.2	5.9	5.2	4.6	3.3	2.8	4.8
Lat	3.5	4.2	4.8	5.6	5.7	5.9	5.8	5.7	5.3	4.9	3.6	3.1	4.9
Lat + 15	3.7	4.3	4.7	5.2	5.0	5.1	5.1	5.2	5.0	5.0	3.8	3.4	4.6
90	3.3	3.5	3.3	3.1	2.5	2.3	2.4	2.9	3.3	3.9	3.3	3.0	3.1
1-Axis (Lat)	4.1	5.0	6.0	7.2	7.5	7.8	7.5	7.3	6.5	6.0	4.2	3.6	6.1
Temp. (°C)	7.7	10.4	16.2	21.6	26.0	30.3	31.9	31.3	28.1	22.5	15.8	10.1	21.0

Tilt	Brownsville, TX Jan	Feb	Mar	Apr	May	Jun	Latitude 25.90°N July	Aug	Sept	Oct	Nov	Dec	Year
Lat − 15	3.2	4.0	4.8	5.4	5.7	6.2	6.4	6.0	5.4	4.9	3.9	3.1	4.9
Lat	3.6	4.3	5.0	5.3	5.4	5.7	5.9	5.8	5.5	5.3	4.4	3.6	5.0
Lat + 15	3.8	4.4	4.9	5.0	4.8	5.0	5.2	5.3	5.3	5.4	4.6	3.8	4.8
90	3.0	3.2	3.0	2.3	1.8	1.6	1.7	2.1	2.9	3.7	3.6	3.1	2.7
1-Axis (Lat)	4.2	5.2	6.1	6.6	6.8	7.6	7.9	7.7	7.0	6.6	5.3	4.1	6.3
Temp. (°C)	20.5	22.3	25.8	28.9	31.0	32.8	34.1	34.2	32.4	29.6	25.7	22.1	28.3

Tilt	El Paso, TX Jan	Feb	Mar	Apr	May	Jun	Latitude 31.80°N July	Aug	Sept	Oct	Nov	Dec	Year
Lat − 15	4.6	5.6	6.6	7.4	7.8	7.7	7.2	6.9	6.4	5.9	4.9	4.4	6.3
Lat	5.3	6.2	7.0	7.3	7.3	7.1	6.7	6.7	6.6	6.4	5.7	5.1	6.5
Lat + 15	5.8	6.5	6.9	6.8	6.4	6.0	5.8	6.1	6.3	6.6	6.1	5.6	6.2
90	5.1	5.2	4.6	3.5	2.5	2.1	2.2	2.8	3.8	4.9	5.2	5.1	3.9
1-Axis (Lat)	6.5	7.9	9.2	10.1	10.4	10.2	9.2	9.0	8.7	8.2	7.0	6.3	8.6
Temp. (°C)	13.4	16.8	21.1	25.9	30.6	35.8	35.6	34.2	30.6	25.8	19.1	14.2	25.3

Radiation data measured in kWh/m^2-day, average daily maximum temperature (°C).

Tilt	Fort Worth, TX Jan	Feb	Mar	Apr	May	Jun	Latitude 32.83°N July	Aug	Sept	Oct	Nov	Dec	Year
Lat − 15	3.8	4.5	5.3	5.9	6.2	6.7	6.9	6.5	5.7	5.0	4.0	3.5	5.3
Lat	4.3	4.9	5.5	5.7	5.8	6.2	6.4	6.3	5.8	5.4	4.5	4.1	5.4
Lat + 15	4.6	5.1	5.4	5.3	5.2	5.3	5.7	5.6	5.7	5.5	4.8	4.4	5.2
90	4.0	4.0	3.6	2.9	2.3	2.1	2.3	2.8	3.4	4.1	4.1	4.0	3.3
1-Axis (Lat)	5.1	6.0	6.9	7.4	7.7	8.4	8.8	8.5	7.4	6.7	5.5	4.9	6.9
Temp. (°C)	12.3	14.9	19.9	24.6	28.3	33.3	35.8	35.7	31.0	25.8	19.3	14.2	24.6

Tilt	Midland, TX Jan	Feb	Mar	Apr	May	Jun	Latitude 31.93°N July	Aug	Sept	Oct	Nov	Dec	Year
Lat − 15	4.3	5.1	6.2	6.8	7.0	7.1	6.9	6.6	5.9	5.5	4.6	4.1	5.8
Lat	5.0	5.7	6.5	6.7	6.5	6.5	6.4	6.4	6.0	6.0	5.3	4.8	6.0
Lat + 15	5.4	5.9	6.4	6.2	5.8	5.6	5.5	5.8	5.8	6.1	5.6	5.2	5.8
90	4.7	4.7	4.2	3.2	2.4	2.0	2.2	2.7	3.5	4.5	4.8	4.7	3.6
1-Axis (Lat)	6.1	7.1	8.4	9.0	9.0	9.1	9.0	8.7	7.8	7.6	6.5	5.8	7.9
Temp. (°C)	13.6	16.4	21.8	26.6	30.8	34.1	35.2	34.5	29.7	25.2	19.0	14.7	25.1

Tilt	San Antonio, TX Jan	Feb	Mar	Apr	May	Jun	Latitude 29.53°N July	Aug	Sept	Oct	Nov	Dec	Year
Lat − 15	3.7	4.5	5.2	5.7	5.9	6.5	6.7	6.6	5.8	5.1	4.1	3.5	5.3
Lat	4.3	4.9	5.4	5.6	5.6	6.0	6.3	6.3	5.9	5.5	4.6	4.1	5.4
Lat + 15	4.5	5.0	5.3	5.2	5.0	5.2	5.5	5.8	5.7	5.6	4.9	4.4	5.2
90	3.8	3.8	3.4	2.7	2.1	1.9	2.0	2.6	3.3	4.0	4.0	3.8	3.1
1-Axis (Lat)	5.1	6.0	6.8	7.0	7.1	7.9	8.4	8.3	7.5	6.9	5.6	4.8	6.8
Temp. (°C)	16.0	18.7	23.1	26.8	29.6	33.2	35.0	35.2	31.8	27.6	22.2	17.5	26.4

Tilt	Salt Lake City, UT Jan	Feb	Mar	Apr	May	Jun	Latitude 40.77°N July	Aug	Sept	Oct	Nov	Dec	Year
Lat − 15	2.9	4.0	5.0	5.9	6.6	7.2	7.3	7.0	6.3	5.0	3.3	2.5	5.2
Lat	3.2	4.3	5.2	5.8	6.2	6.6	6.7	6.7	6.4	5.4	3.7	2.9	5.3
Lat + 15	3.4	4.4	5.1	5.4	5.5	5.6	5.8	6.1	6.1	5.5	3.9	3.1	5.0
90	3.2	3.9	3.9	3.5	3.0	2.8	2.9	3.6	4.3	4.5	3.5	2.9	3.5
1-Axis (Lat)	3.7	5.1	6.5	7.7	8.7	9.6	9.8	9.4	8.6	6.8	4.3	3.3	7.0
Temp. (°C)	2.4	6.4	11.2	16.3	22.2	28.2	33.4	31.9	26.2	18.9	10.4	3.2	17.6

Tilt	Sterling, VA Jan	Feb	Mar	Apr	May	Jun	Latitude 38.95°N July	Aug	Sept	Oct	Nov	Dec	Year
Lat − 15	3.1	3.9	4.7	5.4	5.8	6.1	6.0	5.7	5.0	4.3	3.2	2.7	4.7
Lat	3.5	4.2	4.8	5.3	5.5	5.7	5.6	5.5	5.1	4.6	3.6	3.1	4.7
Lat + 15	3.7	4.3	4.7	4.9	4.8	4.9	4.9	5.0	4.9	4.6	3.7	3.3	4.5
90	3.4	3.7	3.5	3.1	2.6	2.5	2.6	2.9	3.3	3.7	3.3	3.0	3.1
1-Axis (Lat)	4.0	5.0	6.0	6.8	7.1	7.4	7.2	6.9	6.3	5.6	4.2	3.5	5.8
Temp. (°C)	4.5	6.6	12.6	18.4	23.5	28.2	30.6	29.7	26.1	19.8	13.7	7.2	18.4

Radiation data measured in kWh/m^2-day, average daily maximum temperature (°C).

Tilt	Burlington, VT Jan	Feb	Mar	Apr	May	Jun	Latitude 44.47°N July	Aug	Sept	Oct	Nov	Dec	Year
Lat − 15	2.6	3.6	4.5	5.0	5.6	5.9	6.1	5.6	4.7	3.5	2.2	1.9	4.3
Lat	2.9	3.9	4.7	4.9	5.3	5.4	5.6	5.3	4.8	3.7	2.4	2.1	4.3
Lat + 15	3.1	4.1	4.6	4.5	4.6	4.7	4.9	4.8	4.5	3.7	2.4	2.2	4.0
90	3.0	3.8	3.9	3.2	2.9	2.7	2.9	3.2	3.4	3.1	2.2	2.2	3.0
1-Axis (Lat)	3.3	4.6	5.8	6.3	7.0	7.4	7.7	7.1	6.0	4.4	2.6	2.4	5.4
Temp. (°C)	−3.8	−2.5	4.1	12.0	19.6	24.3	27.3	25.5	20.6	13.9	6.7	−0.9	12.2

Tilt	Seattle, WA Jan	Feb	Mar	Apr	May	Jun	Latitude 47.45°N July	Aug	Sept	Oct	Nov	Dec	Year
Lat − 15	1.5	2.3	3.5	4.6	5.4	5.7	6.1	5.6	4.7	3.0	1.7	1.3	3.8
Lat	1.6	2.5	3.6	4.4	5.1	5.2	5.7	5.4	4.7	3.2	1.8	1.4	3.7
Lat + 15	1.7	2.5	3.5	4.1	4.5	4.5	4.9	4.9	4.5	3.2	1.8	1.4	3.5
90	1.5	2.2	2.8	3.0	3.0	2.8	3.1	3.4	3.4	2.7	1.7	1.3	2.6
1-Axis (Lat)	1.8	2.8	4.3	5.5	6.7	7.0	7.9	7.2	5.9	3.7	2.0	1.5	4.7
Temp. (°C)	7.2	9.7	11.5	14.0	17.7	21.1	24.0	24.0	20.7	15.4	10.3	7.3	15.2

Tilt	Madison, WI Jan	Feb	Mar	Apr	May	Jun	Latitude 43.13°N July	Aug	Sept	Oct	Nov	Dec	Year
Lat − 15	3.0	3.9	4.5	5.1	5.8	6.2	6.2	5.7	4.8	3.8	2.5	2.3	4.5
Lat	3.4	4.3	4.7	5.0	5.5	5.7	5.8	5.5	4.8	4.0	2.8	2.6	4.5
Lat + 15	3.6	4.4	4.6	4.6	4.8	4.9	5.0	5.0	4.6	4.0	2.9	2.8	4.3
90	3.5	4.0	3.7	3.2	2.9	2.8	2.9	3.2	3.4	3.3	2.6	2.7	3.2
1-Axis (Lat)	3.9	5.0	5.8	6.4	7.3	7.8	7.7	7.1	6.0	4.8	3.2	3.0	5.7
Temp. (°C)	−4.0	−1.1	5.3	13.7	20.5	25.7	28.0	26.4	21.9	15.5	6.7	−1.2	13.1

Tilt	Lander, WY Jan	Feb	Mar	Apr	May	Jun	Latitude 42.82°N July	Aug	Sept	Oct	Nov	Dec	Year
Lat − 15	3.7	4.8	5.7	6.2	6.5	7.0	7.0	6.8	6.1	5.1	3.8	3.4	5.5
Lat	4.3	5.3	6.0	6.1	6.1	6.4	6.5	6.5	6.2	5.5	4.3	4.0	5.6
Lat + 15	4.6	5.6	5.9	5.7	5.4	5.5	5.6	5.9	6.0	5.6	4.6	4.3	5.4
90	4.5	5.1	4.8	4.0	3.2	2.9	3.1	3.6	4.3	4.7	4.3	4.3	4.1
1-Axis (Lat)[a]	5.0	6.5	7.6	8.3	8.5	9.4	9.5	9.3	8.3	6.9	5.2	4.6	7.4
Temp. (°C)	−0.4	2.8	7.6	13.2	18.9	25.2	30.1	28.8	22.4	15.4	5.8	0.1	14.2

[a] The 1-axis tracker is south-facing at a tilt angle equal to its latitude.

Source: NREL (1994). *Solar Radiation Data Manual for Flat-Plate and Concentrating Collectors.*

APPENDIX F

MAPS OF SOLAR INSOLATION

Renewable and Efficient Electric Power Systems. By Gilbert M. Masters
ISBN 0-471-28060-7

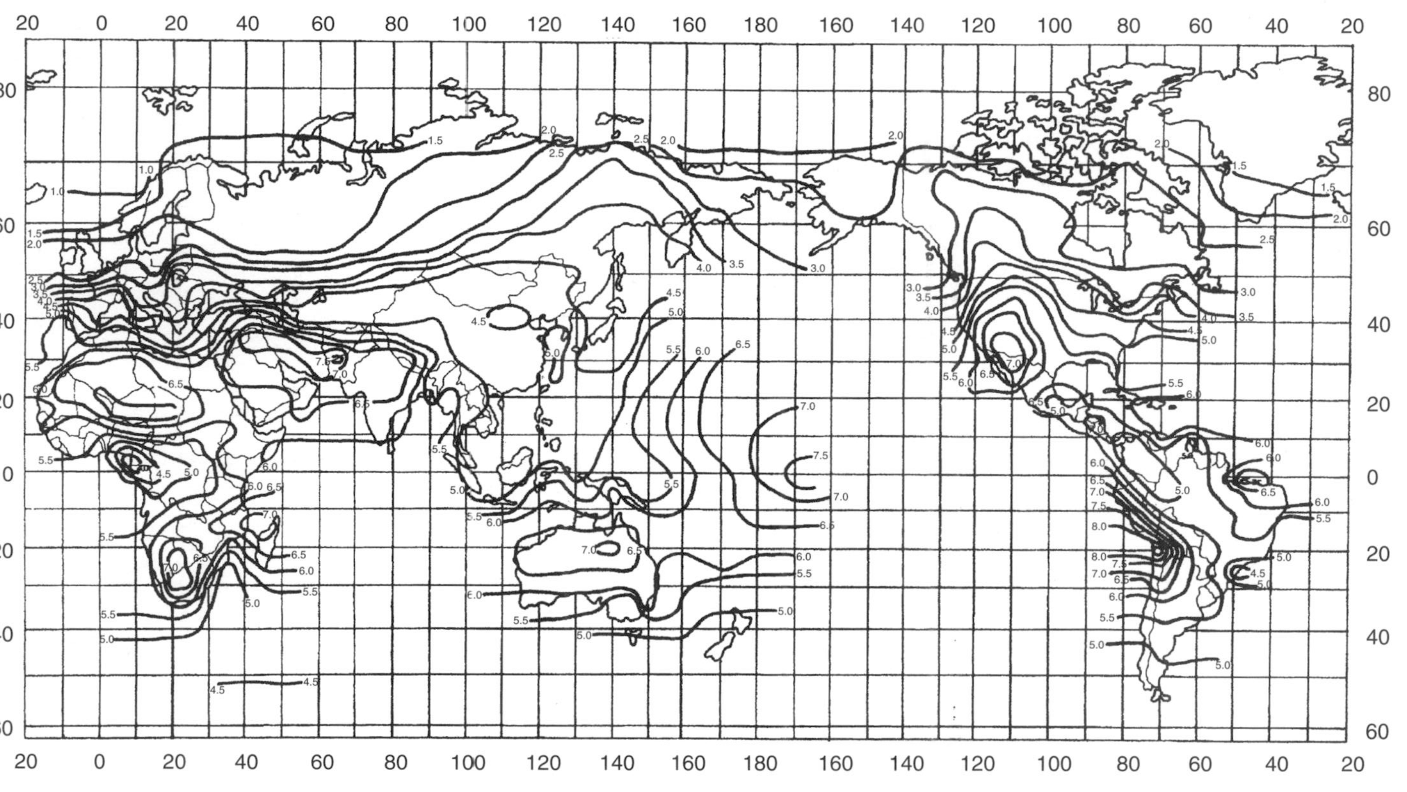

AUTUMN - Tilt angle equals the latitude angle
Daily total solar radiation incident on a tilted surface in $kWH/m^2/day$

Figure Source: Thomas, 1987

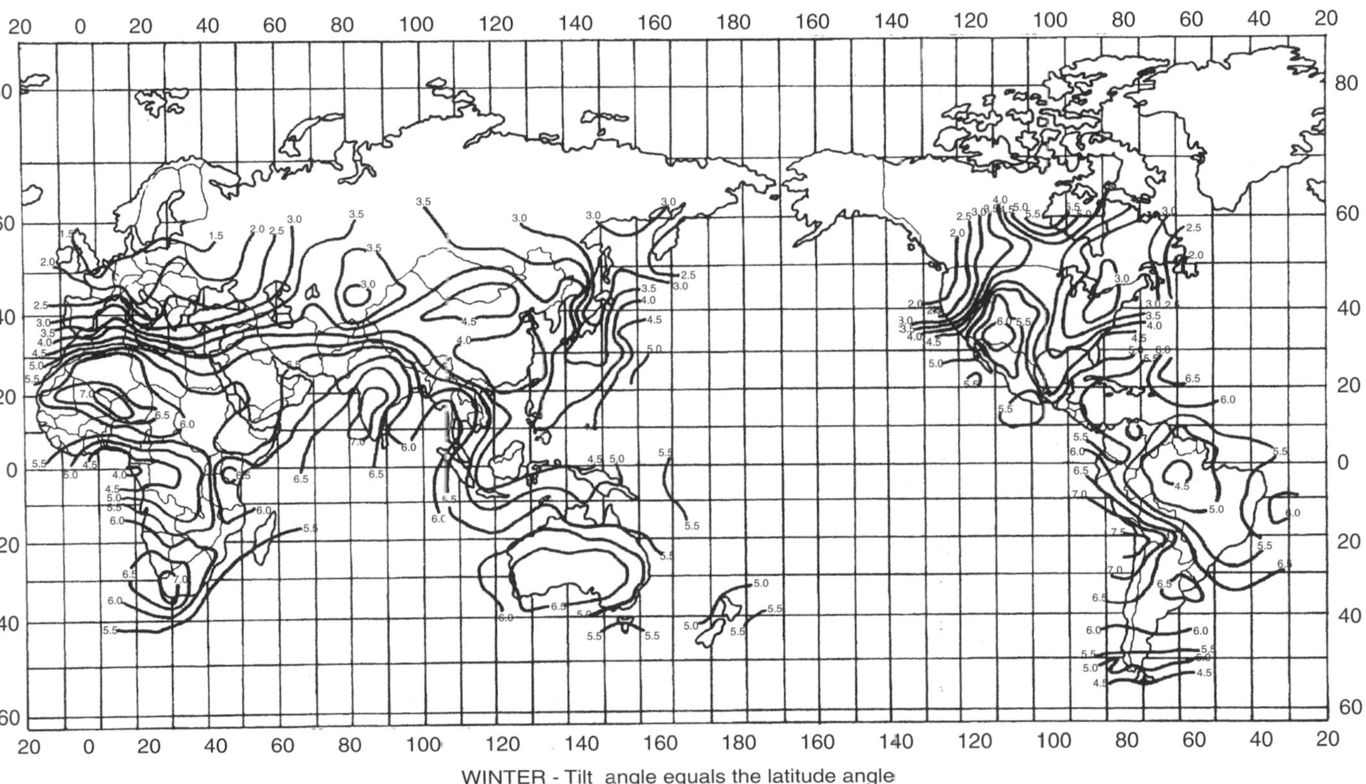

WINTER - Tilt angle equals the latitude angle

Daily total solar radiation incident on a tilted surface in kWH/m^2/day

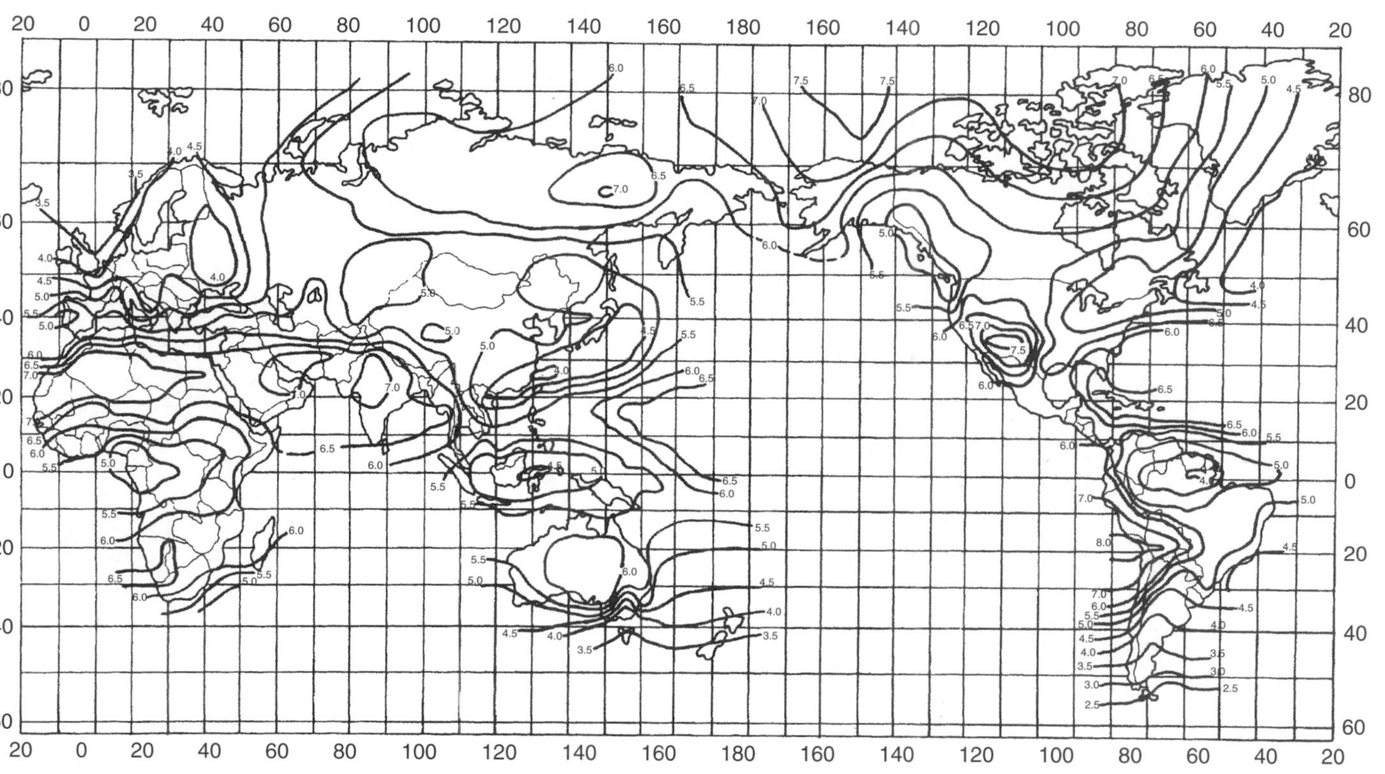

SPRING - Tilt angle equals the latitude angle

Daily total solar radiation incident on a tilted surface in $kWH/m^2/day$

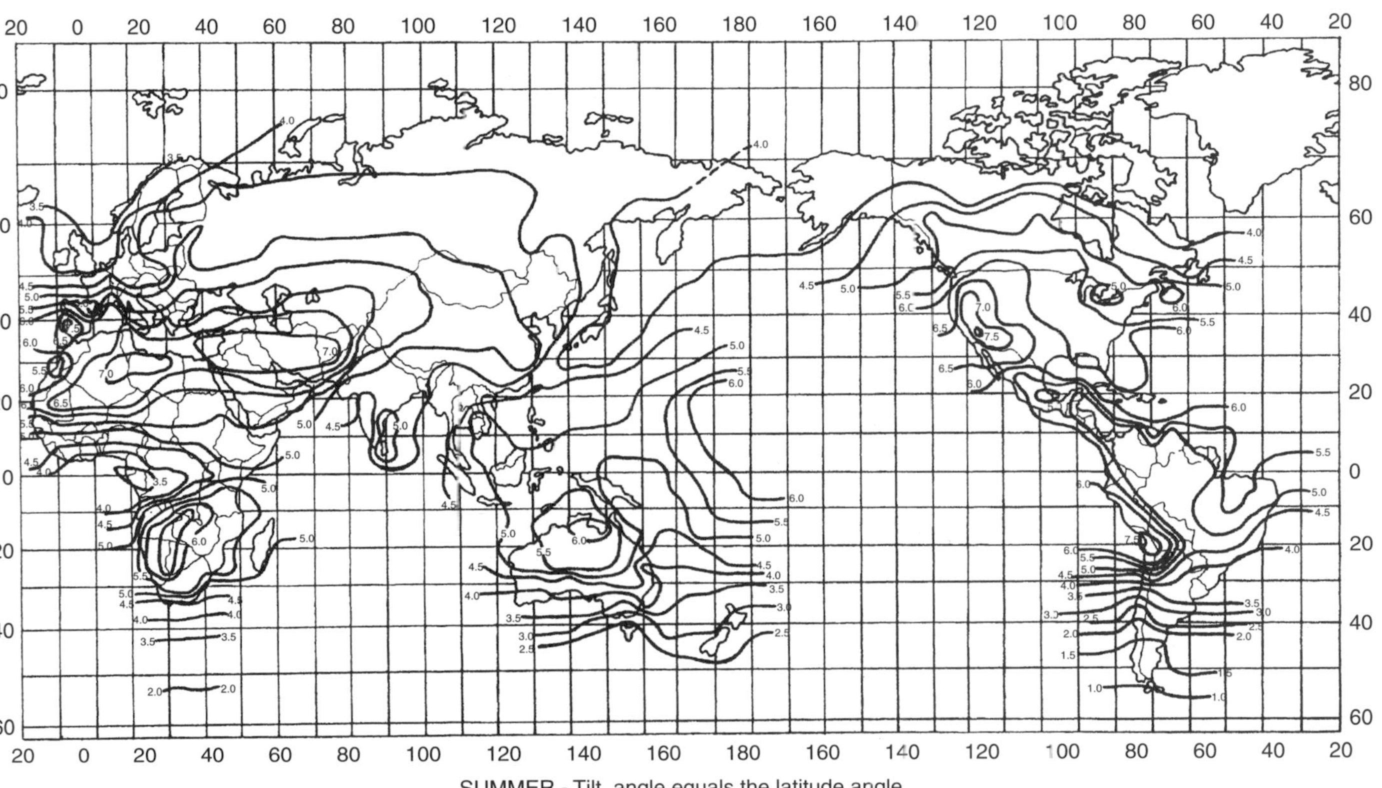

SUMMER - Tilt angle equals the latitude angle

Daily total solar radiation incident on a tilted surface in kWH/m^2/day

INDEX

Renewable and Efficient Electric Power Systems. By Gilbert M. Masters
ISBN 0-471-28060-7